CELLULAR SIGNAL PROCESSING

SECOND EDITION

An Introduction to the Molecular Mechanisms of Signal Transduction

CELLULAR SIGNAL PROCESSING

SECOND EDITION

An Introduction to the Molecular Mechanisms of Signal Transduction

Friedrich Marks
Ursula Klingmüller
Karin Müller-Decker

Garland Science
Taylor & Francis Group

NEW YORK AND LONDON

Vice President: Denise Schanck
Senior Editor: Summers Scholl
Senior Editorial Assistant: Katie Laurentiev
Senior Digital Project Editor: Natasha Wolfe
Typesetter: Thomson Digital
Copy Editor: Chris Purdon
Proofreader: Susan Wood
Cover Designer: Matt McClements, Blink Studio, Ltd.
Illustrator: Oxford Designers & Illustrators
Indexer: Simon Yapp, at Indexing Specialists

ISBN 978-0-8153-4534-3

Library of Congress Cataloging-in-Publication Data

Names: Marks, Friedrich, 1936- author. | Klingmüller,Ursula, author. | Müller-Decker, Karin, author.
Title: Cellular signal processing: an introduction to the molecular mechanisms of signal transduction / Friedrich Marks, Ursula Klingmüller, Karin Müller-Decker.
Description: Second edition. | New York, NY: Garland Science, 2017.
Identifiers: LCCN 2016058713 | ISBN 9780815345343
Subjects: LCSH: Cellular signal transduction.
Classification: LCC QP517.C45 M37 2017 | DDC 571.7/4—dc23
LC record available at https://lccn.loc.gov/2016058713

10 0779252 2

Published by Garland Science, Taylor & Francis Group, LLC, an informa business, 711 Third Avenue, New York, NY 10017, USA, and 3 Park Square, Milton Park, Abingdon, OX14 4RN, UK.

Printed in Great Britain by Ashford Colour Press Ltd
15 14 13 12 11 10 9 8 7 6 5 4 3 2 1

Garland Science
Taylor & Francis Group

NEW YORK AND LONDON

Visit our website at http://www.garlandscience.com

For Lilli, who opened my eyes to the wonderful
world of binarily encoded patterns. F.M.

Preface

All Life is problem solving
  Sir Karl R. Popper

All cells process signals: they associate an exogenous stimulus into a "meaning"—that is, they are able to distinguish between good and bad in order to respond in an adequate manner. Even for most primitive unicellular organisms, signal processing (often called signal transduction) is essential for survival in an unpredictable environment. It is understood that a wide variety of diseases including cancer, diabetes, dementia, psychoses, and cardiovascular disorders are caused or at least promoted by disturbances in cellular signal processing, and most therapeutic drugs, as well as many narcotics, target signaling pathways. Therefore, signal transduction is a subject that ranks among the most rapidly developing fields within biomedical science.

This book deals with the biochemical mechanisms of cellular signal processing. It is intended for undergraduate students with a good knowledge of basic biochemistry and cell biology, and for graduate students in biology, biochemistry, and pharmacology, as well as medical students. The book emerged from our scientific work at the German Cancer Research Center and from a series of lectures we have given at the University of Heidelberg for many years. Although cell biology textbooks provide excellent introductions into the field of signal transduction, our students consistently requested a more comprehensive treatment of the subject that, in particular, addressed evolutionary aspects and questions of medical interest. Moreover, we identified the need for a textbook that would offer a more complete overview of the field than currently provided by available books. As such, it may be useful not only for the student but also for the scientist.

It is currently impossible to depict and describe all the molecular interactions that occur when a cell receives a signal. In an effort to navigate the middle ground between incomprehensibility and superficiality, we have confined ourselves to describing more or less linear signaling cascades that illustrate mechanistic principles of major signaling pathways, although systems biology is currently integrating these cascades into networks. One of the hallmarks of this textbook is its presentation of a unifying concept of signal transduction. Cell signaling is traced to a few simple biochemical switching reactions that link logical operations—mainly performed by protein–protein interactions—with energy-delivering processes of cellular metabolism.

As with the first edition, Chapter 1 deals with the ability of proteins to form a "neural" network that, together with the genome, resembles a cellular brain. In this network, each protein acts as a "molecular neuron" that, according to the laws of chemical thermodynamics and mathematical logic, transforms input signals (environmental and systemic stimuli) into output signals (cellular behavior facilitating adaptation). According to the second law of thermodynamics, a data-processing apparatus needs to be supplied with energy. For the protein network, this energy is derived from a series of exergonic biochemical reactions such as redox processes, GTP and

ATP hydrolysis, protein phosphorylation, and discharge of the membrane potential. This biochemical arsenal is explained in Chapter 2. As discussed in Chapter 3, the biochemistry of cellular signal transduction has a long evolutionary history. It was initiated billions of years ago and, similar to metabolic and genetic mechanisms, has been conserved during evolution. These basic themes are discussed in more detail in Chapters 4 through 16. Finally, Chapter 17 places cellular signal processing in the context of systems biology and describes the formation and function of networks.

In response to our readers, we have updated and reorganized many sections of the text accordingly. New coverage includes intrinsically disordered proteins, signaling by non-coding RNAs, new evolutionary concepts of signal transduction, inflammasomes, regulated necrosis, autophagy, and optogenetics. In addition, the lists of Further Reading found at the end of each chapter have been entirely revised, providing an up-to-date guide for exploring the more specialized literature. To this end, we have focused on the most recent review articles, rather than original publications. Many of the questions raised today will certainly be answered in the near future, whereas other problems not yet recognized will become apparent. Nevertheless, our hope for this new edition of the book, as for the previous one, is that it will be an indispensable resource for students and scientists who are exploring and working in one of the most exciting fields in the life sciences.

Acknowledgments

Academic lecturing is a process of giving and receiving, and we are grateful to our students for their attentive and critical interest and helpful feedback on the first edition of the book. On this second edition, we are particularly grateful to Dennis Bray (University of Cambridge), Stefan Rose-John (University of Kiel), Felix Wieland (University of Heidelberg) and our colleagues from the German Cancer Research Center Heidelberg: Tobias Dick, Sven Diederichs, Karsten Gülow, Doris Mayer, and Marcel Schilling for critically reviewing individual chapters in their areas of interest. We also thank the following reviewers for their suggestions in preparing this new edition: J. Chloe Bulinski (Columbia University), Ahmet Carhan (Yildirim Beyazit University), Joseph Eichberg (University of Houston), Laura Frost (Point Park University), Eric Gauthier (Laurentian University), Kathy Gillen (Kenyon College), Pascal Lafontant (DePauw University), Boon Chuan Low (National University of Singapore), Victoria Moore (Elon University), Eric Price (University of Arizona College of Medicine), and Giovanni Zifarelli (Italian National Research Council). We kindly acknowledge those who helped on the first edition: Ingrid Hofmann, Peter Krammer, Frank Rösl (all from the German Cancer Research Center Heidelberg), Mark A. Lemmon (University of Pennsylvania School of Medicine), Shigeki Miyamoto (University of Wisconsin, Madison), Bradley J. Stith (University of Colorado, Denver), Wei-Jen Tang (University of Chicago), and Alexey Veraksa (University of Massachusetts). In particular, we would like to thank Summers Scholl, Katie Laurentiev, and Natasha Wolfe at Garland Science for their editorial and production support and Oxford Designers & Illustrators for their work on the figures. The editorial team deserves highest appreciation for their professionalism, helpfulness, and patience.

Instructor Resources Website

The images from *Cellular Signal Processing*, Second Edition, are available on the Instructor Site in two convenient formats: PowerPoint® and JPEG. They have been optimized for display on a computer. The resources may be browsed by individual chapter and there is a search engine. Figures are searchable by figure number, figure name, or by keywords used in the figure legend from the book.

Accessible from www.garlandscience.com, the Instructor's Resource Site requires registration and access is available only to qualified instructors. To access the Instructor Resource site, please contact your local sales representative or email science@garland.com.

Brief Contents

Detailed Contents

The "Brain of the Cell": Data Processing by Protein Networks

1

The term signal transduction has become established for the molecular mechanisms by which cells process information transmitted by exogenous or endogenous stimuli. The goal of signal transduction is to find the response that optimally safeguards survival.

In this chapter we shall address the following questions:

- What is the nature of the cell's information-processing system?

- According to which principles does the system work, and what is it able to do?

- Is the computer a suitable metaphor for cellular signal processing, or would it be described better in neurological terms?

1.1 Metaphors and reductionism: temptations and limitations

Living beings are connected inseparably with their environment. Appropriately, apart from many other qualities, organisms are data-processing or cognitive systems. On the basis of a program of inherited and acquired information, they attribute a distinct "meaning" to sensory impressions and other exogenous stimuli that may be understood as "input signals." This enables them to calculate an "output signal," that is, to adjust their behavior, which includes genetic readout, metabolism, reproduction, shape, and motility, to the given environmental situation. This ability, which is a condition of survival, is an essential characteristic of life, from primeval cells to humans.

The processing of large and complex amounts of data requires a **network of switching devices** that, as a minimum, make Yes–No decisions according to the basic laws of mathematical logic and are able to adapt and learn. In biology, such a network is represented most impressively by the brain with its billions of interacting cells working as interconnected switching elements. However, the reception and interpretation of signals is not restricted to animals possessing a brain but is a general property of organisms, even of the most primitive bacteria. What is the data processing equipment of a single cell or, so to speak, the "**brain of the cell**"? It must be a network of interacting molecules, each resembling a molecular "neuron," able to identify the meaning of an input signal and to "calculate" an output signal. This job is done primarily by proteins, interacting with other macromolecules such as nucleic acids of the genomic data bank. It is compelling to assume that as far as data processing is concerned both a network of cells (such as a brain) and a network of proteins (such as a cell) are working according to the same principles. After all, both have the same evolutionary background.

Figure 1.1 Data-processing networks at different levels of complexity Technical and biological data-processing (cognitive) networks exist at different levels of complexity. Even a single protein molecule may be understood as a data-processing network with the individual molecular domains that make up its three-dimensional conformation acting as switching elements. Meaningful data processing requires these switches to interact as directed by the laws of mathematical logic.

Whether an analogous conclusion can be drawn for the next higher level of complexity, for example, for populations of interacting individuals (**Figure 1.1**), remains a matter of worldview (*Weltanschauung*). For some people such a claim would have an unbearable biologistic taste, at least when it is extended to human society, whereas others may remain firm in their conviction of a unifying principle that—emerging in the course of evolution—dominates living nature.

Still another heuristically tempting metaphor is wandering through the popular and even the scientific literature: the data-processing network of microprocessors called a computer. Today it has become fashionable to compare cells, brains, or even organisms with computers and, vice versa, to denote electronic networks capable of learning as "neural." However, this concept, although it has the advantage of clarity, rapidly leads into a cul-de-sac. Biological data processing—even if based on the same mathematical principles—is certainly more ingenious and sophisticated than today's computer technology, and it is still far from clear whether it can ever be interpreted and imitated by a technological approach. So one should treat the computer metaphor with caution, being aware of its narrow limits.

May a similar objection not be raised about the phrase "brain of the cell?"After all, we are still a long way from understanding the mode of operation of a real brain, which some people believe (perhaps in an exaggerated opinion of themselves) to be the most complex data-processing device in the universe. When compared with the computer metaphor, however, the "brain of the cell" concept has a clear advantage: it emphasizes the strong resemblance of biological data processing at levels of different complexity. As a consequence, the investigation of cellular data processing is expected to inspire brain research and vice versa.

Currently, both molecular brain research and molecular cell biology are still more or less in a phase of hunting and gathering. We are collecting and describing processes, trying to trace them back in a strictly reductionistic manner to elementary chemical and physical events. This approach has been extremely successful and is as indispensable as it is highly satisfying; it is also of tremendous importance for medical applications. Certainly the results obtained thus far provide an estimate of the complexity of biological data processing. However, the expectation that such a method will lead to a deeper understanding of cellular behavior appears to be as illusory as a neurobiologist's anticipation that detailed knowledge of molecular and cellular interactions in the brain would be sufficient to comprehend a systemic property such as consciousness or even mind, or the conviction of nineteenth century physicists that knowledge of the locations and impulses of all elementary particles would enable a complete prediction of the universe's future.

The reason for this is that complex systems such as living organisms are always much more than the sum of their parts. Life is the result of irreversible historical processes called evolution and development that continuously create new properties, which cannot be predicted just by counting and categorizing molecules. Only quite recently has molecular biology, hitherto the playground of reductionists, become aware of this problem. As a consequence, strong efforts are being made to extend molecular research into the area known as systems biology (see Chapter 17). By a combined approach of bioinformatics, biophysics, bio-engineering, biochemistry, and molecular biology (including genome- and proteome-wide screening studies), one is trying to translate the language of molecular structures and interactions into the language of complex systems behavior and the other way around. While there is no doubt that this methodology will dominate the bioscience field in the twenty-first century, we have to concede that currently such approaches are still in their infancy.

Thus, lacking a novel intellectual (or even mathematical) concept of the complexity of living systems, we have to be satisfied by reviewing the present results of our hunting and gathering of molecules and mechanisms, an endeavor that is not trivial. In fact, at present the flood of data is rising impressively and alarmingly. However, one should take care not to embrace new illusions; currently the physiological functions of no more than 20% of all signal-processing proteins are known. Taking as an example a special switching reaction such as protein phosphorylation, we know that at least one-third, or more than 10,000, of the cellular proteins undergo this chemical modification, but we understand the physiological relevance of less than 20% of the phosphorylations occurring *in vivo*, and this is still in a reductionistic rather than a systemic manner. Thus, our data collection is far from complete, and many conclusions drawn today may have a rapid use-by date.

Nevertheless, certain basic principles and a couple of biochemical reactions upon which cellular data processing is based have emerged. They clearly show that the signal-transducing network primarily consists of proteins communicating with each other via quite a limited set of noncovalent interactions. These interactions are connected with a dozen or so biochemical reactions delivering the energy required for data processing, thus maintaining signal intensity and giving the flow of data a definite direction. The biochemistry of these reactions is surprisingly simple. In fact, complexity arises from an almost unlimited number of combinations of a few basic elements rather than from a huge number of complicated structures and reactions.

Summary

- Data processing is a general property of organisms, being most efficiently developed in neuronal networks such as the brain. A network of proteins interacting with each other and with the genome is understood as the "brain" of a *single* cell.

- Both neuronal and molecular networks are assumed to work according to the same rules, enabling information processing on the basis of mathematical logic.

- Currently the investigation of biological data processing follows a strictly reductionistic approach of hunting and gathering, which results in the construction of (highly artificial) linear cascades of signal-transducing biochemical reactions. To arrive at a more in-depth understanding, this approach must be broadened by a comprehensive theory of (biological) complexity.

1.2 Information and signals

The ability to respond appropriately to environmental stimuli, called sensitivity (*Reizbarkeit*) in the older literature, belongs to the basic properties of life according to the maxim "the lower the degree of uncertainty, the higher the chance to survive." Therefore—and to repeat what has already been stated—organisms, from the simplest prokaryotes to humans, are cognitive, or data-processing and learning, systems.

On the molecular level, cognition is based on the unique ability of proteins to process data according to logical principles. Their tremendous functional versatility, their unsurpassed structural flexibility, and their incomparable capability to interact with each other and with other molecules make them efficient switching elements

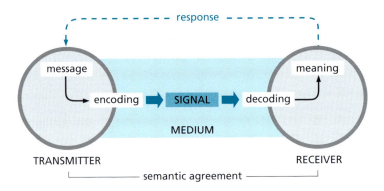

Figure 1.2 Communication by signal transmission and signal transduction A transmitter encodes a message into a signal, which is transmitted to a receiver along a physicochemical medium. Having decoded the meaning of the signal, the receiver transmits a response signal provided a semantic agreement exists between both partners; that is, strictly speaking, both master the same language. In biological systems, receiver and transmitter may be individuals, cells, or protein molecules. The decoding of signals by a cell is usually called signal transduction.

of an information-processing molecular network. Proteins are at the same time enzymes and constructive elements of the cell.

In other words, the network of enzyme-catalyzed metabolic reactions, as well as the processes regulating cellular architecture and motility, are inseparably connected with the data-processing network. In a cell, nothing happens that is not under the control of this network. This system represents a state of extreme order that—according to the second law of thermodynamics—can be generated and maintained only by continuous energy consumption and on the basis of detailed information that has to be processed by the cell.

Information is understood as the content of news exchanged between partners—let us call them transmitters and receivers—aimed at reducing uncertainty and disorder. Such an exchange requires information to be represented by symbols such as letters and words, that can be transmitted via any physicochemical medium, for example, sound waves or printed paper (**Figure 1.2**). Physicochemical entities used for the exchange of information are called **signals**.

Information technology distinguishes between analog and digital signals. Analog signals are omnipresent in music and spoken language. They are characterized by a continuous encoding of information: within fixed limits, the signal may take on any value and may be of any length. In contrast, digital signals are generated by discontinuous encoding of information in the form of short pulses, the intensities of which take on only a few values. The simplest example of a digital signal is binary encoding with only two values. In addition, information can be encoded in the frequency, that is, the interval between the individual signal pulses. A classic example is provided by the Morse alphabet, which also clearly demonstrates the interchangeability of analog and digital signaling systems. Living systems continuously change between the two systems.

In information technology, signals are mostly electromagnetic impulses and waves, or sound waves, whereas inter- and intracellular communication preferentially makes use of molecules and ions (**Table 1.1**). As a consequence, cells process data primarily by use of biochemical reactions.

Frequently the symbolic systems are changed in the course of signal transduction: for instance, when a sensory impression is transformed into an electrochemical nerve impulse. Such transformations resemble the translation of languages. They require a **code**, that is, an instruction of translation by which one symbolic system is related to another one: for instance, the system of German language to the system of English language, the system of visual impressions to the system of nerve impulses, or the system of nucleic acid structure to the system of protein structure.

Information may be encoded in both the intensities and the temporal sequence of signals, called **amplitude- and frequency-dependent modulation**. As an example, a black-and-white sketch is based mainly on amplitude modulation, while a color

Table 1.1 **Biological signals**

Environmental stimuli: sound, light, temperature, touch, taste, and smell; molecules, xenobiotics, toxic substances, other stress factors
Between organisms: pheromones, gamones, sound, sight, touch, taste, smell
Between cells: systemic mediators such as hormones; local mediators such as tissue hormones, cytokines, lipid mediators, neurotransmitters, and nitrogen monoxide; cell surface proteins such as cell adhesion molecules
Within cells: second messengers such as cyclic nucleotides, diacylglycerol, inositol phosphates, phosphoinositides, Ca^{2+}; interaction domains of proteins
Within protein molecules: conformational changes

picture is based on both amplitude- and frequency-modulated light signals. An analogous difference is that between noise and music. These examples illustrate that, as compared with amplitude-dependent modulation, frequency-dependent modulation enables the transmission of much more information, a fact upon which information technology critically depends. Both types of modulation are also found in biological systems. The prototype of a frequency- modulated signal is the nerve impulse. However, many (perhaps most) biochemical reactions are oscillating as well and may generate or be regulated by frequency-modulated signals. Thus, frequency-dependent modulation of endogenous signals is expected to be the rule rather than the exception, and like the brain, a cell may be understood as an ever-oscillating system.

While classical information theory deals primarily with the techniques of encoding and transmission of news, in biology the meaning of signals has top priority. Since information can be encoded in any physicochemical medium, the meaning is medium-independent ("the medium is not the message"). Instead, it is read into the signal on the basis of an inherited and acquired pre-information or program. By convention, in biology this process of signal decoding by the recipient is called **signal transduction**, although signal processing would be a more appropriate term.

From the principle of medium-independent signaling it follows that signals resemble keywords and are principally ambiguous (**Figure 1.3**): for different receivers, the same signal may have different meanings. This is an important point, since one can be tempted to seek the meaning of endogenous biological signals in their chemical structures, although it is exclusively assigned by the receiver. As an example, let us take the signaling molecule cyclic adenosine 3′,5′-monophosphate (cAMP). It is interpreted by many bacteria as a request to activate certain genes (details in Section 3.3.2), whereas the slime mold *Dictyostelium discoideum* responds to the same signal by aggregation into multicellular structures and development into a spore capsule. Numerous and completely different effects of cAMP occur in vertebrate cells. Another instructive example is provided by the vertebrate hormone estradiol that, depending on species and tissue, exhibits quite different effects (see Figure 1.3). Of course, the reverse holds true as well: different signals may evoke the same response. For example, in the liver, the degradation of glycogen is induced by at least seven different hormones.

Although based mainly on biochemical interactions, inter- and intracellular signal transduction follows the same principles as language, where signals (words and sentences) contain conceptual information only for those who are proficient with the language or

Figure 1.3 Signals are ambiguous (the medium is not the message) As the scent of a medium-sized animal has a deterrent effect on small animals but attracts a larger animal, biological signals have completely different meanings depending on the species, the tissue type, and the physiological context, that is, on the program of signal decoding.

cAMP
- prokaryotes ———————— transcriptional co-activator
- *D. discoideum* ———————— attractant and hormone
- vertebrates ———————— second messenger with various meanings

estradiol
- uterus ———————— proliferation, differentiation
- vagina ———————— epithelial keratinization
- brain ———————— control of hormone biosynthesis
- oviduct ———————— synthesis of ovalbumin and other proteins
- liver ———————— synthesis of vitellogenin

who possess the program of decoding. A comparison between language and chemical signal transduction immediately clarifies the analogy between the systems (**Figure 1.4**). Moreover, it tells us that there is a direct causal relationship between the cellular and the social levels of complexity: on the one hand language is the result of elementary molecular events occurring in neurons (bottom-up effect); on the other hand, emotionally charged sentences inevitably have an impact on the biochemistry of cells, including the activity of genes (top-down effect). Such a reciprocal resemblance is a characteristic feature of living systems. It will be discussed in more detail in Chapter 17.

Summary

- Organisms are cognitive or information-processing systems. On the basis of inherited and acquired knowledge (the program), they relate a certain meaning to an environmental stimulus that is transmitted as a signal, or a distinct modulation of a physicochemical medium.

- Signals per se are meaningless. Their interpretation depends solely on the receiver's program and the physiological context. Therefore, a signal is ambiguous: it may have different meanings depending on the recipients and the given situation.

Figure 1.4 Resemblance between acoustic and chemical signal transmission In both cases the speaker's data bank (memory) constructs an image of its environment that is equipped with a meaning and transmitted as a signal (molecule or word) to a listener, who decodes the signal, interpreting its meaning by means of his signal-processing apparatus, that is, asking his memory store.

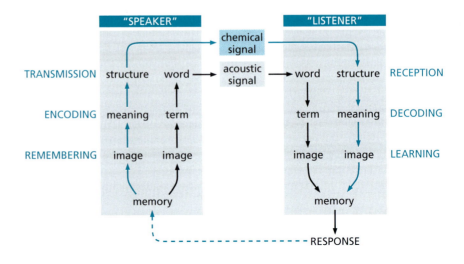

1.3 Proteins are binary switches

Information-processing systems are made up of interconnected switching elements or elementary calculators that transform an input signal into an output signal according to the laws of mathematical logic. In the simplest form, a switch represents the binary values 0 (OFF) and 1 (ON) (**Figure 1.5**). It is a special property of proteins to exist *at least* in two different states: an enzyme, for example, is either inactive (state 0) or active (state 1). The states differ in their three-dimensional structure or conformation. Any protein molecule may be considered, therefore, to be an elementary signal-processing unit. Input signals are influences that change the equilibrium between 0 and 1 and thus the functional state. As a rule such changes are caused by chemical ligands, but depending on the type of protein, changes may also be caused by other stimuli such as pressure, light, and temperature. The output signal is the effect resulting from the altered function of the protein, which in turn is a result of an altered conformation. For instance, an enzyme transforms the input signal "substrate concentration" into the output signal "product concentration."

For most enzymes, substrate concentration is not the only input signal but their activities are controlled and fine-tuned by numerous activators and inhibitors that act as additional input signals. This regulatory principle also holds true for other proteins. How such additional input signals manipulate the protein switch is explained by the theory of **allostery**. It is based on the concept of regulatory protein domains, which are not identical with the active center but, nevertheless, influence protein function via long-range effects on the protein's three-dimensional structure, the conformation. In its most concise form (as originally formulated by Monod, Wyman, and Changeux in the early 1960s), the theory assumes a protein to exist in two different conformations, being in equilibrium and differing in their activities as well as their affinities for regulatory factors, which we may call input signals S (**Figure 1.6**). To make clear that the allosteric effect is a binary switching event, we shall replace the historical terms T (tense) and R (relaxed) by 0 (=T) and 1 (=R).

Let us assume S to be a stimulatory signal or, in pharmacological terms, an **agonist**. In the absence of S, the protein may be assumed to be mostly in conformation 0: the switch would be in the OFF position. In the terminology of chemical thermodynamics, conformation 0 is said to have a low standard free energy. This means that the transition from 0 to 1 is an endergonic reaction that does not occur spontaneously but requires a supply of energy. When S interacts with the regulatory domain of the protein, the equilibrium is shifted to conformation 1: the switch is turned into the ON position, provided the signal molecule S has a higher affinity for 1 than for 0 and the energy released upon binding of S is sufficient to push the conformational change.

The theory also explains the effect of an inhibitory signal or **antagonist**. It has a higher affinity for 0 than for 1 and arrests the switch in the OFF position. Frequently one is dealing with mixed agonists–antagonists that, depending on the concentration, act either as activators or as inhibitors.

Figure 1.5 Proteins as binary switches Most proteins exist in at least two different functional states (here called states 0 and 1) that are represented by different conformations. In state 0 the protein is assumed to be inactive (in the closed or off position) because in this conformation the active site is blocked by an intramolecular interaction. An "input signal" is any influence that transforms or switches one state into the other. The conformational change thus induced releases the protein from autoinhibition, resulting in an "output signal" generated by the active site (another input signal may trigger the reverse 1 → 0 switch). This elementary data-processing unit is an oversimplification because in the cell proteins may exist in much more than two different states or conformations (see, for instance, Section 1.8.5).

Figure 1.6 Allosteric switching of protein function A protein with spatially separated regulatory and functional (catalytic) domains is thought to exist in at least two conformations, 0 and 1, that are in an allosteric equilibrium. An agonistic input signal stabilizes conformation 1, thus shifting the equilibrium downward. In contrast, an antagonistic input signal would stabilize conformation 0, shifting the equilibrium upward (not shown).

Agonistic and antagonistic signaling factors are either small molecules and environmental stimuli or proteins that interact with their target protein via specific domains. These domains are amino acid sequences that are able to recognize and to bind complementary sequences of their own (homotypically) or of a different structure (heterotypically). They will be discussed in more detail in Section 1.8. Such protein–protein interactions occur according to the principles of allosteric regulation.

The interaction of a signaling molecule with a protein may be either covalent or noncovalent. Protein phosphorylation and other post-translational modifications represent covalent binding, while noncovalent binding is typical for the interaction of a hormone with its receptor or for most protein–protein interactions.

To return the switch back to 0, the agonistic signal S has to be removed from the binding equilibrium, a process known as signal extinction. A dilution or chemical inactivation of S is sufficient to reverse a noncovalent interaction, whereas a covalent modification requires a chemical cleavage, which as a rule has to be catalyzed by a specific enzyme (**Figure 1.7**).

When a protein consists of several subunits, the theory of allostery predicts **cooperativity**. Accordingly, the interaction of one subunit with the signaling molecule either facilitates (positive cooperativity) or hinders (negative cooperativity) the

Figure 1.7 Interactions of signaling molecules with proteins (A) Binding of cyclic AMP (cAMP) as an example of a noncovalent interaction. The cAMP signal is extinguished by dilution and by enzymatic degradation, driving the reaction in the reverse direction. In an analogous manner, proteins interact with each other. (B) Phosphorylation as an example of a covalent interaction or post-translational modification. Both the forward and reverse reactions require enzymatic catalysis.

binding of additional signaling molecules, resulting in conformational changes of the other subunits. A cooperative effect is indicated by a sigmoidal curve when the activity of the protein, or the output signal (for instance, measured as enzymatic reaction rate), is drawn in relation to the ligand concentration, or the input signal S. Signal-transducing systems with a sigmoidal characteristic are resistant to low irregular signal intensities called **noise**, which is feared as a disruptive factor in information technology. In other words, cooperative proteins possess a built-in noise filter (**Figure 1.8**). In contrast, proteins without quaternary structure mostly exhibit linear or hyperbolic kinetics, and thus they lack a noise filter. This may be one major reason that signal-transducing proteins mostly consist of several subunits forming oligomeric structures and that, in addition, they form complexes with other signaling proteins. Such signal-processing particles have been called **"signalosomes"** (examples will be found throughout this book). Moreover, a system with a sigmoidal characteristic operates like an on–off switch, whereas systems with linear or hyperbolic kinetics rather work as accelerators or brakes (see Section 1.9).

Most signaling proteins form not only oligomers but, moreover, are expressed in several subtypes or **isoforms** that result from individual genes, from genes with alternative promoters, and from post-transcriptional modifications such as alternative splicing of pre-mRNA. Such isoforms may be restricted to individual tissues and cell types and may exhibit different functions. In addition, they provide the signal-processing apparatus with a high degree of **redundancy**, which in technology is considered to be a feature of superior design since it considerably diminishes the susceptibility to faults (see also Chapter 17).

The theory of allostery was originally developed to explain the cooperative behavior of proteins with quaternary structures. However, it is not restricted to such proteins (as is frequently stated in textbooks) but applies to any protein with separated regulatory and catalytic domains, though in the absence of a quaternary structure the allosteric interaction is not easily recognized from the kinetics but has to be proven by costly structural analyses.

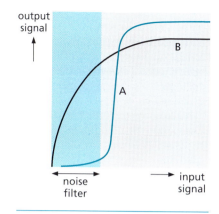

Figure 1.8 Noise suppression by a sigmoidal characteristic of signal transmission With sigmoidal kinetics (curve A), a significant output signal is produced only when the input signal surpasses a certain value. Such a system has a built-in noise filter and resembles a switch with first trigger pressure. In contrast, a system with hyperbolic characteristics (curve B) responds to very weak input signals including accidental noise. It resembles a rotary switch or an accelerator pedal.

Summary

- Proteins may be looked upon as switching elements of a data-processing molecular network.

- Switching is due to a conformational change induced by an input signal. As a result, the protein function representing the output signal is modulated.

- The conformational change results from an intra- or intermolecular allosteric interaction between a regulatory domain (receiving the input signal) and a functional domain (transmitting the output signal).

- In proteins with a quaternary structure, cooperative allosteric interactions provide a noise filter.

1.4 Signal-transducing proteins are "nanoneurons"

The cells of the brain are specialized for information processing. The same holds true for the proteins of cellular signal transduction. They differ from metabolic enzymes or architectural proteins as neurons differ from liver or skin cells. For such signal-transducing proteins, the concept of working as simple binary switches describes a borderline case, since most of them are controlled by much more than one input signal.

In fact, in their capability to convert numerous input signals (or allosteric effectors) into an output signal or even digital into analog signals, such proteins—albeit on a lower level of complexity—resemble neurons, where a large number of dendritic

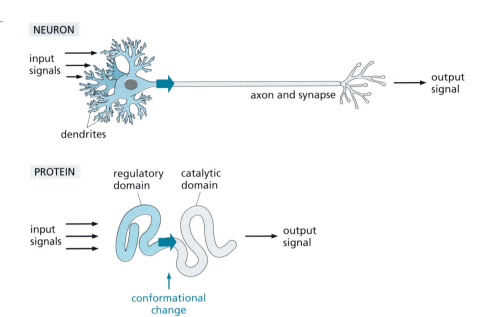

Figure 1.9 Proteins as "nanoneurons" A neuron converts numerous input signals (nerve impulses) received at the dendrites into an output signal transmitted at the axonal synapse. In an analogous manner, a protein receives input signals at its regulatory or interaction domain that, through a conformational change, become converted into an output signal released by a catalytic or functional domain.

synapses serve as signaling input terminals and the axonal synapse acts as an output terminal. In signal-transducing proteins, such synaptic contacts are replaced by structural domains. As a rule, the input terminals are binding sites or interaction domains for allosteric regulators, whereas the functional domain (for instance, the catalytic center of an enzyme) serves as the signaling output terminal (**Figure 1.9**).

Since function strictly depends on the three-dimensional structure of a protein, the conversion of input signals into the output signal occurs through (mostly allosteric) conformational changes acting in either an additive, synergistic (more than additive), or antagonistic manner. As will be explained in Section 1.8, regulatory protein domains react not only with input signals from the outside but also with each other. These intramolecular effects are an important condition of signal conversion because they determine the algorithm of the protein calculator. Their elucidation ranks among the most important and most difficult tasks of current biochemistry and structural biology.

Summary

- In general, signal-transducing proteins have more than one receiver site for regulatory input signals.

- In their ability to integrate various input signals and to calculate an output signal, they resemble neurons, albeit at a lower scale of complexity.

1.5 Proteins form logical gates and "neural networks"

Switches—even simple binary ones—may be used to carry out fundamental logical operations of the type NOT, AND, OR, and NOR according to the rules of **Boolean algebra**. These operations are sufficient to process *any* kind of logical information. The corresponding circuits, called logical gates, are the elementary units of technical data-processing networks that, as already mentioned, resemble the nervous system to a certain extent. Because these technical devices are able to adapt and to learn, they are called neural networks, with the idea that someday they may imitate living systems, including intelligent behavior (artificial intelligence). Such a concept was formulated initially in the 1940s, when it was shown theoretically that a switching device of a few simplified and strongly idealized neurons, called "McCulloch–Pitts neurons," could perform logical operations. When they are understood as cellular "nanoneurons", proteins are also expected to operate as logical gates and, like molecular McCulloch–Pitts neurons, assemble to form a data-processing network, or the "brain of the cell" (**Figure 1.10**).

It should be noted that like the concept of binary protein switches, the rigid Yes–No decisions of Boolean logic represent a borderline case that cannot fully describe biological data processing. Instead, so-called **fuzzy logic**, based on the mathematical theory of fuzzy sets, is more suited to explain decision and regulatory processes occurring in response to the incomplete, imprecise, and even contradictory information an organism typically receives from its environment.

For the brain, the actual state of function of its neuronal network resembles what is called random access memory (RAM) in computer technology. The corresponding memory store of a cell is the functional state of its protein network. Like a brain, it uses its genetic and acquired program to construct an image of the surroundings (for a unicellular organism or for a sensory cell, it is the environment, and for a tissue cell, it would be the internal milieu of the body). This enables the cell to calculate a response necessary for survival. Therefore, the idea of a protein network representing something like the "brain of the cell" appears to be not too far-fetched, though one should refrain as much as possible from anthropomorphic metaphors, keeping in mind that a single cell does not "think." Moreover, the images constructed by the protein network are necessarily incomplete, because only those signals for which the network is phylogenetically and ontogenetically programmed are recognized and processed (though this also holds true for our brain!).

A more questionable or even dangerous metaphor is that of the "computer of the cell." Albeit it certainly has the advantage of clarity and for this reason alone it is used at times in this book: for instance, when referring to switching elements or random access memories. Although information technology tries very hard to imitate biological data processing by neural networks, the results are still rather meager since certain principles of biological systems either are not fully understood or are realized only with difficulty. Thus the hardware of protein networks, in contrast to customary electronic networks, is not irreversibly wired but self-organizes continuously into ever-changing patterns depending on the task and the physiological context. This self-patterning is a direct consequence of a characteristic property of proteins, which is the ability to interact with each other and with other signaling molecules at any place in the cell. Thus, data-processing protein networks are continuously reorganized both spatially and temporally.

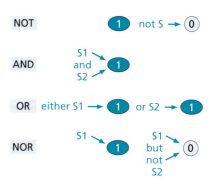

Figure 1.10 Proteins as logical gates The active protein (producing an output signal) is depicted in teal (state 1), and the inactive protein is shown in gray (state 0). A protein working as a NOT gate does not transmit an output when it receives an input signal S. An AND gate produces an output signal only when it receives two input signals S1 and S2 simultaneously (so-called coincidence detector). An OR gate responds either to S1 or to S2. To produce an output signal, a NOR gate must receive only signal S1 but not signal S2.

Summary

- Interacting proteins form logical NOT, AND, OR, and NOR gates, enabling the cell to process any kind of information.

- When the extreme flexibility and complexity of cellular data processing is considered, the computer metaphor appears to be questionable. It would be more appropriate to speak of the "brain of the cell."

1.6 Privacy versus publicity

Signals that transmit news may have a private character, where they address a discrete partner, or they may be directed at the public (**Figure 1.11**). This principle also holds true for inter- and intracellular signal transduction. As shown in **Figure 1.12**, signaling between cells may range from highly public to highly private. Thus, the endocrine system works via public long-range effects, while the nervous system operates mainly via private short-range (para- and autocrine) effects. The highest level

Figure 1.11 Private versus public signaling (Adapted from Sempe JJ [1968] Volltreffer. Deutscher Taschenbuch-Verlag.)

of privacy is reached when cells communicate via membrane-bound signaling molecules (juxtacrine signaling).

In an analogous way, proteins may interact with each other in a widespread or in a strictly targeted manner. Targeted short-range effects that maintain privacy are due to interaction domains, working like electronic contacts, while for widespread long-range effects, an enzymatically active protein produces a large amount of diffusible messenger molecules that interact, either covalently or noncovalently, with a high number of remote partner proteins (publicity). A characteristic example of long-range effects is provided by second messengers such as cAMP or calcium ions, which are released in large amounts and may diffuse throughout the cytoplasm upon stimulation of a few transmembrane receptors (see Chapter 3).

Figure 1.12 Different pathways of signal transmission between cells and molecules Long-range signaling between cells is called endocrine when the signal molecule is transported by the bloodstream (typical for hormones), paracrine when the signal diffuses between neighboring cells across the extracellular matrix (typical for neurotransmitters and many so-called tissue hormones or local mediators), and autocrine when the signal re-acts on the transmitter cell (typical for neurotransmission and developmental processes). Short-range signaling through a direct contact between cells via membrane-bound signaling proteins is called juxtacrine (typical for developmental processes). While endocrine signaling has a public character, para-, auto-, and juxtacrine signals are more private. Endocrine signals find their intracellular counterparts in signal molecules (such as second messengers), which are produced by an enzymatically active protein providing intermolecular long-range signaling by diffusing through the cytoplasm. The more private para and autocrine signaling resembles interactions that depend on a close encounter between the proteins. Protein phosphorylation provides an illustrative example. A direct signaling contact between proteins resembles juxtacrine signaling. The signals are symbolized by teal triangles.

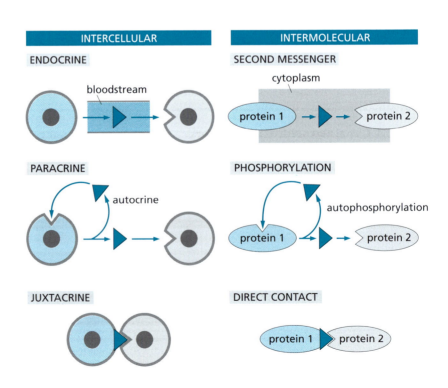

INTERCELLULAR	INTERMOLECULAR
ENDOCRINE	SECOND MESSENGER
bloodstream	cytoplasm
	protein 1 → protein 2
PARACRINE	PHOSPHORYLATION
autocrine	autophosphorylation
	protein 1 → protein 2
JUXTACRINE	DIRECT CONTACT
	protein 1 protein 2

Depending on the context, signaling through second messengers may have also a more private character. In fact, short-range interactions in subcellular networks (spatial privacy) are facilitated by potent mechanisms of signal extinction that restrict a signal to a very limited area of the cell. Moreover, certain proteins that are specialized to act as adaptors or scaffolds may bring interacting proteins into contact or dock them only at specific cellular sites (see Section 1.8; for some illustrative examples see Sections 4.6.1 and 11.4).

Another hallmark of biological signaling is temporal **transitoriness**. A neuronal signal, for instance, is short-lived in order to avoid overstraining the system. Long-lasting effects require, therefore, an oscillating signal that, at the same time, enables frequency-dependent modulation. This is a typical feature of signaling in the nervous system. Intracellular signal processing by protein networks is dominated by the same rule. Here a wide variety of inactivating mechanisms guarantees transitoriness. Moreover, signal-generating and signal-extinguishing proteins frequently are coupled to modular complexes that fulfill the conditions of molecular oscillators (see Section 1.9.3). Therefore, a wavelike pattern and a frequency-dependent modulation of intracellular signals may be assumed to be the rule rather than the exception (for details, see Sections 4.6.1, 11.6, and 14.5.4).

Summary

- By interacting with each other and with other cellular constituents, proteins form a neural network of logical gates. Network formation is based on the property of proteins to communicate with each other by both short- and long-range interactions (thus resembling organisms).

- Short-range effects are due to direct contacts (using specific contact or binding domains), while long-range interactions are transmitted by signaling molecules. Thus, subcellular signal processing resembles both widespread hormonal signaling and targeted neuronal signaling.

- Both neuronal and subcellular signal processing are dominated by the same principles: spatial privacy, temporal transitoriness, redundancy of the components, and frequency-dependent modulation of the signals (**Sidebar 1.1**).

Sidebar 1.1 Trivial and nontrivial calculators
The physicist Heinz von Foerster has clearly distinguished the differences between trivial and nontrivial calculating machines. A trivial machine such as an abacus uses a fixed algorithm, always converting a given input value into the same output value: it is reliable and predictable but unable to adapt and to learn. In contrast, a nontrivial machine uses an adaptable algorithm depending on the previous operations.

It is able to learn and has a memory. In the extreme case this means, however, that a nontrivial machine reacts chaotically, unpredictably, and unreliably. Since living systems have to survive in an unpredictable environment, they need to be adaptable and be able to learn. Their data-processing apparatus requires, therefore, a certain degree of nontriviality to enable the organism to balance on the narrow ridge between reliability and chaos and, thus, to make meaningful decisions. This holds true for proteins: each "arithmetical operation" leaves a track in the molecular conformation, which influences the forthcoming operations and renders proteins adaptable if they possess a structural memory. Since one never knows the previous operations with certainty, it is impossible to precisely predict the function of a protein. This nontrivial behavior of proteins is the basis of all biological data processing, culminating in the abilities of the human brain. Von Foerster has emphasized that any kind of education—but in particular military drill—is aimed at a reduction of nontriviality.

1.7 Cross talking and network formation

The nervous system is an extremely complex network of cells communicating via a large variety of para-, auto-, and juxtacrine interactions. This organization, being mandatory for a data-processing system, finds its counterpart at the subcellular level in the network of signal-transducing proteins, albeit with a lower degree of complexity.

Previously, only linear pathways of signal transduction connecting the cellular periphery with the metabolic and genetic apparatus could be analyzed. It became clear, however, that the picture thus obtained resembled an extreme oversimplification because in reality such pathways were found to be interconnected by a phenomenon called "signaling cross talk." In fact, this term is nothing but a synonym for signal processing by a protein network. It is now generally accepted that any signal received by a cell activates a substantial part of the data-processing protein network rather than stimulating just a simple sequence of a few biochemical reactions, a concept that, for an in-depth investigation, imperatively requires novel approaches such as that of systems biology (see Chapter 17). The situation closely resembles data processing in the brain, where a single sensory input induces an extended (diffuse) activation pattern rather than stimulating just a small group of specialized neurons (**Figure 1.13**).

Impressive microscopic photographs and three-dimensional models showing the beauty and complexity of the neuronal network in the brain have given a vivid impression of the cross talk between nerve cells. For molecules, such a visualization is not yet possible. Instead we have to fall back on more indirect evidence. We know, for instance, that cytoskeletal and membrane structures play a key role in support of the signal-processing protein network. Thus, the complexity of such hardware, or scaffolding, may reflect the complexity of the "cellular brain" to a certain extent (**Figure 1.14**).

As far as teaching is concerned, signaling cross talk creates a problem because for the sake of clarity it is impossible to depict and describe all the interactions that occur when a cell receives a signal (**Figure 1.15**). So in textbooks—including this one—still more or less linear, artificial signaling cascades are presented. However, the reader should always keep in mind that this is an oversimplification, illustrating mechanistic principles rather than the complex reality.

1.7.1 Network biology

A new discipline of theoretical biology known as network biology is emerging that deals with the rules and laws dominating the formation and function of molecu-

Figure 1.13 Diffuse signal processing by the human brain In the brain, the processing of even rather simple sensory inputs results in extended and diffuse excitation patterns involving a large number of neurons and developing on the basis of apparently chaotic background activity. At the molecular level, an analogous situation has to be assumed for cellular signal processing. The excitation patterns were visualized by a method measuring the blood circulation. Albeit providing more details, advanced techniques of imaging principally have confirmed this result. (Adapted with modifications from Fischbach GD [1992] *Sci Am* 267:48–57.)

hearing words

reading words

speaking words

generating words

(A)

1 μm

(B)

1 μm

Figure 1.14 The cytoskeleton: hardware support of the data-processing protein network The photographs show the extremely complex structure of cytoskeletal fibers in the neighborhood of a mitochondrion (A) and in the border zone between cytoplasm and nucleus (B). (A, adapted from Darnell JE, Lodish HF & Baltimore D [1990] Molecular Cell Biology, 2nd ed. Scientific American Books. B, adapted from Penman S [1995] *Proc Natl Acad Sci USA* 92:5251–5253.)

lar networks. Combining molecular biology, biophysics, and mathematics, this research has focused on five types of network: metabolic, genetic, transcription factor, protein interaction, and protein phosphorylation. These approaches are strongly facilitated by novel high-throughput techniques that use cDNA microarrays, protein chips, and other automated methods to allow genome- and proteome-wide screening. Moreover, a conceptual framework has been developed to characterize different types of network (see Chapter 17).

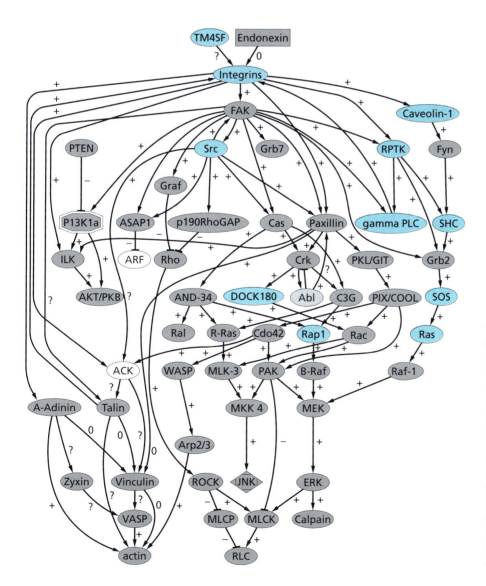

Figure 1.15 The integrin signaling network The figure shows the complex signal processing network activated by a single membrane receptor (integrin) that interacts with two different extracellular signals. It must be emphasized that the sketch depicts only a small fraction of what is expected to happen in reality. The situation is exemplary for most signal-processing events. (Adapted from Martin KH, Slack JK, Boerner SA, Martin CC & Parsons JT. Integrin signaling pathway. *Sci Signal* (connection map in the database of cell signaling).)

Using simple organisms such as yeast, one can generate computerized graphs that show an illustrative, albeit rough, survey of intermolecular crosstalk and allow the discovery of hitherto-unknown interactions, at least on paper. There is no doubt that a more systematic understanding of these networks is urgently needed. However, network biology is still in its infancy, particularly because of the difficulty in uncovering the physiological role of the interactions described. Therefore, one must wait and see whether this new discipline will significantly broaden our understanding of cellular phenomena, especially of signal processing, or—remaining on a descriptive level—if it will only provide a new mathematical formalism. In the following section, as well as in Chapter 17, the reader will be able to glimpse the problems network biology is facing.

1.7.2 Some remarks on the complexity of the "cellular brain"

When compared with the size of a genome, the number of genes encoding signaling proteins is rather modest. Thus, analysis of the human genome has shown that, of the approximately 30,000 genes, about 1500 encode signal receptors, 500 encode protein kinases, 150 encode protein phosphatases, and 1500 encode transcription factors, just to mention some of the most common switching elements of the data-processing protein network. Most of these genes contain several (up to 200) exons, which means that transcribed pre-mRNA probably undergoes alternative splicing. As a rule, the proteins encoded by such splice variants exhibit different properties such as ligand affinities, intracellular distributions, enzymatic activities, and stabilities. Arbitrarily assuming at least three different splice variants are produced per gene, one arrives at 4500 different signal receptors, 1500 protein kinases, 450 phosphatases, and at least 4500 transcription factors.

Each of these proteins is expected to undergo covalent post-translational modifications and so its properties are changed further. Let us again assume that three different modifications per protein occur on average (which certainly is an understatement). These modifications produce $2^3 = 8$ different states, since each modification exists in two states, either present or absent. This calculation increases the number of receptor variants to 36,000, kinase variants to 12,000, phosphatase variants to 3600, and transcription factor variants to 36,000, thus greatly outnumbering the total number of genes. The degree of complexity is furthermore enlarged dramatically by the understanding that most signal-processing proteins exist as homo- and hetero-oligomers rather than as monomers and that the number of functional states of a network increases exponentially with the number of the components. In fact, in a single cell the number of dynamic states generated by the interactions between signal-processing proteins must be astronomically high.

It must be emphasized, however, that in a given tissue cell only a fraction of these variants becomes realized. The pattern of active signaling proteins thus generated changes continuously, depending on the state of differentiation and the actual physiological context. This is what we have called the rearrangement of the hardware components and actual state of the random access memory. The overwhelming complexity of the "cellular brain" indicates that our efforts to understand life are still in their infancy.

Summary

- The data-processing network of the cell is based on chemical interactions or crosstalking between the components. Therefore, a single signal received by a cell evokes a diffuse excitation pattern rather than activating one or a few signaling pathways.

- As yet, the extreme complexity and flexibility of such networks prevent a more in-depth analysis of their physiological functions.

1.8 Interaction domains: how the network is plugged together

Proteins are assembled into highly organized signal-transducing complexes and data-processing networks. This occurs on demand in a fully reversible manner. How is this achieved? The original concept of proteins as freely diffusible molecules encountering each other more or less by chance is considered to be outdated today. Owing to the extremely high concentration of macromolecules (known as macromolecular crowding), the interior of a cell needs to be viewed as a semi-solid medium in which free diffusion would be far too slow for life processes. In addition, neither a precise arrangement nor a controlled interaction of the components would be possible. However, both are mandatory for cellular data processing. Thus, protein interactions must be brought about by other more specific and more targeted processes.

1.8.1 Protein modules

Complex formation is a direct consequence of a characteristic structural feature known as the **modularity of proteins**. Proteins are composed of defined amino acid sequences or modules according to a unit construction system. Generally one may distinguish between functional and interaction modules. A functional module is, for example, the catalytic domain of an enzyme, whereas interaction modules serve as contact sites for all kinds of allosteric regulators—including proteins with complementary contact modules, diffusible signaling molecules such as second messengers, and systemic and environmental factors—as well as for cellular structures. Interaction modules guarantee relatively stable (though mostly noncovalent) binding between the partner molecules. In signal-transducing proteins, the interaction modules are identical with the regulatory domains that serve as terminals of input signals, while functional modules are the catalytic domains that generate output signals (see Figures 1.5, 1.6, and 1.8). As a rule, signal transducing proteins have many more interaction modules than other proteins such as metabolic enzymes. In the course of evolution, the protein modules were recombined into new patterns by genetic rearrangements. Accordingly, innovations in signal processing (resulting in novel properties of the organism) are brought about by unique combinations of existing building blocks rather than by invention of entirely new protein structures. This mechanism of evolvability is a major driving force of complexity. By means of such a combinatorial strategy, cells establish the astronomically large number of protein contacts by using relatively few interaction domains.

The pattern of protein domains represents something like a signaling syntax, and the highly organized multimodular structures and multiprotein complexes thus formed resemble microprocessors to a certain extent. Nevertheless, and in contrast to a computer, proteins are not firmly interconnected but are assembled and disassembled upon demand by energy-consuming reactions. This reversible wiring is promoted, in particular, by covalent post-translational modifications of protein modules, generating ad hoc new sites for inter- and intramolecular contacts that immediately become untied when they are no longer needed. By this means, new interaction patterns are continuously formed between proteins, just as one might expect for an adaptable and educable neural network. The cooperation of the individual domains in such patterns lends protein interactions the necessary precision.

The principles of domain interactions are depicted in **Figure 1.16**, while in **Table 1.2**, selected interaction domains are summarized and arranged according to the recognition sites of their interaction partners. The first group includes domains that recognize short peptide sequences. The domains of the second group interact with peptide sequences containing covalent post-translational modifications due for instance to phosphorylation, acetylation, methylation, and ubiquitylation. Interaction domains not listed in the table include binding sites for extra- and intracellular signal molecules, found in receptors and in many other signal transducing proteins, as

Figure 1.16 Interaction domains The figure shows interaction domains of proteins that are binding sites for other protein molecules carrying a specific recognition signal. To the left, an interaction domain is depicted that recognizes a heterotypical signal, for instance, a post-translational protein modification. In the middle, a heterotypical interaction is shown occurring in the same protein molecule. The scheme to the right shows a homotypical interaction between two identical interaction domains. Other interaction domains recognize low molecular mass messenger molecules, specific nucleic acid sequences, and membrane lipids. Receptors exhibit an enormous variety of interaction domains for intercellular signals and environmental stimuli.

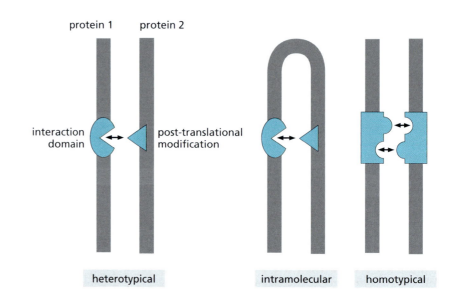

well as for nucleotide sequences, found in transcription factors and other DNA- or RNA-binding proteins.

Domains interacting with post-translational modifications (Figure 1.16) recognize the modified amino acid together with a relatively short amino acid sequence in the immediate neighborhood. These flanking areas may differ from protein to protein, lending the interactions additional specificity, especially since the cognate interaction domains are also variable. Thus, a protein with a distinct recognition site does not simply interact with any other protein possessing the corresponding interaction domain but strictly selects its partner molecule.

This selectivity of interactions is impressively demonstrated by receptor tyrosine kinases. Upon ligand binding, these dimeric transmembrane proteins undergo trans-autophosphorylation of several Tyr residues in their cytoplasmic domains, thus generating contact sites for proteins with SH2 and PTB domains (see Table 1.2). Each of those proteins has a special SH2 or PTB variant that recognizes only one of the various phospho-Tyr residues on the receptor. In other words, each interacting protein finds its correct place on the receptor protein (a detailed discussion of this subject is found in Section 7.1.2). Conversely, the same recognition site may react with several *different* interaction domains. For instance, SH2 domains can interact with phospho-Tyr residues and sometimes also with SH3 domains, and PDZ domains may interact with each other (homotypically) as well as with carboxy-terminal amino acid sequences (see Table 1.2 and Section 7.1.4).

Figure 1.16 distinguishes between domains that interact with nonrelated domains (heterotypically) and those that interact with their own type (homotypically). Examples of domains interacting exclusively in a homotypical manner are the death domains DD and DED. They play a key role in apoptotic signaling (see Section 13.2.2). Another group of interaction domains recognizes membrane components such as phospholipids. Such interactions enable reversible binding of proteins to membranes and generation of signal-processing protein complexes in the neighborhood of membrane receptors. Many membrane lipids are, in addition, signaling molecules or second messengers that control the function of the interacting protein.

1.8.2 Intramolecular interactions

Interactions between regulatory domains may also occur within the same protein molecule, with various consequences for protein function. Thus, one interaction domain may either suppress or promote the function of another, like an antagonist or

Table 1.2 A selection of interaction domains

Interaction partner	Interaction domain
Group 1: the recognition site is a short amino acid sequence	
Pro-x-x-Pro	SH3
Arg-x-x-Lys	SH3
Pro-Pro-x-Tyr	WW
x-y-z-Val-COOH or -Leu-COOH	PDZ
Group 2: the recognition site is a short amino acid sequence containing a covalent modification	
phospho-Tyr	SH2, PTB
phospho-Ser	14-3-3, WW, MH2, etc.
phospho-Thr	FHA, WD 40
acetyl-Lys	bromo
methyl-Lys	chromo
ubiquityl-Lys, polyubiquityl-Lys	UIM, UBA, CUE, UEV, PAZ, NZF, etc.
Group 3: the recognition site is a membrane component	
diacylglycerol	C1
phosphatidylserine	C2
phosphatidic acid	C2
phosphoinositides, Gβγ subunits	PH
3-phosphorylated phosphoinositides	FYVE
3-phosphorylated phosphoinositides	PX
Group 4: the interaction domain is identical with the recognition domain (homotypical interaction)	
death domain (DD)	DD
death effector domain (DED)	DED
CARD	CARD
PDZ	PDZ
SAM	SAM

an agonist, or two different interaction domains may become occupied sequentially to induce a specific function. It is easily conceivable that such interactions provide logical gates.

Frequently, intramolecular interactions stabilize special protein conformations. This provides a general mechanism to keep a signaling protein in an inactive conformation (or vice versa in an active one). Such an intrasteric blockade is broken up by an input signal that either competes with one of the domains or removes the interacting residue (if it is a covalent modification such as a phosphate group) or induces a conformational change by binding to a third domain (**Figure 1.17**). One immediately recognizes that the lifting of an intramolecular blockade resembles a reversible switching process such as those depicted in Figures 1.5 and 1.6. Examples of this central theme of cellular signal processing are found throughout this book (see Sections 2.6.1 and 7.1.2).

1.8.3 Docking domains

A dogma of biochemistry states that to become recognized by an enzyme, a substrate has to bind stereospecifically to the catalytic center. However, in contrast to metabolic enzymes, most signal-transducing proteins are able to interact with a

Figure 1.17 Conformational change by lifting an intramolecular interaction Owing to an interaction between domains 1 and 2, a protein is kept in a closed conformation, which in this example is assumed to be equivalent to an inactive state (domain 1 represents an active center). There are at least three ways (representing input signals) in which the closed conformation may become unfolded so that the protein becomes activated: enzymatic removal of a critical residue, generally a covalent amino acid modification, in domain 2; interaction with a competing protein exhibiting a binding domain for domain 2; or covalent or noncovalent allosteric interaction of a signaling molecule S with a regulatory domain 3. S may be a low molecular mass factor such as a second messenger, a membrane lipid, a protein, a nucleic acid, or a covalent modification such as phosphorylation.

wide variety of substrates. This is typical, for example, for the majority of protein kinases. For such proteins the conventional mechanism of substrate recognition would create a problem of specificity. Therefore, they frequently exhibit additional substrate binding sites that with a high degree of selectivity recognize a docking motif of the substrate protein that is distinct from the site interacting with the catalytic center. Such docking domains not only confer specificity but also help the protein to find its substrate. In other words, they may have a stimulatory effect on the signaling event. An example is discussed in Section 11.1.

Like other interaction motifs, certain docking domains may be generated as needed by covalent post-translational modifications of amino acid residues. Such a mechanism renders the process of enzyme–substrate encounter as an adjustable target of input signals and logical AND gates. Examples are provided by the so-called priming of protein kinases and substrates of ubiquitin ligases by phosphorylation (see Sections 2.8.1, 4.6.4, and 9.4.4).

1.8.4 Adaptor and scaffold proteins

Since the assembly of multiprotein complexes at specific cellular sites would require proteins with many different docking domains, transferring the job of targeting and assembling to a third party would dramatically increase the system's degree of freedom. This might have been the evolutionary pressure for the development of adaptor and scaffold proteins, which dominate signal processing, particularly in eukaryotes (**Figure 1.18**). Adaptor proteins are composed mainly of interaction domains and frequently lack functional domains. As "molecular plugs," they fit together other proteins to form functional units. By use of a single adaptor, a cell may interconnect several receptor types with the same intracellular effector mechanism (see, for example, Sections 3.4.4 and 6.3).

From adaptor proteins it is only a short step to the scaffold(ing) proteins that fulfill the function of a platform in that they arrange other proteins into specific switching patterns, thus enabling highly selective interactions. With the reservation that differently from technical devices scaffold proteins have a flexible and "soft" structure (see below) these multiadaptors can be compared most closely with the sockets of electronic devices. They are extremely important for signal processing since they establish new but reversible connections between signaling pathways, at the same time giving them temporal and spatial specificity. Examples of scaffold proteins are found throughout the book (see Section 11.1).

Figure 1.18 Adaptor and scaffold proteins By means of interaction domains, adaptor and scaffold proteins plug together other proteins to form functional units and multiprotein complexes.

1.8.5 Intrinsically disordered proteins

A dogma that has been kept up for a long time is that a native or functional protein has a well-defined and fixed three-dimensional structure (folding geometry) that is fully determined by the amino acid sequence.

However, as will be explained later (Sections 2.5.1 and 8.5), the native state—defined as the state of lowest energy—only in a few cases is formed spontaneously in the course of protein synthesis. Moreover, it is by no means always identical with the functional state. So, during biosynthesis many proteins need the assistance of helper proteins, the chaperones, to acquire their functional conformation. This has turned out to be a borderline case since the concept of rigid protein conformations has been generally challenged by the discovery of intrinsically disordered proteins that partly or completely lack a defined three-dimensional structure. Such proteins are functionally active and must not be confused with denatured proteins prone to be eliminated as cellular waste. In eukaryotes up to 50% of all proteins may be intrinsically disordered. They are, in particular, involved in cellular signaling. Examples for proteins with a rigid ordered structure are, on the other hand, metabolic enzymes. The percentage of intrinsically disordered proteins increases with the complexity of the organism. Their major role is molecular recognition and complex formation. Both processes are greatly facilitated by disorder ensuring maximal structural and functional flexibility. Many intrinsically disordered proteins exhibit promiscuity: through multiple disordered domains they are able to interact with a wide variety of quite different partner molecules. An example is provided by the transcriptional co-activator CBP/p300 (see Section 2.7 and Section 8.2.2, Figure 8.7). This multidomain protein can recruit up to 20 different proteins in signal-processing complexes. Their property to function as versatile scaffolds and hubs makes intrinsically disordered proteins indispensable for the formation of signaling networks.

Most intrinsically disordered proteins acquire an ordered functional three-dimensional structure upon contact with a partner molecule, a process called **induced folding** or disorder-to-order transition (**Figure 1.19**). This phenomenon challenges the conventional lock-and-key principle of signaling interactions since it indicates that such interactions may follow a "fuzzy logic" rather than a rigid Yes-or-No pattern. According to the second law of thermodynamics, any disorder-to-order transition requires energy. Thus, conventional chaperone effects frequently are coupled with ATP hydrolysis. Differently, the induced folding of most intrinsically disordered proteins is fueled solely by the free energy released by the association with the partner molecule. Since this energy is quite low, the interaction is—albeit highly specific—weak and quickly reversible. Such a combination of low binding affinity with high specificity is essential for the formation of signaling complexes that has to occur rapidly and transitorily at a short timescale down to milliseconds. Again the reader may recall the analogy between the nervous system and the signal-processing protein network of a cell.

Figure 1.19 Induced folding of an intrinsically disordered protein By interaction with a binding partner the intrinsically disordered protein or protein domain (teal) reversibly acquires an ordered three-dimensional structure. (Adapted from Dyson JH & Wright PE [2005] *Nature Rev Mol Cell Biol* 6:197–208.)

1.8.6 The complexity conundrum

The modular composition of proteins is a fundamental feature of biological evolution. Many interaction domains are, therefore, already found in prokaryotes, while others have developed later (such as, for example, the SH2 domains simultaneously with the appearance of tyrosine phosphorylation in multicellular animals). In the post-genomic era, this evolutionary conservatism provides a basis for a census of signaling proteins in that by computerized genome-wide analysis the total number of proteins with related domains and putatively related functions (such as protein kinases, G-proteins, AAA+ proteins, transmembrane proteins) can be determined. Such studies demonstrate that the collection of eukaryotic interaction domains is predetermined in simple uni- and multicellular organisms such as yeast and the nematode *Caenorhabditis elegans* and has been enlarged only marginally in humans. Dramatically enlarged, however, is the number of *combinations* of such domains in a protein molecule, that is, the complexity of modular protein structure. For example, sequences encoding the SH3 domain have been found in 31 yeast, 132 nematode (*C. elegans*), and 273 insect (*Drosophila*) genes but in 894 human genes. In fact, the principle of (almost) free mixing of protein modules enables biological evolution to constantly construct new data-processing circuits, thus climbing the ladder of complexity, although the number of basic units remains rather limited. Moreover, in the course of evolution, an interaction domain may have split into a wide variety of isoforms that differ in their properties such as binding specificity.

This tendency of combinatorial evolution, which also becomes apparent in other processes such as the alternative splicing of pre-mRNA, may explain what has been called the complexity conundrum: why the size of the human genome (20,000 genes) has about the same size as that of the mouse and differs so unexpectedly (not to say disappointingly) little from that of the tiny and apparently primitive worm *C. elegans* (19,000 genes), the small insect *Drosophila melanogaster* (14,000 genes), or even a weed such as *Arabidopsis thaliana* (26,000 genes). Note that the emergence of molecular complexity from a limited number of characters resembles the evolution of a language.

Summary

- The hardware basis of cellular data processing is the modularity of protein structure, consisting of specific amino acid sequences that fulfill the role of communicative and functional domains. In the course of evolution, these modules have been combined in patterns of ever-increasing complexity.

- Signal-transducing proteins differ from metabolic enzymes and structural proteins because they have a particularly high number of interaction domains that receive input signals or enable other kinds of specific interactions (such as substrate and cellular targeting and intramolecular interactions).

- Interaction domains are either intrinsic to protein structure or formed on demand by induced folding and post-translational modifications of amino acid residues.

- Adaptor and scaffold proteins are specialized for assembling proteins and other factors into signal transducing complexes. Owing to a high degree of intrinsic disorder they are particularly versatile and flexible.

1.9 Generation of signaling patterns

In this section we examine some mechanisms used by signal-processing proteins to generate a wide variety of signaling patterns with important consequences, particularly for the type and pattern of the output signal and the stability of the signaling system. The output signal may resemble a pulse or it gradually increases or decreases like a sound growing louder or becoming softer, and exhibits either a continuous or an oscillatory pattern. Stability is understood as resistance to perturbations and as the ability to respond over a prolonged period of time without running the risk of becoming desensitized by overstraining. In technology such properties are considered to be a feature of high performance. For biological systems they are a matter of survival.

Perturbation insensitivity or **robustness** can be improved by adaptation, noise filtering, redundant components and mechanisms (see also Chapter 17), and temporal stability of signaling by oscillatory behavior. While the benefits of redundancy have already been addressed in Section 1.3, noise filtering, adaptation, and oscillatory signaling require a somewhat more detailed treatment.

1.9.1 Switches and regulators

Mathematical analysis of biochemical reactions, carried out on the basis of the fundamental laws of chemical kinetics (see Chapter 17), supplies a series of simple network components or circuits, the functions of which are described by a set of nonlinear differential equations. Examples of these circuits are depicted in **Figure 1.20**. In cells such elementary units are combined into highly sophisticated devices that, if one likes such a metaphor, may be looked upon as biochemical microprocessors.

Provided a protein exists in two states ON (1) and OFF (0), which are converted by a signal S (an allosteric effector), the relationship between signal intensity and response is linear or hyperbolic for a monomeric protein and sigmoidal for a protein consisting of allosterically cooperating subunits (see Figure 1.20A). Sigmoidal behavior resembles that of a switch that requires some pressure to overcome its built-in resistance to accidental triggering. This resistance is equivalent to a **noise filter**. In contrast, linear or hyperbolic behavior resembles that of a sliding regulator such as a throttle.

Switch-like devices with noise filters are also established by **positive feedback**, for instance, when the signal-receiving protein activates itself or stimulates an activating factor. In the beginning the autostimulation increases slowly and a rise of signal intensity has no significant consequence. However, when the signal intensity surpasses a critical value (symbolized by the arrows in Figures 1.20B and 1.20C), the response suddenly reaches its maximal value and the protein activity (that is, the output signal) becomes independent of the input signal S. The result is

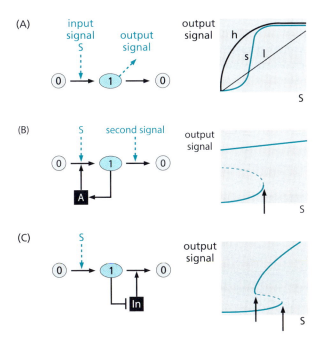

Figure 1.20 Simple biochemical switching elements An input signal S transfers a protein from an inactive state 0 (gray circles) into an active state 1 (teal ellipses) that transmits an output signal and returns to state 0. (A) Simple switch with—depending on the structure of the protein and the reaction type—linear (l), hyperbolic (h), or sigmoidal (s) characteristics. (B) State 1 activates an activator A that promotes the activation of 0 by S. This switch has a discontinuous characteristic and the protein remains active unless it is inactivated by a second signal. (C) State 1 inactivates an inhibitor In that promotes the re-switching of 1 to 0. Again the characteristic of this switch is discontinuous, but in contrast to case B, it can re-switch to 0 by its own.

a discontinuous signal–response curve (see Figure 1.20B). In principle, the process is irreversible unless an additional signal switches the protein back to the 0 state. Autophosphorylation of the insulin receptor serves as an example. It is induced by the insulin signal and results in an insulin-independent long-term activation of the receptor until it is abolished by a phosphoprotein phosphatase activated by a second signal (details in Chapter 7). Without countercontrol, positive feedback systems tend to develop in an explosive and catastrophic manner.

A discontinuous function resembling a (reversible) toggle switch is obtained when the positive feedback is due to the abolition of an inhibitory effect, that is, when the signal-activated protein inhibits or destroys an inhibitor (see Figure 1.20C). Such a case, resembling the arithmetical instruction "**minus times minus is equal to plus**" (double negation), is particularly abundant in biological systems because it works with pre-formed components and, therefore, guarantees a particularly quick and robust response. Moreover, it again provides a highly efficient noise filter. Examples of this regulatory principle are found throughout this book (for a particularly clear one, see Section 5.9).

1.9.2 Oscillators

Oscillating systems open up the possibility of frequency-dependent modulation of signals. When the input signal S not only activates the protein but also stimulates a factor that sets the protein switch back to 0, the signal–response curve shows a **dampened oscillation** resembling an **adaptation by negative feedback** (**Figure1.21A**). In Section 3.4.4 a classical example, the chemotactic reaction of bacteria, is described. Another prominent case is the adaptation of sensory organs (see Chapter 15).

Adaptation contributes to robustness. Moreover, it is a fundamental prerequisite to learning and a critical condition for a signal-receiving device to respond over a wide range of signal intensity with constant sensitivity, thus being able to track down and pursue the source of the signal.

To generate an **undampened oscillation**, a combination of positive and negative feedback (or a twofold negative feedback) is required. To this end, at least three proteins have to interact (**Figure 1.21B**).

Such systems convert into oscillators at intermediate signal intensities. They are suited best for long-term signaling. An example is the oscillating cAMP signal that induces cell aggregation and differentiation of the slime mold *D. discoideum*. cAMP stimulates, via a cellular receptor, its own production (positive feedback), while at higher concentrations it inactivates its receptor and simultaneously promotes its own degradation by stimulating a cAMP phosphodiesterase (negative feedback). Another example is provided by the oscillations of intracellular Ca^{2+} signals, representing

Figure 1.21 Two biochemical oscillators As in Figure 1.20, the activation (0 → 1) and deactivation (1 → 0) of a signal-controlled protein is depicted. Shown are two systems generating either a dampened or an undampened oscillation. (A) The signal S simultaneously activates an inhibitor In that re-switches 1 to 0. This results in a dampened oscillation of the output signal. (B) In addition to the situation depicted in (A) the protein activated by the signal S stimulates its activation via an activator A (positive feedback). As a result, an undampened oscillation of the output signal may develop within a certain intensity range of the input signal S.

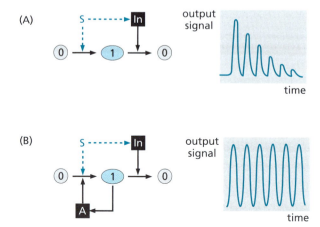

one of the most impressive cases of an oscillating signal (details in Section 14.5.4). Here the positive component is a Ca^{2+}-induced Ca^{2+} release from the endoplasmic reticulum, whereas the negative component consists of an inhibition of this release and an activation of Ca^{2+} pumps at higher Ca^{2+} concentrations. When such systems consist of many units (for instance, slime mold cells or Ca^{2+}-transporting membrane channels) and the inhibitory field is considerably more extended than the stimulatory field, the output signal may develop into a time independent **spatial pattern**. In the reverse case, when the inhibitory field is the smaller one, the output signal spreads temporally and spatially as a **wave**. It is easily conceivable that such systems play an essential role in morphogenesis, with *Dictyostelium* differentiation being a well-known example. Moreover, a steadily increasing body of evidence indicates that wavelike patterning and frequency modulation of signals are general hallmarks of subcellular data processing. Thus a cell may be considered to represent a "musical" rather than only a trivial "chemical" system!

Summary

- By combining negative and positive feedback loops, interacting proteins may self-organize into complexes resembling technical devices that transform input signals into special patterns of output signals.

- Circuits resembling double negation according to the rule "minus times minus is equal to plus" are particularly abundant. Of special interest are oscillating systems, since they enable frequency-dependent encoding of signals.

Further reading

Sections 1.1 and 1.2

Bray D (2009) Wetware. A Computer in Every Living Cell. Yale University Press, New Haven and London.

Sections 1.3 and 1.4

Changeux JP & Edelstein SJ (2005) Allosteric mechanisms of signal transduction. *Science* 308, 1424–1428.

Lavoie H et al. (2014) Dimerization-induced allostery in protein kinase regulation. *Trends Biochem Sci.* 39:475–468.

Lim WA (2002) The molecular logic of signaling proteins: building allosteric switches from simple binding domains. *Curr Opin Struct Biol* 12:61–68.

Sections 1.7 and 1.8

Bhattacharyya RP, Remenyi A, Yeh BJ et al. (2006) Domains, motifs, and scaffolds: the role of molecular interactions in evolution and wiring of cell signaling circuits. *Annu Rev Biochem* 75:655–680.

Bienz M (2014) Signalosome assembly by domains undergoing dynamic head-to-tail polymerization. *Trends Biochem Sci* 39:487–495.

Flynn DG (2001) Adaptor proteins. *Oncogene* 20:6270–6272.

Papin JA, Hunter T, Palsson BO et al. (2005) Reconstruction of cellular signaling networks and analysis of their properties. *Nat Rev Mol Cell Biol* 6:99–111.

Pawson T & Nash P (2003) Assembly of cell regulatory systems through protein interaction domains. *Science* 300:445–452.

Pufall MA & Graves BJ (2002) Autoinhibitory domains: modular effectors of cellular regulation. *Annu Rev Cell Dev Biol* 18:422–458.

Seet BT, Dikic I, Zhou MM et al. (2006) Reading protein modifications with interaction domains. *Nat Rev Mol Cell Biol* 7:473–483.

Tompa P (2012) Intrinsically disordered proteins: a 10-year recap. *Trends Biochem Sci* 37:509–516.

Uversky VN (2011) Intrinsically disordered proteins from A to Z. *Int J Biochem Cell Biol* 43:1090–1103.

Wright PE & Dyson HJ (2015) Intrinsically disordered proteins in cellular signalling and recognition. *Nat Rev Mol Cell Biol* 16:18–29.

Zhu X, Gerstein M & Snyder M (2007) Getting connected: analysis and principles of biological networks. *Genes Dev* 21:1010–1024.

Section 1.9

Ferrell JE & Ha SH (2014) Ultrasensitivity part III: cascades, bistable switches, and oscillators. *Trends Biochem Sci* 39:612–618.

Housden BE & Perrimon N (2014) Spatial and temporal organization of signaling pathways. *Trends Biochem Sci* 39:457–464.

Tyson JJ, Chen KC & Novak B (2003) Sniffers, buzzers, toggles and blinkers: dynamics of regulatory and signaling pathways in the cell. *Curr Opin Cell Biol* 15:221–223.

Supplying the Network with Energy: Basic Biochemistry of Signal Transduction

2

As discussed in Chapter 1, cellular signal transduction is first of all due to biochemical reactions that mediate the interactions within the protein network while at the same time supplying the energy required for data processing. This chapter describes the most abundant and important switching reactions, and will provide the reader with an "Ariadne's thread" to help find a path through the maze of details discussed throughout the rest of the book.

2.1 No order without work

The goal of data processing is to minimize the level of uncertainty. This is synonymous with creating order or a higher level of organization, which thermodynamically resembles a lowering of entropy in the system (**Sidebar 2.1**). According to the second law of thermodynamics, such a task requires energy. Data-processing systems, thus, depend on work, or on a permanent energy supply (**Figure 2.1**): a computer has to be continuously supplied with electricity and a brain with metabolic energy. The same condition applies to the signal-processing protein network of the cell. In fact, the overall theory is that signal transduction reflects a fundamental principle of life: namely, that a process of self-organization (the emergence of order) must always be coupled with another process that provides energy by producing disorganization or chaos. But what is going on at the molecular level?

We return to the concept of a protein resembling a simple binary switch that can be switched on or off (being aware of the fact that this is a borderline case) (**Figure 2.2**). Such a switching process resembles a chemical reaction that is not in equilibrium. This means that when the forward reaction is exergonic (releasing free energy and thus occurring spontaneously), the reverse reaction necessarily must be endergonic

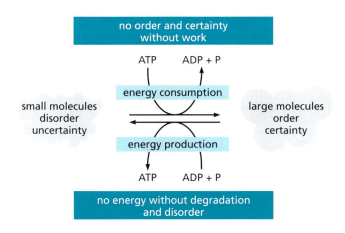

Figure 2.1 **Energy processing by living systems** Biological energy processing is a direct consequence of the second law of thermodynamics. To synthesize large molecules (such as sugars, lipids, proteins, and nucleic acids), maintain a highly ordered state, and minimize the degree of uncertainty, an organism consumes energy, first and foremost by hydrolyzing ATP. ATP is produced (energy is stored) by degrading large molecules (taken up as food) and generating disorder in the environment. Therefore, organisms are called "dissipative systems."

Sidebar 2.1 A living cell resembles a charged battery
The second law of thermodynamics states that, in a *closed* system (such as the universe), disorder measured as entropy increases continuously. In contrast, cells and organisms are *open* systems that, for a certain period of time (the lifespan), are able to struggle against this verdict, albeit at the expense of their environment. This phenomenon is explained by chemical thermodynamics. A chemical reaction either delivers or requires energy. Only a certain portion of this energy, traditionally called "free energy," can be used for work—transformed into chemical, mechanical, electrical, or electromagnetic energy—whereas the rest is lost as heat, increasing the disorder. The energetic turnover of a reaction occurring at constant temperature and constant pressure (as in cells) is described by the **Gibbs–Helmholtz equation**:

$$\Delta H = \Delta G + T\Delta S$$

where ΔH is the change of total energy, ΔG is the change of free energy, ΔS is the change of entropy, and T is the absolute temperature. The term $T\Delta S$ represents the "bound" energy that unavoidably is processed as heat. According to $T\Delta S = \Delta H - \Delta G$, a reaction occurs spontaneously at negative values of ΔG (indicating a loss of free energy by the system), because then the entropy and thus the degree of disorder increases. A positive value of ΔG means, on the other hand, that the reaction needs the supply of free energy because the entropy decreases. Such a reaction does not occur spontaneously.

Since the dimension of G and H is kilocalories per mole, the energy status of a reaction depends on the concentration of the molecules involved. Thus the ΔG value of ATP hydrolysis is about −12 kcal/mol under normal cellular conditions (low ADP concentration) but becomes much less negative, even approaching 0, when the ratio of ATP to ADP concentrations decreases in extreme stress situations, exhausting the cell's energy store.

In fact, depending on the conditions, ATP hydrolysis and any other reaction has the tendency to reach a thermodynamic equilibrium. In this situation ΔG is equal to 0. This means that the reaction has suffered from "heat death" and has become useless as an energy-delivering process. Therefore, the biochemical status of a living cell represents a system far out of thermodynamic equilibrium, thus disposing of a large amount of free energy that is used to maintain a state of high order. Such a state requires a continuous supply of free energy from the environment. This free energy is hidden in chemical bonds (or, in plants, derived from sunlight), for instance in food, where it is released by metabolism and stored in energy-rich cellular compounds such as ATP and macromolecules. In other words, a living cell resembles a battery that continuously has to be recharged at the expense of its environment, whereas in a dead cell the battery has been irreversibly discharged. Systems that maintain a high degree of organization by producing disorder in their environment are called "dissipative." For his pioneering work on non-equilibrium thermodynamics and dissipative systems, Ilya Prigogine was awarded the 1977 Nobel Prize in Chemistry.

Figure 2.2 Switching element consisting of two interacting proteins An interaction between the proteins A and B takes place only in state 1. Input signals enable the switching process (in both directions), while the activity of the AB complex resembles the output signal. According to the rules of chemical thermodynamics, one of the two reactions is exergonic with the reverse reaction being necessarily endergonic (requiring an energy supply). As a rule, exergonic reactions pass off spontaneously, though frequently very slowly, due to a high activation energy. In such a case the input signal has a catalytic effect; that is, it lowers the activation energy. When the reaction is endergonic, the input signaling is coupled with an energy-delivering process.

(needing an energy supply), or vice versa. Therefore, a protein switch must be coupled with an energy-delivering reaction, because otherwise it could be switched into one direction only but not back. Moreover, in a network of switching units, the input signal becomes dispersed rapidly and fades away; therefore, it has to be permanently amplified. Finally, the targeting of the signal requires energy derived from irreversible, strongly exergonic reactions. We have to conclude, therefore, that all cellular switching processes are coupled with biochemical reactions delivering a

Figure 2.3 Most frequent noncovalent and covalent switching reactions of cellular signal processing The reactions summarized here lead to noncovalent changes or covalent post-translational modifications (marked by asterisks) of signal-transducing proteins. They depend on or are coupled with energy-supplying processes such as ATP hydrolysis (gray box) or redox reactions.

large amount of energy. These are, in particular, redox reactions, the hydrolysis of nucleoside triphosphates (NTPs) and other energy-rich compounds, the degradation of macromolecules, and ion flows along electrochemical gradients across membranes (**Figure 2.3**).

2.1.1 Post-translational modifications

Frequently the energy-supplying biochemical reactions coupled with signal transduction generate covalent changes of the signal-transducing protein. Such changes are called post-translational modifications. They are fully reversible with the forward and the reverse reactions catalyzed by enzymes. As summarized in Figure 2.3, the most frequent ones are the following:

- oxidations and nitrosylations of thiol groups, and glutathionylation

- phosphorylations of His, Ser, Thr, and Tyr residues, catalyzed by protein kinases

- ubiquitylations of Lys residues, catalyzed by ubiquitin transferases

- acetylations of Lys and Arg residues, catalyzed by acetyltransferases

- methylations of Lys and Arg residues catalyzed by methyltransferases

- ADP-ribosylations of Arg, Asp and Glu residues, catalyzed by ADP-ribosyltransferases

- partial proteolysis, catalyzed by proteases

The biochemical and functional principles of these modifications are explained in the following sections. Additional covalent reactions of proteins involved in signaling include methylations of glutamate/aspartate carboxyl group hydroxylations, and reversible binding of *N*-acetylglucosamine. They are discussed in Sections 3.4.4, 8.2.4, 8.7, and 9.4.3, respectively. As far as signal transduction is concerned, covalent modifications of proteins serve at least two purposes (**Figure 2.4**).

(1) The conformational and functional change caused by the switching process in the effector protein (protein B in Figure 2.2) is "frozen in" to a certain extent, until the modification is reversed by corresponding enzymes (such as oxidoreductases, phosphoprotein phosphatases, ubiquitin proteases, deacetylases, or ADP-ribosylhydrolases) or sometimes nonenzymatically. Such enzymes are also targets of input signals. Thus, compared with a noncovalent conformational change, covalent post-translational modifications provide additional possibilities to feed input signals into the switching process and, in particular, to control this process temporally.

(2) The covalent change provides a contact site for noncovalent interaction with other proteins and nonprotein molecules, provided these partners possess an appropriate interaction domain, like a lock fitting a key. Such interactions are

Figure 2.4 Two consequences of covalent post-translational modification The modification M may change the protein conformation in such a way that, in the case of an enzyme, a catalytic domain (CD) becomes uncovered, allowing the access of a substrate (pathway 1), or generates a contact site for a partner molecule with a corresponding interaction domain (pathway 2). This mechanism plays a key role in the formation of signal-transducing protein complexes and for the intracellular localization of proteins. Frequently a single modification has both effects. By the same token, a post-translational modification may have an inhibitory (antagonistic) effect when it stabilizes the closed conformation of an enzyme (not shown). The scheme may be generalized. Apart from the catalytic domain of an enzyme, any interaction domain (for instance, a binding site for an allosteric effector or a cytoskeletal component) may be either uncovered or hidden by the conformational change triggered by a post-translational modification.

required for the formation and intracellular targeting of signal-processing complexes and networks of macromolecules. Thus, covalent modification also controls a switching process spatially.

The principle of post-translational modifications has been illustrated by a **writer, reader, and eraser terminology**: the message is "written" by the modifying enzyme, "read" by the molecule interacting with the modified protein, and "erased" by the enzyme removing the modification.

Summary

- Because it is an entropy-lowering process, cellular signal processing depends on a permanent energy supply.

- The energy is derived from metabolic redox reactions, the hydrolysis of nucleoside trisphosphates (ATP, GTP) and other energy-rich molecules such as NAD^+, controlled protein degradation, and a discharge of the membrane potential.

- The enzymes catalyzing these processes are targets of input signals.

- Every signaling event is coupled with at least one of these energy-supplying biochemical reactions. Frequently this coupling results in a reversible covalent change (post-translational modification) of the signaling protein.

2.2 Redox and nitrosylation switches: a balance on a narrow ridge

We begin our examination of energy-supplying reactions with redox processes because they provide the primary energy source of an aerobic cell. Ironically, their role beyond metabolism has been recognized only recently: the redox potential, defined as the (electrochemically measured) energy released either by oxidation or upon withdrawal of electrons or separation of hydrogen, appears to have been used for signal transduction from the very early days of evolution. In fact, proteins respond to changes of the redox potential by conformational rearrangements, provided they possess structures monitoring changes of the redox potential. These redox sensors are the sulfhydryl groups of cysteine residues that are oxidized to disulfide and sulfenic acid groups. This simple reaction, which may occur even nonenzymatically, probably represents one of the most ancient post-translational modifications.

A second group of redox sensors involved in *bacterial* redox signaling are iron ions bound in so-called sulfur clusters. These are protein cofactors that consist of inorganic iron sulfide entities of the composition [Fe2nS2n] complexed by Cys (and His) residues of the protein. The iron ions can exist in the redox states Fe^{2+} and Fe^{3+}. Such clusters are found in bacterial transcription factors controlling the production of enzymes that deal with the potentially hazardous effects of oxygen (see Section 3.6.5), but also in metabolic redox enzymes including ferredoxin (an electron carrier in photosynthesis), bacterial nitrogenase (see Section 3.4.2), xanthine oxidase, and the cytochrome bc1 complex. While sulfhydryl groups are oxidized via a two-electron transition, iron–sulfur clusters are specialized on one-electron transitions as they are induced by free radicals.

To become a binary switch, the redox reaction must be forced under the control of enzymes. Moreover, it has to be reversible: the oxidation must be reversed by a reduction. Such a redox switch is shown in **Figure 2.5**. Here a signal-transducing interaction between two proteins occurs only when one of the partners is reversibly oxidized. The oxidases and reductases catalyzing these reactions are the sensors of input signals.

The oxidants generated by oxidases are **reactive oxygen species (ROS)**, whereas the reductases use peptides and proteins with thiol groups as well as reduced cofactors such as NADH and NADPH as reductants. ROS is a collective term for intermediates arising during the reduction of molecular oxygen to water. These intermediates are generated by one-electron transitions and are superoxide anion radicals, hydrogen peroxide, and hydroxyl radicals:

$$O_2$$
$$\downarrow + e^-$$
superoxide anion radical $O_2^{-\bullet}$
$$\downarrow + e^-$$
hydrogen peroxide H_2O_2
$$\downarrow + e^-$$
hydrogen peroxide $OH^{\bullet}$
$$\downarrow + e^-$$
water H_2O

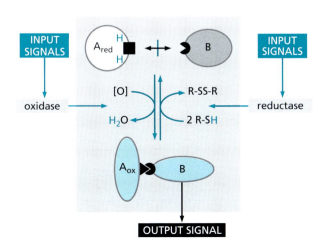

Figure 2.5 Redox switch Protein A interacts with and changes the conformation and function of protein B only when its SH groups have first been oxidized. Oxidants are reactive oxygen species [O] generated as intermediates during the enzymatic or nonenzymatic reduction of molecular oxygen. The corresponding enzymes (oxidases and oxygenases) are controlled by input signals. The same holds true for reductases that set the switch back to 0 by using peptides with thiol groups (R-SH) as reductants. The switching energy is derived from the oxidations. Like any protein switch, this one may also work in the opposite direction, that is, when A interacts with B only in a reduced form.

An additional ROS is singlet oxygen, $^1O_2$, generated during photosynthesis and as a byproduct of enzymatic oxidations such as those of unsaturated fatty acids. One-electron transitions are typically catalyzed by oxidases and oxygenases containing flavin nucleotides, heme, or non-heme iron as prosthetic groups. These families of enzymes include major ROS generators such as the NADPH oxidases (NOX) of the plasma membrane and the xanthine oxidase of the Golgi apparatus, the respiratory-chain complexes of the mitochondrial matrix, as well as sulfhydryl oxidases, and cyclo- and lipoxygenases of unsaturated fatty acid metabolism. While mitochondria are a major source of ROS, other cell organelles such as endoplasmatic reticulum (containing cytochrome P450 oxygenases) and peroxisomes may contribute to ROS production as well.

Since ROS are more aggressive than molecular oxygen, their use is a risky venture. To neutralize their harmful effects, cells possess an arsenal of enzymes, reductants, and radical scavengers that are essential for survival in an oxygen atmosphere. When this defense line is overrun, a situation called **oxidative stress** emerges, which leads eventually to cell death. Oxidative stress is considered to be one of the major causes of aging and diseases connected with aging, such as dementia, atherosclerosis, and cancer.

ROS are used by specialized defense cells such as phagocytes to kill microbes. Moreover, early in evolution, cells learned to harness ROS generated in subtoxic concentrations for a controlled post-translational modification of proteins and thus for signal processing. As mentioned previously, this modification affects the sulf-hydryl groups of cysteine residues, which are reversibly oxidized to disulfide and sulfenic acid groups. Disulfide formation may occur intramolecularly or with other partners, in particular, glutathione and thioredoxin (**Figure 2.6**). In the latter case the protein is said to be glutathionylated or thioredoxinylated.

The ROS that typically oxidizes sulfhydryl groups is **hydrogen peroxide**. It is the most abundant and most important messenger in redox signaling. In contrast, superox-ide anion radicals are less suited for signaling: their diffusion across membranes is limited by their negative charge, their intracellular lifetime is much shorter than that of H_2O_2, and they preferentially interact with iron–sulfur clusters rather than with sulfhydryl groups. Finally, the extremely dangerous hydroxyl radicals are not used as messengers but destroyed immediately by radical scavengers and antioxidants. Whether an SH group is oxidized by H_2O_2 depends on its sterical accessibility and its acidity, which is determined by electrostatic effects of neighboring amino acid residues. As expected for a signaling event, the oxidation is highly selective. In fact, frequently only one of the many cysteine residues of a protein is attacked by the oxidant.

Figure 2.6 Biochemistry of redox signaling: the forward reaction In the course of various enzymatic or nonenzymatic reactions, reactive oxygen species (ROS) are generated that oxidize cysteinyl thiol groups of proteins to sulfenic acid groups or to intra- and intermolecular (mixed, here with glutathione) disulfide groups.

Figure 2.7 Biochemistry of redox signaling: the reverse reaction The protein modified by the redox signal (here with an intramolecular disulfide bridge) is reduced by nicotinamide adenine dinucleotide phosphate (NADPH) with glutathione (GSH) or thioredoxin (Trx) acting as an electron transmitter. As a rule, the reaction with glutathione occurs spontaneously and relatively slowly, whereas thioredoxin is a dithiol-oxidoreductase catalyzing a fast redox reaction. The flavoprotein enzymes glutathione reductase and thioredoxin reductase regenerate the reduced forms of glutathione and thioredoxin at the expense of NADPH.

The reductases that reset the sulfhydryl–disulfide switch to 0 mainly use glutathione and thioredoxin as reductants. **Glutathione** is a Glu-Cys-Gly tripeptide found in cells in millimolar concentrations. It is rather resistant to ROS but is oxidized to a dimeric disulfide by signaling proteins with disulfide groups. This oxidation is catalyzed by the enzyme glutaredoxin, again via a disulfide intermediate, whereas glutathione reductases are responsible for the reduction of oxidized glutathione by NADPH. The much more effective **thioredoxins** are ubiquitous 12 kDa dithiolox-idoreductases. Although mechanistically resembling glutaredoxins, they do not catalyze glutathione oxidation but catalyze their own oxidation to intramolecular disulfides. These disulfides are reduced again by specific reductases, with NADPH acting as an electron donor (**Figure 2.7**).

Redox signaling has been found in both pro- and eukaryotes and is probably more abundant than hitherto assumed. The proteins modified by this reaction include major signal transducers such as transcription factors (see, for instance, Section 3.6.5), protein kinases, and protein phosphatases (see Sections 2.6 and 7.1). A modification of redox signaling is the **nitrosylation** of cysteinyl–SH groups in proteins. In this case nitric oxide (NO) is the reactive agent generated by an enzymatic oxidation of arginine (for details see Section 16.5). NO is both an inter- and intracellular signal molecule. Downstream, it activates cytoplasmic guanylate cyclases by noncovalent binding and modifies other proteins by covalent nitrosylation forming cysteinyl–S–NO groups. When NO is pathologically overproduced, the nitrosylation occurs nonspecifically. The **nitrosative stress** thus evoked is as harmful for the cell as oxidative stress. Under normal conditions, however, NO production is strictly controlled and nitrosylation is exclusively used for signal-controlled post-translational modifications of proteins. Today more than 100 proteins are known, the functions of which are selectively regulated along this pathway. Even though nitrosylation is, strictly speaking, a nonenzymatic reaction, it specifically attacks only selected cysteine residues. The reason is the same as that for the selectivity of redox signals: intramolecular interactions modifying the nucleophilic reactivity (acidity) as well as the sterical accessibility of an SH group. An additional cause of selectivity is targeted interactions of NO synthases with the substrate proteins of nitrosylation.

Enzymes catalyzing denitrosylation set the nitrosylation switch to 0. Again glutathione and thioredoxin play a central role in this process. Thioredoxin transfers NO from the substrate protein to one of its cysteine residues and from there to glutathione where subsequently it is reduced and released by thioredoxin–S–NO or glutathione–S–NO reductase. This trans-nitrosylation of one protein by another seems to work in the reverse direction as well, with nitrosylated thioredoxin and glutathione functioning as NO donors for protein nitrosylation.

Summary

- An ancient source of energy required for signal processing is redox reactions.

- Redox signaling is mediated by reactive oxygen species (ROS) generated along various metabolic pathways (or nonenzymatically).

- By oxidizing sulfhydryl groups, ROS, in particular H_2O_2, post-translationally modulate protein structure and function. The process is reversed by reductases that use peptides (glutathione) or proteins (thioredoxins) with sulfhydryl groups as reductants and NADPH as an electron donor.

- A related post-transcriptional protein modification with signaling character is the nitrosylation by NO of sulfhydryl groups.

- An overactivation of these signaling reactions results in oxidative or nitrosative stress with manifold pathological consequences.

2.3 Switches operated by enzymes that hydrolyze energy-rich compounds

From the very beginning, cells have learned to store the energy obtained from oxidative reactions and other elementary metabolic processes in metabolites containing energy-rich chemical bonds, first and foremost in the phosphoric acid anhydride bond of nucleoside triphosphates, NTPs (see **Sidebar 2.2**). Most *targeted* cellular processes that result in a lowering of entropy—that is, those that lead to a state of higher order—are linked to the energy-supplying hydrolysis of NTPs such as ATP and GTP. These include the biosynthesis of macromolecules, the generation of ion gradients across cell membranes (resulting in the membrane potential), transport reactions and movements, as well as the great majority of signal-transducing reactions.

The phosphoric acid anhydride bond of NTPs is thermodynamically labile but, for the reasons explained in Sidebar 2.2, kinetically stable; that is, under the conditions of a living cell, its hydrolytic cleavage, although strongly exergonic, does not occur spontaneously but has to be catalyzed. This is achieved by two types of enzyme: nucleoside triphosphatases (NTPases) and kinases. NTPases transfer the γ-phosphate residue of NTP to water, producing free inorganic phosphate (P_i). Kinases transfer the γ-phosphate residue to nucleophilic residues (XH) of other molecules, which become phosphorylated. In a second step, catalyzed by a phosphatase or proceeding spontaneously, the phosphate residue is transferred to water, yielding phosphoric acid:

NTPase reaction:	$NTP + H_2O \rightarrow NDP + P_i$
kinase reaction:	$NTP + RXH \rightarrow NDP + RXP$
phosphatase reaction:	$RXP + H_2O \rightarrow RXH + P_i$

Guanosine and adenosine triphosphatases (GTPases and ATPases) and protein kinases/phosphatases are particularly abundant switching devices of the signal-processing protein network. While the NTPases induce noncovalent conformational changes, the kinases/phosphatases catalyze covalent phosphorylations/ dephosphorylations of the corresponding effector protein. The activities of these enzymes are regulated by an enormous variety of noncovalent interactions with other proteins, low-molecular-mass factors, and environmental stimuli as well as by post-translational modifications. These are the input signals driving the switching process either from 0 to 1 or from 1 to 0.

Sidebar 2.2 The unique phosphoric acid
In the data-processing protein network, most switching reactions are linked directly or indirectly to the strongly exergonic hydrolysis of NTPs and the resulting phosphorylation of proteins. What is the distinctive feature of phosphoric acid that makes it so indispensable for life processes and essentially irreplaceable by any other structure? The phosphoric ester bond was most likely present in living systems at the very beginning of evolution. The discovery of ribozymes, RNA with enzymatic activity, led to the hypothesis that the most ancient biomolecules are ribonucleic acids rather than proteins as previously postulated (see Section 3.2). In fact, RNA exhibits several properties essential for life such as catalytic activity, the ability to copy itself, and the capability to store information in its linear structure. The latter requires sufficient chemical stability as the information would otherwise be lost rapidly and could not be transferred reliably to descendants.

For these purposes the phosphoric ester bond is particularly suitable, because of its additional negative charge, which a comparable sulfuric acid ester would be lacking. This charge renders RNA an extremely hydrophilic substance that cannot be lost across cellular membranes and protects the ester bond by electrostatic repulsion from nucleophilic attack such as hydrolysis (**Figure 2.8**).

Nevertheless, RNA is still relatively sensitive because hydrolysis proceeds via a cyclic 2′,3′-phosphate as an energetically favored intermediate. This fact may have provided the selective pressure for the development of DNA, which cannot form such an intermediate and, as a data carrier, is about 100 times more stable than RNA. In parallel with the evolution of DNA, RNAs were replaced as enzymes by proteins, probably because protein structures are much more flexible and variable than nucleic acid structures.

The synthesis of a macromolecule such as RNA is accompanied by a lowering of entropy, thus requiring an energy supply. This may be the reason for the evolutionary "invention" of NTPs. Their phosphoric acid anhydride bond is energy-rich and thermodynamically unstable. Nevertheless, it is kinetically stable since the hydrolysis requires a large amount of activation energy. This is again due to the negative charge, which protects not only the ester but also the anhydride bond from nucleophilic attack. Under the conditions of a cell, the hydrolysis of NTPs has to be enzymatically catalyzed. This peculiarity makes the NTPs, first and foremost, ATP, the universally applicable energy suppliers for the majority of endergonic biochemical reactions. As a rule such reactions proceed via *short-lived* intermediates where the enzyme is phosphorylated at suitable amino acid residues such as His, Asp, and Cys, with ATP as a phosphate donor. From this point, it was probably only a small evolutionary step towards the application of prokaryotic protein phosphorylation as a more persistent modulation of protein structure and function and thus as an energy-dependent switching reaction for cellular data processing.

Simultaneously with the development of protein phosphorylation cells may have used ATPase and GTPase reactions, the membrane potential, redox processes, and degradation of macromolecules to obtain the energy for data processing.

Since the energy stored in the membrane potential and in macromolecules ultimately stems from the hydrolysis of NTPs, it may be stated without exaggeration that cellular data processing is mainly based on the unique chemical properties of phosphoric acid.

Figure 2.8 Phosphate group: self-protection against hydrolysis Through electrostatic repulsion, a cloud of negative charge at neutral pH shields the phosphate group from nucleophilic attack by polarized water molecules. Therefore, the hydrolytic reaction, although exergonic, is blocked since it requires a high activation energy. To occur under the conditions of a living cell, it must be sped up by enzymatic catalysis. R symbolizes an organic residue in the case of a phosphate ester group (as in nucleic acids) or a second phosphate group in the case of a phosphoric acid anhydride group (as in ATP).

Summary

- Most of the energy driving cellular signal processing is derived from the hydrolysis of ATP and GTP. These reactions are catalyzed by nucleoside triphosphatases (NTPases) and protein kinases/phosphatases.

- The energy thus obtained is used for reversible conformational changes or post-translational modifications of signaling proteins.

2.4 GTPase or G-protein switch

One of the simplest and most abundant signal-transducing reactions is based on the hydrolysis of GTP catalyzed by GTPases. This reaction supplies as much free energy as ATP hydrolysis. In the late 1960s it was observed that the activation of the cellular adrenaline receptor required GTP (see Section 4.3). The corresponding GTP-binding protein was identified as a GTPase interacting with the receptor. Since then numerous regulatory GTPases or G-proteins have been found to belong to the standard equipment of the data-processing protein network. As prototypes of switching elements using the energy derived from NTP hydrolysis, G-proteins are involved in almost all signal-processing mechanisms of eukaryotic cells. Evolutionarily, they originate in prokaryotic precursors (see Section 3.6.4).

A distinction needs to be made between "small" and "large" G-proteins. Small G-proteins are monomers of about 200 amino acids. According to their prototype, the p21 Ras protein, they are grouped into the so-called Ras superfamily. Large G-proteins are heterotrimeric αβγ complexes with the Ras-related α-subunit harboring the GTPase function. Other G-proteins include the ribosomal elongation factors, the signal-recognition particles of the endoplasmic reticulum, and the dynamins. The different G-protein families are discussed in detail in Chapters 4 and 10.

Regulatory GTPases have many functions that, at a first glance, seem to have almost nothing to do with each other. However, all these effects target cellular processes by making them irreversible through entropy reduction brought about by GTP hydrolysis.

2.4.1 G-proteins are amplifiers, rectifiers, and organizers

G-proteins are signal-transmitting and signal-transforming switches. They induce reversible conformational changes rather than covalent modifications of proteins. In an exemplary fashion G-proteins demonstrate two basic features of signal transduction: one, connection of signaling processes with an energy-supplying biochemical reaction and two, spatial and temporal control of signaling. A G-protein has bound GDP in state 0 and GTP in state 1, in a 1:1 ratio. In the GTP-loaded state the G-protein interacts with downstream effector proteins, changing their conformation and function in a time-dependent manner (**Figure 2.9**) or acting as an organizer, targeting them to defined sites such as the cell membrane where they come in contact with other proteins forming signal-transducing complexes (**Figure 2.10**). These functional changes and the altered intracellular localization of the effector proteins resemble the output signals of the G-protein switch.

Activation, or the exchange of GDP for GTP, has to be accelerated enzymatically and is thus an input terminal for signals. The same holds true for the reverse reaction,

Figure 2.9 GTPase switch as a controlled timer When it is loaded with GDP, the G-protein or GTPase (G) cannot interact with an effector protein. Only when GDP is exchanged for GTP (catalyzed by guanine nucleotide exchange factor, GEF) is the conformation changed allosterically, now enabling contact with a corresponding binding domain of the effector protein, which becomes altered structurally and functionally. The switch is set back to 0 by the endogenous GTPase activity of G (stimulated by the regulatory GTPase activating protein, GAP). Input signals control the activity of GEF or GAP, whereas the functional change of the effector protein as induced by G resembles the output signal.

Another effect of G is to bring its effector protein in contact with other components of the signal-processing network (see Figure 2.11).

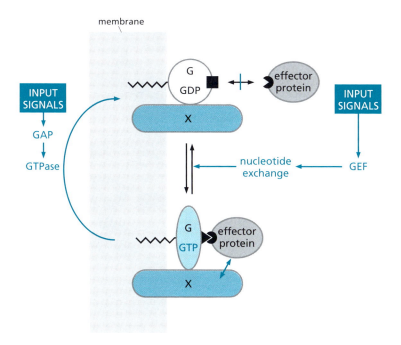

membrane

INPUT SIGNALS

GAP

GTPase

INPUT SIGNALS

GEF

nucleotide exchange

G GDP

effector protein

X

G GTP

effector protein

X

Figure 2.10 GTPase switch as a spatial organizer Upon GDP–GTP exchange, the GTPase G anchored to a cellular structure (for instance, to the plasma membrane through a lipid anchor, shown as a zigzag line) undergoes a conformational change allowing the binding of the effector protein, which is thus brought in contact with another protein X (teal). When the interaction of G does not induce an energy-dependent conformational change of the effector protein, the activated GTPase continuously recruits effector proteins and passes them on to other proteins until it inactivates itself by the GAP-stimulated GTPase activity. The energy required for this organizing process is supplied by interactions between the effector proteins and protein X. A defect of the GTPase function, for instance due to a gene mutation, resembles a switch that cannot set back to 0. It may have fatal consequences, as will be explained later for the GTPase Ras.

GTP hydrolysis. Being GTPases, G-proteins are able to switch back to position 0 on their own. However, their GTPase activity (and with that the duration of the active state) is modulated by other input signals within wide limits. Thus G-proteins are not only amplifiers and rectifiers but also adjustable timers.

2.4.2 Switching mechanism of G-proteins

All regulatory GTPases share an evolutionary ancient core of about 200 amino acids. It is called the **G-domain** and resembles the small G-protein Ras in size and basic structure. The G-domain distinguishes G-proteins from other GTP hydrolyzing enzymes such as β-tubulin or phosphoenolpyruvate carboxykinase. By means of X-ray analysis and directed mutations, the architecture and mode of function of the G-domain could be elucidated for several G-proteins, beginning with the prokaryotic ribosomal translation factor EF-Tu.

The basic structural elements of the domain are six anti-parallel β-sheets with five α-helices grouped in between. Five of the β-sheets are placed in a plain surrounded by the helices. The catalytic center is located in the amino (N)-terminal area of this structure, containing a highly specific GTP/GDP. A characteristic structural element of this site is the P-loop between β-sheet 1 and helix 1. Here the negatively charged phosphate residues (present as Mg^{2+} complexes) of the nucleotide are fixed by circularly arranged amino acids via electrostatic and hydrogen bonds (PM-motif) while the guanine base fits an adjacent hydrophobic pocket, the G-motif. Together PM- and G-motifs build up the catalytic center. The water molecule executing the nucleophilic attack on the phosphoric acid anhydride bond of GTP is polarized by a highly conserved Thr or Gln residue and placed into position in such a way that it can split off the terminal phosphate residue of GTP in the course of an SN2 reaction.

Both GDP–GTP exchange and GTP hydrolysis are accompanied by far-reaching conformational changes of the G-protein, affecting in particular two structural elements of the G-domain, switches 1 and 2 (**Figure 2.11**). These elements contain highly conserved amino acid residues (Thr and Gly) forming hydrogen bridges with the γ-phosphate residue of GTP. When these bridges are pulled down in the course of the GTPase reaction, the switches fold down into another conformation similar to a relaxing string (**Figure 2.12**). These conformational changes of the switch regions are transferred onto the effector protein.

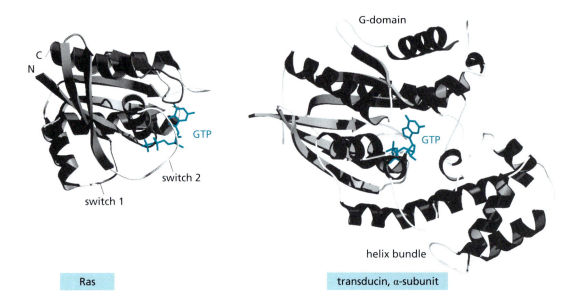

Figure 2.11 Structure of the small G-protein Ras as compared with the α-subunit of the trimeric G-protein transducin Ras exhibits the typical architecture of a G-domain. In the transducin subunit an additional helix bundle is inserted into the G-domain.

Figure 2.12 Effect of GDP–GTP exchange on structure of the G-domain The small G-protein Arf is shown as a GDP complex (left) or a GTP complex (right). The nucleotide exchange results in a considerable rearrangement of the switch regions.

2.4.3 Guanine nucleotide exchange factors are inducers of G-protein signaling

At G-proteins the exchange of GDP for GTP is thermodynamically favored by high GTP excess in the cell. Nevertheless, because of the extremely strong binding of GDP (and GTP) to the G-domain, it requires so much activation energy that it takes hours to occur spontaneously. Therefore, it has to be catalyzed enzymatically (for the same reason, G-proteins do not exist in cells in a guanine nucleotide-free form). The corresponding enzymes are called guanine nucleotide exchange factors (GEFs). Each G-protein has one or more GEFs (the human genome encodes more than 1000 GEFs). With the nucleotide-liganded G-protein, the GEF forms a ternary transition complex where the binding between nucleotide and G-domain is up to 10^6 times weaker as in the absence of GEF, resulting in a dramatic lowering of the activation energy (**Figure 2.13**). In this state the GDP–GTP exchange can proceed in a fraction of a second with the surplus GTP displacing the GEF from the G-protein.

GEFs weaken the affinity of G-proteins to GDP/GTP by deforming the nucleotide binding site. This catalytic mechanism has been studied in detail for mSOS, a GEF of the small G-protein Ras. mSOS blocks the phosphate binding site in switch 2 of the

Figure 2.13 Catalytic effect of GDP–GTP exchange factor GEF (schematically) Though it is exergonic, the transition from G-GDP to G-GTP is kinetically blocked by the high activation energy. The latter is lowered by GEF through the formation of a transition complex with G, GDP, and GTP; that is, GEF has the characteristic catalytic effect of an enzyme.

G-protein and simultaneously pushes aside switch 1 by an extended hairpin loop, resulting in the release of GDP from the binding pocket. Since GEFs have different structures, this mechanism may be representative in principle but certainly not in detail. Thus the GEFs of the large trimeric G-proteins are receptors with seven transmembrane domains and are unable for steric reasons to activate the GDP–GTP exchange at the GTPase-active α-subunit. Instead the exchange reaction may be assisted by the β-subunit. For G-proteins GEFs are the major transducers of input signals. To this end they contain, in addition to the catalytic domain, a wide variety of interaction domains that control substrate recognition, cellular localization (for instance membrane binding or transport into the nucleus) and cross talk with other signaling pathways.

2.4.4 GTPase-activating proteins are terminators of G-protein signaling

The GTPase reaction terminates G-protein signaling. It critically depends on a basic amino acid residue (mostly Arg) in the catalytic center of the G-protein, neutralizing the negative charge of the phosphate residue and thus facilitating the nucleophilic attack of water on GTP. In addition, this residue brings the water-binding Thr or Gln residue in position for the hydrolytic reaction. Small G-proteins lack the critical Arg residue; therefore, their G-domain is incomplete and the GTPase activity is very low. The lacking Arg residue is supplied by another protein called GTPase-activating protein (GAP). It is localized at the tip of a loop structure that, as an "arginine finger," thrusts into the catalytic center of the G-protein (**Figure 2.14**). Strictly speaking, small G-proteins are heterodimeric GTPases consisting of a catalytic subunit and a regulatory subunit, (GAP). In addition to GEFs, GAPs are targets of input signals controlling G-protein activity.

A reliable rule of biology is that there is no rule without exception. In fact, the GAPs of the small G-proteins of the Ran and Rap families supply an asparagine instead of an arginine residue to stabilize the correct orientation of water and GTP in the catalytic center of the G-protein.

In contrast to the small G-proteins, the heterotrimeric G-proteins have a complete G-domain. Owing to several inserts, in particular a bundle of six helices containing the critical Arg residue (see Figure 2.14), this G-domain is about three times as large as that of a small G-protein. As a consequence, the α-subunits of large G-proteins exhibit up to 10^4-fold higher GTPase activity than small G-proteins. Notwithstanding, nature did not miss the chance to use the GTPase reaction as a terminal for input signals. In fact, for many large G-proteins, so-called regulators of G-protein signaling (RGS) have been found that stimulate the GTPase reaction in a signal-dependent manner, though by a mechanism differing from that of GAPs (details in Section 4.3.2).

large G-protein α-subunit

small G-protein

Figure 2.14 Mechanism of action of GTPase-activating protein GAP GAP functions as a subunit of small G-proteins in that it provides an Arg residue required for the GTPase reaction. In the G-domain of the α-subunits of large G-proteins, this Arg residue is already present.

Summary

- G-proteins are GTPases characterized by a unique G-domain in the catalytic center. They function as molecular switches: binding and hydrolysis of GTP results in a conformational change that is transferred to interacting effector proteins. As a result, the activity of the effector protein is changed or the formation of signaling multiprotein complexes is triggered.

- There are two types of G-protein: monomeric (small) and heterotrimeric (large). For both types, GTP loading is catalyzed by enzymes called guanine nucleotide exchange factors (GEFs).

- While large G-proteins exhibit intrinsic GTPase activity, small G-proteins require interaction with regulatory subunits called GTPase-activating proteins (GAPs). The GTPase activity of many large G-proteins is also controlled by accessory proteins. These as well as GEFs and GAPs are receptors of input signals.

2.5 ATPase switches

Similar to G-proteins, two families of ATPases are specialized to induce conformational changes in other proteins and to regulate the association and dissociation of (signal-transducing) protein complexes. These families include the chaperones and the AAA+ proteins. Both types are responsible for protein quality control and for targeted cell physiological processes. Other ATPases work as membrane transporters and nanotechnical motor proteins, causing mechanical alterations of the cell structure.

2.5.1 Molecular chaperones or how to make proteins fit for work

Molecular chaperones represent a large group of mostly unrelated proteins that assist with protein folding and protein complex assembly. They are essential for the maintenance of the highly ordered state of a living cell. The classic theory of protein folding starts from the assumption that the *native* state of a protein represents the most favorable thermodynamic conformation of the polypeptide chain, which arranges spontaneously from the genetically determined primary structure (Anfinsen theory; the 1972 Nobel Prize in Chemistry awarded to Christian B. Anfinsen, Stanford Moore, and William H. Stein). In an aqueous environment, the energy for the native folding stems from the tendency of proteins to hide the hydrophobic side chains of the amino acids in their interior; that is, to adopt a low-energy conformation. However, the native state is by no means always identical with the *functional* state (see Section 1.8.5). Moreover, a spontaneous folding into the native conformation has been observed only for small proteins under nonphysiological conditions. Larger proteins tend to form stable, incompletely folded intermediates during polypeptide biosynthesis at the ribosome and to nonspecifically aggregate because of the very high protein concentration of 300–400 mg/ml (called macromolecular crowding), thus rendering the cytosol a semi-solid medium. Moreover, a wide variety of stress factors including elevated temperature and starvation cause protein denaturation and misassembly (for this reason some chaperones are known as heat-shock proteins). Disturbances of correct protein folding and assembly impair cellular functions and may have severe cytotoxic effects with fatal consequences (for example, Alzheimer's dementia; see Section 13.1.3) Therefore, it is essential to prevent such accidents. This is the task of molecular chaperones.

Since chaperones generate order, their effects depend on energy-supplying reactions, primarily ATP hydrolysis. To this end they either have intrinsic ATPase activity or couple with ATPases. [Here a narrow definition of chaperones as proteins that modify protein conformations in a noncovalent ATP-consuming reaction is preferred. Other authors also consider enzymes that act on the tertiary structure of proteins as chaperones; for example, protein disulfide-isomerases and peptidyl-prolyl *cis/trans* isomerases (see Section 8.5), which catalyze covalent modifications and do not exhibit ATPase activity.]

Two types of ATP-consuming chaperone, heat-shock proteins, **HSP70** and **HSP90**, belong to the basic equipment of all cells and rank among the most abundant cytosolic proteins (though variants are also found in mitochondria and in the endoplasmic reticulum). They are expressed in numerous isoforms and are essential for survival. These chaperones cooperate with each other and not only protect proteins from stress but also have important additional functions. The HSP70 chaperone (found in both pro- and eukaryotes) promotes the folding of growing polypeptide chains. In conjunction with HSP90, it plays a central role in cellular signal processing by inducing the formation of functional protein conformations that are thermodynamically unstable. Examples include the activation of steroid hormone receptors (this topic is discussed in more detail in Section 8.5) and the activation loop phosphorylation of protein kinases (see Section 2.6.1). As is typical for most signaling proteins, the functional forms of HSP70 and 90 are dimers. The function of the chaperones culminates in the **chaperonins.** These are large cylindrical complexes made up by stacked rings of chaperone aggregates. A hole inside the complex called an "Anfinsen cage" protects the ATP-consuming folding process of the substrate or client protein from disturbing influences.

Chaperones recognize their client proteins by linear hydrophobic amino acid sequences that are typically exposed by immature or denatured proteins. For complete folding, the client protein undergoes several cycles of binding to and dissociation from the chaperone driven by ATP hydrolysis. Like G-proteins, chaperones may be looked upon as binary switches (**Figure 2.15**). In state 0 they are loaded with ADP. The switching event 0 → 1 involves the exchange of ADP for ATP and the binding of the client protein. The latter activates the chaperone's ATPase function, thus releasing the energy required for the conformational change of the client protein. During step 1 → 0 the modified client protein dissociates from the chaperone, which returns to the initial state. Depending on the chaperone species this mechanism may vary in detail. Although various models have been developed (see Figure 2.15B), the precise mechanism of chaperone action *in vivo* is still not fully understood.

The output signal of the chaperone switch is the altered function of the client protein: for example, enzymatic activity or the tendency to correctly interact with signaling

Figure 2.15 Chaperone switch (A) The ADP-loaded chaperone (Chap) interacts noncovalently with a client protein (CP). Following ADP–ATP exchange, the chaperone changes its conformation and induces a conformational change of the client protein. Upon ATP hydrolysis, the complex dissociates and the chaperone returns into the ADP-loaded conformation. Input signals affect both the nucleotide exchange and the ATPase reaction. The altered conformation (and function) of the client protein resembles the output signal. It should be remembered that not all proteins classified as chaperones are ATPases. (B) "Nutcracker" model: the chaperone dimer (in this example, heat-shock protein HSP90) forms a clamp that, upon nucleotide exchange, changes into a more closed conformation, breaking up the conformation of the client protein.

ligands. When the ligands are other proteins, the chaperone promotes the formation of functional complexes. In fact, the generation of multiprotein structures such as chaperonins, nucleosomes, and proteasomes is controlled by a series of highly specific chaperones.

Both the ADP–ATP exchange and the ATPase reaction are controlled by regulatory proteins or **co-chaperones** interacting with chaperones via specific domains. The ATPase activity of HSP70 chaperones is, for instance, stimulated by proteins called HSP40 and proteins with a J-domain (named after a bacterial co-chaperone DnaJ), while the ATPase activity of HSP90 is negatively regulated by Cdc37-type proteins (the name refers to a yeast protein involved in cell cycle control). Cdc37 and its homologs are multifunctional scaffolds that bring HSP90 in contact with a wide variety of client proteins.

The ADP–ATP exchange is catalyzed by a wide variety of nucleotide exchange factors that are under the control of signaling pathways. Moreover, it is inhibited by an HSP70-interacting protein (HIP), while an adaptor protein (HOP) brings about the contact between HSP70 and HSP90. Via such accessory proteins, input signals are transferred to the chaperone switch. This multi-stage regulation is strongly reminiscent of the control of G-protein function by GEFs and GAPs.

2.5.2 AAA+ proteins or formation and disruption of macromolecular complexes

AAA+ proteins (frequently referred to as triple-A plus proteins) are ATPases Associated with various cellular activities. As indicated by the name, they constitute an extremely diverse family of pro- and eukaryotic proteins. The only structural feature these proteins have in common is an ATPase domain of about 240 amino acids called the AAA motif. This module contains AAA-specific sequences in addition to the so-called Walker A and B motifs that are typical for most ATPases. As far as their functions are concerned, fluid transitions exist between AAA+ proteins and chaperones, and some AAA+ proteins are counted among the heat-shock proteins.

An important function of AAA+ proteins is to unfold native proteins in order to make them accessible for proteolytic degradation (**unfoldase activity**). Since the folded native protein represents a thermodynamically favored state, unfolding requires ATP energy. Moreover, AAA+ proteins function as energy-consuming disassembly machines for thermodynamically stable protein–protein and protein–nucleic acid complexes. Therefore, AAA+ proteins play an important role in promoting the reversibility and repetition of cellular processes. An example is the AAA+ protein **NSF**, which dissociates the SNARE protein complex of secretory vesicle fusion, making it reusable (see Section 10.4.2).

Other types of membrane fusion are also controlled by AAA+ proteins, for instance, in the course of peroxisome formation or during the post-mitotic reconstruction of the Golgi apparatus and the endoplasmic reticulum. An AAA+ protein, called **katanin**, dissolves tubulin complexes, thus depolymerizing microtubules. Katanin is concentrated in centrosomes, playing an important role in the separation of chromosomes during cell division.

Formation of the initiation complex of DNA replication is regulated by no fewer than 10 different AAA+ proteins, which play a similar role as G-proteins in mRNA translation (see Chapter 8). A major player in this process is the DNA helicase MCM that winds up the DNA double strand. Replication factor C, another heterooligomeric AAA+ protein, forms a complex with proliferating cell nuclear antigen (PCNA, an experimentally used indicator of cell division) and DNA polymerase Δ at the replication fork. This complex displaces the polymerase α/primase complex, thus starting the elongation of the new DNA strand (for details on DNA replication, the reader is referred to textbooks of molecular biology).

A key function in cellular signaling is attributed to the **AAA+ proteases**. These enzymes contain, in addition to the AAA+ motif, a protease catalytic domain, located either on the same or on a separate polypeptide chain. Most AAA+ proteases exist as hexameric ring or barrel structures with a central pore harboring both unfoldase and protease activities. The most prominent example of such highly organized AAA+ proteases is the **proteasomes.** These proteolytic nanomachines cooperate with chaperones in removing damaged proteins that can no longer be renatured and are potentially harmful to the cell. Moreover, they terminate or promote signaling by eliminating stimulatory or inhibitory signaling proteins. Proteasomes are hexagonal multiprotein complexes consisting of six catalytic subunits enclosed by two hexamers of regulatory subunits (for a schematic representation, see Section 2.8). The unfoldase activity is located in the regulatory subunits that deliver the unfolded protein to the protease in the central cavity of the catalytic subunits. Unfolding is required for threading the substrate protein into the narrow proteasome pore and facilitating proteolysis.

The catalytic domains of the AAA+ unfoldase seem to be mobile polypeptide loops facing the central pore. By hydrolyzing one ATP molecule, they shift from an UP position to a DOWN position, thus working like a crowbar that breaks up the intramolecular interactions of the substrate polypeptide chain. Running such a machine is an expensive undertaking for the cell: depending on the stability of its conformation, the complete unfolding of an average protein may cost more than 500 ATP molecules! Nevertheless, the elimination of potentially harmful products as well as the precise termination of signaling seems to be worth the price.

For a protein to be degraded in proteasomes, it must be labeled by a polyubiquitin tag interacting with specific binding sites at the surface of the 19S proteasomal subunits. This tagging, catalyzed by E3 ubiquitin ligases, is the major target of signals controlling protein degradation (see Section 2.8).

Proteasomes are molecular machines running on six (or more) cylinders, and the same holds true for most other AAA+ proteins catalyzing the unfolding and unwinding of polypeptide and DNA chains and the dissociation of complexes. Mechanical force is generated by ATP hydrolysis, which leads to the destruction of intramolecular interactions and drives motion. In fact, a particularly clear analogy to nanotechnology is provided by the AAA+ proteins of the **dynein** subfamily. These are motor proteins that under ATP consumption move along microtubule "tracks" of the cytoskeleton, transporting cargo such as organelles (see Section 10.3.5) and, during cell division, chromosomes. Dyneins are also components of eukaryotic flagellae and ciliae.

Like G-proteins and chaperones, the AAA+ proteins may be looked upon as binary switches with the effect on the protein or DNA substrate being the output signal (**Figure 2.16**). Again the input signals most probably control the ADP–ATP exchange

Figure 2.16 AAA+ protein switch The switch widely resembles the GTPase switch (Figure 2.9). Again the altered conformation of the effector protein resembles the output signal.

as well as the ATPase reaction. How this occurs and which accessory proteins are involved are still widely unknown.

2.5.3 The generation of motion by motor proteins and membrane transporters

The dyneins are an example of motor proteins. Generally speaking, motor proteins are ATPases transforming the chemical energy derived from ATP hydrolysis into motion. By this means they change the cellular architecture and shape as well as the intracellular distribution of organelles and macromolecules and generate cell movements. To this end they interact with the actin and microtubule cytoskeleton that support the cell and serve as tracks for material to be transported.

During ATP hydrolysis, a motor protein undergoes a reversible conformational change, thereby bending and transferring mechanical energy to the cytoskeletal fiber. These single steps add up to a movement that results either in a shift of the interacting fiber or in a wandering of the motor protein along the fiber (for the purpose of transporting cargo) or in a pressure or tension between both partners (**Figure 2.17**). In addition to the dyneins, ATPases acting as motor proteins include the kinesins and the myosins. While kinesins, like dyneins, interact with microtubules, myosins prefer actin fibers. Motor proteins are major targets of signaling pathways.

Kinesins

For the relatively large eukaryotic cell, an ATP-driven transport of cargo-filled vesicles and of macromolecular complexes is essential. The alternative would be free diffusion, which is neither targeted nor nearly fast enough when one considers, for instance, the enormous distances in nerve fibers. Therefore, the transport is carried out by kinesins, which are specialized in migrating along microtubule tracks and carrying cargo. For this purpose they hydrolyze one molecule of ATP per step; that is, the costs of intracellular transport are high. Kinesins play an important role in cellular data processing because they rearrange complexes of signaling proteins, thus adapting the "hardware construction" to the current demands. In addition, kinesins are components of the mitotic spindle apparatus.

Kinesins constitute a large and heterogeneous protein family including more than 100 members. They have in common an ATPase or motor domain in which the P-loops and switch regions 1 and 2 closely resemble the active centers of G-proteins and of myosins. These three protein families are assumed to have developed from an ancestral NTPase, with kinesins and myosins becoming molecular motors and G-proteins becoming molecular switches. In contrast, dyneins, like AAA+ proteins, belong to another phylogenetic tree.

Figure 2.17 Modes of operation of motor proteins (A) The motor protein (kinesin, myosin, or dynein) is anchored to a larger cell structure and moves the interacting cytoskeletal filament. (B) The cytoskeletal filament is anchored: the motor protein moves along the filament and may transport cargo. This mode of operation is typical for the interaction of kinesins and dyneins with microtubules. When both the motor protein and filament are anchored, a tension develops, resulting in either muscle contraction by an actin–myosin interaction or beating of flagella by a microtubule–dynein interaction. The gear-wheel drive serves as an illustration; eukaryotic motor proteins do not contain rotating structural elements but perform kinking and step-by-step movements (Figure 2.18).

Figure 2.18 Basic structure and mode of movement of a kinesin tetramer The upper panel shows the basic structure of kinesin in a schematic manner. The motor protein consists of two heavy chains, held together by coiled-coil dimerization domains, and two light chains (teal) noncovalently bound to the C-terminals of the heavy chains. In the lower panel the mode of movement is depicted schematically. (A) With its ADP-loaded head 1, the kinesin contacts a tubulin subunit of a microtubule MT. (B) Exchange of ADP for ATP causes a conformational change resulting in a spreading of the two heads at the neck region. This allows contact between head 2 and the next binding site on the microtubule, resembling a single step forward. (C) Subsequently ATP becomes hydrolyzed and head 1 catches up with head 2. For further details, see text.

Figure 2.18 depicts the basic structure of a conventional kinesin.*It is a tetrameric protein consisting of two heavy and two light chains each. A characteristic feature of the heavy chain is the N-terminal globular motor domain containing the binding site for microtubules, and a carboxy (C)-terminal globular tail domain with the binding sites for the light chain and other adaptor proteins. The central part of the molecule is an extended helical dimerization domain forming a coiled-coil structure with a partner protein of its kind. The helical regions of this domain are interrupted by hinge sequences providing the conformational elasticity required for the movement but also forming a rebend hairpin loop that structurally locks in the inactive form of the kinesin. The light kinesin chains contain interaction domains establishing the contact with the cargo, thus acting as adaptor proteins. Cargo vesicles are bound to the kinesin by so-called cargo receptors integrated into the vesicular membrane.

The stepwise moving pattern of kinesins is due to two reactions resembling the switching reactions of AAA+ and G-proteins: the exchange of ADP for ATP leads to a conformational change that is reversed by the subsequent ATP hydrolysis (see Figure 2.18). Both these events as well as the loading with cargo are controlled by input signals; however, these are still fairly unknown. An example of microtubule-directed cargo transport controlled by kinesin and dynamin phosphorylation is discussed in Section 10.3.5.

Myosins

Myosins constitute a protein family that in size and heterogeneity is comparable to the kinesin family. Together with their cytoskeletal partner F-actin they form contractile fibers enabling various types of movement such as muscle contraction,

---

*In addition to these conventional kinesins, a nonconventional type exists that is characterized by the motor domain being placed either C-terminally or in the middle of the polypeptide chain. On microtubule tracks the nonconventional kinesins migrate opposite to the conventional kinesins, that is, toward the microtubule-organizing center.

separation of cells during cytokinesis, and membrane extrusions and invaginations (for instance, amoeboid crawling, phagocytosis and pinocytosis, and formation of microvilli). Whether or not myosins are also involved in cargo transport is an open question. In contrast to the kinesins they are unable to perform stepwise movements, although both types of motor protein are structurally related, forming tetramers of two light chains and two heavy chains each. The heavy chains containing the ATPase- and actin-binding domains in the N-terminal region dimerize by an interaction of extended coiled-coil structures, whereas the myosin light chains have an autoregulatory function.

Like G-proteins and kinesins, myosins undergo a two-step switching reaction: there is an exchange of ADP–ATP, which induces the conformational change required for the movement, followed by ATP hydrolysis, which returns the protein to the initial state. Both steps are stimulated by actin and blocked by regulatory proteins, the activities of which are controlled by input signals such as protein phosphorylation and Ca^{2+} ions (for details see Section 14.6.2). These regulatory proteins include the myosin light chains and, in the case of the conventional muscle myosin, tropomyosin, both of which hinder the interaction between myosin and F-actin.

In addition to the muscle myosins, a large number of nonconventional myosins have been found. They are regulated by light chains and have numerous domains for interactions with other proteins and membrane constituents, so they are firmly linked into the signal-processing protein network. The mechanisms under which these interactions work are unknown.

Membrane transporters

Another large family of regulatory ATPases is specialized for transporting substances across membranes, mostly against a concentration gradient. These proteins are characterized by an ATP-Binding Cassette and are called **ABC transporters**. Switching between an ATP- and an ADP-bound state, they mainly promote the excretion of molecules trapped in a conformational cage by the cell. ABC transporters are discussed in more detail in Section 3.3.1.

Summary

- ATPases involved in cellular signal processing include chaperones, AAA+ proteins, and motor proteins.

- Chaperones help other proteins to acquire and maintain their functional conformations and to assemble functional complexes, in particular during de novo synthesis and signal transduction. In stress situations they may protect proteins from denaturation. Most chaperones are ATPases and use the energy derived from ATP hydrolysis for conformational changes. Some chaperones such as the heat-shock protein HSP90 play a particular role in inducing and stabilizing the active conformations of signaling proteins. Like G-proteins, chaperones are associated with regulatory proteins, modulating both ADP–ATP exchange and ATPase activity. Such proteins are receivers of input signals.

- AAA+ proteins are ATPases characterized by a unique ATP-binding motif. They are found in both pro- and eukaryotes, where they work as molecular machines. Their major functions include unfolding of proteins for degradation in proteasomes, disassembly of (signal-transducing) protein complexes, protein disaggregation, cellular movements, and modulation of cellular signal transduction. Prominent examples of AAA^+ proteins are proteasomes and dyneins.

- Motor proteins are a large and heterogeneous family of ATPases specialized for transforming, by conformational changes, the chemical energy derived from ATP hydrolysis into mechanical energy. By interacting with cytoskeletal anchor proteins they are able to move, thereby transporting cargo including signaling protein complexes, and to generate mechanical tension, resulting in changes of cell shape and cellular movements.

2.6 Protein phosphorylation

Reversible phosphorylation, the most abundant post-translational modification of protein structure, has been estimated to affect at least a third of all cellular proteins. Phosphorylation also represents the major switching reaction of cellular data processing, with step 0 → 1 catalyzed by protein kinases and step 1 → 0 catalyzed by phosphoprotein phosphatases (**Figure 2.19**). There are three phosphate donors and energy suppliers: ATP, GTP, and phosphoenolpyruvate, the last used mainly by prokaryotes (see Section 3.3.2). Targets of phosphorylation are the amino acid residues Ser, Thr, and Tyr (which deliver phosphoric acid esters), His (which delivers an energy-rich, unstable phosphoric acid amide), Asp (which delivers a mixed acid anhydride), and Cys (which delivers a labile phosphoric acid thioester) (**Figure 2.20**). Phosphorylation of Lys, Arg, and Glu residues may also be possible, but convincing proof of these processes is still lacking. Protein kinases do not phosphorylate free amino acids. Instead, the phosphorylatable amino acid residue (the phosphorylation site) has to be embedded in a short amino acid sequence of a protein. This so-called **consensus phosphorylation sequence** is recognized specifically by an individual protein kinase type. By convention the amino acid residues that are situated C-terminally of the phosphorylation site are numbered with a positive prefix and the residues that are situated N-terminally are numbered with a negative prefix. Depending on whether a consensus phosphorylation sequence is found only in a few or in many proteins, protein phosphorylation may be either a strongly selective event with a high degree of "privacy" or a more "public" reaction addressing a large number of different targets. However, even those protein kinases that are not highly selective because their cognate consensus phosphorylation sequence is abundant among proteins are able to select individual substrates by interacting with additional binding or docking sites, targeting subunits and scaffold proteins (see Sections 4.6. and 11.1 for more details).

The mechanisms of action of protein kinases and phosphatases as well as the impact of phosphorylation on protein structure are discussed in the following sections. Individual phosphorylations and questions regarding the evolution of protein phosphorylation are treated later.

Figure 2.19 Protein phosphorylation switch The switch consists of a protein kinase (PK), a protein phosphatase, and a substrate protein. In the absence of ATP, the protein kinase is inactive (in analogy to other switching elements here it is formally shown as ADP-loaded). The active kinase transfers a phosphate residue (P) from ATP to the substrate protein, the conformation and function of which are thus changed, resembling the output signal. A phosphoprotein phosphatase sets the switch back to 0. Input signals control the activation of the kinase (summarized in Figure 2.26) and of the phosphatase.

Figure 2.20 Biochemistry of protein phosphorylation (A) Reversible phosphorylation of a group XH on a protein as catalyzed by a protein kinase and a protein phosphatase. (B) Amino acid residues shown to become phosphorylated (black circles show the phosphate acceptor site X; teal circles symbolize the phosphoryl group PO_3^{2-}, somewhat imprecisely biochemists call the transfer of this group between molecules "phospho=transfer reaction").

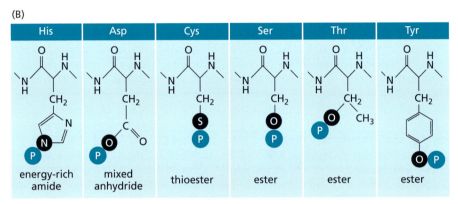

2.6.1 Protein kinases are the workhorses of signal transduction

Protein kinases constitute one of the largest enzyme families known. In the human genome, 518 kinase genes (1.7% of all genes, third-largest gene family) and more than 100 inactive pseudogenes have been found. The corresponding numbers for other species are 130 (2.1%, largest gene family) for the yeast *Saccharomyces cerevisiae*, 454 (2.4%, second-largest gene family) for the fly *Drosophila melanogaster*, and 1049 (4.1%, largest gene family) for the herb *Arabidopsis thaliana*. Also in prokaryotes, protein kinase genes cover a significant percentage of the genome.

There are two large kinase families: the mostly prokaryotic **histidine autokinases** and the so-called **eukaryotic protein kinases**. Eukaryotic kinases, which are also found in prokaryotes, catalyze the phosphorylation of either Ser and Thr or of Tyr residues, or in a very few cases of all three amino acids. They are not related to the histidine autokinases. In addition, several minor families of protein kinases are known that have been called **atypical** because in their structures they differ considerably from the two major families.

Histidine autokinases: structural elements

The catalytic core of a histidine autokinase covers approximately 160 amino acids containing several conserved sequences that are characteristic for this kinase family. They form a binding site for the ATP–Mg^{2+}–substrate complex. This site is characterized by a typical structural element, the so-called Bergerat fold, and harbors the amino acid residues required for the phosphotransfer reaction. The conserved sequences are constituents of an α, β sandwich consisting of five planarly arranged anti-parallel β-sheets superimposed by a layer of three α-helices (**Figure 2.21**). This molecular architecture distinguishes the histidine autokinases from other protein kinases, relating them to chaperone HSP90, type 2 topoisomerase/gyrase (catalyzing the formation of superhelices from circular double-stranded DNA), and MutL recombinase (an enzyme for DNA repair). Together these enzymes constitute the **GHKL family** (from Gyrase, HSP90, Kinase, and MutL). They all generate states of higher order, thereby hydrolyzing ATP.

A special type of His autokinases is the EI enzymes of the bacterial phosphotransferase systems, which use phosphoenolpyruvate instead of ATP as a phosphate donor. The prokaryotic protein kinases are discussed in detail in Section 3.3.2.

Figure 2.21 **Catalytic cores of protein kinases** Shown are the catalytic centers of (A) a bacterial His autokinase (EnvZ of *E. coli*) and (B) a eukaryotic protein kinase (cAMP-dependent protein kinase of mice).

Eukaryotic protein kinases: structural elements

Eukaryotic protein kinases are by far the largest and probably the most ancient kinase family. Their catalytic domain, covering about 250 amino acids, has another basic structure than that of His autokinases. Characteristic building blocks are 12 highly conserved subdomains, allowing an identification of the corresponding genes. These subdomains contain several invariant amino acid residues that are essential for the binding of the ATP–Mg^{2+} (or Mn^{2+}) complex and of the substrate protein, the phosphotransfer reaction, and the regulation of enzymatic activity. In the 1990s the three-dimensional structure was first elucidated by X-ray crystallography for cyclic adenosine 3′, 5′-monophosphate (cAMP)-dependent protein kinase A (PKA). Further studies on numerous other kinases proved this structure to be representative for eukaryotic protein kinases (see Figure 2.21). The predominant structural features are two lobes connected by a hingelike sequence, subdomain V, and forming the catalytic cleft. This structure opens and closes like a mouth with the ATP binding site localized inside and the binding site for the substrate protein outside the oral aperture.

The N-terminal "upper jaw" (N-lobe) with subdomains I–IV is composed of five anti-parallel β-sheets and harbors the ATP binding site, whereas the larger C-terminal "lower jaw" (C-lobe) with subdomains VIA–XI consists mainly of α-helices that are responsible for binding the substrate protein and for the phosphotransfer reaction. Like GTP in G-proteins, ATP is fixed by hydrogen bridges and electrostatic bonds between its nontransferable α- and β-phosphate groups and corresponding amino acids (in subdomains I–IV) as well as by a hydrophobic pocket (in subdomain I) fitting the adenine residue. Subdomain I contains a glycine-rich sequence that is essential for binding the phosphoryl residues. In analogy to G-proteins and other phosphate-transferring enzymes, this sequence is called the P-loop.

In subdomain VIB (in the "lower jaw"), a catalytic loop structure is localized that is characterized by two invariant amino acid residues. One of them is an Asp residue essential for the phosphotransfer reaction; the other is a basic residue (Lys or Arg) neutralizing the negative charge of the γ-phosphate group of ATP, thus facilitating nucleophilic attack by the substrate protein. For the phosphotransfer reaction, the triphosphate group of ATP must be arranged linearly. This is brought about by an Mg^{2+} ion bound as a salt by acidic amino acid residues in subdomain VII.

Subdomain VIII stabilizes the architecture of the "lower jaw" by means of a highly conserved Ala-Pro-Glu sequence. It also harbors a binding site for the substrate protein as well as an **activation loop** (also known as **T-loop** because it contains a strategic Thr residue). The three-dimensional structure of the catalytic cleft is common

Figure 2.22 Different mechanisms of action for His autokinases and eukaryotic protein kinases (A) Upon autophosphorylation, His autokinases (HK) provide phosphate donors for an autophosphorylation of proteins called response regulators (RR) at Asp residues. (B) Eukaryotic protein kinases (PK) catalyze the phosphorylation of substrate proteins (SP) by ATP (step 2). To acquire full enzymatic activity, many of them have to undergo autophosphorylation first (or become phosphorylated by another kinase, step 1). However, in contrast to His autokinases, phosphorylation occurs at hydroxy-amino acids such as Ser, Thr, and Tyr (ROH). These phosphoryl groups (P) cannot be transferred to other proteins but instead change the conformation of the kinase, facilitating interaction with ATP and the substrate protein (SP). To terminate eukaryotic protein phosphorylation, a protein phosphatase is needed to inactivate the (auto)phosphorylated kinase.

to all eukaryotic protein kinases. The specificity for selective phosphorylation consensus sequences is due to subtle variations in the amino acid sequences of the substrate-binding sites resulting in altered charge and hydrophobicity. Tyr-specific kinases differ from Ser/Thr-specific enzymes by a deeper catalytic cleft that is able to accommodate the more voluminous aromatic side chain of the tyrosine residue.

Different mechanisms of action

The different structures of His autokinases and eukaryotic protein kinases are mirrored by two fundamentally different mechanisms of action (**Figure 2.22**). Upon activation by an input signal, His autokinases autophosphorylate each other in a process called trans-autophosphorylation. The phosphorylation site (H box) is localized *outside* the catalytic center in a dimerization domain consisting of two anti-parallel helices, known as the HPt domain (for details see Section 3.3). The amide bond of the phosphorylated histidine residue is unstable, that is, energy-rich. Therefore, in a kind of pingpong mechanism, the phosphoryl group is immediately transferred to other proteins that are called response regulators. These are protein autokinases that use the autophosphorylated His kinases as phosphate donors for an autophosphorylation of aspartate residues. Thus, His autophosphorylation resembles a *short-lived transition state* of a phosphotransfer reaction between two proteins.

Eukaryotic protein kinases catalyze the phosphorylation of other proteins, though most of them also undergo (trans-)autophosphorylation of the activation loop *within* the catalytic center. This regulatory autophosphorylation is not a short-lived transition state of a phosphotransfer reaction; rather, it is a *stable post-translational modification* activating the enzyme by an allosteric conformational change.

In many protein kinases the activation loop closes the oral aperture and hinders contact between ATP and substrate protein (see Figure 2.21). When the kinase receives an input signal, the loop turns aside, allowing access to the catalytic center. For more sustained activation the loop has to be phosphorylated at one strategic Thr and/or Tyr residue. This is accomplished by either autophosphorylation or by another protein kinase (see Figures 2.22 and 2.23). Most—but not all—kinases require such **activation loop phosphorylation** to become fully active.

Since most kinases are intrasterically blocked by inhibitory motifs, activation loop phosphorylation occurs only in the presence of input signals that release the enzyme from the intramolecular restraints and open the catalytic cleft between the N- and C-lobes. But even then in many kinases the activation loop may not be accessible for

inhibitory sequence
(for example, pseudosubstrate)

INPUT SIGNALS

Ca²⁺ (calmodulin)
cAMP
cGMP
DAG
phospholipids
ceramide
5-AMP
RNA
DNA
G-proteins
receptors
other regulatory proteins
growth factors, insulin
protein kinases
protein phosphatases

catalytic domain

CONFORMATIONAL CHANGE

substrate

chaperone
HSP90

activation loop
phosphorylation

INPUT SIGNALS

Figure 2.23 Activation of protein kinases In an inactive protein kinase, the catalytic domain is blocked by an inhibitory peptide sequence. Interaction with an activator (a selection is shown in the teal box) induces a conformational change by which the inhibitory sequence becomes displaced and the catalytic domain is opened for the interaction with the substrate protein and ATP. To gain full activity, many protein kinases require, in addition, a phosphorylation of the activation loop in the catalytic center, which frequently requires the assistance of a chaperone such as HSP90 (see Figures 2.21 and 2.22). Such protein kinases are logical gates with at least three signaling inputs.

phosphorylation but must be uncovered by an additional conformational change induced by a chaperone ATPase such as HSP90. In addition, this chaperone protects protein kinases from dephosphorylation, thus prolonging their active state.

Frequently the inhibitory motifs stabilizing the inactive kinase conformation are **pseudosubstrate domains**. These are polypeptide sequences that resemble the phosphorylation consensus sequence of a substrate protein and are bound by the catalytic center. However, since they lack the phosphoacceptor amino acid (which is replaced by a nonphosphorylatable residue, for instance, Ser by Ala), these sequences are not phosphorylated but act as competitive inhibitors. The interaction of the kinase with an input signal induces a conformational change displacing the inhibitory domain from the catalytic center (**Figure 2.23**). The pseudosubstrate sequence may be either a part of the kinase molecule (as in the case of protein kinase C, Section 4.6.3) or localized on separate subunits (as in the case of protein kinase A, Section 4.6.1).

Apart from pseudosubstrate sequences, other inhibitory structures may bring about intrasteric auto-inhibition. An example is provided by domains containing a phosphorylated tyrosine residue interacting specifically with certain phospho-Tyr binding domains. Activation of this kinase type requires an enzymatic dephosphorylation of the critical Tyr residue or a competitive interaction with a protein having a phospho-Tyr binding site. In most of those kinases, an additional phosphorylation of the activation loop is mandatory (for example, see tyrosine kinase Src, Section 7.1.2).

Kinases that require both the displacement of an auto-inhibitory domain from the catalytic center and an activation loop phosphorylation to be turned on have, so to speak, a double safety lock and two (or three if they require the assistance of a chaperone) separate input terminals for signaling. Thus, they represent logical AND gates or coincidence detectors of at least two different input signals. In addition, many protein kinases contain further input terminals, giving them the properties of more complex logical switches or molecular nanoneurons. Owing to such safety locks, most protein kinases are inactive in the absence of input signals. There are, however, exceptions to this rule: for instance, glycogen synthase kinase 3, which is constitutively active and becomes *inhibited* by input signals (see Section 9.4.4). For a binary switching event it is insignificant, however, whether the event occurs by inhibition or by activation. What matters is that the switch has been turned.

2.6.2 Effects of phosphorylation on proteins

We shall now turn to the question of how phosphorylation changes the function of a protein. Exhibiting a pK value around 7, the phosphate group carries two negative charges in neutral medium, and, in addition, is able to form hydrogen bonds via its four oxygen atoms. We have to expect, therefore, rather pronounced effects on protein structure. Apart from this, the negative charge of the phosphate group could, solely by **electrostatic repulsion**, hinder negatively charged substrates from approaching the active center of an enzyme. This rather infrequent mechanism has been found for the bacterial enzyme isocitrate dehydrogenase kinase, which inhibits isocitrate dehydrogenase by phosphorylating a Ser residue in the catalytic core. A similar inhibition could be achieved experimentally by replacing this Ser residue by a negatively charged Glu or Asp residue but not by neutral residues, demonstrating an electrostatic mechanism. Moreover, X-ray crystallography did not provide any indication for a conformational change.

As a rule, however, phosphorylation leads to considerable **changes of protein conformation** (though it must not be overlooked that only a few proteins have been studied thoroughly in this respect). When, for instance, muscle glycogenphosphorylase, a homodimeric enzyme, is phosphorylated by phosphorylase kinase at a single Ser residue, a long-range allosteric rearrangement (extending over a distance of 35 Å) takes place, opening the active center for the substrate. In the homologous yeast enzyme an allosteric inhibitor, glucose 6-phosphate, is displaced by such a rearrangement.

Apart from such allosteric long-range effects, a phosphate group causes conformational changes in the immediate vicinity as well. As an example, we have already discussed the effect on protein kinases of activation loop phosphorylation. Here electrostatic effects between the phosphate group and negatively charged amino acid residues help to open the active center for the substrates. Indeed, a comparable activation is achieved when the phosphate-acceptor amino acid is replaced by a negatively charged Glu or Asp residue. Some kinases are already equipped with such residues at the critical position in the activation loop and do not require phosphorylation to be activated. Sometimes phosphorylation does not exhibit any measurable effect on protein function, an occurrence that is known as **silent phosphorylation**.

Phosphoamino acid-binding domains: controlling complex formation, degradation, and sequestration

Many proteins contain contact domains that recognize and bind to amino acid sequences with phosphorylated Ser, Thr, or Tyr residues. Therefore, another important function of protein phosphorylation is to generate the complementary domains, thus bringing about **reversible contacts** between proteins (or between domains within the same protein molecule). Such contacts are essential for the ad hoc assembly of signal-processing complexes and networks (for more details see Sections 2.13 and 7.1.2). Binding sites for sequences with phosphorylated Tyr residues are the SH2 and the PTB domains, whereas phospho-Ser residues are recognized by 14-3-3, WW, and MH2 domains, and phospho-Thr by FHA and WD40 domains. These domains are found in a wide variety of proteins that function in a vast range of cellular events.

Leaving aside the formation of signaling multiprotein complexes, among these events protein degradation and sequestration are of uppermost importance for signal transduction. In both processes, protein phosphorylation plays a key role. Proteolytic protein degradation may influence signaling in both a positive and a negative way depending on whether it leads to release of an active fragment from an inactive precursor protein or to complete destruction. A major pathway of proteolysis makes use of ubiquitylation for targeting the protein to proteasomes. As explained in Section 2.8, ubiquitylation is catalyzed by three enzymes, with the **ubiquitin ligases** (enzymes E3) lending substrate specificity. Many ubiquitin ligases recognize

PROTEIN PHOSPHORYLATION

14-3-3
proteins
↓
sequestration

ubiquitin–
proteasome
pathway
↓
degradation

↓
complex
formation

↓
activation
inactivation

Figure 2.24 Consequences of protein phosphorylation

their substrate proteins only when these have become phosphorylated. Therefore, protein phosphorylation is a common trigger for ubiquitylation and degradation.

The same holds true for protein sequestration. The key players in this game are the **14-3-3 proteins**, which recognize their partner proteins by specific amino acid sequences carrying phosphorylated Ser residues. Like protein degradation, sequestration may have both negative and positive effects on signaling because it either reserves the protein for a certain period of time or facilitates its integration into a signaling complex. The 14-3-3 proteins are treated in more detail in Section 10.1.4. The major consequences of protein phosphorylation are summarized in **Figure 2.24**.

2.6.3 Protein kinase families of vertebrates:

The whole set of protein and lipid kinases encoded by the genome of a given organism is called **kinome**. The human kinome comprises 518 protein kinases. Four hundred and seventy-eight of them exhibit a distinct sequence homology of their catalytic subunits and are called **eukaryotic protein kinases (ePKs)**. The remaining 40 members of the kinome lack this sequence homology and are called **atypical protein kinases (aPKs)**. On the basis of structural and functional similarities, the eukaryotic protein kinases are subdivided into several families that can be arranged in a phylogenetic tree (**Figure 2.25**). The major branches of this tree represent the following kinase families.

AGC family

Sixty-one kinases are arranged in this group comprising the prototypical representatives: protein kinases A, G, and C. It includes the following enzymes:

- cAMP-activated protein kinases, PKA (see Section 4.6.1)
- cyclic guanosine monophosphate (cGMP)-activated protein kinases G, PKG (see Section 4.6.2)
- diacylglycerol- (and Ca^{2+}-) activated protein kinases cPKC and nPKC (see Section 4.6.3)
- phosphoinositide-activated protein kinases PKB/Akt, PDK1, atypical PKC, and PKC-related PRK (see Sections 4.6.3 and 4.6.6)
- Gβγ-activated protein kinases for G-protein-coupled receptors (GPCR kinases, Sections 5.7 and 5.8)
- ribosomal S6 kinases S6K and RSK (see Section 9.4.2)

As can be seen, many AGC kinases are activated by low-molecular-mass second messengers such as cAMP, cGMP, diacylglycerol, Ca^{2+} ions, and phosphoinositides as well as by Gβγ-subunits of trimeric G-proteins.

Moreover, practically all of them require activation loop phosphorylation of a conserved Thr residue, which is catalyzed by the kinase itself (autophosphorylation) or by an upstream kinase, frequently the phosphoinositide-dependent kinase PDK1

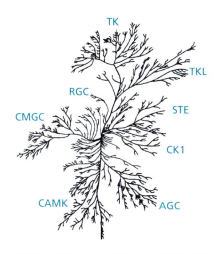

Figure 2.25 The phylogenetic tree of eukaryotic protein kinases Each endpoint of a branch represents an individual protein kinase. The kinase families are depicted in teal. (Adapted from Chartier M, Chenard T, Barker J & Najmanovich R (2013) Kinome Reader: a stand-alone and web-accessible tool to annotate the human protein kinase tree. *PeerJ* 1:e126.)

(which activates itself by autophosphorylation). For maximal activation most AGC kinases have to be phosphorylated, in addition, on a Ser residue in a C-terminal hydrophobic sequence.

Ca^{2+}/calmodulin-dependent kinases (CAMK family)

For activation, these enzymes require Ca^{2+} ions that interact with a regulatory subunit, the Ca^{2+}-binding protein calmodulin (CaM). Among many other enzymes, the CaM kinase family includes the following:

- phosphorylase kinase of glycogen metabolism (see Section 5.3.1)
- myosin light-chain kinase of the actomyosin system (see Section 14.6.2)
- elongation factor 2 kinase (also known as CaM kinase III) of ribosomes (see Section 9.3.4)
- multifunctional CaM kinases I, II, and IV (see Section 14.6.3)
- distantly related CaM kinase kinases (see Section 14.6.3)

The CaM kinase family also includes structurally and phylogenetically related kinases that do not contain calmodulin as a subunit. Examples include protein kinase D (see Section 4.6.5) and 5′-AMP-activated protein kinase (see Section 9.4.3). It should be noted that the conventional term CaM kinases is somewhat misleading because these enzymes do not phosphorylate their activator CaM.

CMGC kinases

These phylogenetically related protein kinases are proline-directed: they catalyze the phosphorylation of Ser and Thr residues in the neighborhood of Pro residues. The acronym CMGC is derived from prototypical members of this family:

- cyclin-dependent protein kinases CDK of cell cycle regulation (see Section 12.4)
- MAP kinases (see Section 11.2)
- glycogen synthase kinases GSK (see Section 9.4.4)
- CDK-like kinases

Ste family

The kinases of this family are relatives of the yeast enzyme Ste 11 and others that when deleted cause sterility in yeast (see Section 11.4). Most of them are constituents of MAP kinase modules functioning as MAP2 and MAP3 kinases (see Sections 11.2 and 11.3).

Further serine/threonine-specific kinase families of vertebrates

From the phylogenetic point of view, these enzymes cannot be classified as belonging to one of the kinase families. Among others, this group includes the following:

- "casein kinases" CK1 and CK2 (see Section 9.4.5)
- kinases WEE1 and MYT1 of cell cycle regulation (see Section 12.4)
- kinases Nek/Nrk, Plk, and Aurora of mitosis control (see Section 12.12)
- kinases IKK of the NFκB modules (see Section 11.7)
- receptor kinases of the TGFβ/activin receptor family (see Section 6.2)

The last are found in a so-called Tyr kinase-like kinase (TKL) family, the members of which are bona fide Ser/Thr kinases, albeit exhibiting a Tyr kinase-related structure. Other kinases belonging to this family are the MAP3 kinases Raf and MLK (see Section 11.3), the IRAKs (see Section 6.3.2) and the LIM kinases (see Section 10.3.3).

A distinct branch of the kinome tree comprises the receptor guanylate cyclases (RGCs). These are pseudokinases containing an inactive kinase domain. They catalyze the formation of cyclic GMP from GTP but do not phosphorylate proteins (see Section 4.5). Beside the receptor guanylate cyclases almost 50 pseudokinases are encoded by the human genome. They are assumed to be involved in the control of active protein kinases (for an example see Section 11.5).

Protein tyrosine kinases (TK family)

Although in mammalian cells only about 2% of phosphoproteins are phosphorylated at Tyr (as compared with 85% phosphorylated at Ser and 13% phosphorylated at Thr), tyrosine phosphorylation plays a pivotal role in cellular signal processing. The 84 human Tyr-specific protein kinases are classified in cytoplasmic (soluble) and receptor tyrosine kinases. They will be discussed in detail in Sections 7.1 and 7.2. The so-called bacterial Tyr kinases (BY-kinases) of prokaryotes are not related to the eukaryotic enzymes but constitute a protein kinase family of its own (see Section 3.5.2).

Some kinases catalyze the phosphorylation of both Tyr and Thr: that is, they are said to be **dual-specific**. The most prominent dual-specific kinases are those of the MEK and MAP2K families, which are highly specific activators of MAP kinases (see Section 11.2). They belong to the Ste11 family.

Representatives of atypical kinases are, the phosphatidylinositol 3-kinases (see Section 4.4.3) and the related kinases mTOR (see Section 9.4.1), the DNA-activated kinases DNA-PK and ATM/ATR (see Section 12.8.1), and the His kinase-related pyruvate dehydrogenase kinases of mitochondria (see Section 3.5.1).

2.6.4 Phosphoprotein phosphatases

Protein kinases switch their substrate proteins on or off by phosphorylation. To return the switch to the 0 position, the substrate protein must be dephosphorylated. This occurs by hydrolysis catalyzed by (phospho)protein phosphatases. Depending on structural parameters, the protein phosphatase family is classified in several nonrelated subfamilies:

- Ser/Thr-specific phosphoprotein phosphatases (PPP)
- Ser/Thr-specific phosphoprotein phosphatases requiring Mg^{2+} (PPM)
- conventional phosphotyrosine phosphatases (PTP)
- small phosphotyrosine phosphatases
- Thr/Tyr-specific Cdc25 phosphatases
- prokaryotic phosphohistidine and phosphoaspartate phosphatases
- prokaryotic protein kinases/phosphatases

PPP family

The enzymes of the PhosphoProtein Phosphatase family, including the subfamilies PPP1, PP2A, PPP2B, and PPP5, have a catalytic core of about 280 amino acids that is characterized by three short sequences ranking among the most conserved structures of eukaryotic enzymes. The core consists of a β-sandwich formed by two planar β-sheets facing each other and being surrounded by several α-helices. In the upper part of the sandwich, two catalytically essential metal ions (either Fe^{3+}, Zn^{2+}, or Mn^{2+}) are bound to Asp and Glu residues. They polarize adjacent water molecules, thus activating them for a nucleophilic attack on the phosphoric ester bond of the substrate protein.

Except for the PPP5 subfamily, the PPP enzymes are composed according to a unit construction system: a few catalytic subunits combine, mostly by noncovalent binding, with a wide variety of regulatory subunits. The latter target the enzymes to distinct cellular sites, control substrate specificity, and modulate enzymatic activity (**Figure 2.26**). The eukaryotic phosphatase PP1 of the PPP1 subfamily has been

Figure 2.26 Different regulation of Ser/Thr phosphatases as compared with Tyr phosphatases In Ser/Thr phosphatases, a catalytic domain may combine noncovalently with various regulatory domains. In Tyr phosphatases, both catalytic and regulatory domains share the same polypeptide chain.

estimated, for instance, to interact with no fewer than 50 different regulatory proteins. Prominent examples of regulatory subunits are those that render PP1 specific for either myosin light chains (see Section 14.6.2) or, alternatively, enzymes of glycogen metabolism (see Section 5.3.1).

The PPP family is very old. Homologous enzymes are found in archaea and bacteria, perhaps playing a role in two-component signal transduction (see Section 3.3). The prokaryotic enzymes also hydrolyze phospho-Tyr bonds.

PPM family

These phosphoprotein phosphatases need exogenous Mg^{2+} ions (therefore 'M'). Although their amino acid sequences are quite different, the three-dimensional architecture of the catalytic core is similar to that of PPP enzymes. Obviously we are dealing with phylogenetic convergence. Whether or not the unit construction principle holds true also for PPM enzymes remains an open question. Like PPPs, PPMs also have a prokaryotic ancestry.

Tyrosine-directed and dual-specific protein phosphatases

Within this group, the **conventional phosphotyrosine phosphatases (PTPs)** constitute by far the most variable group of eukaryotic protein phosphatases (for more details see Section 7.4). In both structure and mechanism of action, they differ considerably from PPP and PPM enzymes. The catalytic core of PTPs covers 230 amino acids, forming a central group of seven anti-parallel β-sheets surrounded by α-helices. The phosphoric acid ester bond of the substrate protein is split by a nucleophilic attack of a cysteine residue, rather than by metal ion-activated water, and generates a labile phosphoric acid thioester as an intermediate that is hydrolyzed in a second step. The catalytically active cysteinyl–SH groups of these enzymes are oxidized quite easily to intramolecular or mixed disulfide or sulfenic acid groups, which are reduced again enzymatically. The activation of many PTPs is therefore assumed to be reversibly regulated by **redox signals** (see Sections 2.2 and 7.1). Another important feature of PTPs is that, in contrast to the Ser/Thr-specific phosphatases, the regulatory domains are integrated into the enzyme molecule rather than being localized on separate subunits (see Figure 2.26).

Some bacteria express conventional PTPs obtained from eukaryotes through horizontal gene transfer. These enzymes are potent virulence factors (see Section 3.5.3). The **small PTPs** seem to be restricted mainly to eukaryotes. They have molecular masses around 18 kDa and are not related to conventional PTPs, although they also hydrolyze substrates via a thioester intermediate. The dual-specific PTPs of the **Cdc25 family** are probably also restricted to eukaryotes. They are absolutely specific for the cell cycle-controlling protein kinases of the CDK family, activating these by dephosphorylation of two vicinal Tyr and Thr residues. Again the reaction involves a thioester intermediate, though no genetic relationship to other PTPs exists.

Prokaryotic histidine and aspartate kinases/phosphatases

These ancient enzymes are bifunctional autokinases/protein phosphatases undergoing auto-inactivation by autodephosphorylation. This mechanism again demonstrates that such prokaryotic protein phosphorylations represent transition states of phosphotransferase reactions. However, in some bacterial species, special phospho-His phosphatases such as SixA and phospho-Asp phosphatases such as CheZ and SpoOE have been found (see Section 3.6.3). These enzymes are highly specific. Since they have to be present in stoichiometric amounts and work only when the Asp-directed autophosphatase activity of their substrate kinases is intact, they are probably allosteric activators rather than genuine phosphatases, functioning in a manner analogous to that of GAPs in that they increase the endogenous phosphatase activities of His and Asp autokinases.

Other prokaryotic enzymes exhibiting both kinase and phosphatase activities are HPr kinase/phosphatase and 6-isocitrate dehydrogenase kinase/phosphatase (see

Section 3.5.3). The combination of both functions in one enzyme appears to be an evolutionarily ancient model that was rejected by eukaryotes. In fact, the separation of kinase and phosphatase activities offers many more possibilities for regulation and network formation. A striking analogy exists between bifunctional kinases/phosphatases and G-proteins in that, via ATP or GTP hydrolysis, both modulate the function of substrate proteins and transfer the released phosphate group onto water.

For vertebrate cells the analysis of the overall protein phosphorylation status (measured by phosphoamino acid analysis upon labeling with ^{32}P, or by mass spectrometry) yields 90% Ser, 10% Thr, and 0.05% Tyr phosphorylation, although the Ser:Thr:Tyr ratio in proteins is about 3:2:1 on average. These steady-state values are due to a combined action of protein kinases and protein phosphatases with the latter showing a bias towards phospho-Thr and in particular phospho-Tyr dephosphorylation.

2.6.5 Phosphorylation cascades and kinase modules

Phosphorylation reactions are often connected in series, with one kinase activating the next. Such reaction cascades or phosphorelays serve signal amplification, noise suppression, temporal sharpening of signals, frequency modulation, and cross talking with other signaling pathways. Frequently the kinases that form a cascade are assembled in modules stabilized by scaffolding proteins. Numerous examples of these cascades are presented in the subsequent chapters.

The most prominent and probably most versatile cascade type is the **MAP kinase cascade**, a central relay and distributor device of eukaryotic signal processing. It will be discussed in detail in Chapter 11. MAP kinase stands for Mitogen-Activated Protein kinase, a historical term because this kinase type was discovered by studies on mitogenic stimulation of cells. Since then it has become clear that MAP kinases are involved in many more cellular processes.

MAP kinase modules are assembled by three protein kinases connected in series. These are the MAP kinases proper; their regulator kinases are named MAP kinase kinases or MAP2 kinases, and the sensor kinases for input signals are called MAP kinase kinase kinases or MAP3 kinases (**Figure 2.27**). These elements of a MAP kinase

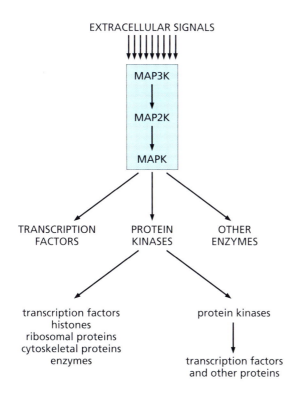

Figure 2.27 MAP kinase modules: universal relay stations Like a ganglion, a MAP kinase module (boxed) processes many different input signals and transmits a large number of output signals, thus influencing almost any cellular process.

module are assembled on scaffolding proteins in such a way that the analogy with integrated electronic circuits is quite striking.

MAP kinase modules are components of numerous signaling pathways connecting almost every receptor type with the metabolic machinery and the genome. The evolutionary success of the module is probably due to the fact that such a device works as a highly efficient noise filter, delivering an extremely sharp signal as a prerequisite for frequency modulation. Indeed, MAP kinase modules are integrated in positive and negative feedback loops in such a way that the conditions for generation of oscillating signals are fulfilled, though experimental proof is still lacking. A frequency-modulated code could easily explain why, depending on the conditions, a single MAP kinase module may exert contrary effects: for example, stimulation of cell proliferation and survival upon short-term activation versus an antimitotic effect or even induction of cell death upon long-term activation. Possibly the survival signal simply has another frequency than the death signal.

Summary

- Phosphorylation ranks among the most ancient and most abundant post-translational modifications of proteins. It is catalyzed by protein kinases that use ATP (or in a few cases GTP or phosphoenolpyruvate) as a phosphate donor. Protein phosphatases render the modification reversible.

- In addition to some minor families of atypical protein kinases, two unrelated superfamilies of protein kinases are known: His autokinases (predominantly found in prokaryotes) and kinases phosphorylating Ser, Thr, and Tyr residues.

- Protein kinases are signal-operated switching elements. In the absence of input signals, most protein kinases are blocked by auto-inhibition. Many kinases require, in addition, phosphorylation of the activation loop (a structure in the catalytic center) to become active.

- Frequently several protein kinases are combined into multiprotein modules; the eukaryotic MAP kinase module being the most prominent example.

- Phosphorylation has three major effects on proteins: electrostatic repulsion of negatively charged partner molecules, conformational changes resulting in altered activity, and generation of contact sites for interacting proteins.

- Like kinases, protein phosphatases are also signal-operated switching elements. Ser/Thr-specific phosphatases exist in a few catalytic domains that noncovalently couple with a large number of regulatory subunits. In Tyr-specific phosphatases, the regulatory and catalytic domains share the same polypeptide chain.

2.7 Protein acetylation and methylation: tools of gene regulation and more

Apart from phosphorylation, acetylation and methylation of proteins represent further post-translational modifications with signal character. Targets are the terminal amino groups of lysine and arginine residues.

Protein acetylation

Protein acetyltion is catalyzed by acetyltransferases that obtain the acetyl residue from acetyl-coenzyme A to the ε-amino group of lysine or the guanidino group of

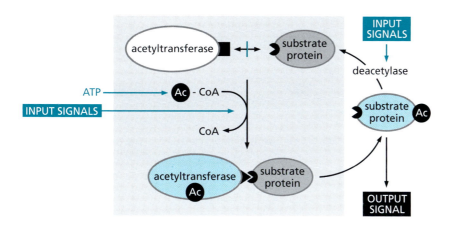

Figure 2.28 Acetylation switch The switching event consists of acetylation of a substrate protein on Lys or Arg residues. The acetate donor is acetyl-coenzyme A (Ac-CoA). The reaction is catalyzed by an acetyltransferase, whereas a deacetylase removes the acetyl residue from the substrate protein. Both enzymes are controlled by input signals. The conformational and functional change induced by acetylation in the substrate protein is the output signal. The switching energy is delivered by splitting the energy-rich thioester bond of Ac-CoA. Reloading of CoA with acetate is an ATP-consuming process.

arginine residues. Corresponding deacetylases hydrolyze the amide bond and thus reverse the switching reaction (**Figure 2.28**).

Both types of enzyme exist in many isoforms that, like protein kinases and phosphatases, differ from each other by their regulatory domains; that is, they respond to different input signals and exhibit different substrate specificities. The deacetylases are, moreover, subdivided into two families with quite different mechanisms of action: hydrolases, which transfer the acetyl residue onto water, and ADP-ribosyltransferases, which transfer the acetyl residue onto ADP-ribose derived from NAD$^+$ (more details are found in Section 8.2.3).

The acetylation of histones has been investigated most thoroughly since it plays a key role in gene activation, with histone acetyltransferases acting as transcriptional co-activators and histone deacetylases as co-repressors (for more details see Sections 8.2.2 and 8.2.3). However, protein acetylation is by no means restricted to histones but appears to be of similar importance as protein phosphorylation. This holds true for numerous transcription factors and related gene-regulatory proteins (for example, NFκB in Section 11.7 and p53 in Section 12.8.3). Their intracellular distribution between cytoplasm and nucleus, their DNA-binding efficiency, and their interactions with other proteins are frequently modulated by acetylation, mostly in combination with phosphorylation. The catalyzing enzymes are called factor acetyltransferases. Their best-known representatives are CBP (CREB Binding Protein) and the homologous p300 (molecular mass of 300 kDa; CREB is the cyclic AMP response element binding protein). The **CBP/p300 acetyltransferases** are prototypes of transcriptional co-activators. Owing to a large number of interactions (see the multidomain scheme in **Figure 2.29**), on chromatin they are able to form extended signal-transducing protein complexes which bundle up various signaling pathways at distinct gene loci. Such sites are marked by gene-specific transcription factors that are components of the multiprotein complex recruited by CBP/p300 (for more details see Chapter 8). Protein acetylation seems to be as ancient as phosphorylation since it is found in bacteria and archaea. For instance, the chemotaxis protein CheY of *Escherichia coli* (see Section 3.4.4) is controlled by phosphorylation as well as by acetylation.

Protein methylation

Protein methylation is catalyzed by methyltransferases that transfer one or more methyl groups from *S*-adenosylmethionine to the terminal amino groups of Lys and Arg residues in polypeptides. *S*-adenosylmethionine is an energy-rich compound generated from methionine and ATP. Its splitting delivers the switching energy of protein methylation. For demethylation the stable bond between methyl and amino group is broken up by oxidation releasing the methyl residue as formaldehyde. This reaction is catalyzed by two types of enzyme: amine oxidases using flavine ad-

Figure 2.29 Acetyltransferase CBP/p300: an integrator of many signaling pathways The figure shows, in a simplified manner, the multidomain structure of the 2441 amino acid residue protein together with the interaction partners. HAT, histone acetyltransferase domain; Bromo, domain for interaction with acetylated proteins; black rectangles, other interaction domains. Further details are found in Sections 8.2.2 and 8.2.5. As in other figures throughout this book, the protein is depicted as a linear structure with the interaction domains symbolized as larger rectangles. Proteins are always shown with the amino terminus to the left and the carboxy terminus to the right.

enine dinucleotide as a cofactor, and Fe^{2+} containing mixed-function oxygenases promoting the reaction of molecular oxygen with the methyl group and with the cofactor α-ketoglutarate. In addition, methylated arginine residues are demethylated yielding citrulline and methylamine (**Figure 2.30**). Protein demethylation is a highly dynamic process that can occur very fast, even in an oscillating pattern. This renders protein methylation a fully reversible post-translational modification that represents a signaling switch comparable to acetylation and phosphorylation. In fact, the number of protein methyltransferases (approximately 40 in humans) and protein demethylases (approximately 30 in humans) is comparable to that of Tyr-kinases and Tyr-phosphatases. The individual enzymes of protein methylation vary in substrate specificity and tissue distribution.

Prototypical for protein methylation is the methylation of histones that like acetylation is essential for the structure and function of chromatin and for gene regulation

Figure 2.30 Biochemistry of protein demethylation (A) Oxidative demethylation of methyllysine catalyzed by a mixed function oxygenase (1) or an amine oxidase (2). As a cosubstrate the mixed function oxygenase requires α-ketoglutarate, which is oxidized to succinate and CO_2 (not shown), the cofactor of the amine oxidase is flavine adenine dinucleotide (FAD). In both cases the methyl residue is released as formaldehyde. Along analogous pathways N-methylarginine is demethylated. (B) Demethylation of a methylarginine residue catalyzed by a peptidylarginine deiminase; arginine is transformed into citrulline and the methyl residue is released as methylamine.

(see Section 8.2.4). Since, however, more than 2000 potential methylation sites have been detected by mass spectrometry in non-histone proteins, the signaling function of protein methylation must be more widespread. For instance, various transcription factors undergo methylation concomitantly with other post-translational modifications such as phosphorylation and acetylation. Depending on target site methylation may either promote or antagonize the effects of other modifications. This may be illustrated by two examples. **FOXO transcription factors** stimulate proapoptotic genes. This effect is inhibited by protein phosphorylation catalyzed by protein kinase B/Akt causing degradation of FOXO (see Section 4.6.5 and Figure 4.42). The phosphorylation is specifically prevented by lysine methylation in the PKB/Akt target sequence of FOXO. The transcription factor **p53** is a central control element of cell proliferation, apoptosis and the response to DNA damage (see Section 12.8.3). It undergoes a variety of post-translational modifications. While methylation of multiple arginine residues promotes cell cycle arrest, demethylation of these sites induces apoptosis. Moreover, monomethylation of a specific lysine residue inhibits p53-controlled transcription, whereas dimethylation of the same site has a stimulatory effect. These and many other examples demonstrate that protein methylation plays an important role in the fine-tuning of signaling.

Summary

Reversible protein acetylation and methylation at Lys and Arg residues represent widespread post-translational modifications that are catalyzed by numerous enzymes. By affecting histones and transcription factors—but also many other proteins—they provide key reactions of transcriptional activation. In particular, histone acetyltransferases and deacetylases are the standard co-activators and co-repressors of gene transcription.

2.8 Protein ubiquitylation: more than a signal of protein degradation

Post-translational modifications of proteins are achieved not only by addition of low-molecular-mass groups such as phosphate or acetate but also by covalent binding of peptides. The prototype of such a peptide is ubiquitin, a highly conserved 8.5 kDa molecule containing 76 amino acids and found in all eukaryotic cells. Its binding to proteins represents an abundant signaling reaction comparable to phosphorylation and acetylation. In fact, in normally growing yeast cells up to 20% of all proteins have been found to be ubiquitin-conjugated and the same may hold true for animal cells. Since ubiquitylation is a highly dynamic process, this number represents a steady-state rather than the absolute number of proteins ubiquitylated in any possible situation.

A ubiquitylating enzyme complex is a module consisting of three components: the enzymes E1 (ubiquitin-activating enzyme), E2 (ubiquitin-conjugating enzyme), and E3 (ubiquitin ligase). While E1 takes care of ubiquitin activation, E2 catalyzes the ubiquitylation of the substrate protein, which is specifically presented by E3 (**Figure 2.31**).

Ubiquitin—which if not immediately available may be proteolytically released from a precursor protein—is activated by binding via its C-terminal glycine residue to a Cys residue of E1. This formation of an energy-rich thioester bond requires ATP hydrolysis and provides the energy source of the ubiquitylation switch. From E1 the ubiquitin residue is taken over by E2—again transiently bound as a thioester—and transferred to a substrate protein where its C-terminal glycine residue forms a so-called isopeptide bond with the ε-amino group of a lysine residue.

Since the ubiquitin residue is removed from the substrate protein by specific proteases or isopeptidases, ubiquitylation is, like phosphorylation, acetylation, and

Figure 2.31 Ubiquitylation switch In principle the switch resembles the phosphorylation switch. In an ATP-consuming reaction, the ubiquitin transferase complex transfers ubiquitin (UB) to a Lys residue of a substrate protein (see Figure 2.32) where it may be removed again by a deubiquitinase. Both ubiquitylation and de-ubiquitylation are controlled by input signals. The conformational and functional change induced by ubiquitylation in the substrate protein is the output signal. The switching energy is delivered by the hydrolysis of ATP or, to be precise, of pyrophosphate (PPi). This scheme applies to ubiquitin-like modifiers as well.

Figure 2.32 Ubiquitin transferase reaction Ubiquitin transferase is a complex of three cooperating enzymes, E1, E2, and E3. (A) Mono-ubiquitylation. In cells, ubiquitin (UB) is directly available or proteolytically released from a precursor protein. E1, the ubiquitin-activating enzyme, binds UB as an energy-rich thioester via a Gly–Cys bond. The reaction requires the hydrolysis of one molecule of ATP, generating a labile UB-AMP intermediate. From E1 the ubiquitin-conjugating enzyme E2 takes over the UB residue, binding it again as a high-energy thioester and transferring it to a substrate protein, where it forms an isopeptide bond between its C-terminal glycine and the ε-amino group of a Lys residue. The substrate protein is presented by the ubiquitin ligase E3 acting as an adaptor or scaffold protein. E3 plays a key role since it renders the ubiquitylation substrate-specific and is controlled by input signals. (B) Poly-ubiquitylation. Several UB residues become linked via isopeptide bonds. Depending on the linking site (Lys48 or Lys63), different downstream reactions result. Ubiquitin-like modifiers such as SUMO are activated and transferred in an analogous manner.

methylation, a reversible switching reaction (**Figure 2.32**). A substrate protein may be either **mono-ubiquitylated** by a single ubiquitin residue or **poly-ubiquitylated** by a polymerized ubiquitin chain. This chain consists of many ubiquitin residues connected by isopeptide bonds between the carboxy-terminal glycine residues and lysine residues. Since ubiquitin has seven lysine residues to be ubiquitylated, a wide variety of linear and branched polyubiquitin chains are constructed, rendering ubiquitylation the probably most versatile of all post-translational protein modifications. The mechanisms of ubiquitin chain assembly are still not known. Moreover, the reason a single protein can be monoubiquitylated in one case but polyubiquitylated in another case is not entirely clear. Possibly each reaction is catalyzed by a specific set of enzymes, or the activity of one enzyme is modulated by accessory factors. In fact, E3 ubiquitin ligases interact with several regulatory subunits that may be exchanged, depending on the input signals. For the discovery of ubiquitin-dependent protein degradation, Aaron Ciechanover, Avram Hershko, and Irwin Rose were honored with the 2004 Nobel Prize in Chemistry.

2.8.1 Ubiquitin E3 ligases transduce the ubiquitylation signal

By bringing together E2 enzymes with selected target proteins and a wide variety of regulatory factors, the ligase E3 determines the substrate specificity of ubiquitylation. While human E2s are encoded by about 40 genes, more than 600 different E3 genes have been found. The variety of ubiquitin ligases exhibiting different substrate specificities, control mechanisms, and tissue distributions thus rivals that of protein kinases. In addition, the significance of these enzymes is underscored by the finding that quite a number of them are tumor suppressors or may mutate to dominant-negative fragments that act as oncoproteins. E3 ubiquitin ligases recognize specific binding sites in their substrate proteins. Well-known recognition sequences are the destruction motifs or "degrons." Since substrate recognition depends on other signaling reactions (such as protein phosphorylation) E3 ubiquitin ligases are the signal-transducing components proper of the ubiquitin transferase complex. Characteristic domains for the interaction with E2 ubiquitin-conjugating enzymes define two E3 families: HECT domain ligases and RING domain ligases.

Ubiquitin ligases with HECT domains

The C-terminal HECT domain characterizing these enzymes contains a strategic cysteine (Cys) residue where the ubiquitin obtained from the ubiquitin-conjugating enzyme E2 is bound transiently as a thioester and transferred subsequently to the substrate bound at the N-terminal domain of the ligase.

A prototypical enzyme of this family is E6-AP (**Figure 2.33**) that is induced by the papilloma virus E6 (HECT stands for Homologous to E6-AP C-Terminus). By priming the transcription factor p53 for degradation, this enzyme allows passage of the DNA structure checkpoint (see Section 12.9) and prevents apoptosis of damaged cells to ensure virus replication. Another HECT domain ligase triggers the destruction of the SMAD transcription factors that mediate the effects of TGFβ and related tissue hormones (see Section 6.2).

Ubiquitin ligases with RING domains

The RING domain (named after a Really Interesting New Gene) contains a group of eight cysteine and histidine residues arranged in a finger-like structure, the "RING finger," that is stabilized by two zinc ions. As the active center of the E3 ligase, this domain is responsible for bringing about the specific interaction between the E2 enzyme and the substrate of ubiquitylation. Thus, in contrast to HECT domain ligases, the RING domain enzymes do not covalently bind the ubiquitin residue. RING domain enzymes constitute the largest family of E3 ubiquitin ligases. Representatives

Figure 2.33 E3 ubiquitin ligases E3 ubiquitin ligases play the role of scaffold proteins that bring together the ubiquitin-conjugating enzyme E2 (shown in color) with selected substrate proteins. Substrate-binding domains/subunits of E3 are depicted in black. E2 catalyzes the transfer of the ubiquitin residue (UB) from the ubiquitin-activating enzyme E1 to the substrate (gray). The figure shows some representative complexes of E2 with HECT and RING domain enzymes, that is, the major families of E3 ubiquitin ligases. HECT, RING, ROC1, and Apc11 are E3 domains/subunits with E2-binding RING domains; cullin, SKP1, elongins, and Apc proteins (except Apc11) are scaffold and adaptor proteins. As indicated APC/C may change between two substrate-binding subunits, Cdc20 and Cdh1 (see Section 12.10).

are MDM2, a negative regulator of cell cycle arrest and apoptosis at the DNA structure checkpoints (see Section 12.8.3), as well as the TNF receptor-associated factors (TNFRs) involved in cytokine signaling (see Section 6.3.1).

An example of a relatively simple RING ubiquitin ligase is provided by Cbl (named after the Casitas B-cell lymphoma of mice, see Section 7.1.1). Here the RING domain and the substrate-binding domain share the same polypeptide chain. This simple structure appears to be an exception rather than the rule, since in most other RING domain ligases the two domains are located on separate subunits.

RING domain ligases have a tendency to form large multiprotein complexes held together by scaffold proteins of the **cullin** family (comprising seven members in humans) and related adaptor proteins. The regulation of such complexes seems to be quite sophisticated. For instance, the function of the RING domain ligases is enhanced by reversible covalent binding of the ubiquitin-like peptide NEDD8 (Neuronal precursor cell-Expressed Developmentally Down-regulated protein 8) at a specific Lys residue of cullin. In contrast to poly-ubiquitylation, "**neddylation**" does not trigger proteolytic degradation.

A clear advantage of the multisubunit structure of most RING-domain ligases is that it offers the chance to choose between different substrate-binding or regulatory subunits, thus conveying a particularly high degree of versatility. Prominent examples of the more complex RING ubiquitin ligases are SCF, VBC, and APC/C.

SCF-type enzymes are composed of at least four subunits: an adaptor protein S̲KP1, the scaffold protein C̲ullin, the RING domain ligase Roc1/Rbx1, and a substrate-binding F̲-box protein (see Figure 2.33). An **F-box*** is a sequence motif of about 50 amino acids that, from yeast to humans (it is estimated that up to 100 F-box genes may exist), is found in various proteins, involved in protein–protein interactions. In humans, F-box proteins differing in their substrate specificities have been identified as subunits of E3 ubiquitin ligases. By choosing between different F-box proteins, an SCF-type ligase may tag different sets of proteins for destruction, for instance, during cell cycle progression.

To become recognized by an F-box, the degron of the substrate protein must be phosphorylated (**Figure 2.34**). This is because the F-box contains a phosphoamino acid-docking domain such as WD40 (a propeller-shaped phospho-Thr-specific sequence of about 40 amino acids ending with tryptophan [W] and aspartate [D]). This substrate phosphorylation provides a particularly clear example of the reversible ad hoc generation of a docking motif and explains why ubiquitylation frequently goes along with phosphorylation, thus representing a logical AND gate.

Cooperating in such a way with various protein kinases, SCF-type ligases prime numerous proteins for degradation in proteasomes. These substrates include the cyclins of the G1 phase, CDK phosphatases (Cdc25), CDK inhibitors (such as p21^{KIP} and p27^{KIP}, and the inactivating kinase WEE1), transcription factors such as E2F, β-catenin and the hedgehog-controlled factor Ci/Gli (see Section 5.9), Notch (see Section 13.1.1), and the inhibitor IκB of the transcription factor NFκB (see Section 11.7). When the variety of SCF substrates is considered, it becomes clear that these ligases play a key role in signal termination and adaptation.

SCF-type ubiquitin ligases are particularly abundant in plants (in *A. thaliana*, about 700 F-box genes have been found). Some of these enzymes have been identified as receptors for the phytohormones auxin and gibberellic acid that, upon ligand

---

*The name F-box goes back to **cyclin F**, which, however, is not a ubiquitin ligase. Instead, it binds to cyclin B–CDK1, probably transporting this complex into the nucleus. In fact, cyclin F has several nuclear localization sequences that cyclin B–CDK1 is lacking. More about cyclins and CDKs is found in Chapter 12.

Figure 2.34 Cooperation between phosphorylation and ubiquitylation: degradation of cyclin E In the active cyclin E–CDK2 complex, cyclin E is phosphorylated at a Thr residue (P). The phosphorylated sequence is recognized by the WD40 domain of an SCF-type E3 ubiquitin ligase. Upon binding by SCF, cyclin E becomes poly-ubiquitylated and, in turn, undergoes degradation in proteasomes. CDK2 is not ubiquitylated and not degraded.

binding, trigger the proteolytic degradation of transcriptional repressors, thus releasing hormone-responsive gene transcription from inhibition (see Section 13.1).

Another subfamily of RING ubiquitin ligases are the **VBC complexes**. They got their name from the ligase VHL and the subunits elongin B and elongin C. The ubiquitin ligase VHL is a tumor suppressor deleted in Von Hippel–Lindau disease (see Section 8.7) and in kidney carcinomas. Elongin B and C were described originally as components of the elongation complex of transcription. Both are adaptor proteins that, together with cullin, organize the VBC complex by binding the E2 enzyme and a substrate-binding protein (see Figure 2.33). The interaction domain of the latter is a **SOCS box** that is structurally and functionally closely related to the F-box (the name SOCS refers to the Suppressors Of Cytokine Signaling that also contain this motif; see Section 7.2.1). VBC ubiquitin ligases, particularly VHL, are involved in the cellular response to hypoxic stress (see Section 8.7). Human cells express about 50 different SOCS box proteins that combine with ubiquitin ligases, providing substrate specificity and versatility.

APC/C is a ubiquitin ligase that plays an essential role in controlling mitosis (see Section 12.10). This rather complex enzyme is composed of at least 12 different subunits. Subunits Apc11 (a RING domain protein) and Apc2 (a cullin analog) are essential for ubiquitin ligase activity, while subunits Apc3, -6, and -8 are SKP1-like adaptors with mostly unknown functions. Cdc20 and Cdh1 act as two different substrate-binding proteins that are exchanged during mitosis (see Figure 2.34). The functions of the other subunits are largely unknown.

2.8.2 Ubiquitin-binding domains

The highly variable patterns of ubiquitylation have been compared with ZIP codes that are used to sort proteins to different destinations. To make sense, a ZIP code has to be deciphered. In fact, like phosphorylated proteins, ubiquitylated proteins make contact with other "reader" proteins by interacting with specific recognition and binding domains.

More than 10 different types of such domain, ranging from 20 to 150 amino acids and exhibiting quite different structures, have been identified in a wide variety of proteins. Examples are provided by the ubiquitin-associating domain (UBA) and the ubiquitin-interacting motif (UIM) as well as the domain for coupling of ubiquitin conjugation to endoplasmic reticulum degradation (CUE). The binding of ubiquitylated proteins to such domains is rather weak and, therefore, readily reversible. It has been suggested that such low-affinity interactions enable the cell to quickly respond to input signals by assembling and disassembling a network of interacting proteins with built-in "dynamic instability." Nevertheless, the precise physiological role of most ubiquitin-binding domains is still poorly understood. Exceptions are the domains found in subunits of proteasomes that bring about the interaction with ubiquitylated proteins to be degraded.

2.8.3 Cellular functions of ubiquitylation

The ubiquitylation of proteins has numerous and diverse consequences. Best known is its role in **proteolytic degradation of proteins in proteasomes**. To be delivered to the proteasome, proteins have to be labeled by a polyubiquitin tag linked via Lys48. Ubiquitin-binding domains showing some preference for this type of polyubiquitin chain are found in proteins of the 19S subunits of the proteasome and in adaptor proteins (**Figure 2.35**). These adaptors bind to proteasomes via ubiquitin-like domains. For the polypeptide chain to be threaded into the central hole of the proteasome, the ubiquitin chain has to be cut off by Lys48 specific deubiquitylating proteases that are components of the 19S regulatory subunits. The ubiquitin chain thus released is not further degraded but is recycled. After de-ubiquitylation, the protein to be degraded is unfolded by AAA+ proteins that are also found in the regulatory subunits of proteasomes (see Section 2.5.2). Frequently, the efficiency of ubiquitylation-induced proteolysis depends on phosphorylation or other post-translational modifications of the substrate protein. Thus, the combination of such modifications with ubiquitylation resembles a logical AND gate.

Ubiquitin-mediated protein degradation is one of the most important events in cellular signal processing, being involved in both signal generation—by inactivating inhibitory proteins (see, for example, Section 11.7)—and signal extinction. For

Figure 2.35 Delivery of polyubiquitylated proteins to the proteasome Substrate proteins (SP) to be degraded are labeled by poly-ubiquitylation and delivered to the 26S proteasome (consisting of a 20S catalytic subunit and two 19S regulatory subunits) either directly (right) or through an adaptor protein (AP, left).

instance, **transcription factors** are labeled for rapid degradation by ubiquitin ligases that are components of the transcription-inducing protein complex. Such a suicide mechanism prevents excessive and uncontrolled gene activation and, thus, allows a precise adaptation of the genetic readout to environmental conditions, albeit in quite a costly manner. In plants the ubiquitylation and proteolytic destruction of transcriptional repressors has been found to be even under direct hormonal control, in that E3 ubiquitin ligases are receptors of phytohormones such as auxin and gibberellic acid (see Section 13.1).

Another example is provided by the **cell cycle**, each phase of which is controlled by cyclin-dependent protein kinases. As discussed in Chapter 12, cyclins, which are the regulatory subunits of these kinases, are produced and degraded in a strictly defined order during the cell cycle, a process that is indispensable for the clockwork-like course of cell division. For degradation the cyclins have to be labeled by ubiquitylation. Poly-ubiquitylation plays a similar role in **DNA repair**. DNA damage leads to an interruption of the cell cycle to give the cell an opportunity to repair the defect or—if this turns out to be impossible—to commit suicide by apoptosis. Stop signals include both ubiquitin-mediated degradation of protein phosphatase Cdc25A, an enzyme that promotes the cell's transition into the phase of DNA replication (see Section 12.4), and stabilization of the anti-proliferative transcription factor p53 by inhibition of ubiquitin-mediated proteolysis (see Section 12.8.3).

A special case is the ubiquitin-controlled **processing of antigens** by antigen-presenting cells. Here specialized immune proteasomes dissect an antigenic protein into fragments of 8–10 amino acids which, together with the major histocompatibility complex, are exposed at the cell surface, stimulating cytotoxic T-lymphocytes.

In contrast to poly-ubiquitylation via Lys48, mono-ubiquitylation or poly-ubiquitylation via Lys residue 63 of ubiquitin does not label a protein for proteasome binding and degradation but, like phosphorylation, for interactions with other proteins. Examples such as ubiquitylation of histone 2B and the transcription factor NFκB are discussed in Sections 8.2.6 and 11.7.

In a substrate protein, several Lys residues may be mono-ubiquitylated. This type of multi-ubiquitylation plays an important role in **endocytosis**, as shown in exemplary fashion by the down-regulation of membrane receptors, through which cells protect themselves from overstimulation. Receptor down-regulation provides a major mechanism of adaptation to exogenous signals (see Section 7.1.1). In the course of down-regulation, the activated receptors are ubiquitylated and subsequently concentrated in endocytotic vesicles. These are split off from the membrane, transporting their cargo to endo- and lysosomes, where it is recycled or degraded (details in Section 5.8.1).

Disturbances of this mechanism may have serious consequences, as demonstrated, for instance, by Cbl and the breast cancer protein BRCA. Both are E3 ubiquitin ligases and induce the degradation of many signaling proteins that, when excessively expressed and activated, would promote tumor growth. An example is provided by Tyr-phosphorylated proteins such as autophosphorylated growth factor receptors. Cbl recognizes the phospho-Tyr site through a corresponding contact domain (SH2 domain). As a consequence the target protein becomes ubiquitylated and is taken up by endocytosis. Cbl thus triggers the adaptive down-regulation of such receptors and of cytoplasmic Tyr kinases (see Section 7.1.1). Deletion of the RING domain transforms Cbl into an oncoprotein. While it cannot contact anymore the E2 ubiquitin-conjugating enzyme, it still binds its substrate proteins, thus acting as a dominant-negative mutant that competitively inhibits the intact ubiquitin ligase derived from the nonmutated allele. In contrast, nonmutated BRCA1 is a tumor suppressor protein; its mono-allelic deletion causes a hereditary predisposition for breast cancer. Another tumor suppressor protein, the genetic inactivation of which is causally related to familiar breast cancer, is BRCA2. Showing no homology to BRCA1, BRCA2 is involved in the regulation of cell proliferation in response to DNA damage (see Section 12.9).

2.8.4 Ubiquitin-like modifiers

Ubiquitin stands for a larger family of protein-modifying polypeptides called ubiquitin-like modifiers (UBL). Their activation and binding mechanisms as well as their three-dimensional structures (but not necessarily their amino acid sequences) resemble those of ubiquitin. The most prominent representative is the protein **SUMO** (Small Ubiquitin-like MOdifier), which exists in several isoforms and is found in all eukaryotic cells. Several SUMO-specific E3 ubiquitin ligases [for instance, protein inhibitor of activated STAT (PIAS; see Section 7.2.1) or Ran-binding protein RanBP2 (see Section 10.5)] and desumoylating proteases (called ubiquitin-like protein-specific proteases), as well as a rather large number of proteins the functions of which are regulated by sumoylation, are known. Substrates of sumoylation include transcription factors such as steroid hormone receptors, c-Jun, STAT, Myb, p53, and others, as well as the GAP of the small G-protein Ran controlling the transport of macromolecules across the nuclear membrane. The enzymes of sumoylation/desumoylation are components of the signal-processing protein network and are regulated, for instance, by protein phosphorylation, which may silence sumoylation signals by inhibiting SUMO ligases and stimulating SUMO proteases.

In contrast to ubiquitylation, sumoylation has never been found to mark a protein for proteolytic degradation but exclusively for interactions with other proteins. In fact, sumoylation may even *prevent* protein degradation by competing with ubiquitylation, thus prolonging the effectiveness of peptide and protein signals such as transcription factors. In addition, sumoylation has a direct effect on the activity of certain transcription factors such as the heat-shock factors HSF1 and 2 (not to be confused with heat-shock proteins), which are stimulated upon temperature stress.

Other ubiquitin-like modifiers are also involved in reactions to stress such as starvation and infections. One of those modifiers is the starvation-activated autophagy regulator **Atg8** that interacts with the membrane lipid phosphatidylethanolamine rather than with proteins, a reaction that is essential for the formation of autophagosomes (for more details on autophagy see Section 9.4.3). This provides another analogy to phosphorylation, which, under certain circumstances, may affect not only proteins but also phospholipids. Interestingly, ubiquitin-like modifiers are also used to control the ubiquitylation reaction proper. An example is neddylation, the modification of certain E3 ubiquitin ligases by the modifier Nedd8 (see Section 2.8.1).

2.8.5 Signal termination by deubiquitylating proteases

Among the 561 human protease genes, around 80 encode enzymes specialized for de-ubiquitylation of proteins, rendering ubiquitylation a reversible switching reaction. Additional proteases are responsible for the removal of ubiquitin-like modifiers such as SUMO and Nedd8. Considering the more than 500 E3 ubiquitin ligases, the enzymatic equipment of ubiquitylation clearly rivals that of protein phosphorylation, comprising in humans 518 kinase and 120 phosphatase genes. Deubiquitylating enzymes (DUBs) constitute a highly diverse group of (mainly) cysteine proteases with molecular masses from 188 to over 3000. Each of them has a limited set of substrates connecting it with distinct physiological functions. Specificity is related to the type of ubiquitin chain, the site of ubiquitylation, and the nature of the substrate protein. In contrast to many other proteases, DUBs are not produced as zymogens but become activated by binding the ubiquitylated substrate. Such a mechanism conveys a high degree of selectivity and prevents accidental degradation of other proteins. In addition to signal termination, the protection of E3 ubiquitin ligases from spurious self-ubiquitylation is a major function of DUBs. For both reasons, they are frequently found to be associated with the ligases.

Summary

- Ubiquitylation is an ATP-consuming post-translational modification that is restricted to eukaryotes. In its versatility and abundance, it rivals or even outdoes protein phosphorylation. It is catalyzed by a complex of three enzymes, E1, E2, and E3, with the ubiquitin ligases E3 determining the substrate specificity of the reaction.

- E3 ligases are signal-operated switching elements. More than 600 different mammalian E3 genes render ubiquitylation specific by coupling E2 ubiquitin-conjugating enzymes with selected substrate proteins. Depending on E2-binding domains, one distinguishes between HECT domain and RING domain ligases. In many ligases, the substrate-binding domain is localized on a separate subunit bound to the E2-binding RING domain subunit by adaptor and scaffold proteins. The role played by E3 ubiquitin ligases in signal transduction rivals that of protein kinases.

- Ubiquitylated proteins recognize their partners by specific ubiquitin-interacting domains.

- Cells make use of ubiquitylation for labeling proteins for degradation in proteasomes, for other protein–protein interactions (network formation), and for the control of protein activities. Proteasomal degradation is a major process of signal termination. It is induced by poly-ubiquitylation, mostly via Lys48 of ubiquitin, whereas the other effects depend on poly-ubiquitylation via other Lys residues or on mono-ubiquitylation. De-ubiquitylating proteases render the reaction reversible.

- Ubiquitin is the prototype of a series of related peptides used for post-translational protein modification. In contrast to ubiquitin, these ubiquitin-like modifiers do not label proteins for proteolytic degradation but serve other purposes in signal transduction.

2.9 Mono- and poly(ADP-ribosylation)

Two types of post-translational modification, mono(ADP-ribosylation) and poly(ADP-ribosylation), use the free energy released by the metabolic breakdown of NAD^+ for signal transduction.

2.9.1 Mono(ADP-ribosylation)

This is a post-translational modification for which an ADP-ribosyl residue released from NAD^+ is transferred to a protein and most frequently linked with an Arg residue via a covalent *N-glycosidic bond* (**Figure 2.36**). The reaction is catalyzed by mono(ADP-ribosyl) transferases and reversed by ADP-ribosylhydrolases. These enzymes have been found in pro- and eukaryotes as well as in bacteriophages. In the human genome, the corresponding genes belong to the approximately 200 genes that were acquired from bacteria. The input signals controlling the enzymes are not yet known.

For some viruses and bacteria, ADP-ribosyltransferases serve as virulence factors. By this means bacteriophages reprogram the bacterial RNA polymerase in such a way that it preferentially transcribes viral genes. The ADP-ribosyltransferases of diphtheria, pertussis, and cholera bacteria are highly active toxins since they intervene strongly with G-protein signaling (more details in Section 4.3.3) or damage the cytoskeleton by ADP-ribosylation of actin. Other bacterial ADP-ribosyltransferases are probably not related to the toxins but control the activities of metabolic enzymes such as the nitrogenases of soil bacteria.

Eukaryotic ADP-ribosyltransferases are distributed over two nonrelated families, extracellular and intracellular enzymes. Extracellular transferases are bound to cell

Figure 2.36 Mono(ADP-ribosylation) Via mono(ADP-ribosyl) transferase, the nicotinamide residue of the redox coenzyme nicotinamide adenine dinucleotide (NAD$^+$) is replaced by a protein, which becomes linked to the ADP-ribose residue via an Arg residue. The hydrolysis of this bond is catalyzed by an ADP-ribosylhydrolase. Both enzymes are controlled by still elusive input signals. The energy for the switching process is supplied by the hydrolysis of the N-glycosidic bond of NAD$^+$. Note that ADP-ribose may be bound also to Glu and Asp residues via an O-glycosidic bond.

membranes by a glycosylphosphatidylinositol anchor (see Section 4.3.1). They are found particularly on cells of the immune system. Little is known about their substrate proteins and physiological role.

Intracellular ADP-ribosyltransferases are components of the signal-processing network, involved (among others) in the temporal limitation and extinction of signals. The βγ-subunits of trimeric G-proteins are inhibited (see Section 4.3), and in photoreceptor cells, signal transduction from the trimeric G-protein transducin to cGMP-phosphodiesterase is blocked by ADP-ribosylation of the inhibitory subunit of the phosphodiesterase (see Section 15.5). ADP-ribosylation suppresses, moreover, the polymerization of cytoskeletal proteins such as actin and desmin to filaments as well as the activity of certain metabolic enzymes such as glyceraldehyde-3-phosphate dehydrogenase and the activity of transcription factors such as NFκB (see Section 11.7). Other processes including the transcription of certain genes and the unfolded protein response (see Section 9.4) are stimulated by mono(ADP-ribosylation).

A special type of ADP-ribosyltransferases, the **Sir enzymes** or sirtuins, are discussed in Section 8.2.3. These enzymes are both NAD$^+$-consuming histone deacetylases transferring the ADP-ribosyl residue to acetyl residues and are catalysts of ADP-ribosylation of proteins.

2.9.2 Poly(ADP-ribosylation)

Mono- and poly(ADP-ribosylation)—also known as "MARylation" and "PARylation"—are chemically closely related but biochemically different protein modifications with signaling character. For poly(ADP-ribosylation) the ADP-ribosyl residues derived from NAD$^+$ are linked via *esterglycosidic bonds* to Asp and Glu residues of the substrate protein. Subsequently, additional ADP-ribosyl residues are added to the initial residue, forming linear and branched poly(ADP-ribosyl) chains of up to 200 units (**Figure 2.37**). The poly(ADP-ribosyl)polymerases (PARPs) catalyzing these reactions are highly conserved enzymes found in archaea and eukaryotes. Mammals express seven isoforms. Like mono(ADP-ribosylation), poly(ADP-ribosylation) is a reversible switching reaction: the ADP-ribosyl polymer is rapidly degraded by poly(ADP-ribose) glycohydrolases, and the initial residue is removed from the protein by ADP-ribosyl protein lyase. Substrate proteins of PARPs are the enzymes themselves (automodification), various transcription factors, proteins involved in DNA replication, cell death mitosis, and histones.

Poly(ADP-ribosylation) of histones belongs to the reactions used by the cell to alter chromatin structure in such a way that gene transcription becomes possible

Figure 2.37 **Poly(ADP-ribosylation)** An ADP-ribosyl residue derived from NAD+ is linked via an ester-glycosidic bond to an Asp or Glu residue of the substrate protein, and additional ADP-ribosyl residues are added to the initial residue forming linear or branched poly(ADP-ribosyl) chains. The reaction is catalyzed by poly(ADP-ribosyl) transferase. Poly(ADP-ribosyl) glycohydrolase depolymerizes the chain and ADP-ribosyl protein lyase hydrolyzes the bond with the protein.

(see Section 8.2.6). The histone-specific PARPs 1–3 have a DNA-binding domain with a two-fold zinc-finger structure (as found also in many transcription factors) and are strongly activated by damaged DNA, for instance, upon irradiation or contact with mutagenic chemicals. These and related observations (for example, in PARP-knockout mice) point to an important role of poly(ADP-ribosylation) in **DNA repair**. Other PARP isoforms have other regulatory domains rather than DNA-binding sites and a different subcellular distribution; therefore, they are assumed to fulfill functions different from those of histone-specific PARPs.

Summary

- Mono- and poly(ADP-ribosylation) are two types of signal-operated post-translational protein modification deriving energy from NAD+ hydrolysis. They are found in both pro- and eukaryotes and, although chemically closely related, are catalyzed by different enzymes. Bacterial ADP-ribosyltransferases are potent virulence factors.

- Poly(ADP-ribosylation) is used for histone modification in the course of eukaryotic transcription and becomes especially activated upon DNA damage.

2.10 Ion channel switches: how to make use of electrical charge

Ion channel switches consist of signaling proteins that use the energy stored in the electric charge of cell membranes. This membrane potential is, in fact, an important energy source for signal processing, rivaling the hydrolysis of energy-rich bonds in NTPs, acetyl-coenzyme A, and NAD+.

2.10.1 Membrane potential is a major energy provider for cellular data processing

The membrane potential is defined as the electrostatic charge caused by an imbalanced distribution of sodium, potassium, calcium, and chloride ions across cell membranes. This "capacitor" can be discharged, with its energy being used for signal transduction, and recharged at the expense of metabolic energy. In other words, cellular membranes function like electric batteries feeding a computer.

The key to understanding the membrane potential is the **semipermeability** of biomembranes. In aqueous solution, each ion binds several water molecules electrostatically: it is surrounded by a hydrate cover. For such hydrated ions, the lipid bilayer of cellular membranes constitutes a barrier that cannot be passed even under an extreme gradient of charge and concentration. Thanks to protein complexes working as pores, cell membranes are, nevertheless, permeable to ions. However, such pores allow only the small ions mentioned to pass, whereas larger charge carriers such as amino acids and nucleotides, let alone electrically charged macromolecules such as proteins and nucleic acids, are held back. This semipermeability causes an imbalanced charge distribution.

To make this clear, one may imagine a model cell filled with a 100 mM solution of K^+ ions and containing potassium channels in its membrane. The positive charge is assumed to be matched by a corresponding amount of negatively charged macromolecules; in the beginning, the interior of the cell is electrically neutral. When such a cell is put into an aqueous solution, K^+ ions pass through the membrane channels along the concentration gradient while the negative counterions cannot leave the cell because they are too large. As a result, a charge difference (negative inside, positive outside) develops at the membrane, hindering the K^+ ion flow, which ends when the electrostatic power matches the chemical power of the ion gradient. This state or equilibrium, also called the Donnan equilibrium, is described by the **Nernst equation**:

$$P_K = \frac{RT}{zF} \ln \frac{[K^+]_o}{[K^+]_i}$$

where P_K is the equilibrium potential of K^+ ions in volts, R is the gas constant, T is the absolute temperature, z is the ion valence (1 for K^+), F is the Faraday constant, and $[K^+]_o$ and $[K^+]_i$ are the K^+ concentrations outside and inside the cell, respectively. If the cell contains additional ion species as well as the corresponding membrane channels, a Donnan potential may be calculated for each type. Under physiological conditions, the following outside:inside ion distributions are established: K^+ 2:1, Na^+ 2:1, and Cl^- 1:2. The resulting membrane potential is about −20 mV.

Despite this charge, a net flow of ions is impossible because the system is *equilibrated*. In other words, under these conditions the membrane potential cannot be used as an energy supplier for signal transduction and other purposes. Apart from this, the high osmotic pressure difference would cause an influx of water (through corresponding membrane pores), which may lead to a disruption of the cell membrane. To prevent this, the cell establishes a *disequilibrium* of the ion distribution to some extent approaching the following concentration ratios (again outside:inside): K^+ 20:1, Na^+ 1:7, and Cl^- 1:25 (actual concentrations are found in **Figure 2.38**). By this means, the membrane capacitor is charged up to a voltage of approximately −60 mV and can be used as an energy provider for coupled reactions, in particular signal transduction.

The charge energy is derived from ATP hydrolysis driving a membrane-bound ion transport system, the Na^+/K^+-dependent ATPase, which, for each hydrolyzed ATP molecule, pumps out three Na^+ ions and allows two K^+ ions to enter the cell. As a result one positive charge is removed from the cell for each pumping step. Consequently, the ATPase has been called an electrogenic ion pump. In addition, certain (not all) cell types possess chloride pumps to regulate the membrane potential, whereas

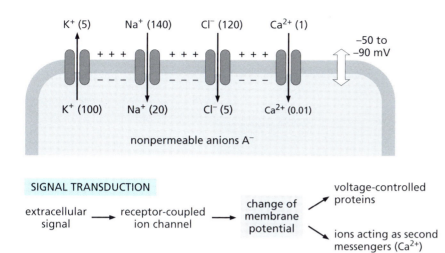

K^+ (5) Na^+ (140) Cl^- (120) Ca^{2+} (1)

−50 to −90 mV

K^+ (100) Na^+ (20) Cl^- (5) Ca^{2+} (0.01)

nonpermeable anions A^-

SIGNAL TRANSDUCTION

extracellular signal → receptor-coupled ion channel → change of membrane potential → voltage-controlled proteins / ions acting as second messengers (Ca^{2+})

Figure 2.38 Membrane potential as an energy supply of signaling reactions The figure shows the distribution of ions across the plasma membrane and the resulting resting potential as brought about by ATP-consuming ion pumps (Na^+/K^+- and Ca^{2+}-dependent ATPases) and ion exchangers. The numbers represent the average ion concentrations in millimoles per liter. The actual values differ between cell types. Moreover, depending on the resting potential and ion concentration, the gating of chloride channels may cause a charge-driven efflux rather than a concentration-driven influx (as shown here) of Cl^- ions. Signals alter the ion distribution by controlling ion channels (or ion pumps). This may result in either an increase (hyperpolarization) or decrease (depolarization) of the negative membrane potential. The resulting change of the membrane potential is recognized as a signal by voltage-controlled proteins, mostly ion channels. In contrast to K^+, Na^+, and Cl^-, Ca^{2+} entering the cell via the corresponding channels in most cells, does not alter the membrane potential to a significant extent but functions as a second messenger.

Ca^{2+}-dependent ATPases or calcium pumps play only a minor role in charge control owing to the comparably low Ca^{2+} concentration in cells.

To maintain the membrane potential, the ion pumps have to work permanently, since cell membranes are leaky for K^+ and Cl^- ions because corresponding ion channels remain open all the time. Therefore, inhibitors of the Na^+/K^+-dependent ATPase such as the active agents of foxglove (*Digitalis purpurea*) may cause a breakdown of the membrane potential, thus paralyzing the cell.

By controlling the activity of the ion pumps, the cell is able to alter specifically the membrane potential. This objective is also achieved by a controlled gating of ion channels, thus allowing an influx or efflux of ions depending on the concentration and charge gradients. As a rule, the membrane potential becomes more negative upon the opening of K^+ channels: following the concentration gradient K^+ ions *leave* the cell causing **hyperpolarization**. The gating of Na^+ channels has the opposite effect: because Na^+ ions *enter* the cell along the concentration gradient the interior of the cell becomes less negative. The cell is said to undergo **depolarization**. In contrast to sodium channels, chloride channels may have either a hyper- or a depolarizing effect depending on the concentration gradient of the ion and on the resting potential.* Thus, the effect of chloride channel gating depends on the cell type and the physiological context. Ion channels exist in a large variety of types that are operated either by changes of the membrane potential or by ligand binding (see Chapter 14).

2.10.2 Action potential is the standard signal of nerve and muscle cells

By acting on ion pumps and ion channels, hyper- or depolarizing signals either hinder or promote processes that depend on the membrane potential. A prototype of such a process is the action potential. Action potentials are waves of depolarization migrating with a speed of up to 120 m/s across cell membranes. These most rapid of all endogenous signals are produced by only a few cell types such as neurons, muscle cells, egg cells, and some others. Such cells have in common an **excitable plasma membrane**. In this membrane type, depolarization-operated Na^+ channels are arranged in such a way that they activate each other and stimulate cooperating K^+ channels which terminate the depolarization.

*The number of ions to be transported for depolarization or hyperpolarization is surprisingly small. According to the equation charge (in coulombs) = capacity (in microfarads per square centimeter) × voltage (in millivolts), and if a cell surface of 10^{-4} cm^2 is assumed, only 5×10^{-17} mol of K^+ ions have to be transferred to change the membrane potential to 50 mV. Compared with the intracellular K^+ content of 10–11 mol, this amount is practically negligible unless the ion flux lasts too long.

	active	inactive
Na⁺ channel	D	H
K⁺ channel	H	D
Cl⁻ channel	H,D	H,D
Na⁺/K⁺ pump	H	D
Cl⁻ pump	H,D	H,D

Figure 2.39 Action potential When a depolarization exceeds a threshold potential (here assumed to be –30 mV) in excitable cells, an action potential of varying length may develop (here depicted for a nerve cell). Its induction is facilitated by rising (depolarizing) and impeded by lowering (hyperpolarizing) the resting potential. Ion channels and ion pumps that change the resting potential when they are activated or inactivated (D, depolarizing effect; H, hyperpolarizing effect) are listed. Whether an influx or efflux of chloride ions induces hyper-or depolarization depends on the ratio between the resting potential and the intracellular chloride concentration.

Depending on the tissue, an action potential lasts between 0.1 ms (skeletal muscle) and 400 ms (heart muscle). Action potentials are generated according to an all-or-nothing rule, provided the depolarization of the cell membrane exceeds a threshold value, and they proceed, like the spark of a fuse, without dampening. The value of the resting potential (the membrane potential of the nonstimulated cell) in relationship to the threshold potential determines the probability of an action potential becoming induced: the induction is impeded by hyperpolarization but facilitated by depolarization of the resting cell (**Figure 2.39**).

To trigger an action potential, depolarization is induced by the gating of Na⁺ channels or nonspecific cation channels. Below the threshold potential, each channel responds independently. Above the threshold potential, the channels cooperate and they activate each other. The result is a strong influx of sodium ions, shifting the membrane potential to positive values for a short time (**Figure 2.40**). Simultaneously, leak channels for potassium ions are closed.

The Na⁺ influx is terminated by a K⁺ efflux. To this end, voltage-dependent K⁺ channels are opened, though with some delay. To give the K⁺ channels time for rapid repolarization, sodium channels are desensitized, becoming reactive again only after a defined refractory period (see Sections 14.2 and 14.4.1). This sequence of events determines the frequency of successive action potentials and, at least in neurons, lets the signal spread in only one direction (for instance, along the axon towards the

Figure 2.40 Ion channel switch An input signal (a chemical ligand or a change of the membrane potential) controls the gating of an ion channel in the cell membrane. The resulting change of the cytoplasmic ion concentration induces a conformational and functional change of an effector protein, generating the output signal. Depending on the ion type, one distinguishes between pH-dependent effects (by protons), a direct modulation by ions of protein function (in particular by Ca²⁺), and voltage-dependent alterations. The energy needed for signal transduction is supplied by ion gradients. These batteries become recharged by ATP-consuming ion pumps.

synapse) because during its refractory period a channel cannot respond to a signal transmitted by the next channel (see Section 14.2). Again an analogy with a fuse that cannot burn down backwards is obvious (though in contrast to a fuse, the cell/primes membrane potential regenerates itself!).

Since action potentials follow an all-or-nothing rule, they cannot be modulated in their intensity but only in their frequency; thus they represent prototypes of **oscillating signals**. Depending on the tissue and species, this frequency varies between less than 1 per second (heart muscle) and more than 1000 per second (skeletal muscle). The major effects of action potentials are muscle contraction and neurotransmitter secretion. In both cases Ca^{2+} ions are the intracellular key signals being released into the cytoplasm through Ca^{2+} channels that open upon depolarization (see Section 14.5).

2.10.3 Ion channels

Ion channels are switches driven by the energy of the membrane potential (see Figure 2.40). They consist of proteins that penetrate the cell membrane, forming a permeable pore that is lined by hydrophilic amino acid residues. For this purpose at least 4×2 transmembrane domains must be arranged in a circular pattern. Such transmembrane domains, mostly α-helices, are either parts of one polypeptide chain or distributed over several protein subunits. To open the pore—a process called gating—the transmembrane domains are assumed to twist like an iris aperture.

Depending on the type of input signals, a distinction is made between voltage- and ligand-controlled channels. Voltage-controlled cation channels contain transmembrane domains acting as voltage sensors because their conformation responds to changes of the membrane potential. Ligand-controlled channels harbor binding sites for extracellular or intracellular signal molecules. Some ion channels can be voltage- as well as ligand-controlled. A special type of ion channel responds to pressure.

In its simplest form, the gating of an ion channel resembles a binary switch in that the channel is either closed (value 0) or open (value 1). In reality, however, an adjustable time factor and a large number of interacting switches are added to allow transformation of the digital signal into an analog one. In addition, many channels possess a built-in switching-off device, turning the channel into a desensitized state upon prolonged activation. This protective mechanism strikingly resembles that of G-protein and protein phosphorylation switches, which also undergo auto-inhibition.

Ion channels let pass either Na^+, K^+, Ca^{2+}, or Cl^- ions. The **cation/anion selectivity** is determined by the electrical charge of the channel: positive for anions and negative for cations. The electrostatic interactions between channel and ion are rather weak since otherwise ion transport would be too slow (whereas in reality it approaches the rate of free diffusion). Therefore, highly charged amino acid residues such as carboxy and amino groups are avoided. Instead, the channel is lined by carbonyl groups (cation channel) or amide groups (anion channel) working as electrical dipoles. Cation channels may be either ion-nonspecific or ion-specific. Most ligand-controlled cation channels are nonspecific—that is they allow the transport of Na^+, K^+, and Ca^{2+} ions even though with some preferences. In contrast, voltage-controlled channels are generally highly ion-specific. In addition to the transmembrane domains, they have a short P- (permeability) loop that dives into (but does not penetrate) the lipid bilayer of the membrane. [These P-loops must not be confused with the P (phosphorylation) loops of G-proteins and other phosphotransferases.] It contains several highly conserved amino acid residues and serves as a selectivity filter. Here the carbonyl or amide groups lining the pore compete with water for binding the ions. As a result the ions that are surrounded by water molecules are dehydrated since otherwise they could not pass the channel. To this end the interaction between ion and P-loop must supply a sufficient amount of energy to strip the hydrate cover. The more voluminous K^+ ions require less dehydration energy (80 kcal/mol) than the smaller Na^+ ions (98 kcal/mol). This difference together with the channel's diameter may explain the ion selectivity of voltage-controlled channels. Thus, the negative charge of the K^+ filter is sufficient to strip K^+ ions but not Na^+ ions, whereas an Na^+-selective channel, though it is able to dehydrate both ion types, is too narrow for K^+ ions to pass. In fact, voltage-dependent Na^+ channels are more than 10,000 times more permeable to Na^+ than to K^+, and vice versa.

Ion channel proteins constitute several families originating from ancient prokaryotic genes (details regarding the evolution of ion channels are found Section 3.6.7). The largest family, called the **S5–S6 family**, covers all passive and voltage-controlled cation channels, though some of them are controlled in addition by ligands. The term S5–S6 refers to a common structural feature, the transmembrane segments (or helices) 5 and 6 of the voltage-dependent K^+ channel of vertebrates (see Section 14.4.1). This channel is composed of two subunits with six transmembrane helices (S1–S6) and a P-loop localized between S5 and S6. Helix S4 is the voltage sensor.

A simpler construction is shown by another K^+ channel of higher eukaryotes: the inwardly rectifying K_{ir} channel, which consists of four subunits with two transmembrane helices each, resembling S5 and S6, and a P-loop in between (see Section 14.4.2). This structure is quite similar to that of the prokaryotic K^+ channel KcsA. Still simpler are the S5–S6 channels lacking a P-loop, such as the epithelial Na^+ channel ENaC (see Section 14.3), the ATP-receptor channel P2X (see Section 16.6), the cAMP- and cGMP-controlled cation channels of sensory cells (see Section 14.5.7), and various neuropeptide receptors of nonvertebrates. Starting from these simplest forms, the more complex ion channels are thought to have developed by gene doubling and gene enlargement. The subunits of the S5–S6 channels always form a single pore.

The **anion channels** that transport Cl^- ions, in particular, belong to six different families. Their complex structures are not related to each other and to those of cation channels. Some of them are chloride/proton exchangers. A special type of anion channel is CFTR, which belongs to the large family of ABC transporters. This ATP-dependent channel plays an important role in transepithelial electrolyte transport and is discussed together with other anion channels in detail in Section 14.7.

Other ion channel families cover the majority of **ligand-controlled channels**. By fusion of a S5–S6 channel protein with a prokaryotic glutamate binding protein, a special channel type, the ionotropic glutamate receptors, has developed. These are nonspecific cation channels that play a central role in the brain (see Section 16.4). Receptor ion channels for other neurotransmitters constitute a separate family with the nicotinic acetylcholine receptor being the prototype (see Sections 16.2 and 16.3). They are either nonspecific cation channels or anion channels made up of five subunits with four transmembrane domains each. A key role in cellular signal processing is played by the Ca^{2+} channels of the endoplasmic reticulum. They exist as tetramers, with six transmembrane domains and a P-loop-like structure each, and are under the control of intracellular signal molecules such as Ca^{2+}, inositol 1,4,5-trisphosphate, and others (see Section 14.5.3).

Summary

- The electrical charge of cell membranes (negative inside, positive outside) is due to semipermeability of the lipid bilayer and to ATP-driven ion pumps that generate disequilibrium of Na^+, K^+, Ca^{2+}, and Cl^- ions. By coupling the influx and efflux of these ions through special ion channels with signal-transducing processes, cells use the energy stored in the "membrane capacitor" for signal processing.

- Excitable cells (particularly nerve and muscle cells) are able to generate action potentials, self-propagating waves of complete membrane depolarization that represent the fastest intercellular signal known. To generate an action potential, a particular arrangement of cooperating voltage-dependent ion channels is needed. Action potentials are frequency-modulated signals.

- Ion channels are membrane proteins with at least four transmembrane domains that form a central pore for the transport of either cations (Na^+, K^+, Ca^{2+}) or anions (mainly Cl^-). Most ion channels are either ligand- or voltage-operated (although some more specialized channels also respond to other stimuli such as pressure). In contrast to ligand-dependent channels, voltage-operated channels are ion-selective due to subtle differences in pore diameter and in the ability to dehydrate ions.

- All eukaryotic ion channels seem to have a prokaryotic ancestry. Major families are S5–S6 channels, including all voltage-dependent channels, and ligand-dependent channels related to the nicotinic acetylcholine receptor.

2.11 Proteolysis switch: protein degradation provides messenger molecules and energy

The energy required for signal transduction is supplied not only by the hydrolysis of energy-rich metabolites and by the membrane potential but also, in certain cases, by the hydrolytic degradation of macromolecules, in particular, proteins. Prominent examples include the proteolytic cascades regulating blood clotting, nonspecific defense (complement system), and programmed cell death (apoptosis). Such a mechanism kills two birds with one stone in that, besides energy, a functionally active fragment of the corresponding protein is released. This holds true also for signal transduction across cell membranes, which makes use of protein degradation to gain generation of both energy and signaling peptides that act as second messengers, resembling an output signal (**Figure 2.41**). An example of this type of signaling is the Notch system (see Section 13.1). Of course, signal transduction by proteolysis is a borderline case and one may hesitate to call this kind of reaction a switch. After all, it is not easy to set such a switch back to 0, as this requires a costly and time-consuming de novo synthesis of proteins.

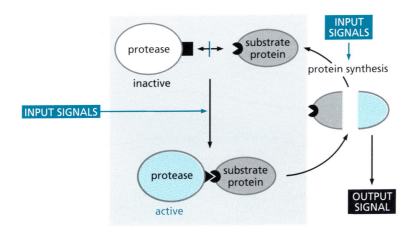

Figure 2.41 Proteolysis switch Here the input signal activates a protease that releases an active fragment from an inactive precursor (substrate) protein. The fragment produces the output signal. The switching energy stems from hydrolysis of the peptide bond.

Summary

The energy stored in the peptide bond can be used to drive signal transduction, and the peptide fragment generated by limited proteolysis may function as a second messenger.

2.12 Receptors: how energy-supplying reactions are combined with signal transduction

Receptors are proteins that recognize extracellular signals. They act as the sensory organs or, if one prefers a technical metaphor, the users' interface of the cell. Extracellular signals may be endogenous messengers such as hormones, neurotransmitters, key metabolic compounds, or environmental stimuli.* The existence of receptors was proposed at the end of the 19th century. British physiologist J. N. Langley postulated that, in order to become effective, pharmacological drugs had to be bound in the organism by "receptive substances" (endogenous signals such as hormones and neurotransmitters were still unknown at this time). At about the same time, Paul Ehrlich proposed that antigens, but also other substances, were bound at the cell surface by receptors which he called "*Seitenketten*" (side chains), expecting them to recognize the *molecular shape* of the ligand according to the lock–key principle. This concept of a stereospecific structural complementarity between ligand and receptor is still the basis of modern receptor theory (**Figure 2.42**). Receptors are prototypical signal-transducing proteins. As such, they are subject to the same rules and laws as any other protein of the data-processing network. This includes allosteric regulation by noncovalent interactions, mainly with their ligands, and post-translational modifications for control and fine-tuning of receptor activity, oligomerization, and communication with other proteins via interaction domains.

A receptor transforms an exogenous input signal into an intracellular signal that is fed into the data-processing protein network. Therefore, a receptor protein consists of a highly specific recognition or discriminator domain and an effector domain. The interaction of the recognition domain with the input signal is always noncovalent and thus fully reversible. It induces an allosteric activation of the effector domain, which then can interact with downstream signaling elements. Since most receptors are complex proteins with a quaternary structure, this effect is cooperative, with sigmoidal kinetics. In other words, receptors are switches with a built-in noise filter.

---

*Such "sensory" receptors must be distinguished from those proteins which specifically interact with *intracellular* signal molecules such as second messengers. Frequently, these proteins also are called "receptors" (see, for instance, Section 14.5.3), although "acceptors" would be the more appropriate term.

Figure 2.42 Principles of receptor function Like all signal-transducing proteins, receptors are composed of a receiver or discriminator domain, recognizing an input signal, and a transmitter or effector domain, generating an output signal. The receptor protein exists in an allosteric equilibrium between an inactive (upper) and an active (lower) conformation. Input signals stabilize either the active or the inactive form; that is, they exhibit either agonistic or antagonistic effects.

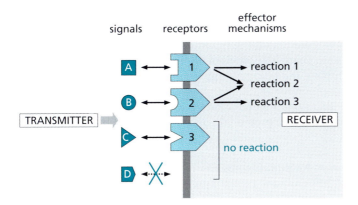

Figure 2.43 Lock–key principle of receptor function Signaling molecules fit the binding site of the receptorlike a key fits a lock. Depending on the effector coupling of its receptor (teal), a signal (A or B) may induce different reactions (1 or 3) or two different signals (A and B) may evoke the same reaction (2). Receptors without effector coupling (3) and signals without a corresponding receptor (D) have no effect.

While recognizing the input signal with high selectivity, a receptor may interact with various downstream pathways that sometimes even have opposite effects (see Figure 2.42). However, a single signaling pathway may also be controlled by various receptors (**Figure 2.43**). In addition, most receptors exist in several isoforms, combining a more or less identical discriminator domain with different effector domains. Taken together, this explains why the meaning of a signal is ambiguous and critically depends on the cell type and the physiological context (**Figure 2.44**).

Most drugs and toxins interact with cellular receptors of endogenous or environmental signals. They are called agonists when they imitate and antagonists when they counteract the effects of the natural ligands. In many cases, it depends on the dose whether a drug or poison has an agonistic or an antagonistic effect. Such effects are most easily explained by the theory of allostery, which predicts that receptor proteins exist in at least two conformations, an active one and an inactive one. Agonists stabilize the active form, pushing the allosteric equilibrium in the corresponding direction, whereas antagonists do the same with the inactive form (see Figure 2.42). Upon overstimulation, most receptors reach a refractory state that does not depend any longer on the presence of a ligand. This desensitization represents a built-in mechanism of protection.

Since signal processing by receptors requires—as any other signaling process—energy, receptors are coupled with all kinds of energy-supplying switching reactions. The **receptors of lipophilic ligands**, which can easily pass the cell membrane—for instance, the steroid hormones—are often found in the cytoplasm or nucleus. They

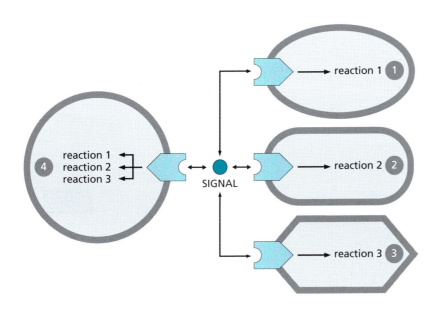

Figure 2.44 Ambiguity of signals As one lock and key may open the doors to different rooms, one signal may evoke different reactions in different cell types (cells 1, 2, and 3), provided its receptor (teal) exhibits different effector couplings. Different effector couplings may also exist within one cell (cell 4).

constitute a family of ligand-controlled or nuclear transcription factors that derive their signaling energy from protein–DNA interactions and chaperone-catalyzed ATP hydrolysis (see Section 8.3). Other cytoplasmic receptors such as the NO-binding soluble guanylate cyclases do not function as transcription factors (see Section 4.5). Here the energy stems from GTP hydrolysis. Finally, plants express hormone receptors (for auxin and gibberellic acid) with a built-in E3 ubiquitin ligase activity that is stimulated upon ligand binding, triggering gene expression by inducing the exergonic proteolytic degradation of transcriptional repressors (see Section 13.1.1).

The **receptors of hydrophilic ligands**, which are unable to penetrate the cell membrane, are transmembrane proteins, exposing their discriminator domains outside and their effector domains inside the cell. The effector domains are coupled with various switching proteins such as GTPases, protein kinases or phosphatases, ion channels, and others (**Figure 2.45**). Depending on the receptor type, this coupling is either covalent, with the effector protein being a domain of the receptor protein, or occurs via a noncovalent interaction between both partners. The three most abundant types of membrane receptor are shown in Figure 2.45.

(1) **GTPase-coupled receptors** all have a characteristic structure of seven transmembrane domains and activate via large heterotrimeric G-proteins enzymes that produce second messengers (see Sections 4.3 and 4.4). Although the structural principle is of prokaryotic origin, these receptors are restricted to eukaryotes and constitute the largest receptor family in animals (not in plants; see Chapter 5). Receptor and G-protein are always separate.

(2) **Protein kinase or phosphatase-coupled receptors** exist at least as dimers, activating each other by trans-autophosphorylation or -dephosphorylation. There are three types of kinase-coupled receptor: histidine kinase-coupled receptors, found mainly in prokaryotes (see Section 3.3); serine/threonine kinase-coupled receptors, abundant in all organisms (see Chapter 6); and tyrosine kinase-coupled receptors, found as yet only in animal cells (see Chapter 7). Receptor and protein kinase are either separate proteins or parts of one polypeptide chain.

(3) **Ion channel-coupled receptors** have one or more ligand binding sites and are found in pro- as well as in eukaryotes (see Section 3.6.7 and Chapter 16).

Special groups not shown in Figure 2.45 are **guanylate cyclase-coupled receptors**, which upon ligand interaction catalyze synthesis of the second messenger cGMP (see Section 4.5), and **protease-** and **ubiquitin E3 ligase-coupled receptors**, which release peptide second messengers or trigger the degradation of signaling proteins (see Section 13.1).

Figure 2.45 Three most abundant types of transmembrane receptor (A) Prototype of a transmembrane receptor consisting of an extracellular discriminator domain (teal), a transmembrane domain, and an intracellular effector domain. (B) The three most abundant types are the protein kinase (PK)-coupled receptor, forming dimers usually with two transmembrane domains; the G-protein-coupled receptor, with seven transmembrane domains and an associated trimeric G-protein (αβγ) being the effector domain; and the ion channel-coupled receptor.

The human genome harbors approximately 1500 genes encoding putative receptor proteins. It has been estimated that receptors can occupy up to 25% of the cell surface which in an idealized average cell (10 μm radius) would amount to 300 μm². Given a radius of 5 nm, a receptor protein would occupy 75 nm² of this area. This means that a cell can expose at its surface 4 million receptor molecules at most, which is a value widely matched by experimental data.

From the 1500 receptor genes, only a few are expressed in a given tissue. Nonetheless, a cell possesses a surprisingly large number of possibilities to respond. Even if only 1% of the genes is active—that is, 15 different receptor types are expressed—a cell could theoretically discriminate between $2^{15} = 32{,}678$ different combinations of ligands (each receptor is assumed to exist only in two states, liganded or not liganded). This model calculation impressively demonstrates that the performance of a signal-processing apparatus depends on the combinations between, rather than on the actual number of, switching events. In the following chapters we shall explore additional examples of this principle.

Summary

- As sensory devices, receptors connect the environment with the data-processing protein network of the cell. Most receptors are transmembrane proteins, but special receptors are found also in the cytoplasm and in cellular organelles.

- Receptors specifically recognize exogenous signals according to the lock–key principle. To deliver the signal to the intracellular network, receptors are coupled with energy-supplying biochemical reactions. The corresponding enzymatic domains and ion channels are associated with the receptor protein either covalently or noncovalently.

2.13 Brief summary of experimental standard methods for investigation of signaling pathways

Signal transduction occurs in highly complex networks of interacting proteins. For the sake of clarity and to approach the problem experimentally, these networks are *arbitrarily* dissected into individual linear cascades of interconnected partner molecules. The standard procedure to analyze such a signaling cascade is to manipulate a selected step with inhibitors or activators and to measure the effect on what are assumed to be the subsequent steps. This may be done by several experimental settings, as shown schematically in **Figure 2.46** and with the help of an example in **Figure 2.47**.

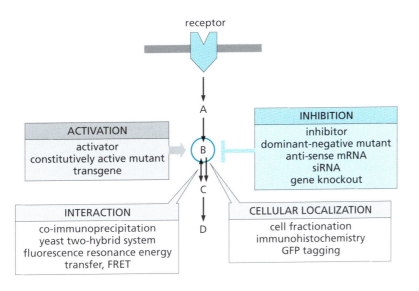

Figure 2.46 Analysis of a signaling cascade For this hypothetical reaction cascade starting with a receptor protein B is assumed to interact with the downstream component C and, in turn, with D. To prove this, the effect on C and D of either inhibition or activation of B is measured. Methods to localize B in the cell or to prove a direct interaction between B and C are also mentioned. They will be discussed in more detail in Section 5.2.

Figure 2.47 Experimental activation and inhibition of protein kinase C
(A) Diagram of the protein kinase C molecule with regulatory and catalytic domains. (B) Protein kinase C is activated relatively specifically by the phorbol ester TPA, a plant toxin (see Section 4.6.3) that binds to the regulatory domain mimicking the effect of the second messenger diacylglycerol. A constitutively active kinase mutant is obtained by inactivation or deletion of the autoinhibitory pseudosubstrate domain. (C) The fungal toxin staurosporine (from actinomycetes) interacts with the catalytic domain but inhibits, in addition to protein kinase C, various other protein kinases. More specific is calphostin C (from *Cladosporium* species), which interacts with the regulatory domain. A dominant negative mutant is obtained by deletion of the catalytic domain.

The simplest approach is to employ chemical activators or inhibitors (**pharmacological perturbation**). A large number of pharmacologically active drugs, as well as natural and synthetic toxins, affect signal-processing proteins. However, the specificities of such compounds do not always meet the expectations because they frequently interact with domains that, because of the modular protein structure, are found in more than one protein. A better, though more expensive, approach is to mutate the corresponding protein itself into a permanently active or an inactive form and then let this mutant compete with the cellular wild-type protein (**genetic perturbation**).

A constitutively active form is generally obtained by a genetic knockout of a regulatory domain that exhibits an auto-inhibitory effect (as, for instance, the pseudosubstrate domain in protein kinases), whereas an inactivation of the effector domain results in a dominant-negative mutant that acts as a competitive inhibitor. By means of targeted (amino-acid-specific) mutations, the effect measured may be related to structural parameters. Other procedures make use of anti-sense mRNA and siRNA (see Section 9.5) as inhibitors. Mutated proteins and other inhibitors and activators mostly are tested on cell cultures. However, targeted mutations as well as the transgene and knockout procedures also allow studies on living model organisms.

When dominant-negative mutants are used, the investigator imitates nature. In fact, physiological protein variants resembling dominant-negative mutants are frequently employed by cells to control signaling pathways. As a rule, they are produced via alternative gene promoters or by alternative splicing of pre-mRNA. Inversely, constitutively active proteins may arise when auto-inhibitory domains (such as the pseudosubstrate sequences of protein kinases) are enzymatically removed. In contrast, constitutively active *gene* mutations are mostly pathological derailments that may cause diseases like cancer.

The experimental standard methods are by no means free of problems and pitfalls. Moreover, in their conventional forms neither the pharmacological nor the genetic approach allow an exact spatial and temporal control of the perturbation and are not precise enough to follow up signaling in real time. This obstacle is widely overcome by the novel method of **optogenetics** where an individual signaling process can be switched on and off by light, a non-invasive perturbation. This technique requires the transfection of a cell with a photosensory protein or a signaling protein tagged with a light-sensitive domain derived from a photosensory protein (for more details see Section 3.4.5, Sidebar 3.4; Section 5.2.1, Sidebar 5.2, and Section 16.4.1, Sidebar 16.2). A major problem of this and related techniques is that transfection of cells with manipulated protein may cause undesirable side effects.

For a mechanistic elucidation, a precise knowledge of the three-dimensional protein structure is mandatory. Here the methods of choice are **X-ray crystallography** and other sophisticated physical techniques such as **mass spectrometry**. Thus, the combination of various methods is required to arrive at the final goal: to understand in detail the molecular mechanism of a given process. The proof of physiological significance requires, in addition, studies on appropriate model organisms and time-consuming clinical and epidemiological evaluations.

Summary

The standard procedure to analyze a sequence (cascade) of interconnected signaling reactions is to inhibit or stimulate a selected step and measure the effect on what are assumed to be the subsequent steps. This can be done by biochemical or pharmacological means or by genetic manipulation either *in vitro* or *in vivo*. Further mechanistic elucidation requires a detailed knowledge of molecular interactions and structures.

2.14 Model organisms for investigation of cellular signal processing

Although the complexity of cellular data processing has steadily increased during the course of evolution, the basic mechanisms entered the stage at the very beginning and have continued up to the present. Therefore, experiments with simpler organisms provide a key for understanding events going on in vertebrate cells and in humans. In addition to prokaryotes such as *E. coli*, the following model organisms have been most frequently used for the investigation of cellular signal transduction:

- yeast
- the slime mold *Dictyostelium discoideum*
- the nematode *Caenorhabditis elegans*
- the fruit fly *D. melanogaster*
- the plant *A. thaliana*
- several vertebrates such as the zebrafish, frog *Xenopus laevis*, the rat, and the mouse

As the simplest eukaryotes, **yeasts** are particularly easy to handle and to study and, most importantly, can be genetically manipulated almost as easily as bacteria. Yeast mutants have advanced the investigation of basic signaling mechanisms in a crucial manner, and upon searching for analogies and homologies, observations on yeast cells have led quite often to pioneering discoveries in vertebrates. Most studies were done using the bakers' yeast *S. cerevisiae* and the fission yeast *Schizosaccharomyces pombe* (used in Africa for brewing the millet beer pombe). *S. cerevisiae* possesses about 6000 genes, and *S. pombe* has about 5000. Being unicellular organisms, yeasts are unsuited for investigation of problems related to multicellularity and tissue formation.

In this respect, **D. discoideum** offers particularly attractive possibilities. Under favorable environmental conditions, the cells of this slime mold are living as free unicellular amoebae. Upon stress, however, they assemble into aggregates of up to 100,000 cells, which go for prey catching as a whole (therefore also called social amoebae) or, upon prolonged stress, differentiate into complex-structured spore carriers. This organism standing at the border between uni- and multicellularity is best suited for studies on chemotaxis, cell aggregation, cell motility, and cell differentiation especially since it can be genetically manipulated almost as easily as yeast.

The nematode **C. elegans** ranks among the simplest and, at the same time, best-investigated metazoan. It is built up from exactly 959 somatic cells (plus a few thousand germ cells), the development and differentiation of which are precisely determined and can be studied in detail. Again the organism is easy to manipulate genetically, thus offering a significant experimental advantage. *C. elegans* is a standard model for studies on cell proliferation and programmed cell death and on intercellular interactions during embryonic development. In the course of such studies, surprising parallels to the situation in vertebrate cells have been revealed. With about 19,000 genes, the genome of the worm is already rather complex.

The tiny fly **D. melanogaster** is considered a pet of geneticists and developmental biologists. It is easy to breed, has a very short generation time (of about 10 days), and is prone to gene mutations. The organization of the genetic material in giant chromosomes considerably facilitates cytogenetic and molecular genetic studies. So *Drosophila* is by far the best-investigated higher invertebrate. As far as the mechanisms of signal transduction are concerned, striking similarities to mammalian cells exist. In fact, many fundamental concepts on cellular data processing have been established based on studies on *Drosophila*. With about 14,000 genes, the genome of the fly is surprisingly small, at least when compared with that of the worm.

The plant **A. thaliana**, belonging to the family Cruciferaceae, is an abundant weed. The ease with which this plant can be bred in the laboratory has made it the botanical counterpart of *Drosophila*. Moreover, *Arabidopsis* is the first plant in which the genome has been completely sequenced. With 26,000 genes it is almost as large as the human genome. Studies on signaling have revealed, in addition to analogies, a number of unexpected differences between plants and animals.

The mechanisms of signal processing in **vertebrates**, especially in mammalian cells, differ from species to species only in minor variants. That means that studies on mice or rats are generally also meaningful for humans.

Cell cultures enjoy great popularity since they are simply organized, easy to manipulate, and usually inexpensive. However, cultivated cells behave differently than cells in an intact tissue, a reservation that holds true in particular for the very popular transformed cell lines, which are purely artificial products. Therefore, it is mandatory to verify the physiological relevance of a result obtained *in vitro* by studies on intact tissues or the whole organism. With this reservation, cell cultures are indispensable for the investigation of cellular data processing.

Summary

Because of the evolutionary conservation of cellular signal processing, the basic mechanisms can be investigated by use of simpler experimental models that provide the possibility of easy genetic manipulation. These include yeast, slime molds, nematodes, and insects as well as cell cultures. To prove the significance of the data thus obtained, *in vivo* experiments with vertebrates as well as (for humans) epidemiological and clinical studies are mandatory.

Further reading

Section 2.1

Minsky A, Shimoni E & Frenkiel-Krispin D (2002) Stress, order and survival. *Nat Rev Mol Cell Biol* 3:50–60.

Schrödinger E (1946) What is life? Cambridge University Press.

Section 2.2

Holmström KM & Finkel T (2014) Cellular mechanisms and physiological consequences of redox-dependent signalling. *Nat Rev Mol Cell Biol* 15:411–421.

Section 2.4

Hall A. (ed) (2000) GTPases. Oxford University Press.

Gasper R et al. (2009) It takes two to tango: regulation of G-proteins by dimerization. *Nat Rev Mol Cell Biol* 10:423–429.

Loirand G, Sauzeau V & Pacaud P (2013) Small G proteins in the cardiovascular system: physiological and pathological aspects. *Physiol Rev* 93:1659–1720.

Ligeti E, Welti S & Scheffzek K (2012) Inhibition and termination of physiological response by GTPase activating proteins. *Physiol Rev* 92:237–272.

Westheimer FH (1987) Why nature chose phosphates. *Science* 235:1173–1178.

Section 2.5

Hancock WO (2014) Bidirectional cargo transport: moving beyond tug of war. *Nat Rev Mol Cell Biol* 15:615–628.

Hartman MA et al. (2011) Principles of unconventional myosin function and targeting. *Annu Rev Cell Dev Biol* 27:133–155.

Pavin N & Tolić-Nørrelykke M (2013) Dynein, microtubule and cargo: a ménage à trois. *Biochem Soc Trans* 41:1731–1735.

Sauer RT & Baker TA (2011) AAA+ proteases: ATP-fueled machines of protein destruction. *Annu Rev Biochem* 80:587–612.

Section 2.6

Endicott JA, Noble MEM & Johnson LN (2012) The structural basis for control of eukaryotic protein kinases. *Annu Rev Biochem* 81:587–613.

Gallego M & Virshup DM (2005) Protein serine/threonine phosphatases: life, death, and sleeping. *Curr Opin Cell Biol* 17:197–202.

Hanks SK & Hunter T (1995) The eukaryotic protein kinase superfamily: kinase (catalytical) domain structure and classification. *FASEB J* 9:576–595.

Lee H et al. (2015) Mining the function of protein tyrosine phosphatases in health and disease. *Sem Cell Dev Biol* 37:66–72.

Manning G, Whyte DB, Martinez R et al. (2002) The protein kinase complement of the human genome. *Science* 298:1912–1934.

Taylor SS & Kornev AP (2011) Protein kinases: evolution of dynamic regulatory proteins. *Trends Biochem Sci* 36:65–77.

Section 2.7

Biggar KK & Li SSC (2015) Non-histone protein methylation as a regulator of cellular signalling and function. *Nat Rev Mol Cell Biol* 16:5–17.

Verdin E & Ott M (2014) 50 years of protein acetylation: from gene regulation to epigenetics, metabolism and beyond. *Nat Rev Mol Cell Biol* 16:258–264.

Section 2.8

DeMartino GN & Gillette TG (2007) Proteasomes: machines for all reasons *Cell* 129:659–662.

Flotho A & Melchior F (2013) Sumoylation: a regulatory protein modification in health and disease. *Annu Rev Biochem* 82:357–385.

Husnjak K & Dikic I (2012) Ubiquitin-binding proteins: decoders of ubiquitin-mediated cell functions. *Annu Rev Biochem* 81:291–322.

Kim YE et al. (2013) Molecular chaperone functions in protein folding and proteostasis. *Annu Rev Biochem* 82:323–355.

Komander D & Rape M (2012) The ubiquitin code. *Annu Rev Biochem* 81:203–229.

Van der Veen AG & Ploegh HL (2012) Ubiquitin-like proteins. *Annu Rev Biochem* 81:323–357.

Varshavsky A (2012) The ubiquitin system, an immense realm. *Annu Rev Biochem* 81:167–176.

Section 2.9

Barkauskaite E et al. (2013) The recognition and removal of cellular poly(ADP-ribose) signals. *FEBS J* 280:3491–3507.

Feijs KLH, Verheugd P & Lüscher B (2013) Expanding functions of intravellular resident mono-ADP-ribose in cell physiology. *FEBS J* 280:3519–3592.

Li N & Chen J (2014) ADP-ribosylation: activation, recognition, and removal. *Mol Cells* 37:9–16.

Section 2.10

Fenno l, Yizhar O & Deisseroth K (2011) The development and application of optogenetics. *Annu Rev Neurosci* 34:389–412.

Evolution of Cellular Data Processing

3

Being a condition of survival, the ability to adapt to an ever-changing environment has accompanied life from the very first day. This means that even the most ancient and primitive cells must have been able to recognize and rate relevant environmental signals. As explained in Chapter 1, this is done by a data-processing protein network that, particularly for unicellular organisms such as prokaryotes, fully resembles a nervous system, justifying the metaphor of a cellular "nanobrain." Investigation of prokaryotic signal transduction has shown that, in its basic biochemical mechanisms, such a primeval network does not differ fundamentally from signal processing networks in eukaryotes and the mammalian nervous system, thus nourishing the concept of a common evolutionary ancestry that had entered the stage in the beginning of life. A closer look at today's prokaryotes gives us some idea of how this might have occurred in the past. It immediately makes clear that the data-processing network must have evolved in parallel to the metabolic network.

3.1 Evolution of biological signal processing

Organisms are classified in three domains: bacteria and archaea (or archaebacteria), together constituting the prokaryotes, and eukaryotes. Eukaryotes differ from prokaryotes by having a nuclear membrane and an intracellular compartmentalization in structures such as endoplasmic reticulum, Golgi apparatus, mitochondria, and chloroplasts. Genetic studies have revealed that eukaryotes are related more closely to archaea than to bacteria (**Figure 3.1**). This was a somewhat surprising discovery because archaea were formerly thought to be the primary organisms, in particular since they are settlers of extreme biotopes. The high degree of horizontal gene transfer between the three domains was another unexpected finding.

Once considered to be primitive, prokaryotes are now acknowledged as unexpectedly complex and highly organized forms of life. Their amazing ability to settle any possible ecological niche gives the reason for their overwhelming evolutionary success. Leaving aside adaptation by gene mutations, the cellular conditions of this ability are as follows:

- rapid and reversible adaptation of gene expression and, thus, of the protein pattern (or "proteome") to the actual environmental conditions
- precise control of behavior such as motility, metabolism, and transport processes
- establishment of social interactions aimed at genetic recombination and division of labor in multicellular structures

To perform such adaptive responses, the cell must be a cognitive system. What are the primary environmental signals that a prokaryotic "nanobrain" has to decipher for the sake of survival? To exist in an unpredictable environment requires a permanent

Figure 3.1 Phylogenetic three-domain tree Broken horizontal lines symbolize horizontal gene transfer. For other details see text.

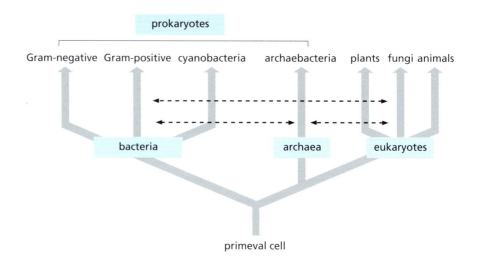

search for optimal conditions of life, or in other words, minimization of stress. It is conceivable, therefore, that the most elementary forms of signal processing are aimed at coping with stress situations such as starvation, disturbances of osmotic pressure, intoxication, dangerous radiation, and attack by predators. A cell unable to track down energy resources and blind to danger has no chance of surviving. Thus, the processing of *environmental signals* stands at the very beginning (**Figure 3.2**). Social contacts or even the formation of multicellular colonies, including specialization of cells by differentiation aimed at a division of labor, may facilitate survival even more. Both require the evolution of an *intercellular* signaling system. It must be emphasized that, on the level of signal decoding, a cell does not make a distinction between environmental and intercellular signals: both are processed by the same network, provided the cell possesses the corresponding sensors or receptors.

A characteristic example of an ancient intracellular signaling reaction dealing with environmental stress is the **stringent response**. This is a primeval bacterial emergency reaction that saves energy. It leads to an almost complete stop of ribosome production and protein synthesis and, as a consequence, of cell proliferation. These are extremely expensive processes devouring almost 50% of the cellular energy. The stringent response is triggered by worsening life conditions, particularly starvation. Owing to a lack of certain amino acids, uncharged transfer RNA (tRNA) accumulates at the ribosomes. As an immediate consequence, the ribosome-associated enzyme of the Rel family, also known as stringent factor, becomes activated. Rel enzymes are guanylate kinases that catalyze the ATP-dependent pyrophosphorylation of GTP in position 3', yielding ppGppp, which subsequently is dephosphorylated to guanosine 3',5'-bis(diphosphate) (ppGpp) by a specific phosphatase (**Figure 3.3**).

Figure 3.2 Evolution of biological signal processing All cells are able to adjust their behavior and pattern of gene expression to exogenous (environmental and intercellular) signals provided these carry a meaning such as food, energy, danger, and partner cells. The meaning is decoded by the cellular apparatus of data processing. The intercellular signaling systems of multicellular organisms have evolved from intercellular communication that prokaryotes require for sexual, symbiotic, and parasitic interactions as well as for the formation of multicellular collectives.

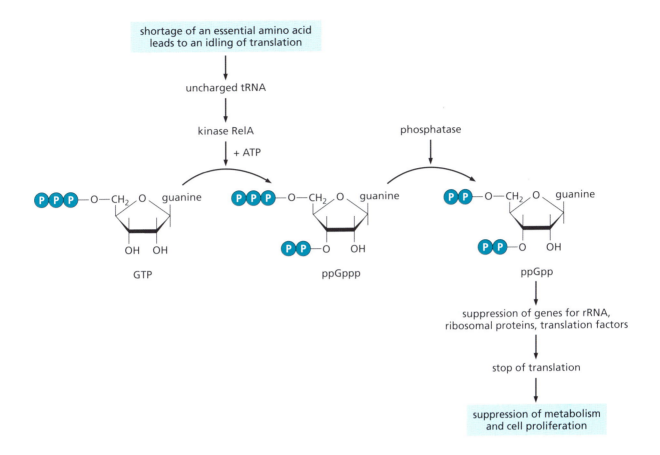

ppGpp has the function of an intracellular emergency signal or "**alarmone**," binding to the catalytic β-subunit of RNA polymerase and inhibiting, in particular, the de novo synthesis of ribosomal RNA, ribosomal proteins, translation factors, and tRNA. Cells containing a defective *rel* gene (*rel*axed mutants) are unable to adapt, literally growing to death in deficiency situations. Evolutionarily advanced species such as myxobacteria use the stringent response for a more sophisticated survival strategy: the formation of multicellular spore capsules (see Section 3.5.2).

The enzymatic apparatus of ppGpp synthesis, including RelA-homologous enzymes, has also been found in **plant chloroplasts**, which are thought to have evolved from bacteria. Like the bacterial system, the plant system is stimulated under stress situations, here in particular by wounding. Eukaryotes possess, in addition, other highly efficient means to turn down translation upon amino acid deficiency, operating by mechanisms that differ entirely from the stringent response (see Sections 9.3 and 9.4).

Another primeval system coping with environmental stress uses RNA molecules as signaling entities. It is discussed in the following section.

Figure 3.3 Signaling of the stringent response in *E. coli* Stress situations such as a shortage of certain amino acids cause an idling of translation, resulting in activation of the ribosomal kinase RelA. The latter transfers a pyrophosphoryl group from ATP to the 3'-position of 5'-GTP. ppGppp thus generated becomes dephosphorylated enzymatically, yielding the "alarmone" ppGpp that acts as a transcriptional suppressor of genes encoding RNA and proteins required for translation. The result is an almost complete stop of protein synthesis. P, phosphate group.

Summary

- By processing a wide variety of exogenous stimuli prokaryotes easily adapt to environmental conditions. The most elementary forms of signal processing are aimed at coping with emergency situations (stress).

- An example is provided by the stringent response, an adaptation of cellular protein synthesis to environmental stress. The response is based on the generation of intracellular messengers (alarmones such as ppGpp) that suppress RNA polymerase activity.

- Social interactions, cell differentiation and formation of multicellular structures depend on the evolution of *intercellular* signaling systems. Environmental and intercellular signals are processed by the same protein–DNA network.

3.2 The RNA world

RNA has unique properties: it can store genetic infomation, replicate itself, catalyze chemical reactions (see Section 2.4 and Sidebar 2.2) and transduce signals. This has led to the hypothesis that the most ancient form of life was an RNA world which later developed into the modern protein and DNA world. Of course, other concepts such as that proteins and RNA had emerged simultaneously cannot be excluded.

Ribozymes

Ribozymes are RNA molecules that catalyze—like enzymes—chemical reactions. Their catalytic properties depend on specific three-dimensional structures. According to the RNA world hypothesis, the most ancient ribozymes must have been an RNA polymerase for RNA replication and a ribonuclease for RNA degradation. Prominent examples of today's ribozymes are the RNAs of the large ribosomal subunit (23S RNA in prokaryotes and 28S RNA in eukaryotes) that catalyze the formation of peptidyl bonds for protein synthesis, and the small nuclear RNAs catalyzing pre-messenger RNA (mRNA) splicing. Both the ribosomal and the small nuclear RNAs are components of larger ribonucloprotein particles where the proteins fulfil the role of RNA chaperones rather than that of enzymes. Some ribozymes that are controlled by low-molecular-mass ligands act as riboswitches. For the discovery of ribozymes the 1989 Nobel Prize in Chemistry was awarded to Sidney Altman and Thomas R. Cech.

Riboswitches

A riboswitch is a non-coding regulatory segment of a mRNA located within the 3′-terminal untranslated region (3′-UTR). It is characterized by a complex tertiary RNA structure (**Figure 3.4**). Riboswitches are signal-transducing elements: upon specific binding of a small molecule (the input signal) the translating activity of the mRNA (the output signal) is changed. Like a receptor (see Section 2.12) the riboswitch is divided into a discriminator domain—here called **aptamer**—and an effector domain—here called **expression platform**, and the signal-tranducing process consists of a change in the RNA conformation. Ligands as input signals interacting with aptamer sequences are metal ions, amino acids, enzymatic cofactors such as *S*-adenosyl-methionin, tetrahydrofolate, thiamin pyrophosphate, and flavin mononucleotide, as well as glucosamine-6-phosphate, purines, and second messengers such as cyclic-di-GMP and cyclic-di-AMP (see Section 3.5). Riboswitches are almost exclusively restricted to prokaryotes. They belong to the oldest genetic control elements. Riboswitches have two major effects (**Figure 3.5**): (1) they interrupt the transcription of operons by exposing upon ligand binding a hairpin structure that functions as a transcription terminator triggering the decomposition of the DNA–RNA–RNApolymerase complex; (2) they inhibit mRNA translation by blocking the ribosome binding site.

Small regulatory RNA

A large proportion of the prokaryotic and the eukaryotic genome is transcribed into RNA that cannot be translated into protein. This non-coding RNA (ncRNA) comprises RNA with housekeeping functions, such as tRNA and ribosomal RNA, and RNA with regulatory functions. In prokaryotes, the RNA with regulatory functions consists mainly of molecules with a length of 50–300 nucleotides, called small RNA (sRNA). Like riboswitches, sRNAs are signal-transducing entities. Their formation is controlled by a wide variety of transcription factors. In a given species the num-

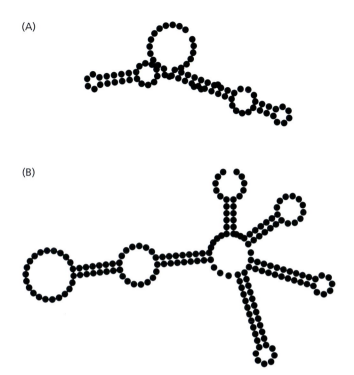

Figure 3.4 Structure of riboswitches The figure shows the cyclic-di-GMP (A) and the lysine (B) riboswitch. Note the sophisticated tertiary structure with several hairpin loops due to intramolecular base-pairing. (Adapted from the Rfm database.)

ber of individual sRNAs rivals that of individual protein transcription factors. The major function of sRNA is to modulate mRNA translation. Thus, sRNAs are powerful regulators of protein synthesis or—in other words—of posttranscriptional control of the genetic readout. To this end sRNA interacts specifically with complementary nucleotide sequences of target mRNA (antisense interaction). sRNAs are critically involved in the regulation of prokaryotic physiology, in particular in stress situations such as starvation, DNA damage, heat, UV-light, and oxidative stress but also in quorum sensing (sRNA suppresses quorum sensing when the cell density is insufficient to form colonies). In fact, sRNAs enable the cell to quickly cope with dangerous environmental situations. One distinguishes between two forms of sRNA. *Cis*-encoded sRNA is a perfect antisense transcript that is synthesized from the DNA strand just opposite (complementary) to the gene of its target mRNA. The complete complementarity ensures a strong and specific sRNA–mRNA interaction with a high degree of privacy. In contrast, *trans*-encoded sRNA is encoded somewhere else in the genome. The interaction of *cis*-encoded sRNA with mRNA results

Figure 3.5 Genetic control by riboswitches (A) The mRNA with riboswitch controls its own transcription. In absence of the ligand (teal circle) the transcription of the downstream operon is permitted. Binding of the ligand to the aptamer domain of the riboswitch causes a conformational change that leads to the formation of an intrinsic terminator structure characterized by a specific hairpin loop. This results in a breakdown of the DNA-RNA-RNApolymerase complex and transcription is stopped. (B) The mRNA with riboswitch controls its own translation. The structural change caused by the binding of the ligand to the aptamer sequence leads to a blockade of the ribosome binding site (RBS) and translation is stopped. (Adapted from Kim J & Breaker RR [2008] *Biol Cell* 100:1–11.)

Figure 3.6 Control of mRNA translation by small RNAs Positive regulation (left): the ribosome binding site (RBS) of mRNA is blocked by intramolecular base-pairing. This blockade is lifted by a complementary sRNA. Negative regulation (right): sRNA inhibits translation by binding to antisense sequences of mRNA.

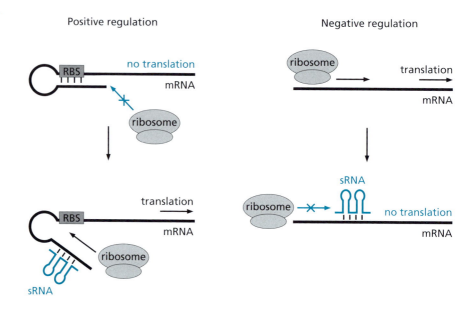

in a suppression of mRNA translation and may end up in mRNA degradation. The same holds true for *trans*-encoded sRNA. However, since the complementarity with target sequences is incomplete, *trans*-encoded sRNA can interact with multiple mRNAs, mainly by binding to or near to the ribosome binding site of the target mRNA. This interaction may not only lead to an inhibition but also to an *activation* of mRNA translation. In the latter case the binding of target mRNA to ribosomes is prevented by an autoinhibitory sequence found on the same RNA strand, and the interaction with sRNA lifts this blockade (**Figure 3.6**). Evidently such a mechanism resembles the blockade of protein kinases by pseudosubstrate domains (see Section 2.6.1).

The weak interaction of *trans*-encoded sRNA has to be enforced—at least in Gram-negative bacteria—by chaperones of the **HfQ protein** family. The first member of this family was identified in *Escherichia coli* as <u>H</u>ost <u>f</u>actor <u>Q</u>, which promotes the replication of bacteriophage Qβ. But this is only one of its numerous effects. A major function of HfQ is to stabilize the sRNA–mRNA complex by remodeling RNA structure and by bringing sRNA and mRNA in close contact. Moreover, the RNA–HfQ complex recruits additional proteins such as endoribonuclease E, required for the final degradation of target mRNA. HfQ is a large ring-shaped protein consisting of six identical subunits and binding sites for sRNA and mRNA. It belongs to the large family of **Lsm proteins**. Like HfQ, Lsm proteins also stabilize—as ring-shaped RNA chaperones—three-dimensional RNA structures. Lsm proteins are found in all organisms. They have a long evolutionary history. Perhaps the most ancient Lsm proteins may have assisted RNA molecules to develop enzymatic activity as primeval ribozymes. In higher organisms Lsm proteins are, for instance, components of spliceosomes where they bind small nuclear RNA which is required for alternative splicing of pre-mRNA (see Sidebar 8.3). Another function of Lsm proteins in eukaryotes is the formation of complexes with small nucleolar RNA. These complexes are required for the maturation of tRNA and rRNA in the nucleolus.

In eukaryotic cells the interaction of regulatory sRNA with mRNA is controlled by another RNA-binding protein family, the **Argonaut proteins** (see Section 9.5), which are not related to HfQ/Lsm proteins. Argonaut proteins are also found in prokaryotes. Here one of their functions is to protect the organism from invasive RNA and DNA which both become degraded upon binding to Argonaut.

According to our present knowledge it has to be concluded that although translational control by small RNA exists both in prokaryotes and in eukaryotes, the mechanisms involved are different.

The adaptive immune system of prokaryotes

To combat infections by viruses (bacteriophages) or contamination with foreign plasmid DNA, prokaryotes have developed non-specific as well as adaptive defense mechanisms that resemble innate and acquired immunity. Non-specific defense is aimed at an inhibition of virus attachment, destruction of invading nucleic acid, or death of infected cells (abortive-infection). Adaptive defense is a "learning system" known as the **CRISPR-Cas** mechanism. It uses a special type of small RNA molecule as intracellular messengers. CRISPR is the acronym for Clustered Regularly Interspaced Short Palindromic Repeats, while Cas stands for CRISPR-associated genes. While CRISPR represents the memory system of adaptive immunity, Cas proteins are used for programming the memory store and as weapons against the invader.

The CRISPR locus of prokaryotic DNA consists of an array of identical repeats (21–48 base pairs) intercalated with "spacer" sequences (26–72 base pairs). The latter are derived from former invaders so that the spacer pattern of CRISPR represents a time record of previous infections. Spacer incorporation into CRISPR is catalyzed by specific Cas proteins. The CRISPR locus is transcribed as a whole into a long precursor RNA which subsequently is processed into small RNA molecules (crRNA), each containing a spacer sequence flanked by sequences derived from the adjacent repeats. The spacer sequences are used to recognize corresponding viral DNA sequences via Watson–Crick base-pairing, while the repeat sequences serve as docking sites for specific Cas nucleases that destroy the invader DNA. CRISPR-Cas loci are found in the DNA of more than 50% of prokaryotes hitherto analyzed. Depending on the species, one can distinguish between a variety of CRISR-Cas systems. To a certain extent the CRISPR-Cas mechanism resembles RNA interference in eukaryotes (see Section 9.5), although the mechanisms are quite different.

In the laboratory, CRISPR-Cas has become the most powerful method of **genome editing**, in particular with **Cas9** as the effector enzyme. Cas9 is an endonuclease introducing a single double-strand break in target DNA. The cleavage triggers endogenous DNA repair reactions that normally result in non-homologous end-joining, causing a deletion of some nucleotides. This may interrupt the coding sequence of a gene and result in a loss-of-function mutation. The pioneering idea was to guide Cas9 with a synthetic crRNA (called synthetic guide RNA, sgRNA) to any DNA sequence of interest. In fact, plasmids carrying Cas9 and sgRNA genes can be expressed in practically all cell types and organisms. Moreover, in the presence of a small exogenous DNA template that matches the cleavage region of the target DNA, homology-directed repair would occur resulting in a controlled exchange of a single nucleotide. A battery of genetically modified Cas9 enzymes has become available serving specific deletion, replacement, and modification of genes.

Summary

- RNA molecules are perhaps the most ancient signaling entities.

- Ribozymes are RNAs with enzymatic activity.

- Riboswitches are sensor sequences of mRNAs that specifically bind small molecules thus modulating transcription and translation.

- Regulatory small RNAs (sRNAs) are either specific inhibitors or stimulators of mRNA translation. Frequently they require the assistance of protein chaperones. sRNAs play a major role in bacterial stress responses.

3.3 Signal-controlled membrane transport: the ancient way to communicate

The controlled uptake of food and the excretion of (toxic) metabolic end products stand at the very beginning of any communication between an organism and its environment. Since the lipid bilayer of the plasma membrane is impermeable for most substances, even the most primeval cells were forced to develop regulative transport mechanisms based on the function of special membrane proteins.

There are four types of transport process:

(1) **Passive transport**, *along* a concentration gradient, is made possible by membrane channels or carrier proteins (uniporters) and therefore has been called "facilitated diffusion." Examples of uniporters are the glycerol transporter of *E. coli* and the glucose transporters GluT1 and GluT4 in liver and muscle or fat tissue, respectively (see Section 7.1.3).

(2) **Primary active transport**, *against* a concentration gradient, requires energy derived from enzymatic hydrolysis of ATP; that is, membrane-bound transport proteins have intrinsic ATPase activity or associate with ATPases. These transporters are discussed in more detail in Section 3.3.1.

(3) **Secondary active transport**, *against* a concentration gradient, derives the energy from an ion flow occurring along a concentration gradient and coupled with the transport process. Examples are the lactose transporter LacY driven by a proton gradient in *E. coli* and the Na^+–Ca^{2+} exchanger powered by an Na^+-gradient (see Section 14.5.1).

(4) **Group translocation** consists of a targeted transport process coupled with an energy-delivering chemical modification of the molecule to be transported. In most cases this is achieved by phosphorylation (the translocation of a phosphate group) requiring the hydrolysis of an energy-rich phosphoric acid derivative such as ATP or phosphoenolpyruvate (PEP). This process is discussed in detail in Section 3.3.2.

Even a very simple prokaryote such as *Mycoplasma genitalium* (harboring only 470 genes) expresses one passive transporter, 13 ATP-powered transporters, and one ion-gradient-powered transporter as well as two group translocation systems. For *E. coli*—although with about 4300 genes it is by no means the most complex prokaryote (the soil bacterium *Sinorhizobium meliloti* has 6300 genes, as many as yeast)—the corresponding numbers are 7, 194, 74, and 22, respectively.

3.3.1 ATP-powered membrane transporters

The proteins of this group are ATPases catalyzing the *targeted* transmembrane transport of ions or larger molecules, frequently against a concentration gradient.

Ion pumps

The bacterial **F-type transporters** use ATP hydrolysis to generate a proton gradient across the cell membrane, driving numerous active transport processes. The reverse reaction results in ATP production. In fact, the ATP synthases of mitochondria and chloroplasts belong to the family of F-type transporters. The **P-type transporters** represent another family of ATP-driven ion pumps. Examples are provided by the Na^+/K^+-dependent ATPases (sodium pumps) and the Ca^{2+}-dependent ATPases (calcium pumps). **V-type transporters** found in the membrane of intracellular vacuoles are restricted to archaea and eukaryotes.

ABC transporters

The largest and most versatile subfamily of ATP-powered membrane transporters is the ABC transporters, named after the ATP-Binding Cassette, a particular ATP binding site in the ATPase domain. Characteristic structural features of this highly

Figure 3.7 Bacterial ABC transporter of type MsbA The protein is composed of two identical subunits with six transmembrane domains (cylinders) and one intracellular ATP-binding site each. ATP hydrolysis causes a conformational change, promoting the excretion of toxic substances (teal ball) trapped in a cage. The transporter becomes reloaded upon ADP–ATP exchange.

conserved cassette are the ATP-binding Walker A motif or P-loop and the Mg^{2+}-binding Walker B motif (ATP is bound as a Mg^{2+} complex). Both motifs are common to nucleoside-triphosphatases and kinases (see Section 2.4).

ABC transporters are modular proteins consisting of four domains: two transmembrane domains, frequently with six transmembrane helices each, and two cytoplasmic domains with one ATP binding site each (**Figure 3.7**). These domains may exist as separate subunits or may be integrated in one or two polypeptide chains. The transport channel is established by a circular arrangement of the transmembrane helices.

In *E. coli* the ABC transporters constitute a large gene family (74 genes). They transport a wide variety of substances, from simple ions, amino acids, and sugars to peptides and proteins, and each transporter is specific for the uptake or excretion of a distinct substance. The uptake of material is facilitated by selective binding proteins mostly found in the periplasm of Gram-negative bacteria. They function as a kind of receptor and interact with the ABC transporters.

One of the best-studied ABC transporters is the **MsbA** protein of *E. coli*. It is the only transporter that is absolutely essential for the bacterial cell. MsbA is a multi-drug resistance protein, which exists in more than 30 homologs. These proteins are responsible for the excretion of lipophilic and toxic (foreign) substances such as drugs. An adaptive overexpression of MsbA is one of the mechanisms that renders bacteria resistant to antibiotics. Resistant bacteria are assumed to be the reason for approximately 60% of iatrogenic diseases.

The MsbA protein is composed of two polypeptide chains with six transmembrane helices and one ATP binding site each. It probably functions as a "flippase," trapping the substrate in a cage at the inner side of the membrane and releasing it to the outside upon a conformational change brought about by ATP hydrolysis (see Figure 3.7).

The human homolog of MsbA is the multi-drug resistance protein **MDR1**, also called P-glycoprotein. Its adaptive overexpression in tumor cells is one of the mechanisms leading to the feared resistance against cytostatics during cancer chemotherapy. MDR1 is the prototype of a large family of eukaryotic ABC transporters with several functions. Another human ABC protein, the cystic fibrosis transmembrane conductance regulator or **CFTR**, is strictly speaking not a transporter but a hormone-controlled chloride channel regulating the function of mucous membranes (for more details see Section 14.7.3). A defective *CFTR* gene is the cause of cystic fibrosis or mucoviscidosis, which is one of the most common human hereditary diseases. In both MDR1 and CFTR, the four domains of an ABC transporter share a single polypeptide chain.

3.3.2 Group translocation systems: primeval forms of data-processing networks combining food uptake with gene regulation

The best-known representatives of this family of membrane transporters are the PEP (phosphoenolpyruvate)-powered **phosphotransferase systems**, abbreviated as PTS. They combine membrane transport with the regulation of gene transcription.

Prokaryotes are capable of switching their genes on and off in a specific and signal-controlled mode in order to adapt to the actual conditions of the biotope. As far as the molecular mechanism is concerned, there is no fundamental difference between prokaryotic and eukaryotic gene transcription. In both cases, an initiation complex containing RNA polymerase and a series of regulatory proteins assembles at the gene at sites that have been marked by transcription factors. This is the key event and primary target of signaling (for details see Chapter 8). A characteristic feature of prokaryotes is that genes controlling an individual metabolic process (structural genes) are combined into functional units known as operons. Operons are transcribed as a whole into a polycistronic RNA and are under the control of both stimulatory and inhibitory gene sequences called promoters, enhancers, and silencers. These are the binding sites of transcription factors. Stimulatory transcription factors facilitate and inhibitory factors (repressors) inhibit the assembly of an active RNA polymerase complex. The major signaling events controlling transcription factor and repressor activity in both pro- and eukaryotes are noncovalent ligand binding and covalent protein phosphorylation. Moreover, repressor activity is frequently regulated by corresponding metabolic products acting via feedback loops or by special signaling reactions. The active form of a transcription factor is always a homo- or heterodimer, enabling a cooperative effect with an intrinsic noise filter (see Section 1.3). Transcriptional control supports the idea formulated in Chapter 1 that signal processing is based on combinatorial processes. A cell contains many more genes than transcription factors. Only 5% of the 4290 genes of *E. coli* encode gene-regulatory proteins. How, then, is *specific* gene regulation achieved? The answer is that the majority of bacterial gene promoters are regulated by a set of different transcription factors rather than by a single regulatory protein. Because of this combinatorial principle, a wide variety of environmental stimuli may affect a large number of genes by use of only a rather limited number of regulatory proteins. The cooperation between activators and repressors renders these systems logical gates (see Section 1.5).

PTSs manage the uptake of a variety of sugars and sugar alcohols by substrate phosphorylation, using PEP as the phosphate donor and energy provider (**Figure 3.8**). Phosphorylation is the starting reaction of substrate metabolism such as glycolysis.

Figure 3.8 Phosphotransferase system (PTS) of *E. coli* The system couples the glycolytic pathway with glucose transport and regulatory reactions. Consisting of the cytoplasmic proteins EI, HPr, and EIIA and the membrane-bound channel complex EIIB/EIIC, it stepwise transfers a phosphate residue (P, teal circle) from one of the two phosphoenolpyruvate (PEP) molecules derived from glucose 6-phosphate to glucose, which is taken up by the cell as glucose 6-phosphate. The second PEP molecule is used for ATP generation. When activated by glucose, the system changes cell motility and inhibits the uptake and metabolism of alternative nutrients (catabolite repression). The phosphorylated amino acids of the PTS proteins are depicted in italic type.

The process is embedded in a network of interacting proteins that adjust behavior and genetic readout to the food supply. PTSs, showing the inseparable connection between metabolism and signal processing, indicate that even on a rather simple level the cellular response to an exogenous stimulus has to be treated as a systemic answer of the whole organism rather than as an isolated biochemical reaction.

PTSs are abundant in bacteria but have not yet been found in archaebacteria and eukaryotes. Even among bacteria they are differentially distributed. *E. coli* expresses 22 PTSs with different substrate specificities, while the 10 times smaller genome of *Mycoplasma genitalium* contains only two PTS genes (for the transport of glucose and fructose) and *Mycobacterium tuberculosis* has none.

PTSs are multiprotein complexes consisting of highly conserved and variable parts. A prototype is the glucose-transporting PTS^{Glc} of *E. coli*. Its conserved parts are the two cytoplasmic enzymes EI and HPr (Histidine-containing phosphocarrier Protein), while the variable part is the transport complex EII. EII is composed of three proteins: the membrane transporter EIIC and the associated proteins EIIA and EIIB (see Figure 3.8). EI and HPr are separate cytoplasmic proteins, whereas the three EII subunits may either share one polypeptide chain (EIIABC; for instance, in the *N*-acetylglucosamine translocator of *E. coli*) or be partly separated (EIIA + EIIBC; for instance, in the glucose translocator of *E. coli*). EIIC contains eight trans-membrane domains and as a dimer constitutes the transmembrane channel.

The homodimeric enzyme EI catalyzes the hydrolysis of PEP, transiently binding the phosphate residue at a histidine residue (His189 in *E. coli*); thus it is acting as a histidine autokinase. From there the residue migrates via HPr, EIIA, and EIIB to glucose, which as glucose 6-phosphate passes the membrane channel EIIC. In both HPr and EIIA the phosphate residue is also bound to His residues; however, in EIIB, which is the glucose phosphorylase proper, the phosphate residue interacts with a cysteine residue.

Together these phosphorylations are thermodynamically unstable, short-lived transition states of phosphotransferase reactions. [Note that the phosphoamide bond of phospho-His contains so much energy that the reaction ATP + protein ↔ ADP + protein-P is not shifted to the left only because the product (protein-P) becomes rapidly dephosphorylated by the subsequent reactions.] The more glucose available, the higher the rate of the cell's catabolic metabolism and the intracellular PEP level. Under such conditions the PTS machinery is running at top speed, phosphorylating glucose at a high rate and, thus, generating a rapid draining of phosphate. As a consequence, the steady-state concentrations of *phosphorylated* PTS proteins are low, whereas they rise upon glucose deficiency. The degree of phosphorylation of PTS components is used for a coupling of the transport process with accessory cellular functions. In fact, PTS represents the backbone of a simple but highly efficient signal-processing protein network enabling the cell to perfectly adapt to environmental conditions.

At a high glucose level, the now underphosphorylated EI inhibits the chemotaxis protein CheA, changing the motility pattern of the cell in such a way that it is able to approach the source of food (positive chemotaxis, see Section 3.4.4). Under the same conditions, underphosphorylated EIIB inhibits the transcriptional repressor Mlc, thus ensuring continuously high activity of the genes encoding the enzymes of glucose transport and metabolism. Moreover, as long as sufficient glucose is available, underphosphorylated EIIA binds and inactivates the transporters for lactose and other sugars, preventing the costly uptake of less attractive nutrients for which the metabolic enzymes are not available, since the corresponding genes, such as the *lac* operon, are also repressed by EIIA. Only when the glucose supply is beginning to run low does this inhibition, called **catabolite repression**, become gradually canceled, enabling the cell to exploit alternative sources of food. The trigger for this genetic and metabolic changeover is the increase in PTS phosphorylation, by which

Sidebar 3.1 Prokaryotic adenylate cyclases Adenylate or adenylyl cyclases catalyze the formation of cAMP from ATP according to

$$ATP \rightarrow \text{cyclic } 3',5'\text{-AMP} + \text{pyrophosphate (PP}_i)$$

(the endergonic reaction is shifted to the right by the subsequent hydrolysis of pyrophosphate). Adenylate cyclases are ancient enzymes found in bacteria, archaea, and eukaryotes. Depending on the species, bacteria may express up to 40 different isoforms.

On the basis of primary sequence data, nucleotide cyclases are subdivided into six classes. Among these, three major families of prokaryotic adenylate cyclases are distinguished. Family 1 consists of adenylate cyclases of Gram-negative bacteria such as *E. coli* with a special function in catabolite repression. Family 2 contains adenylate cyclases of archaea and toxic adenylate cyclases of certain pathogenic bacteria, such as *Bacillus anthracis, Bordetella pertussis, Pseudomonas aeruginosa*, and *Bacillus bronchiseptica*. These enzymes are released together with other toxins into the host cell, causing a pathological overproduction of cAMP, for example, during anthrax and whooping cough. Family 3 consists of "eukaryotic" adenylate cyclases, some of which are found also in prokaryotes (for instance, in mycobacteria) but not in archaea (for details see Section 4.4.1).

Guanylate cyclases that transform GTP into cyclic 3',5'-GMP seem to be entirely restricted to eukaryotes. Bacteria (but not archaea and eukaryotes) possess diguanylate cyclases that transform two GTP molecules into the second messenger cyclic di-GMP (see Section 3.5.2).

a signal is released that changes the transcriptional pattern. In Gram-negative bacteria such as *E. coli*, this signal is **cyclic adenosine 3',5'-monophosphate (cAMP).** It is generated from ATP when highly phosphorylated EIIA activates the membrane-bound enzyme adenylate cyclase (see **Sidebar 3.1** and also Section 4.4.1 for a reaction scheme).

The function of cAMP is that of an intracellular signaling molecule or second messenger. To this end it binds to a cAMP receptor protein, CRP (also called Catabolite gene Activator Protein, CAP), a transcripton factor that upon cAMP-binding stimulates a series of catabolite-repressed gene complexes such as the *lac* operon (**Figure 3.9**).

Being an ancient control mechanism, cAMP signaling is by no means restricted to *E. coli*-like bacteria; many other prokaryotes express a wide variety of adenylate cyclase isoforms. This holds true even for most primitive species such as mycobacteria that have been found to produce cAMP in unusually large amounts. The genome of *Mycobacterium tuberculosis*, for instance, encodes no less than 17 nucleotide cyclases as well as several cAMP phosphodiesterases and other cAMP binding proteins, in particular putative transcription factors. Evidence is accumulating that highly pathogenic *Mycobacteria* (causing tuberculosis, leprosy, and other diseases) use cAMP signaling for proliferation and subversion of host defense mechanisms. Therefore, the corresponding proteins are considered to provide potential targets

Figure 3.9 Activation of CRP by cAMP in *E. coli* Upon binding of 4 cAMP molecules (teal) the three-dimensional structure of CRP is changed in such a way that it can interact specifically with DNA. This process is prototypical for signal-dependent activation of transcription factors. Based on X-ray crystallography. (Adapted from Green J et al. [2014] *Curr Opin Microbiol* 18:1–7.)

for novel therapeutic approaches, in particular against bacterial strains that have become resistant to conventional drug therapy.

The PTS mechanism depicted in Figure 3.8 applies also to Gram-positive bacteria. However, for catabolite repression they pursue another strategy, because many of them—such as *Bacillus subtilis*—neither express a CRP-homologous protein nor produce cAMP. Instead, the catabolite-repressible genes are blocked by a transcriptional repressor that is removed upon glucose deficiency.

The key reaction is a phosphorylation of HPr at a serine (Ser) residue that competes with the EI-catalyzed His phosphorylation of HPr (which is still essential for the sugar uptake). The Ser phosphorylation requires ATP and is catalyzed by an HPr kinase/phosphatase. The latter is activated by a high intracellular level of ATP (and a low level of inorganic phosphate) as well as by intermediates of glycolysis; that is, under conditions of maximal glucose supply. In its structure and bifunctional activity (being both a kinase and a phosphatase), this primeval enzyme differs from all known protein kinases and phosphatases.

In contrast to the labile phospho-His bond, the ester bond of phospho-Ser is kinetically stable. Therefore, Ser-phosphorylated HPr is not a short-lived intermediate but can act as a transcriptional co-repressor by binding the so-called **catabolite control protein A (CcpA)**. This complex interacts with gene regulatory sequences named catabolite responsive elements, *cre*, and inhibits the transcription of catabolite-repressible operons (**Figure 3.10**). Upon glucose or ATP deficiency (or at a high level of intracellular inorganic phosphate), the kinase activity of the HPr kinase/phosphatase is turned down, whereas the phosphatase activity of the enzyme remains unchanged, now catalyzing the dephosphorylation of HPr. This leads to a dissociation of the HPr–CcpA complex and to an activation of the catabolite-repressed genes. Ser

Figure 3.10 Catabolite repression in Gram-negative as compared with Gram-positive bacteria In Gram-negative species, the enzyme EIIA is underphosphorylated at high glucose concentration, inhibiting cAMP formation and thus the activation of cAMP-dependent (catabolite-repressed) genes. In Gram-positive bacteria, the PEP-dependent His phosphorylation competes with an ATP-dependent Ser phosphorylation of HPr. The Ser-phosphorylated HPr together with the protein CcpA inhibits transcription of catabolite-repressed genes.

phosphorylation of HPr not only adjusts gene transcription to the food supply but also controls PTS by negative feedback, since the competing His phosphorylation, and thus the sugar uptake, is hindered.

Phosphotransferase system and evolution of signal transduction

PTS and catabolite repression are components of a data-processing protein network that regulates access to different nutrients in a highly economical manner. These components also cause the intake and metabolism of the particular carbon source to become adjusted at an optimal level that does not overtax the metabolic capacities of the cell. This enables bacteria to adapt perfectly to environmental conditions, albeit within their genetically programmed limits.

Of particular interest are the evolutionary aspects of the system. The network makes use of reversible protein phosphorylation, and the origin of this signaling reaction from short-lived enzymatic transition states becomes especially evident here. Moreover, with cAMP the prototype of a second messenger enters the stage for the first time. By transducing the effects of exogenous signals to the genome, this molecule has become an evolutionary model of success (though its function has changed from a cofactor of a gene-regulatory protein to an activator of protein kinases and ion channels that indirectly control transcription; see Sections 4.4.1 and 4.6.1). Further models of success are the two ways by which the activity of a transcription factor is controlled: either by ligand binding (cAMP–CRP) or by protein phosphorylation (Ser phosphorylation of HPr). These major principles of signaling to the genome have survived evolution up to today's most advanced organisms.

From one-component to multicomponent systems

The screening of domain libraries obtained by gene sequencing has shown that the majority of prokaryotic signal transducers combine just a single input with a single output domain, thus representing the simplest modular structure. Such proteins function as genuine binary switches. To distinguish them from the more complex two-component systems (see Section 3.4), these signaling devices are also called one-component systems. Most of them are transcriptional regulators with a DNA-binding and transactivating site as an output domain; some also have enzymatic functions. The input domains mostly recognize small intracellular molecules such as cAMP and metabolic products or are phosphorylated. The cAMP receptor protein CRP and the kinase substrate HPr provide examples of one-component systems.

One-component systems are probably the most ancient signaling proteins of all. The abundance of these simple transducers in prokaryotes indicates that, in the earlier days of evolution, linear signaling cascades with a rather moderate degree of cross talk may have predominated. One-component systems then seem to have evolved into more complex signal-transducing devices such as two-component systems (see Section 3.4) and multidomain proteins found in today's advanced bacteria and, in particular, in eukaryotes. Here a signaling protein with various input modules is able to communicate with numerous partner proteins, forming a multicomponent system and enabling intense cross talk between individual signaling cascades.

Summary

- The controlled transport of material across cell membranes represents the most ancient form of communication between a cell and its environment. Transport against a concentration gradient must be powered by coupled ion flows or ATP hydrolysis. ATPase-coupled transporters that use ATP energy include the F, P, and V types, working as ion pumps, and ABC transporters (carrying an ATP-binding cassette) that are involved in food intake, control of electrolyte concentrations, and waste disposal, including the excretion of toxic substances.

- The phosphotransferase system (PTS) and the machinery of catabolite repression connect membrane transport with gene regulation. This network enables the cell to adjust the uptake and metabolism of a particular carbon source to both the environmental situation and its metabolic capacities. In evolution, PTS provides an early example of the signaling function of protein phosphorylation—the signaling function of cAMP, the control of transcription factor activity by either ligand binding (cAMP and cAMP receptor protein, CRP), or protein phosphorylation (His-containing phosphocarrier protein, HPr). CRP and HPr are prototypes of one-component systems, the most ancient and most abundant signal-transducing devices in prokaryotes.

3.4 Sensor-dependent signal processing: two-component systems

Up to a certain extent, membrane-bound transport systems are the cellular organs for food intake and waste excretion. In PTS, food intake is coupled with a simple but highly efficient network of signal processing, allowing an adaptation to the particular source of food. Nevertheless, and despite their selectivity, transport processes of this type are "blind" to the environment. PTS controls the motility of the cell by modulating the activity of the chemotaxis protein CheA, but it does not enable a behavior of prey-seeking, a tracking down of optimal biotopes, or even more important, an escape reaction in the case of danger. In fact, such blind survival programs are sufficient only for very simple prokaryotes living in a constant biotope, whereas more complex and free-living species require specific **sensor proteins** working independently of food intake and metabolism. These proteins enable the cell to adjust both its genetic readout and its behavior not only to the food supply but also to many other environmental influences.

Since most environmental factors do not pass or must not be allowed to penetrate the cell membrane, they cannot be recognized by the cytoplasmic one-component systems. Consequently, sensor proteins had to be placed at the cell surface. This was done by an evolutionary trick: to distribute input and output domains onto two different proteins, with one taking on the role of a membrane-bound sensor, and the other that of a cytoplasmic effector. In other words, a two-component system was made from a one-component system. In addition, an effective mechanism of signal transduction between both partners was designed.

Two-component systems dominate the communication between bacteria and environment. Their sensor proteins are receptors for environmental stimuli, the meaning of which can be decoded by the data-processing network of the cell, an ability that is essential for survival. The effector proteins of the sensors are called **response regulators**. They transmit the signal from the sensors to the metabolic and genetic apparatus of the cell. With a few exceptions, sensor genes and their cognate response regulator genes are organized in operons.

Between sensor and response regulator, the signal is transduced by protein phosphorylation, in particular by His phosphorylation, which plays a similar key role as for PTS (**Figure 3.11**). In fact, all sensor proteins are **His autokinases** or at least associated with His autokinases. When the sensors are integrated in the plasma membrane, such two-component systems represent a prototype of a transmembrane signaling device operating by post-translational protein modification. But two-component systems are also used for intracellular signaling. In this case the sensor domain of a membrane-bound His autokinase is facing the cytoplasmic side

Figure 3.11 Prototype of a two-component system with a transmembrane sensor The scheme shows a dimeric sensor protein (component I) with four transmembrane domains and two cytoplasmic His autokinase domains (teal) as well as the corresponding response regulator protein (component II). Upon activation, the sensor undergoes trans-autophosphorylation on a cytoplasmic His residue with ATP as a phosphate donor. The phosphate residue (P) is transferred to an Asp residue in the receiver (or input) domain of the response regulator. By the resulting long-range conformational change, the effector (or output) domain of the response regulator becomes activated. Some response regulators thus control the behavior, while others stimulate the genetic readout (aiming at metabolic adaptation) of the cell. From the receiver domain, the phosphate group is removed either by autodephosphorylation, by the sensor protein (exhibiting both His kinase and Asp phosphatase activities), or by separate phosphatases. Here, *activation* by phosphorylation is shown. However, depending on the system, dephosphorylation may also stimulate downstream signaling (with the phosphorylated response regulator being the inactive form). Sensors with two transmembrane domains per monomer are particularly abundant. However, there are also sensor His kinases with no transmembrane helices or up to 20 transmembrane helices per monomer (for examples see Figures 3.10 and 3.11).

(such as in the case of the energy sensor Aer; see Figure 3.17 in Section 3.4.4), or the His autokinase even lacks transmembrane domains (for an example, see Section 3.4.2). In fact, about a third of all His autokinases known seem to be cytoplasmic rather than membrane-bound enzymes. They are assumed to process endogenous signals such as metabolite concentrations, redox status, and acidity. However, little is known about their role in metabolic regulation. We are much better informed about the functions of membrane-bound His kinases operating as sensors for environmental stimuli. Here signal transduction occurs as shown in Figure 3.11.

When the sensor receives an environmental signal, it undergoes His autophosphorylation at its cytoplasmic domain with ATP as a phosphate donor. Since sensors are homodimers, the autophosphorylation occurs in *trans*: that is, the monomers phosphorylate each other. As in the PTS, the His-phosphorylated sensor protein represents a short-lived energy-rich transition state with a very high turnover rate (it is estimated that less than 1% of the His autokinases of a cell are phosphorylated at a given time) that serves as a phosphate donor for the phosphorylation of the response regulator. This occurs at a highly conserved aspartate (Asp) residue and changes the conformation and the activity of the response regulator. Since the response regulator catalyzes this phosphorylation itself, it might be defined as an Asp autokinase. However, apart from the fact that the term kinase is linked to ATP (or GTP) serving as a phosphate donor, the response regulator protein does not contain the typical protein kinase domains.

Though thermodynamically labile, the acyl–phosphate bond of phospho-Asp becomes stabilized by interactions within the polypeptide structure since its energy is used for the conformational change of the response regulator. Therefore, the dephosphorylation of response regulators requires enzymatic catalysis, contributed either by the response regulator itself, by the corresponding sensor His kinase (being a bifunctional kinase/phosphatase), or by a special response regulator phosphatase such as CheZ in bacterial chemotaxis (see Section 3.4.4) or SpoOE in spore formation (see Section 3.6.3). Depending on the type of response regulator, the lifespan of the phosphorylated state may vary between seconds and hours. Stimuli acting as input signals may control both the His autokinase activity and the response regulator phosphatase activity of the sensor.

For the cell, two-component systems provide extremely efficient means to adjust the genetic activity to the environment. To this end, in their amino (N)-terminal (mostly extracellular) domains the sensor His kinases are equipped with receptor domains recognizing a huge variety of environmental stimuli. The structural variability of these receptor domains reflects the variability of the signals, whereas the cytoplasmic kinase domain is rather conserved.

Response regulators are signal-propagating proteins feeding the input signal into the data-processing protein–DNA network of the cell. Most of them are transcription factors exhibiting the characteristic architecture of a variable DNA-binding and transactivating domain combined with a conserved phosphoacceptor domain. Other response regulators may interact with proteins positioned downstream in a signaling cascade (see, for example, CheY in Figure 3.18 in Section 3.4.4), or the phospho-acceptor domains may be coupled with other signal-transducing domains, such as the catalytic centers of enzymes controlling the production of the intracellular messenger cyclic di-GMP (GGDEF and EAL; see Section 3.5.2) or post-translational protein modification by methylation (CheB; see Section 3.4.4). As far as their mode of function is concerned, response regulators seem to come close to digital (binary) switches (see Section 1.3). In the following sections, the mechanisms of action and cellular functions of two-component systems are explained and illustrated with selected examples.

3.4.1 To shrink or to burst: adaptation to osmotic pressure

For an organism living in an aqueous milieu, adaptation of cell volume and inside pressure to the osmolarity of the environment is a basic requirement of survival, since otherwise a cell would either burst or shrink and dry up. The osmoregulator of *E. coli* is the simplest of all two-component systems, demonstrating in an exemplary manner the characteristic features of these signaltransducing devices. It is composed of the osmosensor EnvZ (Env, envelope) and the response regulator OmpR (Omp, outer membrane protein), a transcription factor. The system controls the genes encoding the porins OmpF and OmpC (**Sidebar 3.2**). These proteins constitute pores in the cell wall or outer membrane and regulate the transmembrane diffusion of hydrophilic and negatively charged molecules.

While the protein OmpF produces wide pores and is expressed at low osmotic pressure, OmpC forms narrow pores and is produced at high osmotic pressure, for instance, in the intestinal milieu, with its high concentration of dissolved substances of low molecular mass. Under such a condition of high osmolarity, the His autokinase activity of the sensor EnvZ is stimulated by a mechanism that is not

---

Sidebar 3.2 Outer membrane channels The family of **porin proteins** forms water-filled channels with a large diameter (1 nm) in the cell wall of Gram-negative bacteria as well as in the outer membrane of chloroplasts and mitochondria. In bacteria these channels serve the specific and unspecific exchange of material between environment and periplasm. The cell adjusts the expression of porins to the particular demands. For bacterial porins the exclusion limit of permeable molecules is at around 600 Da, with the transport following the concentration gradient passively. Though porins respond to changes of the membrane potential, the physiological role of this effect is still debated.

The pore of a porin consists of a β-barrel mostly composed of 16–22 anti-parallel β-sheets. A central bottle-neck formed by several loops connecting the β-sheets determines the substrate specificity of the channel. In the bacterial cell wall the porins are arranged as homotrimers. Parasitic bacteria may perforate the host cell membrane by means of porin-like toxins.

Another family of outer membrane channels in Gram-negative bacteria is the **trans-periplasmic channel tunnels** of the type TolC. They have only one pore that again is formed by a β-barrel. Via a long tunnel domain they bridge the periplasm, directly interacting with transport proteins of the inner cell membrane. TolC channel tunnels are specialized for the export of proteins (for instance, bacterial toxins) and other molecules (for instance, substances toxic for the cell).

Figure 3.12 Two-component system of osmoregulation (*E. coli*) (A) Adaptation of pore size in outer membrane by osmolarity-controlled de novo synthesis of the pore proteins OmpC and OmpF. High osmolarity activates the sensor EnvZ in the cell membrane. (B) Control of *ompC/ompF* gene transcription by a two-component system consisting of the sensor His kinase EnvZ and the response regulator OmpR (P, phosphate). Dark gray vertical arrows symbolize inhibition or activation of gene transcription.

yet fully understood. This results in an Asp phosphorylation of OmpR, which then induces transcription of the *ompC* gene while at the same time inhibiting *ompF* transcription (**Figure 3.12**). EnvZ is, thus, the prototype of a sensor responding to hyperosmotic stress. It is a protein of 450 amino acids exhibiting the characteristic architecture of a sensor His kinase, which always exists as a homodimer with one monomer phosphorylating the other one (trans-autophosphorylation, see **Figure 3.13**). Each monomer is anchored in the membrane by two N-terminal transmembrane helices flanking the signal recognition domain (this is the most abundant sensor type, but other forms with more than two transmembrane helices also occur; see Section 3.4.3).

In the cytoplasmic part of all sensor kinase dimers, a bundle of four helices constitutes a dimerization domain containing the His residues to be phosphorylated (so-called H-box) and the binding sites for the response regulator. The catalytic domain of the His kinase is localized in the carboxy (C)-terminal domain and oriented in such a way that trans-autophosphorylation of the subunits becomes possible (see Figure 3.13). The dimerization domain of EnvZ also harbors the active center of an Asp-specific protein phosphatase that catalyzes the dephosphorylation of the response regulator, thus terminating signal transduction.

As EnvZ represents a typical sensor kinase, OmpR, a 27 kDa protein, is the prototype of a response regulator (and of a signal-transducing protein in general), consisting of an N-terminal regulatory or receiver domain and a C-terminal effector or transmitter domain. The conserved receiver domain contains the characteristic structural elements of a response regulator including the essential Asp residue. Its trans-autophosphorylation with His-phosphorylated EnvZ as a phosphate donor is again facilitated by a homodimerization of OmpR and induces a conformational change enabling the dimer to interact with the regulatory sequences of the *ompC* and *ompF* genes.

Like EnvZ, OmpR also exhibits Asp-specific protein phosphatase activity: it is able not only to activate but also to inactivate itself, thus, together with EnvZ, controlling the lifespan of the active state. So, the response regulator has the qualities of a

Figure 3.13 **Sensor His kinase EnvZ of *E. coli*** (A) Simplified domain structure of an EnvZ monomer with the strategic His residue 243 undergoing trans-autophosphorylation and the catalytic domain binding ATP. (B) Membrane anchoring and schematic folding geometry of an EnvZ homodimer. (C) Simplified scheme as used in other figures. The arrows symbolize His autophosphorylation.

switching element with a built-in timer. Such devices play an important role in the transformation of analog into digital signals and vice versa.

3.4.2 Nitrogen fixation and hydrogen utilization

Enzymatic reduction of molecular nitrogen to ammonia represents a fundamental biochemical process essential for life on earth. It is managed by certain bacteria and archaea such as the rhizobia living as symbionts in leguminosae. Requiring the hydrolysis of 12 ATP per one N_2, the reduction of nitrogen ranks among the most expensive (that is, energy-consuming) biochemical reactions and is therefore under strict control:

$$N_2 + 6H^+ + 6e^- + 12ATP + 12H_2O \rightarrow 2NH_3 + 12ADP + 12P_i$$

The nitrogenases catalyzing the process are mostly molybdenum- (or vanadium-) containing iron–sulfur proteins with a complex quaternary structure. They become expressed only when nutritious nitrogen sources, such as amino acids and ammonia salts, are not available.

Nitrogenase expression is controlled by a two-component system consisting of the sensor NtrB and the response regulator NtrC (Ntr = nitrogen). NtrB is a prototype of a *cytoplasmic* sensor His kinase lacking transmembrane domains (other examples of sensor His autokinases without transmembrane domains, such as HoxJ, KinA, and CheA, are described later). The protein monitors the intracellular state of nitrogen metabolism. In the presence of good nitrogen sources, NtrB is blocked by an inhibitor protein in order to avoid an unnecessary and costly nitrogen reduction. With a deficiency of good nitrogen sources, NtrB phosphorylates the response regulator, which in turn stimulates the transcription of a gene encoding another transcription factor, NifA (Nif = nitrogen fixation). As a master key, NifA controls the whole collection of genes required for nitrogen fixation including that of nitrogenase (**Figure 3.14**).

Nitrogenases are extremely oxygen-sensitive enzymes. Nevertheless, the cell needs oxygen for ATP generation, which is then used as energy supply for nitrogen reduction. To adapt to an optimal oxygen concentration (microaerobic conditions),

Figure 3.14 Signaling pathways for the regulation of nitrogen fixation The figure shows the cytoplasmic NtrB/NtrC two-component system inducing transcription of genes of nitrogen fixation (left) and the membrane-bound FixL/FixJ two component system (right) serving as an oxygen sensor active at low oxygen concentration but inactive when potentially damaging oxygen levels are encountered. The cytoplasmic NifL system (bottom right) is expressed by some species to adjust gene transcription to the cellular redox state. P, phosphate. For further details see text.

therefore, rhizobia possess an additional two-component system consisting of the sensor His kinase FixL and the response regulator FixJ (Fix = [nitrogen] fixation). FixL represents a sensor with four transmembrane domains. It harbors a structurally unusual heme molecule bound by a PAS domain (**Sidebar 3.3**). This heme serves as the oxygen sensor: at low oxygen concentrations, the system is active and the phosphorylated response regulator induces the transcription of the *nifA* gene, whereas at higher oxygen concentrations, oxygen occupies the Fe^{2+} of heme, thus inhibiting the His kinase activity of FixL (see Figure 3.14). By this means an uneconomical synthesis of nitrogenase is avoided under conditions where the enzyme is running the risk of being inactivated by oxygen.

Sidebar 3.3 PAS and GAF domains As sensory domains recognizing a wide variety of input signals, these ubiquitous protein interaction motifs play a major role in both pro- and eukaryotic signal processing. Among the prokaryotic signal transducers, various His autokinases, adenylate cyclases, phosphodiesterases, ATPases, and diguanylate cyclases (see Section 3.4.2) contain such domains functioning as regulatory elements.

PAS domains consisting of 100–300 amino acids are widely distributed among pro- and eukaryotes. The name PAS refers to three proteins where these domains were discovered originally:

● Period protein of *Drosophila*, involved, like homologous vertebrate proteins, in the regulation of genes controlling diurnal rhythms

● ARNT protein (Aryl hydrocarbon Receptor Nuclear Translocator) of vertebrates; among others, it controls

(together with the aryl hydrocarbon receptor) genes encoding detoxifying enzymes (see Section 8.4.3)

● Sim protein, a developmental regulator of *Drosophila*

In eukaryotes, PAS domains mainly establish homo- and heterotypical interactions between proteins. They are found in many signal-transducing proteins including protein kinases, receptors, and ion channels, as well as in phytochromes and other photoreceptor proteins of plants. The considerably smaller PAS domains of prokaryotes are distantly related to the eukaryotic domains. They are found mainly in signaling proteins such as the oxygen sensor FixL, the hydrogen sensor HoxJ, the sporulation-controlling sensor KinA, the aerotaxis-controlling sensor Aer, and the light sensor PYP.

GAF domains are named after the three different proteins in which they were first found: cGMP-phosphodiesterase, adenylate cyclase (of the cyanobacterium *Anabaena*), and FhlA, a transcription factor of *E. coli*. They rank among the most ancient, widespread, and abundant small-molecule binding protein motifs and are found in organisms ranging from cyanobacteria to humans. Known ligands are the cyclic nucleotides cAMP and cGMP (that is not found in prokaryotes), while other ligands have yet to be identified.

GAF domains are distinct from cyclic nucleotide binding domains found in the bacterial transcription factor CRP and in eukaryotic effector proteins such as cAMP- and cGMP-activated protein kinases, cyclic nucleotide-gated ion channels, and the Rap exchange factor EPAC (see Section 4.6.1). The regulatory domains of eukaryotic cGMP-controlled phosphodiesterases (see Section 4.4.2) are GAF domains. The *Anabaena* adenylate cyclase is superactivated by positive feedback through an interaction of cAMP with GAF domains.

Another protective mechanism has been found, in particular in *Klebsiella* species. Here the operon encoding NifA contains an additional gene encoding the regulator protein NifL. By means of an associated flavin adenine dinucleotide (FAD), NifL monitors the redox state of the cell: in the presence of oxygen, the cofactor is oxidized to FAD and NifL represses transcription of the genes of nitrogen fixation (see Figure 3.14). The reduced form NifL-FADH$_2$, indicating low oxygen pressure, does not exhibit this repressive effect. In contrast to the heme of FixL, here the redox cofactor is not a component of a sensor His kinase.

Hydrogen utilization

Many prokaryotes are able to cleave hydrogen, which accumulates as a byproduct of the nitrogenase reaction or is produced by anaerobic fermentation, into two protons and two electrons, resembling the oxidation of H$_2$ to water. This ancient reaction, which originally may have emerged in an oxygen-free primeval atmosphere, is catalyzed by hydrogenases and used as an energy supply in a wide variety of bacteria and archaea where the expression of hydrogenases is controlled by a two-component system. The sensor domain of the corresponding His kinase (for instance, the cytoplasmic HoxJ in *Ralstonia eutropha*, with Hox standing for H-oxidation) is via a PAS domain associated with a regulatory hydrogenase (HoxBC), an iron–nickel–sulfur protein acting as the H$_2$ sensor. The system's response regulator (here HoxA) represses transcription of the hydrogenase genes when phosphorylated. Hydrogen inhibits the His kinase of the sensor, thus promoting the accumulation of nonphosphorylated response regulator that induces gene transcription. This example demonstrates that the phosphorylation of a response regulator may be both a positive and a negative signal, a situation analogous to osmoregulation in yeast (see Section 3.6.1) and chemotaxis of *E. coli* (see Section 3.4.4).

3.4.3 Phosphorelays serve as devices for fine-tuning and cross talk

A rather baroque type of two-component system, transducing the signal along cascades of several interconnected phosphorylations, is represented by phosphorelays. In such relays each sensor His kinase delivers the phosphate residue to an Asp residue of a separate receiver domain, from where it migrates again to a His residue of a so-called HPt (His-containing Phosphotransfer) domain and finally to an Asp residue of a response regulator. HPt domains, which exhibit bundles of four helices, superficially resemble dimerization domains of sensor His kinases (see Figure 3.13). The domains of a phosphorelay are either separate proteins or subdomains of one polypeptide chain. Like the PTS cascade, phosphorelays provide additional locations for cross talk with the signal-processing protein network, as well as probably more efficient noise suppression.

The phosphorelays most thoroughly studied are those controlling the sporulation of bacteria of the species *Bacillus* and *Clostridium*. Sporulation is triggered by environmental stress signals such as starvation. These signals are transduced by several sensor His kinases (for instance, the five membrane-bound or cytoplasmic sensor kinases KinA–E of *Bacillus subtilis*) to a highly conserved phosphorelay controlling the

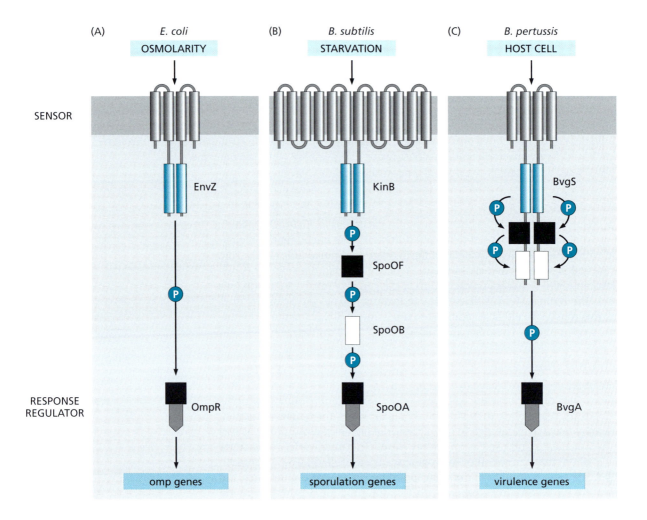

SENSOR

RESPONSE
REGULATOR

Figure 3.15 Selected two-component systems of bacteria
(A) Simple system of osmolarity control, shown for comparison (see also Figure 3.8). (B, C) More complex phosphorelay systems. Teal cylinder, His kinase domain; black square, receiver domain (site of Asp phosphorylation) of a response regulator; gray wedge-shaped symbol, effector domain of a response regulator; white rectangle, HPt domain.

transcription of sporulation genes (**Figure 3.15**). The variety of stress signals is reflected by a corresponding variability of the sensors' extracellular receiver domains.

KinB, exhibiting 12 transmembrane domains per dimer, is a prototype of a group of sensors that functionally resemble the G-protein-coupled receptors of eukaryotes (see Chapter 5); like those, they bind their ligands in the cavity formed by the gobletlike structure of multiple transmembrane helices. Such sensors may contain up to 40 transmembrane domains per dimer. Examples include the quorum sensors monitoring population signals in bacterial colonies (see Section 3.5.1). Another well-known phosphorelay regulates the transcription of the virulence genes of *Bordetella pertussis*. Here all components of the relay including the sensor are found on a single polypeptide chain.

3.4.4 The "nanobrain" of prokaryotes or learning by doing

A brain interprets sensory impressions and calculates a behavioral pattern aiming at adaptation. The data-processing protein network of bacteria does the same, albeit at a lower scale. The major type of prokaryotic behavior consists of controlled changes of motility. Thanks to the inspiring observations made by the German botanist Wilhelm Pfeffer at the end of the 19th century, it became clear that bacteria respond to beneficial and hazardous stimuli by seeking and flight movements, respectively. Such targeted movements are called taxes, and depending on the environmental stimulus, one distinguishes between chemotaxis, phototaxis, thermotaxis, osmotaxis, magnetotaxis, galvanotaxis (electricity), and thigmotaxis (touch). Representing a classical example of quasi-intelligent behavior, chemotaxis of *E. coli* ranks among the best-investigated physiological reactions of prokaryotes and is considered to be at present the most thoroughly studied mechanism in all biological data processing.

Chemotactic behavior

Prokaryotes qualified for taxis move by means of long flagella performing rotating or oscillating motions. In the absence of a stimulus, the cell executes a random walk, which in the case of *E. coli* consists of forward movements (swimming phases) that become interrupted after about 1 s each by a tumbling phase lasting for about 0.1 s. During each tumbling phase, the cell randomly orients into a new direction for swimming. This motility pattern results in an aimless three-dimensional zigzag course. For swimming the flagella rotate counterclockwise, bundling to some kind of a propeller, whereas for tumbling the rotation is switched into a clockwise mode, resulting in a disintegration of the flagellar bundle (in other species, the mode of rotation can be the other way around). Possessing only a forward and a neutral gear, the flagellar motor of *E. coli* is rather simple. Other species such as the soil bacterium *Sinorhizobium meliloti* are able to activate their flagella separately and stepwise, thus changing their speed, as well as to move in curves and to switch into a reverse gear.

The flagellar apparatus of bacteria is a complex device composed of many different proteins. Being an "electric motor," it obtains its energy from ion gradients across membranes. The rotation is brought about by sequential, possibly electrostatic, interactions between rotor and stator proteins that are arranged in a circular pattern (**Figure 3.16**). The molecular mechanism for changing the direction of rotation is not yet fully understood.

When a bacterial cell picks up a stimulus, it changes the random walk into a targeted seeking or escape movement depending on whether the stimulus is an attractant (such as food) or a repellent (such as a poison). To this end, the swimming phases become prolonged when oriented correctly or shortened when oriented in the wrong direction. The cell is thus able to measure the concentration gradient of a substance and promptly adjust its behavior to the result obtained. A bacterium is too small to monitor a spatial concentration gradient, for example between front and back. Instead, prokaryotes recognize *temporal* differences of concentration. To do this the cell must remember where it was before. In other words, targeted movements require learning and memory. Therefore, the data-processing protein apparatus of tactic behavior not only controls the flagellar machine but also must be able to learn and to adapt in a very short time, enabling the bacterial cell to make its

(A)

counterclockwise rotation:
SWIMMING

clockwise rotation:
TUMBLING

(B)

flagellum

bearing   outer membrane

27 nm

inner
membrane

rotor (fli proteins)

peptidoglycan
layer

stator (mot proteins)

CheY-P

2 gears, 6000 rpm,
5 hp per kg, fuel: protons

Figure 3.16 Motility apparatus of *E. coli* (A) Two modes of movement. Depending on the direction of rotation, the 5–7 helical flagella either form a bundle, working like a propeller, or fall apart. (B) Schematic representation of the flagellar motor and its interaction with phospho-CheY (CheY-P).

decisions in a fraction of a second. As one might expect, bacterial chemotaxis can also be misused, for instance, by myxobacteria and myxamoebae, which produce attractive signals to lure *E. coli*, their favored prey.

Processing of chemotactic signals

Tactic signals are processed by a highly conserved two-component system that in its construction differs somewhat from other two-component systems. The major difference is that sensor and His kinase are separate proteins and the His kinase interacts with two different response regulators that affect behavior rather than gene transcription by controlling the flagellar motor and the cellular memory. The separation of sensor protein and His kinase is highly economical since it allows a single signal-processing apparatus to be stimulated by a wide variety of environmental signals, thus enabling an integrated response of the cell.

Apart from the lack of intrinsic His kinase activity, the sensors of tactic signals resemble conventional sensor His kinases in that they exist as dimers with mostly two pairs of transmembrane helices with a variable signal recognition site at the outside (for Gram-negative bacteria, the periplasmic side) and a conserved cytoplasmic domain. These chemoreceptors are known as **methyl-accepting chemotaxis proteins (MCPs)** since they become methylated at several carboxyl groups in the course of a chemotactic response. The degree of change in methylation that occurs during signal processing and subsequently results in an inhibition of receptor activity resembles a bacterial short-term memory store. This feature is discussed in more detail later in the chapter.

E. coli expresses five different MCPs responding to certain sugars, amino acids, dipeptides, and oxygen as well as to changes in pH and temperature. To be recognized by an MCP, sugar molecules have to bind first to specific receptor proteins in the periplasm (**Figure 3.17**). The oxygen-sensing MCP called Aer binds the redox cofactor FAD at its cytoplasmic side via PAS domains. FAD is coupled to an electron transport chain and signals, by its redox state, either a surplus or a shortage of oxygen. Aer enables a more general **energy taxis** such as seeking for energy-rich food, the metabolism of which activates the cellular electron transport chains. By this means the cell may find a biotope with an optimal ratio between oxygen and oxidizable metabolic substrates. Other examples of energy-tactic sensors are the Ser receptor Tsr (see Figure 3.17) and the stress sensor ArcB of *E. coli*. Tsr does not contain a redox

Figure 3.17 Chemotaxis sensors of *E. coli* Depicted are the five membrane-integrated sensor proteins Tsr, Tar, Trg, Tap, and Aer, together with their preferred ligands. M, G, R, and P are periplasmic binding proteins for various sugars and peptides. The Che proteins, including His autokinases and response regulators, are the common downstream effectors of the sensors, controlling both rotation of the flagellar motor and, in a negative feedback loop, methylation of the sensor proteins. Ser, serine; Asp, aspartate; Gal, galactose; Mal, maltose; Rib, ribose; FAD, flavin adenine dinucleotide.

chemotactic signal

MCP

CheR
methyltransferase

CheW   CheW

ATP   ATP

CheA
His kinase

CheZ
protein phosphatase

P   P

CheY

CheB
demethylase

flagellar
motor

Figure 3.18 Chemotactic two-component system of *E. coli* The autophosphorylated His kinase CheA (teal), coupled by the adaptor protein CheW to the membrane-bound sensor (MCP), is the phosphate (P) donor for Asp phosphorylation of the response regulators CheY (controlling the flagellar motor) and CheB (demethylating the sensor). Opponents are the CheY-specific protein phosphatase CheZ and the methyltransferase CheR. Both are constitutively active. The phosphorylated CheY switches the flagellar motor towards clockwise rotation. Teal arrows indicate phosphotransfer reactions.

cofactor but probably responds by a conformational change to pH fluctuations generated by redox reactions. ArcB contains a reactive dithiol group in a PAS domain, which under aerobic conditions becomes oxidized to an intramolecular disulfide group (see Section 3.6.6).

While Aer represents an indirect oxygen sensor, other species such as *B. subtilis* and haloarchaea express direct oxygen sensors coupled with a heme moiety. All in all, the sensor array of *E. coli* appears to be rather meager, since there are bacteria that may express up to 10 times more different sensor proteins.

The apparatus of tactic data processing is highly conserved (**Figure 3.18**). It consists of the so-called chemotaxis proteins, two of which are response regulators:

- CheA, a cytoplasmic His autokinase
- CheW, an adaptor protein linking the sensor protein with CheA
- CheY, the response regulator controlling the flagellar motor
- CheB, the response regulator controlling adaptation, which is a demethylase for the sensor protein
- CheZ, an Asp-specific protein phosphatase for signal termination, supporting the weak autophosphatase activity of CheY
- CheR, a methyltransferase catalyzing methylation of the sensor protein by *S*-adenosylmethionine

It should be noted that the term "chemotaxis protein" is somewhat misleading, because the system processes not only chemotactic but all kinds of tactic signals. In fact, separation of the sensor domain from the His autokinase domain CheA has the advantage that a large number of sensors can be coupled via the adaptor CheW with the same signalprocessing network. We shall encounter this principle of domain separation again when discussing eukaryotic signal transduction.

Some models have been developed to explain the mechanism of signal transduction from the sensor to the His kinase CheA. They postulate either a rotation or a shift of the sensor's subunits. There is, for instance, experimental evidence indicating that a tiny pistonlike movement (of 0.16 nm) of the outer transmembrane segments of the

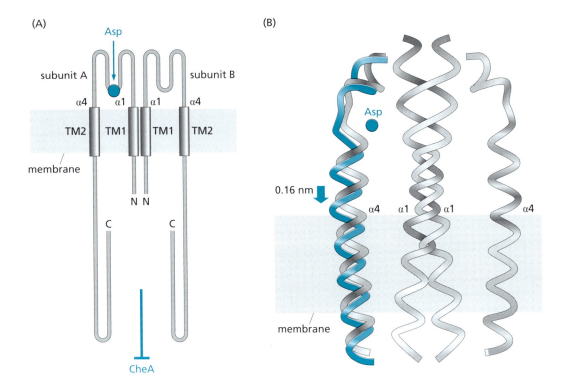

sensor is sufficient to inhibit the His kinase activity of CheA (**Figure 3.19**). During such a minimal shift, the interactions between the side chains of neighboring helices would be preserved, providing the sensor with the properties of an elastic spring that automatically returns into the relaxed position upon termination of the signal. However, how a tolerable signal-to-noise ratio is maintained under such conditions remains an open question. Apart from this, it should be kept in mind that CheA is controlled not only by (chemo)taxis sensors but also by PTS (see Figure 3.8).

The structure of His kinase CheA differs clearly from that of sensor His kinases such as EnvZ (**Figures 3.20 and 3.21**). Though CheA also exists as a dimer, the His phosphorylation occurs in N-terminal HPt domains rather than in the helix bundle of the dimerization domains. Moreover, each CheA monomer contains a regulatory domain for interaction with the adaptor protein CheW and the sensor as well as a substrate-binding domain for interaction with the response regulators.

Intelligent behavior by adaptation

In *E. coli* the activity of the His kinase CheA and thus the degree of phosphorylation of the Che proteins is high in the *absence* of attractants (or in the *presence* of repellents). In other words, the interaction of an MCP sensor with nutrients inhibits the Che signaling system. Owing to the Asp phosphorylation, the response regulator CheY loses contact with CheA, instead interacting as an allosteric regulator with the switching protein FliM of the flagellar motor. This event causes the motor to rotate clockwise, into neutral gear, resulting in a tumbling of the cell. Conversely, the binding of an attractant inhibits CheA. As a consequence, dephosphorylation of CheY by CheZ gains the upper hand, interrupting the interaction between CheY and FliM and switching the motor into counterclockwise rotation, into forward gear. [Note that in other species (such as *B. subtilis*), attractants may *stimulate* and repellents *inhibit* CheA activity, providing an instructive example for the ambiguity of signaling.]

In parallel, the degree of sensor methylation becomes changed: while the methyltransferase CheR is permanently active, the demethylase CheB is stimulated by Asp phosphorylation. As a result, the degree of methylation becomes diminished in

(A)

(B)

(C)

Figure 3.20 Histidine kinase CheA (A) Domain structure of CheA. (B) Schematic folding geometry of the homodimer. Autophosphorylated His residues are shown as teal boxes. Catalytic domains with ATP-binding site are shaded teal. (C) Molecular model of the CheA dimer based on X-ray crystallography.

Figure 3.21 Active centers of some two-component proteins (Adapted from West AH & Stock AM [2001] *Trends Biochem Sci* 26:369–376.)

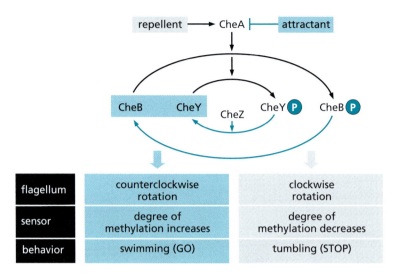

Figure 3.22 Chemotactic responses of *E. coli* Attractants inhibit and repellents stimulate the CheA-catalyzed phosphorylation of response regulators CheB and CheY, resulting in the opposite responses shown in the boxes. In the absence of CheA activity (lack of a repellent or contact with an attractant), CheB undergoes slow autodephosphorylation, whereas CheY is dephosphorylated rapidly by the phosphatase CheZ. As a consequence, the methylated state is more persistent than the phosphorylated state, providing a short-term memory store.

neutral gear, during the tumbling phase, and increases upon switching into forward gear, during the swimming phase (**Figure 3.22**).

The stepwise methylation of several (four or five) methyl groups gradually desensitizes the sensor, probably because the piston movement of the transmembrane segments becomes more and more blocked. This leads to reactivation of the His kinase CheA and thus to termination of the swimming phase. The system is switched back to position 0 and can now respond to the next higher attractant concentration, again with maximal sensitivity. In other words, the system has become adapted.

Our sensory organs are subject to similar adaptation, albeit following an entirely different mechanism. Such adaptive processes guarantee that the sensory apparatus remains fully sensitive over a wide range of signal intensity (at least five orders of magnitude). Thus, a targeted movement of bacteria is possible: without adaptation, the cell would become maximally excited at a minimal signal intensity, inevitably letting it go over the top.

Mainly because of the high phosphatase activity of CheZ, the degree of CheY phosphorylation changes within milliseconds, whereas sensor methylation is maintained for seconds. This difference is due to both the continuous activity of the methyltransferase CheR and the relatively slow auto-dephosphorylation of the phosphorylated demethylase CheB, which is *not* accelerated by CheZ. Owing to this sluggishness, the sensor methylation plays the role of a molecular short-term memory: the degree of methylation was high when the cell was in an area of high attractant concentration *immediately before*, and vice versa. It is easily understood that adaptation of the data-processing apparatus and learning or memory are two sides of the same coin; for example, I am eager to learn as long as my curiosity has not become satisfied or as long as I have not become adapted!

The prokaryotic "nanobrain"

Sensors and chemotaxis proteins are integrated into stable complexes (**Figure 3.23**). This fixed wiring guarantees a particularly rapid signal processing as well as a selective adaptation to individual stimuli. In the cell membrane, many thousands of such complexes are combined into large formations localized predominantly in the anterior pole of the cell. This sensor field may be formed by hexagonal lattices consisting of six triads of sensor dimers, each with associated CheW and CheA proteins (**Figure 3.24**). Such lattices are assumed to exhibit a high degree of cooperativity. This means that allosteric changes of conformation induced by the ligands in a few sensors spread across a larger distance. Such an arrangement is a potent signal amplifier

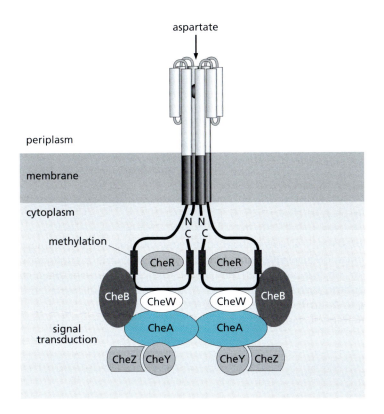

Figure 3.23 Chemotactic signaling complex of *E. coli* Schematic representation of the aspartate receptor Tar and the associated Che proteins. Helical domains of Tar are symbolized by cylinders. (Adapted with modifications from Blair DF [1995] *Annu Rev Microbiol* 49:489–522.)

and allows the transformation of digital input signals into analogous output signals. The result is a gradual behavior of the cell paired with high sensitivity. In principle, this situation resembles sensory signal processing in higher organisms. The (chemo-) taxis system has been described, therefore, as a prokaryotic "nanobrain." Indeed, the phosphorylation and methylation pattern of the taxis proteins, or the excitation pattern of the "nanoneurons," somehow portrays the actual environmental situation just as the excitation pattern of brain neurons does. It should be kept in mind, however, that the taxis system itself is embedded into a larger data-processing protein network, and thus it represents only a part of the bacterial "nanobrain."

Finally, it should be noted that although signal processing operates at the level of the single cell, chemotactic behavior is a systemic phenomenon involving a large number of cells that resemble a school of fishes. As we shall see in Section 3.5.1, the cells communicate by means of so-called quorum signals. These are hormone-like metabolites that also interact with sensors coupled to the chemotaxis system of signal transduction. Thus, in a particularly clear manner, bacterial chemotaxis illustrates a major challenge of today's research: to understand the function of a data-processing network and the emergence of new properties from complex systems, one has to proceed from studying single molecules or single cells to an investigation of large populations of interacting partners. That is, one must go beyond the reductionist approach (see Chapter 17).

Bacterial IQ

A closer look at the prokaryotic cell, with its clear conditions, strongly supports the metaphor introduced in Chapter 1 of the data-processing protein network as the "brain of the cell." Indeed, bacterial signal transduction widely fulfills basic requirements of a neural network, such as performing logical operations by parallel data processing using nontrivial algorithms that are able to adjust. Are bacteria really intelligent?

A partial affirmative answer to this question is given in the definition of the chemotactic response as quasiintelligent behavior. But what is intelligence? Primarily it may be

Figure 3.24 Architecture of a sensor–CheA complex in the inner membrane of *E. coli* (A) CheA dimer interacting, via its regulatory domains R (Figure 3.16), with two CheW adaptors. On average, three sensor molecules are bound to each CheW molecule via their extended (hairy) cytoplasmic domains. Several thousand such complexes constitute the bacterial nanobrain. (B) Model of a lattice of chemotactic receptor complexes, shown looking onto the plasma membrane (receptors are in black, CheW adaptor proteins are in gray, and CheA His kinases are in teal). The sensory field of the cell membrane is a lattice formed by a large number of such hexagonal structures. (A, adapted with modifications from Stock J & Levit M [2000] *Curr Biol* 10:R11–R14. B, adapted with modifications from Shimizu TS et al. [2000] *Nat Cell Biol* 2:792.)

understood as an organism's potential to deal with environmental (and endogenous) challenges in a rational and flexible way. On the molecular level, the number of signaling proteins relative to the size of the genome may be taken as a rough measure of intelligence. This cellular IQ seems to be a matter of lifestyle: the higher the environmental challenges, the higher the IQ value. Analysis of a large number of prokaryotic genes has shown a clear tendency for high IQ values in free-living microbes that encounter an ever-changing environment, whereas the price of living in luxury and idleness is stultification. Thus, the majority of symbiotic and parasitic microbes get by on an absolute minimum of signal transducers, frequently being devoid of two-component systems including that of chemotaxis. On this scale *E. coli*, the pet of microbiologists, turns out to be rather stupid. For instance, compared with free-living Gram-negative bacteria, which sometimes express more than 50 different chemotactic MCPs, the five MCPs of *E. coli* look rather poor, and the same holds true for other signaltransducing proteins (for example, 30 *E. coli* His autokinases versus up to 200 types in other species). This shows the limited value of studies relying just on one or a few "dumb" species.

It must be emphasized that our definition of intelligence is somewhat narrow since it does not include what is known as associative memory. This wonderful property enables us to extract common parameters from countless details that help us to recognize and categorize objects, subjects, and systems as related even when they are not identical. The absence of associative memory is a serious problem of today's computer technology, as demonstrated by daily experience. Whether prokaryotes in particular, or single cells in general, possess an equivalent of associative memory remains to be shown.

3.4.5 Phototaxis and the "invention" of rhodopsin

Phototaxis is defined as a targeted movement directed by light. It has been studied in more detail for **halobacteria**. These are a subclass of archaea inhabiting extreme biotopes, that are, for instance, high in salinity, such as the Dead Sea (salt concentration 4.3 M compared with 0.6 M in ocean water). Halobacteria are able to utilize sunlight as an energy supply for ATP production. Light, particularly at short wavelengths, is also a deadly hazard. To deal with this problem, halobacteria have developed phototactic systems that recognize short-wavelength light as a repellent and long-wavelength light as an attractant. This enables the cells to actively find optimal environmental conditions.

The molecular system decoding light signals resembles that of chemotaxis, though the sensors are, of course, quite different. Halobacteria express two types of light sensors or halobacterial transducers (Htrs): one (HtrI) for long-wavelength light and the other (HtrII) for short-wavelength light. Both resemble chemotaxis sensors as they interact with CheA and the other chemotaxis proteins upon illumination. However, being unable to respond to light on their own, the Htr proteins form complexes with two light-sensitive proteins, the sensory rhodopsins SRI and SRII. When the SRI/HtrI complex becomes stimulated, for instance by orange light, the cell responds with seeking behavior, whereas an escape reaction is induced by activation of the SRII/HtrII complex, for instance by blue light. In other words, prokaryotes equipped with such a signal-processing system are able to "see" colors.

Rhodopsins are proteins with seven transmembrane domains arranged in a goblet-shaped pattern. At the outside, this structure has bound retinal (vitamin A aldehyde) as an imide (Schiff base) at the ε-amino group of a lysine residue. In the absence of light, the retinal is locked in the *all-trans* configuration and the nitrogen of the imide bond is protonated, carrying a positive charge. Upon illumination, retinal isomerizes into the 13-*cis* form, resulting in an allosteric rearrangement of the transducer protein Htr that causes an activation of the phototactic system.

In addition to the phototactic sensor rhodopsins (SR type), other microbial rhodopsins are known. These participate in the transformation of light into metabolic energy. In fact, these bacterio- and halorhodopsins (BR and HR) play the role of light-powered ion pumps serving the generation of ATP and the control of the cellular pH value. Upon illumination, BR produces a proton gradient across the cell membrane by transferring, through a conformational rearrangement, the proton bound to the imide nitrogen from the inside to the outside of the cell. By powering a membrane-bound ATP synthase, the proton gradient then transforms light energy into chemical energy (**Figure 3.25 and Figure 3.26**). HR uses the positive charge of the protonated Schiff base to bind chloride ions electrostatically. Here the proton does not dissociate from the illuminated protein but, again due to a light-dependent conformational rearrangement, the N–H dipole is folded down, releasing the chloride ion into the interior of the cell. This transport process regulates intracellular acidity.

Halobacteria perfectly adjust to environmental conditions. When there is ample food and oxygen (that is, the oxygen signal is received by heme-containing aerotaxis sensor HtrVIII and transduced to the Che complex), only light protection by the HtrII/SRII system is needed. Consistently, the energy-supplying HRs and BRs become expressed only upon oxygen shortage in the water, as caused, for instance, by an increase of the salt concentration due to evaporation.

Evolution of rhodopsins

Rhodopsins are universal light sensors that are classified in two families: microbial rhodopsins (type 1) and rhodopsins of the animal eye (type 2). Whether both types developed independently or are the result of lateral gene transfer is still a matter of debate. Microbial type 1 rhodopsins are by no means restricted to haloarchaea

Figure 3.25 Reaction cycle of archaebacterial rhodopsin In the inset the reactions occurring in the cell membrane are shown: light-dependent *cis–trans* isomerization of retinal bound as a Schiff base at a lysine residue of rhodopsin.

As a consequence, the conformation of the membrane-bound rhodopsin is changed in such a way that a proton bound to the Schiff base is released into the extracellular space. The reflux of the protons following an electrical charge gradient powers a membrane-bound ATP synthase. In halorhodopsin, the protons are not released but facilitate the influx of chloride ions.

but have been found in a wide variety of prokaryotes as well as in unicellular eukaryotes, green algae, and fungi. Of special interest are the **channelrhodopsins** of green algae. These light-gated ion channels permeable for small cations such as Na^+, K^+, Ca^{2+} and protons are sensors for phototactic movements. Channelrhodopsins and derivatives thereof are widely used as experimental tools in neuroscience (see Chapter 16, Sidebar 16.2).

The molecular architecture of rhodopsin, with its seven transmembrane helices and an extracellular ligand binding site, is one of the most successful standard models of evolution. It serves as a receptor for retinal and light as well as for a wide variety of other signaling molecules, though the receptors of higher eukaryotes are probably not descendants of the microbial rhodopsins (see Section 5.1).

Light sensors of cyanobacteria

Other prokaryotes using light as an energy supply are the cyanobacteria or blue-green algae. They are considered to be the evolutionary "inventors" of chlorophyll-

Figure 3.26 Phototaxis of haloarchaea Different rhodopsins (teal) are shown integrated into the plasma membrane and serving energy-delivering ion transport (center) and controlling, as complexes with corresponding sensors, phototactic movements. Note that in the situation depicted by the right figure seeking behavior is triggered by the HtrI/SRI complex, whereas an escape reaction is induced by the HtrII/SRII complex. For further details see text.

Figure 3.27 Light-sensitive reaction of cyanobacteria Light-dependent isomerization of the tetrapyrrole pigments in the phycochrome proteins is shown. The light-activated Pfr form activates two-component systems of phototaxis and chromatic adaptation. Absorption maxima of the pigments are given.

dependent photosynthesis, being the ancestors of plant chloroplasts. In cyanobacteria, phototaxis is controlled by the phycochromes phycoerythrin and phycocyanin rather than by rhodopsins. Moreover, by means of these light-sensitive proteins, the cell is able to utilize light with wavelengths for photosynthesis to which chlorophyll is relatively insensitive.

The light-sensitive pigments of phycochromes are two tetrapyrrole derivatives, phycoerythrobilin (for short-wavelength light) and phycocyanobilin (for long-wavelength light), which are structurally related to bile pigments. Light induces an isomerization of the pigment molecules, acting as an input signal for the phototaxis system (**Figure 3.27**). In addition, as in rhodopsin, a proton is released and fed into the photosynthetic complex. Apart from the Che apparatus, another two-component system is used by the cell to adjust the expression of phycochromes to the light conditions, a response called chromatic adaptation. The phycochromes are the evolutionary ancestors of the phytochromes serving light adaptation in plants (**Sidebar 3.4** and Section 3.6.1).

Sidebar 3.4. Photosensory proteins
Photosensory proteins are receptors that use light as an input signal. They contain low-molecular-mass chromophores which upon absorption of a photon undergo a structural rearrangement. The conformational change thus induced in the protein triggers downstream signaling events such as protein phosphorylation, ion transport, second messenger generation or inactivation, and gene activation.

Six families of photosensory proteins are known.

- **Rhodopsins** using retinal as a chromophore (see Sections 3.4.5 and 15.5).

- **Phytochromes** with tetrapyrrole derivatives as chromophores. They are found in cyanobacteria (phycochromes) and plants (see Section 3.4.5, **Figure 3.27**). In cyanobacteria the light signals are processed by a "degenerate" two-component system with a His kinase-related Ser/Thr kinase.

- **Cryptochromes** using flavin adenine dinucleotide as a chromophore. Cryptochromes are found in prokaryotes and throughout the eukaryotic kingdom. They play a role in the control of day-night rhythms (see Section 9.4.5).

- **LOV domain proteins** are found in plants, fungi and bacteria. The chromophore is flavin adenine mononucleotide reversibly bound to an N-terminal LOV domain (Light-Oxygen-Voltage domain). An example is provided by the phototropins that in plants control light-dependent periodic movements (phototropism), and—via effects on gene transcription—morphogenetic processes such as growth, flowering and germination. In plants the LOV domain of the sensor interacts with a Ser/Thr kinase which undergoes autophosphorylation and becomes active upon illumination. In some fungi LOV proteins control the response to daily and seasonal light periods.

- **Blue light sensors using flavin adenine dinucleotide as chromophore (BLUF proteins)** are more or less restricted to prokaryotes. In certain bacteria they induce the inactivation of the second messenger cyclic di-GMP (see Sidebar 3.5) by stimulating a corresponding phosphodiesterase. Another example is provided by certain bacteria and flagellates where BLUF proteins regulate phototaxis by controlling the activity of adenylate cyclase.

- **Xanthopsins.** The chromophore of these prokaryotic proteins is *p*-hydroxycinnamic acid. The prototype, photoactive yellow protein, provides a standard model for studies of the effects of light on protein conformation. The major physiological function is light protection since activation by short wavelenght light induces an escape reflex.

Photosensory proteins are important experimental tools for the methods of optogenetics (see Chapter 16, Sidebar 16.2).

Bacterial signal transduction: more elaborate than previously assumed

The sequencing of several hundred prokaryotic genomes has provided a wealth of data showing that prokaryotic signal processing is a theme with many variations. Thus, depending on the species, putative sensor domains of transmembrane proteins are fused with His autokinases, adenylate cyclases, serine/threonine (Ser/Thr)-specific protein kinases, protein phosphatases, diguanylate cyclases (a novel type of second-messenger generating enzymes; see Sidebar 3.4), and DNA-binding domains. These interactions indicate the existence of transmembrane signaling mechanisms that go beyond the conventional two-component system, albeit in most cases the extracellular input signals as well as the downstream effector proteins of such receptorlike proteins are not known. Moreover, such enzymatically active domains are integrated in various cytoplasmic proteins, which by the presence of typical protein and DNA interaction domains are characterized as potential transducers in an intracellular signaling network. These data, which are derived from genomic screening, need to be validated by biochemical studies. Nevertheless, they show that the prokaryotic cell has mastered the task of constructing new protein species by recombining a limited set of modules in a virtuoso manner, thus laying the foundations for most of the signaling processes found in mammalian cells.

Summary

- Two-component systems provide the standard model of receptor-coupled signal processing in prokaryotes. The sensor expresses His autokinase activity or couples with His autokinases. The effector protein, called response regulator, usually is a transcription factor. About two-thirds of all sensors are transmembrane proteins processing environmental stimuli; the rest are found in the cytoplasm processing intracellular signals.

- Upon activation by a signal, the sensor undergoes ATP-dependent His autophosphorylation, immediately transmitting the phosphate residue to a regulatory Asp residue of the response regulator. Phosphatase activities render the signaling process reversible.

- Phosphorelays resemble two-component systems but consist of more than two partners. Such phosphorylation cascades offer additional possibilities for control, network formation (cross talk), and signal fine-tuning.

- By means of a universal two-component system consisting of individual sensors and a common apparatus of signal transduction (Che complex), prokaryotes are able to adjust their movement pattern to a wide variety of chemical and physical stimuli acting as either attractants or repellents. Adaptation enabling targeted, quasi-intelligent movement is due to sensor desensitization through reversible protein methylation. The phosphorylation and methylation patterns represent a short-term memory store of the cell.

- In light-sensitive halobacteria the phototactic sensors of two-component systems are coupled with rhodopsin proteins containing *all-trans*-retinal as a light-sensitive pigment. Upon illumination, the retinal is isomerized to the 13-*cis* form and the chemotaxis system is activated, regulating the movement pattern and the adaptive behavior of the cell. Other microbial rhodopsin types function as light activated ion pumps, ion channels and use light as an energy source.

3.5 From vagabonds to societies: "bacterial hormones"

If hormones are understood as mediators of *inter*cellular communication in metazoans, a term such as "bacterial hormones" appears to be nonsensical at first glance. However, most prokaryotes are by no means loners but are capable of amazing social performances, thereby acting like multicellular organisms. This process begins with the formation of characteristically shaped colonies and culminates in complex, differentiated structures that consist of millions of cells. Examples are provided by the chainlike colonies of certain cyanobacteria, the predatory spheres of myxococci, the fruiting bodies of myxobacteria, the hunting packs of *Proteus mirabilis*, and the so-called biofilms. Frequently, such structures are aimed at the division of labor through cell differentiation.

3.5.1 Quorum sensing and auto-inducers

For aggregation and differentiation, cells communicate by signals. Among such **social signals**, the so-called quorum sensing has been investigated in detail. Quorum sensing monitors the population density of a collection of cells to find out whether or not it makes sense to start an expensive genetic program—such as sporulation or biofilm formation—which requires a larger group of cells to become effective. The principle of quorum sensing is simple: since cells continuously release the social signal, its concentration is a direct function of cell density, and the target genes become activated above a distinct threshold concentration.

Pioneering experiments were performed on *Vibrio fischeri*. This Gram-negative symbiont inhabits the light-producing organs of certain fishes and cephalopods, causing bioluminescence. In return for this the host protects the settlers from enemies and provides food. For the host, light production makes sense only when the intensity of bioluminescence exceeds a certain value. However, the prokaryote produces light only at a sufficiently high population density (at least 10^{10} cells/mL, compared with a density of a few hundred cells per milliliter in sea water). Therefore, bioluminescence is triggered by a population-dependent signal that is released by the cells. This auto-inducer stimulates neighboring cells to transcribe the *lux* genes that encode bioluminescence enzymes such as luciferases (more information on such enzymes is found in Section 5.2) and enzymes catalyzing the production of the auto-inducer.

Auto-inducers are not restricted to luminous bacteria but have been found in many other species. As far as Gram-negative bacteria are concerned, auto-inducers are almost exclusively **acyl-homoserine lactones** (**Figure 3.28**), differing from each other by the acyl residue and exhibiting species-specific effects. This provides quorum sensing with a high degree of privacy. Acyl-homoserine lactones are lipophilic

Vibrio fischeri   *Vibrio harveyi*   *Staphylococcus aureus*

Figure 3.28 Some auto-inducers of quorum sensing Shown are an acyl-homoserine derivative (left), a furanosylborate diester (middle), and a cyclic octapeptide (right) from different microorganisms.

enough to penetrate cellular membranes with increasing cell density, thereby "soaking" the whole population.* The corresponding cellular receptors are transcription factors that by convention are called LuxR in reference to the bioluminescence system, although they are by no means involved only in light production. They may control many other population-dependent functions including the formation of spores, colonies, and biofilms as well as of virulence factors (these are illness-causing toxins and proteins used by pathogenic microorganisms for the invasion of host cells). Upon binding of a specific auto-inducer, the LuxR factors dimerize and either induce or repress the transcription of individual operons (**Figure 3.29**). This mechanism of action closely resembles that of steroid hormones, which in higher eukaryotes interact with so-called nuclear receptors (see Section 8.3).

Gram-positive bacteria do not produce acyl-homoserine lactones. Instead their auto-inducers are short, partly **modified peptides** released from larger precursor proteins (an example, the cyclic octapeptide of *Staphylococcus aureus*, is shown in Figure 3.28). In some species this signaling system is completed by another one using **furanosylborate diesters** as auto-inducers (see Figure 3.28). This provides an interesting and rare example of the utilization of boron in a biological system. Evidently the arsenal of "bacterial hormones" is much more elaborate than hitherto

---

*Free diffusion requiring a minimum of water solubility is only one way to distribute social signals within a cell population. Highly hydrophobic molecules have been shown to be packed in vesicles that are pinched off from the outer membrane of certain Gram-negative bacteria to become fused with the membrane of a neighboring cell.

Figure 3.29 Mechanisms of action of auto-inducers: two major pathways along which extracellular signals control gene transcription Depending on the chemical structure, auto-inducers (black dots) activate transcription factors either by direct interaction (left, Gram-negative bacteria) or via a receptor-controlled phosphorylation cascade (right, two-component system of Gram-positive bacteria). LuxR, transcription factors first studied in the bioluminescence system; RR, response regulator; P, phosphate. For other details see text.

assumed, since novel factors have been found that differ chemically from modified peptides, acyl-homoserine lactones, and furanosylborate diesters. They include derivatives of oligosaccharides, dihydroxypentadienones, butyrolactones, quinolones, and others, with many of them exhibiting quite exotic structures.

In contrast to acyl-homoserine factors the auto-inducers of Gram-positive species are unable to penetrate cellular membranes. Instead, their effects are mediated by highly specific two-component systems controlling the transcription of corresponding genes (see Figure 3.29). The analogy to peptide hormones of higher eukaryotes that interact with protein kinase-coupled receptors is striking. Thus, auto-inducers control the activity of transcription factors either directly by binding or indirectly by phosphorylation. These two mechanisms have been preserved in the course of evolution and still represent the major signaling routes leading to gene regulation (see Section 8.3.1).

Prokaryotes, but also algae and flowering plants, produce inhibitors of quorum sensing, thus outcompeting and keeping in check populations of competitor, parasitic, and symbiotic bacteria. There is some evidence that human cells also use anti-quorum strategies to fight bacterial infections. For instance, cells from bronchial epithelium have been shown to inactivate the auto-inducer controlling the virulence genes in *Pseudomonas aeruginosa*. It is clear that the investigation of such defense mechanisms and the potential clinical application of synthetic inhibitors offer highly attractive possibilities in medicine.

3.5.2 Biofilms and fruiting bodies: benefits of multicellularity

Population-dependent signals control not only parasitism and symbiosis but also the aggregation of bacteria into multicellular, differentiated structures such as biofilms and fruiting bodies. Biofilms are large bacterial aggregates growing on surfaces. In these complex structures the cells are embedded like tissue cells into an extracellular polysaccharide matrix that is produced by the cell collective under the control of the signaling molecule cyclic di-GMP (Sidebar 3.5). Common examples are tooth plaques, the slimy coat of stream pebbles, mats of bacteria lining water pipes, and the bacterial colonies in the ruminant stomach. The discovery of biofilms has revolutionized the widely held idea of prokaryotes as "primitive" unicellular organisms. Indeed, rather than single cells living as vagabonds, biofilms are the predominant form of prokaryotic life. As indicated by fossils, the history of such multicellular structures reaches back to the early days of evolution.

Biofilms are highly organized three-dimensional structures with fluid channels running through a spongelike body. The microbes inhabiting a biofilm may belong to one or to several different species. The individual cell types are organized in colonies that communicate with each other and divide labor, for instance, breaking-up of food, protection from stress factors, and parasexual exchange of genetic material. The degree of cell differentiation, expressed as a pattern of genetic activity, depends on the position in the biofilm.

Like metazoans, such highly organized forms of prokaryotic life can exist only when there is a permanent exchange of intercellular signals. Here the auto-inducers of quorum sensing play a key role. Because of their resistance to antibiotics and disinfectants, biofilms are a major cause of acute bacterial intoxications and chronic infections. The investigation of the signaling mechanisms is, therefore, of practical value.

One of the most complex physiological processes in the prokaryotic world is the sophisticated survival strategy of **myxobacteria.** In emergency situations they develop fruiting bodies, consisting of thousands of cells that have become differentiated to spores. This type of cell differentiation is triggered by the ancient prokaryotic stress reaction of the stringent response, resulting in an almost complete

Sidebar 3.5 Cyclic di-GMP, a bacterial "lifestyle molecule" Cyclic diguanosine mono-phosphate (c-di-GMP) is produced from GTP by diguanylate cyclases and degraded to inactive linear 5'-phosphoguanylyl-3'-5'-guanosine (pGpG) by cognate phosphodiesterases (**Figure 3.30**).

These enzymes exist in many variants and are abundant throughout the prokaryotic kingdom. For instance, bacteria with a "high IQ" may express more than 50 isoforms of cyclase. Both enzyme types may become activated (or inhibited) by extracellular signals, probably via transmembrane receptors, a situation which is related closely to the eukaryotic cyclic AMP-system. In fact, c-di-GMP is considered now as one of the most versatile second messenger molecules in prokaryotes. Only recently, another prokaryotic second messenger has been found. This is **cyclic-di-AMP.** Its functions seem to resemble that of c-di-GMP.

Among the various—proposed or established—functions of c-di-GMP, the regulation of bacterial "lifestyle" is the most prominent one: an increase of the c-di-GMP level (caused, for instance, by certain autoinducers or environmental stress) leads to a transition of the motile into the sessile form. i.e. biofilm formation, whereas a decrease has the reverse effect and may promote virulence. Such global physiological consequences indicate that a wide variety of cellular proteins, including those regulating gene transcription and DNA replication, are under the control of c-di-GMP and c-di-AMP. In fact, major targets of both cyclic nucleotides are specific riboswitches that—depending on the species—regulate multiple genes.

Considering the role in virulence as well as the high resistance of biofilms to conventional antibacterial drugs, the signaling cascades regulated by cyclic dinucleotides are emerging as attractive targets for novel therapeutic strategies. There are only a few examples of these messenger molecules in eukaryotes. Thus, in the slime mold *Dictyostelium discoideum* c-di-GMP is one of the factors that prompt swarming amoeboid cells to become sessile forming a multicellular spore capsule (**Figure 3.31**), and in mammalian cells cyclic dinucleotides help to trigger the innate immune response (see Chapter 6) by activating a protein called "stimulator of interferon genes" (STING). Here they may stem from invading bacteria, but, in addition, mammalian cells are able to produce cyclic dinucleotides by themselves upon contact with foreign (bacterial!) DNA.

Figure 3.30 Signal-controlled biosynthesis and inactivation of cyclic di-GMP

(A) aggregation phase

Myxobacteria *Stigmatella aurantiaca*

Myxamoebae *Dictyostelium discoideum*

(B) differentiation phase

0.1 mm

2 mm

Figure 3.31 Aggregation and differentiation of myxobacteria as compared with myxamoebae In both cases, apparently amorphous cellular aggregates (A) develop into fruiting bodies (B). (Myxobacteria adapted from J.A. Shapiro, *Sci. Am.* 82–89, June 1988. Myxamoebae (left) courtesy of Dr. Rupert Mutzel, Free University Berlin (right) adapted from R. Phillips et al., Physical Biology of the Cell. New York and London: Garland Publishing, 2013.)

stop of cell proliferation (see Section 3.1). In addition, stressed myxobacteria release proteases into the extracellular space that hydrolyze extracellular proteins. The amino acids thus produced serve both as food and as quorum signals (called A-signals) that are processed by two-component systems, enabling the cell to monitor whether enough cells are available to generate a fruiting body. When this is the case, the cells emit aggregation signals of still-unknown structure, inducing chemotactic responses. The cell that happens to be the strongest transmitter develops into an aggregation focus, gathering together some 100,000 other cells. The aggregate then differentiates into a fruiting body. The differentiation is controlled by a morphogenetic signal protein (C-signal) exposed at the cell surface, interacting with sensors at the surface of neighboring cells. This juxtacrine signal transduction strongly resembles events regulating the embryonic development of vertebrates (for examples, see Sections 7.1.4, 7.3, and 16.1). By this means, *Myxococcus xanthus* produces fruiting bodies and also differentiates alternatively into another multicellular structure, the so-called predatory sphere, serving the collective catching of prey.

In their mobility mode, myxobacteria differ from less complex species such as *E. coli* in that they have no flagella but execute sliding movements resembling that of amoebae. They possess two genetically fixed programs of mobility, one for single cells and the other for groups of cells sliding like slugs on a slimy secretion. Both programs are probably coordinated by a signaling cascade that formerly had been found only in animals. In this cascade, a transmembrane tyrosine kinase (MasK of *M. xanthus*) interacts with a GTPase (MglA) that, among all prokaryotic G-proteins, is most similar to the small G-proteins of the eukaryotic Ras family. These findings strongly indicate a fluent transition between prokaryotic and eukaryotic mechanisms of signal processing.

Myxobacteria developed approximately 2 billion years ago, representing the most ancient multicellular organisms. Independently, multicellularity was also "invented" by eukaryotic myxamoebae or slime molds. The modes of motility, aggregation, and differentiation, most thoroughly studied for *Dictyostelium discoideum*, are quite similar to those of myxobacteria, though they employ other mechanisms. This provides a fascinating example of evolutionary convergence (see Figure 3.31).

There is a blurred dividing line between the auto-inducers of quorum sensing and bacterial **pheromones**. These cell-derived attractants are small peptides controlling various aspects of social life. In *Enterococcus faecalis*, for instance, they induce the exposure of surface proteins for "sexual" contacts (serving the exchange of genetic material). This function resembles that of plant or animal sexual hormones. In other species pheromones control the uptake of free DNA or the invasion of tissues by pathogenic bacteria. The effects of the prokaryotic pheromones known thus far are mediated by two-component systems.

With **programmed cell death**, another aspect of the social life of bacteria has been detected only recently. At first glance this seems to be a paradoxical behaviour for a unicellular organism. However, for a cell population coping with stress, such as starvation or virus infection, an altruistic self-sacrifice of cells would make sense because it may help other members to survive. Programmed cell death in bacteria is controlled by the so-called **holins**, namely proteins that in structure and function resemble the pro- and antiapoptotic factors of the Bcl-2 family in eukaryotes (see Chapter 13). A major executor enzyme is peptidoglycan hydrolase, which disintegrates the outer cell membrane.

Summary

- The formation and function of bacterial colonies and complex multicellular structures (such as biofilms and fruiting bodies) are controlled by intercellular signals including factors that monitor the population density (quorum sensing).

- In Gram-negative species, such hormone-like auto-inducers are acyl-homoserine lactones able to penetrate the cell membrane and to interact directly with transcription factors.

- The auto-inducers of Gram-positive species are modified peptides, furanosylborate diesters, and other compounds. Unable to pass the membrane, they instead interact with sensors of corresponding two-component systems that transduce the signal through Asp phosphorylation of transcription factors, or response regulators.

- Thus, intercellular signaling by "hormones" and the major principles of transcriptional regulation—either ligand binding or phosphorylation of transcription factors—are of prokaryotic origin.

3.6 From bacteria to humans: evolution of signaling mechanisms

Since cells cannot survive without communicating with the environment, the mechanisms of cellular signal processing must have developed in parallel with the mechanisms of food intake, energy metabolism, and reproduction. Indeed, there are fluid transitions between metabolic and signaling reactions, and even the most primitive prokaryotes possess a complex arsenal of signal-processing proteins, while more advanced species communicate via hormone-like factors. A comparison of different species clearly indicates that practically all eukaryotic mechanisms of signal transduction must have emerged from prokaryotic precursors. Examples are provided by membrane transporters, second messengers, protein kinase-coupled and rhodopsin-like receptors, and the pathways of transcription factor control. In the following sections, the discussion of evolutionary aspects is expanded to include standard biochemical devices of signal transduction such as kinases, phosphatases, GTPases, proteases, redox enzymes, and ion channels.

3.6.1 "Prokaryotic" protein phosphorylation in eukaryotes

Phosphorylation is a key reaction of energy metabolism, and at the same time is the most versatile and perhaps most ancient mechanism for post-translational control of protein function. Most probably, protein phosphorylation was applied by the precursors of archaea and bacteria. In the course of evolution, several large families of protein kinases and phosphatases have emerged (see Section 2.6 for an overview).

The His-specific autokinases of prokaryotic two-component systems represent a transition between the phosphotransferases of energy metabolism and signal-processing protein kinases. In the course of prokaryotic evolution, they have become models of success: 1.5% of the genome of *E. coli* includes 30 genes encoding His kinases and 32 genes encoding response regulators. *Nostoc punctiformis*, *Myxococcus xanthus*, and cyanobacteria express more than 150 different two-component systems. Only very simple bacteria such as mycoplasma can survive without two-component systems. They probably have lost the corresponding genes because of their special life conditions (see remarks on the "bacterial IQ" in Section 3.4.4).

Histidine kinases and two-component systems are also expressed by archaebacteria, slime molds, fungi, and plants. It is assumed that they had been taken over from bacteria by horizontal gene transfer. Apart from the mitochondrial genome, no such genes have been found in animals. All eukaryotic two-component systems are constructed as phosphorelays, thus providing more possibilities for interactions and signaling cross talk.

Only assumptions can be made concerning the reason why, on the way to eukaryotes, bacterial His phosphorylation has fallen into oblivion, being replaced by Ser/Thr/Tyr phosphorylation. Because the phosphoric ester bonds are much more stable than the phospho-His and phospho-Asp bonds, they certainly provide a better memory effect, more precise regulation, and a superior signal-to-noise ratio, albeit at the price of speed.

Yeast

The bakers' yeast *Saccharomyces cerevisiae* has only one His kinase gene and three response regulator genes encoding a system for **osmolarity control** (**Figure 3.32**). A striking feature of this system is its connection with a "modern" Mitogen-Activated Protein (MAP) kinase module (see Section 2.6.5), which has not yet been found in prokaryotes. In fact, yeast cells express several MAP kinase modules with different functions (see Section 11.1). By phosphorylating the corresponding transcription factors, one of these modules stimulates genes encoding proteins that protect the cell from osmotic stress, for instance, enzymes of glycerol production. Under normal conditions this module is inhibited by the phosphorylated response regulator SsK1. Hypertonic stress blocks the associated sensor His kinase Sln1, resulting in dephosphorylation of the response regulator and, in turn, activation of the MAP kinase module. Via the interconnected HPt protein Ypd1, the His kinase Sln1 can phosphorylate a second response regulator, Skn7, which does not control a MAP kinase module but functions as a transcription factor of additional stress genes. An analogous system has been found in the fission yeast *Schizosaccharomyces pombe*.

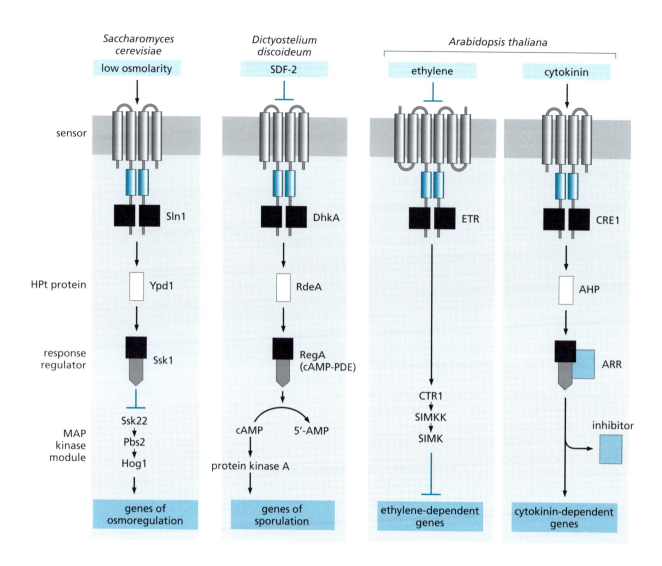

Figure 3.32 "Two-component" phosphorelays of eukaryotes Teal cylinder, His autokinase domain; black square, Asp-containing receiver domain; white rectangle, HPt domain; gray wedge-shaped symbol, effector domain of response regulator.

Dictyostelium discoideum

Containing 20 His kinase and phosphorelay genes, this slime mold is currently known to be the eukaryote equipped with the most two-component systems. Among these systems, however, only one is understood in more detail in that it has been found to be involved in the coordination of cell differentiation by controlling the production of the signal molecule cyclic AMP. For *Dictyostelium*, cAMP has two functions: it is a chemotactic attractant for the aggregation of single cells to a multicellular structure and it is a morphogenetic signal inducing the differentiation of spore capsules. These effects of cAMP are mediated by G-protein-coupled receptors (see Chapter 5). The action of cAMP depends on its concentration: at low levels it is an attractant, at high levels a morphogen. The cAMP level is controlled by two enzymes, an adenylate cyclase catalyzing biosynthesis from ATP and a phosphodiesterase catalyzing hydrolysis to inactive 5'-AMP. cAMP phosphodiesterase (cAMP-PDE) is a C-terminal subdomain of a response regulator, RegA, and becomes activated by Asp phosphorylation of RegA. This phosphorylation is stimulated by two sensor His kinases, DhkA and DhkB (*Dictyostelium* histidine kinases), that are inhibited upon binding of their ligands; Slime mold Differentiation Factor SDF-2; and discadenin, a cytokinin derivative. As a result, the interconnected HPt protein RdeA and the response regulator RegA become dephosphorylated, the phosphodiesterase activity is suppressed, and the level of cAMP increases (see Figure 3.32). Another sensor His kinase of *Dictyostelium* is probably involved in the adaptation to osmolarity.

Plants

Plants express a limited number of two-component systems. For instance, 11 His kinase genes and 16 response regulator genes have been found in the genome of *Arabidopsis thaliana* (compared with more than a thousand genes encoding "eukaryotic" protein kinases). Some of these systems mediate the response to osmotic stress and to the phytohormones cytokinin and ethylene as well as light adaptation by phytochromes.

In structure and function, the **osmosensor Athk1** (*A. thaliana* histidine kinase 1) resembles the corresponding sensor Sln1 of yeast. In both cases the His kinase activity is depressed by hypertonic stress. Athk1 is found predominantly in root cells.

The gaseous plant hormone **ethylene** controls processes of senescence such as fruit ripening and the fall of leaves. *Arabidopsis* expresses at least five different ethylene receptors, which are tissue-specifically distributed. These are sensor His kinases that inhibit the expression of ethylene-inducible genes when activated. Ethylene blocks the His kinase activity. Like the osmosensor system of yeast, the signaling cascade includes a MAP kinase module with the Ser/Thr kinase CTR1 acting as a MAP3 kinase (see Figure 3.32). As functional antagonists of ethylene, **cytokinins** promote cell division and retard senescence. In *Arabidopsis*, two sensor His kinases have been identified as cytokinin receptors. Upon phosphorylation, the corresponding response regulators become released from a complex with an inhibitor protein (see Figure 3.32). For other pathways of plant hormone signaling, see Sidebar 6.1 and Section 13.3.

As mentioned in Sidebar 3.4, the photosensory phytochromes of plants control a "degenerate" two-component system with a His kinase-related Ser/Thr kinase coupled to the sensor. By means of this system, the transcription of genes encoding proteins of light adaptation is controlled. "Degenerate" His kinases have also been found in prokaryotes such as *B. subtilis*, indicating that the change of amino acid specificity is not a result of eukaryotic evolution.

Animals

The genomes of animals sequenced thus far do not contain His kinase and response regulator genes. An exception is mitochondria, the matrix of which contains two Ser/Thr kinases exhibiting some sequence homology with His kinases. These enzymes are subunits of the large dehydrogenase complexes metabolizing pyruvate and other α-keto acids. The dehydrogenase activity is suppressed by phosphorylation. This reaction is stimulated by protein deficiency, indicating that it provides a protective mechanism against a counterproductive degradation of amino acids.

The occurrence of degenerate His kinases in mitochondria is explained by the evolution of these organelles from endosymbiotic prokaryotes. It should be noted that in contrast to the labile His phosphorylation, the Ser/Thr phosphorylations catalyzed by degenerate His kinases are stable post-translational modifications.

Since conventional two-component systems are not found in animals, they provide interesting targets for novel antibacterial drugs, which are under development.

3.6.2 "Eukaryotic" protein phosphorylation in prokaryotes

At the transition from prokaryotes to eukaryotes, the "prokaryotic" His kinases were widely replaced by Ser/Thr and Tyr kinases. Nevertheless, the adjective "eukaryotic" commonly used for these enzymes is somewhat misleading, since they are by no means restricted to eukaryotes but are found, together with the corresponding protein phosphatases, in the majority of prokaryotes as well. In fact, in 29 out of 35 prokaryotic genomes sequenced, corresponding genes have been identified (for His kinases this ratio is 26:35). Even mycoplasma lacking His kinases express "eukaryotic"

Ser/Thr kinases. Obviously these enzymes have a long evolutionary history, probably going back to a primeval form that emerged before the separation of bacteria, archaea, and eukaryotes. In contrast, His kinases seem to be genuine bacterial "inventions" taken over by other species through horizontal gene transfer. Such a concept does not contradict the fact that some microbes are devoid of either bacterial His kinases or "eukaryotic" kinases and phosphatases. As a rule, such species are highly adapted pathogenic organisms, such as mycoplasma and rickettsia, as well as inhabitants of isolated biotopes with constant environmental conditions. It is assumed that the corresponding genes devoted to the processing of environmental stimuli turned out to be unnecessary and were eliminated.

The functions of "eukaryotic" protein phosphorylation in prokaryotes are understood only in a few cases. Examples are the kinase Pkn1 of *Myxococcus xanthus*, which is involved in sporulation, and the kinase AfsK, which phosphorylates the transcription factor AfsR and thus regulates the biosynthesis of antibiotics in *Streptomyces coelicolor*.

Kinases unrelated to both His kinases and eukaryotic kinases are called atypical. Examples of atypical kinases in prokaryotes are provided by bacterial Tyr kinases (**Sidebar 3.6**) and the bifunctional protein kinases/phosphatases (see Section 3.6.3).

3.6.3 Evolution of protein phosphatases

Among the protein phosphatases, the Ser/Thr-specific enzymes of the PPP family (see Section 2.6.4) seem to be phylogenetically most ancient. Like Ser/Thr kinases, they probably were introduced by the common primeval precursor of pro- and eukaryotes. In contrast, the bacterial enzymes of the PPM family (see Section 2.6.4) are thought to have been acquired from eukaryotes by horizontal gene transfer since they have not yet been found in archaea. A similar assumption is made concerning Tyr-specific phosphatases, identified in a few, mostly pathogenic, bacterial species where they play a role as virulence factors (**Sidebar 3.7**). No genes of the dual-specific Cdc25 phosphatases regulating the cell cycle have yet been found in prokaryotes.

The phospho-His and phospho-Asp residues generated by two-component systems are hydrolyzed either nonenzymatically or by the intrinsic phosphatase activity of the bifunctional His kinases and response regulators. There are, in addition, Asp-specific phosphatases such as CheZ (chemotaxis), SpoOE (sporulation of *B. subtilis*), RapA and B (sporulation), and the His-specific phosphatase SixA (adaptation

---

Sidebar 3.6: Bacterial protein-tyrosine kinases

Until recently, Tyr phosphorylation was thought to be a rare event or even absent in bacteria. This view has completely changed, since new techniques of protein analysis have revealed a large number of Tyr-phosphorylated proteins and of the corresponding kinases in prokaryotes. Structurally most bacterial Tyr kinases are not related to their eukaryotic counterparts but form an enzyme family of their own (called BY kinases) not yet found in eukaryotes. On the other hand, a few "eukaryotic" Tyr kinases have been found in several prokaryotic species indicating a long evolutionary history of such enzymes similar to that of the "eukaryotic" Ser/Thr kinases.

BY kinases appear to be transmembrane enzymes that undergo autophosphorylation upon activation. Their extracellular domain might provide binding sites for intercellular signaling molecules that are still elusive, however. Interestingly, such a mechanism of action strongly reminds one of eukaryotic receptor tyrosine kinases (see Chapter 7). Thus, the same principle of transmembrane signaling may have evolved in two different enzyme families. Another example is provided by receptorlike Ser/Thr kinases that are abundant in plants and are found also in some prokaryotes.

A steadily increasing body of evidence indicates an important role of BY kinases in bacterial physiology, for instance in DNA replication, stress response, cell wall formation, and virulence. It is hoped that in the near future these enzymes and their substrate proteins will become targets of new antimicrobial strategies.

Sidebar 3.7 Phosphotyrosine phosphatases as virulence factors: God's scourge of the Middle Ages Some bacteria produce Tyr-specific protein phosphatases of the conventional subfamily (see Section 2.6.4). They are assumed to have acquired this ability from eukaryotes by horizontal gene transfer. The bacterial phosphatases are extremely potent virulence factors. The enzyme YoPH of plague bacteria (*Yersinia*), together with some additional toxins, is thought to be responsible for the depopulation of large areas of Europe and Asia in the Middle Ages. YoPH is injected by the bacteria into the host cell and destroys focal adhesions—the contacts to other cells and to the extracellular matrix—thus inducing cell death by apoptosis. A comparably destructive effect, in this case on the cytoskeleton, is exhibited by the Tyr-specific phosphatases STP of *Salmonella typhimurium* and MPtpB of *Mycobacterium tuberculosis*. Other examples of the fatal consequences of a disturbance by bacteria of cellular data processing are provided by the cholera, pertussis, and diphtheria toxins attacking G-proteins and by the pathogenic adenylate cyclases of anthrax. They will be discussed in more detail in Section 4.3.3.

of *E. coli* to oxygen deficiency). These enzymes are monospecific: they have only one substrate protein each and are considered to be the result of highly specialized development. CheZ dephosphorylates exclusively the response regulator CheY, and SixA affects only the HPt domains of the sensor ArcB, the cytoplasmic part of which is a typical phosphorelay consisting of a His kinase, an Asp kinase, and an HPt domain.

Other extreme specialists are the bifunctional bacterial Ser/Thr kinases/phosphatases, represented by the isocitrate dehydrogenase kinase/phosphatase (AceK) of *E. coli* and the HPr-kinase/phosphatase of *B. subtilis*. AceK is of historical interest since it was the first protein kinase found in prokaryotes. The enzyme inhibits isocitrate dehydrogenase by phosphorylating a strategic Ser residue in the catalytic center, thus inhibiting a starter reaction of the citrate cycle and shifting the isocitrate metabolism to the glyoxylate pathway:

$$NADP^+ + isocitrate \rightarrow NADPH + \alpha\text{-ketoglutarate} + H^+ + CO_2$$

In the kinase, the same catalytic center is responsible for both phosphorylation and dephosphorylation. Dephosphorylation occurs through a re-transfer of the protein-bound phosphate residue to ADP. The subsequent hydrolysis of the ATP thus formed renders the reaction exergonic. In other words, the kinase/phosphatase is also an ATPase. Such a mechanism appears to be unique and is not a precedent in evolution. It also holds true for HPr kinase/phosphatase which is, however, devoid of ATPase activity. Obviously both enzymes represent evolutionary impasses.

Beside protein phosphorylation, other post-translational modifications are involved in bacterial signal processing. The methylation of carboxyl groups has already been mentioned as a mechanism to store memory in the course of (chemo)taxis. Other modifications include N-acetylation and N-methylation, glycosylation, nitrosylation, ADP-ribosylation, and attachment of ubiquitin-like proteins. They are important for a wide variety of physiological effects including gene expression, stress response, and infectivity. Mostly the corresponding enzymes as well as the gross molecular mechanisms resemble those found in eukaryotes indicating a long evolutionary history.

As far as prokaryotes are concerned, it is beyond the scope of this book to treat these modifications in detail. Instead, the reader is referred to the list of publications at the end of the chapter.

3.6.4 Evolution of regulatory GTPases

In addition to protein kinases and phosphatases, GTPases constitute a particularly important family of signaltransducing proteins. There are two major types of GTPases: the FtsZ/tubulin family and the G-protein family. Both seem to have evolutionary histories comparable to that of protein kinases, since the corresponding genes have been found in all prokaryotes studied so far.

The prokaryotic **FtsZ proteins** derive their name from a group of *E. coli* proteins involved in cell division. In their three-dimensional architecture and their functions they closely resemble the eukaryotic β-tubulins, though very little sequence homology exists. FtsZ proteins control the directed polymerization of cytoplasmic filaments or protofilaments in prokaryotes, whereas β-tubulins regulate the formation of microtubules in eukaryotic cells. Protofilaments are components of the Z-ring that separates the daughter cells during cell division, while microtubules underpin the structure of the mitotic spindle as well as cilia and flagella. As expected, FtsZ-related proteins have also been found in mitochondria and chloroplasts.

The **G-proteins** are not related to FtsZ proteins and β-tubulins. As biochemical switches they regulate a wide variety of cellular processes, causing spatial and temporal coordination and irreversibility (see Section 2.4). First the translation of mRNA has to be mentioned, which is controlled by the GTPase subfamily of initiation and elongation factors. The prokaryotic translation factors are summarized in **Table 3.1** (for details see Section 9.1). This mechanism is highly conservative and common to all forms of life. Equally abundant are G-proteins such as the signal recognition particles and their receptors, which manage the transport of newly synthesized polypeptide chains across membranes (see Section 9.2). They have prokaryotic counterparts such as the G-proteins Ffh and FtsY of *E. coli*.

Additional prokaryotic G-proteins seem to be involved in ribosome formation, namely the Era (*E. coli* ras-like) proteins, facilitating the maturation of ribosomal RNA; the EngA proteins and the ribosome small subunit-dependent GTPases RsgA promoting the assembly of large and small ribosomal subunits, respectively; and the ribosome associated Obg proteins, which may function as stress sensors to measure the fall of the cellular GTP level during a stringent response.

Homologous GTPases playing a role in ribosome biogenesis are also found in other organisms. Gene mutations have shown that most of the prokaryotic G-proteins are essential.

Members of the Ras superfamily of small G-proteins have also been found in prokaryotes where they might play a role in signal processing. An example is the GTPase MglA of myxobacteria (see Section 3.5.2). This protein is a member of an enzyme family that seems to be restricted to prokaryotes, whereas another prokaryotic GTPase type called RUP has been postulated to provide the ancestor of eukaryotic Ras and Arf GTPses.

The search for receptor-coupled heterotrimeric G-proteins in prokaryotes has been in vain. This supports the conclusion that prokaryotic transmembrane signaling is strongly dominated by the two-component mechanism.

Table 3.1 Prokaryotic translation factors

Factor	Function
IF1	dissociation of the ribosome, binding of initiator tRNA
IF2	GTPase, formation of the 70S initiation complex
IF3	ATPase, binding and scanning of mRNA
EF-Tu	GTPase, binding of aminoacyl tRNA at the A-site
EF-Ts	GDP–GTP exchange factor of EF-Tu
EF-G	GTPase, translocation of peptidyl-tRNA
RF1	release of the completed polypeptide
RF2	release of the completed polypeptide
RF3	GTPase, termination

Taken together, the prokaryotic G-protein mechanism appears to be simple when compared with the elaborate G-protein networks in eukaryotic cells. In fact, as explained in Chapter 10, the majority of eukaryotic G-proteins regulate the traffic between intracellular compartments, which are absent from prokaryotes.

3.6.5 Sensors of oxidative stress

Life under an oxygen atmosphere is a risky venture. Organisms may profit from aerobic conditions that enable highly efficient energy production. However, they are also endangered by the aggressive chemical properties of oxygen and its metabolic products (see Section 2.2). To deal with oxidative stress, aerobic bacteria have developed simple but efficient protective mechanisms by which they recognize dangerous reactive oxygen species (ROS) and initiate countermeasures. In this situation, the mechanistic principle of signal processing operates, in which certain signaling proteins become post-translationally modified by oxidation of iron-sulfide groups or cysteinyl–SH groups.

To make this clear, let us examine two ROS-controlled transcription factors of *E. coli*: the superoxide anion radical sensor SoxR and the H_2O_2 sensor OxyR. **SoxR** contains as a redox cofactor a $[Fe_2-S_2]$ iron sulfide entity complexed by cysteinyl residues. This so-called iron sulfur cluster (which is found in many redox enzymes) typically undergoes single electron transitions when reacting with free radicals such as superoxide anion radicals. As a result the conformation of SoxR is changed facilitating the binding to DNA. In turn, genes that encode enzymes degrading the highly toxic superoxide anion radical are activated. The prototype of such enzymes is superoxide dismutase catalyzing the reaction $O^{2-\bullet} + 2H^+ \rightarrow O_2 + H_2O_2$. The activation of SoxR is reversible because the factor is reduced by NADH/NADPH. The transcription factor **OxyR** takes care of the elimination of hydrogen peroxide that is generated along this and other metabolic pathways. OxyR represents the second family of ROS sensors that contain as a redox center oxidable cysteinyl residues instead of an iron sulfur cluster. Such proteins are less reactive with free radicals but are especially suited to react with peroxides that oxidize the thiol groups of Cys to sulfenic acid and disulfide groups (see Section 2.2). As for SoxR this modification is reversible becoming reduced by NADH/ NADPH via glutathione as an intermediate reductant. In the oxidized state, OxyR induces the activity of genes controlling peroxide metabolism. Another example of a bacterial redox sensor containing a reactive dithiol configuration is the His kinase ArcB. It will be discussed in more detail in Section 3.6.6.

Saccharomyces cerevisiae possesses a similar protective system; however, the sensor and transcription factors are separated. The yeast peroxide sensor is **Orp1**, a protein that in contrast to OxyR is not a transcription factor but exhibits the properties of a peroxidase, which is an enzyme that catalyzes the oxidation of substrates by hydrogen peroxide. In the course of this reaction, two cysteine residues of Orp1 become oxidized, forming a disulfide bridge or sulfenic acid groups that immediately become reduced again by another SH protein, **Yap1**. Yap1 is a transcription factor controlling several genes taking care of antioxidant production. It is inactivated by thioredoxin-catalyzed reduction. An adaptor protein, Ybp1, brings the sensor Orp1 in close contact with Yap1. Compared with the bacterial OxyR, the yeast system impressively demonstrates the evolutionary change from a simple peroxide sensor to an H_2O_2-sensitive complex of signal processing consisting of a sensor, an adaptor, and an effector. As discussed in more detail in Section 2.2, in eukaryotes, ROS indeed play a double role as both stress factors that have to be eliminated and intracellular signaling molecules. An example for ROS signaling in higher eukaryotes is found in Section 7.1.

3.6.6 Signal-controlled proteolysis

In eukaryotes, controlled degradation as well as proteolytic activation of signal-transducing proteins is a major regulatory mechanism. As a rule, proteins are

Sidebar 3.8 Sigma factors The RNA polymerase of *E. coli* consists of a core and a regulatory subunit called σ-factor.

The core is a heterotetrameric protein composed of two α-subunits, one β-subunit, and one β′-subunit. It is able to bind to DNA and to catalyze RNA biosynthesis but is unable to recognize the promoter sequences of individual genes. Such selectivity is conferred by the σ-factor. *E. coli* possesses seven different α-subunits exhibiting different promoter specificities. Everyday needs are satisfied by the σ^{70}-subunit, a protein of 63 amino acids. The additional six σ-factors become activated during stress situations and control the transcription of "emergency genes" such as those encoding heat-shock proteins (see Section 2.5.1). σ^{70} has two binding sites for DNA, with the C-terminal site recognizing double-stranded regions and the N-terminal site recognizing single-stranded regions. The corresponding interaction sequences of DNA are a TTGACA hexamer at −35 bp (35 base pairs upstream of the transcription start) and a TATAAT hexamer, the **TATA-box**, localized in a double-stranded area as well as at −10 bp, where the double strand becomes unwound into single strands. This unwinding, also called melting, is a prerequisite for strong binding between DNA and the RNA polymerase complex. σ^{70} interacts with these two sites only as a complex with the core enzyme due to an auto-inhibitory conformation of the σ-factor where the N-terminal sequence blocks the binding site for the DNA sequence at −35 bp. This inactive form becomes unfolded to the active conformation upon binding to the core enzyme.

marked for degradation by ubiquitylation to become hydrolyzed subsequently in proteasomes. This principle finds its precursor in prokaryotes. An example is the activation of the **alternative sigma factor σ^{S}** of *E. coli* (**Sidebar 3.8**). This factor becomes up-regulated only in stress situations such as starvation, high temperature, or osmotic crisis, and directs the RNA polymerase to genes specialized for stress protection. Under stress-free conditions, σ^{S} is kept at a low level by permanent proteolytic degradation. To be recognized by the protease ClpXP (a bacterial counterpart of the proteasome), the σ factor needs to be labeled by binding the recognition factor **RssB** (note the striking analogy to ubiquitylation). RssB, a protein, is active only when phosphorylated, thus providing an input terminal for stress signals that are mediated by a two-component system with the His kinase **ArcB** as a sensor and RssB as a response regulator. ArcB is an energy sensor that, under aerobic conditions, exists in an oxidized disulfide form. Upon energy shortage indicated by oxygen deficiency, ArcB is reduced to the dithiol form and its His kinase activity is suppressed. As a consequence, the α factor becomes resistant to proteolytic degradation. Such a mechanism, which instead of inducing de novo synthesis only alters the steady-state concentration of a regulatory protein, has the advantage of responding very rapidly to changes in the environment. This principle is expensive but obviously essential for coping with stress situations.

The release of active signaling proteins from membrane-bound precursors by intramembrane proteolysis (see Chapter 13) has also been introduced by prokaryotes. An example is the **alternative sigma factor σ^{K}** directing RNA polymerase to sporulation genes. σ^{K} becomes released from a precursor protein when *B. subtilis* receives environmental signals inducing sporulation. This signaling reaction is a perfect counterpart of the animal notch system (see Section 13.1.1).

3.6.7 Prokaryotic ion channels

Ion channels are absolutely essential for metabolism, equalization of osmotic pressure, and regulation of intracellular acidity. In parallel they were employed for data processing from the very beginning of life. An instructive example is provided by Ca^{2+} ions. Life depends on phosphate compounds such as nucleotides and nucleic acids. Therefore, it is mandatory for the cell to keep the cytoplasmic calcium concentration at a level that is much lower than in the extracellular world (e.g. sea water or body fluids), since otherwise insoluble Ca-phosphate complexes would precipitate and paralyze the cell. This control is managed by an interplay between extrusion and influx of Ca^{2+} via membrane transporters and ion channels. The strong constraint of cellular calcium homeostasis has offered a unique possibility for signaling, requiring the evolution of proteins the activity of which is under the control of Ca^{2+}. Thus Ca^{2+} has become perhaps the most versatile chemical messenger (see Chapter 14.5) in both pro- and eukaryotes. In prokaryotes many proteins have been found that structurally resemble eukaryotic calcium-binding proteins. However, only little is known about their physiological functions. Moreover, the investigation of prokaryotic calcium channels is still in its infancy. However, there is an increasing body of evidence indicating a key role of calcium in emergency and survival situations such as heat-shock response, biofilm formation, sporulation etc.

Clearly the evolutionary history of ion channels dates back as far as that of protein kinases and GTPases. Only the simplest prokaryotes, such as some mycoplasma, seem to be devoid of the corresponding genes, leaving open the question whether they had been lost by adaptive processes. Research on prokaryotic ion channels has made an important and significant contribution to our understanding of these signal-propagating devices, in particular since several channel proteins could be crystallized and studied by X-ray analaysis. The overall result emerging from these studies is that all basic structures and mechanisms of action of ion channels were introduced by prokaryotes.

A common structural feature of ion channels is a series of transmembrane helices (at least four) forming a central pore that, through its width and electrostatic charge, determines the channel's ion selectivity (see Section 2.10.3). Prokaryotic ion channels that have been investigated in great detail include the mechanosensitive channel MscL, several potassium and sodium channels, and the chloride channel ClC. Each of these channels is a prototype of a large and evolutionary extremely successful family of channel proteins. Representatives of prokaryotic cation channels are shown in **Figure 3.33**.

Mechanosensitive channels

These channels are abundant in both prokaryotes and eukaryotes. Their major role is that of an excessive pressure valve protecting cells from disruption by hypo-osmotic shock. Gating of the channels is triggered by the horizontal tension of the plasma membrane due to a swelling of the cell. This tension causes an irislike twisting of the channel's transmembrane domains, leading to a widening of the central pore (**Figure 3.34**). Such a response is particularly essential for free-living organisms that are permanently endangered by a hypo-osmotic situation, for instance, rain.

Lacking a ligand-binding site as well as an ion selectivity filter and a voltage sensor, **MscL** (L stands for large conductance) is one of the simplest ion channels. It represents the primeval form of the so-called S5–S6 channels (see Figure 3.33 and Section 2.10.3) and might be a precursor of the ligand-controlled ion channels of higher eukaryotes. In prokaryotes, however, no ligand-controlled channels (except a glutamate-dependent potassium channel GluR0) have been found.

While MscL channels occur predominantly in bacteria, a second family of mechanosensitive channels is distributed more widely. These are the structurally highly

Figure 3.33 Subunits of prokaryotic ion channels of the S5–S6 family: precursors of eukaryotic channels Mechanosensitive channel MscL is the prototype of the following eukaryotic ion channels: ATP-receptor cation channel P2X (Section 16.6); epithelial sodium channel ENaC (Section 14.3); and several neuropeptide receptor ion channels of invertebrates. Potassium channel KcsA resembles the inwardly rectifying potassium channel K_{ir} of eukaryotes. In other eukaryotic K^+ channels, two KcsA-like subunits (TWIK) or one KcsA-like subunit and one Kchlike subunit (TOK) share one polypeptide chain each (Section 14.4). Potassium channel Kch resembles the cyclic nucleotide-gated ion channels of higher eukaryotes (Section 14.5.7). Voltage-dependent sodium and potassium channels NaChBac and KvAP may be precursors of voltage-dependent Na^+, K^+, and Ca^{2+} channels of eukaryotes (see Sections 14.2, 14.4.1, and 14.5.2). Glutamate-dependent potassium channel GluR0 is probably the precursor of the ionotropic glutamate receptor cation channels of vertebrates (Section 16.4). Gray cylinders, transmembrane domains and P-loops (P); teal cylinder, transmembrane helix with voltage sensor; dark gray domains, extracellular Glu-binding site of GluR0. Numbering S1–S6 of the transmembrane domains refers to the voltage-dependent K^+ channel of eukaryotes. The number of subunits constituting a channel is shown beneath the name.

MscL 5   KcsA 4   Kch 4   NaChBac, KvAP 4   GluR0 4

cytoplasm

Figure 3.34 Model of the mechanosensitive channel MscL The channel pore is depicted in a three-dimensional schematic manner (bottleneck structures are shown in teal). Transmembrane helices are represented by black lines. The channel is gated by a horizontal tension of the membrane caused by the swelling of the cell in a hypo-osmotic environment.

variable **MscS channels** (S stands for small conductance). The corresponding channel of *E. coli* consists of seven symmetrically arranged subunits with three transmembrane helices each and becomes gated by both mechanical stress and membrane depolarization, thus representing a primeval form of voltage-dependent ion channels.

Potassium and sodium channels

A prokaryotic prototype of these channels is **KcsA** (K$^+$ channel of streptomyces A) isolated from *Streptomyces lividans* (homologous channels of *E. coli* are KefB and KefC). X-ray analysis has revealed a subunit structure that is characterized by two transmembrane helices and a P-loop rendering the channel selective for K$^+$ ions (the problem of channel selectivity is discussed in more detail in Section 2.10.3). Each channel consists of four subunits representing the primeval form of all ion-selective potassium channels of the S5–S6 family (see Figure 3.33). The KcsA channel opens at low pH, probably due to an interaction of protons with the extracellular domains. As in the case of the MscL channel, this gating is assumed to be caused by an irislike twisting of the transmembrane helices (**Figure 3.35**).

The structure of KcsA resembles that of inwardly rectifying potassium channels of vertebrates (see Section 14.4.2). A special species of this channel type is the potassium channel **MthK** of *Methanobacterium thermoautotrophicum* (see Figure 3.35). It

(A)

Figure 3.35 Schematic representation of architecture and gating of the KcsA channel
(A) Arrangement of transmembrane helices S5 and S6 as well as the P-loop (P) in the plasma membrane (cross section showing two of the four subunits of the channel). (B) Gating of the central pore by an irislike twisting of the transmembrane helices.

(B)

Figure 3.36 Gating of the Ca²⁺-dependent K⁺ channel MthK The insert shows the schematic structure of one of the four channel subunits with the transmembrane segments S5 and S6 and the P-loop (P) forming the selectivity filter. The extended C-terminal domain (teal) contains Ca²⁺ binding sites. Below a space-filling model of the channel based on X-ray crystallography is shown in a closed (left) and in an open form (right). Channel gating is due to a conformational change of the ring of C-terminal domains (teal) that is induced by Ca²⁺ binding. (Modified from Schumacher M & Adelman JP [2002] *Nature* 417:501–502.)

contains an extended cytoplasmic C-terminal domain that binds Ca^{2+} ions resulting in channel gating. X-ray crystallography has provided a deep insight into channel structure and the gating process (**Figure 3.36**). Functionally MthK resembles the mammalian Ca^{2+}-gated potassium channels BK and SK that, however, have a more complex structure (see Section 14.4.1).

Voltage-dependent cation channels of vertebrates are typically composed of four subunits with six transmembrane domains and a P-loop each whereby transmembrane segment S4 acts as a voltage sensor (see Section 14.1). An analogous structure has been found for prokaryotic cation channels, such as the potassium channel **Kch** of *E. coli*. While Kch does not significantly respond to voltage, other bacterial channels do: the **voltage-dependent Na⁺-channel** of *Bacillus halodurans* (see Figure 3.33) and the archaeal potassium channel **KvAP** of *Aeropyrum pernix*. Certainly, these channels are evolutionary precursors of vertebrate voltage-dependent ion channels. Their four subunits resemble those of Kch; however, transmembrane segment S4 together with a part of S3 acts as a voltage sensor (see Figure 14.3). The elucidation of the KvAP structure has revolutionized our concepts of ion channel gating (see Section 14.1). While the function of the potassium channel is not yet understood, the sodium channel has been proposed to be a part of the flagellar motor, which in this species is supplied by energy derived from an Na⁺ gradient across the cell membrane. It is possible that these channels also participate in regulation of the very negative ion potential (ranging between −100 and −150 mV) of bacterial cells. Cooperating with mechanosensitive channels they may, in addition, protect the cell from osmotic stress. The subunit and domain architecture of voltage-dependent ion channels is also found in cyclic nucleotide-gated channels that play a key role in vertebrate sensory signal processing (see Chapter 15). The ancestor of these channels might be represented by the cyclic nucleotide-gated potassium channel **MloK1** of *Mezorhizobium loti*, a plant symbiont.

A **glutamate-dependent potassium channel GluR0**, which has been found in some bacterial species, has evolved from the fusion of a KscA-like channel protein with a periplasmic binding protein for glutamate. GluR0 is considered to represent the evolutionary precursor of the ionotropic Glu receptors of higher eukaryotes (see Section 16.4).

Figure 3.37 Membrane topology of chloride and water channels (A) Schematic representation of a chloride channel subunit showing 18 helices (gray cylinders). Note the internal symmetry caused by an anti-parallel repeat of the first nine helices (see also Section 14.7 for more details). (B) Subunit of an aquaporin water channel showing six transmembrane helices and two P-loops. (C) Channels composed of two and four identical subunits, respectively, exhibiting either two or four pores.

Chloride channels

Like the S5–S6 channels, the chloride or anion channels of the ClC type (see Section 14.7.1) also originate from prokaryotic precursors, though not all prokaryotes have the corresponding genes. X-ray analysis of the channel protein **EcClC** of *E. coli* has revealed an unexpectedly complex structure. This channel is composed of two identical subunits with 18 strongly twisted helices each, 17 of them penetrating the cell membrane partly or completely. In contrast to cation channels, each subunit of EcClC forms a pore; that is, the channel has two pores (**Figure 3.37**). This structure seems to be prototypical for all chloride channels of the ClC type found as yet (see Section 14.7). The function of the prokaryotic chloride channels is unclear. They may participate in regulation of pH value and cell volume. Because of the small volume of a bacterial cell (10^{-12} mL on average), the gating of ion channels must be precisely terminated; otherwise, intolerable changes of the ion concentration and osmotic pressure would occur. How this control is achieved is still an open question.

Water channels

A large and ubiquitous family of membrane channels is specialized for the transport of water and small polyalcohols, which on their own cannot penetrate the lipid bilayer. These channels are called **aquaporins** (Aqp, only for water) and **glyceroaquaporins** (Glp, for water and for polyalcohols such as glycerol). Both types are widely distributed among prokaryotes and participate in osmoregulation and volume control, respectively. Human cells express 10 different aquaporins. The channels AqpZ and GlpF of *E. coli* have been studied in detail. Their structure seems to be representative for all (glycero)aquaporins. Each channel is composed of four subunits, each of which contains four transmembrane helices and two P-loops and forms one pore; that is, the complete channel has four pores (see Figure 3.34). Bottlenecks and charged amino acid residues guarantee the selectivity of the pores. Hydrated ions are too large to pass through the pore, and the charge of the pore is insufficient for stripping the ion from water molecules. Therefore, water channels are impassable for ions and other charged molecules.

Summary

- While Ser/Thr-specific protein kinases have been found in all species studied thus far, His autokinase-based two component systems are restricted to prokaryotes and, to a smaller extent, slime molds, fungi, and plants.

- Two unrelated families of Tyr-specific protein kinases are known, the bacterial BY kinases and the eukaryotic tyrosine kinases. Both families have evolved independently.

- Protein phosphatases are phylogenetically as ancient as protein kinases. His autokinases are bifunctional, exhibiting both kinase and phosphatase activity. In addition, monofunctional His-, Asp-, and Ser/Thr-specific phosphatases have been found in prokaryotes. Some bacterial Tyr phosphatases acquired from animals by horizontal gene transfer play a fatal role as virulence factors.

- The most ancient G-proteins are translation factors and proteins that catalyze the transmembrane transport of newly synthesized polypeptide chains. Small G proteins of the Ras superfamily are also found in prokaryotes.

- Redox signaling through reversible oxidation of SH groups in proteins is an ancient principle of regulation. In prokaryotes it may have developed from mechanisms that protected the organism from oxidative stress.

- Mechanisms resembling both the ubiquitin-controlled degradation of signaling proteins and the proteolytic release of active factors from protein precursors are also found in prokaryotes. In bacteria such signaling events play an important role in stress situations.

- All ion and water channels of eukaryotes can be traced back to prokaryotic precursors. In prokaryotes such channels are involved in cell volume control, regulation of the osmotic pressure, and maintenance of the membrane potential.

Further reading

Section 3.1

Boutte CC & Crosson S (2013) Bacterial lifestyle shapes the regulation of stringent response activation. *Trends Microbiol* 21:174–180.

Section 3.2

Cech TR & Steitz JA (2014) The noncoding RNA revolution – trashing old rules to forge new ones. *Cell* 157:77–94.

Gottesmann S & Storz G (2011) Bacterial small RNA regulators: versatile roles and rapidly evolving variations. *Cold Spr Harbour Persp Biol* 22011;3:a003798.

Hur JK, Olovnikov I & Aravin AA (2014) Prokaryotic Argonautes defend genomes against invasive DNA. *Trends Biochem Sci* 39:257–259.

Sauer E (2013) Structure and RNA-binding properties of the bacterial Lsm protein Hfq. *RNA Biol* 10: 610–618.

Section 3.3

Davidson AL & Chen J (2004) ATP-binding cassette transporters in bacteria. *Annu Rev Biochem* 73:241–268.

Görke B & Stülke J (2008) Carbon catabolite repression in bacteria: many ways to make the most out of nutrients. *Nature Rev Microbiol* 6:613–624.

Green et al. (2014) Cyclic AMP and bacterial cyclic-AMP receptor protein revisited: adaptation for different ecological niches. *Curr Opin Microbiol* 18:1–7.

Kremling A et al. (2015) Understanding carbohydrate catabolite repression in *Escherichia coli* using quantitative models. *Trends Microbiol* 23:99–109.

Section 3.4

Beier D & Gross R (2006) Regulation of bacterial virulence by two-component systems. *Curr Opin Microbiol* 9:143–152.

Briegel A et al. (2014) New insights into bacterial chemoreceptor array structure and assembly from electron cryotomography. *Biochemistry* 53:157–185.

Casino P, Rubio V & Marina A (2010) The mechanism of signal transduction by two-component systems. *Curr Opin Struct Biol* 20:763–771.

Galperin MJ (2005) A census of membrane-bound and intracellular signal transduction proteins in bacteria: Bacterial IQ, extroverts and introverts. *BMC Microbiol* 5:35–54.

Spudich JL et al. (2014) Mechanism divergence in microbial rhodopsins. *Biochim Biophys Acta* 1837:546–552.

Ulrich LE, Koonin EV & Zhulin IB (2005) One-component systems dominate signal transduction in prokaryotes. *Trends Microbiol* 13:52–56.

Zhao X, Zeth K & Liu J (2014) Molecular archirecture of the bacterial flagellar motor in cells. *Biochemistry* 53:4342–4333.

Section 3.5

Boyd CD & O'Toole GA (2012) Second messenger regulation of biofilm formation: breakthrough in the understanding c-di-GMP effector systems. *Annu Rev Cell Develop Biol* 28:439–462.

Boyle KE et al. (2013) Social evolution in biofilms. *Curr Opin Microbiol* 16:207–212.

Claessen D et al. (2014) Bacterial solutions to multicellularity: a tale of biofilms, filaments and fruiting bodies. *Nature Rev Microbiol* 12:115–124.

Corrigan RM & Gründling A (2013) Cyclic-di-AMP: another second messenger enters the fray. *Nature Rev Microbiol* 11:513–524.

de la Fuente-Núñez C et al. (2013) Bacterial biofilm development as a multicellular adaptation: antibiotic resistance and new therapeutic strategies. *Curr Opin Microbiol* 16:580–589.

Römling U, Galperin MY & Gomelsky M (2013) Cyclic-di-GMP: the first 25 years of a universal bacterial second messenger. *Microbiol Mol. Biol Rev* 77:1–52.

Schuster M et al. (2013) Acyl-homoserine lactone quorum sensing: from evolution to application. *Annu Rev Microbiol* 67:43–63.

Shapiro JA (1988) Bacteria as multicellular organisms. *Scientific Am* 256:82–89.

Sharp P (2013) Cyclic di-nucleotide signalling enters the eukaryotic domain. *LUBMP Life* 65:897–903.

Section 3.6

Anantharaman V, Iyer LM & Aravind L (2007) Comparative genomics of protists: new insights into the evolution of eukaryotic signal transduction and gene regulation. *Annu Rev Microbiol* 61:453–475.

Attwood BV (2013) Histidine kinases from bacteria to humans. *Biochem Soc Trans* 41: 1023–1028.

Bayles KW (2014) Bacterial programmed cell death: making sense of a paradox. *Nature Rev Microbiol* 12:63–68.

Booth IR et al. (2015) The evolution of bacterial mechanosensitive channels. *Cell Calcium* 57:140–150.

Capra EJ & Laub MT (2012) The evolution of two-component signal transduction systems. *Annu Rev Microbiol* 66:325–347.

Cain JA, Solis N & Corwell SJ (2014) Beyond gene expression: the impact of protein posttranslational modifications in bacteria. *J Proteomics* 97:265–286.

Chao JD, Wong D & Av-Gay Y (2014) Microbial protein-tyrosine kinases. *J Biol Chem* 289:9463–9472.

Delcour AH (2013) Electrophysiology of bacteria. *Annu Rev Microbiol* 67:179–197.

Dominguez DC, Guragain M & Patrauchan M (2015) Calcium-binding proteins and calcium signaling in prokaryotes. *Cell Calcium* 57:151–165.

Grangeasse C, Nessler S & Mijakovic I (2012) Bacterial tyrosine kinases: evoution, biological function, and structural insights. *Phil Trans Roy Soc Biol Sci* 367:2640–2655.

Keshishian EA & Rashotte AM (2015) Plant cytokinin signalling. *Essays Biochem* 58:13–27

Lanouette S et al. (2014) The functional diversity of protein lysine methylation. *Mol Syst Biol* 10:724.

Larrieu A & Vernoux T (2015) Comparison of plant hormone signaling systems. *Essays Biochem* 58:165–181.

Maupin-Furlow JA (2014) Prokaryotic ubiquitin-like protein modifications. *Annu Rev Microbiol* 68:155–175

Plattner H & Verkhratsky A (2015) The ancient routs of calcium signaling evolutionary tree. *Cell Calcium* 57:151–165.

Podgornaia AI & Laub MT (2013) Determinants of specificity in two-component signal transduction. *Curr Opin Microbiol* 16:156–162.

Wilson ME et al. (2013) MscS-like mechanosensitive channels in plants and microbes. *Biochemistry* 52:5708–5722.

Wuichet K & Sogaard-Andersen L (2015) Evolution and diversity of the Ras superfamily of small GTPases in prokaryotes. *Genome Biol Evol* 7:57–70.

Basic Equipment: G-Proteins, Second Messengers, and Protein Kinases

4

4.1 Life is stress

As explained in Chapter 3, the molecular "nanobrain" of prokaryotes processes signals mainly by means of protein switches operated by reversible protein phosphorylation. The data-processing apparatus of higher cells essentially follows the same principles. In bacteria, one major task of cellular data processing is to cope with environmental stress to discriminate between good and bad influences. The same holds true for more advanced organisms, albeit on a much higher level of complexity.

About 80 years ago the term "stress" was coined by Hans Selye to characterize the various adaptive reactions of the human organism to external and internal stimuli. Today the word stress has a negative overtone since it is understood mostly as a disease-causing alarm reaction. However, when the term is used in a broader sense to characterize *any* type of physiological and psychic mobilization, even the transition from the sleeping to the waking state is considered stress (as probably most readers would confirm).

In our body such a transition from rest to activity leads to stimulation of the three major data-processing networks: the endocrine system, the nervous system, and—in more extreme cases—the nonspecific and specific immune system. These systems produce a wide variety of signaling molecules such as hormones and local mediators including neurotransmitters and antibodies, thus communicating with each other and with their target organs. No matter whether we are concerned with ourselves or with the environment, these factors are always forming interconnected causality chains leading from the macro-world of our environment (or our imaginations) into the nano-world of cellular molecules and back. It is essential to illustrate this principle with the help of a concrete example such as the particular stress situation depicted in **Figure 4.1**.

4.1.1 The fight-or-flight response

A stress response starts with a sensory impression (leaving aside imaginary stress situations). The environmental signals are transformed by sensory cells into nerve impulses, a process described in detail in Chapter 15. The brain decodes these impulses with the help of stored information and, in the case depicted in Figure 4.1, recognizes the situation as a threat. As a consequence, an appropriate response is selected that, depending on the situation and the psychic constitution, may be either "fight" (consisting of an active change of the environmental condition, either forcibly or verbally) or "flight" (consisting of an adaptation to the situation). In either case motion is required, though it may be of the mouth and vocal cords only, so the skeletal muscles have to be activated by the motor neurons via the neurotransmitter acetylcholine as a stimulatory signal (**Figure 4.2**; details are found in Sections 14.6.2 and 16.2).

fight-or-flight response

↓

alertness
activation of muscles
mobilization of energy reserves

Figure 4.1 A stress situation
The tiger should be taken as a metaphor, since the situation depicted here symbolizes many types of stress including the alarm clock in the morning, political debates, marital crisis, and academic examination.

Figure 4.2 Processing of stress signals by the neuronal and endocrine network When the central nervous system recognizes the stress situation, it induces a fight-or-flight reaction (Figure 4.1). Immediately, the skeletal muscles (left) and the sympathetic branch of the vegetative nervous system are activated by means of acetylcholine. By releasing the stress hormones and neurotransmitters adrenaline and noradrenaline, the adrenal medulla and the sympathetic nerves guarantee an overall stimulation of the organism, including mobilization of biochemical energy reserves to supply the energy for muscle contraction. Long-term adaptation is brought about by the endocrine stress axis (right). Here certain neurosecretory cells of the hypothalamus release the peptide hormone CRH (Corticotropin Releasing Hormone), which induces the production of ACTH (AdrenoCorticoTropic Hormone, also known as corticotropin) by the adenohypophysis. In the adrenal cortex ACTH stimulates the biosynthesis of glucocorticoid-type stress hormones.

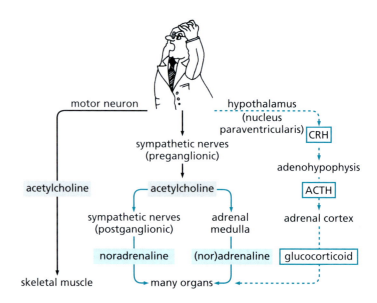

As a result the energy reserve of the muscle cell, in the form of glycogen, rapidly becomes exhausted and the blood sugar drops, indicating that it is time now to open further energy stores such as liver glycogen and fat reserves. To this end the autonomous nervous system transmits stress signals to the corresponding target organs. These signals include the sympathetic transmitter noradrenaline and the neurosecretions noradrenaline and adrenaline that are released from the adrenal medulla. Via intracellular signaling cascades, these signals induce the metabolic degradation of fat and glycogen to fatty acids and glucose, respectively, which are distributed in the bloodstream and are used by the target organs (for example, muscle cells) for energy production (for details, see Section 5.3).

In parallel, dramatic changes in blood circulation occur. To supply muscles and nerves sufficiently, the blood supply is increased for these tissues but decreased for other organs through effects on the vascular smooth muscles. Moreover, to raise the flow rate, the heartbeat becomes faster and stronger. In addition, breathing is deepened for a better oxygen supply and intestinal peristalsis is inhibited, since digestion is a matter of relaxation rather than of activity (**Figure 4.3**). For these and many other effects, essentially *one* signal is needed, namely (nor)adrenaline, the endogenous stress mediator *par excellence*.

If the stress situation is perpetuated and the energy reserves cannot be refilled fast enough, proteins are "burned" for energy generation. To this end they are fed into the pathways of carbohydrate metabolism. Under normal conditions this

Figure 4.3 Effects of sympathetic and parasympathetic stimulation (Adapted from S. Silbernagl & A. Despopoulos [1991] Taschenatlas der Physiologie. With permission from Thieme-Verlag, 1991.)

process, called gluconeogenesis, is suppressed so as not to waste precious material, but it is activated in emergency situations. The triggering signals are the glucocorticoid hormones of the adrenal cortex. Their biosynthesis is induced along a sequence of orders called the endocrine stress axis, originating in the brain (see Figure 4.2). It is easily seen that long-lasting activation of this axis may have deleterious consequences.

The stress response clearly demonstrates how, on the way from the tiger to the molecules, signals are exchanged and transformed until the sensory impression has evoked a response from the whole organism. On the molecular level, three stages of adaptation are evident:

(1) After a moment of shock, the skeletal muscles begin to contract and the peripheral blood flow is changed (visible as turning pale or blushing). Both responses are triggered by the most rapid of all endogenous signals, a change of the membrane potential in nerve and muscle cells mediated by the neurotransmitter acetylcholine.

(2) With some delay, but still within a few minutes, the fat and glycogen reserves are opened. The enzymatic apparatus required is already present, waiting for a stimulation by a stress signal such as (nor)adrenaline.

(3) In contrast, long-term adaptation requires de novo synthesis of proteins such as the enzymes of gluconeogenesis, achieved by stimulation of gene expression mainly by glucocorticoids. This may take several hours.

4.1.2 Vegetative nervous system

The short- and medium-term responses are mediated by the **vegetative** or **autonomous nervous system** that in vertebrates controls practically all organ functions and the internal milieu of the body, (**Sidebar 4.1**) while at the same time cooperating with the endocrine system and the central nervous system. This cooperation implies that the vegetative nervous system, although frequently reacting via local feedback loops (in a reflectory manner), is by no means "autonomous" but, in fact, is permanently under the control of environmental, hormonal, and psychogenic factors (see Figure 4.2).

The contact sites between central and vegetative nervous systems are synapses in ganglions arranged in the spinal marrow. Long axons (nerve fibers) originating from the postsynaptic ganglionic neurons, approach the target organs. These organs include heart muscle; smooth muscles controlling peripheral blood flow, peristalsis, urinary bladder contraction, functions of genital organs (erection, ejaculation, uterine contraction) and respiration; metabolically active organs such as liver and fat tissue; and exocrine glands.

Sidebar 4.1 Internal milieu The term "*milieu interieur*" was coined by the French physiologist Claude Bernard at the end of the nineteenth century to describe the state of the extracellular milieu in multicellular organisms. The interior milieu is thought to replace the external milieu of unicellular organisms originally living in sea water. Important parameters of the internal milieu are osmolarity (including Ca^{2+} concentration), blood pressure, pH, temperature, and concentration of energy-supplying molecules (such as glucose, fatty acids, and oxygen) as well as of metabolic waste products (such as carbon dioxide). As a rule, the organism tolerates only very small fluctuations of such parameters. Therefore, it needs expensive regulatory mechanisms to guarantee the constancy or homeostasis of the internal milieu. Major control devices are the nervous system, the endocrine–paracrine complex, and the immune system. These systems are not isolated from each other but are consistently cross talking, forming a network that connects the environment with the signal-processing apparatus of the tissue cells, since these are normally blind to environmental stimuli.

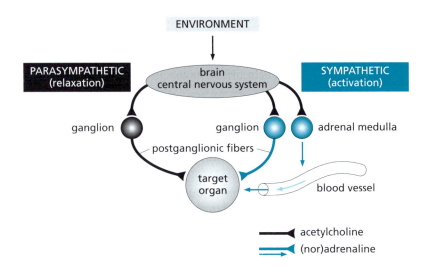

Figure 4.4 Organization of the vegetative nervous system The two branches of the vegetative nervous system, sympathetic and parasympathetic, are controlled by the central nervous system via spinal ganglions. By releasing the neurotransmitter noradrenaline (sympathetic, teal) and acetylcholine (parasympathetic, black), the postganglionic fibers regulate the function of almost any organ. The adrenal medulla represents a neurosecretory endocrine gland, evolved from a sympathetic ganglion. It releases adrenaline and noradrenaline into the blood.

Anatomically and functionally there are two branches of the vegetative nervous system, known as **sympathetic** and **parasympathetic**. They operate in an antagonistic way; that is, the vegetative nervous system controls organ function according to a two-rein principle. The sympathetic branch assumes the excitation role and the parasympathetic branch assumes the steadying role (see Figure 4.3).

Vegetative neurons do their jobs mainly by use of two transmitters: acetylcholine and noradrenaline. For fine regulation these are joined by a collection of co-transmitters and neuromodulators. Acetylcholine is released by the preganglionic neurons, which transmit the signal from the central to the vegetative nervous system, and by the postganglionic parasympathetic neurons, whereas noradrenaline is the neurotransmitter of the postganglionic sympathetic neurons, which transmit the signal from the nervous system to the target organs (exceptions to this rule are eccrine sweat glands, the periosteum, and blood vessels of skeletal muscles where sympathetic neurons use acetylcholine instead of noradrenaline). Noradrenaline and adrenaline (also known as norepinephrine and epinephrine) are also released into the bloodstream by the adrenal medulla, acting as stress hormones. The adrenal medulla is a sympathetic ganglion that has become anatomically remodeled to a neurosecretory gland (**Figure 4.4**).

The fact that, for its various functions, the vegetative nervous system gets by with only two major signaling molecules demonstrates very clearly that the medium is not the message. Whether and how a signal is decoded by the receiver cell is dependent solely on the inherited and acquired programming of the data-processing network. This fact had been established early by pharmacological and toxicological experiments. British pharmacologist Sir Henry Hallett Dale (1875–1968; awarded the 1936 Nobel Prize in Physiology or Medicine) is regarded as a pioneer in this field. As early as 1906 he observed that extracts from adrenal medulla (adrenaline and noradrenaline were still unknown at that time) evoked antagonistic sympathetic effects on blood pressure and uterine contraction depending on whether or not the experimental animals had been pretreated with extracts from ergot (*Claviceps purpurea*). Today we know that ergot alkaloids suppress the α-adrenergic effects of (nor)adrenaline, such as uterine contraction, leaving untouched the antagonistic β-adrenergic effects such as uterine dilation. Similar observations were made for acetylcholine: here the poison of the deadly nightshade (*Atropa belladonna*) inhibits selectively the muscarine-like parasympathetic effects but not the nicotine-like effects in the ganglions of the spinal marrow and on skeletal muscles, whereas the South American arrow poison curare exhibits just the reverse effect.

Summary

- Environmental or psychogenic stimuli provoke a systemic reaction known as stress response, consisting of a general activation that may culminate in a fight-or-flight syndrome.

- On the molecular level of endogenous signaling, the response comprises three major stages: a rapid nervous stimulation of muscular activity mediated by acetylcholine, followed by a more delayed mobilization of glycogen and fat reserves due to enzyme activation triggered by (nor)adrenaline, and finally long-term adaptation due to gene transcription induced by glucocorticoids and other hormones.

- The vegetative nervous system controls the functions of most organs by using only two neurotransmitters: noradrenaline as a sympathetic signal and acetylcholine as a parasympathetic signal. Acetylcholine is, in addition, the transmitter in the spinal ganglions that connect the vegetative with the central nervous system.

- The stress hormone adrenaline is released by the adrenal medulla, a sympathetic ganglion anatomically remodeled into a neuroendocrine gland.

4.2 Discovery of intracellular signal transduction

Early investigations with organ extracts and later studies with acetylcholine and (nor)adrenaline had shown that the same signaling substance might have different or even antagonistic effects depending on the tissue and the physiological context. This principle is now firmly established for the majority of hormones, local mediators, and neurotransmitters. There was only one plausible explanation for such a phenomenon: a cell can interpret the meaning of the same signal in different ways. Therefore, different receptor types were postulated, namely, α-and β-adrenergic receptors for (nor)adrenaline and muscarinic and nicotinic receptors for acetylcholine. The existence of these receptors was supported by a wealth of experimental data (these receptors have been isolated and analyzed in detail; see Chapters 5 and 16). Since acetylcholine and (nor)adrenaline are not lipophilic enough to pass the cell membrane, their receptors were expected to be localized at the cell surface. But how was the signal transduced to the functional apparatus inside the cell? The first and pioneering answer to this question was provided by experiments focusing on an adrenaline-regulated metabolic effect that could be analyzed by the biochemical methods available at that time. This effect was the stress-induced degradation of glycogen. In fact, for the first time these studies gave a hint as to how cellular signal transduction might work.

Glycogen, a polymer of glucose mainly linked via 1,4-α-glycosidic bonds, represents an instantly available energy supply for the body and is found in most tissues, particularly in liver and skeletal muscle. On demand, it becomes degraded via glucose 1-phosphate and glucose 6-phosphate to glucose, which is released by the liver into the bloodstream and oxidatively metabolized for ATP generation by other tissues (**Figure 4.5**; more details in Section 5.3.1). The course of this fundamental metabolic process was elucidated in the mid-twentieth century, with Carl Ferdinand Cori and Gerty Theresa Cori at Washington University in St Louis playing a pioneering role (they were awarded the 1947 Nobel Prize in Physiology or Medicine, together with Bernardo A. Houssay). The key reaction of glycogenolysis was found to be a phosphorolytic splitting of the glycosidic bond, catalyzed by the enzyme glycogen phosphorylase and yielding glucose 1-phosphate (see Figure 4.5). Glycogen phosphorylase was found to exist in two forms, an enzymatically active a-form and an inactive b-form. In the 1960s, Edmund H. Fischer

Figure 4.5 Biochemistry of glycogenolysis Glycogen is phosphorolytically degraded to glucose 1-phosphate, which is isomerized to glucose 6-phosphate by phosphoglucomutase. Glucose 6-phosphatase catalyzes the dephosphorylation of glucose 6-phosphate yielding glucose. Glycogen phosphorylase, the key enzyme of signal-induced glycogen degradation, exists in an inactive b-form and an active a-form. The a-form is generated from the b-form by reversible enzymatic phosphorylation of Ser residues.

and Edwin Krebs demonstrated that, in contrast to the b-form, the a-form was phosphorylated at a strategic serine residue, a reversible reaction catalyzed by an enzyme called glycogen phosphorylase kinase competing with a corresponding protein phosphatase (see Figure 4.5). This pioneering discovery of regulatory protein phosphorylation was honored by the 1992 Nobel Prize in Physiology or Medicine.

The depolymerization of glycogen is stimulated by adrenaline, glucagon, and other sugar-mobilizing hormones. In fact phosphorylase kinase becomes phosphorylated and more active in the presence of such hormones. A crucial discovery was made by Earl W. Sutherland (1925–1974), a former coworker of the Coris, by demonstrating that the hormones could activate phosphorylase kinase not only in the intact tissue but also in liver homogenate. For the first time a hormonal effect could be studied *in vitro*; the door had been opened for a detailed biochemical analysis performed in a classical manner. In the course of these studies, the cellular membrane fraction separated by centrifugation was found to contain an Mg^{2+}-dependent enzyme X that was stimulated in the presence of adrenaline, producing from ATP a heat-stable factor of low molecular mass. This factor was able to stimulate the enzymes of glycogenolysis that were enriched in the soluble cell fraction. Referring to adrenaline as the "first messenger," this factor was called a "second messenger" and was considered to be the sought-after signal, which bridged the gap between the hormone receptor at the cell surface and the enzymatic apparatus of glycogen metabolism inside the cell. The chemical identification of the second messenger was facilitated by an individual mixture of intuition and chance. It was determined that the factor had to be a nucleotide consisting of equal portions of adenine, ribose, and phosphoric acid. However, an exact structural analysis was impossible with the methods available at that time because—despite tremendous efforts—only tiny amounts of the factor could be isolated.

An inquiry addressed to the nucleotide expert Leon A. Heppel (National Institutes of Health) about the nature of the second messenger went unanswered at the time. However, according to a rumor, Heppel was in the habit—said to be quite widespread among scientists—of collecting incoming mail in three-months portions before responding. Thus Sutherland's letter was in the same stack of paper as a letter from David Lipkin in St Louis, which reported a novel nucleotide that was obtained upon treatment of ATP with barium hydroxide (**Figure 4.6**). The nucleotide was cyclic adenosine 3′,5′-monophosphate (cAMP), which ultimately turned out to be identical with Sutherland's second messenger. The adrenaline-activated enzyme X in the liver cell membrane, catalyzing the formation of cAMP from ATP, was then called adenylate cyclase (see Section 4.4.1). For his epoch-making discovery, Sutherland was awarded the 1971 Nobel Prize in Physiology or Medicine.

Figure 4.6 Chemical generation of cyclic AMP from ATP

Summary

The first pathway of intracellular signal transduction was discovered during the investigation of stress-induced glycogen degradation in the liver. Upon interaction with the membrane-bound receptor adrenaline, the first messenger, was found to evoke the intracellular release of cAMP, the second messenger, produced from ATP by adenylate cyclase. In turn, cAMP stimulated the activation of glycogen phosphorylase by protein phosphorylation, catalyzed by a specific protein kinase and reversed by a protein phosphatase.

4.3 Trimeric G-proteins: coupling of receptors with the protein network

With the discovery of the second messenger mechanism and regulatory protein phosphorylation, Sutherland had identified two central principles of intracellular signal transduction. However, two questions still remained: how the adrenaline receptor activated adenylate cyclase, and how cAMP stimulated the phosphorylation of glycogen phosphorylase. A key observation was that the activation of adenylate cyclase required not only ATP as a substrate but, quite unexpectedly, also GTP, which became hydrolyzed to GDP and inorganic phosphate in the course of cAMP formation. A GTP-hydrolyzing enzyme was found to be interconnected between receptor and adenylate cyclase. This result led to the discovery of the regulatory guanosine triphosphatases (GTPases or G-proteins), which was acknowledged by the 1994 Nobel Prize in Physiology or Medicine, awarded to Martin Rodbell and Alfred G. Gilman. GTP hydrolysis is one of the major switching reactions for transmembrane signaling.

4.3.1 Transmembrane signaling by trimeric G-proteins

The GTPase discovered by Rodbell, Gilman, and coworkers proved to be a protein complex consisting of the three subunits $G\alpha$, $G\beta$, and $G\gamma$. It represents the prototype of the heterotrimeric or "large" G-proteins, the only function of which is to couple a particular class of membrane receptors to the signal-processing network of the cell. The GTPase activity is restricted to the $G\alpha$-subunit, being the G-protein proper. As typical for G-proteins, the interaction of $G\alpha$ with effector proteins such as adenylate cyclase is initiated by the exchange of bound GDP for GTP, catalyzed by a GDP–GTP exchange factor (GEF). The GEFs of the trimeric G-proteins are transmembrane receptors that are characterized by seven trans-membrane domains resembling the topography of rhodopsin. These so-called G-protein-coupled receptors (GPCRs), representing the largest of all receptor families in animals, are discussed in detail in Chapter 5.

The binding of GTP catalyzed by the receptor causes a conformational change of the $G\alpha$-subunit resulting in its dissociation from the $\beta\gamma$-dimer, which does not dissociate further (**Figure 4.7**). Upon dissociation, both the Gα- and Gβγ-subunits transmit input signals from the receptors to interacting effector proteins (see Figure 4.7). The major effector proteins of Gα subunits are the following:

- adenylate cyclases
- phosphatidylinositol (PI)-specific phospholipase C (PLC), type β
- cGMP-specific phosphodiesterase in retinal cells
- the GEF of the small G-protein Rho

Typical Gβγ-effector proteins include the following:
- some isoforms of adenylate cyclase
- PLCβ
- phosphatidylinositol 3-kinase (PI3K), type γ (see Section 4.4.3)
- β-adrenergic receptor kinase (see Section 5.7)
- $G_{i,o}$-regulated potassium channels (see Section 14.4.2)
- voltage-controlled calcium channels in presynaptic membranes of neurons (see Section 14.5.2)
- protein kinase Ste20, involved in pheromone signaling of yeast and probably related kinases of vertebrates (see Section 11.3)

Which of these proteins interacts and whether it becomes activated or inactivated depends on the particular isoforms of the Gα- or Gβγ-subunits in the G-protein (see Figure 4.7). G-proteins possess a built-in mechanism of signal termination: within a short time the Gα-subunit undergoes auto-inactivation due to its GTPase activity. As a consequence the trimeric G-protein reassociates and the βγ-dimer becomes inactivated as well.

Figure 4.7 Trimeric G-proteins: variability and mode of operation The upper part of the figure shows how G-protein subunits are anchored at the inner side of the cell membrane by tetradecanoyl (or palmitoyl) and farnesyl (or geranylgeranyl) residues (see also Figure 4.8). An activated receptor—of the heptahelical type only—acts as a GDP–GTP exchange factor (GEF) and induces dissociation into α- and βγ-subunits, which both interact with downstream effector proteins. These proteins are listed in the gray boxes, in the case of the Gα-subunits together with the related G-protein types (teal, inhibition; black, activation; the lists are not complete but include the major effector types). The GTPase activity of Gα renders the process reversible. The average molecular masses of the subunits are 31–52 kDa for Gα, 37 kDa for Gβ, and 8 kDa for Gγ.

Membrane anchoring

A characteristic structural feature of large G-proteins is that their subunits are anchored through covalently bound lipid residues at the inner side of the plasma membrane in immediate proximity to the receptor (see Figure 4.7). Membrane anchoring of signal-transducing proteins is not restricted to G-proteins but represents a general principle of organization in data-processing protein networks. For this purpose, cells employ several biochemical reactions resulting in post-translational protein modifications (lipidations) that enable membrane associations of different stabilities. The principle of anchoring by covalently bound lipids is briefly summarized in **Sidebar 4.2**.

Sidebar 4.2 Lipid anchors Through covalently bound long-chain lipids acting as hydrophobic "anchors," proteins may associate with membranes. This is an important condition for the *ad hoc* assembly of signal-transducing complexes, in particular in the vicinity of transmembrane receptors. The lipid anchor can be hidden inside the protein molecule, becoming reversibly exposed only by a signal-induced conformational change (examples are discussed in Sections 10.4.3 and 14.6.1). By this means, signals may control a reversible membrane anchoring of proteins. The following lipid anchors have been described (**Figure 4.8**).

(1) Amides of tetradecanoic acid (myristoyl anchor). Here the acyl residue activated by coenzyme A is bound by an amino (N)-terminal amino group, mostly of a glycine (Gly) residue in the sequence Gly-x-x-x-Ser. Frequently, positively charged amino acid residues

are found in the immediate vicinity. They promote association with the negatively charged phospholipids of membranes. This modification seems to be irreversible since as yet no enzymes have been found that hydrolyze the amide bond. Examples of proteins anchored by myristoyl residues are the α-subunits of trimeric G-proteins, the small G-protein Arf (see Section 10.4.3), and the tyrosine (Tyr) kinases of the Src family (see Section 7.2).

(2) Thioesters of palmitoic acid. The acyl residue is bound to the SH group of a (usually carboxy (C)-terminal) cysteine (Cys) residue. In contrast to the amide bond, the thioester bond can be hydrolyzed by the cell, so the modification is reversible. Palmitoylated proteins include α-subunits of trimeric G-proteins, the small G-protein Ras (see Figure 4.7 and Section 10.1), and some cytoplasmic protein kinases.

Figure 4.8 Lipid anchors
(A) Enzymatic prenylation, including the steps prenylation, removal of AAX, and carboxyterminal methylation (SAM, S-adenosylmethionine).
(B) Glycosylphosphatidylinositol anchor at the cell surface. The inositol ring is depicted in a simplified form. (C) Twofold anchoring of the small G-protein Ras at the inner side of the plasma membrane. P, phosphate group.

(3) Thioethers of oligoprenols such as farnesol (C_{15}) and geranylgeranol (C_{20}). The prenyl residues are bound to SH groups of C-terminal cysteine residues. The reaction starts from prenyl pyrophosphates and is catalyzed by protein prenyltransferases recognizing the consensus sequence protein-Cys-A-A-X (so-called CAAX box, with A symbolizing an aliphatic amino acid and X a C-terminal serine, threonine, methionine, leucine, alanine, or glutamine residue). After the prenylation of the Cys residue, the AAX tripeptide is removed by a prenylprotein peptidase, and the free carboxyl group of Cys becomes methylated by a highly specific prenylprotein carboxymethyltransferase that uses S-adenosylmethionine as a methyl donor (see Figure 4.8). Examples of proteins anchored by a prenyl residue are the γ-subunits of trimeric G-proteins, various small G-proteins such as Ras, and kinases phosphorylating GPCRs. Since the oncogenic effect of mutated Ras depends on prenylation, inhibitors of the enzymes involved are considered to be potential anti-cancer drugs.

(4) Amides with glycosylphosphatidylinositol. This rather complex structure called GPI anchor (see Figure 4.8) is used by cells to reversibly fix proteins at the *outer* surface of the cell membrane. Examples are acetylcholinesterase and cell-surface antigens.

Subfamilies

Like most signal-transducing proteins trimeric G-proteins exist in various isoforms. Theoretically, the 16 Gα, 5 Gβ, and 12 Gγ species expressed by mammalian cells together with numerous splice variants might combine to form more than 1000 individual G-proteins. Even if not all of them are realized, the degree of variability is almost unlimited considering the more than 1000 isoforms of GPCRs (see Section 5.2).[*] On the basis of sequence homologies of the α-subunits and their sensitivity to certain bacterial toxins, the following subfamilies have been defined (see Figure 4.7).

(1) **G_s family** (s = stimulatory): the α-subunit *stimulates* adenylate cyclase. This family also includes the olfactory G_{olf} proteins expressed only in the olfactory epithelium. G_s-proteins are selectively stimulated by cholera toxin (see Section 4.3.3).

(2) **$G_{i,0}$ family** (i = inhibitory; 0 refers to a subfamily in brain): the α-subunit *inhibits* adenylate cyclase. This family also includes the transducins or G_t-proteins of the retina and the G_{gust}-proteins of the taste cells. G_t- and G_{gust}-proteins activate cGMP-phosphodiesterase and phospholipase Cβ, respectively. With the exception of a G_z-subgroup, these G-proteins are selectively inhibited by pertussis toxin (see Section 4.3.3).

(3) **$G_{q,11}$ family**: the α-subunit stimulates phospholipase Cβ. This family includes several subgroups and is insensitive to cholera and pertussis toxin.

(4) **$G_{12,13}$ family**: the α-subunit activates, among others, some GEFs of the Rho family of small G-proteins (see Section 10.3.1).

Many G-proteins are expressed ubiquitously, whereas others are restricted to certain cell types and tissues: for instance, G_{t1} are found in retinal rod cells and G_{t2} in retinal cone cells. Most abundant are the $G_{i,0}$-proteins. They provide the major source of Gβγ-subunits, which (compared with the α-subunits) require considerably higher concentrations to become effective.

AGS proteins: receptor-independent activation of trimeric G-proteins

The only reliable rule of cellular signal transduction is that there is no rule without an exception. Thus, the concept of trimeric G-proteins being stimulated exclusively

---

[*]The nematode *Caenorhabditis elegans* has 20 genes encoding 17 G_i types. In contrast, in plants (at least in *Arabidopsis thaliana*) only a few signaling G-proteins have been found. Here signal processing seems to be based more on protein phosphorylation.

by receptors has been modified by the discovery of AGS proteins (<u>A</u>ctivators of <u>G</u>-protein <u>S</u>ignaling), which activate trimeric G-proteins independently of receptors. Three families have been identified, each operating through a particular mechanism.

(1) AGS1, representing the first family, is a *bona fide* nucleotide exchange factor that resembles, particularly for $G_{i,0}$-proteins, a GPCR. Interestingly, it is a small G-protein, belonging to the Ras superfamily (see Chapter 10).

(2) AGS proteins of the second family act as guanine nucleotide dissociation inhibitors by binding to and stabilizing the GDP form of Gα via a G-protein regulatory motif. They promote the release of active Gβγ from the trimeric complex, thus selectively stimulating Gβγ-controlled signaling while blocking Gα effects.

(3) AGS proteins of the third family form complexes with Gβγ. Their role in signaling is not yet known.

The wide distribution of these proteins indicates important functions in G-protein-dependent signal transduction that are not yet fully understood. Clearly, they bypass receptor-dependent pathways and possibly control G-protein subunits in cellular compartments that are devoid of receptors. An increasing body of evidence points to a role of AGS proteins in several aspects of neurotransmission. The discovery of AGS proteins indicates that G-protein-dependent signaling is considerably more complex and connected with other signaling pathways to a much higher degree than hitherto assumed.

4.3.2 Regulators of G-protein signaling: GTPase-activating proteins of trimeric G-proteins

Trimeric G-proteins undergo auto-inactivation due to the intrinsic GTPase activity of the α-subunits. In many cases, however, this reaction is not fast enough. For instance, auto-inactivation of the G-protein transducin that mediates light processing in the retina would take around 20 s, whereas an image fades almost immediately when we close our eyes. Therefore, cells strongly stimulate the intrinsic GTPase activity of their G-proteins. This function may be taken over by effector proteins via negative feedback (**Figure 4.9**). So, phospholipase Cβ is activated by the α-subunit of $G_{q,11}$-proteins, and at the same time stimulates the GTPase activity of that subunit. The same holds true for transducin and its effector protein cyclic GMP-phosphodiesterase.

Even more effective and versatile are special GTPase-activating proteins called **RGS** (<u>R</u>egulators of <u>G</u>-protein <u>S</u>ignaling). They constitute a large protein family found in all eukaryotic cells from yeast to humans. In yeast they counteract the effects of the mating pheromones as well as differentiation of the fermentation phenotype, both of which are mediated by GPCRs (see Section 5.2). Moreover, the deficiency of special RGS proteins leads to defects of reproduction and development in the mold

Figure 4.9 Regulation of trimeric G-protein activity (A) Dual control of Gα activity by a G-protein-coupled receptor (GPCR) acting as GEF and by a GTPase-activating effector (phospholipase Cβ, cGMP phosphodiesterase) and an RGS protein. Some RGS proteins may be also effectors of G-proteins. (B) Sharpening of the Gα signal by an RGS protein. (C) p115 Rho-GEF as an example of a multifunctional RGS protein with various interaction domains (DH, DBL homology domain, see Section 10.3.1; PH, pleckstrin homology domain; PIP$_2$, phosphatidylinositol 3,4-bisphosphate). More information on Rho and Rho-GEF is found in Section 10.3.

Aspergillus nidulans and to malformations of the nervous system in the nematode *Caenorhabditis elegans.* Such observations clearly prove a fundamental role of RGS in eukaryotes.

RGS proteins are quite heterogeneous. The only structural feature they have in common is the RGS domain that, upon interacting with a complementary domain of $G\alpha$-subunits, stimulates GTPase activity. This effect may be very specific; for instance, the isoform RGS9 reacts only with transducin. The RGS domain is not related to the GAP (GTPase-Activating Protein) domain interacting with small G-proteins.

Apart from an RGS domain, most RGS proteins have additional interaction domains and, therefore, couple GTPase activation with other effects, thus promoting signaling cross talk. Examples include the RGS protein **AKAP-2** (A-Kinase Anchor Protein 2) and the RGS of $G_{12,13}$ proteins. AKAP-2 anchors protein kinases A (PKA) and C (PKC) and other signal-transducing proteins at their cellular sites of action (see also Section 4.6.1), while the $G_{12,13}$ RGS is not only a GTPase activator but, in addition, a $G_{12,13}$-effector protein that stimulates as a GDP–GTP exchange factor the small G-protein Rho (see Figure 4.9).

Many RGS proteins owe a special property to this multifunctionality: they facilitate not only the switching-off but also the switching-on mechanism of G-proteins, that is, the effect of the receptor on GDP–GTP exchange. By this means the $G\alpha$-signal is "sharpened" (see Figure 4.9B), becoming less sensitive to accidental noise and suited for a more rapid and possibly frequency-modulated transmission.

4.3.3 Diseases caused by disturbances of trimeric G-protein signaling

Considering the central position of trimeric G-proteins in signal transduction across cell membranes, it is conceivable that deregulation of G-protein signaling may have serious consequences. A classic example is cholera, caused by the bacterium *Vibrio cholerae.* The major symptom of this disease is an extreme loss of fluid across the intestinal epithelium: instead of normally 2 L, 20–30 L is excreted per day, resulting in high mortality. The **cholera toxin** is an oligomeric protein interacting with gangliosides at the cell surface. As a result, a subunit expressing ADP-ribosyltransferase activity is released into the cell. There this enzyme specifically catalyzes the ADP-ribosylation of an Arg residue in the active center of $G\alpha_s$-subunits, inhibiting their intrinsic GTPase activity. This effect leads to a continuous activation of the G-protein and its effector adenylate cyclase, and the cell is swamped with cAMP, which stimulates the export of fluid across the cell membrane (a detailed discussion of the mechanism is found in Section 14.7.3). To exert its effect, cholera toxin has to cooperate with a cellular protein, the **ADP-ribosylation factor (Arf)**. This is is a small G-protein normally regulating the formation of secretory vesicles (see Section 10.4.3). Cholera toxin is highly specific and, therefore, experimentally used to distinguish effects of G_s-proteins from those of other G-proteins. The heat-labile enterotoxin of pathogenic *Escherichia coli* strains has a similar effect as cholera toxin, while the heat-stable enterotoxin of the same organism stimulates an overproduction of cyclic GMP (cGMP) with similar consequences (see Section 14.7.3).

Another bacterial protein disturbing G-protein signaling is **pertussis toxin**, a component of a whole toxin cocktail of *Bordetella pertussis,* the bacterium causing whooping cough. Like cholera toxin, pertussis toxin is an ADP-ribosyltransferase; it selectively attacks a Cys residue in $G\alpha_{i,0}$-subunits thus interrupting the interaction with the receptor. As a result $G_{i,0}$ effects are suppressed. In cells expressing G_s-proteins, this leads to an overactivation of adenylate cyclase (which normally is inhibited along the $G_{i,0}$ pathway) and, as in the case of cholera toxin, a pathogenic accumulation of cAMP. This effect is strongly augmented by a toxic adenylate cyclase of *B. pertussis.* In addition, all signaling reactions mediated by Gβγ-subunits, which are predominantly released from $G_{i,0}$-proteins, become blocked.

Examples of other diseases related to disturbances of trimeric G-protein signaling include the following:

- a form of night blindness due to a genetic defect of the α-subunit of transducin

- McCune–Albright syndrome, characterized by defects of hormonal regulation. It is caused by a gain-of-function mutation of $G\alpha_s$ that correlates with massive disturbances of G_s-controlled hormone production

- pseudohypoparathyroidism, caused by a loss-of-function mutation of $G\alpha_s$, as a result the effects of parathyroid hormone (that interacts with a G_s-coupled receptor) are suppressed leading to premature puberty with obesity, growth retardation, and skeletal deformities

Summary

- Trimeric (large) G-proteins are major devices of transmembrane signaling. Upon interaction with an activated receptor, which functions as a guanine nucleotide exchange factor (GEF), G-proteins dissociate into α- and βγ-subunits, both of which are signal transmitters that interact with second-messenger-producing enzymes or ion channels. Depending on the downstream coupling, several subfamilies of trimeric G-proteins are distinguished. In addition to transmembrane receptors, certain trimeric G-proteins are also activated by AGS proteins.

- Trimeric G-proteins may undergo auto-inactivation via their intrinsic GTPase activities. Frequently, this process is stimulated in addition by regulators of G-protein signaling (RGS proteins). Owing to their interaction domains, RGS proteins enable cross talk of trimeric G-proteins with a wide variety of other signaling reactions.

- The α-subunits of trimeric G_s- and $G_{i,0}$-proteins are attacked in a highly selective manner by bacterial toxins exhibiting ADP-ribosyltransferase activity. While cholera toxin renders $G\alpha_s$-subunits constitutively active by inactivating the GTPase activity, pertussis toxin suppresses $G_{i,0}$ signaling by interrupting the receptor/G-protein interaction. Both effects are used for a characterization of G-protein functions.

4.4 Downstream of G-proteins: enzymes producing second messengers

Trimeric G-proteins couple a large number of membrane receptors to the data-processing protein network of the cell. They do this mainly by controlling the activities of enzymes that catalyze the generation of second messengers (**Figure 4.10**). These effector enzymes of trimeric G-proteins are:

- adenylate cyclases, producing cAMP

- PI3Ks, producing phosphatidylinositol 3,4-bis- and 3,4,5-trisphosphate (PIP_2 and PIP_3)

- PLCs producing diacylglycerol (DAG) and inositol 1,4,5-trisphosphate ($InsP_3$)

Additional effects of trimeric G-proteins are discussed later in the chapter.

4.4.1 Adenylate cyclases of mammalian cells

Among the effector enzymes of trimeric G-proteins, the adenylate (or adenylyl) cyclases (ACs) were discovered first. They catalyze the hydrolysis of ATP to cAMP and pyrophosphate. The latter is immediately hydrolyzed by a cooperating

Figure 4.10 Second messenger generation controlled by trimeric G-proteins: an overview Upon activation by membrane-bound receptors, the G-protein subunits change the activity of several enzymes. As a result, cellular levels of the second messengers shown in the teal box are up- or downregulated.

pyrophosphatase to two molecules of inorganic phosphate, rendering the overall reaction exergonic.

As explained in Chapter 3, adenylate cyclases are ancient enzymes found in almost all cells from the simplest prokaryotes to humans. In the course of evolution they have developed a bewildering diversity of structures (**Sidebar 4.3**). The adenylate cyclases of vertebrates are large proteins (up to 140 kDa). Their overall structure, consisting of 2—6 transmembrane domains and a catalytic center with the two ATP-binding sites on the cytoplasmic side (**Figure 4.11A**), is reminiscent of that of a membrane transporter with an ATP-binding cassette (ABC transporter; see Section 3.3.1), although the two classes of proteins are only distantly related.

The mammalian organism expresses at least ten isoforms that are tissue-specifically distributed. The cyclases AC1–AC9 are integrated as dimers or oligomers in the plasma membrane, where they form signal-transducing complexes with G-proteins, receptors, and ion channels. In contrast, AC10 is a cytoplasmic enzyme (soluble AC) that lacks the transmembrane domains characteristic for the membrane-bound adenylate cyclases. Moreover, AC10 is controlled by bicarbonate ions rather than by G-proteins. The enzyme is ubiquitously expressed and plays a role in bone formation and neuronal metabolism.

As shown in Figure 4.11, all membrane-bound isoforms are activated by the α-subunits of trimeric G_s-proteins; however, they differ in their sensitivity to other signals such as $G\alpha_i$- and $G\beta\gamma$-subunits and Ca^{2+} ions (as calmodulin complexes). In fact,

Figure 4.11 Mammalian transmembrane adenylate cyclases (A) Membrane topology of the two cyclase domains C1 and C2 including the ATP-binding site. In reality, the active enzyme is a dimer or oligomer. Forskolin is a plant-derived adenylate cyclase activator. (B) List of nine adenylate cyclase (AC) isoforms with regulatory factors: (+) stimulation, (−) inhibition, (0) no effect, (?) doubtful or not tested.

	$G\alpha_s$	$G\beta\gamma$	$G\alpha_i$	Ca^{2+}/calmodulin
AC1	+	−	−	+
AC2	+	+	0	0
AC3	+	−	−	+
AC4	+	+	0	0
AC5	+	0	−	0
AC6	+	0	−	?
AC7	+	+	0	0
AC8	+	−	−	+
AC9	+	?	?	0

Sidebar 4.3 "Eukaryotic" adenylate and guanylate cyclases In addition to the adenylate cyclases of Gram-negative bacteria and the toxic adenylate cyclases of certain pathogenic microbes discussed in Section 3.3.2, the "eukaryotic" adenylate cyclases constitute a separate family. Because "eukaryotic" cyclases have been found also in prokaryotes, the term "adenylate cyclases type III" is now preferred. The common structural element of these enzymes is a catalytic cyclase homology domain. In the membrane-bound adenylate cyclases of metazoans, two such domains are combined with 2 × 6 transmembrane domains (**Figure 4.12**), a membrane topology resembling that of ABC transporters, although those are not closely related to cyclases. Soluble AC10 of metazoans lacks these transmembrane domains. The adenylate cyclases of prokaryotes and unicellular eukaryotes have many additional interaction domains. The bewildering variability of these domains indicates an extended cross talk of adenylate cyclases with other signaling pathways and provides a particularly instructive example of the combinatorial possibilities of protein modules (see Figure 4.12). Cyclase homology domains are also found in guanylate cyclases, which are counted as type III cyclases although they differ completely from adenylate cyclases in other structural parameters.

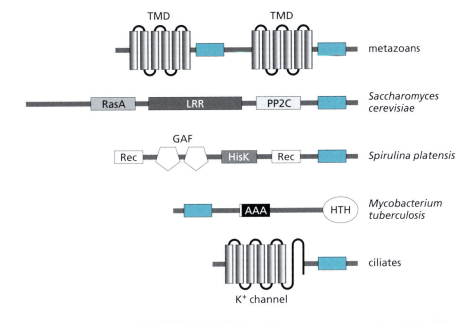

Figure 4.12 Variability of adenylate cyclases: some examples Depending on the species, the cyclase homology domain (teal) is combined with various interaction domains. TMD, hexahelical transmembrane domain of metazoan adenylate cyclases (Figure 4.11); RasA, binding domain for the small G-protein Ras of *Saccharomyces cerevisiae*; LRR, leucine-rich repeats; PP2C, protein phosphatase 2C domain; HisK and Rec, His autokinase and Asp receiver domains of a two-component system; GAF, interaction domain for various small signaling molecules (see Sections 3.3.2 and 4.4.2); AAA, triple-A domain with ATPase function (see Section 2.5.2); HTH, helix–turn–helix motif of a DNA-binding site. In ciliates the cyclase domain is linked to the hexahelical transmembrane structure of a voltage-dependent potassium channel.

the individual isoforms of adenylate cyclases provide particularly clear examples of logical gates. AC1 constitutes an AND gate or coincidence detector for signals acting via $G\alpha_s$ AND Ca^{2+}/calmodulin and a BUT NOT gate with $G\alpha_{i,0}$-activating signals. On the other hand, AC2 may be activated by G_s- OR $G_{i,0}$-proteins (via $G\beta\gamma$) since it is not inhibited by $G\alpha_i$. These differences in regulation have important physiological consequences, as shown by the following examples.

● Isoforms AC5 and AC6, expressed predominantly in heart muscle cells, are activated via the G_s pathway by the sympathetic signal noradrenaline and inhibited via the $G_{i,0}$ pathway by the parasympathetic signal acetylcholine. By this means heart activity is either enhanced or reduced (for details, see Section 5.5). In addition, the influx of Ca^{2+} accompanying each heart muscle contraction causes an inhibition of AC5 and perhaps also of AC6, establishing a negative feedback loop.

- In the central nervous system, Ca^{2+} controls cellular processes thought to be involved in learning and memory (see Section 16.4). The predominant adenylate cyclases in brain neurons are AC1 and AC8, both of which are activated by Ca^{2+}/calmodulin, and AC9, which is inhibited by calcineurin, a Ca^{2+}-dependent protein phosphatase (see Section 14.6.4). These adenylate cyclases are strongly expressed during memory training and, therefore, may play a key role in learning processes, in particular by mediating calcium- and cAMP-dependent gene expression required for strengthening of synaptic contacts.

- In neurons, but also in heart muscle cells, the calcium-sensitive adenylate cyclases are found in close proximity to voltage- and ligand-controlled Ca^{2+} channels such as the L-type channel and the NMDA receptor (see Sections 14.5.2 and 16.4, respectively), ensuring particularly rapid interaction.

In fact, adenylate cyclases are scaffolds recruiting a wide variety of signal-transducing proteins that are under the control of cAMP or regulate the cAMP level. In the so-called microdomains (see next Section, Sidebar 5.4) surrounding such complexes a steep gradient of cAMP is observed indicating that cAMP production occurs in a strong field of cAMP breakdown (see Section 4.4.2). Thus, cAMP provides an excellent example for the "privacy" and the transitoriness of intracellular signaling (see Section 1.6).

Contrary to many other signal-transducing proteins, almost no drugs and only a few poisons are known to act on adenylate cyclases. An exception frequently used for experimental purposes is the diterpene **forskolin** (see Figure 4.11) from the plant *Coleus forskohlii*, which activates all isoforms except AC9 and the soluble AC10. The cyclases bind forskolin at a pseudosubstrate site that in the absence of forskolin (or a stimulatory G-protein) blocks the active center. An endogenous factor fitting this site has been postulated but not yet identified. In Indian folk medicine, extracts from *C. forskohlii* roots have been used against asthma, heart disease, and thyroid insufficiency.

4.4.2 Phosphodiesterases and the importance of signal extinction

The efficacy of a signal-processing apparatus critically depends on signal extinction, since otherwise the system would sink rapidly in a flood of data, becoming paralyzed. Signal extinction may be a signal by itself, since for data processing it is irrelevant whether a binary process proceeds from 0 to 1 or from 1 to 0, or whether a switch is turned to the left or to the right. Therefore, for the cell, signal generation and signal extinction are equivalent events that are both controlled by input signals. This is demonstrated clearly by the cyclic nucleotides: in the liver, the *generation* of cAMP induces glycogenolysis, whereas in the retina, the *degradation* of cGMP causes an impulse in the visual nerve.

The termination of cAMP and cGMP signals is achieved by phosphodiesterases catalyzing their hydrolysis to inactive nucleoside 5′-monophosphates. By controlling (together with the cyclases) the cellular level of cyclic nucleotides, the phosphodiesterases thus adopt a key position in signal processing (**Figure 4.13**). Pharmaceuticals interacting with phosphodiesterase activities are widely used in the clinic, and considerable efforts are made to expand this field.

In mammalian cells phosphodiesterases are encoded by no less than 20 genes, and by alternative splicing and related mechanisms more than 50 variants may be produced. These isozymes are grouped in 11 subfamilies on the basis of sequence homologies, tissue distribution, and substrate specificity. The fact that there are considerably more phosphodiesterases than cyclases emphasizes the importance of these enzymes in signal transduction. One distinguishes between cAMP-specific

Figure 4.13 Generation and termination of the cAMP/cGMP signal The enzymes involved, cyclases and phosphodiesterases, are expressed in numerous isoforms controlled by various input signals.

phosphodiesterases (subfamilies 4, 7, and 8), cGMP-specific phosphodiesterases (subfamilies 5, 6, and 9), and enzymes that hydrolyze both cyclic nucleotides (subfamilies 1, 2, 3, 10, and 11).

As is typical for signal-transducing proteins, phosphodiesterases exhibit a modular structure and mostly exist as homodimers. The catalytic domain common to all isoforms contains two Zn^{2+} ions. In contrast to the catalytic domain, the regulatory domains are highly variable, reflecting the equally high variability of input signals. In fact, the phosphodiesterases are regulated as strictly as the cyclases. Like protein kinases they are intrasterically inhibited in the absence of an input signal, because an auto-inhibitory sequence blocks the catalytic center. This inhibitory sequence may be a part of the enzyme molecule or may exist as a separate subunit as, for example, in the case of the phosphodiesterase isoform PDE6 in the retina. Phosphodiesterase-activating input signals overcome the auto-inhibition by inducing an allosteric change of protein conformation. With the exception of PDE6, which seems to be the only phosphodiesterase controlled by a trimeric G-protein (transducin), such signals are Ca^{2+} ions, cGMP, and protein phosphorylation.

Here are some examples:

- PDE1 is widely distributed and stimulated by Ca^{2+}/calmodulin, thus connecting Ca^{2+}-dependent and cyclic nucleotide-dependent signaling.

- PDE2 is allosterically activated by cGMP. The cGMP-binding site in the regulatory domain is known as **GAF domain,** belonging to one of the largest and most ancient domain families for interaction of proteins with smaller molecules such as cGMP (see Sidebar 3.3 in Section 3.4.2). Its structure differs from that of the cGMP-binding site in the catalytic center. PDE2 plays an important role in the hormonal regulation of blood volume and blood pressure. In the adrenal cortex, the pituitary gland hormone ACTH (adrenocorticotropic hormone) activates adenylate cyclase via a G_s-protein-coupled receptor, and the cAMP generated in turn promotes the production and release of the mineralocorticoid aldosterone. By acting on the kidney, the latter causes an increase of blood volume and blood pressure. These effects are counteracted by the natriuretic peptide ANP released from atrial heart cells. As will be discussed later, the ANP receptor in the adrenal cortex is a receptor guanylate cyclase producing cGMP that activates PDE2. PDE2 degrades cellular cAMP and as a result the release of aldosterone ceases.

- PDE3 and PDE4 are widely distributed. Both are activated by phosphorylation catalyzed by the cAMP-dependent protein kinase A. Accordingly, they are negative feedback inhibitors in cAMP-controlled pathways such as glycogenolysis and lipolysis. Another protein kinase that activates PDE3 is PKB/Akt. This enzyme is a component of signaling cascades stimulated by insulin, leptin, and other hormones (see Section 4.6.6). In contrast to PDE2, PDE3 is competitively inhibited rather than stimulated by cGMP. PDE3 and PDE4 are of particular clinical interest since both are targets of pharmaceuticals that inhibit the enzymes thus evoking cardiotonic, vasodilatory, and thrombolytic effects.

- PDE5 is allosterically activated by cGMP. Its major function is the regulation of vascular tone, being an antagonist of nitrogen monoxide, NO. Using cGMP as a second messenger NO induces vasodilatation (see Section 16.5). The result is an increased blood flow that causes, among other effects, penile erection. This response is dampened by PDE5 catalyzing degradation of cGMP. PDE5 has gained considerable attention since inhibitors such as Viagra (sildenafil), which prolong the relaxation phase of vascular smooth muscles, are widely used as erection-promoting drugs.

- PDE6 is more or less restricted to retina cells where it occupies a central position in visual signal processing (see Chapter 15).

With the exception of PDE8 and PDE9, all PDE isoforms are inhibited by methylxanthines such as caffeine, theophylline, and isobutylmethylxanthine which occupy the cyclic nucleotide binding domains. Many of the effects of these drugs are thus explained. However, additional cellular targets have been identified, in particular for caffeine (see Section 16.9.1).

4.4.3 Phosphatidylinositol kinases

Cyclic nucleotides are not the only second messengers released along G-protein-controlled pathways. In fact, various intracellular signaling molecules are produced from membrane phospholipids, with phosphatidylinositols (PIs) being a particularly versatile source. Although PIs comprise only about 10% of membrane phospholipids, they are of great importanc for signal processing. There are two mechanisms by which membrane-bound signaling molecules are generated from these phospholipids: phosphorylation of the inositol ring and hydrolysis (see Section 4.4.4). Phosphorylation is catalyzed by specific phospholipid kinases that use ATP as a substrate and attack positions 3, 4, and 5 of the inositol ring, thus yielding phosphatidylinositol phosphates (PIPs), also known as phosphoinositides (**Figure 4.14**). PI phosphorylation is reversed by a series of lipid phosphatases. As a result of these enzymatic reactions in mammalian cells, at least seven types of PIPs have been found that differ in the position and degree of phosphorylation. These relationships are summarized in **Figure 4.15**.

A characteristic property of these lipids is that they are able to generate specialized membrane domains where they may recruit signal-transducing proteins via specific interaction domains. These proteins thus become concentrated and activated, forming signal-transducing complexes (signalosomes) at specific sites. Examples of membrane-targeting protein domains interacting with PIPs are:

- **PH domain** (Pleckstrin Homology domain; pleckstrin is a functionally ill-defined protein kinase C substrate protein in blood platelets). This domain is the most abundant PIP-binding motif in human cells. It is absolutely specific for 3-phosphorylated PIPs which are stereospecifically bound. Because of this

Figure 4.14 Points of attack of phosphatidylinositol kinases Membrane-bound phosphatidylinositols are a source of various signaling molecules generated by signal-controlled phospholipid kinases that catalyze the ATP-dependent phosphorylation of the inositol ring at the positions shown. Frequently phosphatidylinositols contain an arachidonic acid residue in position 2 of the glycerol moiety. The inositol ring is shown in a simplified form (A). A stereochemically correct form is shown in part B.

property, proteins with PH domains are major targets of signaling pathways that lead to an activation of phosphatidylinositol 3-kinases. PH domains exist in various isoforms. Only a minority of them recognize PIPs, while the majority are engaged in protein–protein interactions. Many G-proteins, for instance, bind to PH domains of their partner molecules.

Figure 4.15 Biosynthesis of phosphorylated phosphatidylinositols The six-membered ring of inositol (here shown in a simplified form) is esterified by phosphatidic acid (PA) and phosphoric acid (P, teal). The phosphorylation is catalyzed by various lipid kinases and results in phosphatidylinositol mono-, bis-, and trisphosphates (PIP, PIP_2, and PIP_3). Monophosphates are generated by PI3 kinases, PI4 kinases, and PI5 kinases, respectively. The monophosphates are further phosphorylated by type I PIP kinases (positions 4 and 5); type II PIP kinases (position 4); and type III PIP kinases (position 5). Phosphatases that exhibit corresponding specificities render the phosphorylations reversible (not shown).

- **FYVE domain** (named after the four proteins Fab1, YOTP, Vac1, EEA1 where this domain was originally discovered). This domain binds selectively to PI 3-monophosphate and is particularly enriched in endosomal proteins.

- **PX domain** (<u>PhoX</u> homology domain; the name refers to the enzyme NAD<u>PH</u> <u>ox</u>idase) interacting with PI 3-mono-, PI 3,5-bis-, and PI 3,4,5-trisphosphates.

A common structural feature of such domains is a binding pocket lined by basic amino acids residues that electrostatically interact with the phosphate groups of PIPs. Specificity is provided by additional amino acid residues. Beside PIP-specific domains membrane-targeting protein motifs are known that recognize other lipids. Examples are found in Sections 4.6.3 and 14.6.1. More unspecific membrane targeting is achieved, in addition, by a binding of lipid residues to proteins (see Section 4.3.1).

Membrane targeting plays an essential role for the recruitment of signal-processing protein complexes to transmembrane receptors as well as for protein sorting within the different membrane compartments of the cell. Owing to strict compartmentalization of lipid kinases and phosphatases, the PIPs are distributed specifically rather than randomly among the cellular membranes. So PI 3-monophosphate is found predominantly in endosomes recruiting proteins with FYVE domains, whereas the PI 4-monophosphate prefers the Golgi apparatus, and the bisphosphates are enriched in the plasma membrane. The energy-consuming lipid phosphorylations generate biochemical gradients that together with protein phosphorylation, GTP hydrolysis, and other switching reactions make cellular processes irreversible, pushing them in one direction. This becomes particularly evident in strictly targeted exo- and endocytotic vesicle transport, which is regulated by PIPs (for details see Section 10.4.4).

Phospholipases C degrade PIPs to diacylglycerols and inositol phosphates, with the 3,4,5-trisphosphate ($InsP_3$) being the most prominent example, whereas almost nothing is known about the putative functions of other inositol phosphates.

Phosphatidylinositol 3-kinases (PI3Ks)

Among the phospholipid kinases, the phosphatidylinositol 3-kinases are most prominent and best investigated because they occupy a key position in cellular signal processing. The major reaction catalyzed by these enzymes is the phosphorylation of PI 4-mono- and 4,5-bisphosphates, produced by position 4- and 5-specific kinases. The products are phosphatidylinositol 3,4-bis- and 3,4,5-trisphosphate (PIP_2 and PIP_3). Both stay in membranes, acting as docking sites for proteins with PH domains. The enzymes of the PI3K family are, therefore, of great importance for reversible association of proteins with membranes and for the formation of signal-transducing protein complexes in the vicinity of membrane-bound receptors.

On the basis of structural features, the mammalian phosphatidylinositol 3-kinases have been subdivided into three families called PI3K-I, -II, and -III. The PI3K-I family, including the four isozymes PI3Kα, -β, -γ, and -δ, has been investigated in more detail. They have a C-terminal lipid kinase domain in common as well as a C2 domain that facilitates membrane association in a Ca^{2+}-dependent manner (for details see Section 4.6.3)*.

Reversible recruitment to membranes is a critical step in PI3K activation because these cytoplasmic enzymes would not otherwise have access to their phospholipid

---

*Although showing almost no homology with eukaryotic protein kinases, the catalytic domains of PI3Ks exhibit both lipid kinase and protein kinase activity. Therefore, these enzymes, together with the PI3K-related kinases (see Section 12.8), are classified as **atypical protein kinases**. In fact, in cell culture PI3Ks may undergo autophosphorylation and catalyze the phosphorylation of other proteins.

Figure 4.16 Domain structure and regulation of phosphatidylinositol 3-kinases Interaction of an α, β, or δ isoenzyme with a p85 adaptor is shown (AdBD, adaptor-binding domain; RasBD, Ras-binding domain; PRO, proline-rich motif for interaction with SH3 domains; IRS, insulin receptor substrate; GAB, Grb2-associated binder). In addition, the domain structure of the γ isoform, including the two sites for interaction with Gβγ-subunits, is depicted. Wortmannin is a fungicide from *Penicillium wortmannii*. As a semispecific inhibitor of some PI3K isoforms, it has found broad—and sometimes rather uncritical—application in research.

substrates. Membrane attachment is supported by additional structural elements of PI3Ks. These are binding sites for membrane-bound proteins such as the small G-protein Ras, Gβγ-subunits, and certain adaptor proteins (**Figure 4.16**).

For instance, **PI3Kγ** contains two Gβγ-binding sites (see Figure 4.16) and therefore is primarily activated by signals processed by $G_{i,0}$-protein-coupled receptors (remember that $G_{i,0}$-proteins are the major source of Gβγ-subunits). **PI3K**α, -β, and -δ are under the control of Tyr kinase-coupled receptors including those of insulin, growth factors, cytokines, antigens, and components of the extracellular matrix (Chapter 7). Activation requires membrane recruitment of PI3K by Ras that is stimulated along such pathways (**Figure 4.17**).

For an interaction with Ras the Ras-binding site of PI3K must be exposed by a conformational change induced by a binding to the receptor. Receptor binding requires adaptor proteins forming complexes with PI3Ks. These adaptors contain binding sites (SH2 domains) for proteins with phosphorylated Tyr residues such as

Figure 4.17 Signaling pathways leading to activation of phosphatidylinositol 3-kinases PI3Ks catalyze the phosphorylation of membrane-bound phosphatidylinositol 4-monophosphate and 4,5 bisphosphate at position 3 of the inositol residue (here depicted in a simplified form as a six-membered ring). The isoform PI3Kγ is activated by Gβγ- subunits derived mainly from $G_{i,0}$-proteins coupled to heptahelical receptors, whereas the isoforms PI3Kα, -β, and -δ are activated by tyrosine kinase-coupled transmembrane receptors in cooperation with active Ras-GTP. Phosphorylated sites are symbolized by colored dots.

autophosphorylated receptors (see Section 7.1.2). In Figure 4.16 the domain structure of the p85 adaptor associating with PI3Kα is depicted. Beside an SH2 domain the protein exhibits an SH3 domain as well as a proline-rich domain as a docking site for other signal-transducing proteins associating to signalosomes.

It is easily conceivable that PI3Ks are multifunctional relay stations for a large number of extracellular signals, in particular for those stimulating cell growth and cell proliferation. In fact, a major function of the PI3K signaling pathway is to coordinate the rates of cell growth and cell proliferation and to adjust them to the energy and food supply to ensure the survival of the cell. Cell proliferation critically depends on cell growth, since cell division requires a certain cellular size.

The antagonists of PI3Ks are the **phospholipid phosphatases of the PTEN family** (the name refers to a Phosphatase-deleting mutation of chromosome TEN). These enzymes are tumor suppressor proteins that are deleted in many neoplastic diseases in humans. The tumor-suppressing effect is easily understood: PTENs counteract the proliferative and survival-promoting effects of PI3K signaling by reversing the phosphorylation of PIs in position 3. Accordingly, tumor cells with deleted PTEN respond more sensitively to mitogenic signals and are rescued from apoptosis. Since PTENs belong to the superfamily of protein phosphatases, they will be discussed together with similar enzymes in Section 7.4.

4.4.4 Phosphatidylinositol-specific phospholipases C

In addition to phosphorylation, enzymatic hydrolysis of phosphatidylinositols provides a pathway for second messenger production. The hydrolysis is catalyzed by phospholipases that are grouped in several families according to the preferred splitting site in the phospholipid molecule (**Figure 4.18**).

Phospholipases C (PLCs) are the producers of the second messengers diacylglycerol (DAG) and inositol 1,4,5-trisphosphate (InsP$_3$). While DAG stays in the membrane serving as a docking site and activator of protein kinase C and some other enzymes (see Section 4.6.3), InsP$_3$ is released into the cytoplasm, where it gates Ca^{2+} channels in the endoplasmic reticulum (see Section 14.5.3). Thus, signals stimulating PLC activity increase the cytoplasmic Ca^{2+} level with far-reaching physiological

Figure 4.18 Hydrolysis of a phosphatidylinositol phosphate by phospholipases Membrane-bound phosphorylated phosphatidylinositols such as the 4,5-bisphosphate shown are a source of various signaling molecules generated by signal-controlled phospholipases. These include phospholipase A$_2$, producing arachidonic acid (the precursor of a wide variety of eicosanoids, see Section 4.4.5); phospholipase C, generating the second messengers diacylglycerol (DAG) and inositol 1,4,5-trisphosphate (InsP$_3$); and phospholipase D, catalyzing the formation of phosphatidic acid.

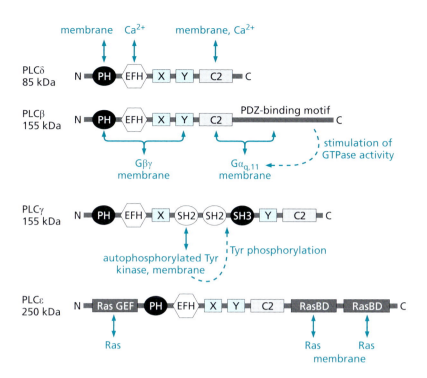

Figure 4.19 Phosphatidylinositol-specific phospholipases C Shown are the domain structures, interacting partners (teal), and sites of membrane recruitment of the four PLC subfamilies of mammals. PLCβ interacts with and becomes activated by Gα$_{q,11}$- and Gβγ-subunits and, in turn, stimulates the GTPase activity of Gα$_{q,11}$. Through its SH2 domains PLCγ binds to autophosphorylated Tyr kinases such as the receptors of insulin and growth factors and becomes activated by phosphorylation. In their domain structures PLCη largely resembles PLCβ, while PLCζ resembles a truncated PLCδ lacking the PH domain (not shown). X, Y (teal), catalytic domains; EFH, EF-hand domain; RasBD, Ras-binding domain, for PDZ domain see Sidebar 7.2. Two-headed arrows indicate binding.

consequences. Particularly in neurons, PLCs have an additional important function: by degrading phosphatidylinositol 3,4-bisphosphate, they control the function of a class of nonspecific cation channels known as TRP channels, some of which are inhibited, while others are activated by the phospholipid. As we shall see in Section 14.5.6 and Chapter 15, TRP channels play a key role in sensory signal transduction, fertilization, and calcium homeostasis.

Phospholipase C enzyme family

Phosphatidylinositol-specific PLCs are Ca^{2+}-dependent cytosolic enzymes. Mammalian cells express at least 13 isoforms as well as a series of splice variants. On the basis of structural similarities, they are divided into the subfamilies PLCβ (four isoforms), PLCγ (two isoforms), PLCδ (three isoforms), PLCε (one isoform), PLCζ (one isoform), and PLCη (two isoforms). PLCα, originally thought to be a separate isoform, is now considered to be a degradation product of PLCδ1. In simple eukaryotes such as yeast, only type δ is found; therefore, it is considered to represent the primeval form of PLC.

As shown in **Figure 4.19**, PLCs share a catalytic domain separated into two areas X and Y, and an EF-hand domain for Ca^{2+} binding (see Section 14.6.1). The active conformation of the catalytic domain is. stabilized by Ca^{2+} ions forming electrostatic bonds with acidic amino acid residues.

As cytoplasmic enzymes, PLCs are confronted by the same problem as the P13Ks: they must be recruited to the cell membrane to come in contact with their membrane-bound substrates. This recruitment is a key event in enzyme activation. It is achieved by PH and C2 domains found in all isoforms and—depending on the isoform—by mechanisms based on an interaction with membrane-anchored activators. These interactions reflect the integration of PLCs in a complex network of signaling pathways.

● PLCβ is recruited and activated by the membrane-anchored α-subunits of G$_{q,11}$-proteins as well as by Gβγ-subunits (**Figure 4.20**). This means that the second messengers DAG and InsP$_3$ are released not only along the G$_{q,11}$-pathways but also through an activation of G$_{i,0}$-proteins, which are the major source of Gβγ-subunits in the cell. On the other hand, PLCβ stimulates the GTPase activity

Figure 4.20 Signaling pathways leading to activation of C-type phospholipases Individual PLC isoforms are activated by the membrane receptors and ion channels shown to the left. They all catalyze the hydrolysis of phosphatidylinositol 4,5-bisphosphate, resulting in the formation of DAG and InsP$_3$ as well as in an altered function of TRP cation channels (phosphorylated sites are symbolized by teal dots).

of G$\alpha_{q,11}$, and thus inhibits its own activation acting as a negative feedback regulator. Among the isoforms, PLCβ1 and -3 are widely distributed, whereas PLCβ2 is restricted to the hematopoietic system and PLCβ4 to retina and nerve cells.

- PLCγ is activated by Tyr phosphorylation rather than by G-proteins. Tyr phosphorylation is catalyzed by either receptor or cytoplasmic Tyr kinases (see Figure 4.20). For this purpose the PLC interacts via its SH2 domains (see Figure 4.19) with the autophosphorylated kinase. The mechanism of activation resembles that of PI3Kα, -β, and -δ, but does not seem to require the assistance of adaptor proteins.

- PLCδ responds quite sensitively to Ca^{2+}. Thus, an increase in the cytoplasmic Ca^{2+} concentration as induced by various signals is probably sufficient for activation. Membrane binding is achieved by the C2 domain interacting with phosphatidylserine and Ca^{2+} and by the PH domain binding to PIP$_3$.

- PLCε plays a special role, because it is both an activator (acting as a GEF) and an effector protein of the small G-protein Ras. It becomes activated along the Ras signaling pathways. Since Ras is membrane-anchored, the interaction results in membrane recruitment of PLCε, resembling the effect of Ras on PI3K. The role of PLCε in signal transduction is discussed in more detail in Section 10.1.4.

- PLCζ, an extremely Ca^{2+}-sensitive enzyme, is found only in mammalian sperm. In the course of fertilization it is injected by the sperm into the egg cell and releases InsP$_3$ and Ca^{2+}, thus stimulating as a "kiss of life" metabolism and mitosis. Loss of function mutations cause male infertility. Research on PLCζ is of considerable medical interest: the enzyme is expected to facilitate artificial fertilization and the production of embryonic stem cells. However, how this phospholipase becomes activated is not known yet.

- PLCη is expressed almost exclusively in the central nervous system. Since it is the most Ca^{2+}-sensitive PLC, the enzyme is assumed to act as neuronal calcium sensor that is able to respond to minute changes in the cytoplasmic Ca^{2+} concentration. Its physiological functions are still a matter of conjecture.

4.4.5 Lipid mediators are intra- and extracellular signals

As we have seen, membrane lipids are a rich source of signaling molecules. The lipid bilayer of cell membranes is composed mainly of glycerophospholipids and sphingomyelins. Both components have a dual function: as building materials and as a source of signaling molecules. Thus, glycerophospholipids are the starting products for the formation of DAG, inositol phosphates, PIPs, arachidonic acid and its many derivatives, and lysophospholipids, whereas ceramides and sphingosines are generated from sphingomyelins. Since the production of these lipid-derived mediators is controlled by extracellular signals, they are considered to be second messengers. However, many of them also function as *inter*cellular messengers, being released from cells and interacting with specific cellular receptors located either at the cell surface or inside the cell. This dual function is facilitated by the physico-chemical properties of lipids, enabling many of them to pass cellular membranes in both directions. As is typical for signaling molecules, the levels of lipid messengers are extremely low in resting cells but increase rapidly upon stimulation by exogenous and endogenous signals. Effective mechanisms of signal termination, such as enzymatic degradation, guarantee that this increase is transient. Because of the high interconnectivity of lipid metabolism, signaling by lipid mediators occurs in a complex network of interactions that controls a vast number of cellular processes. Moreover, owing to their physicochemical properties lipid mediators are enriched in cellular membranes. This provides an additional level of complexity as far as subcellular localization, metabolism and signaling mechanisms are concerned.

Sphingosines and ceramides are cellular stress signals and local mediators of cell death

Stress as well as a series of endogenous mediators activate **sphingomyelinases** (SMases), thus inducing a hydrolytic degradation of sphingomyelin to ceramide and choline phosphate. SMases are widely distributed from bacteria to humans. There are three enzyme families, the members of which differ in terms of their pH dependency. As far as signal transduction is concerned, the *neutral* SMases are of special interest because they are membrane-bound proteins interacting with receptors. *Acid* SMases are required for lipid catabolism in lysosomes and, when expressed at the cell surface and released from cells, for the extracellular generation of ceramides. *Alkaline* SMases are mainly found in the gastrointestinal tract and are considered to be digestive enzymes.

Ceramides are hydrolyzed by **ceramidases** to sphingosines and long-chain fatty acids. Like SMases, ceramidases exist in several isoforms and their activities probably are controlled by signals. *Acid* ceramidases are confined to lysosomes, functioning as catabolic enzymes, *alkaline* ceramidases are concentrated in intracellular membranes, and *neutral* ceramidases are transmembrane enzymes with their catalytic domains facing the extracellular space. They are assumed to control the level of ceramides and sphingosines in blood plasma and other body fluids.

Sphingomyelinases and ceramidases become active in emergency situations such as temperature stress, oxidative stress, genotoxic stress, and lack of growth factors, as well as in the presence of endogenous stress signals such as the cytokines interleukin-1, tumor necrosis factor-α, and Fas ligand (for more information about these factors see Sections 6.3 and 13.2.2). In addition, SMases are stimulated by a series of GPCRs. According to recent observations, the contact between receptor and SMase is brought about by a special adaptor protein FAN (Factor Associated with N-sphingomyelinase). This is shown in **Figure 4.21** for the cannabinoid receptor as an example, but holds true for cytokine receptors as well. Thus, ceramides and sphingosines are *bona fide* second messengers.

As a rule, ceramides and sphingosines inhibit cell proliferation and, depending on the cell type (at least *in vitro*), induce terminal differentiation, senescence, or apoptosis.

Figure 4.21 Release of intracellular lipid messengers by G-protein-coupled receptor signaling A transmembrane receptor (here as an example, the cannabinoid receptor CB1, see Section 16.7) activates, via adaptor protein FAN, the enzyme sphingomyelinase, thus generating membrane-bound ceramide and sphingosine, which is phosphorylated to soluble sphingosine- 1-phosphate. Alternatively, the receptor may open a conventional G-protein-controlled signaling pathway.

These effects may be due to structural changes of cellular membranes. Ceramides promote the formation of lipid rafts (see Section 5.8) where, for instance, the pro-apoptotic "death receptor" Fas accumulates, forming death-inducing signaling complexes (see Section 13.2.2). Moreover, they may participate in the permeabilization of the mitochondrial membrane, leading to cell death (see Section 13.2.5). Although effector proteins have been identified, no ceramide-specific interaction motif has been found yet in proteins. The ceramide effector proteins include both **protein phosphatases** and **protein kinases**. The phosphatases belong to the families PP1 and PP2A (see Section 2.6.4 for a classification of phosphatases) and mediate the anti-proliferative and pro-apoptotic effects of ceramides in that they dephosphorylate the transcription factor Rb, thus stopping the cell cycle (see Section 12.6), and the anti-apoptotic proteins PKB/Akt and Bcl2, which both become inactivated. A ceramide-activated protein kinase is the atypical protein kinase Cζ (see Section 4.6.3) that, depending on the cell type, is integrated in signaling pathways leading to either cell proliferation or cell death.

Lysophospholipids and phosphatidic acid are local modulators of survival and immune responses

Sphingomyelins are sources of additional signaling molecules. These are **ceramide-1-phosphates (C1Ps)** and **sphingosine-1-phosphates (S1Ps)**. Both are functional opponents of ceramides and sphingosines: as typical survival signals, they promote cell proliferation and cell motility while inhibiting apoptosis.

C1Ps and S1Ps are generated by ceramide- and sphingosine-kinases (**Figure 4.22**). These lipid kinases use ATP as a phosphate donor and are found both inside and outside cells. To become active they have to be phosphorylated by mitogenic MAP kinases such as ERK or by proteinkinase C or D. Thus, a wide variety of signals, all stimulating these kinases via their membrane receptors, trigger the formation of C1Ps and S1Ps.

S1Ps are lysophospholipids

Lysophospholipids are sphingomyelins lacking the amide-bound fatty acid residue, or glycerophospholipids lacking the fatty acid residue at the sn2 position of glycerol. In sphingomyelins specific deacetylases split the amide bond, while in glycerophospholipids the sn2 ester bond is hydrolyzed by phospholipase A$_2$ (**Figure 4.23**). These enzymes are stimulated by many hormones and other intercellular messengers. Other lysophospholipid mediators are sphingosylphosphorylcholine (SPC) and ly-sophosphatidic acid (LPA). S1Ps, SPC, and LPA are released from cells and bind to specific receptors, thus acting as auto- and paracrine local signals.

The corresponding membrane receptors are GPCRs. They are expressed by mammalian cells in three LPA-specific, five S1P-specific isoforms, and at least one SPC-specific type. Depending on the isoform, these receptors couple with $G_{i,0}$, $G_{q,11}$, or $G_{12,13}$ proteins. Along the corresponding signaling cascades, S1P, SPC, and LPA evoke a wide variety of effects on cell proliferation and cell survival. Moreover, these lysophospholipids control the dynamics of the actin cytoskeleton and thus cell architecture and motility. To this end they activate the small G-proteins Rho and Rac along $G_{q,11}$- and $G_{12,13}$-triggered pathways (for details see Section 10.3). In parallel, Ca^{2+} ions become released by the $G_{q,11}$–PLCβ–InsP$_3$ cascade, activating the actomyosin complex and, thus, cell motility as well as secretory processes and smooth muscle contraction (S1P also directly stimulates Ca^{2+} release from the endoplasmic reticulum, albeit via an ill-defined mechanism [see **Figure 4.24** and Section 14.5.3]). The effects of lysophospholipids on cell motility and architecture are critical for embryonic development, in particular for the vascular and immune systems. In fact, mice with an inactive (knocked-out) S1P receptor gene die before birth, since their network of blood capillaries has not developed properly.

Equally important is the function of lysophospholipids in defense and emergency reactions. Thus, they augment and integrate the effects of proinflammatory mediators, and act as chemotactic signals in inflammatory reactions and in the course of wound healing. In the immune system, S1P and LPA are major regulators of development, chemokine responsiveness, and recirculation of lymphocytes from lymph nodes and thymus to the periphery. Newly developed immunosuppressive drugs (such as FTY720) are S1P1 receptor ligands inducing receptor down-modulation.

Figure 4.22 Biosynthesis of signaling molecules from sphingolipids Putative functions of the compounds are shown in the gray boxes.

phosphatidylcholine

choline

phospholipase D

phospholipase A₂

phosphatidic acid

lysophosphatidylcholine

choline

phospholipase A₂

lysophospholipase D

lysophosphatidic acid

Figure 4.23 Biosynthesis of lysophosphatidic acid Pathways leading to intra- and extracellular formation of lysophosphatidic acid are shown. Phospholipases A_2 and D are signal-activated intracellular enzymes, whereas lysophospholipase D and its substrate lysophosphatidylcholine are found in the blood plasma.

LPA exhibits a peculiarity in that it not only interacts with membrane-bound receptors but also, like a steroid hormone, is able to penetrate the cell membrane, binding to ligand-activated transcription factors: the so-called peroxisome proliferator-activated receptors (PPARs) that are involved in the regulation of lipid metabolism (see Section 8.4.1). Such dual receptor interactions have also been reported for other lipid mediators, such as the eicosanoids.

Figure 4.24 Sphingosine-1-phosphate: an intra- and extracellular messenger Upon receptor-mediated activation (here by the autophosphorylated PDGF receptor tyrosine kinase), sphingosine kinase phosphorylates sphingosine. The sphingosine-1-phosphate thus generated may act as an intracellular messenger, gating a special class of Ca^{2+} channels in the endoplasmic reticulum (ER), or may leave the cell as a paracrine or autocrine tissue hormone interacting with a GPCR. Through the activated receptor, various intracellular signaling cascades become stimulated, leading (together with Ca^{2+}) to the physiological responses shown in the box. Arrestin is a scaffold protein relieving the G-protein coupling of heptahelical receptors and coupling them to other signaling pathways such as the MAP kinase cascades (explained in Section 5.8).

Figure 4.25 Phosphatidic acid as an intracellular messenger Input signals interact with membrane receptors and stimulate (via the small G-proteins Arf and Rho and via protein kinase C) phospholipase D, which hydrolyzes membrane-bound phosphatidylcholine to phosphatidic acid and choline. Phosphatidic acid unfolds its cellular effects either directly or after hydrolysis to lysophosphatidic acid catalyzed by phospholipase A$_2$.

Phosphatidic acid and platelet-activating factor

Another highly efficient membrane-bound messenger molecule with numerous functions is **phosphatidic acid (PA)**, the biosynthetic precursor of LPA. However, in contrast to LPA and S1P, PA is most probably not released by cells but acts exclusively as an intracellular signal. PA is liberated from phospholipids (in particular phosphatidylcholine and phosphatidylethanolamine) by **phospholipase D (PLD)** that hydrolyzes the choline-phosphate bond (**Figure 4.25**). Isoforms of this signal-transducing enzyme are found in the plasma membrane and in intracellular membranes. PLD contains several domains for interactions with membrane phospholipids (in particular PIP$_2$), protein kinase C, calcium ions, and small G-proteins of the Arf and Rho families that recruit PLD to membranes (see Sections 10.4.3 and 10.3). To gain full activity, PLD has to be phosphorylated by either protein kinase Cα (see Section 4.6.3) or the Rho-activated kinase PRK (see Section 10.3). Along these pathways PLD is stimulated by a wide variety of signals to produce phosphatidic acid.

Like lysophospholipids, phosphatidic acid is primarily a survival signal that is released under various cellular stress situations. In its functions it largely resembles LPA (see Figure 4.25). Since LPA may be generated from phosphatidic acid, it is sometimes difficult to distinguish between genuine and secondary (LPA-mediated) effects of phosphatidic acid. Several signaling proteins have been found to be activated by phosphatidic acid. These include the small G-protein Arf and the protein kinase RAF1 as well as the protein kinase mTOR. As will be shown in Section 9.4.1, mTOR stimulates protein synthesis and cell growth. As such it is a major player in exercise-promoted muscle growth, and, indeed, phosphatidic acid is sold as an anabolic food supplement for body-builders.

Finally, **platelet-activating factor**, an unusual glycerophospholipid with an ether bond (**Figure 4.26**) has to be mentioned. Platelet-activating factor is a highly active local mediator that is released, in particular from white blood cells, in response to inflammatory stimuli. In an auto- and paracrine manner it induces the aggregation

Figure 4.26 Platelet-activating factor PAF

of thrombocytes and promotes other reactions involved in inflammation and wound healing. Its receptor is a GPCR that couples with $G_{i,0}$- and $G_{q,11}$-proteins.

Lipid mediators derived from unsaturated fatty acids

Polyunsaturated fatty acids such as linoleic acid (C_{18}, two double bonds), α-linolenic acid (C_{18}, three double bonds), and arachidonic acid (C_{20}, four double bonds) are characteristic components of phospholipids. For humans and many animal species, linoleic and α-linolenic acids are essential: they cannot be synthesized but have to be taken up through a vegetable diet. In contrast, our body can synthesize arachidonic acid from linoleic acid by enzymatic lipid-chain elongation. While in earlier times the major function attributed to these compounds was the control of membrane viscosity, they are now known to play a key role in intercellular communication, particularly by serving as precursors of a very large and extremely versatile family of signaling lipids formed primarily along oxidative pathways.

The characteristic structural feature of polyunsaturated fatty acids found in cells is that the double bonds are *cis*-configured and not conjugated: they are separated from each other by a methylene group. Lipids containing such 1,4-*cis,cis*-pentadiene structures are highly oxygen-sensitive, becoming rapidly autoxidized to products of bewildering variety. Biological evolution has benefited from this reactivity by forcing the individual reactions under the control of enzymatic catalysis. As a result, cells now possess a very large collection of lipid messengers called **oxylipins**. Depending on whether C_{18} or C_{20} fatty acids are the precursors, one distinguishes between the subfamilies of **octadecanoids** and **eicosanoids**. While in plants linoleic and linolenic acid-derived octadecanoids predominate, in animals the eicosanoids are the most abundant oxylipins, with arachidonic acid being the major source. The eicosanoid families currently known are summarized in **Table 4.1**.

To become transformed into oxylipins, the polyunsaturated fatty acids have to be released from membrane phospholipids where they are bound primarily in the *sn*-2

Table 4.1 Eicosanoids[a]

Type	Biosynthesis	Major functions
prostaglandins	COX[a], prostaglandin synthases	smooth muscle contraction, pain sensation secretory processes, luteinization immunosuppression
thromboxanes	COX, thromboxane synthases	platelet aggregation, aorta constriction
prostacyclins	COX, prostacyclin synthases	thromboxane antagonists
HPETE and HETE	LOX, cytochrome P450 monooxygenases	inflammation, blood pressure control, kidney function neurotransmitter?
leukotrienes	LOX, leukotriene synthases	bronchoconstriction, leukotaxis
lipoxins	LOX	anti-inflammatory, resolution of inflammation
15-epi-lipoxins	acetylated COX[b]	anti-inflammatory, resolution of inflammation
hepoxilins	LOX	Ca^{2+} release, neuromodulation?
trioxilins	LOX	unknown
epoxyeicosatrienoic acids	cytochrome P450 monooxygenases	vasodilatation, cardioprotection, kidney protection, brain protection, kidney function
isoprostanes	nonenzymatic	stress signals
anandamide	not clear	neurotransmitter

[a]COX, cyclooxygenase; LOX, lipoxygenase; HPETE, hydroperoxyeicosatetraenoic acid; HETE, hydroxyeicosatetraenoic acid.
[b]Generated by the reaction of COX with acetylsalicylic acid (aspirin).

position of glycerol. Mostly (but not exclusively) this reaction is catalyzed by A_2**-type phospholipases (PLA$_2$)** (see Figure 4.18), which in a Ca^{2+}-dependent step are recruited via their C2 domains to membranes. PLA_2 are activated by phosphorylation catalyzed by MAP kinases, in particular ERK. In other words, the very large number of signals processed by MAP kinase modules may release polyunsaturated fatty acids and, thus, stimulate the synthesis of octadecanoids and eicosanoids provided Ca^{2+} has been made available by another signal (note that the two signals form a logical AND gate). This means that formation of these lipid mediators is a general event in cell physiology. In fact, eicosanoids and octadecanoids are released by a wide variety of environmental as well as hormonal, immunological, and neural stimuli.

Free unsaturated fatty acids are extremely short-lived because they are immediately reincorporated into phospholipids or are oxidized to eicosanoids or octadecanoids. Moreover, they undergo rapid autoxidation. Nevertheless, free arachidonic acid has turned out to be a signaling molecule on its own. Two of its effects, the inhibition of Ras–GAP and the activation of Ca^{2+} channels, are discussed in Sections 10.1.3 and 14.5.5, respectively.

Short lifespans also hold true for octadecanoids and eicosanoids, because they also become autoxidized, metabolically inactivated, or incorporated into phospholipids. As a result, their range mostly is limited to the tissue or even the local cell group of origin. They are considered, therefore, to act as local mediators (also called tissue hormones or autocoids), a property they share with other lipid messengers such as S1Ps and lysophospholipids.

Prostaglandins and thromboxanes

Pharmacological effects of eicosanoids such as stimulation of uterine contraction were discovered in the 1930s and by mistake were attributed to an elusive prostate hormone called "prostaglandin." Although since then every tissue has been found to be able to produce such active agents, the name prostaglandins had become irreversibly established for a group of prototypical eicosanoids. The importance of this field is emphasized by the 1982 Nobel Prize in Physiology or Medicine, awarded to Sune K. Bergström, Bengt I. Samuelsson, and John R. Vane for their fundamental research on the biochemistry and physiology of prostaglandins and related compounds.

Characteristic structural elements of prostaglandins are two double bonds and a five-membered ring the structure of which characterizes individual prostaglandin types (see Figure 4.26). Together with the related thromboxanes, prostaglandins constitute the prostanoid family of eicosanoids.

Prostaglandins arise by a two-fold addition of molecular oxygen to arachidonic acid, resulting in an unstable cyclic endoperoxide called prostaglandin H_2 that serves further enzymes as a substrate (**Figure 4.27**). Prostaglandin H_2 synthesis is catalyzed by **cyclooxygenases**. These hemoproteins combine two enzymatic activities: an oxygenase with a peroxidase.

Mammalian cells express two cyclooxygenase isoforms. Cyclooxygenase 1 (COX-1) is a constitutive enzyme satisfying the cell's basic needs for prostanoids. In contrast, cyclooxygenase 2 (COX-2) becomes transiently expressed in most tissues only in situations of increased prostanoid requirement such as infections, wounding, and tumor growth. Both COX-1 and COX-2 are inhibited by non-steroidal anti-inflammatory drugs (to be distinguished from anti-inflammatory steroids such as glucocorticoids). The prototype of such drugs is acetylsalicylic acid, better known as aspirin. Its well-known effects substantiate the central role eicosanoids play in inflammatory processes and pain (see Section 15.3). The applicability of aspirin and related drugs is hampered by side-effects such as gastric and intestinal bleeding that are due to an inhibition of COX-1 rather than of COX-2. Therefore, novel drugs selectively inhibiting COX-2 were expected to represent a breakthrough in the treatment of chronic

Figure 4.27 Receptor-mediated synthesis of prostanoids
Phospholipase A_2, the rate-limiting key enzyme of eicosanoid formation, is activated by Ca^{2+} and phosphorylation (catalyzed by MAP kinases such as ERK) and induces the release of arachidonic acid from membrane phospholipids. By combining the activities of an oxygenase with that of a peroxidase, cyclooxygenase oxidizes arachidonic acid to prostaglandin H_2 which is the common precursor of prostaglandins, prostacyclin, and thromboxanes. Cyclooxygenases are inhibited by aspirin and related drugs. The pathways along which membrane receptors activate Ras are explained elsewhere (see Sections 5.8 and 10.1). The individual prostaglandin types (named prostaglandins A, B, D, E, F, and J) differ from each other by the structure of the oxidized five-member ring system. Types D, E, and F are generated by individual prostaglandin synthases, whereas types A, B, and J are secondary products of types D and E. Prostacyclin synthase and thromboxane synthase convert prostaglandin H_2 into prostacyclin and thromboxane, respectively. The suffix 2 means that a prostanoid carries two double bonds in the side chains; that is, it is derived from arachidonic acid. Prostanoids with one or three double bonds also are found in cells. They are derived from other, less abundant polyunsaturated C20 fatty acids.

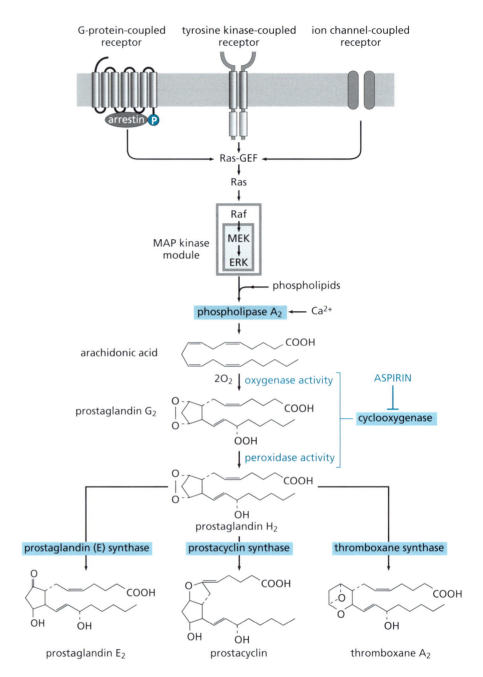

inflammatory diseases such as rheumatoid arthritis. Moreover, an overexpression of COX-2 correlates with or even is causally related to tumor development in most tissues indicating that such drugs might be useful in **cancer prevention**. In fact, aspirin and related drugs were found to diminish tumor incidence, in particular in colon and esophagus, when applied over a long period. Unfortunately, severe and sometimes deadly side effects, in particular on the cardiovascular system, have brought the COX-2 inhibitors into disrepute and dramatically dampened the initial enthusiasm.

Another well-known therapeutic effect of aspirin, **prevention of stroke and myocardial infarction**, is due to COX-1 inhibition causing an attenuation of blood clotting. Here the critical step is the aggregation of platelets at sites of blood vessel damage. This reaction, which is induced by a variety of factors, is enormously amplified by an autocrine mechanism involving **thromboxane A_2**, a prostanoid produced in platelets (thrombocytes) from prostaglandin H_2 by thromboxane synthase, an enzyme strongly expressed in these cells. The endogenous antagonist of thromboxane A_2 is **prostacyclin** (also known as prostaglandin I_2), generated from prostaglandin H_2 by a prostacyclin synthase that is particularly abundant in endothelial cells. The biosynthesis of both prostanoids starts with COX-1 and is inhibited, therefore, by aspirin and related drugs. However, while endothelial cells overcome this blockade by activating the transcription of the *COX-1* gene, platelets cannot adapt because they lack a nucleus. Thus, repeated intake of aspirin results in a lasting inhibition of platelet aggregation and blood clotting.

The cellular effects of prostanoid signals are primarily mediated by heptahelical transmembrane receptors coupled to G_s-, $G_{i,0}$-, and $G_{q,11}$-proteins. The thromboxane receptor, for instance, causes an increase of cytoplasmic Ca^{2+} along the $G_{q,11}$–PLCβ–InsP$_3$ pathway that is the signal for typical thromboxane effects such as platelet aggregation and, in addition, contraction of vascular smooth muscles, in particular those around the aorta. In contrast, the antagonistic effects of prostacyclin are mediated via the G_s–adenylate cyclase–cAMP pathway.

15-Deoxy-$\Delta^{12,14}$-prostaglandin J$_2$, an endogenous metabolite of prostaglandin D$_2$, provides an exception to the rule in that it does not interact with a transmembrane receptor but binds directly to, and thus activates, the nuclear transcription factor PPARγ (see Section 8.4.1). This mechanism of action resembles that of a steroid hormone. One may speculate that, besides interacting with membrane receptors, other prostanoids might also be ligands of transcription factors because they are lipophilic enough to penetrate the cell membrane.

Other eicosanoids

Apart from cyclooxygenases, other enzymes catalyze the formation of eicosanoids from arachidonic acid. Several **lipoxygenases** add molecular oxygen stereoselectively to the 1,4-*cis,cis*-pentadiene structures, causing hydroperoxidation together with a shift and a *cis–trans* isomerization of the double bond attacked. The products are hydroperoxyeicosatetraenoic acids (HPETEs), which become rapidly reduced to the corresponding hydroxy compounds (HETEs). HPETEs may also be rearranged into epoxides that, in turn, are either hydrolyzed or added to the tripeptide glutathione. Along this enzymatic pathway 5-HPETE is transformed into **leukotrienes,** a group of physiologically active intercellular messengers. Leukotriene B$_4$ (**Figure 4.28**) is one of the most potent chemotactic attractants for leukocytes and plays an important role in inflammatory processes. Via Ca^{2+} release mediated by a $G_{q,11}$-protein-coupled transmembrane receptor and the PLCβ–InsP$_3$ signaling cascade, it facilitates cell motility.

The glutathione adduct leukotriene C$_4$ evokes, via a $G_{q,11}$-protein-coupled transmembrane receptor and Ca^{2+}, a contraction of airway smooth muscles. Its release— for instance, in the course of an allergic reaction—is considered to be a major cause

Figure 4.28 Generation of eicosanoids by lipoxygenase-catalyzed oxidation of arachidonic acid Teal arrows indicate the points of attack of oxygen catalyzed by individual lipoxygenases of mammals. Leukotrienes B$_4$ and C$_4$ are physiologically active eicosanoids generated along the 5-lipoxygenase pathway.

of bronchial asthma. In this respect leukotriene C_4 cooperates with histamine, the primary mediator of allergy, by positive feedback: each agent promotes the release of the other one. Upon contact with an allergen histamine is released from mast cells and, like leukotriene C_4, causes smooth muscle contraction along the $G_{q,11}$–PLCβ–$InsP_3$–Ca^{2+} pathway.

Compared with the leukotrienes, little is known about the physiological functions of most other lipoxygenase-derived eicosanoids. The same holds true for the receptors aside from 8-HETE, which—similar to deoxyprostaglandin J_2—activates a transcription factor of the PPAR family (in this case isoform PPARα) by direct binding. An analogous effect has been described for leukotriene B_4 and PPARγ.

Additional enzymes metabolizing arachidonic acid to eicosanoids are **cytochrome P450-dependent monooxygenases**. They catalyze the formation of a bewildering variety of hydroxylated and epoxidized products with widely unknown functions.

A special class of arachidonic acid metabolites not generated by oxidation are the arachidonyl amides with ethanolamine and glycine. These so-called **anandamides** (referring to the Sanskrit word for "bless") have found particular interest as endogenous ligands of receptors originally characterized as cellular binding sites of tetrahydrocannabinol, the marijuana drug (see Section 16.7). In addition, they interact with a cation channel, the vanilloid receptor, that mediates pain sensations (see Section 15.3).

Finally, the **isoprostanes** should be mentioned. These eicosanoids are endogenous products of a nonenzymatic oxidation of arachidonic acid. Isoprostanes are thought to be messengers signaling emergency situations such as oxidative stress, since they arise upon tissue damage caused by severe intoxication, for instance after misuse of alcohol, tobacco, and drugs (**Sidebar 4.4**).

Conclusion: lipid mediators are higher-class second messengers

At first glance the various types and functions of lipid mediators present quite a confusing picture. However, there is a common denominator: most lipid mediators represent a special type of second messengers that, upon stimulation of a cell by a primary signal (environmental stimuli, hormones), are released into both the cytoplasm and the extracellular space. While in the cell the lipids function as conventional second messengers, their task outside the cell is to coordinate and modulate (amplify, integrate, or dampen) the tissue's overall response to the primary stimulus. Thus they play a key physiological role.

An example is provided by functional blood circulation. When, for instance, a hormone stimulates the function of a gland, an increase in blood flow is essential to rapidly distribute the glandular products released or when muscle activity triggered by a neurotransmitter an increased supply of oxygen and glucose via blood circulation

Sidebar 4.4. Oxylipins as invertebrate and plant hormones Invertebrates are able to produce and, in contrast to vertebrates, to store eicosanoids. Certain coral species provide probably the richest source of prostaglandins in nature, using them to deter predatory fishes. The structural variability of invertebrate eicosanoids seems to be almost unlimited, whereas our knowledge of their functions is rather fragmentary. Such fragments include the neuromodulatory effect of hepoxilins in the sea snail *Aplysia californica* and a hatching effect of the same compounds on barnacles, as well as the induction by 8-HETE of larval metamorphosis of freshwater and sea water polyps and of egg cell maturation in sea stars. In plants, arachidonic acid is quite rare. Instead, lipid messengers are produced from linoleic and linolenic acid along lipoxygenase-catalyzed pathways. The octadecanoids thus obtained are metabolized to a wide variety of biologically active compounds acting, among others, as stress signals and wound hormones (**Figure 4.29**). Sea algae use the lipoxygenase pathways to produce small hydrocarbons serving the gametes as sexual pheromones.

Figure 4.29 Linolenic acid-derived stress signals of plants By oxidation of α-linolenic acid along the ω3-lipoxygenase pathway, signaling lipids are produced that trigger defense reactions and tissue repair. Jasmonic, isojasmonic, tuberonic, and cucurbic acids represent the large family of jasmonate-type phytohormones. They induce the formation of chemical compounds counteracting attacks by fungi, insects, and microbes. Volatile derivatives, such as jasmonic acid methyl ester, seem to act as long-distance signals between remote parts of a plant. In perfumery, such compounds are used as aromatic essences. Traumatin is released upon wounding, stimulating wound repair.

is required. Another example is the inflammatory response, where various cell types including white blood cells, platelets, endothelial cells, vascular smooth muscle cells, pain-sensing neurons, fibroblasts, epithelial cells, and others cooperate. For the coordination and modulation of this complex interplay, lipid mediators play a critical role. Since lipid mediators assist rather than replace primary stimuli, many of them are not essential, although deficiencies may result in pronounced functional disorders.

Summary

- Adenylate cyclases catalyze the formation of cAMP from ATP whereas phosphodiesterases terminate signaling by catalyzing the hydrolysis of cAMP (and cGMP). These enzymes exist in numerous isoforms that are controlled by a variety of input signals. Phosphodiesterases are important targets of pharmaceutical drugs.

- Cellular membranes are a rich source of signaling molecules most of them acting as second messengers. Their formation is catalyzed by signal-controlled enzymes expressed in various isoforms.

- Phosphatidylinositol 3-kinases (PI3Ks) catalyze the generation of membrane-bound phosphatidylinositol 3,4-bis- and 3,4,5-trisphosphates, which are docking sites and activators of proteins with pleckstrin homology (PH) domains. Phospholipid phosphatases such as PTEN, which antagonize PI3K effects, are tumor suppressor proteins.

- Phosphatidylinositol-specific phospholipases C (PLCs) are Ca^{2+}-dependent enzymes that catalyze the formation of diacylglycerol (DAG, a stimulator of protein kinase C) and inositol 1,4,5 trisphosphate (which increases cytoplasmic calcium). PLC-activating signals are of critical importance for induction of Ca^{2+}-dependent cellular processes.

- The enzymatic hydrolysis of glycerophospholipids and sphingomyelins triggers the release of ceramides, sphingosines and other lipid mediators which - via G protein-coupled receptors - modulate the effects of a wide variety of primary signals involved in the control of cell proliferation, cell death and stress responses.

- Phospholipase A_2, a signal-controlled enzyme activated by Ca^{2+} ions and MAP kinase-catalyzed phosphorylation catalyzes the release of polyunsaturated fatty acids such as arachidonic and α-linolenic that are the sources of additional signaling molecules modulating the effects of primary signals such as hormones and stress factors.

- The formation of polyunsaturated lipid mediators is catalyzed by cyclooxygenases and lipoxygenases. Cyclooxygenases promote the oxidation of arachidonic acid to pro-inflammatory prostanoids and are inhibited by pain relievers such as acetyl salicylic acid. Lipoxygenase products such as leukotrienes are powerful leukotactic and bronchoconstrictive agents.

4.5 Guanylate cyclases are not controlled by trimeric G-proteins

The second cyclic nucleotide functioning as an intracellular messenger is cyclic guanosine 3′,5′-monophosphate (cGMP). It is as widely distributed as cAMP, but the tissue concentrations are much lower with the exception of brain and retina, where high levels are found. cGMP formation from GTP is catalyzed by guanylate (or guanylyl) cyclases. Apart from a similar catalytic domain, guanylate cyclases have little in common with adenylate cyclases. Although guanylate cyclases are not controlled by trimeric G-proteins, they are discussed here in more detail.

There are two families of guanylate cyclases: soluble and receptor types (**Figure 4.30**). The **soluble guanylate cyclases** (sGCs) are cytoplasmic enzymes found in practically all vertebrate tissues. Their active form is an α,β-heterodimer linked by disulfide bridges. In the N-terminal regulatory domain a heme is bound as a prosthetic group, interacting with nitrogen monoxide (NO) in nanomolar concentrations and, with much lower affinity, with carbon monoxide (CO) but unlike most other hemoproteins not with oxygen. Both NO and CO are gaseous intercellular mediators with a wide spectrum of physiological activities (for details see Section 16.5). Binding of NO/CO causes a conformational change resulting in cyclase activation. Physiologically significant functions of sGCs are relaxation of vascular smooth muscles (resulting in increased blood supply), retrograde regulation of synaptic signal transmission, inhibition of platelet aggregation, and immunomodulatory effects.

Receptor guanylate cyclases exhibit quite a different structure. They do not contain heme (and therefore are not activated by NO and CO) but resemble receptor protein kinases in their dimeric structure and the way enzyme activity is regulated. In fact, the cytoplasmic part of these enzymes harbors both a cyclase and a protein kinase homology domain, the latter being inactive, as it lacks an essential aspartate residue in subdomain VI (see Section 2.6.1).

GC-A and GC-B:	receptors of natriuretic peptide hormones
GC-C:	receptor of heat-stable enterotoxin and guanylin
GC-D:	receptor of odorants (pheromones?)
GC-E and GC-F:	guanylate cyclases of retina (ligands?)
GC-G:	?

Figure 4.30 Guanylate cyclases (Left) Receptor guanylate cyclase. The isoforms and corresponding ligands are shown in the box. (Middle) Soluble guanylate cyclase. (Right) Mammalian adenylate cyclase for comparison. Cyclase domains are symbolized by teal ellipses, inactive ATP-binding protein kinase domains of the receptor guanylate cyclase by light gray cylinders, and transmembrane domains by dark gray cylinders.

Nevertheless, this kinase homology domain plays a key role in regulating cyclase activity, because upon phosphorylation by an unknown protein kinase of several Ser and Thr residues, it blocks the access of GTP. This autoinhibition is relieved by an allosteric interaction with ATP that occurs when the ligand is bound to the extracellular receptor domain. The active state of a receptor guanylate cyclase is short-lived because the enzyme becomes rapidly desensitized by dephosphorylation rendering it insensitive to further stimulation. Resensitization requires new phosphorylation. This built-in switch resembling an automatic fuse prevents a prolonged overstimulation of the system. Related safety measures, albeit following other rules, are known for a variety of other receptors (see, for instance, Section 16.2).

In mammals, seven isoforms of receptor guanylate cyclases have been found. Isoforms A and B are widely distributed, serving as receptors for the **natriuretic peptides ANP, BNP, and CNP.** These hormones, produced by atrial cells of the heart, control salt and water homeostasis (osmolarity) and thus blood pressure. A major effect is relaxation of vascular smooth muscles resulting in vascular dilation, increased blood flow, and decreased blood pressure. Receptor isoform C is concentrated in epithelial cells of the small intestine. Its ligands are bacterial **enterotoxin** and the tissue hormone **guanylin**, which both activate the chloride ion channel CFTR along a signaling pathway including a cGMP-dependent protein kinase (details in Section 14.7.3). Expression of isoform D is restricted to a special subgroup of olfactory neurons; it probably serves as a receptor of pheromone-like odorants (see Section 15.4). Isoforms E and F are found only in the retina (and pineal gland) and play a key role in **visual signal processing** (see Section 15.5). Their endogenous ligands, if they exist at all, are not known. The same holds true for the more widely distributed isoform G.

Invertebrates seem to have more guanylate cyclase genes than vertebrates *(C. elegans*, for instance, has 27). They may have quite another structure than the vertebrate enzymes (a form with 2–6 transmembrane domains, resembling the structure of adenylate cyclase, has been found). The functions of the invertebrate enzymes are more or less obscure with the exception of a receptor guanylate cyclase in the membranes of **sea urchin spermatozoa**. Its ligands are peptides (called speract and resact) that are released from the egg cell, stimulating the motility and acrosomal reaction (rendering the sperm cell able to penetrate the egg cell) of sperm cells. The question as to whether such proteins are also expressed by mammalian egg cells is still open. It is easily conceivable that a more in-depth investigation of this subject may have far-reaching consequences for contraception.

Summary

- cGMP is a second messenger resembling cAMP. Its bioformation is catalyzed by guanylate cyclases that are expressed either as soluble cytoplasmic hemoproteins or as transmembrane receptors.

- Soluble guanylate cyclases are specifically activated by NO and CO, whereas receptor guanylate cyclases interact with endo- and paracrine peptide mediators.

4.6 The next level: protein kinases as sensors of second messengers

The second messengers produced along receptor- and G-protein-controlled pathways are input signals of various Ser/Thr-specific protein kinases, namely cAMP for PKA, cGMP for PKG, PIP_2 and PIP_3 for PKB/Akt and phospholipid-dependent kinase 1, and DAG for PKC and PKD. Ca^{2+} ions released by $InsP_3$ are also activators of several protein kinases. These kinases belong to the large family of AGC kinases (see Section 2.6.3) and exhibit some common features:

- they are relatively nonspecific but gain substrate specificity by **anchor proteins** directing the kinases to distinct intracellular sites and to individual substrates

- to become active they require **phosphorylation**, at least of the activation loop and frequently also of a hydrophobic domain adjacent to the C-terminus of the catalytic domain

- with the exception of PKA and PKG, they undergo reversible **membrane binding** to come in contact with their membrane-bound activators (achieved by various membrane-binding domains, not found in PKA and PKG). Because of their low but regulated substrate specificity, AGC kinases are multipurpose enzymes and universal relay stations in the data-processing protein network. As such they distribute, transform, and modulate most of the signals transmitted by membrane-bound receptors.

4.6.1 Protein kinase A, a cAMP sensor

Several receptor proteins of cAMP have been identified in eukaryotes, including, first and foremost, PKA (also found as cAMP-dependent protein kinase, cAPK, in the literature), cation channels of sensory cells (see Section 14.5.7), and the GDP–GTP exchange factor EPAC of the small G-protein Rap (see Section 10.2). To this list the G-protein-coupled cAMP receptor found in the cell membrane of *Dictyostelium discoideum* (for the slime mold, *extracellular* cAMP is a chemotactic aggregation signal) should be added. These proteins have in common a conservative cAMP-binding domain that can be traced back to the predecessor of all cAMP-binding proteins, the transcriptional co-activator CRP of bacteria (see Section 3.3.2). As shown by crystallographic studies, binding of cAMP to those domains induces far-reaching conformational changes of the cAMP effector proteins.

The ubiquitously distributed PKAs, expressed in 11 mammalian isoforms, differ from other AGC kinases by two special features: they undergo autophosphorylation of the activation loop instead of being phosphorylated by a separate kinase and the regulatory and catalytic domains are located on separate polypeptides. In fact, PKA is a tetrameric protein consisting of two cAMP-binding (regulatory) and two kinase (catalytic) subunits each. In the absence of cAMP, a pseudosubstrate sequence in the

Figure 4.31 Domain structure and activation of protein kinase A
(A) In the absence of cAMP, catalytic subunits (C) of PKA form a stable complex with a homodimer of the regulatory subunits (R). In this complex the substrate-binding site of C is blocked by a pseudosubstrate sequence of R (symbolized by two-headed arrows). Upon binding of four molecules of cAMP, the complex dissociates, exposing the catalytic domains of C. cAMP molecules leaving R are immediately hydrolyzed by phosphodiesterases. (B) Domain structure of R and C. The catalytic domain contains binding sites for ATP and substrate protein. DD, dimerization domain; PSD, pseudosubstrate domain; cAMP, cAMP binding domain.

regulatory subunits blocks the catalytic center. The high-affinity binding of two molecules of cAMP per regulatory subunit induces a conformational change resulting in dissociation of the tetramer and release of the active kinase subunits (**Figure 4.31**).

Protein kinase A modules

Though it is a soluble enzyme, PKA interacts via its regulatory subunits with anchor proteins and thus becomes concentrated at distinct sites and in cellular organelles. These proteins are called **AKAPs** (A-Kinase Anchor Proteins).

Mammalian cells express more than 30 isoforms with quite variable interaction domains. The term AKAP is somewhat misleading, because these proteins also anchor other protein kinases and other signal-transducing proteins in addition to PKA. In fact, AKAPs are typical multivalent scaffold proteins connecting signal-transducing proteins to complexes or modules that are found, for instance, in the neighborhood of receptors. Such an arrangement allows individual signal transduction in spite of identical G-protein coupling. This is shown in **Figure 4.32**: although two different signals, 1 and 2, both stimulate the G_s–adenylate cyclase–PKA pathway, they may evoke different cellular responses. Thus, AKAPs provide signaling with a high degree of privacy (see Section 1.6).

Many AKAPs bring together PKA and other kinases with enzymes of signal extinction such as protein phosphatases and cAMP phosphodiesterases. Through such a construction, the PKA signal is dramatically sharpened and may become subject to frequency modulation (for more details see Section 14.5.4). **Figure 4.33** shows the prototype of an AKAP module exhibiting a complex pattern of positive and negative feedback loops. The resemblance to a neural or electronic switching element can hardly be ignored.

Selective anchoring in the vicinity of individual substrate proteins is a crucial condition of PKA specificity, since the phosphorylation consensus sequence Arg-Arg-x-Ser/Thr-x is rather abundant and, in the test tube PKA is all but choosy. The number of PKA substrate proteins is large, including structural proteins, enzymes, and transcription factors. Because of this versatility, PKA is involved in a wide variety of physiological processes extending from simple metabolic reactions up to complex brain functions such as learning and memory fixation. Examples are found throughout this book.

The ability of the cell to equip, by specific localization, a relatively nonselective signaling protein with specificity, efficacy, and privacy is not restricted to PKA but represents a general motif of cellular signal transduction. It is one of the combinatorial tricks enabling the cell to achieve a large number of effects with a rather limited arsenal of tools.

Figure 4.32 "Individualization" of transmembrane signaling by Akinase anchor proteins Different AKAPs couple protein kinase A (PKA) with different G_s-protein-coupled receptors (R1, R2) and different substrate proteins (S1, S2). Thus, different signals may evoke individual effects despite the fact that they activate the same G_s–adenylate cyclase–cAMP pathway.

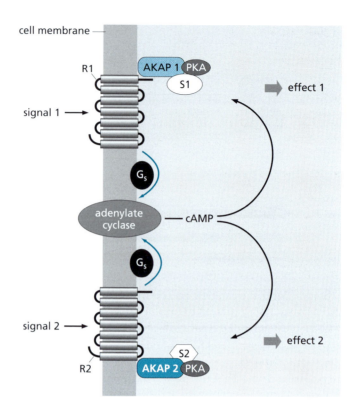

Figure 4.33 Signal-transducing protein module organized by an A-kinase anchor protein In this model, which is based on experimental evidence, the AKAP associated with a membrane brings together protein kinase A (PKA) with another protein kinase (PK; an example is provided by the MAP kinase ERK5), a cAMP–phosphodiesterase (PDE) inhibiting PKA activation, and a protein phosphatase (PPase) inhibiting protein phosphorylation. PDE is assumed to be activated by PKA and inhibited by PK. The module processes at least two different input signals acting on PKA and PK, respectively, and might function as an oscillator, generating a frequency-modulated output signal.

4.6.2 cGMP-activated protein kinases

Three types of intracellular cGMP receptors have been identified: cGMP-activated protein kinases, cGMP-controlled phosphodiesterases, and cGMP-controlled ion channels. In this section only the kinases are treated in detail; information about phosphodiesterases is found in Section 4.4.2, and ion channels are discussed in Section 14.5.2. cGMP-activated protein kinases or PKGs (another abbreviation found in the literature is cGKs) catalyze Ser/Thr phosphorylation. In contrast to PKA, the catalytic and regulatory domains of PKG share a single polypeptide chain. The enzymes form dimers, ensuring activation by trans-autophosphorylation of the activation loop upon interaction with cGMP. Mammalian cells express two isoforms, PKGI and PKGII. PKGI is widely distributed and has a large number of substrate proteins including several transcription factors (**Figure 4.34**). In smooth muscle cells, the enzyme transduces the relaxing effect of the NO signal to the contractile apparatus by counteracting myosin light-chain phosphorylation and $InsP_3$-induced Ca^{2+} release (see Section 16.5). To this end PKGI phosporylates and activates myosin light chain phosphatase and inactivates $InsP_3$ receptor Ca^{2+} channels in the endoplasmic reticulum. Along this pathway the kinase controls blood pressure and penile erection. A well-established function of PKGII is the phosphorylation and activation of the CFTR chloride channel (see Section 14.7.3). A role of PKG in NO-dependent long-term potentiation of synaptic signaling in brain neurons is implied by the high cGMP level in nerve cells. An illustrative example of the role of cGMP in general and of PKGI in particular is provided by the hormonal regulation of blood pressure and kidney function. As shown in **Figure 4.35**, situations leading to an increase of blood pressure stimulate atrial heart cells to release hormones—atrial natriuretic peptides (ANPs)—which in the kidneys induce vasodilatation and water and salt excretion. This effect is mediated by cGMP, produced by an ANP receptor guanylate cyclase, and by PKGI. The major ANP antagonists are the steroid hormone aldosterone and the peptide hormone angiotensin II. The latter is derived from a precursor protein (angiotensinogen) that is cleaved by the proteases renin and angiotensinogen-converting enzyme. Angiotensin II strongly elevates blood pressure by causing vasoconstriction along the $G_{q,11}$–$InsP_3$–Ca^{2+} axis. Moreover, it stimulates the release of aldosterone from the adrenal cortex, where aldosterone production is controlled by

Figure 4.34 Substrates and effects of cGMP-activated protein kinases Black arrows indicate activation; teal bars show inhibition. NOS, NO synthase; CREB, cAMP response element-binding protein; MEK and MEKK, protein kinase components of MAP kinase modules (see Chapter 11).

the pituitary hormone ACTH. ANP counteracts angiotensin II by inducing vasodilatation and switches off the aldosterone effects by inhibiting, along the cGMP–PKG pathway, the release of renin from the kidneys and by suppressing aldosterone synthesis in the adrenal cortex. Here the major effect of cGMP is to activate phosphodiesterase type 2 that hydrolyzes cAMP, the second messenger of ACTH.

4.6.3 Protein kinases C, a family of multipurpose enzymes

The second messenger DAG is an activator of PKCs. These enzymes are found in all eukaryotes from yeast to humans. Mammalian cells contain nine *pkc* genes and express 10 isotypes including two splice variants (PKCβI and -βII). Some of them, such

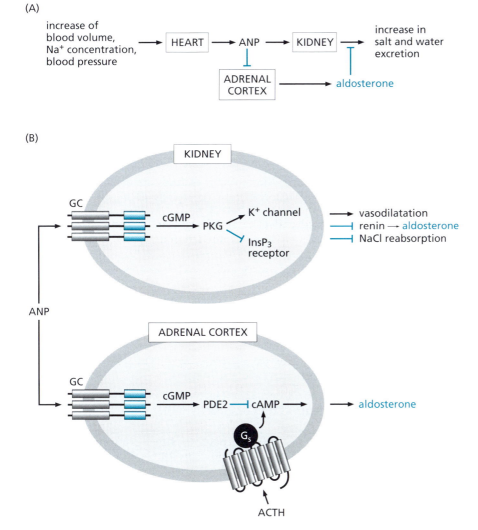

Figure 4.35 Blood pressure control through cGMP signaling (A) Atrial cells respond to an increase of blood volume, blood pressure, and blood osmolarity by releasing natriuretic peptide (ANP), thus enhancing the excretion of water and salt in the kidney. ANP also inhibits the synthesis of aldosterone in the adrenal cortex. (B) In the kidney ANP activates a trimeric receptor guanylate cyclase (GC). The cGMP formed stimulates cGMP-dependent protein kinase (PKG), which phosphorylates a hyperpolarizing potassium channel (resulting in activation) and the InsP3 receptor Ca2+ channel of the endoplasmic reticulum (resulting in inhibition). These effects culminate in increased blood flow due to vasodilatation and inhibition of renin secretion and NaCl reabsorption. Renin is a protease that, together with angiotensin-converting enzyme, releases angiotensin II from a precursor protein. Angiotensin II induces vasoconstriction and promotes the release of aldosterone from the adrenal cortex. In the adrenal cortex, ACTH (adrenocorticotropic hormone) triggers aldosterone synthesis via the Gs–cAMP–protein kinase A pathway. cGMP activates PDE2, thus lowering the cAMP level. This results in a diminished release of aldosterone.

Figure 4.36 Protein kinase C families The 10 mammalian isoforms of protein kinase C are marked by Greek letters. For activation, conventional cPKCs require diacylglycerol (DAG), phosphatidylserine (P-serine), and Ca^{2+}; new nPKCs require only DAG and P-serine (but not Ca^{2+}); and atypical aPKCs require only P-serine. Like nPKCs, PKC-related kinases (PRK or PKN) have an atypical C2 domain exhibiting a low affinity for Ca^{2+}. In addition, PRKs possess HR1 domains that interact with the small G-protein Rho. In the considerably larger Pkc1 of yeast, the interaction domains of PKC and PRK are combined, indicating that this enzyme is an evolutionary precursor of the PKC families. In PKC the C-terminal kinase domain is separated from the N-terminal regulatory domains by a hinge sequence; proteases may split the molecule here, releasing a constitutively active kinase fragment. In fact, protein kinase C was discovered as a proteolytically activated "protein kinase M" (refers to Mg^{2+} dependency) in brain preparations in 1979. Whether the proteolytic activation has a physiological function is not clear. PS (teal box), pseudosubstrate motif.

Pkc1 of *Saccharomyces cerevisiae*

as PKCα, -δ, and -ζ, are found in almost every tissue, whereas others are expressed by certain cell types only; for instance, PKCγ is produced exclusively by neurons.

In contrast to PKA, regulatory and catalytic domains in PKC share one polypeptide chain. Differences in the regulatory domains lead to a definition of three subfamilies (**Figure 4.36**).

(1) **Conventional PKCs (cPKCs).** These enzymes require both DAG and Ca^{2+} ions for activation. Association with membrane-bound DAG is accomplished by two **C1 (Conserved) domains** arranged in a tandem structure. C1 domains are Cys-rich sequences of about 50 amino acids with a DAG-binding pocket stabilized by two zinc ions. The role of DAG is to shield the hydrophilic region of the domain, rendering it hydrophobic enough to interact with membrane phospholipids, in particular with phosphatidylserine. **C2 domains** provide a second binding site for acidic phospholipids, enabling a Ca^{2+}-dependent association of cPKC with membranes in addition to the C1-mediated binding. The Ca^{2+} ions are bound by negatively charged amino acid residues connecting these with the anionic headgroup of phosphatidylserine. As already discussed, C2 domains play an important role in the reversible docking of proteins to membranes. The name protein kinase C refers to the Ca^{2+} dependency of these domains, which were originally discovered in PKC. However, in contrast to C1 domains, C2 domains are not restricted to DAG-interacting proteins such as PKC but are found in many other proteins (see Section 14.6.1).

(2) **New PKCs (nPKCs).** These enzymes are activated by DAG but do not require Ca^{2+}. The reason is an altered C2 domain that lacks acidic amino acid residues needed for Ca^{2+} binding, whereas the C1 domain is intact.

(3) **Atypical PKCs (aPKCs).** These enzymes (which must not be confused with atypical protein kinases!) require neither DAG nor Ca^{2+} for activation, since they lack a C2 domain and the C1 domain exists only as a single copy that cannot bind to DAG but still interacts with phosphatidylserine.

Within the PKC family only atypical PKCι has been identified as an oncoprotein that is overexpressed in various human cancers. Thus, this PKC isoform appears to be a particularly attractive target of anticancer strategies. This holds true also for other PKC isoforms because—even if not oncogenic per se—they occupy central points in cellular signal processing that are involved in several stages of carcinogenesis (see also Section 13.3).

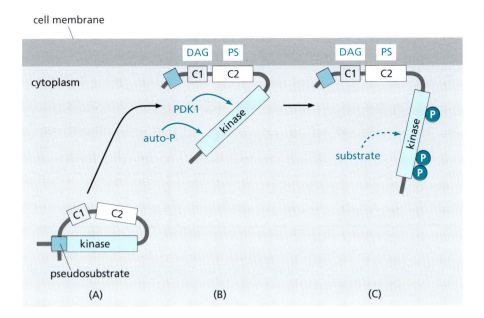

cell membrane

cytoplasm

pseudosubstrate

(A)   (B)   (C)

Figure 4.37 Activation of conventional protein kinase C Activation is a multi-step process. (A) In the inactive enzyme, a pseudosubstrate motif occupies the binding sites for protein kinase PDK1 (which phosphorylates the activation loop of PKC) and for the substrates. (B) The blockade is lifted by binding to membrane-bound diacylglycerol (DAG) and phosphatidylserine (PS). PKC can be transferred into the substrate-binding open conformation by autophosphorylation (auto-P) and phosphorylation by PDK1. (C) In the open conformation, PKC interacts with the substrates ATP and effector protein. P, phosphate group. For the sake of clarity only one of the two C1 domains of cPKC is depicted.

Mechanism of activation

PKC becomes activated in a rather complicated multi-step process. First the enzyme must associate with the inner side of the cell membrane. To this end, membrane-binding domains such as C1 and C2 are indispensable because all PKC isoforms are intrasterically blocked by a pseudosubstrate sequence that is displaced from the catalytic center only when the kinase comes in contact with acidic membrane phospholipids such as phosphatidylserine, causing a conformational change (**Figure 4.37**). This enables another protein kinase to gain access and phosphorylate the activation loop of PKC. The "activation loop kinase" has turned out to be identical with the phospholipid-dependent kinase **PDK1** (see Section 4.6.4). To gain full activity, PKC—like other AGC kinases—needs to be phosphorylated, in addition, at two sites in the substrate-binding domain. This is achieved by autophosphorylation or by the protein kinase TOR (see Section 9.4).

The simple phosphorylation consensus sequence Arg-x-x-Ser/Thr-x-Arg-x recognized by PKC is found in many proteins, making them potential PKC substrates. Thus, in principle PKC is as nonselective as PKA. But again specificity is gained by various anchor proteins. These include the AKAPs (see Section 4.6) as well as receptors of active C-kinases (RACKs) and substrates of inactive C-kinases (STICKs). In addition, anchor proteins have been found that bind individual PKC isotypes and modulate their activities. Such anchor proteins direct PKC not only to substrates but also to regulatory proteins, thus acting as scaffolds of signal-transducing protein complexes. A striking example is provided by **InaD**. In the *Drosophila* eye, this scaffold protein assembles a complex of light-processing proteins including PKC, which here has a negative control function (see Section 15.5).

Beside PKC some other proteins interact with DAG (**Sidebar 4.5**). Moreover, several natural products have been found that specifically imitate the intracellular effects of DAG. The most prominent of these DAG mimics are the phorbol esters. Their effect is indeed unique, since no other substance is known to imitate so specifically a second messenger. Phorbol is a diterpene alcohol found as a diester in certain plants of the spurge family (Euphorbiaceae). The prototype of the phorbol esters is 12-*O*-tetradecanoylphorbol 13-acetate or TPA (frequently found in the literature under the synonym phorbol myristate acetate, PMA; **Figure 4.38**) from *Croton tiglium*. TPA overactivates DAG-controlled pathways since it has a higher binding affinity to C1 domains and is metabolically much more stable than DAG. For this reason TPA has become a preferred tool of experimental cell biology.

Sidebar 4.5. Diacylglycerol effector proteins: more than just protein kinase C High-affinity C1-binding sites for phosphatidylserine and DAG are found not only in PKC but also in other proteins. These include the following:

● protein kinase D (see Section 4 6.5)

● DAG kinases, which catalyze phosphorylation of DAG to phosphatidic acid and thus extinction of the DAG signal (but at the same time generation of phosphatidic acid signals); mammals express nine isoforms, but only DAGKα, -β, and -γ are activated by DAG

● GRP, a GDP–GTP exchange factor of the small G-protein Ras (see Section 10.1.3)

● chimaerins, GTPase-activating proteins of the small G-protein Rac (see Section 10.3.1)

● Munc-13, Mammalian homolog of the protein unco-ordinated 13 of *C. elegans*, restricted to neurons; DAG promotes association with the presynaptic membrane, where Munc-13 acts as a chaperone for syntaxin, thus facilitating neurotransmitter release.

Long before its role as a DAG imitator was discovered, TPA gained great significance as a highly efficient tumor promoter in experimental cancer research (see Section 13.3).

Protein kinase C-related kinases

More distant relatives of PKC are the kinases of the **PRK family** (Protein kinase C-Related Kinases, also known as protein kinases N; see Figure 4.36) occurring in several isoforms in all eukaryotes. Like PKC, these versatile enzymes must associate with membranes to be phosphorylated and activated by PDK1. However, the mechanism of membrane anchoring differs from that of PKC in that it depends on several N-terminal HR1 domains rather than on C1 and C2 domains. HR1 domains specifically interact with the small G-protein Rho, which in its GTP-loaded form reversibly binds to membranes, thus providing a membrane docking site for PRK. In other words, PRKs are Rho effectors: their activation depends on an active Rho signal (for more details on Rho signaling, see Section 10.3). In yeast a primeval Pkc1 has been found. This enzyme combines the properties of PKC and PRK and is considered to represent an evolutionary precursor of both kinase families (see Figure 4.36).

4.6.4 Activation loop kinase PDK1, a master key for activation of AGC kinases

By catalyzing activation loop phosphorylation, the kinase PDK1 (Phospholipid-Dependent Kinase 1) is a key regulator of most kinases of the AGC superfamily. Gene-knockout experiments have shown the kinase to be essential for survival. PDK1 differs from most other protein kinases in that there is only one corresponding gene in the human genome and that no PDK1 isoforms (such as a "PDK2") have been found. The name of the enzyme indicates that it interacts with membrane phospholipids to phosphorylate its substrates recruited to membranes. However, PDK1 activity does not depend on phospholipids since the enzyme is constitutively active because of autophosphorylation of its activation loop. Thus, input signals

Figure 4.38 Phorbol ester TPA

Figure 4.39 Activation loop kinase PDK1 and its substrates PDK1 is reversibly bound at the inner side of the plasma membrane by an interaction of its PH domain with phosphatidylinositol 3,4,5-trisphosphate (PIP$_3$) and catalyzes the activation loop phosphorylation of various kinases of the AGC family shown in the teal boxes. These kinases are recruited to the membrane and prestimulated by membrane-bound activators (teal letters; SPHI, sphingosine) or primed by phosphorylation (catalyzed by MAP kinases and mTOR kinases; see Section 9.4.1). The question as to whether PKA is activated by PDK1 or by autophosphorylation only is still open.

only control the intracellular distribution of PDK1 by generating docking sites at the membrane. These docking sites are 3-phosphorylated PIs such as PIP$_3$ produced along the PI3K signaling pathways. In fact for PIP$_3$ binding, PDK1 contains a PH domain.

PDK1 substrates, in addition to PKC and PRK, include PKB/Akt, ribosomal S6 kinases p90 RSK and p70 S6K (see Section 9.4.2), p21-activated kinase PAK (see Section 11.3), and others (summarized in **Figure 4.39**). To be phosphorylated by PDK1, these substrate kinases have to undergo preactivation to displace the pseudosubstrate sequences from the activation loops. This is achieved by an interaction with membrane-bound activators such as DAG and phosphatidylserine (for PKC), PIP$_3$ (for PKB), Rho-GTP (for PRK), and sphingosine (for PAK) or by phosphorylation (for ribosomal S6 kinases). In other words, the effects of PDK1 depend on both a special conformation and a specific cellular localization of the substrate protein.

4.6.5 Protein kinases D are integrators of phosphatidylinositol 3-kinase, protein kinase C, and G-protein signaling

In the course of PKC research, a novel kinase was encountered that, because of its DAG-binding C1 domains, was thought to be a new PKC isoform and therefore was called PKCμ. However, the enzyme differed so much from PKC that now it is considered to be the representative of a separate protein kinase family. This family is called protein kinase D (PKD) and includes several isoforms.

Like c- and nPKC, PKD has two C1 domains, whereas the auto-inhibitory role of a pseudosubstrate domain is taken over by a **PH domain** interacting with the catalytic center. This PH domain becomes displaced by signals activating PKD, such as the membrane-bound mediators PIP$_2$/PIP$_3$ and Gβγ (**Figure 4.40**). The function of the C1 domains, as in the case of PKC, is to reversibly anchor PKD to membrane-bound DAG, thus fulfilling the condition for activation loop phosphorylation, catalyzed here by DAG-bound a- and nPKC rather than by PDK1. Thus the enzymes of the PKD family are downstream effectors of PKC as well as of PI3K (providing PIP$_2$/PIP$_3$), G-proteins (providing DAG and Gβγ), and Tyr kinase-coupled receptors (providing DAG and activating PI3K). This multiplicity of input signaling predestines the PKDs to function as versatile logical gates. Nevertheless, the physiological function of PKD is still not entirely clear. Some data suggest a role in mitogenic signaling and in the control of the Golgi apparatus.

4.6.6 Protein kinase B/Akt and the processing of survival and insulin signals

A major performer in the signaling cascades controlled by PI3K is protein kinase B (PKB). Since this Ser/Thr-specific enzyme may mutate to the oncoprotein Akt (referring to a spontaneous thymoma of the AKR mouse strain where this mutation was

Figure 4.40 Protein kinase D (A) Domain structure of PKD with interacting factors. (B) Activation by DAG and PKC. Via its second C1 domain, PKD binds to DAG in the membrane. This neutralizes the intrasteric inhibition by the PH (pleckstrin homology) domain (which subsequently may interact with Gβγ or PIP$_3$ localized in the membrane), thus opening the catalytic center for activation loop phosphorylation by PKC and substrate binding.

found for the first time), it is frequently called PKB/Akt or simply Akt. Mammalian cells express three isoforms. Like other AGC kinases, PKB/Akt is activated in several steps, including at least two phosphorylations and a specific interaction with cellular membrane systems. PKB has no pseudosubstrate sequence. Instead, as in PKD, the catalytic center is blocked by an N-terminal PH domain (**Figure 4.41**). This auto-inhibition is relieved when the PH domain associates with the second messengers PIP$_2$/PIP$_3$ at the inner side of the plasma membrane. So, the activation of PKB always requires cooperation with PI3Ks supplying PIP$_2$/PIP$_3$ as input signals (see Figure 4.41). In membrane-bound PKB, the active center is accessible for PDK1-catalyzed activation loop phosphorylation. To open the kinase domain for substrate

Figure 4.41 Domain structure and activation of protein kinase B/ Akt (Top) Domain structure of PKB/ Akt (PH, pleckstrin homology domain; HM, hydrophobic motif). (Bottom) Scheme of multi-step activation. (A) In inactive PKB, the kinase domain is blocked by an intramolecular interaction with PH and HM. (B) Partial unfolding of inactive PKB, through an interaction of PH with membrane-bound phosphatidylinositol 3,4,5-trisphosphate (PIP$_3$), enables activation loop phosphorylation by PDK1 and phosphorylation of HM by an unknown kinase. (C) PKB with fully unfolded substrate-binding site. P, phosphate.

Figure 4.42 Protein kinase B-controlled signaling pathways In a shortened way, the figure shows the activation of PKB by PIP$_3$ generated along the receptor–phosphatidylinositol 3-kinase (PI3K) pathway. The phospholipid phosphatase PTEN antagonizes PKB activation and acts as a tumor suppressor protein. (A) Pro-mitogenic signaling: PKB phosphorylates and inactivates several anti-proliferative factors such as cell cycle inhibitors p21, p27, and MYT1; protein kinase GSK3, suppressing the activity and synthesis of cyclins and mRNA translation; and FOXO transcription factors, inhibiting cyclin transcription and promoting synthesis of the cell cycle inhibitor retinoblastoma protein (see also Section 12.7). (B) Antiapoptotic signaling: PKB phosphorylates and inactivates proapoptotic proteins and FOXO transcription factors that stimulate pro-apoptotic genes (see Section 13.2.4). FasL, Fas ligand. (C–E) Insulin signaling: PKB promotes glycogen synthesis by inactivating GSK3 (C) and recruiting the glucose transporter GluT4 to the membrane (E) and stimulates protein synthesis by inactivating GSK3 and indirectly activating the protein kinase mTOR (D; see Section 9.4.1).

proteins, another protein kinase, most probably mTORC2 (see Section 9.4.1), has to phosphorylate, in addition, a Ser residue in a C-terminal hydrophobic motif of PKB (see Figure 4.41). mTORC2, which is itself a PKB substrate, is a regulator of insulin effects (see Section 9.4.1).

Like the other AGC kinases, PKB is a multipurpose kinase that occupies a key position in vertebrate signal processing. Among the various effects, two functions stand out (**Figure 4.42**): transduction of survival signals by suppression of apoptosis in favor of cell proliferation, and transduction of insulin signals.

Protein kinase B, cell survival, and cancer

The survival effect is based on the phosphorylation by PKB of diverse pro-apoptotic proteins (see Section 13.2.5) that are thus inactivated. These factors include the mitochondrial protein Bad, the enzyme caspase 8, and the **FOXO transcription factors** (Forkhead box, type O). The latter are widely distributed in eukaryotes. Their major function is an adaptive response to stress. To this end they stimulate the transcription of genes encoding pro-apoptotic proteins such as Bim and Fas ligand as well as of genes encoding cell cycle inhibitors such as p21$_{Kip}$ and p27$_{Kip}$ (see Section 12.4), and of the retinoblastoma protein p130 (blocking the cell cycle in the absence of mitogenic signals; see Section 12.6). FOXO factors *inhibit* the transcription of the *cyclin D* gene, thus blocking a major input terminal of mitogenic signals (see Section 12.6). In animal models FOXO activity has been shown to promote longevity. Population studies indicate a similar correlation between longevity and FOXO (in particular a specific polymorphism of the *foxo3* gene) in humans. This life-prolonging effect is most probably due to an increase of cellular stress resistance. By promoting autophagy and protein degradation in proteasomes FOXO enables the cell to survive under stress conditions such as starvation, and to get rid of defective proteins and organelles that might be toxic. Moreover, FOXO suppresses oxidative stress and prevents a premature depletion of the stem cell pool thus promoting tissue renewal. The breakdown of such cytoprotective mechanisms plays a key role in age-related diseases such as neurodegeneration (Alzheimer's, Parkinson's, and Huntington's disease, vascular dementia) muscle atrophy, atherosclerosis, cancer, and others.

Through PKB-catalyzed phosphorylation, FOXO is hindered from entering the nucleus since it is sequestered in the cytoplasm by **14-3-3 proteins** (see Section 10.1.4 for more details on this sequestering mechanism). This inhibitory effect is seen under nutrient-rich conditions, resulting in an increased release of anabolic hormones such as insulin that activate the PI3K–PKB cascade.

PKB suppresses not only de novo synthesis but also activity of the previously-mentioned cell cycle inhibitors. At the same time, the kinase turns off glycogen synthase kinase 3 (GSK3), an anti-proliferative enzyme that inhibits mRNA translation (see Section 9.4.4) and interrupts the cell cycle by phosphorylation of cyclins D and E thus priming them for ubiquitylation and proteasomal degradation. MYT1, another anti-proliferative protein kinase, is also inactivated by PKB. By phosphorylating the cyclin B–CDK1 complex, MYT1 prevents the transition from G2- into M-phase (see Chapter 12 for details on cell cycle regulation). These observations indicate a key role of PKB in the regulation of tissue homeostasis, the steady-state between the birth and decay rates of cells. Thus, it is easily conceivable that a permanent overactivation of PKB due to oncogenic mutation promotes tumor development. The complex activation mode of the enzyme, including at least two "safety locks" (two-fold phosphorylation), may provide a measure against accidental stimulation. Moreover, a protein preventing PKB activation is expected to function as a tumor suppressor. Such a protein is the phospholipid phosphatase PTEN, the antagonist of PI3K (see Section 4.4.3). In fact, the *pten* gene has been found to be inactive in many human cancers

Protein kinase B and insulin action

By promoting major anabolic effects such as glycogen and protein synthesis, PKB occupies a central position in the mechanism of action of insulin (see Figure 4.42). A major PKB substrate related to insulin action is glycogen synthase kinase 3β (GSK3), which becomes inactive upon phosphorylation. Since GSK3 inhibits glycogensynthase (see Section 5.3.1), PKB indirectly stimulates the deposition of glycogen, an insulin effect *par excellence*. To supply glycogen synthase with its substrate glucose, PKB promotes, in addition, the exposure at the cell surface of the glucose transporter GluT4, which facilitates the uptake of glucose from blood into muscle cells, another characteristic insulin effect (for details see Section 7.1.3).

Finally, the anabolic actions of insulin on protein synthesis and cell growth are mediated by PKB: the kinase activates a signaling reaction leading to a stimulation of the protein kinase mTOR that increases mRNA translation, at the same time suppressing the inhibitory effect of GSK3 on ribosomal protein synthesis (see Section 9.4.1).

Summary

- Protein kinase A (PKA) is activated by cAMP. PKA has a wide variety of potential substrate proteins and functions. Specificity is achieved by anchor proteins concentrating the kinase at distinct cellular sites and in the vicinity of selected substrates and other signal-transducing proteins.

- cGMP-dependent protein kinases exhibit various effects including vasodilatation, diuresis, and water transport in mucous epithelia.

- Protein kinase C (PKC) enzymes require activation loop phosphorylation by phospholipid-dependent kinase 1 (PDK1) and attachment to acidic membrane lipids. The subgroup of nPKC needs, in addition, an interaction with membrane-bound diacylglycerol (DAG), and cPKC subgroup requires both DAG and Ca^{2+} ions. Thus, all PKC isoforms operate downstream of PI3K (activating PDK1), and cPKC and nPKC, in addition, operate downstream of PLC (providing DAG and Ca^{2+}). The rather nonselective PKCs gain substrate specificity through anchor proteins.

- PKC-related kinases operate downstream of signaling pathways leading to activation of the small G-protein Rho.

- PDK1 operates downstream of PI3Ks since it requires membrane attachment via 3-phosphorylated phosphatidylinositols (PIs). Its main function is to activate protein kinases of the AGC group by catalyzing activation loop phosphorylation.

- Protein kinases D (PKDs), like PKCs, need DAG for activation. Membrane attachment is facilitated by 3-phosphorylated PIs and Gβγ-subunits, while activation loop phosphorylation is catalyzed by PKC. Thus PKDs are a common downstream target of several signaling pathways.

- Protein kinases of the B/Akt family (PKB/Akt) require 3-phosphorylated PIs for membrane attachment and PDK1 for activation loop phosphorylation; they operate downstream of PI3Ks. PKB/Akt significantly contributes to cell survival (but also to tumorigenesis). A second major function is transduction of anabolic insulin signals.

Further reading

Section 4.3

Blumer JB & Lanier SM (2014) Activators of G protein signaling exhibit broad functionality and define a distinct core signaling triad. *Mol Pharmacol* 85:388–396.

Cabrera-Vera TM, Vanhauwe J, Thomas TO et al. (2003) Insights into G protein structure, function, and regulation. *Endocr Rev* 24:765–781.

Dohlman HG & Thorner JW (2001) Regulation of G protein-initiated signal transduction in yeast. *Annu Rev Biochem* 70:703–754.

Lemmon MA (2008) Membrane recognition by phospholipid binding domains. *Nat Rev Mol Cell Biol* 9:99–11.

McCudden CR, Hains MD, Kimple RJ et al. (2005) G-protein signaling: back to the future. *Cell Mol Life Sci* 62:551–577.

Neves SR, Ram PT & Iyengar R (2002) G protein pathways. *Science* 296:1636–1639.

Riobo NA & Manning DR (2005) Receptors coupled to heterotrimeric G proteins of the G_{12} family. *Trends Pharmacol Sci* 26:146–154.

Wettschureck N & Offermanns S (2005) Mammalian G proteins and their cell type specific functions. *Physiol Rev* 85:1159–1204.

Section 4.4

Airola MV & Hannun YA (2013) Sphingolipid metabolism and neural sphongomyelinases. *Handb Exp Pharmacol* 215:57–76.

Balla T (2013) Phosphoinositides: tiny lipids with giant impact on cell regulation. *Physiol Rev* 93:1019–1137.

Cooper DM F & Tabbasum VG (2014) Adenylate cyclase-centered microdomains. *Biochem J* 462:199–213.

Francis SH, Blount MA & Corbin JD (2011) Mammalian cyclic nucleotide phosphodiesterase: molecular mechanisms and physiological functions. *Physiol Rev* 91:651–690.

Gericke A et al. (2013) PI(4,5)P_2-mediated cell signaling: emerging principles and PTEN as a paradigm for regulatory mechanism. *Adv Exp Med Biol* 991:85–104.

Gomez-Cambronero J (2014) Phospholipase D in cell signaling: from myriad of cell functions to cancer growth and metastasis. *J Biol Chem* 289:22557–22566.

Kadamur G & Ross EM (2013) Mammalian phospholipase C. *Annu Rev Physiol* 75:127–154.

Kühn H, Banthiya S & van Leyen K (2015) Mammalian lipoxygenases and their biological relevance. *Biochim Biophys Acta* 1851:308–330.

Rosen H et al. (2013) Sphingosine-1-phosphate and its receptors: structure, signaling, and influence. *Annu Rev Biochem* 82:627–662.

Seifert R, Schneider RH & Bähre H (2015) From canonical to non-canonical cyclic nucleotides as second messengers: pharmacological implications. *Pharmacol Ther* 148:154–184.

Serhan CN, Chiang N & Dalli J (2015) The resolution code of acute inflammation: Novel pro-resolving lipid mediators in resolution. *Semin Immunol* 27:200–215.

Wang D & Dubois RN (2010) Eicosanoids and cancer. *Nat Rev Cancer* 10:181–193.

Worby CA & Dixon JE (2014) PTEN. *Annu Rev Biochem* 83:641–669.

Section 4.5

Kuhn M (2016) Molecular physiology of membrane guanylyl cyclase receptors. Physiol. Rev. 96: 751–804.

Section 4.6

Ellwanger K & Hausser A (2013) Physiological functions of protein kinase D in vivo. *IUBMB Life* 65:98–107.

Hers I, Vincent EE & Tavare JM (2011) Akt signalling in health and disease. *Cell Sign* 23:1515–1527.

Martin-Liberal J et al. (2014) Targeting protein kinase C in sarcoma. *Biochim Biophys Acta* 1846:547–559.

Martins R, Lithgow GJ & Link W (2015) Long live FOXO: unraveling the role of FOXO proteins and longevity. *Aging Cell* doi: 10 111/accel 12427.

Parker P et al. (2014) Atypical protein kinase C as a human on-cogene and therapeutic target. *Biochem Pharmacol* 88:1–11.

Rosse C et al. (2010) PKC and the control of localized signal dynamics. *Nat Rev Mol Cell Biol* 11:103–112.

Rozengurt E (2011) Protein kinase D signaling: multiple biological functions in health and disease. *Physiology* (*Bethesda*) 26:23–33.

Scheffzek K & Welti S (2012) Pleckstrin homology (PH) domains – versatile modules in protein-protein interaction platforms. *FEBS Lett* 586:2662–2673.

Scott JD, Dessauer CW & Tasken K (2013) Creating order from chaos: cellular regulation by kinase anchoring *Annu Rev Pharmacol Toxicol* 53:187–210.

Steegborn C (2014) Structure, mechanism, and regulation of soluble adenylyl cyclase - similarities and differences to transmembrane adenylyl cyclases. *Biochim Biophys Acta* 1842:2535–2547.

Steinberg SF (2008) Structural basis of protein kinase C isoform function. *Physiol Rev* 88:1341–1378.

Su B & Jacinto E (2011) Mammalian TOR signaling to the AGC kinases. *Crit Rev Biochem Mol Biol* 46:527–547.

Taylor SS & Kornev AP (2011) Protein kinases: evolution of dynamic regulatory proteins. *Trends Biochem Sci* 36:65–77.

Toker A & Marmiroli S (2014) Signaling specificity in the Akt pathway in biology and disease. *Adv Biol Reg* 55:28–38.

Signal Transduction by Receptors with Seven Transmembrane Domains

5

In Chapter 4 we saw that trimeric G-proteins generally receive their input signals from a distinct class of receptors localized in the cell membrane. These are characterized by seven transmembrane domains and are called heptahelical, seven-pass, 7TM, or serpentine receptors. In animal cells they constitute the largest receptor family. They interact with a huge number of intercellular and environmental ligands, and in medicine they are the most important molecular targets of therapeutic measures. In fact, this receptor type is targeted by more than one-third of clinically employed drugs. For their work on heptahelical receptors Robert J. Lefkowitz and Brian Kobilka were awarded the 2012 Nobel Prize in Chemistry.

In this chapter, the basic mode of action of this receptor type will be explained with the help of a few examples. Moreover, it will be shown that such receptors not only activate trimeric G-proteins but transduce exogenous signals to the signal-processing protein network of the cell also along other pathways.

5.1 An evolutionary model of success

Heptahelical receptors represent one of the most successful models of evolution. Since its emergence as archaeal rhodopsin, this particular protein structure has evolved in animals into the largest receptor family and one of the largest gene and protein families overall. While yeast cells possess only three (**Sidebar 5.1**), the nematode *Caenorhabditis elegans* has more than 1000 such genes, representing the largest gene family of this animal. Encoding mainly chemoreceptors for environmental odorants and taste stimuli, these genes guarantee the special way of life of the worm.

In the human genome, around 800 active and several hundred pseudogenes of heptahelical receptors have been found. The active genes constitute the fourth largest gene family and encode 460 olfactory and 342 non-olfactory receptors. Owing to post-transcriptional (such as alternative splicing of pre-mRNA) and post-translational modifications, the actual number of receptors is considerably higher.

It appears that the structural principle of seven interconnected transmembrane domains (resembling the magic number seven!) was "invented" several times in the course of evolution, since no significant sequence homologies exist between the prokaryotic, invertebrate, and mammalian proteins. Such a pronounced evolutionary convergence indicates that this structure is uniquely suited for the interaction with signaling molecules and the transduction of signals across the cell membrane. The circular arrangement of the seven transmembrane domains linked by polypeptide loops starting in an amino (N)-terminal extracellular polypeptide tail of variable length obviously offers nearly unlimited variability to adjust to any possible shape of

Sidebar 5.1 G-protein-coupled receptors in yeast and plants and as viral pathogens In the yeast *Saccharomyces cerevisiae*, only three GPCRs have been found that are related neither to bacterial rhodopsins nor to GPCRs of vertebrates. They are sensors of sexual pheromones and of glucose. The action of **yeast pheromones** ranks among the best-investigated cellular events controlled by GPCRs. The pheromones are peptides produced as a- and α-factors by the haploid gametes (a- and α-cells) in the course of sexual reproduction. In the corresponding partner cell, each pheromone induces readiness for mating, culminating in the fission giving a diploid wild-type cell. To this end a-cells express an α-receptor Ste2 and α-cells express an a-receptor Ste3 (the term Ste refers to sterile mutants). Both are heptahelical proteins coupling with the trimeric G-protein Gpa1. The βγ-subunit of Gpa1 activates the transcription of mating-specific genes via a protein kinase Ste20 and a mitogen-activated protein (MAP) kinase module (**Figure 5.1**). Investigation of this signaling pathway has led to the discovery of the regulators of G-protein signaling (RGS proteins).

The **glucose receptor Gpr1 of yeast** is also G-protein-coupled. It activates a signaling cascade resembling the β-adrenergic signaling pathway in vertebrate cells, though Gpr1 is not related to the β-adrenergic receptor. So, the α-subunit of the Gpr1-coupled G-protein Gpa2 activates adenylate cyclase and the cAMP thus produced induces a fermentation phenotype via a yeast homolog of PKA and a series of protein phosphorylations (see Figure 5.1).

Compared with animals, the family of heptahelical receptors is entirely underdeveloped in plants. Although about 50 GPCR-like genes have been identified in the *Arabidopsis* genome, their sequences are widely unrelated to those of animal GPCR genes and only two of the corresponding proteins (GCR1 and GCR2) have been found to interact with a Gα-subunit. As far as G-proteins are concerned there is little choice, since *Arabidopsis* possesses only one Gα-, one Gβ-, and two Gγ-genes, and practically the same holds true for all other plants the genomes of which have been sequenced.

Moreover, plants lack the typical downstream effectors such as adenylate cyclase and phospholipase Cβ as well as regulatory proteins such as arrestins and GPCR kinases (see Sections 5.7 and 5.8).

The Gα-subunit (called GPA1 for *Arabidopsis*) seems to fulfill several purposes. One of its functions is to control a mitogenic signaling cascade which probably is induced by plant hormones of the brassinosteroid family. However, the brassinosteroid receptor is coupled to a protein kinase rather than to a G-protein (see Section 6.1) and the mechanism of signal transduction remains a contentious issue.

GPA1 is also involved in the control of stomata opening. Stomata are pores at the surface of leaves that serve as the exchange media of water and gases. They open and close depending on environmental conditions. These movements are controlled by ion channels regulating the

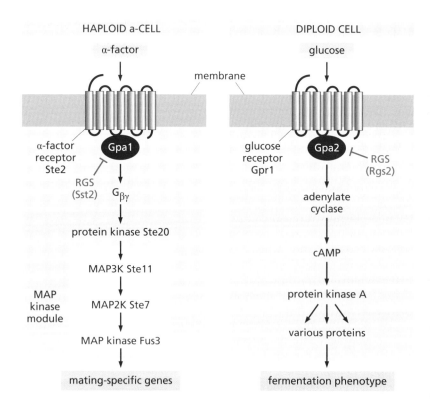

Figure 5.1 GPCR-regulated signaling pathways in yeast (Left) Pheromone signaling pathway of a haploid a-cell. (Right) Fermentation pathway of a diploid cell. Inhibition of G-protein signaling by regulator of G-protein signaling (RGS) proteins, activating intrinsic GTPase activity, is shown in teal.

turgor pressure in the stomata cells. **Abscisic acid**, a multifunctional plant hormone, inhibits one of these channels and thus prevents the gating of stomata. This reaction is mediated by GPA1. In fact, abscisic acid interacts with one of the above- mentioned GPCR-like receptors (GCR2). Another receptor of abscisic acid is a cytoplasmic protein that stimulates a protein kinase activating transcription factors, ion pumps and other effectors.

Several types of human **herpesvirus**, including the cytomegalovirus as well as the tumorigenic forms of Kaposi sarcoma virus HHV8 and Epstein-Barr virus, captured vertebrate GPCR genes (especially those of the chemokine receptor family) in the course of evolution and misuse them as pathogens. Compared with their mammalian counterparts, the viral GPCRs are mutated in such a way that they do not require a ligand; they are constitutively active.

Depending on the virus type, the receptors have been found to activate $G_{i,0}$-, $G_{q,11}$-, $G_{12,13}$-, and G_s-controlled pathways, thus triggering cell proliferation, inhibiting apoptosis, promoting cell migration and angiogenesis, and escaping immune surveillance. Considering the role herpesviruses play in a wide variety of potentially life-threatening diseases, the receptors may turn out to be highly important targets of therapeutic measures.

cell membrane

ligand (11-*cis*-retinal)

Figure 5.2 Arrangement of transmembrane helices in a heptahelical receptor (mammalian rhodopsin) In this schematic representation of a model based on X-ray analysis, the circular arrangement of the transmembrane helices (cylinders) within the protein (gray cloud) is shown. Note that the ligand 11-*cis*-retinal (teal), a small molecule, is bound in the cavity formed by the transmembrane helices (whereas more bulky ligands such as peptides and proteins interact with the extracellular domains of the receptor protein). Beside rhodopsin, the structures and interactions of many other heptahelical receptors have been elucidated in detail.

signaling molecules (**Figure 5.2**). Technology has realized this principle in the combination lock to be opened by a discrete series of numerals. It must be remembered, however, that this story of success holds true for animals only. In plants and prokaryotes the heptahelical receptor type seems to be quite rare (see Sidebar 5.1) or absent and signal transduction is dominated by protein kinase-coupled receptors.

Summary

In animal cells, receptor proteins anchored in the cell membrane by seven transmembrane domains represent the most abundant receptor type encoded by one of the largest gene families. Phylogenetic predecessors are the rhodopsins of archaebacteria.

5.2 G-protein-coupled receptors: structure and mode of operation

The best-known function of most receptors with seven transmembrane domains is to catalyze (as GDP–GTP exchange factors) the loading of trimeric G-proteins with GTP upon binding of the ligand. Thus they are called G-protein-coupled receptors

Figure 5.3 Major signaling pathways downstream of G-protein-coupled receptors: a recapitulation Upon ligand binding, G-protein-coupled receptors (GPCRs) activate the G_s-, $G_{i,o}$-, and $G_{q,11}$-dependent signaling cascades regulating second messenger release and the membrane potential. An individual GPCR may interact either selectively with one or promiscuously with more than one G-protein (see Figure 5.4). The second messengers generated and the downstream actions of the effector proteins (in the teal boxes) are shown to the right. As can be seen, the $G_{i,o}$ family comprises the most versatile G-proteins. Black arrows indicate activation; teal pistils symbolize inhibition. GIRK, $G_{i,o}$-protein-gated inwardly rectifying K+ channel.

(GPCRs). The major signaling cascades originating from these receptor/G-protein couplings have been described in Chapter 4. A schematic overview is seen in **Figure 5.3**.

It should be remembered that there is practically no aspect of cellular life—whether development, survival, motility, metabolism or genetic readout—that is not addressed by these signaling cascades.

Table 5.1, which presents a selection of GPCR ligands acting as input signals and their corresponding receptors, illustrates the variability and versatility of this receptor type. The number of ligands is striking indeed, ranging from ions, nucleotides, amino acids, amines, and lipids to peptides and proteins. It must not be overlooked, however, that many of the GPCRs are still "orphans": their natural ligands are unknown. Elucidation of their physiological functions is a major task in the field of molecular pharmacology.

Table 5.1 G-protein-coupled receptors of mammals: a selection[ab]

Ligand	$G_{q,11}$	$G_{i,o}$	G_s	Alternative receptor coupling
acetylcholine (5)				cation channel
adenosine (4)				
adrenaline (9)				
angiotensin (2)				
bradykinin (2)				
calcitonin and relatives (4)				
calcium ions (1)				
cannabinoid (2)				
chemokine (17)				
cholecystokinin/gastrin (2)				
CRH (2)				
dopamine (5)				
endothelin (2)				
fatty acids (5)				
FSH				
GABA (2)				anion channel
galanin (3)				
glucagon				

glutamate (8)	cation channel
histamine (4)	
leukotriene	
LH	
lysophospholipid	
melanocortin (5)	
melatonin (3)	
neuropeptide Y (6)	
neurotensin (2)	
opioid (4)	
oxytocin	
platelet-activating factor	
prostanoid (15)	
purine (4)	cation channel
rhodopsin (retinal)	
serotonin (13)	cation channel
somatostatin (5)	
sphingosine 1-phosphate	
tachykinin (3)	
thrombin (4)	
TRH (2)	
TSH	
vasopressin (3)	
VIP (3)	

[a]The number of isoforms unequivocally identified is given in parentheses. For a more comprehensive list the reader is referred to Wettschureck & Offermanns (2005) and Civelli et al. (2013).

[b]CRH, corticotropin-releasing hormone; FSH, follicle-stimulating hormone; GABA, γ-aminobutyric acid; LH, luteinizing hormone; TRH, thyrotropin-releasing hormone; TSH, thyroid-stimulating hormone; VIP, vasoactive intestinal peptide.

As can be seen in Table 5.1, almost every ligand interacts with more than one receptor isoform. This **principle of multiple receptors** is reinforced by the principle of **variable G-protein-coupling**. Accordingly, G-protein-coupling is not necessarily restricted to a specific receptor isoform, but the same isoform may interact with different G-proteins.

Many GPCRs exist in several G-protein-specific conformations in equilibrium with each other (**Figure 5.4**). A change of G-protein specificity may be due to post-translational modifications of the receptor protein. So the β2-adrenergic receptor turns from G_s- into G_i-coupling upon phosphorylation by protein kinase A (PKA). Such modifications provide almost unlimited possibilities for tissue-specific fine-tuning of signaling. In view of this variability it is not surprising that the pharmaceutical industry is developing countless synthetic ligands of GPCR. They comprise the majority of all medical drugs.

In the postgenomic era it has become possible to construct phylogenetic trees for individual gene families. By this means, five GPCR families have been identified for human cells, which may have evolved from a single primeval gene by gene doubling, exon shuffling, and other molecular mechanisms of evolution. Only 24 out of almost 1000 GPCR genes cannot be assigned to these families. The GPCR families of man are summarized in **Table 5.2** (among the ligands, only the most frequent ones are

Figure 5.4 Mechanisms of variable G-protein-coupling of GPCRs
(A) Different isoforms of a receptor interact selectively with individual G-proteins. (B) A receptor exists in several conformations, each of which interacts with an individual G-protein. (C) Change of G-protein-coupling by post-translational modification shown for the β_2-adrenergic receptor as an example. Upon feedback phosphorylation (P) by protein kinase A and exchange of the G-protein, the adenylate cyclase-activating becomes an adenylate cyclase-inactivating receptor.

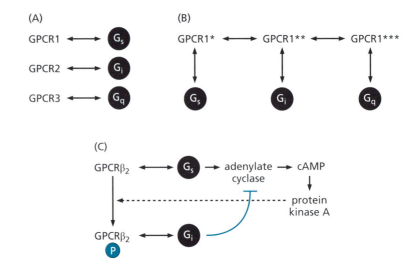

Table 5.2 Human GPCR families[a]

Class A: Rhodopsin family (702)
α-group (89): amine receptors (40) of serotonin, histamine, (nor)adrenaline, dopamine, acetylcholine and chemokines; prostanoid receptors (15), opsins of retina, receptors of adenosine, melatonin, melanocortin, cannabinoids, lysophosphatidic acid
β- and γ-group (95): receptors of peptide hormones and peptide neurotransmitters
δ-group (518): receptors of pituitary gland hormones, purine/pyrimidine receptors, thrombin receptor, odorant receptors (460)

Class B: Secretin receptor family (15)
receptors of secretin, calcitonin, glucagon, parathyroid hormone, corticotropin releasing hormone

Class C: Glutamate receptor family (22)
metabotropic glutamate receptors, metabotropic GABAB receptors, calcium receptors, T1 taste receptors

Adhesion receptor family[b] (33)

Class F: Frizzled/TAS2 family (35)
Wnt receptors (10), T2 taste receptors (24), Smo proteins of Hedgehog signaling[c]

[a]The number of different receptor types is given in parentheses. The table does not contain classes D and E because they comprise non-vertebrate receptors such as the fungal mating pheromone receptors (D) and cAMP receptors of slime molds (E).

[b]These receptors are also known as LNB-7TM proteins, with LN referring to an extraordinarily long N-terminal domain and B to a former classification of GPCR families. The extracellular N-terminal domain contains interaction motifs that are found also in cell adhesion molecules and receptors acting in a juxtacrine manner. It has been proposed, therefore, that LNB-7TM proteins mediate interactions between adjacent cells by binding homotypically to each other or heterotypically to other cell surface proteins.

[c]Most effects of these receptors occur without G-protein-coupling

quoted). The largest family by far is the rhodopsin group, covering about 700 receptors arranged in four subgroups.

5.2.1 A closer look at the architecture and mode of operation of G-protein-coupled receptors

GPCRs are type 1 transmembrane proteins (N-terminus outside, carboxy (C)-terminus inside the cell). Each transmembrane helix of a GPCR is about 25–35 amino acids long and penetrates the plasma membrane. This membrane binding is frequently

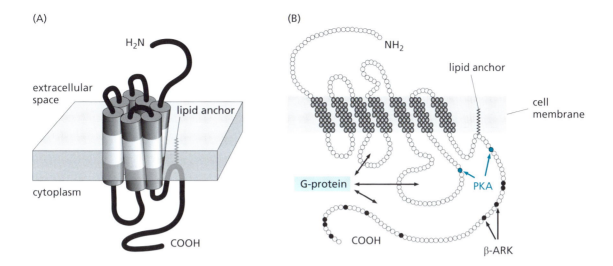

reinforced by a long-chain lipid anchor (**Figure 5.5**). Structural investigations have shown that the interaction of the receptor with the ligand causes a conformational change that induces the GDP–GTP exchange and the dissociation of the G-protein into subunits. As a result, the binding site for the effector protein becomes uncovered in the Gα- or Gβγ-subunit.

How this conformational change proceeds has been investigated in detail by using rhodopsin as a model. It was found that the ligand—here retinal—is bound by a lysine residue of transmembrane helix 7 as a protonated Schiff base, resembling the situation in bacteriorhodopsin (see Section 3.4.5). The positively charged amide group interacts through a "salt bridge" (an electrostatic bond) with a negatively charged glutamate (Glu) residue of transmembrane helix 3. Light induces a *cis–trans* isomerization of retinal, causing the hydrolysis of the Schiff base bond. As a result the salt bridge breaks down and helices 6 and 7 move outward opening a cleft in the cytoplasmic surface of the receptor. This cleft harbors the binding site for the Gα-subunit and the catalytic center for GDP–GTP exchange (**Figure 5.6**). It must be emphasized that the activation of a receptor such as rhodopsin by *hydrolysis* of a covalent bond between ligand and receptor protein represents a rare exception. However, it was found that other GPCRs canonically activated by noncovalent ligand *association* undergo similar conformational changes through specific horizontal movements of the transmembrane helices. In fact, for a switching process, the direction of switching (association versus dissociation) does not matter.

5.2.2 Benefits of receptor oligomerization

Like most receptors, GPCRs exist as homo- and heterodimers (**Figure 5.7**). This dimerization is mostly constitutive but in some cases may be induced also by ligands. Oligomerization has far-reaching consequences for receptor function, altering ligand specificity or modulating G-protein coupling of one monomer by the other one. Below are some examples.

(1) **Change of ligand specificity.** The G_s-coupled T1 taste receptors respond as T1R1–T1R3 heterodimers to certain amino acids mediating the taste sensation "umami," whereas in the combination T1R2–T1R3 they are sweet sensors (see Section 15.1.2).

(2) **Change of sensitivity.** Certain $G_{i,0}$-coupled receptors in brain neurons such as dopamine D2 receptor, somatostatin SSTR5 receptor, and opioid receptors DOP and KOP are much more effective as heterodimers of different isoforms than as homodimers.

Figure 5.5 Constructive principle of G-protein-coupled receptors (A) Three-dimensional scheme. Note the tilted (nonparallel) transmembrane helices (as shown also in Figure 5.2) and the lipid anchor. (B) Schematic "serpent model" of the β-adrenergic receptor including the lipid anchor and the phosphorylation sites of PKA (teal) and β-adrenergic receptor kinase (β-ARK, black), as well as the domains of G-protein interaction in the second and third cytoplasmic loops and in the C-terminal part of the molecule. Phosphorylation mostly inhibits the interaction with the G-protein (however, see Figure 5.4 for an exception). Receptors of this type bind small ligands such as adrenaline or acetylcholine in the central channel formed by the helices, whereas more bulky ligands such as peptides and proteins interact with the extracellular loops and the amino-terminal extension.

Figure 5.6 Mechanism of rhodopsin activation (A) Breakdown of the intramolecular salt bridge between helices 3 and 7 of the opsin protein caused by hydrolysis of the covalent Schiff base bond between opsin and retinal. (B) View of the seven transmembrane helices (cylinders) from below. The conformational change due to breakdown of the salt bridge opens the interaction site for the α-subunit of the G-protein. Not shown are additional interactions between amino acid residues in helices 2 and 7 as well as in helices 3 and 6, which further stabilize the conformation of the inactive receptor and become disrupted upon activation. (B adapted from Bourne HR [1997] *Curr Opin Cell Biol* 9:134.) For other details, see text.

(3) **Function.** The function of the metabotropic B-type receptors of the neurotransmitter γ-aminobutyric acid (GABA) critically depends on the heterodimerization of the isoforms GABAB-R1 and GABAB-R2, which show only 35% sequence homology. The corresponding homodimers are inactive (see Section 16.3).

A special case is the hetero-oligomerization of certain GPCRs with **RAMPs** (Receptor Activity Modifying Proteins). These are single-pass transmembrane proteins with a large extracellular domain. The three RAMP isoforms known have been shown to facilitate the transport of some GPCRs, particularly those of the secretin receptor family (such as receptors for the thyroid gland hormone calcitonin and related peptides), to the cell surface. In addition, they modulate the functions of those GPCRs in such a way that the same receptor acquires different ligand specificity depending on the interacting RAMP isoform. These findings indicate a much greater diversity of GPCR functions than previously imagined. In the context of receptor oligomerization, the reader should remember that oligomeric protein molecules provide a noise filter, which is an important condition of efficient and specific signal processing (see Section 1.3). For details on methodologies for analysis of protein-protein interactions, see **Sidebar 5.2**.

Summary

- Heptahelical receptors are incorporated in the plasma membrane through a circularly arranged group of seven transmembrane domains that may be completed by a long-chain lipid residue. Upon binding of extracellular ligands most of them stimulate associated trimeric G-proteins by catalyzing GDP–GTP exchange and subunit dissociation (G-protein-coupled receptors or GPCRs). By interacting with different receptor isotypes or conformations, most ligands activate more than one G-protein type.

- While in animals GPCRs are the most abundant receptors, plants and yeast express very few GPCRs.

- Through oligomerization of GPCRs, with each other or with receptor activity modifying proteins (RAMPs), a noise filter is established and ligand specificity and sensitivity are changed.

Figure 5.7 Dimerization of G-protein-coupled receptors Dimers are formed by an interaction (A) between extracellular domains (for instance, homodimerization of the β2 adrenergic receptor); (B) between transmembrane domains (found in various GPCRs); or (C) between intracellular domains (for instance heterodimerization of GABAs receptor isoforms through a coiled-coil structure). (D) Heterodimerization of a class 2 GPCR (secretin receptor family) with a receptor activity modifying protein (RAMP) is shown. Depending on the RAMP type, a receptor interacts with different ligands. Each dimer probably activates only one G-protein molecule.

Sidebar 5.2 Analysis of protein–protein interactions
To demonstrate noncovalent interactions between proteins, such as the heterodimerization of receptors, several methods have been developed. They are of utmost importance for research because cellular signal processing is primarily due to interactions of proteins aiming at network formation.

Preliminary indications of a noncovalent binding of two proteins are obtained in the course of protein purification, when, for example, different proteins cannot be separated by chromatography (**co-chromatography**) or when an antiserum against a protein 1 also precipitates a protein 2 (**co-immunoprecipitation**). The molecular-biological method of the **yeast two-hybrid system** is more reliable (**Figure 5.8**).

Figure 5.8 Yeast two-hybrid system for analysis of protein–protein interactions The DNA-binding domain (light gray) and the transactivating domain (dark gray) are (A) separated enzymatically from transcription factor GAL4 of yeast at the sites indicated and then (B) fused with proteins to be tested. If the bait and the prey thus obtained interact with each other (C), linking the functional domains of the transcription factor at a correct distance, the reporter gene of the yeast cell transformed with the fusion products becomes activated. For further details, see text.

Figure 5.9 Experimental proof of protein–protein interactions by fluorescence resonance energy transfer (A) Biological resonance energy transfer between the protein aequorin (containing the fluorophore coelenterazine) and the green fluorescent protein (GFP) in the luminous pacific jellyfish *Aequorea victoria*. The fluorescence of aequorin is excited by the ATPase luciferase (or by Ca^{2+} ions) under oxygen consumption. By Förster resonance energy transfer, the excited aequorin induces fluorescence of GFP. Independently, GFP fluorescence is also induced by UV light. (B, C) Proof of protein–protein interaction by fluorescence resonance energy transfer (FRET). The two proteins expected to interact have been labeled with different fluorophores (F1 and F2), sending out light of different colors, and fluorescence is activated by UV light (B). If the conditions of FRET are fulfilled, the fluorescence intensity of the donor protein decreases to the same degree as the fluorescence intensity of the acceptor protein increases (C).

Here one makes use of the fact that the activity of certain transcription factors depends only on the correct distance between DNA-binding and transactivating domains; it is insignificant what type of polypeptide sequence bridges the gap between the domains. An example is provided by the transcription factor GAL4 of yeast, inducing transcription of the β-galactosidase gene. Here the sequence between the N-terminal DNA-binding domain and the C-terminal transactivating domain can be cut out enzymatically and replaced by a foreign protein. To prove an interaction between two foreign proteins 1 and 2, protein 1 is fused with the DNA-binding domain and protein 2 with the transactivating domain and both fusion products are expressed in yeast cells. Upon interaction, the so-called reporter gene *gal4* becomes activated and β-galactosidase is synthesized. For a quantitative evaluation of enzyme activity, a simple test is at hand: upon hydrolysis of the synthetic substrate *o*-nitrophenyl β-D-galactoside, the product *o*-nitrophenol is generated, which is easily assayed because of its yellow color. When the transactivating domain of GAL4 is fused with all proteins obtained from a cDNA library, the hybrid molecule of protein 1 with the DNA-binding domain can be used as "bait" to fish out the "prey"; that is, all interacting proteins. Although the method is very sensitive, it frequently produces false-positive results due to nonspecific interactions. Moreover, the results obtained do not necessarily reflect the situation in the living cell. Thus, they have to be verified by microscopic techniques such as **immunofluorescence microscopy** or **multicolor fluorescent protein imaging** (which, on the other hand, both suffer from rather poor spatial resolution).

A method to prove protein–protein interactions more precisely is based on a quantum-physical phenomenon discovered by the German physical chemist Theodor Förster in 1946. It is known as **fluorescence resonance energy transfer (FRET**; a variant is called bioluminescence resonance energy transfer or BRET). The method is based on the following principle. Fluorescent protein 1 transfers a certain portion of its fluorescence energy to protein 2 fluorescing at another wavelength, provided the emission spectrum of the donor overlaps with the excitation spectrum of the acceptor and the distance between the proteins is less than 100 A. If this is the case, the fluorescence intensity of protein 1 is diminished exactly as much as the fluorescence intensity of protein 2 increases (**Figure 5.9**). This energy transfer can be measured. Since most proteins do not fluoresce on their own, they have to be labeled with suitable chromophores that either are synthetic compounds or are derived from biological sources.

firefly luciferin

luciferin coelenterazine
of jellyfish aequorins

fluorophore of the green fluorescent protein GFP,
(a tripeptide forming a
p-hydroxybenzylidene-imidazolinone
integrated in a polypeptide)

Figure 5.10 Biological fluorophores

In fact, certain organisms make use of resonance energy transfer to produce bioluminescence. The light energy is obtained by the oxidation of a luciferin, catalyzed by an oxygen- and ATP-consuming luciferase. Luciferins are prosthetic groups of photoproteins (**Figure 5.10**). A well-studied example is provided by the pacific jellyfish *Aequorea victoria*. This animal uses the luciferase-catalyzed oxidation of the luciferin coelenterazine embedded in the photoprotein aequorin (emitting blue light) to stimulate, by resonance energy transfer, a green fluorescent protein (GFP) to emit green-yellow light (see Figure 5.9). The fluorophore of GFP is a cyclic tripeptide integrated in the protein (see Figure 5.10). This system finds many and diverse applications in research. The luciferase reaction may be employed as an extremely sensitive ATP assay (the lower limit is less than 0.1 pmol of ATP). Moreover, since light emission by aequorin is also stimulated by Ca^{2+} ions, it can be used to monitor local changes of cellular Ca^{2+} (see Section 14.5 for an example). Finally, GFP is a highly suitable probe to localize proteins in cells. For this purpose, fusion proteins with

GFP are prepared. These hybrid molecules are expressed in the cell and their fluorescence is stimulated by UV light. It has turned out that the tagging with GFP does not significantly change the intracellular distribution of most proteins.

The most advanced technique to study protein–protein interactions is **optogenetics**. Originally this method was developed for neuroscience (see Section 16.4.1, Sidebar 16.2). It makes use of photosensory proteins obtained from prokaryotes and plants (see Section 3.4.5, Sidebar 3.4). These proteins function as light receptors in signaling cascades. Upon illumination they undergo a conformational change that induces either an ion flux or a specific interaction with downstream binding proteins. An example is provided by the photosensory protein phytochrome PHYB of the plant *Arabidopsis thaliana* that upon exposure to red light reversibly binds to the protein phytochrome-interacting factor PIF. To study the interaction of two proteins 1 and 2 the target cell is genetically manipulated in such a way that it expresses these proteins as chimeras: protein 1 fused with the light-sensitive

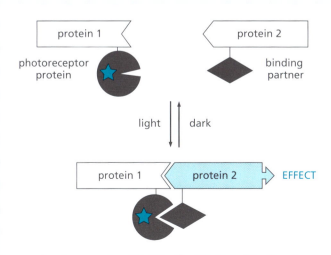

Figure 5.11 Experimental proof of protein–protein interactions by optogenetics To exert an effect (for example an enzymatic reaction), protein 2 has to interact with protein 1. This interaction can be induced by light when protein 1 is labelled with a photoreceptor domain that undergoes a conformational change upon illumination, and protein 2 with a binding partner that interacts with the light-activated photoreceptor domain. The teal asterisk symbolizes the chromophore of the photoreceptor domain.

binding domain of the phytochrome, and protein 2 fused with the corresponding binding domain of PIF (in addition, the target cell has to get to produce the phytochrome chromophore, a phytocyanobilin; see Section 3.4.5). Through light exposure, protein 1 and protein 2 are brought together allowing an interaction (**Figure 5.11**). At first glance, the method appears to be closely related to the yeast two-hybrid system. The great advantage of the optogenetic approach is, however, that the interaction can be induced or interrupted *ad libitum* and with a high degree of precision in the living target cell, in an intact tissue or even in a living organism by a non-invasive signal. Moreover, when the binding partner of the photosensory protein has been equipped with a lipophilic membrane anchor the technique is highly suitable also to study the attachment of proteins to cellular membranes. In an analogous fashion the interaction of proteins with nucleic acids can be investigated in an approach of light-switched gene transcription. Currently optogenetic techniques outdo any other method as far as temporal and spatial resolutions are concerned. It is the only method that allows kinetic studies under nearly real-time conditions at a millisecond scale.

5.3 Adrenergic receptors: sensors of stress and sympathetic signals

To treat the GPCR family in all detail would break up the frame of this book and make little sense, since the basic functions common to all receptors of this type may be explained with the help of a few examples. Particularly good examples are the stress effects and the interplay between sympathetic and parasympathetic systems explained in Chapter 4. In fact, the adrenergic receptors of adrenaline and noradrenaline that process stress and sympathetic effects are GPCRs. Vertebrate cells express these receptors in nine isoforms. In the first half of the twentieth century, pharmacological studies led to the definition of two receptor families, called α- and β-adrenergic. Since then it has become clear that this classification is based on different G-protein couplings: the receptors of the α_1-family (three isotypes) activate the $G_{q,11}$–PLCβ pathway, α_2-receptors (three isotypes) activate $G_{i,0}$-proteins, and β-receptors (three isotypes) are coupled to G_s-proteins stimulating cAMP formation (**Figure 5.12**). The receptor isoforms differ in their sensitivities for adrenaline versus noradrenaline as well as in tissue distribution. β_1-Receptors prefer noradrenaline and β_2-receptors prefer adrenaline, whereas β_3-receptors do not distinguish between the two β_1-Receptors are found predominantly in heart muscle cells; β_2-receptors are particularly abundant in liver, smooth muscle, skeletal muscle, and glandular cells; and β_3- receptors are characteristic of adipose tissue. While α-adrenergic receptors principally mediate sympathetic and stress effects on blood pressure, major functions of β-adrenergic receptors are the stimulation of heart activity and the mobilization of the body's energy reserves.

Figure 5.12 Stress and sympathetic effects mediated by G-protein-coupled adrenergic receptor

α_1-adrenergic	α_2-adrenergic	β-adrenergic
smooth muscle contraction (blood pressure rises) viscous saliva (dry mouth) goose flesh	smooth muscle relaxation (blood pressure falls)	increased heart beat glycogenolysis and lipolysis relaxation of airway muscles inhibition of peristalsis

5.3.1 Biosynthesis and degradation of glycogen: how signals control metabolic processes, part I

To open the endogenous energy reserves in stress situations, β-adrenergic receptors transduce the (nor)adrenaline signal to the metabolic apparatus of the cell, stimulating glycogen and fat metabolism. The investigation of this process represents a milestone in science history. In fact, in the mid-twentieth century it was shown for the first time how a metabolic process, glycogen metabolism, is controlled by a signal-processing protein network. As pointed out in Section 4.2, these investigations culminated in the discovery of regulatory protein phosphorylation and of cAMP as a second messenger of hormonal signals. Moreover, they led to the first verifiable theory of how a signal is transduced from the cellular periphery to the metabolic machinery inside the cell.

Glycogen metabolism

Glycogen, a polymer of glucose molecules mainly linked by 1,4-α-glycosidic bonds, is the primary energy depot of the body. While most cells can accumulate small amounts of glycogen, the major stores are in the liver and the skeletal muscles (only when these stores are filled will the energy-richest food component, fat, be deposited). After a meal, the liver and muscles take up glucose from the blood, the liver by a constitutively expressed membrane transporter (the protein GluT2) and the muscle by a related transporter (GluT4), the expression of which has to be induced by insulin. Inside the cell, glucose is phosphorylated at position 6 by hexokinase. The glucose 6-phosphate thus generated either may be degraded along the glycolytic pathways, destined for ATP formation or it becomes transformed into glucose 1-phosphate and UDP-glucose, destined for incorporation into glycogen. The last step is catalyzed by the enzyme glycogen synthase (**Figure 5.13**).

Figure 5.13 Breakdown and biosynthesis of glycogen P, phosphate residue.

Figure 5.14 Regulation of blood sugar level The glucose level in the blood is an essential parameter of the internal milieu. Normally, it is precisely adjusted to 0.06–0.10% owing to a steady-state between glucose release (mainly from the liver) and glucose consumption by skeletal muscle and other tissues. A deviation from this mean value may result in hyper- or hypoglycemic shock. Therefore, the steady-state is controlled continuously by the two peptide hormones glucagon (promoting glycogen breakdown and glucose release) and insulin (promoting glucose uptake and glycogen formation). Both hormones are synthesized by individual pancreatic cell populations that monitor the rise or fall of the glucose level by means of appropriate sensor proteins. To match the strongly increased glucose consumption in stress situations, glycogen breakdown and glucose release are reinforced by (nor)adrenaline (teal). While glucagon and insulin constitute an autonomous regulatory system (being, of course, under the supervision of the vegetative nervous system), the stress axis is dominated by the central nervous system monitoring the environmental (or emotional) situation.

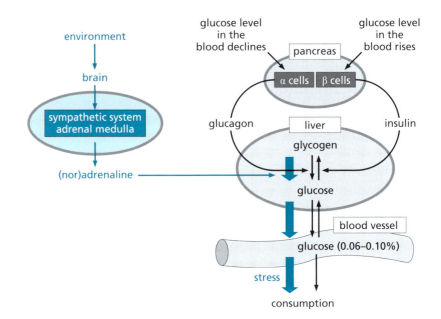

The synthase has an opponent in glycogen phosphorylase, which catalyzes the phosphorolytic breakdown of glycogen to glucose 1-phosphate. The latter is rearranged to the 6-phosphate, which may be either utilized for glycolysis or dephosphorylated to glucose to be released into the bloodstream (**Figure 5.14**).

Both glycogen synthase and glycogen phosphorylase are under the strict control of the signal-processing protein network. They are "switched" by signal protein phosphorylation, albeit with an important difference: phosphorylation *activates* the phosphorylase but *inactivates* the synthase. While the activation of glycogen phosphorylase is specifically catalyzed by a phosphorylase kinase, glycogen synthase may be inactivated by various kinases. In stress situations; however, first and foremost, it is controlled by cAMP-dependent PKA activated along the β-adrenergic route.

The cellular signal-processing network controlling the activity of the glycogen-metabolizing enzymes is under the supervision of numerous hormones and neurotransmitters and, in addition, is influenced by physical activity. The major players in this game are **glucagon**, promoting glycogen degradation, and **insulin**, promoting glycogen formation. Both are peptide hormones released by individual cell types of the pancreatic Langerhans islets. These cells are equipped with glucose-sensing proteins monitoring any rise or fall of the blood sugar level (see also Section 14.4.2). In a competitive manner both hormones adjust the glucose concentration in blood to a value of 0.06–0.10%. When the energy demand increases, for example, under stress, this regulatory feedback loop becomes overtaxed and has to be reinforced by additional hormones and neurotransmitters, first of all (nor)adrenaline (see Figure 5.14).

Although glucagon and (nor)adrenaline interact with different GPCRs, both use the same G_s–cAMP–PKA signaling pathway to stimulate glycogen degradation. In fact, PKA thus activated has two major substrates: glycogen synthase, which is turned off, and the above-mentioned phosphorylase kinase, which is turned on by phosphorylation.

Phosphorylase kinase

Phosphorylase kinase has become famous as the first protein kinase identified. In contrast to PKA, it is a very specific enzyme with only one major substrate, glycogen phosphorylase, which by phosphorylation of a Ser residue is transformed from an inactive b-form into an active a-form (see Chapter 4, Figure 4.5). With four different subunits,

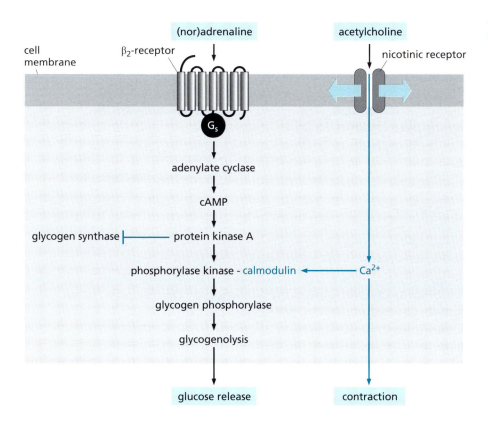

Figure 5.15 Coupling of contraction and glycogen degradation in skeletal muscle Phosphorylase kinase is activated by protein kinase A as well as Ca^{2+}, the signal of muscle contraction. Ca^{2+} enters the cytoplasm through acetylcholine-controlled cation channels (nicotinic receptors) as well as Ca^{2+} channels in the endoplasmic reticulum (not shown; see Section 14.5.3). The calcium sensor is the calmodulin subunit of phosphorylase kinase. The effect of phosphorylation is to increase the Ca^{2+} sensitivity of the enzyme.

phosphorylase kinase has a rather complex quaternary structure. One of these subunits is the Ca^{2+}-sensor protein calmodulin (CaM) that plays an important role especially in skeletal muscle, where it couples the Ca^{2+}-dependent muscle contraction, an energy-consuming process, with glycogen degradation, an energy-supplying process (**Figure 5.15**; more details on muscle contraction are found in Section 14.6.2).

Glycogen synthase kinase 3

PKA activated along the beta-adrenergic pathway is only one among several kinases inhibiting glycogen synthase. Other Ser/Thr-directed protein kinases turning down glycogen synthesis are glycogen synthase kinase 3 (GSK3), the "casein kinases" CK1 and CK2 and the 5′-AMP-dependent kinase AMPK. As discussed in Section 9.4, these kinases are regulated along signaling pathways other than those for PKA.

For an understanding of insulin action, GSK3 is of special interest. In the past, insulin was thought to directly stimulate protein phosphatases, thus reversing the phosphorylation of glycogen synthase and glycogen phosphorylase. Today another mechanism is favored where insulin via phosphatidylinositol 3-kinase activates protein kinase B (PKB)/Akt, which in turn inhibits GSK3 (see Section 4.6.6 for more details on PKB/Akt). As a result, constitutively active protein phosphatases gain the upper hand and dephosphorylate the enzymes of glycogen metabolism. This holds true, in particular, for **protein phosphatase PP1**, a (in principle) nonspecific enzyme that nevertheless is selectively bound to glycogen by regulatory glycogen-targeting subunits (see also Section 2.6.4 for the control of protein phosphatases).

As far as phosphorylation of glycogen synthase is concerned, the situation depicted in Figure 5.15 is somewhat oversimplified. In fact, for the inhibition of glycogen synthase, several protein kinases have to cooperate. In a first step the enzyme becomes phosphorylated by CK1 at a specific serine (Ser) residue. This "priming" provides the condition for additional phosphorylations by PKA. The whole sequence resembles a logical AND gate and has been called **hierarchical phosphorylation**. In a similar manner, GSK3 and AMPK cooperate with protein kinase CK2 (**Figure 5.16**).

Figure 5.16 "Hierarchical phosphorylation" of glycogen synthase Shown are the phosphorylation sites for protein kinase A (PKA), AMP-dependent protein kinase (AMPK), casein kinases CK1 and CK2, and glycogen synthase kinase 3β (GSK3).

CK1 and CK2 are constitutively active regulator kinases, whereas AMPK represents an energy sensor of the cell, becoming active upon a fall of the ATP level and interrupting energy-consuming metabolic reactions such as glycogen synthesis. These three kinases as well as GSK3 are treated in more detail in Section 9.4.

5.3.2 Hormone-stimulated mobilization of fat deposits: how signals control metabolic processes, part II

Besides glucose, long-chain fatty acids are major energy sources of the body. They are stored as glycerol esters in fat cells. The initial step of fat mobilization consists of a hydrolytic degradation of these triglycerides, yielding glycerol and free fatty acids. Like glycogen degradation, this reaction is stimulated by various hormones and catalyzed by a **hormone-sensitive lipase**, an enzyme found predominantly in adipose tissue but also in skeletal muscle and steroid hormone-producing glands. While in adipose and muscle tissue the enzyme is mainly stimulated by (nor)adrenaline or glucagon, in the glands it becomes activated by pituitary hormones such as adrenocorticotropin and releases cholesterol, the precursor of steroids, by hydrolyzing cholesteryl esters. Together these hormones induce lipolysis along the G_s–cAMP–PKA pathway. In fact, as far as the principles of signal transduction are concerned, the hormone-controlled lipolysis represents a simplified version of glycogenolysis. Like glycogen phosphorylase, the hormone-sensitive lipase is activated by protein phosphorylation catalyzed by PKA. In other words, every extracellular signal stimulating cAMP formation along the G_s-adenylate cyclase pathway triggers lipolysis, provided the fat cell expresses the corresponding receptor. Under stress, the mobilization of fat is provoked by (nor)adrenaline interacting with a β3-adrenergic receptor (**Figure 5.17**).

Figure 5.17 Stress hormone-stimulated lipolysis Protein kinase A, activated along the β-receptor–Gs signaling pathway, phosphorylates lipase, which becomes stimulated, and the lipase blocker perilipin, which becomes inactivated. The fatty acids produced thereupon are released into the bloodstream.

However, the phosphorylation of lipase by PKA is not sufficient: to gain access to its substrates, the lipase has to be docked to the fat droplets of adipocytes, where its binding sites normally are blocked by the regulatory protein **perilipin**. To make room for the lipase, perilipin also has to be phosphorylated by PKA. As soon as the hormonal stimulus has ceased, both lipase and perilipin are dephosphorylated again by constitutively active protein phosphatases. This dephosphorylation is augmented by insulin through an activation of PKB/Akt that, in turn, phosphorylates and activates a cAMP phosphodiesterase, thus suppressing the cellular cAMP level.

Insulin not only inhibits lipolysis but also promotes the incorporation of fatty acids into triglycerides. There is a feedback in that fatty acids stimulate insulin release from the pancreas. This effect is controlled by a special class of GPCRs that are activated specifically by saturated and nonsaturated fatty acids ranging from acetic acid to long-chain species. In pancreatic β-cells these **free fatty acid receptors (FFARs)** are strongly expressed but they are also abundant in other tissues such as intestine, adipose tissue, immune cells, and brain. By interacting predominantly with $G_{q,11}$-proteins FFARs stimulate secretory processes through an activation of phospholipase Cβ and, in turn, of InsP3-dependent calcium release (see Figure 5.3). FFARs play an important role in energy metabolism and inflammatory reactions and are attracting much attention as potential targets of novel drugs against obesity, metabolic syndrome, and diabetes.

5.3.3 α-Adrenergic receptors and blood pressure control

Under stress, the blood pressure rises: (nor)adrenaline promotes the contraction of vascular smooth muscles, thus reducing the vascular volume. This effect is mediated by α_1-adrenergic receptors and the $G_{q,11}$ –PLCβ signaling cascade generating $InsP_3$, which leads to a rise of the cytoplasmic Ca^{2+} level by gating Ca^{2+} channels in the endoplasmic reticulum. Ca^{2+} stimulates the protein kinase **MLCK (Myosin Light-Chain Kinase)** that, like phosphorylase kinase, contains the Ca^{2+} sensor protein CaM as a subunit. In turn, the kinase phosphorylates the regulatory myosin light chains, thus lifting their inhibitory effect on the contractile actomyosin complex (for details see Section 14.6.2).

Even during permanent stress, the blood pressure does not increase indefinitely, thus preventing fatal consequences. This self-control is managed by a negative feedback mechanism by which noradrenaline turns down its own release. To this end noradrenaline interacts with an α_2-adrenergic receptor incorporated in the presynaptic membrane, that is, in the membrane of the same sympathetic neuron that releases noradrenaline (**Figure 5.18**). The presynaptic receptor is coupled to a $G_{i,0}$-protein and inhibits noradrenaline release along three pathways:

(1) a $G_{i,0}$-protein-gated inwardly rectifying K^+ channel (see Section 14.4.2) becomes gated, hindering the development of an action potential by hyperpolarization of the neuron

(2) a voltage-dependent Ca^{2+} channel in the presynaptic membrane, which transports the Ca^{2+} required for exocytosis of noradrenaline-filled vesicles, is *inhibited*

(3) the same Ca^{2+} channel is *activated* by PKA-catalyzed phosphorylation; this is the target of the third inhibitory pathway, $G_{i,0}$-dependent suppression of adenylate cyclase activity (for details on Ca^{2+} channels see Section 14.5).

Because of its dual function, noradrenaline is called a mixed $\alpha_1\alpha_2$-agonist. Its structure can be altered synthetically in such a way that pure α_1- and α_2-agonists become

Figure 5.18 α-Adrenergic signal transduction As an example, blood pressure control by regulation of the volume of peripheral blood vessels is shown. In the smooth muscle cell equipped with α_1-adrenergic receptors, contraction is induced along the $G_{q,11}$–PLCβ–$InsP_3$–Ca^{2+} pathway. In the sympathetic neuron innervating the smooth muscle cell, noradrenaline inhibits its own release via a $G_{i,0}$-coupled α_2-adrenergic receptor incorporated in the presynaptic membrane (see Figure 5.19). MLCK, myosin light-chain kinase; DAG, diacylglycerol; ER, endoplasmic reticulum.

Figure 5.19 Blood pressure control by antagonistic α-adrenergic effects (A) α-Adrenergic synapse between a sympathetic neuron and a vascular smooth muscle cell. (B) α₁- and α₂-Adrenergic agonists compared with noradrenaline.

available. This process had led to a broad field of clinical applications, particularly, of course, in blood pressure therapy (**Figure 5.19**).

Summary

- The constancy of the blood sugar level is based on a steady-state between glycogen degradation and glycogen formation. Key enzymes are phorphorylase kinase (for degradation) and glycogen synthase (for formation).

- Glycogen degradation is stimulated by all signals interacting with G_s-coupled receptors. Along this pathway phosphorylase kinase is *activated* by PKA-dependent phosphorylation. Phosphorylase kinase is activated also by Ca^{2+} ions coupling contraction and glycogen degradation in skeletal muscle.

- Glycogen synthase is *inactivated* by phosphorylation catalyzed by various protein kinases. Among these, glycogen synthase kinase 3 (GSK3) becomes switched off by insulin along the phosphatidylinositol 3-kinase–protein kinase B/Akt pathway. This provides a major condition of insulin-dependent glycogen synthesis.

- Lipolysis, the release of free fatty acids from triglycerides, is catalyzed by hormone-controlled lipases. β-Adrenergic and other hormonal stimuli induce lipolysis through PKA-catalyzed phosphorylation of both the lipase that becomes activated and the inhibitor perilipin that becomes inactivated. In the presence of insulin, both proteins are dephosphorylated and lipolysis is inhibited.

- α-Adrenergic receptors play an important role in blood pressure control by the sympathetic nervous system. While postsynaptic α_1-receptors coupled to $G_{q,11}$-proteins provoke an increase of the blood pressure through a Ca^{2+}-dependent contraction of vascular smooth muscles, presynaptic α_2-receptors are components of a negative feedback loop in that they inhibit noradrenaline release from the sympathetic neuron along $G_{i,0}$-controlled pathways.

5.4 Muscarinic acetylcholine receptors are sensors of parasympathetic signals

Acetylcholine is the standard neurotransmitter of the parasympathetic nervous system and, therefore, the major endogenous opponent of noradrenaline. In addition, it transmits contraction-inducing signals at neuromuscular synapses and interneuronal signals in spinal ganglions and in various parts of the central nervous system. In contrast to noradrenaline, acetylcholine interacts with two quite different types of "cholinergic" receptors: while the parasympathetic effects are mediated by a G-protein-coupled type (also called metabotropic type) found in virtually all tissues, the other effects are transmitted by an ion channel-coupled type (also known as ionotropic type; **Figure 5.20**). For historical reasons the GPCRs are called "muscarinic" and the receptor ion channels are known as "nicotinic" acetylcholine receptors. The latter are treated in detail in Section 16.2.

Figure 5.20 Acetylcholine (or cholinergic) receptors Acetylcholine opens cation channels (via nicotinic receptors) and activates $G_{i,0}$- and $G_{q,11}$-controlled signaling cascades (via muscarinic receptors). Nicotinic receptors are specifically blocked by the arrow poison curare, and muscarinic receptors are blocked by atropine, the poison of the deadly nightshade (*Atropa belladonna*). The numbers beneath the muscarinic receptors indicate the five different receptor isotypes M_1–M_5.

Figure 5.21 Parasympathetic effects mediated by muscarinic acetylcholine receptors Depending on the receptor type, acetylcholine may exhibit antagonistic effects, for instance, on smooth muscles.

At the beginning of the twentieth century, when acetylcholine was still unknown, cellular binding sites of muscarine, one of the poisons of the fly agaric mushroom, had been postulated to explain the characteristic toxic effects of this compound. Although muscarine was later found to be an acetylcholine agonist and the binding sites turned out to be acetylcholine receptors, the name **muscarinic (or muscarine) receptors (M-receptors)** had become firmly established (and the same holds true for the nicotinic receptors, originally postulated to explain the pharmacological effects of nicotine).

Mammals possess five genes encoding M-receptors that are tissue-specifically expressed. The M-receptors couple with different G-proteins: types M_1, M_3, and M_5 interact with $G_{q,11}$, and types M_2 and M_4 activate $G_{i,0}$. Some typical parasympathetic effects triggered by M-receptors are shown in **Figure 5.21**. Like adrenergic receptors M-receptors are targets of many pharmaceutical drugs. Of historic interest is the contraction of the eye pupils, since it provides an early example of how a signaling pathway was manipulated. In the past, women used an M-receptor blocker, the plant poison atropine of the deadly nightshade, *Atropa belladonna*, to counteract pupil contraction for a radiant and tempting look.

Summary

- Acetylcholine interacts with both ion-channel-coupled (nicotinic) and G-protein-coupled (muscarinic) receptors.

- Parasympathetic effects are transmitted by M-receptors expressed in five isoforms. They are either $G_{i,0}$- or $G_{q,11}$-coupled and are selectively inhibited by atropine.

5.5 Stress and the heart: the competition between sympathetic and parasympathetic signals

Our treatment of stress and signaling in the vegetative nervous system will be concluded by an instructive example demonstrating the interplay between cholinergic and adrenergic signals, that is, between parasympathetic and sympathetic stimuli, in the regulation of heart activity. In a particularly clear manner it shows how organ function is controlled by a network of antagonistic signaling pathways.

As for most target organs of the vegetative nervous system, the heart is subject to control "by two reins:" upon parasympathetic activation it is slowed down, whereas upon sympathetic activation it beats faster and stronger. Under stress the heart "is pounding up to the neck." The discovery that, upon parasympathetic stimulation, an isolated frog heart released a substance into the culture medium that dampened the frequency of a second heart was a scientific milestone. Using this experimental approach, Otto Loewi (Nobel Prize in Physiology or Medicine for 1936) in Graz demonstrated that neurons transmit chemical messengers instead of communicating exclusively through electrical signals as had been assumed before. Loewi called the signaling substance (today classified as a neurotransmitter) "Vagusstoff," referring to the synonym nervus vagus for parasympathetic nerves. Not until years later was "Vagusstoff" identified as acetylcholine.

Each heartbeat is triggered by an action potential that is generated by the pacemaker cells in the sinoatrial knot and induces in the heart muscle cells an influx of Ca^{2+} through voltage-controlled membrane channels (for details see Section 14.6.2). The *power of contraction* depends on the strength of the Ca^{2+} influx, that is, the length of time in which the Ca^{2+} channels remain open. The opening period is extended by a PKA-catalyzed phosphorylation of the channel protein, which is activated along a β-adrenergic signaling pathway. This provides the mechanism by which the sympathetic nerves and stress, or (nor)adrenaline, increase the *power* of heart contraction (**Figure 5.22**).

The *heart frequency* is controlled by HCN (Hyperpolarization- and Cyclic Nucleotide-activated) ion channels in the pacemaker cells (see Section 14.6.2 for a more in-depth discussion of this subject). They become opened at the end of an action potential (at maximal hyperpolarization) and, being Na^+ channels, have a depolarizing effect. The more active the pacemaker channels, the shorter the intervals between action potentials and the faster the heartbeat. Pacemaker channels belong to the family of cyclic nucleotide-gated cation channels and are sensitized by cAMP binding (see Section 14.5.7). This is the signaling pathway along which sympathetic and stress signals increase the heart frequency (see Figure 5.22).

Acetylcholine activates the inhibitory $G_{i,0}$-cascade via an M_2 receptor, resulting in hyperpolarization, blockade of voltage-dependent Ca^{2+} channels, and inhibition of cAMP production. As a consequence, the heart activity is turned down.

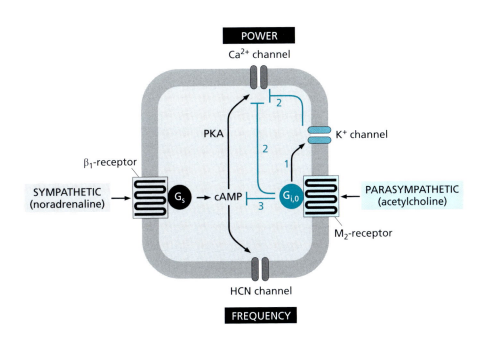

Figure 5.22 Control of heart muscle activity Schematic representation of the heart muscle equipped with a $β_1$- adrenergic receptor for the sympathetic signal (noradrenaline, left) and an M_2-muscarinic receptor for the parasympathetic signal (acetylcholine, right, in teal) as well as with three different ion channels: a voltage-controlled Ca^{2+} channel regulating heart contraction or power (above); a hyperpolarizing K^+ channel (right); and a depolarizing pacemaker channel HCN controlling heart frequency (below). cAMP released by the sympathetic signal activates the Ca^{2+} channel (via PKA-catalyzed phosphorylation) and the HCN channel (directly), thus increasing the contraction power and frequency of heartbeats. This effect is antagonized by the parasympathetic signal that induces hyperpolarization through K^+ channel gating (pathway 1), hinders Ca^{2+} influx by hyperpolarization and a direct effect of $G_{i,0}$ on the Ca^{2+} channel (pathway 2), and inhibits adenylate cyclase (pathway 3). The figure illustrates the regulatory principle without laying claim to anatomical correctness. In fact, HCN channels are restricted to the pacemaker cells in the sinoatrial knot that, through gap junctions, are coupled electrically and metabolically with the rest of the heart.

Summary

- The heart muscle provides a classic example of an organ controlled by antagonistic signals of the vegetative nervous system.

- Via a β-adrenergic receptor, the sympathetic transmitter noradrenaline stimulates both contraction power and frequency by activating voltage-dependent Ca^{2+} channels and cAMP-controlled hyperpolarization- and cyclic nucleotide-activated (HCN) pacemaker channels.

- These effects are antagonized by the parasympathetic transmitter acetylcholine that, along a $G_{i,0}$-controlled pathway, inhibits cAMP formation, blocks the Ca^{2+} channels, and causes hyperpolarization (by gating K^+ channels).

5.6 Protease-activated receptors are major players in blood clotting and inflammation

For most GPCRs, the ligands are separate molecules that interact spontaneously. There are two exceptions: rhodopsins, with their covalently bound ligand retinal, and a special class of receptors that are activated enzymatically by extracellular proteases. In mammals four isoforms of **protease-activated receptors (PARs)** have been found. Originally they had become known as "thrombin receptors" although they react with other proteases as well.

Proteases are not the ligands proper of these receptors. Instead, PARs contain a tethered ligand, a hexapeptide incorporated in the N-terminal extracellular domain. In the resting state this domain is prevented from interacting with the ligand-binding site (localized in the second extracellular peptide loop) by the N-terminal peptide residue of the receptor protein. This intramolecular blockade is abolished by proteases splitting off the auto-inhibitory domain (**Figure 5.23**). Protease-activated receptors exhibit "G-protein promiscuity": they interact with G_q-, $G_{i,0}$-, and $G_{12,13}$-proteins, activating numerous signaling pathways.

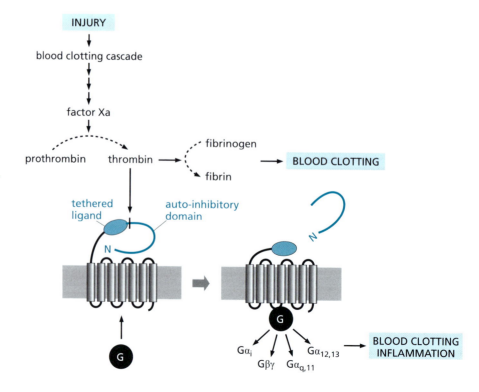

Figure 5.23 Activation of thrombin receptor PAR1 As an example, the proteolytic release of thrombin in the course of blood clotting is shown. Thrombin has two effects: it catalyzes the formation of fibrin from fibrinogen, and it activates PAR1 in various cell types by removing an N-terminal auto-inhibitory domain, thus enabling an interaction of the tethered ligand with the receptor. PAR1 may couple with various G-proteins. For more details, see text.

For the cell the activation of PARs is an expensive venture because it is irreversible: the regeneration of the system requires the synthesis of new receptor protein. Therefore, it makes sense that PARs are employed predominantly in emergency situations where the costs are unimportant. Such an emergency situation is wounding, with blood clotting being an immediate response. In fact, thrombin becomes activated by the blood clotting cascade and has two functions: it catalyzes the proteolysis of fibrinogen to fibrin, thus leading to clot formation, and it activates PARs in cells and tissues that support clotting and perform defense functions.

A key event of blood clotting, including the pathological forms of clotting such as stroke and myocardial infarction is the aggregation of platelets, promoted by an emptying of secretory vesicles containing aggregation-inducing substances such as ATP. The Ca^{2+} ions needed for this exocytotic process are provided by the $G_{q,11}$–PLCβ–InsP$_3$ signaling cascade triggered by a PAR. In fact, thrombin ranks among the most potent inducers of platelet aggregation.

Cells involved in defense reactions are endothelial and epithelial cells, vascular smooth muscle cells, leukocytes, and sensory neurons. Their stimulation evokes the typical symptoms of inflammation such as redness and heat (because of dilation of blood vessels), swelling (because of increased vascular permeability), and pain. In these responses PARs expressed by these cell types play an important role, although it must be emphasized that inflammation is an extraordinarily complex process that is by no means regulated by proteases alone but by many additional factors (see, for instance, Section 7.3.2).

The widening of blood vessels due to a dilation of vascular smooth muscles is induced indirectly by thrombin, since the protease activates, via the $G_{q,11}$ coupled receptor PAR1, the Ca^{2+}-dependent biosynthesis of nitric oxide and some other mediators evoking smooth muscle relaxation (for details see Section 16.5.1). In addition, thrombin stimulates, again through activation of PAR1, the proliferation of endothelial and smooth muscle cells, thus promoting the formation of new blood vessels (angiogenesis). To this end the protease turns up several signaling cascades controlled by PKB/Akt, protein kinase C (PKC), and arrestin (more about arrestin is found in Section 5.8). In parallel the de novo synthesis of the vascular endothelial growth factor (VEGF) is induced, which additionally stimulates these signaling cascades.

Inflammatory pain is caused by an activation of temperature-sensitive heat and pressure receptors in sensory neurons (see Section 15.3). Such neurons are stimulated by several pro-inflammatory mediators including thrombin and related proteases. In fact, neurons as well as neuron-associated astrocytes express several PAR isoforms. In sensory neurons the receptor PAR2 activates along the $G_{q,11}$–PLCβ–diacylglycerol pathway PKC that sensitizes the vanilloid receptor VR1 by phosphorylation. VR1 is a receptor ion channel evoking pain sensations (see Sections 15.3 and 16.7). Certain sensory neurons respond to thrombin and other proteases by a release of pro-inflammatory neuropeptides that cause a "neurogenic inflammation" in adjacent tissues. The cellular events of inflammation are, in addition, dramatically amplified by the Ca^{2+}-dependent release of prostaglandins and related lipid mediators. This response provides the target of most widely used anti-inflammatory drugs such as aspirin and its relatives (see Section 4.4.5).

Cancer development is accompanied frequently by inflammation. Here PARs seem to play a fatal role as tumor-promoting factors in that they induce cell proliferation. To this end they couple with adaptor proteins called arrestins that activate mitogenic signaling cascades (see Section 5.8).

PARs are found not only in peripheral nerves but also in the brain. Brain cells produce small amounts of thrombin that may activate survival, (anti-apoptotic) pathways thus protecting neurons from death due to glucose and oxygen deficiency or oxidative stress, as occurring, for instance, in the course of stroke. In higher concentrations, however, thrombin may promote cell death by inducing Ca^{2+} release. To suppress this pro-apoptotic function of thrombin, neurons produce several antidotes including the protein nexin-1, the strongest thrombin inhibitor known.

Summary

- Protease-activated receptors (PARs) constitute a subfamily of G-protein-coupled receptors predominantly involved in the handling of emergency situations such as wounding and inflammation. They are stimulated by a tethered ligand that is incorporated in the extracellular domain and are activated when the auto-inhibitory N-terminus is removed by proteolysis

- Deregulation of PAR signaling is involved in severe pathological processes such as stroke, myocardial infarction, cancer, and inflammatory diseases.

5.7 Adaptation of G-protein-controlled signaling: a matter of feedback

Rapid signal extinction and adaptation protect the signal-processing apparatus from overloading and collapsing. Moreover, these mechanisms help the apparatus to respond to oscillating signals that are typical within the nervous system. As we have learned from bacterial chemotaxis (see Section 3.4.4), adaptation also ensures maximal sensitivity over a broad range of signal intensity and may function as a molecular short-term memory store.

Signaling pathways controlled by G-proteins are highly adaptive. Let us look at a prototypical example: adaptation of β-adrenergic signal transduction. Principally, adaptation occurs at two levels: extinction of the input signal, and adjustment of the signal-processing apparatus. The standard mechanisms for the extinction of chemical messengers are dilution and metabolic inactivation. At adrenergic synapses, dilution of the noradrenaline released into the synaptic cleft is mainly caused by a recapture mechanism. For this purpose the presynaptic membrane of the neuron is equipped with specific transporter proteins that catalyze the re-uptake of the neurotransmitter (see Section 16.9). In addition, neurons and other cells express enzymes catalyzing the inactivation of noradrenaline. Two of them, the catechol O-methyltransferase COMT and the monoamine oxidase MAO (**Figure 5.24**), are targets of drugs used to treat high blood pressure (owing to α-adrenergic effects) and depression (as is discussed in more detail in Section 16.9, noradrenaline is critically involved in the processing of emotions in the brain).

Adaptation of the signal-processing apparatus takes place primarily at the receptor according to a mechanistic scheme that probably holds true for most GPCRs. A strong input signal leads to a **desensitization** of the receptor within seconds. In the minutes thereafter, the receptor becomes internalized via uptake into endocytotic vesicles, removing it from the membrane. The internalized receptors may be either re-exposed through exocytosis or degraded in lysosomes. The whole sequence of events, called **down-modulation**, is a major reason for tolerance and tachyphylaxis, or addiction to pharmaceutical drugs and narcotics.

The desensitization starts with a phosphorylation of several Ser and threonine (Thr) residues in the cytoplasmic domains of the receptor. It is controlled by two negative feedback loops. In one of these loops, phosphorylation is catalyzed by downstream effector kinases of G-protein-controlled signaling such as PKA (for G_s-signaling) or PKC (for $G_{q,11}$-signaling). Since such kinases may also become activated along other signaling pathways, this type of desensitization is said to be heterologous.

Figure 5.24 Adaptation of β-adrenergic signal transduction The signaling molecule noradrenaline may be inactivated enzymatically by O-methylation, catalyzed by catechol O-methyltransferase, or by the oxidative deamination of the monoamine to an aldehyde group, catalyzed by monoamine oxidase. However, at both α- and β-adrenergic synapses the predominant mechanism of signal extinction consists of the recapture of noradrenaline by the presynaptic neuron. At the level of intracellular signal processing, various mechanisms of adaptation exist. The most important ones are shown here: (1) inactivation of the receptor by phosphorylation, catalyzed by both PKA and a specific GPCR kinase GRK2; (2) interruption of G-protein interaction by β-arrestin; (3) inhibition of adenylate cyclase along an antagonistic signaling pathway; (4) extinction of the cAMP signal by phosphodiesterase; (5) dephosphorylation of PKA substrate proteins; and (6) receptor down-modulation. These pathways are interlinked and connected with additional pathways controlled by other signals, such as those controlling the activities of cAMP phosphodiesterases and protein phosphatases. For the sake of clarity, this signaling cross talk is not shown.

Somewhat more selective is another type of desensitization. Here the receptor phosphorylation is catalyzed by a **G-protein-coupled Receptor Kinase (GRK)** that in mammals is found in seven isoforms belonging to the AGC kinase family. Some GRKs are restricted to individual tissues and are relatively specific: for example, rhodopsin and iodopsin kinases GRK1 and GRK7 are found only in retina cells. In contrast, other GRKs such as the β-adrenergic receptor kinases GRK2 and GRK3 (also known as βARK1 and βARK2) are widely distributed, interacting not only with β-adrenergic receptors but with with most other GPCRs.

To phosphorylate the membrane-integrated receptor, a GRK has to associate with the cell membrane. For this purpose, GRK2 and 3 possess pleckstrin homology (PH) domains that interact with the membrane-bound second messenger phosphatidylinositol 3,4-bisphosphate (PIP$_2$) and with membrane-anchored Gβγ-subunits that both serve as GRK activators. In contrast, other GRKs such as the rhodopsin kinases are constitutively active because they are anchored in the membrane by their own covalently bound lipid residues and do not need the assistance of PIP$_2$ or Gβγ. Such permanent GRK activity is essential for the most rapid light adaptation of the retina (see Section 15.6).

It must be emphasized that GRKs attack exclusively GPCRs activated by the corresponding ligand. In the case of GRK2 and 3 this is easily conceivable, since only an active receptor is able to release the Gβγ-subunits required for GRK activation. Moreover, Gβγ stimulates phosphatidylinositol 3-kinase type γ, which catalyzes the formation of PIP$_2$ (see Section 4.4.3). Since, however, both Gβγ-subunits and PIP$_2$ are generated along a wide variety of other pathways, GPCR desensitization is integrated in a network of signaling cross talk.

Summary

- Via a series of feedback loops, G-protein-dependent signaling is under strict negative control.

- Rapid desensitization is caused by receptor phosphorylation catalyzed by specific G-protein-coupled receptor kinases (GRKs) or other downstream kinases.

- As a result, the contact with the G-protein becomes interrupted and the receptor undergoes internalization through endocytosis. This reversible down-modulation may become irreversible when the receptor protein is degraded in lysosomes.

5.8 Arrestins are multifunctional adaptors for signaling cross talk

To desensitize a receptor, phosphorylation is not sufficient but provides a prologue for the next step, which involves a displacement of the G-protein by an arrestin. Arrestins are eukaryotic proteins that bind specifically to *phosphorylated* GPCRs, thus cooperating with GRKs and other protein kinases. Again one can distinguish between two rhodopsin-specific visual arrestins, found only in retina cells, and two ubiquitously distributed β-arrestins 1 and 2. As in the case of GRKs the prefix β is somewhat misleading since β-arrestins by no means are restricted to β-adrenergic receptors but interact with most other GPCRs. Arrestins are essential proteins since genetically manipulated knockout animals die as embryos.

The interruption of the receptor/G-protein interaction is only one function of arrestins. In addition, they induce receptor internalization and may interlink GPCRs with G-protein-independent signaling pathways. The reason for this versatility is that arrestins, as scaffold proteins, assemble several components of the signal-processing network into functional complexes.

5.8.1 Receptor internalization

As already mentioned, internalization is understood as an uptake of membrane proteins, particularly receptors, into endocytotic vesicles and endosomes. At least in the early phase, this process is reversible, enabling the cell to rapidly and flexibly adjust the number of receptors exposed at the surface to the particular conditions. Internalization starts from dents in the membrane. In these **caveolae** and **coated pits**, receptors accumulate and are internalized according to the mechanism of receptor-induced endocytosis (see Sidebar 5.3). The role of arrestins is to connect as scaffolds the phosphorylated receptor protein with a complex of endocytotic control proteins (**Figure 5.25**). The arrestin-binding site of the receptor is called, therefore, the internalization motif. In addition, arrestins promote endocytosis directly: the separation of the vesicle from the membrane is controlled by the small G-protein Arf6 (see Section 10.4.3) that together with its GDP–GTP exchange factor ARNO (ARf6 Nucleotide binding site Opener) is bound by arrestin and thus concentrated in the coated pit. The ability of arrestins to promote endocytosis is modulated by phosphorylation and ubiquitylation, indicating that these proteins are knots in a more complex signaling network with various feedback and feedforward loops.

The final step of adaptation is the proteolytic degradation of the receptor protein in lysosomes. It represents the cell's ultimate answer to overstimulation. In the subsequent recovery phase, receptors have to be newly synthesized. The individual stages of a membrane receptor's life cycle are summarized in **Figure 5.26**.

Figure 5.25 Internalization of G-protein-coupled receptors by clathrin-dependent endocytosis Activated receptors assembled in coated pits become phosphorylated (P) and bound to the proteins of the endocytotic apparatus (clathrin, Arf6, ARNO) by use of arrestin and the protein AP2 as adaptors. For endocytosis, the invaginated area of the cell membrane is released as a vesicle into the cytoplasm by means of the GTPase dynamin. Afterward the vesicle is stripped of the clathrin coat, which is reutilized. Arf6 is a small G-protein required for vesicle formation (see Section 10.4.3), and ARNO is the corresponding GDP–GTP exchange factor.

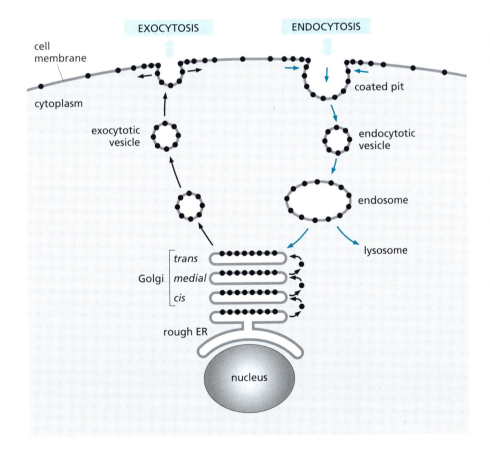

Figure 5.26 Life cycle of transmembrane receptors Receptor proteins (black dots) are synthesized at the rough endoplasmic reticulum (ER), packed into secretory vesicles in the Golgi apparatus, and exposed at the cell surface by exocytosis (left). During adaptation by down-modulation (teal arrows), the membrane receptors assemble in membranous microdomains and membrane invaginations such as clathrin-coated pits, which as endocytotic vesicles become released into the cytoplasm. Several vesicles fuse to endosomes that deliver their content either to the Golgi apparatus for re-utilization or to lysosomes for final degradation.

Sidebar 5.3 Microdomains and endocytosis On membranes, receptors and other membrane proteins are concentrated in microdomains rather than distributed randomly. Such microdomains are due to a demixing of membrane lipids: because of their higher melting temperature, sphingomyelins and cholesterol have the tendency to aggregate to **lipid rafts** swimming in the phospholipid bilayer, which has an oily consistency at body temperature. In these semisolid structures the lateral diffusion of membrane-bound proteins is strongly reduced and the assembly of stable signal-processing protein complexes is facilitated. Such complex formation seems to be supported by intrinsic raft proteins such as caveolins, annexins, and flotillins. How the cell controls the formation and disappearance of lipid rafts is still widely unknown.

A type of lipid raft visible under the microscope is the **caveolae**. These are small dents in the cell surface where the lipid rafts are pulled together and stabilized at the inner side of the membrane by the membrane protein caveolin. Caveolae may be precursors of endocytotic vesicles. Similar membrane invaginations are the **coated pits**. At the inner side of the membrane they are coated by the structural protein **clathrin**, which, owing to its triskelion structure forms basket-like aggregates, stabilizing the invagination. Like caveolae, coated pits are precursors of endocytotic vesicles. In both cases the separation of the vesicle from the membrane is controlled by **dynamin**. This monomeric G-protein aggregates at the neck of the invagination, forming a collar that contracts under GTP hydrolysis (see Figure 5.25). Endocytosis of clathrin-coated pits is the major mechanism by which cells take up macromolecules and distribute them between individual membrane compartments such as plasma membrane, endosomes, lysosomes, Golgi apparatus, and endoplasmic reticulum.

Separation of the vesicles and their transport between cellular compartments is regulated by a special class of adaptor proteins (APs). These proteins form complexes with clathrin and control vesicle trafficking between plasma membrane and endosomes (protein AP2) or endosomes and Golgi apparatus (proteins AP1 and AP3). Before the vesicle fuses with its target membrane, the clathrin coat has to be removed. This **uncoating** is under the control of several proteins, with the chaperone Hsc70 and its co-chaperone auxilin playing key roles. Auxilin is the ATPase-activating protein of Hsc70 (itself an ATPase) that is active only in the ADP-loaded form. Hsc70 changes the conformation of clathrin in such a way that the vesicle coat disintegrates. During this uncoating, the ADP bound to Hsc70 is replaced by ATP (for details on the mechanism of action of chaperone ATPases, see Section 2.5.1).

Extracellular macromolecules to be taken up by the cell are bound by transmembrane receptors that are enriched in lipid rafts and coated pits undergoing endocytosis. This **receptor-mediated endocytosis** (see Section 10.4.1) supplies cells with essential nutrients such as iron and cholesterol. In blood, iron is bound to the transport protein transferrin and taken up via a specific transferrin receptor, while cholesterol is concentrated in low-density lipoprotein (LDL) particles that interact with an LDL receptor. Both receptors are homodimeric transmembrane proteins with one transmembrane domain each per subunit. Receptor-mediated endocytosis is also used by cells to clear active compounds such as antigen–antibody complexes and hormones from the blood. Finally, it represents the standard pathway for the internalization of signal receptors.

5.8.2 Reprogramming of receptors: from G-protein coupling to arrestin coupling

As we have seen, GRKs and arrestins block G-protein-dependent signaling pathways. However, arrestins can do more: upon displacement of the G-protein, they interlink *phosphorylated* GPCRs, which are unable to interact with G-proteins, with signaling cascades, which until recently have been thought to be stimulated by tyrosine (Tyr) kinase-coupled receptors only (see Section 7.1). A typical example is the mitogenic MAP kinase module consisting of the protein kinases RAF1, MEK1, and ERK. The conventional pathway leading to a stimulation of this module starts from Tyr kinase-coupled growth factor receptors and proceeds via the small G-protein Ras that assembles the three kinases of the module to a functional complex at the inner side of the cell membrane (details in Section 10.1.4 and Chapter 11). The organizing role of Ras can be taken over by an arrestin that acts as a scaffold protein for the module. This works only when the arrestin is bound to a phosphorylated GPCR and means that the input signal stimulating the MAP kinase module is identical with the GPCR ligand that induces adaptive receptor phosphorylation and arrestin binding. In a similar manner, the stress-activated MAP kinase modules (see Section 11.1) become stimulated by arrestin-coupled receptors.

Such signaling cascades are activated along still another route. In this case the arrestins recruit cytoplasmic Tyr kinases of the Src family, which phosphorylate Tyr kinase-coupled receptors, such as those of cytokines, antigens, and growth factors, as well as the substrate proteins thereof. This phosphorylation has the same activating effect on the receptor as autophosphorylation, as it is normally induced by the receptor ligand. In other words, Tyr-kinase-coupled receptors (and MAP kinase modules) are not only activated in the conventional way by their genuine ligands, but also by ligands interacting with GPCRs and transforming them into arrestin-coupled receptors. Such cross talk between different signaling cascades has been called **receptor transactivation** (**Figure 5.27**). It enables GPCR ligands to secure a direct grasp on the genome, because a major effect of MAP kinase modules is modulation of gene transcription.

Latest observations have added additional pieces to the arrestin puzzle. Not only β-arrestins interact with GPCRs but also with several cytokine receptors, the Toll-like receptors of bacterial components (see Section 6.3.3), the protease-coupled receptor Notch (see Section 13.1.1) and others. Moreover, arrestins even translocate into the nucleus, probably augmenting gene transcription by supporting the interaction between transcription factors and transcriptional co-activators such as histone acetyltransferases.

5.8.3 Receptor transactivation by ectodomain shedding

GPCRs may also transactivate receptor Tyr kinases, independently from arrestin coupling, by triggering the release of the corresponding ligand. Such a mechanism has been described for the epidermal growth factor (EGF) receptor family. Here an ADAM-type protease (see Section 13.1.2), activated along a still ill-defined GPCR-controlled pathway, detaches EGF or related growth factors from membrane-bound precursor proteins through a mechanism known as ectodomain shedding (see Section 13.1 for more information). The process seems to play a role in hyperproliferative diseases such as tumors and cardiac hypertrophy.

Figure 5.27 Reprogramming of heptahelical receptor by phosphorylation and arrestin (A) Signal S activates, via a GPCR, a G-protein-controlled signaling cascade such as for instance the G_s–cAMP pathway. In the course of adaptation, the receptor becomes phosphorylated by a GPCR kinase. (B) Arrestin has displaced the G-protein from the phosphorylated (P) receptor, thus interrupting G-protein-controlled signaling. Instead, arrestin-controlled signaling becomes active. Owing to the scaffold function of arrestin, the receptor is coupled to the components of MAP kinase modules or Tyr kinases of the Src type that both become activated. (C) Src-type kinase activates a dimeric Tyr kinase-coupled receptor by phosphorylating in the cytoplasmic domains those sites that, upon ligand binding, would undergo trans-autophosphorylation. In turn, the receptor stimulates the MAP kinase modules and activates other signaling pathways. (D) Arrestin translocates into the nucleus, acting as transcriptional co-activator. In summary, one signal S has cleared a variety of signaling routes.

Summary

- Arrestins are proteins playing a key role in receptor desensitization by displacing G-proteins from phosphorylated, desensitized G-protein-coupled receptors (GPCRs).

- They may induce receptor down-modulation through endocytosis, or they may cause receptor reprogramming by acting as scaffold proteins to couple the phosphorylated GPCRs with Tyr kinase-coupled receptors and cytoplasmic Tyr kinases.

- Upon translocation into the nucleus, arrestins may act also as transcriptional co-activators.

5.9 Heptahelical receptors in hedgehog and Wnt signaling pathways

While practically all trimeric G-proteins are activated by heptahelical receptors, not all receptors of this type are coupled to G-proteins. The most prominent examples of such a case are provided by **Smo** and **Fz**, two receptors originally discovered in the *Drosophila* mutants "smoothened" and "frizzled" but since then found in homologous forms in all animals hitherto studied. The input signals activating these receptors are the peptide factors **"hedgehog" (Hh)** for Smo and **"wingless" (Wnt)** for Fz. Both rank among the most important and most abundant morphogens regulating the development of embryonic and the maintenance of adult tissues. To this end both factors control signaling networks of extreme complexity the detailed treatment of which is outside the scope of this textbook. Instead we shall restrict to the so-called "canonical" pathways. These follow a simple mechanistic principle in that hedgehog/Hh or wingless/Wnt induce gene transcription by protecting specific transcription factors from proteolytic degradation occurring along the ubiquitin-proteasome pathway (**Figure 5.28**). This mechanism provides a particularly clear example of the principle of double negation ("minus times minus is equal to plus"), which is a favorite of cellular signal processing. Certainly, it is an expensive way to control gene activity, since in the absence of the hedgehog and Wnt signals the transcription factors are continuously synthesized and destroyed. However, it has the advantage of responding in a rapid, variable, and flexible manner and thus is obviously worth the cost.

5.9.1 Hedgehog signaling

Hedgehog (Hh) is a 19 kDa lipoprotein. It received its name from mutated *Drosophila* larvae that exhibited bristles similar to those on hedgehogs. Owing to gene doubling, three isoforms have evolved in vertebrates. Their names—sonic hedgehog (Shh), indian hedgehog (Ihh), and desert hedgehog (Dhh)—obscure the already idiosyncratic nomenclature used by the *Drosophila* community.

The "canonical" pathway of hedgehog signaling

It must be emphasized that hedgehog does not directly bind to the heptahelical transmembrane protein Smo. Instead, the hedgehog receptor proper is **Ptc ("patched")**, a protein with 12 transmembrane domains structurally resembling bacterial membrane transporters. Ptc is highly conserved among species. It inhibits the activity of Smo and this inhibition is relieved when Ptc specifically binds hedgehog.

Since Ptc does not bind directly to Smo, the mechanism by which Ptc suppresses Smo activity is not entirely clear. A currently favored hypothesis proposes that Ptc, being a putative membrane transporter, releases a cholesterol-like lipophilic signaling molecule such as provitamin D_3 that interacts with a corresponding sensor domain of Smo,

Figure 5.28 Principle of "canonical" hedgehog and Wnt signaling The input signal activates a heptahelical receptor that, in turn, produces a G-protein-independent signal interrupting the continuous degradation of transcription factors along the ubiquitin–proteasome pathway. As a consequence, the transcription of specific genes is upregulated.

Figure 5.29 "Canonical" Hedgehog signaling The receptor of the lipoprotein hedgehog is the protein Ptc (patched), with 12 transmembrane domains resembling the structure of a membrane transporter. (A) In the absence of hedgehog, Ptc inhibits the heptahelical transmembrane protein Smo (smoothened), probably by releasing an inhibitory lipid S such as provitamin D_3. Under these conditions, the cellular level of the transcription factor Ci is very low because Ci is phosphorylated by the protein kinases depicted in the box and partially degraded along the ubiquitin (UB) pathway, yielding fragment CiR. CiR acts as a transcriptional repressor. (B) Hedgehog neutralizes the inhibitory effect of Ptc probably by preventing the release of S. As a result, additional Smo molecules become exposed at the cell surface and aggregate to active complexes, leading to stabilization of Ci along a still ill-defined pathway. (C) Schematic domain structure of Ci and CiR with the DNA-binding site (zinc finger, light gray box, DBD), the transactivating domain (TD) that interacts with the transcriptional co-activator CBP/p300, and the repressor domain (RD). The teal arrows show the site of proteolytic cleavage; black arrows indicate phosphorylation sites of Ci.

thus inhibiting signal transduction. When hedgehog binds to the large extracellular domains of Ptc, the transmission of this inhibitory signal would be interrupted resulting in Smo aggregation and activation (**Figure 5.29**).

Upon Smo activation the transcription factors **Ci** in the case of *Drosophila* (referring to a mutant *Cubitus interruptus*) or **Gli** in the case of vertebrates (referring to Glioblastoma cells, where these factors were found for the first time) become protected from proteolytic degradation and, in turn, the transcription of the corresponding target genes is activated. This transcriptional activation is augmented by an accompanying event, namely a relief from gene repression caused by degradation products of Ci/Gli. These are called **CiR** and **GliR**, respectively. In fact, Ci/Gli carries apart from its DNA-binding domain a transactivating domain at the C-terminus and a repressor domain at the N-terminus. By proteolytic degradation only the transactivating domain is destroyed whereas the DNA- binding site and the repressor domain remain intact. In other words: by proteolysis the stimulatory transcription factors Ci and Gli are transformed into gene repressors (see Figure 5.29).

This proteolysis is controlled by an intricate regulatory network. It occurs only in a complex with other signaling proteins and depends on a hierarchical phosphorylation of Ci/Gli by the kinases PKA, GSK3β, and CK1 (see Section 5.3.1). As is characteristic for many ubiquitylation reactions, phosphorylation generates a docking motif that is recognized by a specific S3 ubiquitin ligase of the SCF family. The latter enzyme labels Ci/Gli for specific degradation by proteasomes (for more details see Section 2.8.1).

A signal generated by activated Smo induces the dissociation of this multiprotein complex and, in turn, the proteolytic degradation of Ci/Gli becomes blocked. Apart from the fact that trimeric G-proteins are not involved, the nature of this signal is not known. It may consist of several protein phosphorylations. An important aspect is that Ci/Gli is prevented from translocation into the nucleus, probably by binding to an inhibitor called SuFu (Suppressor of Fu). This blockade is overcome by SuFu phosphorylation catalyzed by a Ser/Thr-specific protein kinase Fu ("fused"). Both SuFu and Fu are held together by a kinesin-like scaffold protein COS-2 ("coastal 2"). This aggregate is an additional constituent of the signaling multiprotein complex assembled at Smo, and FU becomes activated upon dissociation of this complex upon Smo activation. Thus, two safety locks, permanent proteolysis and control of intracellular translocation, prevent an accidental activation of Ci/Gli.

Recently several "non-canonical"" pathways of hedgehog signaling have been identified in vertebrates. They do not require Gli transcription factors. In one of these pathways Smo interacts with a trimeric G_i-protein thus functioning as a genuine GPCR. For more details the reader is referred to the list of publications at the end of this chapter.

Physiological and pathological hedgehog signaling

Hedgehog controls, among other processes, the formation of body segments in insects and, in vertebrates, the development of the neural tube, limbs, and right–left asymmetry. Disturbances of hedgehog signaling result, therefore, in massive malformations. In adult tissues, hedgehog is involved in the control of homeostasis, the equilibrium between cell gain and cell loss that ensures a constant tissue mass and function. Accordingly, a permanent overactivation of the hedgehog system may result in tumor development. Hedgehog's function as a morphogen depends on a strictly controlled distribution in the developing organism and on steep extracellular concentration gradients. For this pupose several conditions have to be fulfilled. Thus, hedgehog carries a cholesteryl residue at the C-terminus and a palmitoyl residue at the N-terminus. Owing to these modifications, it is sufficiently lipophilic to accumulate in lipid rafts at the surface of its target cells, where it also finds its receptors. On the other hand, hedgehog acts as a paracrine messenger, diffusing from cell to cell. This is made possible through an interaction with strongly hydrophilic proteoglycans that shield the lipid residue, rendering hedgehog water-soluble. Both processes lead to a more unspecific accumulation of hedgehog. A more specific distribution is caused by a distinct expression pattern of proteins that selectively sequester hedgehog, delivering it to the receptor complex on demand. Very recently such proteins have been found in *Drosophila*, known as **iHog** "interference hedgehog" and **Boi** "brother of iHog", while their mammalian counterparts are known as **Cdo** and **Boc**. With a single transmembrane domain and several extracellular immunoglobin- and fibronectin-like domains, they structurally resemble cell adhesion proteins (see Section 7.3.1). The proteins seem to be expressed preferentially in hedgehog target cells, forming a specific pattern at the cell surface. As hedgehog co-receptors they may selectively enrich the morphogen at the cell surface in the immediate neighborhood of Ptc and Smo. Experiments with genetically manipulated animals have revealed an essential role of Cdo and Boc in the development of the central nervous system.

As already mentioned, defects of hedgehog signaling may lead to tumorigenesis. Thus a permanent overactivation of sonic hedgehog signaling correlates with **basal cell carcinoma**, the most frequent type of skin cancer. The reason is a gain-of-function mutation of the receptor Smo. On the contrary, Ptc and SuFu have been identified as tumor suppressor proteins. Loss-of-function mutations of their genes are associated with medulloblastoma (a brain tumor), with prostate carcinoma, and again with basal cell carcinoma. Clinical research focuses on the development of inhibitors of hedgehog signaling some of which may become important for the treatment of certain cancer types.

For their work on *Drosophila* development, including the discovery of the hedgehog system, Edward B. Lewis, Christiane Nüsslein-Volhard, and Eric F. Wieschaus were awarded the Nobel Prize in Physiology and Medicine in 1995.

5.9.2 Wnt signaling

For embryonic development and the maintenance of tissues, the paracrine Wnt factors of vertebrates are as important as hedgehog. Their *Drosophila* prototype, wingless, is a palmitoylated glycoprotein that, among others, controls the correct development of cell patterns in wings and cuticle. For humans, 19 isoforms of homologous Wnt proteins have been described that are tissue-specifically expressed and exhibit different functions. The term Wnt is a combination of Wingless and Int, the chromosomal locus of integration of a murine mammary tumor virus the proliferative effect of which led to the discovery of the Wnt/Int factors as homologs of *Drosophila* wingless. The Wnt signaling network is highly conserved and apparently goes back into evolutionary history to the first appearance of multicellular animals.

Wnt receptors

Specific Wnt receptors are the ubiquitously distributed heptahelical Fz proteins that in man are expressed in 10 different isoforms. Depending on the tissue and Wnt isotype, these receptors may stimulate the $G_{q,11}$–PLCβ signaling cascade, acting as genuine GPCRs. The role of PLCβ activation has been described in detail in Section 4.4. A major effect is the increase of the cytoplasmic Ca^{2+} level with its numerous consequences (see Sections 14.5 and 14.6). Therefore, this response is called the Wnt-Ca^{2+} pathway. Other signaling cascades induced by Wnt/Fz activate gene transcription in a G-protein-independent way. To this end Fz requires the assistance of co-receptors. These are single-pass transmembrane proteins which form a ternary complex with Fz and Wnt. At least five different co-receptors have been found including the receptor Tyr-kinases Ryk and Ror (see Section 7.1) and LDL receptor-related lipoprotein LRP. Probably many different combinations between Fz isoforms and co-receptors are coupled to individual pathways of downstream signaling. Moreover, some other developmental factors (norrin and R-spondin) are known to interact with Fz, confirming the notion of multipurpose usability of signaling proteins. In addition, a variety of paracrine peptides have been detected, which inhibit the binding of Wnts to their receptors. These observations indicate a network of positive and negative signaling that is of bewildering complexity.

The "canonical" pathway of Wnt signaling

Among the G-protein-independent signaling cascades activated by Wnt-receptor complexes, the β-catenin pathways are the most prominent. They are called "canonical". **β-Catenins** (alias "armadillo" in *Drosophila*) are transcriptional co-activators that stimulate the expression of various genes when heterodimerizing with transcription factors of the **TCF/LEF** family (T-Cell Factor/Lymphoid Enhancer-factor-binding Factor). In the absence of β-catenin these factors are inactive. The genes stimulated by TCF/LEF and β-catenin encode proteins controlling cell proliferation such as cyclin D (see Section 12.3), transcription factors such as cJun and cMyc, and growth factors (**Figure 5.30**). The activity of β-catenin is strictly controlled, involving a dual safety lock consisting of controlled intracellular distribution and protein phosphorylation. In fact, two different β-catenin pools are found in the cell. In the first pool β-catenin exists as a complex with cell adhesion molecules of the cadherin family (see Section 7.3 for details). In this complex it is sequestered and thus not available as a transcription factor. The second pool consists of free cytoplasmic β-catenin. In the absence of Wnt this pool is kept extremely small by permanent proteolytic degradation,

Figure 5.30 "Canonical" Wnt–β-catenin signaling (A) In the absence of Wnt, β-catenin is bound to a "destruction complex" consisting of the protein kinases GSK3 and CK1 (activates GSK3) and the scaffold proteins APC and axin. Under these conditions GSK3 catalyzes the phosphorylation of β-catenin thus labeling it for ubiquitylation and subsequent proteolytic degradation. As a result the expression of Wnt-responsive genes is blocked by TCF/LEF. (B) When Wnt binds to its receptor Fz and to the co-receptor LRP the cytoplasmic protein DVL is recruited to Fz where it becomes activated by phosphorylation (by the protein kinase CK2). The phosphorylated DVL inhibits GSK3/CK1. Moreover, axin undergoes proteolytic degradation and the destruction complex disintegrates. β-catenin thus released becomes dephosphorylated and stabilized being now able to neutralize the repressor effect of TCF/LEF so that the WNT-responsive genes can be transcribed.

promoted by a β-catenin destruction complex. The destruction complex contains, in addition to β–catenin, the scaffold proteins **axin** (the name is reminiscent of a characteristic Wnt effect in vertebrate development, duplication of the body axis) and **APC** (referring to the disease Adenomatous Polyposis of the Colon), as well as the protein kinases GSK3 (glycogen synthase kinase 3) and CK1 (casein kinase 1) and the protein phosphatase PP2A (see Figure 5.30). Axin and APC organize and stabilize the complex in such a way that β-catenin undergoes phosphorylation by GSK3 (CK1 is the primer kinase of GSK3, see Section 9.4.4). This phosphorylation labels the factor for degradation along the ubiquitin–proteasome pathway. So, the situation strongly resembles the hedgehog system. In fact, the SCF-type E3 ubiquitin ligase triggering β-catenin degradation seems to be identical with the enzyme inducing partial proteolysis of the transcription factor Ci of the hedgehog system.

Wnt signals induce the dissociation of the destruction complex. As a result, the phosphorylation of β-catenin is stopped and dephosphorylation by PP2A gains the upper hand, leading to an increased level of free β-catenin and, subsequently, to transcriptional activation of the Wnt target genes. The pathway connecting the Wnt-Fz complex with the destruction complex is known only in fragments. A key role is played by the protein **DVL** (the *Drosophila* homolog first discovered is called "dishevelled", DSH). DVL binds to the cytoplasmic region of Fz. Upon activation of Fz by Wnt it becomes phosphorylated by the protein kinase CK2. This triggers—along a still ill-defined pathway—the proteolytic degradation of axin, resulting in complete disassembly of the destruction complex.

As we have seen, the protein kinase GSK3 may be regarded as the key enzyme controlling β-catenin activity. It should not be overlooked, however, that GSK3 has many other physiological functions (see Sections 5.3.1, 9.4.4, 11,6.2, 12.6, and 13.1.3). Thus, its stabilization by Wnt not only leads to an activation of β-catenin-targeted genes but is expected to have many other consequences for the cell.

Wnt signaling and disease

The sophisticated and expensive regulation of Wnt signaling reflects the central role of this pathway in the development and maintenance of tissues. While defects of the signaling cascade result in severe embryonic malformations, Wnt signaling is required throughout life for the homeostatic balance of rapidly self-renewing tissues such as intestinal crypts and hair follicles as well as for the regulation of bone mass, and it has been suggested as a factor in the maintenance and control of the stem cell pool of a tissue. As would be expected, disturbances of Wnt signaling may cause uncontrolled cell proliferation and cancer. Indeed, in various human tumors β-catenin has been found to be mutated to an oncoprotein. Such a mutation renders β-catenin widely resistant to proteolytic degradation. The reason is that it cannot be phosphorylated by GSK3 since the target Ser and Thr residues have become replaced by other amino acids. Frequently, this gain-of-function mutation is accompanied by loss-of-function mutations of axin and APC, which both are tumor suppressor proteins. The deletion of APC, in particular, ranks among the most frequent causes of benign and malignant tumor growth. A germline mutation of the *APC* gene results in **adenomatous polyposis of the colon**, a relatively rare pre-malignant condition characterized by benign tumors in colon epithelium. Owing to the very high number of these polyps, the tendency for malignant development is practically 100%, being only a matter of time. Malignant transformation results from an accumulation of additional genetic defects occurring accidentally (see also Section 4.6.5, Sidebar 4.6, and Section 13.3). A non-hereditary loss-of-function mutation of APC as probably caused by environmental factors is the triggering event of the great majority of spontaneous **colorectal carcinomas** that develop from spontaneous polyps and rank among the most frequent and dangerous forms of cancer. Since such polyps are found in about half of the older population, the deletion of *APC* seems to be by far the most abundant tumorigenic gene defect in man. In fact, in about 90% of colorectal cancers the signaling pathways controlled by β-catenin are up-regulated. Therefore, it is not surprising that the Wnt signaling system is in the forefront of experimental and clinical cancer research.

Summary

- While almost all trimeric G-proteins are coupled to heptahelical receptors, a few members of this receptor family are linked to G-protein-independent pathways.

- Examples are provided by the receptors of the morphogenetic and mitogenic hedgehog and Wnt proteins. These hormone-like factors are widely distributed in eukaryotes.

- Both stabilize and activate certain transcription factors of the Ci/Gli and β-catenin type, respectively.

- Deregulation of the signaling cascades may result in tumor development such as basal cell carcinoma (hedgehog) and large bowel cancer (Wnt).

5.10 Other G-protein-independent heptahelical receptors: lessons from slime molds and Homer

Apart from the hedgehog and WNT systems, other signaling pathways have been described that start with heptahelical receptors but do not (or not exclusively) proceed via trimeric G-proteins. An interesting example is provided by the slime mold *Dictyostelium discoideum*, a social amoeba switching between uni- and multicellularity. Under stress, tens of thousands of cells aggregate to form a multicellular structure that develops into a spore capsule (see Figure 3.31). The signal for both aggregation and differentiation is cAMP, released from a few pacemaker cells and received by adjacent cells. The latter respond in three ways: they move towards the pacemaker cells (chemotactic effect); they start to produce cAMP on their own, thereby attracting further cells (relay effect); and they activate the genes controlling aggregation and differentiation (morphogenetic effect). Thus, for the slime mold, cAMP is both a pheromone and a hormone.

Depending on the phase of differentiation, *Dictyostelium* cells express four different cAMP receptors (CAR1–CAR4) of the GPCR type and possess 11 genes encoding Gα-subunits and one gene each for Gβ- and Gγ-subunits. The rearrangement and activation of the contractile actomyosin complex required for the chemotactic response

Figure 5.31 Two pathways leading to Ca²⁺ release by metabotropic glutamate receptor mGlu1 in neurons The protein Homer bridges the gap between a G-protein-coupled receptor and the InsP₃-activated Ca²⁺ channel in the endoplasmic reticulum, thus inducing channel gating in a G-protein-independent way. Homer signaling is interrupted by complex formation.

is induced by CAR1 and the $G\beta\gamma$-subunits, whereas the other cAMP receptors evoke the obligatory Ca²⁺ influx and most of the gene activations along G-protein-independent pathways, which are still not known in their details.

In contrast to the process occurring in *Dictyostelium*, the mechanism of a G-protein-independent Ca²⁺ mobilization by another type of heptahelical receptors is known in more detail. These receptors are the **metabotropic glutamate receptors** of neurons. In addition to coupling in the conventional manner with $G_{i,0}$- and $G_{q,11}$-proteins, they have been found also to interact directly with those Ca²⁺ channels of the endoplasmic reticulum that are normally controlled by InsP₃. The gap between receptor and channel is bridged by an adaptor protein "**Homer**" interacting with proline-rich binding sites (called "Homer ligands") localized both in the cytoplasmic C-terminus of the receptor protein and in the InsP₃ receptor and ryanodine receptor Ca²⁺ channels (**Figure 5.31**).

This mechanism offers the neuron two possibilities to gate the Ca²⁺ channels in the endoplasmic reticulum: via InsP₃ (or the still-elusive ryanodine receptor ligand; see Section 14.5.3) or via Homer. Upon excitation, neurons produce an isoform (a-type Homer) that interrupts Homer signaling and Ca²⁺ release, thus protecting the cells from an overstimulation by an excessive Ca²⁺ signal that otherwise would result in cell death, for instance as a consequence of stroke and other neurodegenerative diseases (see Section 16.4.2). Moreover, this feedback control has been proposed to play an important role in memory fixation. A steadily increasing body of evidence indicates, on the other hand, that hereditary or acquired defects of Homer signaling are associated with psychoses, epilepsy, drug addiction (particularly alcoholism), and other neuropsychiatric disorders.

Some GPCRs induce the release of the intracellular messenger ceramide along a G-protein-independent signaling pathway. To this end they activate the membrane-bound enzyme **sphingomyelinase** by means of a special adaptor protein, a mechanism that has been discussed in Section 4.4.5.

Summary

- Heptahelical receptors interacting with the chemotactic and morphogenetic signal cAMP in *Dictyostelium discoideum* stimulate gene transcription and mobilize intracellular Ca²⁺ along G-protein-independent pathways.

- Heptahelical metabotropic glutamate receptors in brain neurons activate both G-protein-dependent and -independent signaling. In the independent pathway, the protein Homer interconnects the receptor with Ca²⁺ channels in the endoplasmic reticulum.

Further reading

Sections 5.1, 5.2

Alexander SPH, Mathie A & Peters (2011) Guide to receptors and ion channels. *Brit J Pharmacol* 164:Suppl. 1.

Benovic JL (2012) G-protein-coupled receptors signal victory. *Cell* 151:1148–1150.

Civelli O et al. (2013) G protein-coupled receptor deorphanizations. *Annu Rev Pharmacol Toxicol* 53:127–146.

Ghosh E et al. (2015) Methodological advances: the unsung heroes of the GPCR structural revolution. *Nat Rev Mol Cell Biol* 16:60–81.

Kallal L & Benovic JL (2000) Using green fluorescent proteins to study G-protein-coupled receptor localization and trafficking. *Trends Pharmacol Sci* 21:175–180.

Katrich V et al. (2013) Structure-function of the G protein-coupled receptor superfamily. *Annu Rev Pharmacol Toxicol* 53:531–556.

Lefkowitz RJ (2004) Historical review: a brief history and personal retrospective of seven-transmembrane receptors. *Trends Pharmacol Sci* 25:413–422.

Meyer T & Truel MN (2003) Fluorescence imaging of signaling networks. *Trends Cell Biol* 13:101–106.

DeWire SM, Ahn S, Lefkowitz J et al. (2007) β-Arrestins and cell signaling. *Annu Rev Physiol* 69:483–510.

Piston DW & Kremers GJ (2007) Fluorescent protein FRET: the good, the bad and the ugly. *Trends Biochem Sci* 32: 407–414.

Stagljar I & Fields S (2002) Analysis of membrane protein interactions using yeast-based technologies. *Trends Biochem Sci* 27:559–563.

Tehan BG et al. (2014) Unifying family A GPCR theories of activation. *Pharmacol Therap* 143:51–60.

Wettschureck N & Offermanns S (2005) Mammalian G proteins and their cell type specific functions. *Physiol Rev* 85:1159–1204.

Sections 5.3–5.7

Ferré S et al. (2014) G protein-coupled receptor oligomerization revisited: functional and pharmacological perspectives. *Pharmacol Rev* 66:413–434.

Gieseler F et al. (2013) Proteinase-activated receptors (PARs) – focus on receptor-receptor interactions and their physiological and pathophysiological impact. *Cell Commun Sign* 11:86.

Hara T et al. (2014) Role of free fatty acid receptors in the regulation of energy metabolism. *Biochim Biophys Acta* 1841:1292–1300.

Heng BC, Aubel D & Fussenegger M (2013) An overview of the diverse roles of G-protein coupled receptors (GPCRs) in the pathophysiology of various human diseases. *Biotechn Adv* 31:1676–1694.

Sexton PM, Morfis M, Tilakaratne N et al. (2006) Complexing receptor pharmacology. Modulation of family B G protein-coupled receptors by RAMPs. *Ann NY Acad Sci* 1070:90–104.

Section 5.8

DeWire SM et al. (2007) β-Arrestins and cell signaling. *Annu Rev Physiol* 69:483–510.

Hupfeld CJ & Olefsky JM (2007) Regulation of receptor T tyrosine kinase signaling by GRKs and β-arrestins. *Annu Rev Physiol* 69:561–577.

LeRoy C & Wrana JL (2005) Clathrin- and non-clathrin-mediated endocytotic regulation of cell signalling. *Nat Rev Mol Cell Biol* 6:112–126.

Moore CAC, Milano SK & Benovic JL (2007) Regulation of receptor trafficking by GRKs and arrestins. *Annu Rev Physiol* 69:451–482.

Parton RG & Simons K (2007) The multiple faces of caveolae. *Nat Rev Mol Cell Biol* 8:185–194.

Section 5.9

Briscoe J & Therond PP (2013) The mechanisms of hedgehog signalling and its role in development and disease. *Nat Rev Mol Cell Biol* 14:416–429.

Clevers H & Nusse R (2012) Wnt/β-catenin signaling and disease. *Cell* 149:1192–1205.

Niehrs C (2012) The complex world of WNT receptor signalling. *Nat Rev Mol Cell Biol* 13:767–779.

Robbins DJ, Fei DL & Riobo NA (2012) The hedgehog signal transduction network. *Sci Signal* 5(246):re 6.

Valenta T, Hausmann G & Basler K (2012) The many faces and functions of β-catenin. *EMBO J* 31:2714–2736.

Section 5.10

Jardin I et al. (2013) Homer proteins in Ca^{2+}-entry. *IUBMB Life* 65:497–504.

Signal Transduction by Serine/Threonine Kinase-Coupled Receptors

6

In Chapter 5, heptahelical receptors coupling with trimeric G-proteins have been hailed as an evolutionary model of success originating from archaeal rhodopsins and dominating signal transduction in animals. Something similar holds true for protein kinase-coupled receptors occupying the first rank in prokaryotes and plants and the second rank in animals. In fact, signal transduction across membranes by means of protein phosphorylation is the most abundant method of communication between cells and environment, and the corresponding receptors have a venerable evolutionary history as the most ancient devices of transmembrane signaling. Constituting a large and heterogeneous protein family, they are found in all living beings. This is in clear contrast to G-protein-coupled receptors, which seem to be restricted to eukaryotes, particularly to animals (leaving aside the special case of archaebacterial rhodopsins).

6.1 The principle of oligomerization-driven signal transduction

We have already encountered ancient prototypes of protein kinase-coupled receptors in prokaryotes: the histidine (His) kinase-coupled receptors of two-component systems. Here the kinase activity is mostly intrinsic, harbored in the cytoplasmic domain of the protein, while only a few receptors (particularly the chemotaxis receptors) associate noncovalently with a separate His kinase. These two principles—intrinsic versus associated protein kinase activity—have been preserved throughout evolution, even though during eukaryotic development the His-specific kinases were replaced by serine/threonine- (Ser/Thr-) and tyrosine-(Tyr-) specific kinase activities (**Figure 6.1**).

All protein kinase-coupled receptors are transmembrane proteins possessing one (or, as in the case of His kinase-coupled receptors, more than one) transmembrane domain. The protein kinase-coupled receptors of eukaryotes share a common mechanism of activation. In the absence of a ligand, they are blocked in an inactive conformation. To transduce signals they have to assemble into dimers or higher oligomers. Depending on the receptor type, the role of the ligand is to induce or stabilize a conformational change that promotes oligomerization or stabilizes preformed oligomeric receptor complexes. As a result the intrinsic or associated kinase domains come into a juxtaposition to activate each other. In most cases this activation requires activation loop phosphorylation that is generally achieved by **trans-autophosphorylation**. Depending on the receptor type, this occurs according to two different principles: either the monomers phosphorylate each other (principle of equal partnership) or—rather an exception—one monomer phosphorylates the other one, which transmits the signal downstream (master–slave principle). By phosphorylating effector proteins, the activated kinases then induce

Figure 6.1 Protein kinase-coupled receptors: an overview The receptors exist as oligomers (at least as dimers) of subunits with at least one transmembrane domain each. The colored rectangles symbolize the protein kinase domains (left) and subunits (right).

	with intrinsic protein kinase activity	with associated protein kinase
histidine-specific	prokaryotes unicellular eukaryotes plants	prokaryotes: (chemo)taxis receptors
serine/threonine-specific	plants animals: TGFβ receptor family	animals: receptors of IL1, TNFα, bacterial compounds, morphogens
tyrosine-specific	animals: receptors of insulin, growth factors, morphogens	animals: receptors of cytokines, antigens, extracellular matrix components

intracellular signaling cascades connecting the periphery with the metabolic apparatus and the genome. Frequently the cytoplasmic domain of the receptor protein itself becomes phosphorylated in addition to activation loop phosphorylation (**Figure 6.2**).

This generates docking sites for interacting proteins, an effect that is particularly pronounced for Tyr kinase-coupled receptors. Moreover, phosphorylation may prolong signal transduction across the cell membrane: even when the ligand has dissociated, the receptor or the associated protein kinase, respectively, stays active until it is dephosphorylated by a cognate protein phosphatase. This is in clear contrast to the prokaryotic receptor His kinases, where the phosphorylated form is a short-lived transition state of a phosphotransfer reaction between the receptor and the downstream substrate protein, the response regulator (see Section 3.4). In the following sections, Ser/Thr kinase-coupled receptors are discussed, while Tyr-kinase-coupled are described in Chapter 7.

Figure 6.2 Mode of action of protein kinase-coupled receptors The receptor protein acts as an adaptor linking an extracellular signal with intracellular protein kinases. (A) The inactive receptor dimer (or oligomer) with intrinsic or associated kinase domains (teal) encounters an extracellular signal molecule S. (B) Binding of S causes a conformational change inducing kinase activity by trans-autophosphorylation (or bringing together constitutively active kinase domains). (C) The activated kinase domains may catalyze the phosphorylation of the receptor protein, thus generating docking sites for the recruitment of cellular signaling proteins, which may become phosphorylated by the receptor-associated kinase or otherwise activated. Until signal extinction by dephosphorylation, the receptor–kinase complex may remain active even upon dissociation of S or becomes downregulated by internalization and proteolytic degradation.

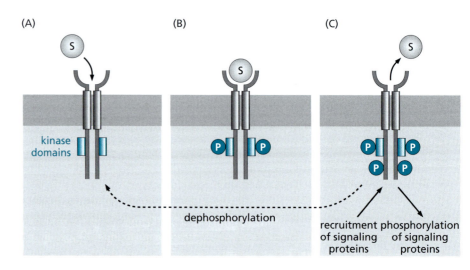

Summary

Protein kinase-coupled receptors of eukaryotes are proteins with one transmembrane domain and either an intrinsic cytoplasmic kinase domain or a docking domain for interaction with a separate kinase. Mutual activation of the kinase domains, mostly by trans-autophosphorylation, requires oligomerization (mostly dimerization).

6.2 Ser/Thr- kinases as receptors: transforming growth factor β receptor family

Transmembrane proteins with intrinsic kinase activity represent the simplest molecular architecture of kinase-coupled receptors. The Ser/Thr-specific forms are by far the most abundant receptor type in plants (see **Sidebar 6.1**), whereas in animals they are restricted to the rather small family of TGFβ receptors. TGFβ stands for **Transforming Growth Factor β**, a historically based term since the factor was discovered by the observation that, together with the unrelated TGFα (which interacts with a receptor with intrinsic Tyr kinase activity; see Section 7.1), it induced the proliferation and neoplastic transformation of cell cultures. However, this is by no means the main function of TGFβ. Instead, the protein has turned out to be a multifunctional tissue hormone regulating many physiological processes from embryogenesis to wound healing. TGFβ ranks among the most potent inducers of extracellular matrix proteins and seems to be the strongest endogenous inhibitor of epithelial, endothelial, and lymphatic cell proliferation.

Beside TGFβ proper, the TGFβ family comprises at least 30 additional factors, mainly exhibiting morphogenetic functions such as the **activins**, the **inhibins**, and the **bone morphogenetic proteins (BMP)**. BMPs are expressed in more than 20 isoforms and, contradicting their name, are involved in more than just bone formation. In fact, gene knockout experiments demonstrate an essential role of TGFβ and related factors in various stages of embryonic development where they cooperate with other morphogenetic signals such as Wnt and hedgehog (see Section 5.9). Particularly exciting is their function as key regulators of stem cell renewal and proliferation.

Functionally and structurally closely related are the morphogenetic factors DPP (referring to the gene locus *decapentaplegic* that regulates the development of

---

Sidebar 6.1 Receptor protein kinases in plants *Arabidopsis thaliana* was the first plant with a completely sequenced genome. It possesses no less than 417 genes (other species such as rice even more than 700) that encode proteins with a single transmembrane domain and a cytoplasmic Ser/Thr kinase domain, a topography resembling that of the TGFβ receptors. As compared, animals are rather low in receptor kinase genes: *C. elegans* contains 3 and 40 (receptor Ser/Thr kinases and receptor Tyr kinases), *Drosophila* 5 and 20, and humans 12 and 58 genes of this type. Although typical Tyr kinase-coupled receptors have not yet been found in plants, the plant receptor kinases may nevertheless phosphorylate both Ser/Thr and Tyr residues, and Tyr autophosphorylation even seems to play a key role in the activation of some plant receptor kinases. This activation includes homo- or heterodimerization of the receptor molecules.

Only a few ligands have yet been identified for plant kinases including the steroid hormone brassinolide, which controls longitudinal plant growth, and surface molecules of pathogenic organisms (microbes, fungi, and insects) which induce defense mechanisms resembling the innate immune response of animals. Structural motifs of the extracellular domains of plant receptor kinases give hints on other ligands such as proteins and carbohydrates of the cell surface as well as components of the extracellular matrix. Physiological processes that have been tentatively connected with plant receptor kinases include pollen recognition, embryogenesis, organ differentiation, development of gametophytes, control of cell volume, and defense of pathogens. As far as cellular signal processing is concerned the complexity of interacting pathways is assumed to rival that found in animals.

15 imaginal discs in *Drosophila* larvae) and corresponding proteins of *Caenorhabditis elegans*, indicating the evolutionary ancientness of this receptor type. The pathways of signal transduction are identical in principle for all these factors and are explained in the following section.

6.2.1 Receptor activation and downstream signaling

Receptors of the TGFβ family are grouped into three types: I, II, and III. All types have a single transmembrane motif, and types I and II also have cytoplasmic Ser/Thr kinase domains. Upon binding of the dimeric ligands (the active forms of TGFβ and its relatives are homodimers, in some cases also heterodimers), each receptor type dimerizes and cooperates with the other types. Signal transduction occurs according to the master–slave principle. The cytoplasmic kinase domain of receptor type II, the master, is constitutively active. It has mainly one substrate: receptor type I, the slave. The job of the ligand is to join both receptor types, thus enabling phosphorylation of type I by type II. To this end, both receptor types have ligand-binding sites. Type I receptors are expressed in several isoforms. These are subsumed under the term **activin-like kinases (ALK1–ALK7)** that are phosphorylated by individual type II receptors.

Phosphorylation results in a stimulation of the type I kinase, which is the actual signal transducer. Deviating from the rule, this phosphorylation does not occur in the activation loop but at multiple Ser and Thr residues in another segment of the kinase molecule. The effect is the same: the active center of the kinase is made accessible for the substrates ATP and protein. One function attributed to the kinase-inactive receptor type III (also known as β-glycan, since chemically it is a membrane-anchored proteoglycan) is to concentrate ligands near the type II receptor. Through this effect, type III receptors are thought to potentiate, for instance, the effects of inhibins. These hormone-like proteins are distantly related to TGFβ. They have become known as endogenous inhibitors of follicle hormone release by the anterior pituitary gland. By blocking the corresponding type II receptors, inhibins may antagonize the

Figure 6.3 Signal transduction by transforming growth factor β receptors (Step 1) Upon binding of the dimeric ligand, receptor types I and II associate; type II activates type I by phosphorylation (teal arrows). (Step 2) Assisted by the membrane-attached adaptor protein SARA, phosphorylated (P) receptor I binds and phosphorylates an R-SMAD protein. (Step 3) The phosphorylated R-SMAD dimerizes with a Co-SMAD protein. (Step 4) The R-SMAD/Co-SMAD complex accumulates in the nucleus and modulates gene transcription. Type I and II receptors exist as homodimers. For the sake of clarity, only the monomeric forms are shown.

effects of another group of TGFβ-related hormones, the activins. The inhibin–activin system plays an important role in the regulation of the menstrual cycle and thus is of interest in reproductive medicine.

In its simple straightforward strategy, the TGFβ system is reminiscent of a bacterial two-component system: the major substrates of the type I receptor kinase are transcription factors. These are known as **SMADs** (the term is composed of Sma, referring to the small body size of *Dauer* larvae of *C. elegans* induced by stress-activated TGFβ signaling, and MAD, referring to Mothers-Against-DPP mutants of *Drosophila*). SMAD proteins are linked with the receptor by the membrane-associated adaptor protein SARA (SMAD Anchor of Receptor Activation; see **Figure 6.3**) but they dissociate upon phosphorylation.

Eight human SMAD isoforms are known. Their different functions gave rise to a distinction between regulatory R-SMADs (isoforms 1–3, 5, and 8), co-regulatory Co-SMADs (isoform 4), and inhibitory I-SMADs (isoforms 6 and 7). Only R-SMADs are phosphorylated by type I receptor. As a result, they form oligomers with the Co-SMAD that enter the nucleus, whereas I-SMADs are negative regulators that promote dephosphorylation of R-SMADs and their degradation along the ubiquitin–proteasome pathway Thus, the receptor–SMAD complex represents a sophisticated system of logical gates.

6.2.2 Transcriptional control by SMADs

R-SMAD/Co-SMAD complexes help to control the transcription of numerous genes and thereby interact with other signaling pathways. For this purpose they bind to specific DNA sequences that are found in many transcriptional enhancers and promoters, which also contain binding sites for other transcription factors such as AP1, ATF2 (see Section 11.6), the nuclear receptors of vitamin D and glucocorticoid (see Section 8.3), and Forkhead activin signal transducers. The task of the SMAD proteins is to assemble transcriptional co-activators (such as the histone acetyltransferase CBP/p300 and the SWI–SNF chromatin remodeling complexes) or co-repressors (such as histone deacetylases) at the initiation sites marked by those transcription factors (**Figure 6.4**). Since the latter are activated along additional signaling pathways and, moreover, SMAD activity is fine-tuned by phosphorylation catalyzed by various protein kinases, TGFβ signaling is integrated in a complex network of signaling cross talk enabling all kinds of logical decisions.

Interaction of SMAD with co-repressors requires the assistance of adaptor proteins that displace co-activators from the complex. These adaptors are called **Ski**,

(A) histone acetylation (activation)

(C) histone deacetylation (repression)

Figure 6.4 Modulatory effect of SMAD and Ski on gene transcription Most R-SMAD proteins have two conserved domains (light gray rectangles) that interact, as a complex with Co-SMAD (not shown), with DNA (albeit some R-SMADs do not bind to DNA) and another transcription factor (TF). (A) Phosphorylated SMAD recruits the co-activator histone acetyltransferase CBP/p300, thus activating the gene marked by TF. (B) Ski displaces the co-activator from the complex. (C) The SMAD–Ski complex recruits a co-repressor, histone deacetylase (HDAC), which slows down gene expression. P symbolizes the phosphorylation of SMAD by the TGFβ receptor.

referring to their discovery as oncoproteins of the Sloan-Kettering retrovirus that mainly infects birds. However, overexpression of Ski has also been described for human tumors. Ski probably facilitates tumor growth by suppressing the anti-proliferative effect of TGFβ in epithelia, which is due to the activation of genes encoding cell cycle inhibitors such as p15^{INK4} (see Section 12.4). Under certain circumstances, Ski may also slow down tumor development, probably by eliminating stimulatory effects of TGFβ on invasive growth and metastasis. Normally Ski seems to be essential for development of the nervous system and skeletal muscles, as indicated by experiments with Ski knockout animals.

In addition to the Ski system, other mechanisms have been discovered that regulate the various effects of TGFβ-like proteins. **BAMBI** (BMP and Activin Membrane-Bound Inhibitor) provides an example. This transmembrane protein exhibits the properties of a BMP receptor but lacks protein kinase activity. Resembling a dominant negative receptor mutant, BAMBI intercepts ligands without transmitting SMAD signals to the genome.

It must be emphasized that TGFβ/BMP signaling has been explained here in a simplified form. Most probably, TGFβ receptors also control signaling pathways without SMADs, thus modulating the activities of mitogen-activated protein (MAP) kinase and nuclear factor (NF) κβ modules. In these pathways, the **TGFβ-Activated protein Kinase TAK1** functions as a relay. TAK1 is a Ser/Thr-specific MAP3 kinase (see Section 11.3) that also provides a link with cytokine signaling (see Section 6.3.2).

Summary

- Hetero-oligomeric receptors of TGFβ and related cytokines constitute the only family of Ser/Thr-coupled transmembrane receptors in animals.

- Ligand binding induces both receptor phosphorylation and, in turn, substrate phosphorylation.

- Major substrate proteins are the SMAD transcription factors, which, in complexes with other transcription factors, stimulate or repress the activity of a wide variety of genes, the majority of which regulate morphogenetic processes.

- In plants, receptorlike Ser/Thr kinases represent one of the largest protein families. Their functions are still widely obscure.

6.3 Cytokine receptors: key players in defense reactions

Protection from microbial and parasitic attacks is an essential and basic condition of survival. For this purpose, organisms make use of an arsenal of defense reactions such as the production of toxic compounds: for instance, reactive oxygen species and antibacterial or antifungal metabolites. This "chemical warfare," known as **innate immunity**, is employed even by simple animals and by plants. In fact, innate immunity provides the first and evolutionarily most ancient line of defense found throughout the metazoan kingdom. In higher animals the major effector cells of the innate immune system are phagocytic white blood cells, as well as epithelial cells of skin, eyes, lung, cervix, vagina, intestine, and other tissues, which function as barriers to microbial invasion. In the past, innate immune surveillance was assumed to function in an entirely nonspecific way. More recently, however, the system was found to be equipped with a sophisticated mechanism of signal transduction that ensures a very selective response to a wide variety of stimuli.

In vertebrates, the innate defense system is complemented by the highly specific **adaptive** (or **acquired**) **immune system**. This late achievement of evolution is based on the generation of tailor-made antibodies and antigen receptors by lymphocytes. In contrast to the immediately responding innate immune system, adaptive immunity reacts with some delay and is characterized by a long-term memory effect. Adaptive and innate immunity are interlinked in that they depend on each other. Thus, phagocytes provide antigenic peptide fragments, and a major task of antibodies is to mark invaders as targets of killer cells of the innate immune system. Note that prokaryotes dispose of another form of adaptive immunity, the CRISPR-Cas system (Section 3.2).

Both the innate and the adaptive immune systems communicate through and are under the control of an intercellular signaling network that makes use of a wide variety of peptide factors, most of which interact with protein kinase-coupled receptors. The evolutionary progress from innate to adaptive immunity is mirrored by the mechanisms of signal transduction: while the factors stimulating and controlling innate immunity use "old-fashioned" Ser/Thr phosphorylation, adaptive immunity is mainly regulated by "modern" Tyr kinase-coupled receptors, in particular those interacting with antigens (see Section 7.2.2). A practical consequence is that the signaling mechanisms controlling innate immunity can be studied in nonvertebrate models such as worms and insects, an approach that is not feasible for adaptive immunity of vertebrates.

Most of the factors that control innate immunity and (in vertebrates) the communication between both innate and adaptive immunity are **cytokines**. Cytokine is a collective term for a large and heterogeneous group of protein and peptide mediators that exhibit autocrine, paracrine, and endocrine effects. While some of them are circulating hormones, most are local mediators or tissue hormones (also called autocoids). Cytokines are key mediators of inflammation and transmit differentiation, proliferation, and death signals, in particular in the hematopoietic and lymphatic systems.

There are distinctions among chemokines, interleukins (IL), interferons, colony-stimulating growth factors (CSF), monokines, and cytokines of the tumor necrosis factor α (TNFα) family. Chemokines (<u>chemo</u>tactic cyto<u>kines</u>) recruit defense cells such as granulocytes, monocytes, macrophages, and lymphocytes to the sites of action. Both the major sources and major targets of interleukins are white blood cells (one subgroup secreted mainly by lymphocytes has become known as lymphokines). Interferons are cytokines specialized for viral defense, while monokines are produced by monocytes and macrophages. Finally, CSFs are multipurpose growth factors secreted by and acting on a variety of cell types, in particular on white blood cells. A particular subgroup called neurokines provides intercellular communication in the nervous rather than in the hematopoietic system.

There are four major families of protein kinase-coupled cytokine receptors characterized by common structural features:

- family I: Tyr kinase-coupled receptors of most interleukins, growth factors G-CSF and GM-CSF, and circulating hormones such as prolactin, leptin, erythropoietin, and thrombopoietin

- family II: Tyr kinase-coupled receptors of interferons and IL10, -19, -20, -22, and -24

- family III: Ser/Thr kinase-coupled receptors of TNFα and related proteins and low-affinity receptor of nerve growth factor*

- family IV: Ser/Thr kinase-coupled receptors of IL1 and microbial surface components

If one defines TGFβ and its relatives as cytokines (as is frequently done), then the TGFβ receptors would constitute a fifth family. However, in contrast to TGFβ

---

*Nerve growth factor also interacts with the high-affinity receptor Trk, which is a receptor Tyr kinase (see Section 7.1.1).

(A)

(B)

Figure 6.5 Basic theme of type III and IV cytokine signal transduction (A) Upon binding of a ligand, the oligomeric cytokine transmembrane receptor undergoes a conformational change. Consequently, interaction domains of the cytoplasmic extension become exposed and bind corresponding adaptor proteins. These recruit cytoplasmic Ser/Thr-specific protein kinases that come in close contact, enabling activation through mutual (auto)phosphorylation. (B) The kinases, thus activated, stimulate pro-apoptotic and/or immune-modulatory and pro-inflammatory signaling devices such as the NFκB modules (left) and the MAP kinase modules (right) primarily aiming at defense reactions (P, phosphorylation).

receptors, the receptors of families I–IV do not harbor intrinsic kinase activity but associate with separate cytoplasmic protein kinases. Still other cytokines, in particular chemokines, interact with G-protein-coupled receptors. Finally, some authors also define growth factors that interact with receptor Tyr kinases as cytokines. We do not follow this classification but discuss this topic separately in Chapter 7.

While family I and II receptors directly bind their cognate Tyr kinases (see Section 7.2), family III and IV receptors use various adaptor proteins that serve as specific links to Ser/Thr kinases (**Figure 6.5**). This situation is reminiscent of the bacterial chemotaxis receptors, where the adaptor protein CheW brings together the His kinase CheA with the sensor proteins. As discussed in Section 3.4.4, such a mechanism enables the cell to couple a single signaling pathway with a variety of receptors and extracellular signals. In the case of Ser/Thr kinase-associated cytokine receptors, this common signaling pathway is the NFκB cascade described in Section 11.7. Most cytokines stimulate, in addition, MAP kinase cascades, particularly those with the stress-activated MAP kinases p38 and JNK (cJun amino (N)-terminal Kinase, see Chapter 11). A particular subgroup is also specialized for transmission of pro-apoptotic signals. Upon stimulation by cytokines a major task of these pathways is to process signals that control the activity of white blood cells, thus triggering the above-mentioned defense reactions. The perceptible outcomes of these signaling events are the various symptoms of inflammation and illness. Thus, many cytokines belong to the large and heterogeneous family of pro-inflammatory mediators. It must be emphasized, however, that, in higher animals, Tyr kinase-coupled receptors also participate in the regulation of inflammatory responses (see Sections 7.2.3 and 7.3.2).

In Figure 6.5 the principle of cytokine signal processing is depicted. In the following discussion we shall present selected examples, each illustrating an individual subtype of signaling. As we shall see, cytokine signaling is a complicated issue. One reason for this is the complexity of the inflammatory response, involving numerous cell types and a wide variety of cellular responses. This situation requires extended signaling cross-talk at both intercellular and intracellular levels. Another reason is that inflammation in the absence of a pathogen would be a dangerous event for the body to be avoided at all costs, given that the hallmarks of inflammation (the activation of aggressive cell types and the production of highly cytotoxic metabolites) are processes that result in tissue destruction. An even more dangerous effect of an acute overproduction of many pro-inflammatory cytokines is shock, the potentially fatal breakdown of blood pressure. In fact, cytokine signaling provides a particularly illustrative example of how signaling cross talk and fine-tuning collaborate to safeguard a risky venture. Understanding these signaling interactions is crucial to any progress in treating some of the most widespread and dangerous diseases.

For the investigation of the innate immune system, Bruce Butler, Ralph M. Steinman, and Jules Hoffmann were awarded the 2011 Nobel Prize in Physiology or Medicine.

6.3.1 Tumor necrosis factor receptors are mediators of inflammation and cell death

The TNF superfamily

On the basis of structural similarities, around 20 different cytokines have been brought together in the so-called TNF superfamily. The name comes from the prototypical representative of this family, **tumor necrosis factor α (TNFα).** TNF-like cytokines are most potent stimulators of the innate immune response. They induce inflammation and may cause cell death by apoptosis and regulated necrosis. They are expressed as homotrimeric transmembrane proteins, particularly by leukocytes and lymphocytes. and exert juxtacrine effects on adjacent cells. However, many of them may also be shed by membrane-bound proteases, in particular those of the ADAM family (see Section 13.1.2), to diffuse freely and act as paracrine signaling molecules. An overproduction of TNFα contributes to chronic inflammatory diseases such as rheumatoid arthritis, asthma, and Crohn's disease (bowel inflammation) and may have fatal consequences such as emaciation (cachexia) of tumor patients.

For TNF-like cytokines more than 40 different receptors have been found in vertebrates. Most of them penetrate the membrane with a single transmembrane domain. The extracellular domains of the receptors are quite similar, reflecting the close relationship of the ligands. Their characteristic structural elements are cysteine rich repeats. The frequently homotrimeric ligands assemble three receptor molecules into a homotrimeric complex, thus gating signaling pathways that may lead to either cell death or cell proliferation (survival). These apparently contradictory effects indicate a sophisticated signal-transducing machinery. To this end the intracellular extensions of the receptors contain domains that interact with different adaptor proteins thus targeting different signaling routes.

The TNFα receptors

For TNFα two different receptors, TNFR1 and TNFR2, have been characterized (**Figure 6.6**). While TNFR1 is expressed by most tissues, TNFR2 is restricted to the cells of the immune system. The major structural difference between the receptors is that TNFR1 contains a cytoplasmic **death domain**, which is not found in TNFR2. This domain is required for the formation of adaptor complexes which induce cell death. TNFR1 shares this ability with five other death receptors that all belong to the TNF receptor superfamily. Since the mechanisms of cellular suicide are treated in more detail in Section 13.2, we shall restrict the discussion of TNFR1-induced cell death to the essentials.

To cause cell death TNFR1 binds—via homotypic death domain interactions—a complex with the adaptor proteins **TRADD** (TNF Receptor-Associated Death Domain protein), **RIPK1** (Receptor-Interacting Protein Kinase 1), and **FADD** (Fas-Associated Death Domain protein) as the major constituents. FADD, in turn, activates caspase 8 and apoptosis. In addition, TNFR1 may stimulate caspase-independent cell death through regulated necrosis. To this end, a complex of the active

Figure 6.6 Signal transduction by the tumor necrosis factor receptors -1 and -2 Trimeric ligand TNFα binds to the trimeric receptor TNFR1 (left) or TNFR2 (right). By its cytoplasmic death domains (black rectangles) TNFR1 recruits the adaptor protein TRADD. When TRADD interacts with FADD and receptor-induced protein kinase RIPK1, caspase 8 and the caspase cascade are activated leading to apoptosis. Upon poly-ubiquitylation (Ub) RIPK1 in a complex with auto-ubiquitylated TRAF 2 interacts with MAP3 kinases and IK-kinases. This results in a stimulation of MAP kinase cascades and the NfκB pathway. As a consequence immunomodulatory and pro-inflammatory genes are activated. The kinase activity of RIPK1 is not required. In addition, TNFR1 can promote regulated necrosis via a complex containing the phosphorylated protein kinases RIPK1 and RIPK3 and their substrate protein MLKL (for details see Section 13.2). TNFR2 (right) lacking death domains only stimulates MAP kinase and NfκB pathways via the TRAF2/RIPK1 complex. Protein kinases associated with the receptor-adaptor complexes are shown in the coloured boxes (P, phosphorylation).

and autophosphorylated kinases RIPK1 and RIPK3 and their substrate protein MLKL is formed.

Cell survival and inflammatory processes are promoted when either TNFR1 or TNFR2 assembles a complex containing RIPK1 and the adaptor protein TRAF2 (TNF Receptor-Associated Factor 2). Provided both proteins have undergone post-translational modification through poly-ubiquitylation, this complex activates MAP kinase and NFκB modules by stimulating the corresponding protein kinases MAPK3 and IKK. Poly-ubiquitylation is under the control of TRAF2, a ubiquitin ligase (**Sidebar 6.2**).

Sidebar 6.2 TRAF adaptors and the dual role of ubiquitylation Mammals express six different TRAF proteins that couple with receptors of the TNF receptor family and of IL1- and Toll-like receptors (see Section 6.3.2) to downstream signaling cascades. TRAFs have a carboxy (C)-terminal TRAF domain for receptor binding and homo- or heterotrimerization as well as (with the exception of TRAF1) a RING domain (**Figure 6.7**). The latter is important for the interaction with protein kinases. RING domains are, furthermore, characteristic functional elements of many E3 ubiquitin ligases (see Section 2.8.1). In fact, TRAF2 can fulfill its adaptor function only when it has catalyzed its own poly-ubiquitylation (note the analogy to protein kinases, most of which have to undergo autophosphorylation to become active). TRAF1 lacking a RING domain seems to function as a competitive inhibitor of TRAF-controlled signaling reactions. In addition to receptors and protein kinases, other signaling proteins are bound by TRAFs, for instance, the caspase inhibitors IAP1 and IAP2 which also exhibit E3 ubiquitin ligase activity (see Section 13.2.4) and the adaptor TRADD.

The signaling pathway connecting TNF receptors with the NFκB module provides an instructive example of the dual effect of ubiquitylation, either being a stimulatory covalent signaling reaction or labeling the protein for degradation in proteasomes. As explained in Section 2.8, for stimulatory ubiquitylation the polyubiquitin residue transferred to the protein is linked via Lys residue 63. Both adaptor proteins, TRAF2 and RIPK1, are modulated in this way. As a result, the TRAF2/RIPK1 complex is enabled to interact with the IK kinases of the NFκB module, recruiting them to the receptor complex. The link is probably provided by the kinase TAK1 together with its regulator proteins TAB1 and TAB2, and the scaffold protein NEMO which all have polyubiquitin-binding domains (see Section 6.3.2).

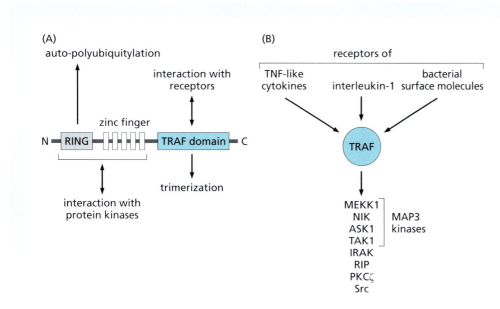

Figure 6.7 Coupling of receptors with cytoplasmic protein kinases by TRAF adaptors (A) Schematic domain structure of TRAFs 2–6. (B) Receptors upstream and protein kinases downstream of TRAF.

To mark a protein for proteolysis, the polyubiquitin residue has to be linked via Lys 48. This reaction plays an important role in NFκB activation since it eliminates the IKK inhibitors. On the other hand, it terminates signal transduction by removing NFκB. Both reactions are catalyzed by two different Lys 48-specific E3 ubiquitin ligases that are not identical with TRAFs (see Section 11.7).

Polyubiquitin chains are degraded by de-ubiquitylating proteases. One of these enzymes, called CYLD, has been identified as a tumor suppressor protein. A loss-of-function mutation of CYLD is the cause of cylindromatosis, a benign neoplastic disease of hair follicles and sebaceous glands. CYLD selectively degrades Lys 63-linked polyubiquitin residues, thus opposing the effects of TRAF2.

6.3.2 Interleukin 1 receptors are master regulators of inflammation

The cytokine **interleukin 1 (IL1)**, a 17 kDa protein expressed in two isoforms, α and β, is a key signal for the coordination of endogenous defense reactions. IL1 is released by leukocytes in response to injury, infections, and inflammatory reactions. It prompts a wide variety of tissues to produce pro-inflammatory tissue hormones (such as prostaglandins and cytokines), leukotactic chemokines, cell adhesion molecules, and hydrolytic enzymes catalyzing tissue degradation; in other words, the whole molecular arsenal of inflammation. Therefore, IL1 is considered to be the major trigger of the **classical symptoms of inflammation**: heat (including fever), redness (due to increased blood flow), swelling (due to increased vascular permeability), pain, and tissue destruction. In fact, IL1 was initially described as an endogenous pyrogen, a factor that, via effects on the thermoregulatory center in the hypothalamus, induced fever.

IL1 activation by inflammosomes

In non-stimulated cells IL1 (and the closely related IL18) is stored as an inactive zymogen. Upon stimulation the active cytokine is released by proteolytic cleavage of this precursor. This process is catalyzed by the protease **caspase 1**, which has been called therefore IL1-Converting Enzyme (ICE). Caspases form a family of proteases that is divided into two subfamilies: apoptotic caspases and inflammatory caspases (for more details see Section 13.2) with caspase 1 being the prototype of an inflammatory caspase. Like IL1, caspase 1 is released from an inactive zymogen (procaspase 1) by proteolytic cleavage. Caspase activation is an autocatalytic process that is catalyzed by the caspase itself. It starts when the procaspase molecules come in close contact (so-called proximity-induced autoactivation). This occurs in cellular structures called **inflammosomes**.

Figure 6.8 Activation and function of a cytoplasmic pathogen receptor protein of the NLR family Binding of a pathogen-associated molecular pattern (PAMP) to the regulatory domain (leucine-rich repeats, LRR) of NLR (here the subtype NLRC4) induces, via a conformational change, oligomerization to a wheel-like structure called inflammosome. This process depends on the binding and probably hydrolysis of ATP (the nucleotide-binding oligomerization domains NOD are shown in light gray). Working like a machine, the oligomer recruits several molecules of procaspase 1 by homotypic interaction of CAR domains (CARD, black circles) and enables the release of active caspase 1, a homodimeric complex, through proximity-induced autoproteolysis. In turn, caspase 1 catalyzes the proteolytic release of IL1 from a precursor zymogen. Finally, the process results in an activation of the inflammatory cascade. The figure represents a simplified scheme of inflammosome assembly the mechanism of which differs from NLR type to NLR type. Frequently additional adaptor proteins facilitate the interaction of subunits and the binding of procaspase 1. Note the close relationship to the activation mechanism for the apopotic cascade (see Figure 13.11)

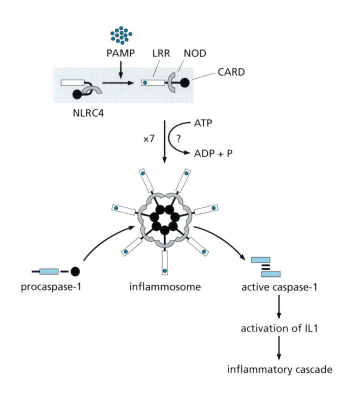

Inflammosomes are protein complexes consisting of seven identical subunits that are arranged as "spokes" in a wheel-like structure (**Figure 6.8**). The formation of such a highly ordered complex requires energy most probably delivered by ATP hydrolysis. In fact, each subunit is supposed to exhibit ATPase activity harbored in a nucleotide-binding domain that also is required for complex formation. The domain is known, therefore, as Nucleotide-binding and Oligomerization Domain, NOD. NOD-containing proteins are expressed in various isoforms called **NOD-Like Receptors (NLRs)**. NLRs are activated to assemble in inflammosomes when they come in contact with invading microbes. To this end they interact with so-called **Pathogen-Associated Molecular Patterns (PAMPs)**. These are surface molecules such as lipopolysaccharides and lipoproteins as well as nucleic acids that are found exclusively in microorganisms. In addition, inflammosome formation is induced by other dangerous stuff including endogenous material released upon cell and tissue damage or deposited in the course of pathological processes. A prominent example of such **Damage-Associated Molecular Patterns** or **DAMPs** is provided by uric acid crystals causing gout through chronic inflammation. Inflammatory processes accompanying Alzheimer's disease may also be due to DAMPs, in this case insoluble protein aggregates. Both PAMPs and DAMPs are specifically recognized by NLRs in the cytoplasm and—as explained in the next section—by receptors at the cell surface. In NLRs, pattern recognition occurs at leucine-rich repeats (LRRs) which contain specific binding sites for PAMPs or DAMPs. Upon binding, the NOD becomes uncovered and oligomerization into an inflammosome occurs. How then is caspase activated? To this end NLRs contain caspase-recruiting domains such as CARD or PYD (see Sidebar 13.5, Section 13.2.5) which homotypically interact with corresponding domains of procaspase 1. As a result, and supported by the wheel-like structure of the inflammosome, the procaspase molecules form a cluster allowing proximity-induced autoproteolysis (see Figure 6.8).

As shown in Chapter 13, programmed cell death (apoptosis) is induced along an analogous pathway except that the apoptotic caspase 9 becomes activated instead of the inflammatory caspase 1 and the "wheel", here called apoptosome, is built up by a special NLR called APAF. APAF responds to mitochondrial proteins rather than to PAMPs or DAMPs, and caspase 9 induces pathways leading to cell death rather than to a release of pro-inflammatory cytokines.

Beside NLRs, other pattern recognition receptors exist in the cytoplasm that are more specialized on monitoring double-stranded RNA as found in viruses. These RNA sensors are RIG1 (derived from a Retinoic acid-Inducible Gene 1) and MDA5 (derived from a Melanoma Differentiation-Associated gene 5). Both activate signaling pathways that lead to the synthesis and release of pro-inflammatory cytokines including antiviral interferon β. A deregulation of cytoplasmic virus defense is thought to be responsible for the catastrophic influenza epidemic in 1918.

Receptor mechanism

Mammalian cells express one IL1 receptor and at least eight related receptors. One of these interacts with closely related cytokine IL18; the ligands of the others—probably members of the IL1 family as is IL18—have yet to be identified. IL18, a potent pro-inflammatory mediator induces T-helper cell differentiation and activates natural killer cells. IL1 receptors are transmembrane proteins. They are expressed in various isoforms that form homo- and heterodimers upon ligand binding. The receptors are characterized by a cytoplasmic **TIR domain** (Toll-like–Interleukin 1 Receptor domain; see **Figure 6.9**) of about 200 amino acids that allows for the binding of adaptor proteins and is essential for signal transduction. The extracellular domain binding IL1 is characterized by three immunoglobulin (Ig)-like domains.

The active form of the IL1 receptor is a heterodimer of two quite similar transmembrane proteins, IL1R and IL1R-associated protein. The conformational change induced by the ligand IL1 enables the receptor to bind the adaptor protein **MyD88** (the name refers to Myeloid cell Differentiation induced by IL1), which for IL1 signaling plays a similar role as the TRAF adaptors for TNF signaling. MyD88 contains an N-terminal death domain and a C-terminal TIR domain. The latter interacts with the corresponding TIR domains of the receptor, thus transmitting the IL1 signal to the MAP kinase and NFκB modules that, in turn, activate the transcription factors of the IL1-responsive genes.

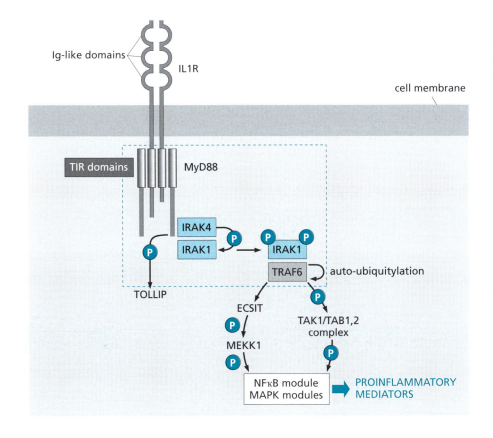

Figure 6.9 Signal transduction by interleukin 1 receptor By its TIR domains (cylinders), the activated receptor heterodimer assembles a complex consisting of adaptor protein MyD88 and protein kinases IRAK1 and IRAK4. Through phosphorylation, IRAK4 induces the degradation of the IRAK1 inhibitor TOLLIP and activates IRAK1. Upon dissociation from the receptor complex, phosphorylated IRAK1 associates with TRAF6, which becomes activated by auto-ubiquitylation and transmits the signal via the adaptor ECSIT to the MEKK1-JNK and MEKK1-ERK modules and via the TAK1/TAB1,2 complex to the p38 MAP kinase and NFκB modules. As a result, pro-inflammatory mediators such as cytokines and arachidonic acid metabolites are synthesized (P, phosphorylation). In the box (teal dotted line) the core of the signal-transducing complex assembing at the receptor is shown.

As with TNFR, the mechanism of signal transduction is rather sophisticated (see Figure 6.9). The major task of MyD88 is to recruit two Ser/Thr-specific protein kinases at the cytoplasmic domain of the receptor. These kinases are called **IRAK1** and **IRAK4** (Interleukin 1 Receptor-Activated Kinases). They resemble the RIP kinases involved in TNF signaling. IRAK4 links IRAK1 with MyD88 and catalyzes the phosphorylation of IRAK1. This enables IRAK1 to dissociate from the receptor and form a complex with the adaptor protein **TRAF6**. Like TRAF2, TRAF6 is a Lys 63-specific ubiquitin ligase (see Sidebar 6.2) that, upon stimulation, activates itself by auto-ubiquitylation. In the cytoplasm, the ubiquitylated TRAF6 then interacts with the MAP3 kinase **TAK1** bound to the regulator proteins TAB1 and TAB2 (see Section 11.2). As a result, TAK1 becomes activated by (auto?)phosphorylation and transmits the signal to the pro-inflammatory p38 MAP kinase and NFκB modules. At the same time, IRAK1 is degraded along the ubiquitin–proteasome pathway. A parallel pathway leads to activation of another MAP3 kinase, MEKK1, resulting in signal transduction to the JNK and ERK MAP kinase modules. This seems to require an additional adaptor protein, **ECSIT** (Evolutionarily Conserved Signaling Intermediate of Toll pathways), linking TRAF6 with MEKK1 and promoting the autophosphorylation of MEKK1 (see Figure 6.9). As in the case of RIPK1, the role of the kinase activity of IRAK1 is not essential for signal transduction but remains to be elucidated.

IL1 signaling is protected from accidental stimulation at different stages of signal transduction and, in addition, by **TOLLIP** (Toll-Interacting Protein), which sequesters inactive IRAK1. Upon activation, simultaneously with the phosphorylation of IRAK1, TOLLIP is also phosphorylated, resulting in its degradation along the ubiquitin–proteasome pathway.

6.3.3 Toll-like receptors are an exposed defense line against microorganisms and viruses

While the cytoplasmic pattern recognition receptors intercept pathogenic organisms and dangerous material inside the cell, another defense line is positioned at the cell surface and in endosomes. The corresponding sensor proteins are called Toll-like receptors (TLR) referring to a *Drosophila* prototype (**Sidebar 6.3** and **Sidebar 6.10**). Similar

---

Sidebar 6.3 Toll-like receptors in *Drosophila*: dual role in embryogenesis and defense Among the TLRs of invertebrates, the prototype ***Toll*** (means "terrific") of *Drosophila* has been studied in the most detail. Its ligand is an endogenous peptide called ***Spätzle*** (referring to the form of the larvae that is reminiscent of a Swabian noodle specialty) that, upon demand, is released by controlled proteolysis from a precursor protein. This occurs either during larval development or as a response to fungal infections. Triggering signals, either morphogens or fungal molecules, activate different proteases such as the enzyme Persephone.

By this means the *Spätzle–Toll* system controls two entirely different physiological processes: dorso-ventral differentiation of the larvae and biosynthesis of fungicidic peptides such as drosomycin. In both cases, gene transcription is stimulated along the NFκB pathways including the same IκB-homologous inhibitor Cactus but differing from each other by the isoforms of the Rel transcription factors (for details see Section 11.7). The corresponding signaling cascades resemble those of vertebrate cells. By means of the adaptor protein dMyD88 (and an additional adaptor Tube found only in insects) the receptor recruits the kinase *Pelle* (homologous to IRAK), dTRAF (homologous to TRAF6), and the adaptor ECSIT, whereas the kinases resembling the vertebrate IκB kinases, MEKK1, and TAK1 have yet to be identified. Other TLRs of insects seem to participate in defense reactions as well.

TLRs are also present in other nonvertebrates lacking an adaptive immune system. In plants (at least in *A. thaliana*) a subgroup of the large family of transmembrane receptors containing a cytoplasmic Ser/Thr kinase domain seem to be involved in defense reactions. Approximately 80 corresponding "receptors" have been characterized as cytoplasmic rather than transmembrane proteins. Their ligands are unknown, but there is some evidence indicating a role in protection from pathogenic organisms such as fungi.

to NLR-type receptors, Toll-like receptors monitor pathogen- and damage-associated molecular patterns (PAMPs and DAMPs). As a result the whole program of inflammation becomes activated. To this end, the receptors control several thousand genes including those of the pro-inflammatory cytokine and chemokine cocktail. In fact, TLRs belong to the most versatile cellular receptors of all. In mammals, TLRs are expressed in at least 13 isotypes. Like the related IL1 receptors, they are active only as homo- or heterodimers. Some of these isoforms are ubiquitously expressed, whereas others are restricted to distinct tissues.

The individual TLR types exhibit different cellular localization and different ligand specificities. Thus, some receptors (such as TLR1, -2, and -4) are expressed at the cell surface, whereas others (such as TLR 3, -7, -8, and-9) are found primarily in endosomal membranes where they interact with PAMPs and DAMPs taken up by endocytosis. As a rule, the surface receptors recognize lipids, peptides, and sugar derivatives, the endosomal receptors nucleic acids. TLR4 has attracted particular interest because it is the cellular receptor of **lipopolysaccharides (LPS)**. These macromolecules are constituents of the cell wall of Gram-negative bacteria including the highly pathogenic species *Salmonella*, *Legionella*, *Bordetella pertussis*, and certain *Escherichia coli* strains. LPS, also known as **endotoxins**, cause severe symptoms of illness ranging from fever to **septic shock** (which still is one of the most frequent causes of death). The reason for this toxicity is an overactivation of MAP kinase and NFκB signaling. This results in a massive release of pro-inflammatory cytokines (such as TNFα and IL1, with the effect enhanced by positive feedback), arachidonic acid metabolites (such as leukotriene C4 and thromboxane A2; see Section 4.4.5), and nitric oxide. These factors, in particular nitric oxide, cause a dramatic and frequently fatal drop in blood pressure (see Section 16.5).

Gram-positive bacteria such as *Bacillus*, *Streptococcus*, and *Staphylococcus* species do not produce LPS. Instead, the innate immune system recognizes other cell wall components such as **lipoteichoic acids**, **lipoproteins**, and **peptidoglycans**. The corresponding cellular receptors are TLR1 and TLR2. TLR2 (and TLR4) also interacts with special cell wall and cell membrane components (lipoglycans and proteins) of mycobacteria and infectious fungi. The flagellar protein **flagellin**, another pathogen-associated molecular pattern of bacteria, binds to TLR5. TLR9, an endosomal receptor, is specific for **CpG-rich DNA** found in a nonmethylated form only in prokaryotes, whereas in eukaryotes it is always methylated (see Section 8.2.8). In endosomes, bacterial DNA is degraded, yielding single-stranded fragments with CpG-rich DNA, which are the TLR9 ligands proper. TLR9 as well as TLR2 and TLR4 in addition recognize surface molecules of infectious protozoans such as *Trypanosoma* and *Plasmodium*. TLRs are also the body's outpost in the battle against viruses (**Sidebar 6.4** and **Figure 6.11**).

Surface receptors such as TLR4 recognize glycoproteins of viral envelopes while the endosomal receptors respond to viral nucleic acids. Thus, double-stranded RNA, a characteristic product of most viral infections, selectively activates TLR3, while TLR7 and -8 are sensors of single-stranded RNA of RNA viruses (for example, HIV and influenza virus) and TLR9 recognizes nonmethylated CpG sequences not only in bacterial DNA but also in in the genomes of DNA viruses (for example, herpes and papilloma virus). Beside the canonical pro-inflammatory signaling pathways, antiviral TLRs trigger a specific response leading to the transcription of the genes of the antiviral interferons α and β.

The interferon genes are specifically up-regulated by interferon response element-binding transcription factors (IRFs) that are expressed in various isoforms and require protein phosphorylation to become active (see Figure 6.10). To this end TLR7, -8, and –9 seem to use the conventional signaling pathway via MyD88, TRAF6 and the IRAKs. As a result interferon synthesis is activated together with NFκB and MAP kinase modules.. The TLR3 signaling pathway is a variation of this theme (see Figure 6.10). The receptor interacts via its TIR domains with the adaptor protein

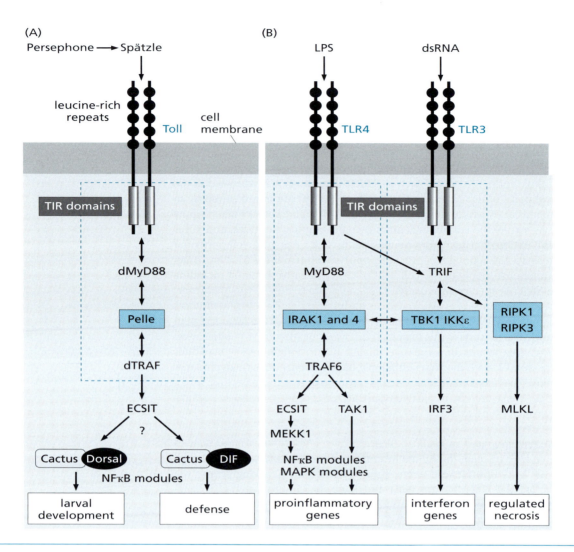

Figure 6.10 Signal transduction by Toll and Toll-like receptors (A) Toll-activated signaling cascades of *Drosophila*. Persephone is one of the proteases releasing the ligand Spätzle from a precursor protein. Dorsal and DIF are two Rel isoforms of the NFκB modules addressing different genes, while Cactus is the IκB-homologous inhibitor becoming degraded upon activation of the modules (see Section 11.8). (B) Two TLR-activated signaling cascades of vertebrates. One of the cascades that is typically induced by TLR4 completely resembles IL1 signaling. It is stimulated by lipopolysaccharide (LPS) and via the MyD88 adaptor it leads to a release of pro-inflammatory mediators. The other cascade, shown here for the receptor TLR3 is stimulated by double-stranded viral RNA (dsRNA). Via the adaptor TRIF it leads to the synthesis of antiviral interferons. Moreover, TRIF may connect the protein kinases RIPK1 and RIPK3 with the receptor. Along this pathway the membrane-damaging protein MLKL is activated and regulated necrosis is triggered. TRIF-dependent signaling may also be activated by TLR4. The protein kinases associated with the receptor/adaptor complexes are shown in the colored boxes.

TRIF (TIR domain inducer of interferon β) rather than with MyD88. TRIF organizes a signaling complex that allows IRF phosphorylations by protein kinases such as TBK1 (TANK-binding kinase 1; TANK, the TRAF associated NFκB activator, is a scaffold protein) and IKKε, which also gate the NFκB pathway. Since the expression of TLRs may be induced by interferons, antiviral signaling amplifies itself by positive feedback (**Sidebar 6.5**).

Like the pro-inflammatory cytokine receptors, TLRs may also induce programmed cell death, in particular regulated necrosis, when adaptors such as TRIF recruit a complex of the protein kinases RIPK1, RIPK3, and their substrate protein MLKL to the receptor (see Figure 6.10).

Sidebar 6.4 Bacterial surface molecules: unique ligands of Toll-like receptors Macromolecules activating TLRs are present in the cell wall of Gram-negative and Gram-positive bacteria as well as in the peptidoglycan layer covering the membranes of both types. Peptidoglycan consists of an oligosaccharide backbone that is composed of disaccharides containing N-acetylgalactosamine and N-acetylmuramic acid with interlinked oligopeptide chains bound to the muramic acid residues (see textbooks of biochemistry and microbiology). The peptidoglycan layer of the cell is anchored in the (inner and outer) membrane by lipoproteins (in Gram-negative species) or lipoteichoic acid (in Gram-positive species). The outer membrane of Gram-negative bacteria is covered to about 75% by LPS, which exhibits a uniquely complex structure not found elsewhere in nature (see Figure 6.11). For this reason, the interaction with TLR4 is absolutely specific.

Figure 6.11 Two bacterial surface molecules acting as ligands of Toll-like receptors (A) **Lipopolysaccharide (LPS).** By its hydroxy-fatty acid residues, the molecule is anchored in the outer membrane of Gram-negative bacteria. The lipid A moiety, mediating the interaction with receptor TLR4, is responsible for the endotoxin effect. The O-antigen is a branched oligosaccharide composed of glucose, galactose, and N-acetylglucosamine. KDO (2-keto-3-deoxyoctulosonic acid), Hept (L-glycero-D-manno-heptose), and 3-hydroxy-fatty acids are characteristic and unique components of LPS. A large number of LPS types are known that differ from species to species in their chemical composition and their endotoxic potency. (B) **Lipoteichoic acid**. This molecule is anchored by a diacylglycerol residue in the cell membrane and by the teichoic acid residue in the peptidoglycan layer of Gram-positive bacteria. Teichoic acid may also be found covalently bound at the outside of the peptidoglycan layer. The chemical composition of lipoteichoic acid varies from species to species (the glycerol may be replaced, for instance, by ribitol). The structures recognized by the TLRs are shown in the colored boxes (P, phosphate residues).

Sidebar 6.5 Antiviral strategies For all organisms the defense against pathogenic viruses is an essential condition of survival. In vertebrates this strategy is based on the functions of the **interferons**. These cytokines exist in two types. Type I comprises the interferons α (23 isoforms) and β that are produced by all cell types in response to viral (and bacterial) infections. Type II is identical with interferon γ, also called immune interferon, because it is exclusively produced by T lymphocytes and natural killer cells. Interferon γ activates macrophages to generate cytotoxic metabolites such as nitric oxide and reactive oxygen species (see Section 2.2) and exhibits immunoregulatory and antitumor activities. The antiviral interferons proper are those of type I. They enable the innate and adaptive immune systems to fight viruses. Interferon receptors couple with cytoplasmic Tyr kinases of the JAK family that phosphorylate and activate the STAT transcription factors (see Section 7.2.2). They control the transcription of more than 100 different genes along this pathway, thus generating an antiviral state that extends to the tissue surrounding virus-infected cells. Key events are those that prompt the infected cells to commit suicide and the surrounding cells to activate signaling reactions that intervene with virus invasion and replication. A powerful strategy is to sequester double-stranded viral RNA or to prevent its translation into protein. While sequestering is achieved by a dynamin-related GTPase Mx, RNA translation is inhibited by the double-stranded RNA-activated protein kinase PKR that phosphorylates and inactivates the translation factor eIF2 (see Section 9.3.1). The genes encoding Mx and PKR are activated by interferons. In addition, interferons stimulate cytotoxic T lymphocytes and natural killer cells and control the immune system at many other points aiming at the production of virus-specific antigens.

Viruses (and bacteria) have developed various strategies to subvert the innate immune system. These include reduction of the CpG sequences in viral DNA, sequestering or proteolytic degradation of host signaling proteins, and production of modified ligands altering or blocking TLR activation. A major target of viral subversion is the protein kinase PKR, which is prevented from activation or deceived by false activators and substrates. In addition, viruses may protect double-stranded RNA from sequestering and destruction. Moreover, by binding to certain proteins, they may render the RNA "invisible" for the cellular RNA sensors TLR3 and RIG1/MDA5, which may also be blocked separately by inhibitory proteins. Viruses also secrete soluble truncated forms of interferon receptors that intercept circulating interferons or they produce inactive forms of IRFs that compete—as dominant negative mutants—with the cellular transcription factors. By inhibiting caspases certain viruses are able to prevent apoptosis of infected cells. As a counter-measure the cells may commit suicide by regulated necrosis which is caspase-independent (for details, see Section 13.2). Viral RNA can also be recognized and destroyed by RNA interference (see Section 9.5). Whether or not this plays a significant role in mammalian virus defense is a matter of debate.

Since cells always try to cope with viral tricks, the situation ends up in an arms race. An understanding of this battle is an essential condition for the clinical control of viral diseases.

Summary

- Receptors of the pro-inflammatory, immunomodulatory, and cell death inducing cytokines are oligomeric transmembrane proteins that couple via specific adaptor proteins with Ser/Thr-specific protein kinases. By this means they trigger nonspecific defense reactions (innate immunity).

- Depending on the receptor type and the adaptor proteins recruited upon activation, the receptors of the tumor necrosis factor (TNF) family may cause cell death by apoptosis or by regulated necrosis. Alternatively, they may stimulate genetic programs of cell survival and for the synthesis of pro-inflammatory mediators.

- Interleukin 1 (IL1) is a master regulator of the inflammatory response. Via its receptor and several adaptor proteins it stimulates specific protein kinases (IRAKs) leading to an activation of pro-inflammatory genetic programs.

- Interleukin 1 is released from an inactive precursor by controlled proteolysis catalyzed by caspase 1. This occurs in inflammasomes. A major constituent of these multiprotein complexes are cytoplasmic receptor proteins that recognize pathogen- and damage-associated molecular patterns.

- Toll-like receptors (TLRs) are membrane receptors of pathogen- and damage-associated molecular patterns. Like TNF, IL1 and cytoplasmic pattern recognition receptors they activate the genetic program of innate immunity.

- Overactivation of TLR signaling may result in endotoxin shock. In insects, TLRs also control morphogenetic processes.

Further reading

Section 6.2

Hartwig T & Wang ZY (2015) The molecular circuits of steroid hormone signalling in plants *Essays Biochem* 58:71–82.

Larrieu A & Vernoux T (2015) Comparison of plant hormone signaling systems. *Essays Biochem* 58:165–181.

Macho AP and Zipfel C (2014) Plant PRRs and the activation of innate immune signaling. *Molec Cell* 54:263–272.

Macho AP, Lozano-Duran R & Zipfel C (2015) Importance of tyrosine phosphorylation in receptor kinase complexes. *Trends Plant Sci* 20:269–272.

Marias MJ, Martin-Malpartida P & Massague J (2015) Structural determinants of Smad function in TGFβ signaling *Trends Biochem Sci* 40:296–308.

Massague J & Just W (2012) Sorting the TGFβ labyrinth. *FEBS Lett* 686:issue 14.

Section 6.3

Akira S, Uematsu S & Takeuchi O (2006) Pathogen recognition and innate immunity. *Cell* 124:783–801.

Bauernfeind F & Hornung V (2013) Of inflammosomes and pathogens - sensing of microbes by the inflammosome. *EMBO Mol Med* 5:814–826.

Beutler B & Rietschel EZ (2003) Innate immune sensing at its roots: the story of endotoxin. *Nat Rev Immunol* 3:169–176.

Boraschi D & Tagliabue A (2013) The interleukin-1 receptor family. *Semin Immunol* 25:394–407.

Cabal-Hierro L & Lazo PS (2012) Signal transduction by tumor necrosis factor receptors *Cell Sign* 24:1297–1305.

Christofferson DE, Li Y & Yuan J (2014) Control of life-or-death decisions by RIP1 Kinase. *Annu Rev Physiol* 76:129–150.

Cohen P (2014) The TLR and IL-1 signalling network at a glance. *J Cell Sci* 127:2383–2390.

DeNardo D (2015) Toll-like receptors: activation, signalling and transcriptional modulation *Cytokine* 74:181–189.

Kawai T & Akira S (2006) Innate immune recognition of viral infection *Nat Immunol* 7:131–137.

Lamkanfi M & Dixit VM (2012) Inflammosomes and their role in health and disease. *Annu Rev Cell Dev Biol* 18:137–161.

Medzhitov R (2008) Origin and physiological roles of inflammation. *Nature* 454:428–435.

Nadiri A, Wolinski MK & Saleh M (2006) The inflammatory caspases: key players in the host response to pathogenic invasion and sepsis. *J Immunol* 177:4239–4245.

O'Neill LA and Bowe AG (2007) The family of five: TIR-domain-containing adaptors in Toll-like receptor signalling. *Nat Rev Immunol* 7:355–364.

Portou MJ et al. (2015) The innate immune system, Toll-like receptors and dermal wound healing: a review. *Vascul Pharmacol* 71:31–36.

Silke J, Rickard JA & Gerlic M (2015) the diverse role of RIP kinases in necroptosis and inflammation. *Nat Immunol* 16:689–697.

Whitfield C & Trent MS (2014) Biosynthesis and export of bacterial lipopolysaccharides. *Annu Rev Biochem* 83:99–128.

Signal Transduction by Tyrosine Kinase- and Protein Phosphatase-Coupled Receptors

7

Protein phosphorylation at tyrosine residues plays a crucial role in cellular signal processing. It is found in all organisms. The corresponding kinases are subdivided in two unrelated families: prokaryotic BY kinases (see Section 3.2) and eukaryotic Tyr kinases. Thus, during evolution Tyr phosphorylation has been "invented" at least twice. As far as eukaryotic Tyr kinases are concerned, one distinguishes between cytoplasmic and transmembrane enzymes. Being sensors of extracellular signals, the latter are called receptor tyrosine kinases (RTKs). The family of mammalian Tyr kinases is composed of 32 cytoplasmic enzymes and 58 receptor tyrosine kinases.

7.1 Receptor tyrosine kinases

Until recently it was proposed that receptor tyrosine kinases have developed in parallel with the evolution of multicellular animals, thus representing the "most modern" cellular sensors. Now, this assumption is met with increasing skepticism since genetic studies and modern methods of protein analysis have revealed many eukaryotic tyrosine kinases (including enzymes similar to receptor tyrosine kinases) in unicellular eukaryotes. While in the beginning of eukaryotic evolution this receptor type may have been specialized for sensing environmental signals, in metazoans this property has been extended to intercellular communication.

The ligands of mammalian receptor tyrosine kinases are hormones such as **insulin** and a large and heterogeneous family of **growth factors** (**Figure 7.1**). The latter regulate embryonic organ development and, in the mature organism, wound repair and other proliferative processes. This physiological function is mirrored by the downstream targets of receptor Tyr kinase signaling: these are first of all mitogen-activated protein (MAP) kinase cascades and the anti-apoptotic phosphatidylinositol 3-kinase–protein kinase B/Akt complex (which is also a key mediator of insulin effects). Both signaling pathways lead to an activation of a wide variety of genes that control cell survival and proliferation (see Chapter 11). There are fluid transitions between Tyr kinase-coupled receptors and cell adhesion proteins that are primarily responsible for the formation and cohesion of tissues. Conversely, such receptors are major targets of defects that lead to chronic metabolic diseases, immunological disorders, embryonic malformations, and, in particular, cancer. Indeed, most of the receptor Tyr kinases have been found to undergo oncogenic mutations.

Receptor tyrosine kinases have one transmembrane domain. Their carboxy (C)-terminal cytoplasmic regions are quite similar, containing the kinase domains and several autophosphorylation sites, whereas the amino (N)-terminal extracellular domains with the ligand-binding sites are highly diverse (**Figure 7.2**). This diversification is mirrored by functional variability, which has led to a classification into about 20 subfamilies. As a rule, each receptor type is expressed in several isoforms that differ in properties and tissue distribution.

Figure 7.1 Receptors with intrinsic Tyr kinase activity Shown are those receptor families (left) for which the ligands are known, together with the corresponding ligands (right). The numbers of receptor isoforms identified to date are given in parentheses.

Ephrin-R (14)	ephrins
PDGF-R (6)	platelet-derived growth factor
	keratinocyte growth factor (KGF)
	colony-stimulating factor (CSF)
	stem cell factor (SCF)
EGF-R (11)	epidermal growth factor family
	transforming growth factor α (TGFα)
FGF-R (4)	fibroblast growth factor family
VEGF-R (3)	vascular-endothelial growth factors
NGFR/Trk (3)	nerve growth factor (NGF), neurotrophins
Ins-R (3)	insulin, insulin-like growth factors (IGF)
HGFR/Met (2)	hepatocyte growth factor (HGF)
Ret	neurotrophic factors from glia cells
Tie/Tek (4)	angiopoietin
Musk	agrin
Axl/Tyro (3)	Gas-6, protein S
Ryk	Wnt-type morphogens
Alk	pleiotrophin, midkine
DDR (2)	collagen

Figure 7.2 Diversification of receptor Tyr kinases Note the variability of the extracellular sensor domains compared with the monotony of the cytoplasmic kinase domains, resembling the situation found for prokaryotic sensor His kinases (Chapter 3). The ligands of receptors Klg, Ros, Ltk/Alk, and Ror are still unknown. Many of the receptors are named after corresponding oncogenes. (Partly adapted from Blume-Jensen P and Hunter T [2001] *Nature* 411:355–365.)

All receptor Tyr kinases either exist as dimers or undergo dimerization upon ligand binding as shown in **Figure 7.3**. To this end the individual isoforms of each receptor type associate into homodimers or heterodimers that are stabilized either by two ligands or by one ligand with two binding sites, one for each receptor subunit. Dimerization accompanies a conformational change that activates the receptor by

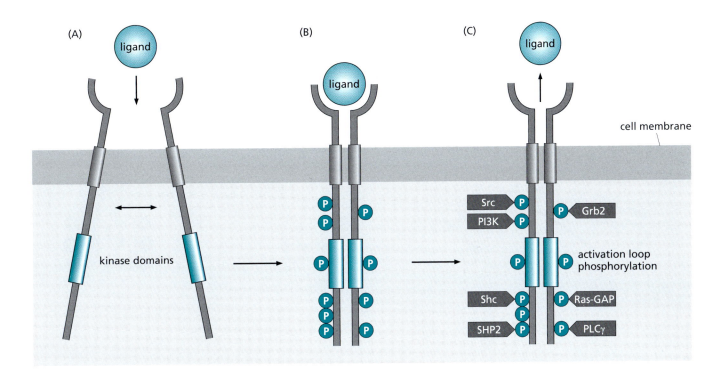

Figure 7.3 Three steps of signal transduction by a receptor Tyr kinase (A) Interaction of the receptor with a bivalent ligand (or two ligand molecules) leads to a conformational change that promotes dimerization, bringing the cytoplasmic Tyr kinase domains in close contact. (B) Trans-autophosphorylation of the activation loops in the kinase domains and of other sites. The phosphate residues are symbolized by P. (C) Site-specific interactions of individual phospho-Tyr residues with some selected proteins carrying Src homology 2 (SH2) or phospho-Tyr-binding (PTB) domains. Src is a cytoplasmic Tyr kinase, Grb2 and Shc are adaptor proteins linking the receptor to MAP kinase pathways, SHP2 is a phospho-Tyr-specific protein phosphatase, PI3K is phosphatidylinositol 3-kinase, Ras-GAP is a GTPase-activating protein, and PLCγ is phospholipase C type γ. Additional proteins undergoing phosphorylation by the receptor are not shown. The autophosphorylated receptor remains active even in the absence of the ligand until it becomes dephosphorylated.

allosteric interactions of the cytoplasmic domains. Currently, this mechanism is best understood for the epidermal growth factor receptors (see Section 7.1.1). For most receptor Tyr kinases, the active state is then stabilized by trans-autophosphorylation of the activation loops. Additional phosphorylations of the cytoplasmic domains generate docking sites for the recruitment of signal-transducing proteins, many of which are Tyr kinase substrates themselves. In other words, receptor Tyr kinases are both enzymes and scaffold proteins that, upon phosphorylation, organize signal-processing protein complexes at the inner side of the cell membrane.

The Tyr-phosphorylated docking sites of the receptors specifically bind proteins with domains that recognize phospho-Tyr residues. Even though an active receptor is phosphorylated at multiple Tyr residues, this binding does not occur randomly but in a highly selective manner (see Figure 7.3). As explained in Section 7.1.2, the reason for this selectivity is found in the variability of phospho-Tyr-binding domains.

7.1.1 Signaling by growth factors

Epidermal growth factor receptors

For a more in-depth understanding of receptor Tyr kinases, we shall have a closer look at the most thoroughly investigated prototype of this family, the epidermal growth factor (EGF) receptor. The EGF receptor was one of the first receptor Tyr kinases characterized and purified. Moreover, its mutated form ranks among the first oncoproteins discovered. In fact, oncogenic v-ErbB protein of avian erythroblastosis virus (which provokes a neoplastic blood-disease in chickens) was found to be identical with a truncated EGF receptor. This defective protein is constitutively active because it lacks the extracellular ligand-binding domain that, in the absence of the ligand, exerts an autoinhibitory effect on the receptor's Tyr kinase activity by preventing receptor dimerization (**Figure 7.4**).

The EGF receptors represent a family of 11 related receptors that are collectively called ErbB1–ErbB11 with ErbB1 being the canonical EGF receptor. Their ligands, if known, are EGF and EGF-like growth factors, a family of 13 hormone-like proteins that evoke a wide variety of cellular responses, particularly cell proliferation but also cell migration, cell differentiation, and cell death. These proteins perform

Figure 7.4 Activation of the EGF receptor ErbB1 (A) Two monomeric ErbB1 proteins in the inactive conformation: the kinase domains (KD) cannot interact with each other. L1 and L2, leucine-rich repeats; CR1 and CR2, cysteine-rich domains. (B) Active dimer stabilized by ligand binding. "Back-to-back" dimerization occurs through an interaction of the CR1 domains: the kinase domains are in close contact, enabling one domain to activate the other one. (C) Active autophosphorylated dimer (P, phospho-Tyr residues). ErbB1 does not need activation loop phosphorylation.

these functions not only in epidermis but in most other epithelia. Which of these effects is realized depends on the cell type, the physiological context, and the ligand and receptor isoforms. Most growth factors of the EGF-family are expressed as transmembrane proteins. To become active as circulating hormones the extracellular domains have to be released by controlled proteolysis (for an analogous example see Section 7.1.4). Gene knockout experiments show that each receptor isoform is important for embryonic development, particularly of epithelia. A constitutive overexpression of ERbB protein (for instance, resulting from gene amplification) or hyperactivity due to oncogenic point mutations, particularly in the ATP-binding domain, are associated with several human malignancies such as cancer of the lung, pancreas, head and neck, and brain. Frequently such defects of the EGF receptor system correlate with a poor prognosis. Therefore, the ErbB system is a preferred target of novel cytotoxic anti-cancer drugs, some of which have been approved for clinical use (see below).

X-ray crystallography has revealed much about the structure of the EGF receptor ErbB1 and its mode of transmembrane signaling. The receptor exists in a ligand-binding and a non-ligand-binding form that are in equilibrium. These forms differ in the conformation of the extracellular domain (see Figure 7.4). In the ligand-binding form, the leucine-rich repeats L1 and L2 of the extracellular domain arrange to a clamp-like structure that accommodates the ligand. Ligand binding facilitates a back-to-back dimerization through an interaction of the cysteine-rich domains CR1, called dimerization arms. In the nonbinding form, the CR1 domain interacts with the CR2 domain and is buried inside the extracellular polypeptide structure, while L1 and L2 are kept at a distance unable to interact with the ligand.

Dimerization provides the condition for an allosteric activation of the cytoplasmic kinase domains. In the resting state, hydrophobic leucine residues block the active center. These keep the activation loop and a strategic α-helix in the N-lobe (see Section 2.6.1) in a position that does not allow ATP binding (the "mouth" of the substrate binding site is closed). With the conformational change induced by dimerization, these structural elements turn into the correct position required for kinase activation (the "mouth" opens). To this end the C-lobe of one monomer interacts with the N-lobe of the other monomer, provided the kinase domains are juxtaposed asymmetrically (see Figure 7.4). The activation mechanism involves a special feature: the EGF receptors belong to those few kinases that do not require activation loop phosphorylation for activation. This is in contrast to other receptor Tyr kinases (such as the insulin receptor) that do need activation loop phosphorylation. Nevertheless, in principle the activation mechanism of ErbB is assumed to be representative for receptor Tyr kinases in general.

A key event of ErbB signaling is the phosphorylation of several Tyr residues in the cytoplasmic tail regions of the receptor protein. By this means, contact sites for proteins with corresponding phospho-Tyr recognition sites (see Section 7.1.2) are generated. This holds true for most receptor tyrosine kinases. In fact, as already mentioned, receptor tyrosine kinases are not only transmembrane signal transducers and enzymes, but also scaffold proteins that organize signal processing protein complexes in the cytoplasm. However, an important property distinguishes receptor tyrosine kinases (and other Tyr kinases) from conventional scaffold proteins with preset binding sites: they are dynamic scaffolds because their binding capacity depends on the degree of Tyr phosphorylation and, therefore, is fully reversible.

As far as the interaction partners of the EGF receptor are concerned, the individual ErbB isoforms exhibit some selectivity. Thus, each ErbB type contributes specifically to signaling. Together, the 11 ErbBs exhibit many potential Tyr phosphorylation sites that may interact with a wide variety of partner proteins. This network of interactions is controlled by 13 different input signals (EGF and its relatives) and regulates the activity of numerous genes. The organization of the ErbB network clearly demonstrates what has been discussed in Chapter 1: signal processing means an excitation of a complex "neural" network rather than a process involving a few individual sequences of biochemical reactions. An in-depth understanding of such a situation demands a systems-based approach because conventional methods rapidly reach their limits.

An interesting variation is found in nerve cells that are confronted with the problem of signaling across long distances. The growth of their axons is stimulated by nerve growth factor interacting with the receptor Tyr kinase Trk at the tip of the nerve fiber. This turns up both local signaling and gene transcription, even though the nucleus may be centimeters or even meters away. To bridge this distance, the ligand–receptor complex is taken up in endocytotic vesicles (endosomes) that, with the help of motor proteins, migrate to the nucleus by a process called **retrograde transport**.

Reactive oxygen species facilitate signaling by receptor tyrosine kinases

Since Tyr phosphorylation occupies a central position in the regulation of tissue formation, tissue maintenance, and tissue regeneration, the cell must keep it under strict control. In this context protein Tyr phosphatases play a particularly important role as the most efficient opponents of Tyr kinases (see Section 7.4). In fact, in most cases, activation of a receptor Tyr kinase is inadequate for efficient signal transduction, unless the phosphatases are switched off. This is achieved by the redox signal H_2O_2, which inactivates Tyr phosphatases by oxidizing catalytically relevant sulfhydryl groups to disulfide and sulfenic acid groups. The thioredoxin cycle renders the reaction reversible (**Figure 7.5** and Section 2.2). How does the cell switch on H_2O_2 production simultaneously with the activation of Tyr kinase-coupled receptors? A major source of hydrogen peroxide is the membrane-bound enzyme NADPH oxidase, which is activated by the small G-protein Rac (see Section 10.3.4). Rac is a target of various signaling cascades including the PI3K pathway controlled by Tyr kinase-coupled receptors. Thus, the release of hydrogen peroxide is a general response to receptor activation. To augment the effect, H_2O_2 simultaneously inhibits the phospholipid phosphatase PTEN, an opponent of PI3Ks.

Receptor inactivation and down-regulation: an extravagant venture

In its autophosphorylated form a Tyr kinase-coupled receptor stays active even after the ligand has dissociated. A standard mechanism of rapid signal termination is,

Figure 7.5 Enhancement by redox signaling of signal transduction by a receptor Tyr kinase The Tyr kinase-coupled receptor activated by a signal molecule S and subsequent autophosphorylation (P) interacts with and activates phosphatidylinositol 3-kinase (PI3K). The phosphatidylinositol 3,4,5-trisphosphate (PIP$_3$) thus generated recruits the small G-protein Rac to the cell membrane, where it stimulates nicotinamide adenine dinucleotide phosphate (NADPH) oxidase activity. The reactive oxygen species (ROS) produced by this enzyme oxidize cysteine sulfhydryl groups (SH) and thus inactivate a protein Tyr phosphatase (PTP) and the phospholipid phosphatase PTEN, which both would inhibit receptor-dependent signaling (teal pistils). By means of thioredoxin (Trx) and NADPH, both enzymes become reactivated. Thus, signaling by receptor Tyr kinases depends on the redox status of the cell.

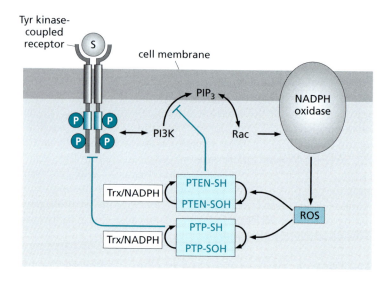

therefore, receptor dephosphorylation by Tyr-specific protein phosphatases. As we have seen, the activity of these is limited, however, by redox signaling activated by the receptor. This situation appears to be rather tricky, indicating a continuous oscillation between phosphorylation and dephosphorylation.

Upon prolonged stimulation, Tyr kinase-coupled receptors become internalized by endocytosis and destroyed inside the cell—similar to the process occurring with G-protein-coupled receptors (GPCRs, see Section 5.7). At least for some Tyr kinase-coupled receptors, this down-regulation is achieved by mono-ubiquitylation and subsequent degradation in lysosomes, or by poly-ubiquitylation followed by proteolysis in proteasomes. The E3 ubiquitin ligase that has been found to play a key role in this process is **Cbl** (the name refers to the Casitas B-cell lymphoma of mice, where Cbl has undergone oncogenic mutation). Its mechanism of action has been investigated in some detail. We shall look more closely at it because Cbl provides a particularly clear example of the function of a ubiquitin ligase in signal transduction.

Cbl-type proteins have been found in multicellular animals; humans express three isoforms. In experimental animals the complete loss of Cbl by gene knockout results in prenatal death, indicating a key role of these enzymes in cellular physiology. A hint to their function is given by the Cbl structure: it includes a Tyr kinase-binding motif containing an SH2 domain as a docking site for Tyr-phosphorylated sequences, a proline-rich domain that is a binding site for SH3 domains, a ubiquitin-associating domain probably interacting with regulatory proteins, and a RING domain that is essential for ubiquitylation (**Figure 7.6** and Section 2.8.1). By means of these domains, Cbl interacts with Tyr-phosphorylated (activated) receptors (and cytoplasmic Tyr kinases as well). The receptors (or the cytoplasmic tyrosine kinases) catalyze a Tyr-phosphorylation of Cbl. As a result its conformation becomes unfolded in such a way that an E2 ubiquitin-conjugating enzyme can bind to the RING domain and ubiquitylate the Tyr kinase-coupled receptor. The viral oncoprotein vCbl of mice is a truncated form lacking the RING domain (see Figure 7.6). Through its Tyr kinase-interacting domain it still binds substrate proteins but is unable to establish contact with the E2 ubiquitin-conjugating enzyme; thus it acts as a dominant negative mutant. The oncogenic effect results from a continuous overactivation of mitogenic and anti-apoptotic signaling pathways stimulated by Tyr kinases and Tyr kinase-coupled receptors that are rescued from degradation.

In reality the control of Cbl activity is even far more complex. Thus, several regulatory proteins (Sts, Sprouty, the small G-protein Cdc42, and others) have been identified that are recruited to the Cbl substrate and specifically counteract ubiquitylation, either by displacing and sequestering Cbl or by promoting its degradation. In

(A)

(B)

Figure 7.6 Domain structure and activation of the E3 ubiquitin ligase Cbl (A) cCbl is one of several human Cbl isoforms. Interaction partners are depicted in teal. The gray box is the Tyr kinase-interacting domain, Pro is a prolinerich domain, and UBA is a ubiquitin-associating domain. Locations that become Tyr-phosphorylated are marked by asterisks. vCbl, the viral oncogene product, is a truncated Cbl acting as a dominant negative mutant. (B) Mechanism of Cbl activation by a Tyr-phosphorylated (P) Tyr kinase (TK). In the inactive conformation of Cbl (left, depicted in a simplified form), the RING domain is blocked. This blockade is lifted by the attachment of TK to the Tyr kinase-interacting domain, in particular to its SH2 domain, and Tyr phosphorylation of Cbl in a "linker domain" (teal P). As a result, an E2 ubiquitin-conjugating enzyme can bind to the RING domain and ubiquitylate TK (right). Tyr kinases such as Src containing an SH3 domain interact, in addition, with the proline-rich domain of Cbl (not shown).

fact, Cbl activated by a substrate Tyr kinase is short-lived since it undergoes self-destruction by auto-ubiquitylation or by an interaction with other ubiquitin ligases. Auto-ubiquitylation means that Cbl and its substrates become down-regulated in a coordinated, reciprocal manner, a phenomenon of self-control that has also been reported for other E3 ubiquitin ligases. Quite an opposite effect is observed when Cbl is down-regulated by another ubiquitin ligase. Because of the high substrate specificity of E3 ligases, such a reaction is selective, meaning it does not include the Cbl substrate that, instead, is protected from ubiquitylation and destruction through Cbl.

Clearly, protein down-regulation is quite an expensive venture, since phosphorylation, ubiquitylation, and in particular the resynthesis of down-regulated proteins require a large amount of ATP energy. Obviously, the precise regulation of cellular signaling is anything but simple and is worth any extravagance!

Receptor Tyr kinases as targets of cancer therapy

Cancer, a characteristic disease of complex multicellular organisms, is caused by somatic or hereditary gene mutations and other persisting defects that disturb cellular signal processing. As stated in Section 13.3, a tumor cell misinterprets the communicative signals of the body, thus to some extent resembling a psychotic person. In the past 10 years, efforts have been made to correct such oncogenic defects of signal transduction. For this approach, receptor Tyr kinases rank among the most attractive targets for two reasons.

(1) They are linked with multicellularity.

(2) Most of them have been found to be targets of oncogenic mutation.

Synthetic Tyr kinase inhibitors are called **tyrphostins** (<u>tyr</u>osine <u>phos</u>phorylation <u>in</u>hibitors). They have been used for the treatment of a variety of cancers. An example is provided by chronic myelogenous leukemia which is characterized by an overactivation of the Tyr kinase BCR-Abl (**Sidebar 7.1**). Here a specific tyrphostin is employed which blocks the ATP-binding site of the kinase. More about tyrphostins and signal transduction therapy of cancer is found in Chapter 13.3.

Sidebar 7.1 The Philadelphia translocation as a cause of leukemia The oncogene of the Abelson leukemia virus of mice encodes a Tyr kinase **vAbl**. The corresponding proto-oncogene *c-abl* is distributed over the entire animal kingdom. It has gained much attention because of the so-called **Philadelphia translocation**, a chromosomal rearrangement considered to be the primary cause of chronic myelogenous leukemia. In this rearrangement the *c-abl* gene is translocated from chromosome 9 to chromosome 22, where it fuses with a truncated form of a gene called *bcr*. The hybrid gene thus generated gives rise to a cAbl/BCR fusion protein exhibiting deregulated cAbl Tyr kinase activity (**Figure 7.7**). **BCR** (Breakpoint Cluster Region) is an atypical Ser/Thr-directed protein kinase containing a Rho guanine nucleotide exchange factor (GEF) domain and a Rac GTPase-activating protein (GAP) domain. Since the small G-proteins Rho and Rac are key regulators of the cytoskeleton (see Section 10.3), BCR may have a similar function.

cAbl is not a receptor Tyr kinase. Instead it exhibits binding domains for DNA and the actin cytoskeleton as well as SH2, SH3, and proline-rich (SH3-interacting) domains and is found in the nucleus and in the cytoplasm, as well as in a lipid-anchored form at the membrane. The kinase activity is strictly autoregulated, mainly by an intramolecular interaction between the SH3 and the Pro-rich domains. As with most Tyr kinases, cAbl must be phosphorylated at a Tyr residue in the activation loop to gain full activity. This occurs either by autophosphorylation, by a Tyr kinase of the Src family (see Section 7.2), or by the DNA damage checkpoint kinase ATR (Ataxia Telangiectasia mutated Rad3-related; see Section 12.8). The active cAbl is a short-lived enzyme that is rapidly inactivated by dephosphorylation and proteolytic degradation. Both autoregulation and degradation are impaired in the oncogenic cAbl-BCR fusion protein. A variety of functions has been attributed to cAbl. First of all, the enzyme transmits proliferative and survival signals, which explains its oncogenic potential. On the other hand, cAbl also

Figure 7.7 Oncogenic activation of Tyr kinase cAbl by the Philadelphia translocation Owing to a reciprocal translocation of fragments of chromosomes 9 and 22, a so-called Philadelphia chromosome is generated in which the *c-abl* gene is fused with the N-terminal fragment of the *bcr* gene. In the constitutively expressed BCR/cAbl fusion protein encoded by the hybrid gene, cAbl has oncogenic potential and is a major cause of chronic myelogenous leukemia. The anti-leukemic drug Gleevec blocks the ATP-binding site of the oncogenic cAbl. Abbreviations: oligo, oligomerization domain; TyrP, domain with phosphorylated Tyr interacting with SH2 domains; CalB, Ca^{2+}-dependent lipid-binding domain; Rac-GAP, domain activating the GTPase activity of Rac; Rho-GEF, domain catalyzing the GDP–GTP exchange of Rho; PRO, proline-rich domain interacting with SH3 domains; NLS, nuclear localization sequence; DNA-BD, DNA-binding domain; actin-BD, actin-binding domain; SH2 and SH3, Src-homology domains 2 and 3.

seems to participate in apoptosis. In fact, in response to DNA damage, its nuclear form becomes stimulated by ATR-catalyzed phosphorylation and subsequently phosphorylates ATR and other stress proteins such as the p53-homologous factor p73 (see Section 12.8). However, since ATR is also activated independently from cAbl, the role of the latter in the cell's DNA damage response is not clear yet.

cAbl is an essential enzyme because corresponding knockout animals are not viable. Indeed, the kinase is thought to play a critical role in the development of the nervous system, controlling the formation of synaptic contacts brought about by targeted growth of nerve fibers or axons. As explained in Section 16.1, this axonal navigation is due to interactions between receptors at the axonal growth cone with stimulatory or inhibitory signaling molecules transmitted by adjacent cells. A major intracellular target of those interactions is the actin cytoskeleton, producing membrane extrusions at the top of the growing nerve fiber. These events critically depend on cAbl activity, as has been shown in particular for the *Drosophila* model. By means of its actin-binding domains, cAbl directly interacts with the cytoskeleton, phosphorylating a series of actin-regulatory proteins and navigation receptors.

7.1.2 Binding domains for phosphotyrosine residues or the art of complex formation

Whether or not a protein becomes bound to an autophosphorylated Tyr kinase-coupled receptor depends on interaction domains that specifically recognize phosphorylated Tyr residues in proteins. These binding sites are called SH2 and PTB domains. They have been found in uni- and multicellular eukaryotes but not in prokaryotes.

Src homology domains

The acronym SH2 means Src Homology 2. It refers to Src, a cytoplasmic Tyr kinase that in a mutated form was discovered as an oncoprotein of a chicken sarcoma virus (see Section 7.2). Src contains sequences that are also found in homologous forms in many other proteins. These Src homology domains are as follows:

- SH1, a domain resembling the kinase domain of Src

- SH2, a domain interacting with sequences containing phospho-Tyr residues

- SH3, a domain interacting with proline-rich sequences with the recognition site -Pro-X-X-Pro

SH2 and SH3 domains have been identified in wide variety of proteins. Such proteins may be enzymes, transcription factors, or adaptors. Some examples are shown in **Figure 7.8**. In prokaryotes SH2 domains have not yet been found.

SH2 and PTB domains recognize sequences with phosphotyrosine residues

An SH2 domain comprises about 100 amino acids. The key structural element is a bundled four-stranded β-sheet flanked by a three-stranded β-sheet and two α-helices (**Figure 7.9**). The four-stranded sheet contains conserved amino acids binding the phospho-Tyr residue, while the loops linking the sheet to the other structures interact with two to six amino acids located C-terminally of the phospho-Tyr residue. These structural elements may differ among SH2 domains and determine the specificity of the interaction, which goes far beyond recognition of the phospho-Tyr residue. On this basis, three major families of SH2 domains have been identified and are named for prototypical proteins:

- Src-type SH2 domains recognize charged residues in positions 1 and 2 as well as a hydrophobic residue in position 3 C-terminally from phospho-Tyr

- phospholipase Cγ-type SH2 domains interact with aliphatic residues in positions 1 to 5

Figure 7.8 A selection of proteins with Src homology domains (A) Enzymes Src, Ras-GAP, PLCγ, Tyr phosphotase SHP2, and transcription factor STAT. SH3, SH2, and SH1, Src homology domains 3, 2, and 1; PTK, protein Tyr kinase domain; PH, pleckstrin homology domain; C2, Ca²⁺/phospholipid-binding domain; GAP, GTPase-activating protein domain; PLC, phospholipase C domain; PTPase, protein Tyr phosphatase domain; DNA-BD, DNA-binding domain. The proteins are not shown to scale. (B) Adaptor proteins without enzymatic activity. Domains that recognize phospho- Tyr residues are shown in color. PTB, phosphotyrosine-binding domain.

- Grb2-type SH2 domains recognize an Asp residue in position 2 (as explained below, Grb2 is an adaptor protein connecting receptor Tyr kinases with downstream effectors)

In addition, other SH2 domains recognize nonphosphorylated Tyr residues, for instance, in the intrasteric inhibition of protein phosphatase SHP1 (**Figure 7.10**).

SH2 domains have two major functions: formation of signal-processing complexes, in particular at autophosphorylated receptors, and intrasteric control of the activity of signal-processing proteins. Frequently both effects are combined. An instructive

Figure 7.9 Structure of an Src-typical SH2 domain The four β-sheets with the central phospho-Tyr (P-Tyr, teal) binding site are seen.

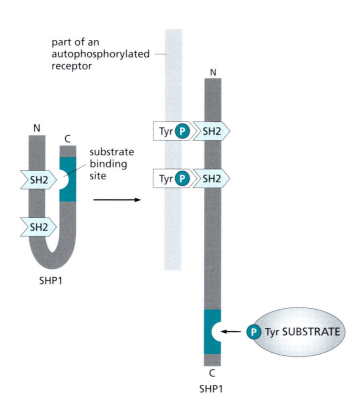

part of an
autophosphorylated
receptor

N

C

substrate
binding
site

SH2

SH2

N

Tyr P SH2

Tyr P SH2

SHP1

P Tyr SUBSTRATE

C

SHP1

Figure 7.10 Role of SH2 domains in activation of the protein Tyr phosphatase SHP1 In inactive SHP1, one of the two tandem SH2 domains shields the substrate binding site (left; no P-Tyr interaction). This blockade is abolished by the conformational change induced by the interaction with two tandem P-Tyr residues of the autophosphorylated receptor (right).

example is provided by the Tyr-specific protein phosphatases of the type SHP1 (SH2-containing protein Tyr Phosphatase1, pronounced "ship") that, among other functions, take care of the dephosphorylation of autophosphorylated receptors, that is signal termination. In a negative feedback loop, the same receptors activate SHP1 by lifting an intrasteric blockade (see Figure 7.10).

In a similar but more sophisticated way, Src-type Tyr kinases are activated. As explained in Section 7.2, these cytoplasmic enzymes are integrated in signaling cascades controlled by Tyr kinase-coupled receptors (**Figure 7.11**). Src activity is controlled by both inhibitory and stimulatory Tyr phosphorylation providing a particularly clear example of the ambiguousness of protein phosphorylation. While a C-terminal phospho-Tyr residue keeps the enzyme in an inactive conformation by interacting with the SH2 domain, phosphorylation of a Tyr residue in the activation loop is required for full activity. Therefore, Src activation proceeds in two steps. For activation the interaction between the SH2 domain and the C-terminal phospho-Tyr residue is first interrupted by a competing Tyr-phosphorylated protein or by enzymatic dephosphorylation. In the second step, activation loop phosphorylation occurs by trans-autophosphorylation. The enzyme is switched off again by dephosphorylation of the activation loop and rephosphorylation of the C-terminal Tyr residue. The latter reaction is catalyzed by a specific C-terminal Src kinase (Csk). Csk is a cytoplasmic enzyme that, to establish contact with Src, interacts with a Csk-binding single-pass transmembrane protein. The binding protein also associates with the SH3 domain of Src and, in turn, is phosphorylated by Src at several Tyr residues. The phospho-Tyr groups thus generated function as anchoring sites for additional proteins with SH2 domains. Thus, the Csk-binding protein is a multifunctional scaffold that organizes via SH2-specific interactions large signal processing complexes at Src-type kinases.

In contrast to SH2 domains, **PTB domains** recognize phospho-Tyr residues together with amino acids located *amino-terminally*, but in some cases they also recognize sequences with nonphosphorylated tyrosine or even without tyrosine. Their structure is characterized by a central seven-stranded β-sandwich and differs considerably from that of SH2 domains. The human genome encodes about 30 proteins with

Figure 7.11 Role of SH2 and SH3 domains in activation of the cytoplasmic protein Tyr kinase Src (A) In the absence of an input signal, Src (membrane-anchored by a tetradecanoyl residue symbolized by the zigzag line) is intrasterically inhibited by two interactions, one between an SH2 domain and a C-terminal phospho-Tyr residue (Tyr527 in humans) and the other between an N-terminal SH3 domain and a proline-rich domain (Pro). The kinase domain is depicted as a teal beam. (B) Src has become unfolded due to competitive interactions of an SH3 protein with the Pro domain and of a Tyr-phosphorylated protein (such as a receptor) with the SH2 domain (for the sake of clarity, the interaction partners are not shown). Alternatively or in addition, phosphatase SHP2 may have dephosphorylated Tyr527 (for an example of this mechanism, see Section 7.2.2). (C) Src has become activated by autophosphorylation of Tyr416 in the activation loop. The activation process is rendered reversible by dephosphorylation of Tyr416 and rephosphorylation of Tyr527 the latter catalyzed by the C-terminal Src kinase (Csk). To become recruited to the membrane-bound Src, the cytoplasmic Csk interacts with a Csk-binding protein, a single-pass transmembrane protein that functions as a scaffold (not shown). The Src oncoprotein lacks Tyr527; therefore, it is partially activated. (D) Molecular model of Src in the inactive conformation (based on X-ray crystallography).

PTB domains. Most of them are adaptor and docking proteins that help to assemble large protein complexes at Tyr-phosphorylated proteins, in particular at autophosphorylated receptor Tyr kinases.

An example of the role of SH2, SH3, and PTB domains in adaptor proteins is provided by receptor-induced activation of the Ras–MAP kinase signaling cascade. For mitogenic stimuli such as growth factors, this pathway represents a signal-transducing main street from the cell periphery to the genome; therefore, it is one of the major

(A)   (B)

Figure 7.12 Role of adaptor proteins in signal transduction by receptor Tyr kinases (A) Membrane-associated signal-transducing complex between autophosphorylated (P) receptor, adaptor Grb2, Ras-GEF mSOS (colored; Pro, proline-rich sequence), membrane-anchored Ras, and the MAP kinase module RAF–MEK–ERK. (B) The same complex but coupled to the receptor by the scaffold protein Shc.

downstream effector systems of receptor Tyr kinases (**Figure 7.12**). Here the adaptor protein **Grb2** (Growth factor receptor-binding protein 2) connects the autophosphorylated receptor, bound via an SH2 domain, with the Ras-GEF "mSOS", bound via an SH3 domain, thus bringing mSOS into the immediate vicinity of the receptor and of membrane-anchored Ras. Through this interaction mSOS, and in turn Ras, become activated. By means of a second SH3 domain (see Figure 7.8), Grb2 may interact with additional proteins such as the docking protein GAB (Grb2-Associated Binding protein, see Section 7.1.3). As a result, large signal-processing protein complexes may be assembled at the receptor. Under certain circumstances (probably at low intensities of the input signal), the receptor does not interact with Grb2 directly but via the scaffold protein **Shc** (SH2 domain-containing protein, pronounced "shic"). Shc binds to the receptor by a PTB domain and to Grb2 by a proline-rich sequence interacting with an SH3 domain. By means of its SH2 domain, Shc may recruit further proteins. Moreover, it becomes phosphorylated at multiple Tyr residues acting as binding sites for other proteins with SH2 domains (see Figure 7.8).

7.1.3 Signal transduction by the insulin receptor, a matter of docking

For most receptor Tyr kinases, a major purpose of autophosphorylation is activation of the enzymatic domain and generation of binding sites for other proteins. By multiple phosphorylations, a bound protein may be rendered into recruitment center for a wide variety of proteins with SH2 and PTB domains, thus completing or even taking over the role of the autophosphorylated receptor protein. Such a protein is called docking protein. Most known docking proteins have a PTB domain by which they become fixed at the autophosphorylated receptor. In addition, many of them are anchored in the membrane by lipid residues, noncovalent interactions with membrane phospholipids, for instance through pleckstrin homology (PH) domains (**Figure 7.13**), or transmembrane domains.

The best-known examples of docking proteins are the **insulin receptor substrates (IRS)**. These are large proteins (130 kDa) that can accommodate many effector proteins. An IRS protein binds through a PTB domain to the autophosphorylated receptors of insulin, insulin-like growth factors, and interleukin 4 (IL4). It is phosphorylated at multiple Tyr residues by the receptor itself or, as in the case of the IL4 receptor lacking an intrinsic Tyr kinase domain, by receptor-associated

Figure 7.13 Docking protein IRS, insulin receptor substrate The insulin receptor consists of four polypeptide subunits held together by disulfide bonds (teal). Upon autophosphorylation, the cytoplasmic domain of the receptor binds to the phospho-Tyr-binding (PTB) domain of the docking protein IRS (interacting P-Tyr, colored circle; P-Tyr in the activation loop of the receptor, black circle). In addition, IRS is attached to the membrane by a pleckstrin homology (PH) domain and becomes phosphorylated by the receptor at multiple Tyr residues (teal circles). Thus it is transformed into a docking center for proteins with SH2 and PTB domains.

Tyr kinases of the JAK family (see Section 7.2.1). IRS is attached to the membrane through a pleckstrin homology (PH) domain (see Figure 7.13). In addition to phospho-acceptor Tyr residues, IRS has many Ser and Thr residues that are phosphorylated by a wide variety of protein kinases, resulting mostly in inhibition of signal transduction. Thus, graded Ser/Thr phosphorylations may fine-tune signaling by IRS and provide a mechanism of feedback control (an example is discussed in Section 9.4.3).

The insulin receptor has a more complex structure than other receptor Tyr kinases. Each monomer consists of two polypeptide chains held together by disulfide bonds (see Figures 7.2 and 7.13). The monomers are linked to each other through disulfide bonds. Thus, in contrast to EGF receptors, the insulin receptor has a preformed dimeric structure that interacts with one ligand molecule. The conformational change resulting from ligand binding triggers a mutual activation loop phosphorylation of the cytoplasmic kinase domains. To evoke the anabolic effects of insulin, the receptor activates a particular set of signaling cascades by using, in addition to IRS, other adaptor and docking proteins such as Shc to stimulate the Ras–RAF–ERK MAP kinase module. Along this mitogenic pathway, insulin-like growth factors (and in cell culture insulin by itself) can induce cell proliferation.

Mechanism of insulin-dependent glucose absorption

Following a meal, insulin is released from pancreatic β-cells and stimulates tissues, first and foremost skeletal muscle and adipose tissue, to absorb glucose from the blood. To this end the hormone activates the **glucose transporter GluT4** and promotes **glycogen synthesis** for glucose storage. GluT4 is a protein with 12 transmembrane domains forming an aqueous membrane channel that regulates glucose influx into muscle and fat cells The transporter shuttles between the cell membrane and the cytoplasm where it is packed in storage vesicles. Insulin induces the translocation of these vesicles to the cell surface, whereas in the absence of the hormone membrane-bound GluT4 rapidly becomes internalized by clathrin-dependent endocytosis.

A key role in the translocation of GluT4 storage vesicles is played by the lipid kinase PI3K and the protein kinase **PKB/Akt**. Upon interaction with insulin the insulin receptor phosphorylates the docking protein IRS, which in turn binds and activates PI3K. PI3K catalyzes the phosphorylation of the membrane lipid phosphatidylinositol 4,5-bisphosphate at position 3. The trisphosphate PIP3 thus generated is required for anchoring of PKB/Akt at the membrane where it is activated by the protein kinase PDK1 (see Section 4.6.6). Substrates of PKB/Akt include the GAPs for two small G-proteins of the Rab and Ral family as well as proteins controlling vesicle fusion with the plasma membrane. As discussed in Sections 10.4.2 and 10.4.4, Rab and Ral proteins are master regulators of intracellular vesicle transport and exocytosis. In the case of GluT4 translocation the GAPs are inactivated by PKB/Akt-catalyzed phosphorylation, a reaction that makes sense since GAPs deactivate small G-proteins. Moreover, PKB/Akt stimulates glycogen synthesis by phosphorylating and inhibiting glycogen synthase kinase 3β (GSK3β) that otherwise would suppress glycogen synthesis (see Section 5.3.1).

An additional signaling pathway leading to GluT4 translocation starts from the small G-protein TC10, a member of the Rho family that interacts with the insulin receptor through a set of special adaptor proteins. PKB/Akt is not involved. The physiological significance of this pathway is not entirely clear.

Another protein kinase inducing GluT4 translocation by inactivating the GAPs of the small G-proteins is the **AMP-activated protein kinase AMPK**. This is of particular interest, since AMPK is a sensor of energy deprivation. It is activated by a drop of the ATP level as it occurs in stress situations such as, for instance, prolonged muscle contraction. In other words: body exercise may have the same effect as insulin. Moreover, it protects the organism from insulin overload which is a major reason of diabetes type 2 (for more details, see Section 9.4.3).

When the insulin level drops, the signaling system is immediately shut off. This is achieved by the protein Tyr phosphatases PTP-1B and SHP1, which catalyze the dephosphorylation of the receptor and the IRS and are in control when the kinases become inactive. To augment this effect, phosphatidylinositol 3,4,5-trisphosphate (PIP$_3$) is inactivated by lipid phosphatases such as PTEN (which dephosphorylates PIP$_3$ at the 3-OH group of the inositol residue) and SHIP2 (SH2-containing Inositol Phosphatase 2, which dephosphorylates PIP$_3$ at the 5-OH group of inositol; not to be confused with SHP2). As a result, PKB/Akt activity is terminated (see Section 4.6.6) and IRS is released from the membrane. In addition, through PKB/Akt a signaling pathway is stimulated along which the insulin receptor and IRS are inactivated through phosphorylation of Ser and Thr residues. The phosphorylation is catalyzed by the ribosomal S6 kinase (p70^{S6K}) that in turn is activated by the protein kinase mTor, a cellular sensor of nutrient and energy supply (see Section 9.4.1). This context is schematically summarized in **Figure 7.14**. In diabetes, insulin signaling is strongly impaired, a subject that is treated in more detail in Section 9.4.3.

Figure 7.14 Major signaling pathways mediating glucose uptake and glycogen synthesis in skeletal muscle Shown to the left is the tetrameric insulin receptor that, upon ligand binding and autophosphorylation, binds (and phosphorylates) the scaffold/docking protein IRS. As a result, protein kinase B/Akt (PKB) becomes activated, promoting the exposure of GluT4 at the cell surface and - by inhibiting glycogen synthase kinase 3β (GSK3β)-glycogen synthesis. As shown to the right GluT4 translocation is also stimulated by the protein kinase AMPK that responds to a drop of the ATP level (increase of AMP), for instance during muscle contraction (exercise). Vice versa its opponent, protein kinase mTOR, is activated by PKB/Akt resulting in an inactivation of the receptor-IRS complex.

Other docking proteins

In addition to IRS, other docking proteins have been identified. Here are some examples:

- **FRS**, Fibroblast growth factor (FGF) Receptor Substrates. Like the closely related IRS, they bind to receptors (mainly those of fibroblast and nerve growth factor) by a PTB domain, whereas membrane attachment is due to a fatty acid anchor.

- **GAB**, Grb2-Associated Binders. They interact with various receptor and receptor-associated Tyr kinases. The membrane-binding site is a PH domain.

- **DOK** (Downstream Of Tyr Kinases) proteins. They preferentially bind Ras-GAP and are therefore negative regulators of Ras signaling. Membrane binding is due to a PH domain.

- **LAT**, Linker for Activation of T cells. This docking protein with a transmembrane domain is associated with the T-cell receptor and plays an important role in T-cell activation (see Section 7.2.2).

- **Cas**, Crk-associated kinase substrate. This docking protein lacks a membrane-binding site but has proline-rich sequences and SH3 domains for interaction with *cytoplasmic* Tyr kinases (see Section 7.2.3).

- **Cbl**. This adaptor and docking protein lacks a membrane binding site but has E3 ubiquitin ligase activity that induces the endocytosis and degradation of Tyr kinase-coupled receptors, Tyr kinases of the Src family, and SH2 proteins bound by phospho-Tyr residues (for more information, see Section 7.2.2).

7.1.4 Ephrin receptors control embryonic development by signaling in two directions

Ephrin receptors constitute the largest subfamily of Tyr kinase-coupled receptors (the name comes from an erythropoietin-producing hepatoma cell line where the proteins were discovered in 1987). Their ligands are the ephrins. One distinguishes between A-type and B-type ephrins. A-type ephrins are proteins bound at the cell surface by a glycosylphosphatidylinositol anchor (see Section 4.3.1). They are easily detached by ADAM proteases, becoming circulating hormone-like factors.

B-type ephrins are genuine transmembrane proteins that interact with the receptors in a juxtacrine manner, limiting their effects to direct cell–cell communication (**Figure 7.15**). A characteristic feature of the ephrin B system is that it transmits signals in both directions: the B-ephrins and the ephrin B receptors are both ligands

Figure 7.15 Signal transduction by the ephrin B system Cell 1 (left) expresses ephrin B; cell 2 (right) expresses the corresponding receptor Tyr kinase. Through a juxtacrine interaction, both partners activate each other. Signal transduction is induced by autophosphorylation of the receptor and by phosphorylation of ephrin B by associated Tyr kinases. As a result, proteins with SH2 and PTB domains become assembled (and possibly phosphorylated) at the intracellular regions of both ephrin B and its receptor (not shown). Through their intracellular carboxy ends (white circles), ephrins and ephrin receptors recruit, in addition, proteins with PDZ domains. The dimeric receptor structure shown here is a borderline case because normally ephrin receptors exist as higher-order oligomers.

and receptors. While the receptors, like other receptor Tyr kinases, are activated by ligand-induced autophosphorylation, the ephrins, which lack intrinsic kinase domains, are phosphorylated by associated Tyr kinases, probably by those of the Src family (see Figure 7.15). The triggering signal for these phosphorylations is the interaction with the ephrin B receptor. Thus, in the ephrin B system, both types of protein kinase-coupled receptor, those with intrinsic and those with associated kinases, are combined. Like other Tyr kinase-coupled receptors, the phosphorylated ephrins and their receptors interact with proteins carrying SH2 and PTB domains. In addition, they are able to bind proteins with PDZ domains (Sidebar 7.2) via C-terminal motifs.

Ephrins and their receptors play an essential role in animal development in that they control **targeted growth processes** in the course of embryogenesis. This includes the directed outgrowth of nerve fibers (a process called axon guidance) together with the formation and stabilization of synapses and the sprouting of blood vessels. Thus, ephrins are essential for cell differentiation and interlinkage formation in both the nervous and vascular systems. The growth of nerve fibers is a strictly targeted process that involves both attractive and repulsive forces. Attraction is supported by juxtacrine binding of ephrins (and other morphogenetic factors; see Section 16.1) to their receptors. Consequently, repulsion seems to depend on the proteolytic release of ephrins from the membrane, as well as on endocytotic down-modulation of ephrins and their receptors.

The cellular effects of ephrins are mirrored by the signaling cascades they activate, which target predominantly the cytoskeleton and influence cell size, adhesion, and motility. In this respect they differ from the pathways activated by other receptor Tyr kinases, which primarily stimulate the mitogenic MAP kinase cascades and the PI3K–PKB/Akt system. The protein **PAR3**, for instance, interacting with ephrins through PDZ domains, activates the small G-proteins Cdc42 and Rac, which are key regulators of cytoskeletal dynamics (see Section 10.3.4). As a result, asymmetrical cell divisions are induced, leading to cell polarity. Another protein with PDZ domains, syntenin, is a scaffold for proteins controlling the formation of intercellular contacts, in particular those of synapses.

Proteins interacting with Tyr-phosphorylated ephrins through SH2 domains also include important regulators of the cytoskeleton such as Grb4, an adaptor protein

---

Sidebar 7.2 PDZ domains PDZ domains rank among the most abundant interaction domains, found in animals, plants, and prokaryotes (the human genome contains about 230 genes encoding PDZ domains). The acronym PDZ refers to three proteins—Postsynaptic density protein 95 and Discs large protein of *Drosophila* and Zona occludens protein of vertebrates—where these domains were found for the first time.

PDZ domains, consisting of 80–90 amino acids, exhibit quite variable sequences. They possess a characteristic three-dimensional structure that enables them to interact heterotypically with both carboxy-terminal tetrapeptide motifs and internal sequences, as well as homotypically with PDZ domains of other proteins, thus participating in the formation of multiprotein complexes. Proteins may contain more than 10 PDZ domains and, in addition, other interaction motifs rendering them to highly versatile scaffold and adaptor proteins. An example is provided by the widely distributed **membrane-associated guanylate kinase homologs (MAGUKs)** that assemble multiprotein signaling complexes rather than catalyzing the phosphorylation of GMP by ATP as genuine guanylate kinases do.

PDZ proteins play an important role in the development of cell polarity, that is, the nonequal distribution of cellular components, particularly in epithelial cells and neurons. The postsynaptic density complex is an example. This assembly of proteins is found beneath the postsynaptic membrane of frequently used neurons and stabilizes the synaptic contact (see Sections 16.1 and 16.4). It contains many proteins with PDZ domains including the prototypical protein PSD95, a MAGUK. Another function of PDZ domains is to organize receptor complexes at cell membranes. This holds true not only for ephrins but equally well for other receptors. The signal-transducing protein Dsh of the Wnt cascade (see Section 5.9), for instance, is a PDZ protein.

involved in the control of the GTPase dynamin, the Rac/Cdc42-activated protein kinase PAK1 (see Section 10.3.4), and the scaffold proteins Dock180 and axin (see Section 5.9), as well as Vav, a GEF of small G-proteins of the Rho family. One negative characteristic of ephrins is that they may promote invasive cancer growth and metastasis as well as (by stimulating angiogenesis) the blood supply of tumors. In fact, ephrins have been found to be overexpressed in neoplastic lesions.

7.1.5 Receptor tyrosine kinases as cell adhesion molecules, or how to kill two birds with one stone

The juxtacrine interactions of the ephrin system provide an example of a role for receptor Tyr kinases in cell adhesion. In fact, some receptor Tyr kinases contain extracellular domains that are characteristic for bona fide cell adhesion molecules (CAMs; see Section 7.3 for more information). Such motifs are, for instance, the Ig- and fibronectin-like domains found in platelet-derived growth factor receptors, FGF receptors and the receptor Axl as well as the cadherin-like domains of the receptor Ret and the discoidin-like domains in DDR (see Figure 7.2). Vice versa, proteins classified as CAMs may function as signal receptors, indicating that there are fluid transitions between the two types of protein. The following examples illustrate the dual function of specific receptor Tyr kinases and their role in embryogenesis.

Ret

The receptor Tyr kinase Ret plays an important role in embryonic development of the nervous system. In cooperation with ephrins, it controls the outgrowth and branching of neurites and the formation and function of synapses. In addition, Ret participates in the differentiation of kidney and sperm cells. In nervous tissue, Ret is activated by **Glial cell-Derived Neurotrophic Factors** (GDNF) that are expressed in the four isoforms GDNF, neuroturin, persephin, and artemin. They interact with Ret only when they have been bound first by **GDNF Family Receptors**, **GFRα**, a class of glycosylphosphatidylinositol-anchored membrane proteins (**Figure 7.16**). The active receptor is a Ret homodimer flanked by two GFRα co-receptors and requires, in addition, heparan sulfate proteoglycans to facilitate the interaction with the GDNF ligand. The primary downstream effectors of this complex are docking proteins such as FRS2 as well as scaffold and adaptor proteins such as Shc and Grb2, gating the usual signal-transducing pathways. A unique feature of Ret is the extracellular cadherin-like domains, which indicate a role in cellular adhesion. In fact, another receptor of GDNFs is the bona fide neuronal cell adhesion molecule N-CAM; like Ret, N-CAM is activated by the GDNF–GFR complex. In contrast to Ret, however, N-CAM is devoid of intrinsic Tyr kinase activity. Instead, it is an example of a kinase-associated receptor that transduces the signal by interacting with the cytoplasmic Tyr kinase Fyn, an Src family member (see Section 7.2). Owing to chromosomal rearrangements, Ret (Rearranged during transfection) may mutate into a constitutively active oncoprotein found in thyroid tumors and multiple endocrine neoplasias. Conversely, given the effects of Ret on nervous tissue, drugs that treat Parkinson's disease and drug addiction may be obtained by developing synthetic Ret agonists.

Fibroblast growth factor receptors

The four isoforms of the FGF receptor interact with 21 different isoforms of FGF that, in addition to their effects as wound hormones, play a key role in early phases of embryogenesis and organ development, particularly of the skeleton. This is drastically demonstrated by malformations caused by an overactivation of FGF receptors due to point mutations, especially of the extracellular Ig-like domains (**Figure 7.17**). These domains are essential for ligand binding, dimerization, and *cis* interactions (interactions within the same membrane plane) with Ig-CAMs (see Section 7.3.1). Whether or not they also contribute to cell adhesion is still not clear. Like Ret, FGF receptors also need heparan sulfate, probably to open the ligand-binding site. This restricts FGF effects to tissue areas exhibiting a high local concentration of this type of proteoglycan.

Figure 7.16 Transmission of glial cell-derived neurotrophic factor signals by receptor Tyr kinase Ret (left) and cell adhesion molecule N-CAM (right) Tyr kinase domains are symbolized by colored rectangles.

Figure 7.17 Malformations of the skeleton caused by point mutations of the fibroblast growth factor receptor (A) The receptor is depicted schematically (TK, colored rectangles, Tyr kinase domains; I, II, and III, IgG-like domains). Colored circles indicate the positions of point mutations in the extracellular and transmembrane domain leading to deregulation and overactivation of the receptor. (B, C) The syndromes shown here are hereditary diseases. They are characterized by a disproportionate growth of extremities and body (frequently dwarf-like) and severe malformations of limbs, skull, and face (craniosynostosis). Achondroplasia obtained by breeding is also the reason for the short legs of dachshunds and Basset hounds. (B and C, from Wilkie AOM, Moriss-Kay GM, Jones EY et al. [1995] *Curr Biol* 5:500–507.)

Axl

The receptor Tyr kinase Axl and its relatives Tyro3 and Mer are strongly expressed by monocytes, macrophages, and cells of the vascular, nervous, and reproductive systems. The extracellular extensions of these receptors contain two Ig and fibronectin-like domains, each possibly interacting with CAMs. As ligands, the proteins Gas6 and S have been described. Gas6 is encoded by a Growth arrest-specific gene, and protein S inhibits the blood-clotting cascade. Like Ret and FGF receptors, receptors of the Axl type are also engaged in embryonic development. In addition, they inhibit the function of antigen-presenting cells. Indeed, upon gene knockout, an extreme hyperplasia of lymphoid organs, accompanied by autoimmune diseases and massive disturbances of the nervous system, has been observed. Oncogenic mutations of Axl have been found in several human neoplasias.

Discoidin domain receptors (DDRs)

The extracellular discoidin-like domains of the discoidin domain receptors are homologous to the lectin (carbohydrate-binding protein) discoidin of the slime mold *Dictyostelium discoideum*. This amoeboid organism expresses such cell adhesion proteins when aggregating to multicellular fruiting bodies (see Section 3.5.2). The ligands of discoidin domain receptors of vertebrates are collagen species. Discoidin domain receptors cooperate with integrins, a class of Tyr kinase associated collagen receptors (see Section 7.2.3). The pathways of downstream signaling and the cellular responses activated by discoidin domain receptors are similar to those of other receptor tyrosine kinases. Defects of discoidin domain receptor signaling have been found to be involved in atherosclerosis and arthritis.

Summary

- Receptor tyrosine kinases are transmembrane proteins with an extracellular ligand binding site and a cytoplasmic kinase domain. They specifically respond to a wide variety of growth factors and to insulin. Moreover, many receptor tyrosine kinases contain extracellular cell–cell interaction motifs. In fact, there are fluent transitions between receptor tyrosine kinases and cell adhesion molecules (CAMs).

- In animals, receptor tyrosine kinases control the formation and maintenance of tissues. They are major targets of oncogenic transformation, and receptor tyrosine kinase inhibitors are used in cancer therapy.

- Receptor tyrosine kinase activation requires dimerization (oligomerization) and - as a rule - trans-autophosphorylation of the kinase activation loops. In addition, other Tyr residues in the cytoplasmic extensions become phosphorylated serving as docking points for proteins with SH2 and PTB domains that specifically interact with phospho-Tyr residues. As a result, signal-processing protein complexes are formed. Thus, receptor tyrosine kinases combine the properties of enzymes with those of highly dynamic scaffold proteins.

- For signal extinction, receptor tyrosine kinases become dephosphorylated catalyzed by Tyr-specific protein phophatases. By inhibiting these enzymes reactive oxygen species augment the effects of receptor tyrosine kinases. Prolonged activation results in down-regulation by (reversible) internalization and (irreversible) proteolysis in lysosomes. This process requires mono-ubiquitylation catalyzed by the ubiquitin ligase Cbl. Dominant negative mutants of Cbl are oncogenic.

7.2 Receptors associated with tyrosine kinases

The advantage of separating a receptor's ligand-binding domain from its effector domain has been discussed in Section 3.4.4: by this "trick," a wide variety of input signals can be processed by the same intracellular apparatus for signal transduction. This occurs, of course, at the expense of a selective targeting of signals. Receptors associated with Tyr kinases function according to the same principles as receptors with intrinsic Tyr kinase activity, that is receptor tyrosine kinases. Ligand binding causes homo- or hetero-oligomerization (or activates preformed oligomers) in such a way that associated cytoplasmic Tyr kinases undergo trans-autophosphorylation.

T-cell antigen receptor
B-cell antigen receptor
Fc-receptor
} Src family, for example, Lck Lyn, Fyn (B cells)

cytokine receptors   Janus kinases (JAK)
receptors of extracellular   focal adhesions-
matrix (integrins)   associated kinases (FAK)

Figure 7.18 Receptors associated with Tyr kinases and their overall functions

Figure 7.19 Cytoplasmic Tyr kinases associated with transmembrane receptors To the left are the names of the kinases; to the right are the receptors that predominantly interact with them. Kinases of the Frk/Brk, Fes/Fer, and ACK subfamilies are found mostly downstream of kinases that directly associate with receptors. The Tyr kinase domain is symbolized by a gray rectangle. C_{14}, tetradecanoyl residue anchoring the kinases of the Src family at the inner side of the cell membrane; BD, interacting domain; FAT, focal adhesion targeting domain; CRIB, Cdc42/Rac-interactive binding domain; NLS, nuclear localization sequence; PH, pleckstrin homology domain; PRO, proline-rich motif that interacts with SH3 domains; RTK, receptor Tyr kinase.

The kinases thus activated phosphorylate effector proteins as well as multiple Tyr residues in the cytoplasmic receptor domains, thus generating contact sites for proteins with SH2 and PTB domains. An overview of this receptor family is presented in **Figure 7.18**. The cytoplasmic Tyr kinases interacting with these receptors or positioned downstream of such interactions are summarized in **Figure 7.19**. Many of them have been found to undergo oncogenic mutations.

The prototype of a non-receptor Tyr kinase is **Src** (pronounced "sark"). Its mutated form encoded by the viral v-*src* gene has become famous as the first oncoprotein discovered in the Rous sarcoma virus in 1970. This chicken retrovirus was characterized by Peyton Rous (Nobel Prize in Physiology or Medicine for 1966) as the first transmissible agent inducing cancer. Further studies led to a theory on the genetic roots of cancer that states that oncogenes in general are gain-of-function mutations of proto-oncogenes that encode signaling proteins involved in normal cell physiology. For this discovery, the Nobel Prize in Physiology or Medicine for 1989 was awarded to J. Michael Bishop and Harold E. Varmus.

Src represents a major subfamily of non-receptor Tyr kinases that include the enzymes Src, Yes, Fyn, Fgr, Lyn, Hck, Lck, and Blk, as well as some closely related enzymes. Common features of these kinases are membrane anchoring by fatty acid residues and a characteristic arrangement of SH2, SH3, and Pro-rich domains as well as a C-terminal phospho-Tyr residue. As explained for Src (see Section 7.1.2), the intramolecular interactions between these domains keep the enzymes in an inactive state, which is unlatched by input signals.

7.2.1 Cytokine receptors associated with tyrosine kinases: the JAK–STAT pathway

Type I and II cytokine receptors (see Section 6.3) associate with cytoplasmic Tyr kinases. In their active form, these receptors are dimers or oligomers of subunits assembled according to a unit construction system. Many subunits are known combining to form many receptors with different ligand specificities. Frequently, the receptors are composed of different subunits as shown for the interleukin 2 receptor in **Figure 7.20**. Other receptors are homodimers that for activation are either "bridged" by one cytokine molecule or interact with one ligand per subunit. Examples are provided by the receptors of erythro- and thrombopoietin, growth hormone, prolactin, and the granulocyte growth factor G-CSF (Colony-Stimulating Factor; this cytokine was discovered by its ability to stimulate the growth of granulocyte colonies in tissue culture). Each receptor subunit has one transmembrane domain. Other structural features are several conserved Cys residues and, only in type I receptors, one or two Trp-Ser-X-Trp-Ser (WSxWS) motifs in the extracellular domains as well as signal-transducing "boxes" of conserved amino acid sequences in the cytoplasmic part.

Through these boxes, cytokine receptors associate with cytoplasmic Tyr kinases of the **JAK** family (the term stands for JAnus Kinase, although another interpretation

Figure 7.20 Signal transduction by the interleukin 2 receptor The receptor is a heterotrimer. Binding of interleukin 2 induces an activation by trans-autophosphorylation of the associated Tyr kinase JAK that, in turn, phosphorylates the receptor (P, phosphate). It generates numerous binding sites (for the sake of clarity only two are shown) for proteins with SH2 and PTB domains and triggering corresponding signaling cascades (not shown). These proteins include the transcription factors STAT. Upon phosphorylation by JAK, STAT dimerizes by a mutual interaction between SH2 domains and P-Tyr groups. The dimer accumulates in the nucleus, inducing the transcription of genes. The domain structure of STAT is depicted below. The teal arrow symbolizes Tyr phosphorylation; the black arrow indicates Ser/Thr phosphorylation; teal double-headed arrows mean interactions. DNA-BD, DNA-binding domain; TAD, transactivating domain; RTK, receptor Tyr kinase.

is said to be "just another kinase"). They exist in four isoforms, JAK1–3 and TYK2 (TYrosine Kinase 2). Through specific N-terminal interaction domains called JAK homology domains, they bind to the receptor subunits and become allosterically activated in such a way that they trans-autophosphorylate their activation loops. Thereupon the activated kinases catalyze the phosphorylation of the receptor (rendering it a docking site for SH2 and PTB proteins) and other substrate proteins, thus opening a series of signaling pathways including MAP kinase cascades and the PI3K–PKB axis. Principally, these signaling responses resemble those evoked by receptor Tyr kinases.

In addition, JAKs gate a route that directly leads from the cell's periphery to the genome and thus is strongly reminiscent of the straightforward strategy of prokaryotic signal transduction. This "two-component system" is called the **JAK–STAT pathway**. STATs (Signal Transducers and Activators of Transcription) are transcription factors that become activated by tyrosine phosphorylation. STAT-like proteins are found throughout the animal kingdom but not in unicellular eukaryotes, fungi, or plants. The mammalian genome encodes seven isoforms that transmit the signals of different cytokines. STATs contain both SH2 and SH3 domains. Through the SH2 domains they bind to the phosphorylated receptor, where they become phosphorylated by the associated JAKs. As a result, they form homo-or heterodimers by mutual interactions between SH2 domains, and phospho-Tyr groups and become enriched in the nucleus. Their target genes are characterized by specific response elements. The most prominent of these response elements is called γ-**interferon activation sequence**, although by no means does it respond solely to γ-interferon but in addition to a wide variety of other cytokines. This versatility is due to numerous sequence variations that are specifically recognized by individual STAT isoforms. Thus, a STAT1 homodimer controls another set of genes than a STAT3 homodimer or a STAT1,2 heterodimer.

The principle of STAT activation is shown in Figure 7.20 for the interleukin 2 (IL2) receptor as an example of a type I cytokine receptor. Other cytokine receptors transmit signals in an analogous way. STATs may also be phosphorylated and activated by other Tyr kinases such as receptor Tyr kinases. In addition, certain GPCRs have been found to activate STATs, possibly along the arrestin–Src signaling pathway (see Section 5.8). Moreover, STAT effects are augmented by Ser phosphorylation of the transactivating domain catalyzed, for instance, by the MAP kinase ERK. Therefore, STATs seem to resemble logical AND gates that gain full activity only when they have received two different signals.

The JAK–STAT system is controlled by sophisticated mechanisms of signal termination. In general, the degree of phosphorylation of the components is kept at a low level by ever working Tyr-specific phosphatases. Moreover, STATs induce the transcription of genes encoding **suppressors of cytokine signaling (SOCSs)**. These proteins block the catalytic domain of JAKs, at the same time triggering ubiquitylation and proteolytic degradation of the receptor and the associated JAK. That is, they play a similar role as arrestins in GPCR signaling. SOCS knockout animals are not viable, giving some idea of how essential this negative feedback control is. The characteristic structural element, called a SOCS box, comprises about 40 amino acids. It is found not only in SOCSs but also in many other proteins, most of which exhibit E3 ubiquitin ligase activity (for more details, see Section 2.8.1). SOCS have gained particular attention since they seem to be involved in obesity. As shown in Section 9.4.3, excessive accumulation of body fat is prevented by adipokinetic hormones such as **leptin**, a cytokine-like protein that interacts with a class I cytokine receptor and activates the JAK–STAT pathway. Chronic stimulation resulting from overnutrition may lead to leptin resistance, due to an accumulation of SOCSs and receptor down-modulation. The consequence would be obesity.

Another family of inhibitors hinders the binding of STATs to DNA. These **protein inhibitors of activated STAT (PIAS)** have functions going far beyond regulation of

STAT activity. Some of them control, for instance, the activity of ion channels, while others are E3 ubiquitin ligases that catalyze the sumoylation of transcription factors such as cJun and p53, which become inhibited.

Finally, STAT forms that lack a transactivating domain are significant as well. They arise from alternative RNA splicing and function as competitive inhibitors resembling dominant negative mutants. In Section 7.2.3, we shall encounter another example of this regulatory principle.

7.2.2 Antigen receptors, a highlight of sophistication

B- and T-lymphocytes express Tyr kinase-coupled receptors that are activated by antigens and control a system of signaling cascades resembling in principle those downstream of receptor Tyr kinases. They are indispensable for the specific immune response because they strictly regulate proliferation, differentiation, and apoptosis of lymphocytes.

Antigen receptors are more complex than the protein kinase-coupled receptors discussed up to now. They are oligomeric transmembrane proteins without intrinsic enzymatic activity composed of a variable antigen-binding part that is exposed at the outside of the cell and an invariant signal-transducing part in the cytoplasmic region. The antigen-binding part consists of proteins of the immunoglobulin family. Owing to the mechanism of gene segment recombination, its variability is almost unlimited (the number of variants amounts to 10^{15}; for details the reader should consult immunology textbooks).

All antigens evoke more or less identical cellular responses. From the beginning, this situation excludes any fixed "wiring" between the sensor and the effector domains of the antigen receptors. In fact, upon ligand binding, all antigen receptors couple with the same invariant signal-transducing complex of various cytoplasmic Tyr kinases and their substrates, thus activating a rather complicated network of signal-processing reactions. This network—which is not yet understood in all details—will be presented here in a summary fashion by use of T-cell activation as an example.

T-cell receptor and its co-receptors

As shown in **Figure 7.21**, the T-cell receptor is composed of six different transmembrane proteins, subunits α, β, γ, δ, ε, and ζ, that are linked noncovalently. Two of them form the α, β-heterodimer, which is the variable part of the receptor. In a juxtacrine manner this heterodimer interacts with the antigen exposed at the surface of an antigen-presenting cell together with the major histocompatibility complex (MHC). This results in a conformational change of the invariant part of the receptor (called CD3 complex*), which consists of a γ, ε heterodimer and a δ, ε heterodimer as well as a ζ, ζ homodimer. As a result these subunits become accessible for an associated Tyr kinase that phosphorylates their cytoplasmic domains at so-called **ITAM sequences** (Immune receptor Tyrosine Activation Motifs) with two Tyr residues each. Since ITAM sequences resemble the autophosphorylated domains of receptor Tyr kinases serving as docking sites for proteins with SH2 and PTB domains, their phosphorylation is absolutely essential for signal transduction.

ITAM phosphorylation is catalyzed by the lymphocyte-specific Tyr kinase **Lck**, a member of the Src subfamily. As a primary target of input signals, Lck plays a key role in antigen-dependent signal transduction. The kinase is—like Src—anchored in the cell membrane by a fatty acid residue and concentrated in the immediate vicinity of the T-cell receptor by a noncovalent binding to **CD4** (in T-helper cells) or

---

*The CD (Cluster of Differentiation) nomenclature is used to classify cell surface molecules and structures, that are recognized by antibodies, in particular in white blood cells. The permanently growing list of CDs comprises already more than 400 entries.

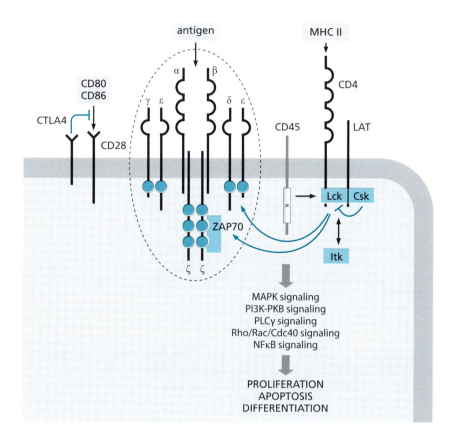

Figure 7.21 Signal transduction by the antigen receptor complex of T cells: the basic events The core of the receptor complex, shown in the oval, is flanked on both sides by co-receptors CTLA4, CD28, and CD4. Colored circles symbolize the ITAM sequences, which upon phosphorylation by the Tyr kinase Lck primarily bind the Tyr kinase ZAP70 via a tandem of SH2 domains. Upon activation by Lck and autophosphorylation, ZAP70 serves as a docking site for proteins with SH2 and PTB domains. CD45 is a protein Tyr phosphatase with two catalytic domains (light gray cylinders) that activates Lck by hydrolyzing a C-terminal phospho-Tyr residue. This reaction is reversed by the Tyr kinase Csk. Lck and its opponent Csk are bound by the transmembrane scaffold protein LAT. Upon Tyr phosphorylation by Lck and ZAP70, LAT binds additional proteins with SH2 and PTB domains including the Csk-related Tyr kinase Itk that further amplifies T-cell receptor signaling.

CD8 (in cytotoxic T cells). These two transmembrane proteins belong to the large family of Ig-CAMs and interact with MHC (see Section 7.3), thus functioning as both co-receptors and adaptors. Another co-receptor required for the contact between the T-cell and antigen-presenting cell is the transmembrane protein **CD28**. Its ligands are the proteins CD80 or CD86 exposed at the surface of antigen-presenting cells. Like CD4/CD8, CD28 also binds Lck that catalyzes Tyr phosphorylation, thus rendering CD28 an additional docking site for proteins with SH2 and PTB domains. This signaling pathway becomes activated mainly during long-lasting immune reactions. CD28 activity is inhibited by the transmembrane protein **CTLA4** that probably sequesters the ligands.

Regulation of tyrosine kinase Lck

Like most other Tyr kinases Lck is not constitutively active but has to be switched on. This occurs in the course of T-cell receptor activation according to the scheme shown for Src in Figure 7.11. As for Src the key event is the release from autoinhibition by the hydrolysis of a C-terminal phospho-Tyr residue. The dephosphorylation is catalyzed by the **Tyr-specific phosphatase CD45**, a transmembrane protein and further component of the T-cell receptor complex. In nonstimulated T cells CD45 undergoes auto-inhibition, probably by dimerization. This blockade is overcome by an allosteric interaction due to receptor and co-receptor activation (more details about this mechanism are found in Section 7.4).

By re-phosphorylating and inactivating Lck, the cytoplasmic Tyr kinase Csk opposes the effect of CD45. Csk is identical with the kinase that inactivates Src by catalyzing C-terminal phosphorylaton (mentioned in Section 7.1.2). As in the case of Src the interaction of Lck with Csk occurs at the surface of membrane-bound adaptor and scaffold proteins. These are **LAT** (Linker of Activated T-cells) and **LIME** (Lck-Interacting MEmbrane protein), two T-cell-specific Csk-binding proteins. Like other Csk-binding proteins, LAT and LIME are multifunctional scaffold proteins. By means of SH2 domains and several phosphorylatable Tyr residues, they recruit large signal-processing protein complexes in the neighborhood of the T-cell receptor. These complexes contain, among others, phospholipase Cγ and the adaptor Grb2 that connects the mitogenic Ras-RAF-MAP kinase cascade with the receptor (see Section 7.1.2).

Csk is a target of immunosuppressive signals. One of these, prostaglandin E2, induces the release of cAMP in T cells by interacting with a G$_S$-protein-coupled receptor. cAMP activates protein kinase A that, in turn, inactivates Csk by Ser/Thr phosphorylation. The recruitment and activation of Csk ensures a negative feedback control of T-cell receptor signaling. Moreover, upon prolonged stimulation, Lck is—like other Tyr kinases of the Src family—degraded in proteasomes. As for receptor tyrosine kinases, this down-regulation is triggered by poly-ubiquitylation catalyzed by E3 ubiquitin ligase Cbl (see Section 7.1.1).

Role of tyrosine kinases ZAP70 and Itk

The signal generated by the T-cell receptor/Lck complex is amplified and further processed by at least two additional Tyr kinases. With their ITAM sequences, the invariable subunits of the T-cell receptor functionally resemble docking proteins. As such, they assemble proteins with SH2 and PTB domains when phosphorylated by Lck. One of these proteins is the Tyr kinase ZAP70 (Zeta-chain Associated Protein kinase of 70 kDa). ZAP70 has two SH2 domains. These domains exactly fit two phospho-Tyr residues of an ITAM sequence, specifically bringing ZAP70 into the immediate vicinity of Lck, which catalyzes activation loop phosphorylation of ZAP70. By subsequent autophosphorylation, the activated ZAP70 renders itself a docking station for other SH2 and PTB proteins. To augment this effect further, ZAP70 phosphorylates the Csk-binding scaffold proteins LAT and LIME, thus generating additional binding sites for SH2 and PTB proteins. In other words, the activation of ZAP70 provides an amplifying mechanism that results in the formation of signal-transducing complexes at the T-cell receptor, a response that is essential for T-cell activation. Consequently, deletion of ZAP70 leads to severe immune deficiencies.

LAT/LIME not only bind and activate Csk but—via additional adaptor proteins—another cytoplasmic Tyr kinase, **Itk** (Interleukin 2-activated tyrosine kinase). As shown by inhibitor experiments, this Csk-related enzyme plays an essential role in T-cell signal processing.

Summarizing these effects, one easily recognizes that the activation of the T-cell receptor triggers an avalanche of Tyr phosphorylations in the signal-processing protein network, which in the cell leads to an extended excitation pattern. As a result, cell proliferation (clonal expansion) and defense competence are stimulated dramatically. Termination of T-cell receptor signaling follows the route that is typical for Tyr kinase-coupled receptors: Tyr-specific protein phosphatases (such as SHP1 and others; see Section 7.4.2) cooperate with the E3 ubiquitin ligase Cbl in inactivating and degrading the tyrosine-phosphorylated components of the receptor complex.

Immunological synapses

When an immune response is prolonged, the T cell develops a special contact zone with the antigen-presenting cell, thus ensuring and reinforcing a long-lasting signal transduction. This "immunological synapse" contains a large aggregate of T-cell receptor complexes and is stabilized by an interaction between **ICAM1** (Immune Cell Adhesion Molecule 1), expressed at the surface of the antigen-presenting cell, and the integrin **LEA** (Lymphocyte Function-associated Antigen), exposed by the T cell. The situation somewhat resembles memory fixation by the formation of strong synaptic contacts in frequently used neurons, where aggregates of neurotransmitter receptors assemble to form postsynaptic densities (see Section 16.4.1).

Other immune receptors

Receptors of the Ig type that resemble the T-cell receptor to some degree are the B-cell receptors, Fc receptors, and myeloid cell receptors. Like the T-cell receptor, these receptors are transmembrane complexes consisting of ligand-binding and signal-transducing subunits. The latter bear cytoplasmic ITAM sequences that are phosphorylated by cytoplasmic Tyr kinases recruited upon receptor activation. Again this is the trigger of a signaling avalanche, which resembles that described for T-cell activation.

Like the T-cell receptor, the **B-cell receptor** consists of variable and invariant parts and processes signals in an analogous manner. The variable part is the membrane-bound IgM, and the invariant part is an Igα,β heterodimer with ITAM sequences. The role played in T cells by the Tyr kinase Lck is taken over here by the closely related Tyr kinases Lyn and Fyn, while ZAP70 is replaced by the related kinase Syk, Itk by the kinase Btk, and the co-stimulator CD28 by the transmembrane protein CD19 (the activation mechanism of which is still unknown).

Fc receptors are expressed primarily by natural killer cells, granulocytes, macrophages, and mast cells. Their major effect is to stimulate the killer activities of those cells. To this end they interact with antibodies that are bound via their variable Fab regions to pathogens. The receptors recognize these antibodies by their invariant Fc or tail region. As a result, the pathogen is bound at the surface of the receptor-bearing cell, which finally destroys the invader. Fcγ receptors, which are the most abundant type, recognize IgG-type antibodies; Fcα receptors are specific for IgA; and Fcε receptors interact with IgE. Fcα and Fcγ receptors are expressed by almost all types of phagocytic white blood cell, where they induce phagocytosis and the release of antimicrobial factors such as reactive oxygen species. In contrast, Fcε receptors are more or less restricted to mast cells and baso- and eosinophilic granulocytes. The mast cell receptors respond to complexes of IgE antibodies with allergens. By recruiting and activating Tyr kinases such as Lyn and Syk, they induce the degranulation of exocytotic vesicles that release histamine and other factors, thus evoking an allergic reaction. The receptors on eosinophilic granulocytes recognize large parasites such as worms coated with IgE antibodies, and they trigger the release of enzymes such as peroxidases and other factors toxic for the invader. Since both phagocytosis and vesicle degranulation require an increase in the cytoplasmic Ca^{2+} concentration, the phospholipase Cγ–InsP$_3$–Ca^{2+} pathway activated by Tyr phosphorylation occupies a central position in Fc receptor signaling.

A few years ago a novel type of immune receptors was discovered and called **TREM** (Triggering Receptors Expressed on Myeloid cells). They are structurally related to Fc receptors. TREMs are expressed in several isoforms and are particularly abundant in monocytes and neutrophils, bone-dissolving osteoclasts, and brain microglia. As their ligands are unknown, a conclusive assessment of their physiological function is lacking. Their major role is thought to be that of amplifiers of inflammatory reactions, for instance, by pushing up the release of cytokines and other pro-inflammatory mediators. To this end they seem to cooperate with TL and NOD-LRR-type receptors, thus effectively lowering the threshold of detection of bacterial and viral pathogens. Their putatively central role in infectious diseases renders TREM receptors a highly interesting subject of clinical research.

7.2.3 Integrins monitor signals from the extracellular matrix

In tissues, cells are embedded in a matrix of proteins and mucopolysaccharides. The contact between cells and matrix is not passive. In fact, the function of a multicellular structure critically depends on a continuous communication between both partners: cells permanently remodel the extracellular medium that, in turn, continuously sends signals to the cells. The ancient form of such interactions is represented by the bacterial biofilms, the cells of which produce an extracellular matrix that controls their behavior (see Section 3.5.2).

Animal cells communicate with the extracellular matrix via the Tyr kinase-coupled receptors of the integrin family (unicellular eukaryotes, fungi, and plants that lack Tyr kinases use other mechanisms). Integrins are heterodimers of two proteins, α and β, exhibiting one transmembrane domain each. Mammalian cells express 18 α- and 8 β-subunits, combining these into at least 24 receptors with different ligand specificities.

Integrin activation

Integrins are receptors. Therefore, the frequently heard term "integrin receptors" is somewhat misleading because it insinuates that integrins are ligands of such receptors. In fact, the ligands of integrins are proteins of the extracellular matrix such as collagen, fibronectin, vitronectin, and laminin as well as proteins of the blood-clotting cascade such as fibrinogen and von Willebrand factor. In addition, cell adhesion molecules of the Ig-CAM and ADAM families exposed by neighboring cells interact with integrins in a juxtacrine mode. A particularly instructive example of such an effect is the egg–sperm interaction, where sperm proteins known as **fertilins** activate integrins of egg cells. The three mammalian fertilin isoforms (fertilin α, fertilin β, and cyritestin) are transmembrane proteins. They are derived from transmembrane proteases of the ADAM family by partial proteolytic degradation of the extracellular domains uncovering integrin binding sites. There is convincing experimental evidence that fertilization depends on the fertilin– integrin interaction, which in animals is achieved by both a fertilin α, β-heterodimer and cyritestin. In humans, however, the genes of fertilin α and cyritestin have degenerated to inactive pseudogenes and it is assumed that the contact between egg cell and sperm is brought about either by a fertilin β,β-homodimer or by another member of the ADAM family (more about ADAM proteases is found in Section 13.1.2).

Most integrins establish contacts with their ligands only upon demand, staying in a resting state unless they receive an additional input signal. Such a signal may be sent by the ligand on its own but can also be transmitted by the signal-processing network of the cell, through what is known as "**inside-out signaling**." The key to understanding this regulation is the structure of the cytoplasmic domains of the α-subunits: by strongly interacting with the β-subunits, they keep the integrin in an inactive conformation characterized by a bending of the extracellular domains that hinders the access of an exogenous ligand (**Figure 7.22**). The inhibition is overcome

Figure 7.22 Model of integrin activation by "inside-out signaling" Inactive integrin differs from active integrin by a "knee-bent" conformation of the ectodomains of the α,β-heterodimer. This structure is due to an inhibitory interaction between the cytoplasmic domains that hinders ligand binding. This inhibition is relieved by an intracellular activator A that is activated by extracellular signals along signal-processing pathways (symbolized by the teal arrow). The nature of A is still somewhat obscure. For some integrin types it is identical with the cytoskeletal protein talin.

by proteins that interact with the cytoplasmic integrin domains triggering what has been called transmembrane domain tilting. These integrin-activating proteins are targets of intracellular signaling cascades, becoming activated by post-translational modifications such as phosphorylation (or dephosphorylation) and partial proteolysis or by binding of second messengers such as phosphoinositides. In most cases these signaling reactions result in the exposure of a PTB domain by which the regulatory protein binds to the integrin β-subunit, displacing the inhibitory α-subunit. Proteins such as talin and α-actinin fulfill the function of intracellular integrin activators by connecting integrins with the actin cytoskeleton. The activation of these proteins is still poorly understood.

Inside-out signaling prevents accidental integrin activation

Thrombocytes, or blood platelets, provide an example of strict control of integrin activity and the importance of inside-out signaling. Their integrin, termed αIIbβ3, is responsible for the interaction of the cell with proteins of the blood-clotting cascade that induce thrombocyte aggregation at the site of clotting. An uncontrolled integrin effect would have fatal consequences such as thrombosis. Therefore, platelet integrins cannot be allowed to bind their ligands unless they are activated by thrombogenic stimuli such as thrombin, adrenaline, or collagen that are released in sufficient amounts or come in contact with platelets only upon injury. Adrenaline and thrombin interact with GPCRs and activate integrin along intracellular pathways of inside-out signaling, whereas collagen interacts with a special integrin subtype to stimulate the integrin αIIbβ3 in an elusive way. Another process requiring a strictly controlled inside-out activation of integrins is **leukocyte invasion** during an inflammatory response; this will be explained in more detail in Section 7.3.2. Finally, the interaction between the adhesion molecule ICAM1 and the **T-cell integrin LFA** should be recalled. As explained in Section 7.2.2, this interaction is required for the formation of immunological synapses between T cells and antigen-presenting cells. It strictly depends on T-cell activation.

The cytoplasmic domains of integrins are themselves substrates of protein kinases. For instance, the phosphorylation of distinct Ser and Thr residues by the MAP kinase ERK stabilizes the inactive conformation. Since ERK is activated by integrin signaling, this provides a means of negative feedback control (**Figure 7.23**).

Figure 7.23 Signal transduction by integrins The input signals cause integrin aggregation at focal adhesion sites. At the cytoplasmic side, the integrin aggregates assemble large complexes of cytoskeletal proteins (teal, shown in the white box) and the Tyr kinase FAK. Upon autophosphorylation, FAK functions as a docking protein for proteins with SH2 and PTB domains. These are summarized beneath the FAK symbol to the right. They lead to an activation of the signaling cascades shown in the black box. Moreover, FAK cooperates with downstream Tyr kinases such as Src and ACK. In addition, FAK controls signaling pathways leading either to activation or to inhibition of the small G-proteins Rac and Cdc42.

Integrin signaling

The activation of integrins has far-reaching consequences for cell proliferation and differentiation. Particularly striking are the effects on the cytoskeleton, becoming visible as changes of cell shape, motility, and adhesion. As explained in Section 10.3, such changes critically depend on the formation of focal adhesions, not to mention special cases such as immunological synapses. At these cellular contact sites, active integrins assemble into large aggregates that anchor the cytoskeleton to structures of the extracellular matrix and to neighboring cells, thus contributing significantly to the integrity and function of a tissue. At their cytoplasmic domains, the integrins accumulate large complexes of signal processing proteins, providing the link with the cytoskeleton. Such complexes are estimated to comprise more than 30 different proteins.

For the transduction of integrin signals, cytoplasmic Tyr kinases play a key role. Since these enzymes are concentrated at the contact sites, they are called **focal adhesion kinases (FAK)**. Two isoforms are known: the ubiquitously distributed FAK1 and the tissue-specifically expressed FAK2. The domain structure of these kinases is shown in **Figure 7.24**. When the FAKs bind to activated integrins by their N-terminal regions, their conformation is changed, enabling a trans-autophosphorylation of Tyr residues in the activation loop and at other sites of the molecule.

As in the case of other receptor-coupled Tyr kinases, contact sites for proteins with SH2 and PTB domains are generated and large signal-transducing complexes are formed around the receptor. Among the proteins recruited, Src and related Tyr kinases are key effectors, phosphorylating further proteins (together with FAK). Thus, as in the case of immunological receptors, the activation of integrins triggers an avalanche of Tyr phosphorylation in the signal transducing protein network.

The activation of G-proteins Rac and Cdc42 is a key event in integrin signaling

Three substrates of Src, namely ACK, Cas, and paxillin, are the origins of signaling pathways leading to an activation of Rac and Cdc42. As explained in Section 10.3, these small G-proteins are master regulators of actin dynamics and focal adhesions.

ACK (Activated Cdc42-associated Kinase) is a cytoplasmic Tyr kinase that is phosphorylated and activated by Src.

Figure 7.24 Focal adhesion kinase and its inhibitor Upon interacting with integrins via its N-terminal region, focal adhesion kinase (FAK) autophosphorylates five Tyr residues (teal arrows), generating binding sites for proteins with SH2 domains including the Tyr kinase Src. A C-terminal focal adhesion targeting domain (FAT) binds the adaptor/docking protein paxillin, linking FAK with focal adhesions. Two proline-rich domains (PRO) provide interaction with SH3 proteins such as Cas. FAK contains, in addition, four Ser/Thr phosphorylation sites (not shown). The catalytic domain is symbolized by the gray rectangle. The box in the lower part of the figure shows a truncated FAK lacking the catalytic domain. This FAK-related nonkinase (FRNK) resembles a dominant negative mutant acting as a competitive inhibitor of FAK.

Being a component of the integrin signaling complex, it binds via a CRIB (Cdc42/Rac-Interactive Binding) domain to the GTP-loaded form of Cdc42, inhibiting its GAP-promoted GTPase activity, thus preventing signal termination (see Figure 7.19). As a result, the signal-transducing efficacy of Cdc42 becomes augmented.

Cas (Crk-associated kinase substrate) is a multifunctional adaptor/docking protein binding through an SH3 domain to autophosphorylated FAK in the immediate vicinity of Src. Upon phosphorylation by Src, it interacts with **Crk**. Crk was originally discovered as a viral oncoprotein and erroneously assumed to be a kinase (therefore the name "<u>C</u>hicken tumor virus <u>r</u>egulator <u>k</u>inase"). Crk represents a family of widely distributed docking proteins with SH2 and SH3 domains as well as proline-rich sequences for an SH3 interaction. In addition, it is phosphorylated at around 15 Tyr residues serving as binding sites for a wide variety of proteins with SH2 and PTB domains including Src (providing positive feedback), PI3K, GAB, and Grb2. By this means, Crk may activate the PI3K/PKB as well as the GAB/Grb2– Ras–MAP kinase cascades, that is, the "survival program" of the cell. This explains the oncogenic potency of mutated Crk. In the context of integrin signaling, the protein **DOCK180** (Dedicator Of CytoKinesis) is an important interaction partner of Crk binding to the SH3 domain. Probably by recruiting a Rac-GEF, it specifically activates the small G-protein Rac.

Paxillin, another starting point of a Rac-activating pathway, is a scaffold protein that links integrins with proteins of the cytoskeleton. Upon multiple Tyr phosphorylations by FAK and Src, it binds and activates proteins with SH2 domains including Src, Crk, DOCK180, the Rac-GEF C3G, and a GEF of Cdc42. Paxillin-associated FAK activates GAPs and attenuates Rac and Cdc42 activities by negative feedback. Thus, the FAK–paxillin complex is an activator of Rac and Cdc42 with a built-in autocontrol mechanism.

Cross talk with other signaling cascades

Like other receptor-coupled Tyr kinases, FAK may stimulate signaling cascades including the Ras–RAF–MAP kinase cascade. Moreover, there is direct cross talk between integrins and receptor Tyr kinases because these are activated by integrins even in the absence of the corresponding ligands. In this case, Src takes over the role of a kinase phosphorylating the activation loops of the receptor kinases. Conversely, growth factor receptors are able to activate signaling pathways controlled by integrins.

Summary

- Most cytoplasmic Tyr kinases exhibit motifs that enable interaction with transmembrane receptors lacking an intrinsic Tyr kinase domain. These are family I and II cytokine receptors, antigen receptors, and integrins.

  The receptor–Tyr kinase complexes transduce signals in a similar way as receptor Tyr kinases.

- The cytokine receptors comprise a large family of homo- and heterodimers or oligomers assembled according to a unit construction system. The associated cytoplasmic Tyr kinases belong to the JAK family. Major JAK substrates are transcription factors of the STAT family.

- Antigen receptors of lymphocytes are stimulated by proteins exposed at the surface of antigen-presenting cells. To this end they interact with co-receptors and form complexes with various downstream Tyr kinases. Structurally and functionally related receptors are Fc receptors of phagocytes, which bind pathogen- and allergen-bound antibodies, and TREM receptors, which amplify inflammatory reactions evoked by bacteria and viruses.

● Integrins are heterodimeric transmembrane proteins that establish contact to other cells and, in particular, to the extracellular matrix. They combine the properties of cell adhesion molecules with those of signal transducers. Upon activation they aggregate (in particular at focal adhesion sites) and recruit cytoskeletal proteins and cytoplasmic Tyr kinases (FAKs) that are essential for the induction of various signaling cascades. This has widespread effects on cell proliferation and cell differentiation and, in particular, on the actin cytoskeleton.

Figure 7.25 Basic types of the most abundant cell adhesion molecules Parts of two cells are shown. The membrane of the lower cell is equipped with individual cell adhesion molecules; the membrane of the upper one has the corresponding ligands. While Ig-CAMs and cadherins interact with their own kind, the ligands of selectins and integrins are membrane-bound carbohydrates and extracellular matrix compounds (ECM), respectively. Integrins and some other Tyr kinase-coupled receptors also interact with individual Ig-CAMs (not shown). CRP, complement regulatory protein.

7.3 Signal transduction by cell adhesion molecules

As we have seen, integrins and some receptor Tyr kinases (such as ephrin receptors, Axl, Ret, DDR, and FGF receptors) are both sensors of extracellular signals and mediators of cellular adhesion, thus supporting the concept of co-evolution of Tyr kinase-coupled receptors and multicellular animals since a selective cohesion of tissue cells is the primary condition of multicellularity. In addition to these receptors, other transmembrane proteins have been defined as cell adhesion molecules (CAMs) because of their ability to interlink cells (**Figure 7.25**). At first glance, such CAMs appeared to be nothing but passive "glue." However, more in-depth studies have revealed that they are able to transduce signals, resembling both receptors and juxtacrine ligands. To discuss CAMs in this chapter is justified because many of them exhibit properties of Tyr kinase-coupled receptors or are ligands of such receptors.

7.3.1 Cellular adhesion molecules combine the effects of intercellular glue with those of signaling receptors

The most abundant CAMs, besides integrins, are cadherins, selectins, and Ig-CAMs. In addition, there are a series of more specialized, tissue-specific adhesion proteins.

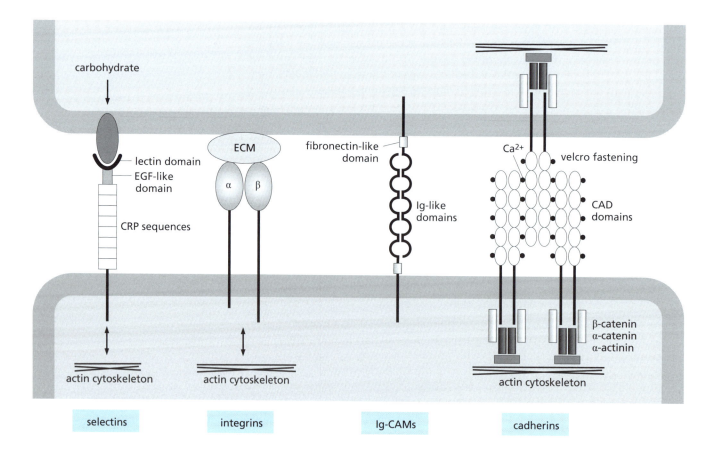

Cadherins

Cadherins are responsible for the cohesion and tensile strength of tissues and are involved in embryogenesis, for instance in gastrulation and morula formation. They are concentrated, in particular, at stable intercellular contact sites such as adherens junctions and desmosomes.

Mammalian cells express more than 80 different cadherins, which are tissue-specifically distributed. Cadherins are their own receptors; that is, they interact homotypically. This interaction is based on typical structural features consisting of five repeats of an extracellular cadherin domain (CAD domain) that comprises 115 amino acids. Binding one Ca^{2+} ion each, the domains of interacting cadherins stick together like a Velcro fastening (*Klettverschluss*). The cytoplasmic domains of most cadherins bind **α- and β-catenins**. These are adaptor proteins that provide contact with the cytoskeleton. In fact, numerous interactions between catenins and the small G-proteins of the Rho family, the major regulators of cytoskeletal dynamics, have been described. As explained in Section 5.9, β-catenin is, in addition, a transcription factor that becomes activated along the Wnt signaling pathway.

Cadherins also activate receptor Tyr kinases independently from the ligands proper and may also become phosphorylated by such receptors. The (neuronal) N-cadherin, for instance, cooperates with one of the FGF receptors to induce the growth of neurites, and the (vasculo-endothelial) VE-cadherin triggers the outgrowth of blood capillaries together with vasculo-endothelial growth factor, a ligand of a receptor Tyr kinase.

A particularly clear example of the close relationship between signal receptors and CAMs is provided by **Ret**, which is both a receptor Tyr kinase and a special cadherin characterized by an extracellular CAD motif (see Figure 7.2). As described in Section 7.1.5, its ligands are neurotrophic factors of glial cells.

Close relatives of the cadherins are the **protocadherins**. These CAMs are mainly expressed by nerve cells. In contrast to cadherins, they do not seem to bind catenins but instead interact with cytoplasmic Tyr kinases such as Fyn.

Selectins

These adhesion proteins use an ancient mechanism of cell–cell contact, the **lectin interaction**, and therefore are also called Lec-CAMs. Lectins are sugar-binding proteins of plants and have become known as phytohemagglutinins because in experiments they cause an aggregation of erythrocytes. However, the lectin interaction, which provides a means of specific intercellular communication, is not restricted to plants but is widely distributed in the animal kingdom as well, although for vertebrates a physiological role has been shown unequivocally only for endothelial cells, leukocytes/lymphocytes, and platelets. In these cases, selectins bring about the adhesion of blood cells to the inner layer of arterial blood vessels in the course of an inflammatory and immune response such as that evoked by injury or infection. This adhesion is a first step in blood coagulation and, as explained in Section 7.3.2, in leukocyte invasion and lymphocyte homing. Like other CAMs, selectins may be considered to be receptors with signaling properties. They interact with and are phosphorylated by cytoplasmic Tyr kinases. However, this is still widely unexplored terrain.

Cell adhesion proteins with immunoglobulin (Ig)-like domains

Proteins with Ig-like domains are derived from one of the largest gene families of vertebrates (comprising 765 genes in humans as compared with 140 genes in *Drosophila* and 64 genes in *Caenorhabditis elegans*). Most of them contain various additional interaction domains characterizing them as components of the signal-processing apparatus. Typical examples are the antigen receptors and the MHC receptors CD4 and CD8 of lymphocytes. Moreover, receptor functions have also been attributed to several hundred **Ig-CAMs**. The canonical feature of these adhesion receptors is approximately seven Ig-like domains in the extracellular region, which in many types are flanked by fibronectin III-like domains (resembling a structure found in the extracellular matrix protein fibronectin). An Ig-like domain consists of about 100 amino acids arranged in several β-sheets and held together by a disulfide bridge.

A subfamily of Ig-CAMs is represented by **junctional adhesion molecules (JAMs)**, concentrated mostly but not exclusively in tight junctions. These cellular contact sites are permeability barriers of endothelial and epithelial cell layers sealing off individual areas or the inner area of the body against each other and the environment, such as in the blood–brain barrier and in kidney and intestinal epithelia. Tight junctions prevent a mingling of apical and basolateral membrane proteins, thus maintaining cell polarization (see Figures 10.17 and 14.9). The junctions are composed of many different proteins. Major components are the claudins, exhibiting four transmembrane domains and interacting with each other in a homotypical juxtacrine mode. Claudins do not belong to the Ig-CAM family.

JAMs seem to control the stability and permeability of the junctions. To this end they dimerize, like other Ig-CAMs, at the cell membrane and bind homotypically to corresponding dimers as well as heterotypically to integrins exposed by adjacent cells. The prototypical tight junction JAM (JAM-A) exhibits all properties of a receptor. Like ephrin receptors, it binds various adaptor proteins with PDZ domains at its intracellular C-terminus, thus recruiting signal-transducing protein complexes. By this means JAM-A modulates both the dynamics of the actin cytoskeleton and the transcription of certain genes.

As discussed in Section 7.1.5, adhesion molecules with intrinsic Tyr kinase activity such as Ret, Axl, DDR, and FGF receptors are critically involved in embryonic tissue development. The same holds true for Ig-CAMs, cadherins, and related proteins. Their role in embryogenesis is vital for the nervous system and the formation of synaptic contacts. This is described in Section 16.1.

7.3.2 How cell adhesion molecules cooperate in the inflammatory response

The inflammatory reaction clearly shows how individual CAMs and receptors work together to bring about a complex physiological response. This response is controlled by a cocktail of pro-inflammatory mediators, also known as "inflammatory soup." The first to occur upon damage or infection is to recruit white blood cells at the site of injury. To this end, leukocytes must be released from the blood vessels through the endothelium into the damaged tissue. This trans-endothelial cell migration or diapedesis is triggered by soluble mediators and accomplished by adhesion molecules. It is tightly regulated because an accidental or incorrectly regulated leukocyte infiltration of healthy tissue would be deleterious, as shown by severe diseases such as rheumatoid arthritis, psoriasis, and asthma.

To attract white blood cells the injured tissue releases "bait": that is, chemotactic factors, such as leukotriene B (see Section 4.4.5) and chemokines. At the same time endothelial cells are made "sticky" for leukocytes by other pro-inflammatory mediators, such as histamine, TNFα, thrombin, and complement factor C5a, and by bacterial components stimulating Toll-like receptors, which induce selectin production. The weak interaction thus established slows down the velocity of the white blood cells in the bloodstream to what is called "leukocyte rolling" over the inner surface of the blood vessel. To firmly fix the cells, both leukocytes and endothelial cells express an Ig-CAM (called PE-CAM1, Platelet-Endothelial Cell Adhesion Molecule 1) that undergoes a homotypic interaction. In addition, leukocyte integrins bind to vasculo-endothelial VE-CAMs and immunological ICAMs, which are enriched in docking structures that develop on the surface of endothelial cells upon contact with leukocytes. For this purpose, the integrins must undergo inside-out activation (as shown in Figure 7.22) by endothelial chemokines. In leukocytes, the CAM–integrin interaction seems to trigger the whole cascade of integrin signaling explained above, including, in particular, the pronounced effects on the cytoskeleton. This strengthens the contact with the endothelial cells and stimulates leukocyte mobility to such an extent that more and more cells assemble at the site of injury and prepare to penetrate the vessel wall. The tight junctions between the endothelial cells are loosened transiently, a response that seems to be brought about by JAMs activated by an interaction with leukocyte integrins. Finally the leukocytes can slip through the endothelial cell layer and leave the blood vessels, invading the injured or infected tissue and becoming visible, at the end, as pus. A closely related process is lymphocyte homing, where lymphocytes escape from blood vessels to enter the lymphatic system and to become enriched in lymph nodes.

7.3.3 AGE and RAGE in diabetes and Alzheimer's dementia

Among the receptorlike cell adhesion molecules, a special type of Ig-CAM is gaining increasing attention in biomedical research. These are the **Receptors of Advanced Glycation End products (RAGEs)**. Their ligands are thought to be involved in the development of severe degenerative diseases including atherosclerosis, Alzheimer's dementia, diabetes, and probably cancer.

Feared long-term effects of diabetes mellitus include disturbances of the blood supply, neuropathies, and a dimming of the eye lens. These symptoms are traced back to tissue defects caused by irreversible protein modifications such as glycation. For this nonenzymatic and nonspecific reaction sugars bind to proteins, establishing Schiff-base linkages with free amino groups. In the course of an Amadori rearrangement, such compounds then become transformed into fructosamine derivatives (**Figure 7.26**), which are deaminated and oxidized, yielding many secondary products. Being chemically very reactive, these may covalently cross-link polypeptide

Figure 7.26 Variability of protein Tyr phosphatases Receptorlike protein Tyr phosphatases (PTPs) and a selection of cytoplasmic PTPs are shown. PRO, proline-rich interaction domains for proteins with SH3 domains; PBM, phospholipid-binding motif; C2, Ca²⁺/phospholipid-binding domain.

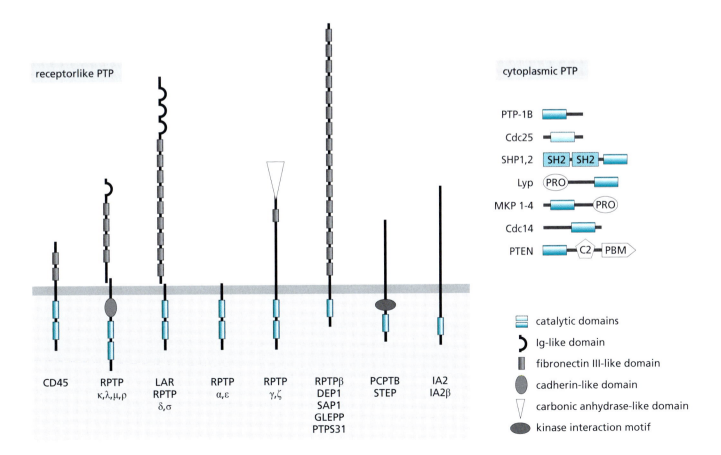

chains, and the so-called **Advanced Glycation End products (AGEs)** are formed. The result is cell and tissue damage. In fact, AGEs are damage-associated molecular patterns (DAMPs, see Chapter 6) that are specifically recognized by corresponding receptors, the RAGEs. RAGEs are found in white blood cells and other tissues and are expressed in response to AGE production. Upon ligand binding they stimulate intracellular signaling cascades such as the MAP kinase and NFκB cascades probably in a similar way as Toll-like and interleukin 1 receptors (see Sections 6.3.2 and 6.3.3). As a result, many genes are activated, in particular those of pro-inflammatory mediators. In other words: activated RAGEs are potent inducers of inflammation and the whole spectrum of the innate immune response. Beside AGEs many other compounds have been found to interact with RAGEs including the leukocyte integrin Mac1 (reacting also with ICAM1 and fibrinogen), certain proinflammatory tissue hormones, the non-histone protein amphoterin, and even bacterial lipopolysaccharides. In some tumors, RAGEs are overexpressed and are assumed to facilitate invasive growth and the formation of metastases. Since RAGEs are also activated by **β-amyloid** (see Section 13.1.3) they may also be involved in Alzheimer's dementia. In fact, high concentrations of AGEs and RAGEs have been measured in the damaged brain areas.

The *physiological* role of RAGEs remains an open question. Possibly, like many other Ig-CAMs, they are important for growth processes during embryogenesis and adolescence, whereas at advanced age the pathological effects become more evident. In the face of the frequent and severe diseases that involve RAGEs, strong efforts are being made in clinical research to bring this receptor type and its cellular effects under control. Several drugs blocking RAGEs are in clinical trial.

Summary

- The most abundant cell adhesion molecules (CAMs), besides integrins, are cadherins, selectins, and immunoglobulin-like (Ig-)CAMs. These transmembrane proteins have dual functions: as a glue establishing intercellular contact sites and as signal-transducing receptors that frequently are associated with Tyr kinases.

- CAMs play a critical role in the formation and maintenance of tissues and in defense reactions (inflammation and wound repair).

- A special type of Ig-CAMs, the receptors for advanced glycation end products (RAGEs), interact with a wide variety of damage-associated molecular patterns and induce an innate immune response. They are thought to play a significant role in degenerative diseases such as diabetes, atherosclerosis, Alzheimer's dementia, and cancer.

7.4 Protein tyrosine phosphatases and phosphatase-coupled receptors

In a mammalian cell, more than a third of all proteins may become reversibly phosphorylated. This permanently fluctuating level of protein phosphorylation resembles a type of short-term memory or, if one prefers the computer metaphor, a random access memory. Biochemically, it represents a steady-state resulting from the competing effects of protein kinases and protein phosphatases.

Generally, the steady-state level of Ser/Thr phosphorylated proteins is much higher than that of Tyr phosphorylated proteins. The reason for this is the high number of Ser/Thr kinases, comprising 428 types in humans as compared with only 90 Tyr kinases. In addition Tyr phosphorylation is controlled by phosphatases in a particularly strict manner. In fact, Tyr phosphorylation occurs in an extended "inhibitory field" of phosphatase activity. Moreover, the phosphatases are frequently activated by the kinases, while the latter stimulate their own activities by autophosphorylation. In other words, a positive feedback of short range is coupled with a negative feedback of long range. Such a situation is described by a set of nonlinear differential equations that give rise to different solutions depending on the initial conditions. Correspondingly, the system may either oscillate, self-organize into a defined pattern, or respond chaotically (see Sections 1.6 and 14.5.4). Oscillation offers the fascinating possibility of frequency-modulated signaling that would enormously increase the efficacy and versatility of cellular data processing. In fact, signal transmission between neurons is based on this principle, and intracellular cAMP and Ca^{2+} signals are also frequency-modulated (see Section 14.5.4). Since the necessary conditions are fulfilled, the same may hold true for Tyr phosphorylation, although convincing experimental proof is still lacking.

7.4.1 Cytoplasmic and receptorlike protein tyrosine phosphatases

The human genome contains 107 genes and some inactive pseudogenes with sequences characteristic for protein Tyr phosphatases. However, only 81 of them encode bona fide protein Tyr phosphatases, whereas the others are genes of phospholipid and RNA phosphatases as well as of some enzymatically inactive proteins. Thus, the number of genuine protein Tyr phosphatases corresponds well with the number of Tyr kinases (90). In reality, however, the number of kinase and phosphatase isoforms is much higher than the number of genes due to alternative splicing of pre-mRNA, alternative gene promoters, and other mechanisms of diversification. Thus, both enzyme families show a high degree of variability, with their individual members exhibiting specific and frequently tissue-restricted functions, which do not really overlap.

Protein Tyr phosphatases are regulated differently than Ser/Thr-specific phosphatases. The latter are mostly small proteins consisting only of a catalytic domain. As explained in Section 2.6.4, they gain specificity by combining *ad libitum* with various regulatory subunits. In contrast, in most protein Tyr phosphatases, the catalytic and regulatory domains are firmly wired, sharing a single polypeptide chain. In other words, they are multidomain proteins. Similar to Tyr kinases, the regulatory domains of Tyr phosphatases seem to have a dual function: on the one hand they control enzymatic activity, and on the other hand they regulate interaction with the substrate and intracellular localization of the Tyr phosphatase. As an example, the regulation of the Tyr phosphatase SHP2 has been discussed (see Figure 7.9).

On the basis of amino acid sequences of the catalytic domains, protein Tyr phosphatases (PTPs) are categorized in several families, which are summarized in **Table 7.1**. Only the so-called classical PTPs are strictly Tyr-specific, while the dual-specific PTPs also dephosphorylate Thr and Ser residues, as well as RNA and phospholipids.

The classical protein Tyr phosphatases are subdivided into cytoplasmic and membrane-bound enzymes. Cytoplasmic Tyr phosphatases are responsible for the strong inhibitory field of Tyr phosphorylation in the cell. Their efficacy is so pronounced that frequently they have to be inactivated transiently to ensure signal transduction from a Tyr kinase-coupled receptor into the cytoplasm or nucleus. This inactivation is brought about by a reversible oxidation by the redox signal H_2O_2, at least as far as PTP-1B and SHP (but also the lipid phosphatase PTEN) are concerned.

With each exhibiting a single transmembrane domain and one or two catalytic domains in the cytoplasmic part, the 21 membrane-bound members of the classical protein Tyr phosphatase family have a topography resembling that of receptor Tyr kinases. The corresponding ligands have been identified in only a few cases.

Thus, pleiotrophin, a neurite growth-promoting cytokine, interacts with PTPζ, inhibiting the phosphatase activity. As a consequence, signaling through Tyr phosphorylation is facilitated and leads to effects on cell adhesion and cytoskeletal dynamics.

Table 7.1 Protein tyrosine phosphatases (PTPs) and related enzymes

Enzyme	Number of human genes	Specificity
Classical PTPs		
receptorlike PTP	21	P-Tyr
cytoplasmic PTP	16	P-Tyr
Dual-specific PTPs		
MAP kinase phosphatases (MKPs)	11	P-Tyr, P-Thr
atypical dual-specific PTPs	19	P-Tyr, P-Thr, phospholipids
PRL	3	P-Tyr
slingshots	3	P-Ser
Cdc14, KAP1,	4	P-Ser, P-Thr
PTEN and myotubularins	21	phospholipids
Others		
Cdc25[a]	3	P-Tyr, P-Thr
LMPTP	1	P-Tyr
EyA[b]	4	P-Tyr, P-Ser

[a]Cdc25 enzymes are not related to other dual-specific PTPs.
[b]EyA enzymes catalyze dephosphorylation through an aspartate rather than a cysteine residue (see also Section 2.6.4).

However, in most cases the ligands are not known and the transmembrane Tyr phosphatases are therefore referred to as **receptorlike**. In many of them the extracellular extensions show structural features typical of cell adhesion molecules, such as Ig- and fibronectin III-like domains (see Figure 7.26), supporting the notion that, like certain receptor Tyr kinases, these phosphatases participate in cell–cell adhesion and interactions of cells with the extracellular matrix. In fact, several receptorlike Tyr phosphatases show high binding affinities for typical matrix components such as proteoglycans and collagen as well as with cell adhesion molecules which are known to play a role in tissue development.

A characteristic structural element of many receptorlike protein Tyr phosphatases is a doubled catalytic domain. Only one partner of this tandem, the membrane-proximal domain, exhibits phosphatase activity, whereas the other one is called a pseudophosphatase domain. It has an auto-inhibitory control function blocking the enzymatically active domain. For steric reasons this is possible only in a dimeric molecule, by a *trans*-interaction. It is assumed, therefore, that activating signals trigger the dissociation of the inactive dimer into two active monomers (**Figure 7.27**) whereas inhibitory signals stabilize the dimeric form. This is in contrast to receptor Tyr kinases, which have to dimerize to become activated by *trans*-autophosphorylation.

7.4.2 Tyrosine phosphatases in health and disease: a short overview

Although protein Tyr phosphatases are extremely important for signal processing, our knowledge of their physiological and pathological functions resembles a patchwork rather than a clear picture. Indeed, as yet only a few enzymes can be associated unequivocally with physiological processes such as cell proliferation, cell adhesion, immune response, and insulin action, all of which are controlled by tyrosine phosphorylation.

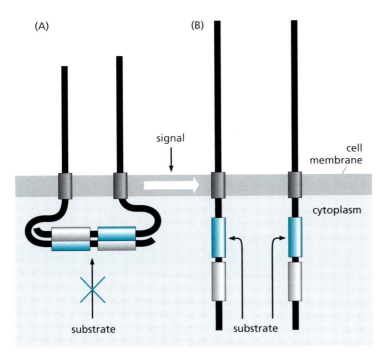

Figure 7.27 Model of the mechanism of activation of receptorlike protein Tyr phosphatases (A) In the dimeric molecule a trans-interaction between the regulatory (light gray cylinders) and catalytically active (colored cylinders) phosphatase domains is assumed to block access of the substrate to the catalytic centers. (B) An input signal has triggered dissociation into monomers, thus exposing the enzymatically active domains.

Cell cycle control

In Section 12.3, we shall examine the dual-specific protein phosphatases of the **Cdc25** subfamily as key regulators of the cell cycle. They catalyze the dephosphorylation of a Thr-Tyr pair in cyclin-dependent kinases, which thus become activated and promote cell cycle progression. In contrast, the phosphatases **Cdc14** and **KAP1** of the dual-specific Tyr phosphatase family *deactivate* cyclin-dependent kinases by dephosphorylating a strategic Thr residue in the activation loop. This reaction is important for the termination of cell cycle phases. Structurally, the Cdc25 phosphatases have ancient precursors in the bacterial rhodanases which are, however, sulfurtransferases rather than protein phosphatases.

Immune response

Another phosphatase for which our knowledge of its physiological role is supported by experimental observations is the receptorlike enzyme **CD45**. Its expression is restricted to hematopoietic cells. As an activator of the receptor-coupled Tyr kinase Lck, CD45 is a master switch in T-cell activation (see Section 7.2.2). Consistently, deletion of the enzyme results in severe disturbances of immune surveillance. There is also evidence indicating a connection between CD45 and multiple sclerosis, an autoimmune disease leading to destruction of nerve cell myelin. While CD45 is involved in lymphocyte *activation*, other Tyr phosphatases expressed in lymphocytes exhibit *inhibitory* effects. Such enzymes include PEP and the related PEST (which inactivate Csk and Lsk by dephosphorylating the activation loops) as well as LMPTP and PTPH1. Another opponent of lymphocyte activation is the classical Tyr phosphatase **SHP1**, which inactivates the Tyr kinases ZAP70 (in T cells) and Syk (in B cells) by activation loop dephosphorylation. As expected, SHP1 knockout mice suffer from immune defects but also from other disorders such as hair follicle development (because of their holey coat they are called "moth-eaten").

Another cytoplasmic Tyr phosphatase that inhibits T-cell activation by catalyzing the dephosphorylation of ZAP70, T-cell receptor, and the ITAM-sequence kinase Lck (see Section 7.2.2) is **Lyp** (lymphoid phosphatase). This enzyme has received attention because a gain-of-function mutation is a risk factor for **autoimmune diseases** such as rheumatoid arthritis, type 1 diabetes, and Graves–Basedow disease, the most common cause of hyperthyroidism and goiter. The Lyp mutation may prevent a sufficient activation of regulatory T cells.

Mitogenic signaling

In contrast to SHP1, the sister enzyme **SHP2** does not inhibit lymphocyte activation. Moreover, it does not seem to antagonize the effects of other Tyr kinases. Instead, it stimulates Ras–RAF–ERK signaling by a still-elusive mechanism, possibly involving a stimulatory dephosphorylation of a Tyr kinase of the Src family. SHP2 may even mutate to an oncoprotein, causing the so-called **Noonan syndrome**. This disorder is characterized by pronounced developmental defects and has been connected with juvenile myelomonocytic leukemia. Since SHP2 knockout animals are not viable, the enzyme certainly plays an essential role in embryogenesis.

Insulin signaling

The cytoplasmic enzyme **PTP-1B** is the prototype of the classical protein Tyr phosphatases. One of its major cellular functions is that of an antagonist of the insulin receptor which, together with the docking protein IRS, becomes dephosphorylated, terminating insulin signaling. In muscle cells of patients who are obese and those with type 2 diabetes, PTP-1B is consistently found to be overexpressed and hyperactive. Moreover, in persons with congenital obesity or type 2 diabetes, the *ptp-1b* gene is frequently mutated. Conversely, PTP-1B knockout mice are particularly insulin-sensitive, even at a high intake of carbohydrate and fat. This indicates that, in these animals, the insulin receptor is not desensitized by excessive insulin as it normally occurs under carbohydrate/fat overload. Therefore, great hopes are placed on clinically applicable inhibitors of PTP-1B; these are not yet on the market, however. Another antagonist of the insulin receptor and, in particular, of the integrin–FAK complex is the receptorlike Tyr phosphatase **LAR** (Leukocyte Antigen Related). The enzyme is expressed in various isoforms and is especially abundant in neurons where it participates in the control of synapse formation and neurite outgrowth.

Cell adhesion, cancer, and other realms of protein tyrosine phosphatases

The receptorlike **RPTPμ** is assumed to stabilize cadherin linkages. In fact, the phosphatase antagonizes Tyr phosphorylation of the cadherin–catenin complexes, which is catalyzed by Src-type and receptor Tyr kinases and destabilizes adherens junctions. Such destabilization of intercellular contact sites has been observed in tumors exhibiting excessive Tyr kinase activity, where it may contribute to metastasis.

Since many Tyr kinases have been found to mutate to oncoproteins, one would expect their opponents, the Tyr phosphatases, to be tumor suppressors. In fact, several classical Tyr phosphatases are candidates for tumor suppressors and are thought to be involved in a variety of malignancies. However, this role has been proven unequivocally for only one of these phosphatases, specifically the receptorlike **DEP1** that is deleted in colorectal carcinomas and some other tumors. DEP1 is thought to antagonize the receptor Tyr kinase Met that is hyperactive in many human neoplastic diseases. Protein Tyr phosphatases may even mutate to oncoproteins. This unexpected finding is understandable when the role of Tyr phosphatases as indirect activators of Tyr kinases, particularly those of the Src type, is considered (see Figure 7.6). One example is the overexpression of **PTPα**, which in colon carcinomas correlates with a hyperactive Src. Another example is the oncogenic mutation of **SHP2** in patients suffering from Noonan syndrome as mentioned above.

A tumor suppressor *par excellence* is the phospholipid phosphatase **PTEN** (Phosphatase- and TENsin-homologous enzyme; tensin is a catalytically inactive Tyr phosphatase). This enzyme specifically removes phosphate groups from the 3-hydroxyl group of inositol residues, thus inactivating the second messengers phosphatidylinositol 3,4-bisphosphate (PIP_2) and 3,4,5-trisphosphate (PIP_3). As a result, the anti-apoptotic PKB/Akt pathway is interrupted (see Section 4.6.6). This explains why loss-of-function mutations of PTEN protect cells from apoptosis. As a result, the mutated cells gain an advantage over their normal neighbors. Such mutations are found in a wide variety of benign and malignant neoplasias. In fact, the genetic inactivation of PTEN ranks among the most frequent events in human cancer development. The molecular event underlying PTEN inactivation is due to a defect of the C2 domain, which results in the mutated PTEN losing its ability to associate with its membrane-bound substrates.

The largest family of phospholipid phosphatases is the **myotubularins**. As indicated by the name, myotubularins of vertebrates are involved in muscle cell differentiation. In fact, mutations of the corresponding genes have been found to correlate with severe neuromuscular diseases such as myotubular myopathy. The substrates of the myotubularins are phosphatidylinositol 3-phosphate and 3,5-bisphosphate rather than PIP_2 and PIP_3. These phospholipids serve as membrane-bound anchoring sites for proteins with lipid interaction domains such as PH, PX, and FYVE (see Sections 2.6.2 and 4.4.3).

Tyr phosphatases with still-unknown functions are the slingshots (named after a *Drosophila* mutant), PRL (phosphatase of regenerating liver), LMPTP, and EyA enzymes. LMPTP, the Low Molecular mass Protein Tyrosine Phosphatase, is a representative of an ancient protein family that is closely related to prokaryotic arsenate reductases.

As mentioned in Section 3.6.3, certain bacteria produce Tyr-directed protein phosphatases as extremely powerful **virulence factors**. The disastrous effects of such enzymes, for instance in the "Black Death" plague epidemics of the Middle Ages, impressively demonstrate the essential role tyrosine phosphorylation plays in human health and survival.

Summary

- Protein Tyr phosphatases constitute a large family of rather diverse cytoplasmic and transmembrane proteins including both Tyr-specific and dual-specific phosphatases but also phospholipid phosphatases. They are multidomain proteins with catalytic and regulatory domains combined in one polypeptide chain. Transmembrane phosphatases structurally resemble receptors and CAMs, although only very few ligands are known.

- Deregulation of tyrosine (and phospholipid) dephosphorylation due to gene mutation, bacterial infection, or other reasons has been linked with numerous diseases.

Further reading

Section 7.1

Barquilla A & Pasquale EB (2015) Eph receptor and ephrins: therapeutic opportunities. *Annu Rev Pharmacol Toxicol* 55:465–487.

Endres FN et al. (2014) Emerging concepts in the regulation of the EGF receptor and other receptor tyrosine kinases. *Trends Biochem Sci* 39:437–445.

Hunter T (2015) The genesis of tyrosine phosphorylation. Cold Spring Harbor Persp. Biol, http://csperspectives.cshlp.Org.

Lemmon MA & Schlessinger J (2010) Cell signaling by receptor tyrosine kinases. *Cell* 141:1117–1134.

Leto D & Saltiel AR (2013) Regulation of glucose transport by insulin: traffic control by GLUT4. *Nat Rev Mol Cell Biol* 13:383–396.

Pendergast A (2002) The Abl family kinases: mechanisms of regulation and signaling. *Adv Cancer Res* 85:51–100.

Rhee SG, Kang SW, Jeong W et al. (2005) Intracellular messenger function of hydrogen peroxide and its regulation by peroxiredoxins. *Curr Opin Cell Biol* 17:183–189.

Sun C & Bernards R (2014) Feedback and redundancy in receptor tyrosine kinase signaling: relevance to cancer therapies. *Trends Biochem Sci* 39:465–474.

Section 7.2

Chakraborty AK & Weiss A (2014) Insights into the initiation of TCR signaling. *Nat Immunol* 15:798–807.

Hordijk PL (2006) Endothelial signalling events during leukocyte transmigration. *FEBS J* 273:4408–4415.

Ingley E (2008) Src family kinases: regulation of their activities, levels and identification of new pathways. *Biochim Biophys Acta* 1784:56–65.

Ivaska J & Heino J (2011) Cooperation between integrins and growth factor receptors in signaling and endocytosis. *Annu Rev Cell Dev Biol* 27:291–320.

Kim C, Ye F & Ginsberg MH (2011) Regulation of integrin activation. *Annu Rev Cell Dev Biol* 27:321–345.

Leitinger B (2011) Transmembrane collagen receptors. *Annu Rev Cell Dev Biol* 27:265–290.

Malissen B et al. (2014) Integrative biology of T cell activation. *Nat Immunol* 15:790–797.

Monteiro RC & van de Winkel JGJ (2003) IgA Fc receptors. *Annu Rev Immunol* 31:177–204.

Pawson T, Gish GD & Nash P (2001) SH2 domains, interaction modules and cellular wiring. *Trends Cell Biol* 11:512–518.

van der Merwe PA & Davis SJ (2002) The immunological synapse—a multitasking system. *Science* 295:1479–1480.

Yaffe MB (2002) Phosphotyrosine-binding domains in signal transduction. *Nat Rev Mol Cell Biol* 3:177–186.

Wang X et al. (2009) Structural biology of shared cytokine receptors. *Annu Rev Immunol* 27:29–60.

Winograd-Katz et al. (2014) The integrin adhesome: from genes and proteins to human disease. *Nat. Rev Mol Cell Biol* 15:273–288.

Wu W et al. (2015) Lipid in T-cell receptor transmembrane signaling. *Progr Biophys Mol Biol* 118:130–138.

Section 7.3

Ibrahim ZA et al. (2013) RAGE and TLRs: relatives, friends or neighbours? *Mol Immunol* 56:739–744.

Juliano RL (2002) Signal transduction by cell adhesion receptors and the cytoskeleton: functions of integrins, cadherins, selectins, and immunoglobulin-superfamily members. *Annu Rev Pharmacol Toxicol* 42:283–323.

Klesney-Tait J, Turnbull IR & Colonna M (2006) The TREM receptor family and signal integration. *Nat Immunol* 7:1266–1273.

Kierdorf K & Fritz G (2013) RAGE regulation and signaling in inflammation and beyond. *J Leukoc Biol* 94:55–68.

Nadiri A, Wolinski MK & Saleh M (2006) The inflammatory caspases: key players in the host response to pathogenic invasion and sepsis. *J Immunol* 177:4239–4245.

Petri B & Bixel MG (2006) Molecular events during leukocyte diapedesis. *FEBS J* 273:4399–4407.

Ramasamy R et al. (2014) Advanced glycation end products and RAGE: a common thread in aging, diabetes, neurodegeneration, and inflammation. *Glycobiology* 15:16R–28R.

Xie J et al. (2013) Cellular signalling of the receptor for advanced glycation end products (RAGE). *Cell Sign* 25:2185–2187.

Zhou FQ, Zhong J & Snider WD (2003) When GDNF meets N-CAM. *Cell* 113:814–815.

Zinn K (2007) Dscam and neuronal uniqueness. *Cell* 129:455–456.

Section 7.4

Ghelis T (2011) Signal transduction by protein tyrosine phosphatases in plants. *Plant Sign Behav* 6: 942–951.

Lee H et al. (2015) Mining the function of protein tyrosine phosphatases in health and disease. *Semin Cell Dev Biol* 37:66–72.

Mohebiany AN et al. (2013) Receptor-type tyrosine phosphatase ligands: looking for the needle in the haystack. *FEBS J* 280:388–400.

Tonks NK (2013) Protein tyrosine phosphatases- from housekeeping enzymes to master regulators of signal transduction. *FEBS J* 280:346–378.

Eukaryotic Gene Transcription: The Ultimate Target of Signal Transduction

8

For all living beings, the ability to adjust the genetic readout to the environmental conditions is an essential condition of survival. In fact, all types of exogenous signal have an impact on the genome: that is, in eukaryotic cells most signaling pathways end in the nucleus. For humans this holds true even for a misplaced word or a panicked thought.

In both pro- and eukaryotes, gene transcription is regulated according to the same principle: transcription-controlling proteins rather than genomic DNA are the targets of input signals. These proteins, the transcription factors, interact with regulatory DNA motifs known as promoter, enhancer, and silencer sequences. By this interaction, genes are specifically marked for the enzyme RNA polymerase, the central player of the game, which catalyzes the formation of RNA from nucleoside triphosphates.

Two general pathways exist for signal transduction from the cell's periphery into the nucleus (**Figure 8.1**): signals that are unable to penetrate the cell membrane are intercepted by transmembrane receptors and give rise to the formation of second messengers or activate other secondary reactions. As a result, intracellular signaling cascades are stimulated, leading to modulation of transcription factor activity by

Figure 8.1 Two standard pathways of transcriptional regulation by exogenous signals The effects of hydrophilic signaling molecules that are unable to penetrate the cell membrane are mediated by second messengers and signaling cascades controlled by covalent modifications (frequently phosphorylations) of transcription factor activity. Most lipophilic signaling molecules penetrating the cell membrane are noncovalent ligands of certain transcription factors called nuclear receptors. In principle these two pathways of transmembrane signaling can be traced back to prokaryotes (see Section 3.5.1, Figure 3.29).

post-translational modification such as protein phosphorylation. Only when the exogenous signaling molecules are lipophilic enough to pass the membrane can they interact directly with transcription factors.

Transcription consists of three phases:

(1) **Initiation.** During this phase a multiprotein complex is formed, composed of RNA polymerase associated with several regulatory proteins. This complex is recruited and noncovalently bound to a specific site of an individual gene called a control or promoter sequence, which has been marked by gene-specific transcription factors. As a consequence, the double-stranded DNA becomes unwound upstream of the coding sequence to enable the polymerase to begin transcription.

(2) **Elongation.** Upon completion of initiation, the regulatory proteins interacting with RNA polymerase in the initiation complex are replaced by corresponding proteins to form an elongation complex. Bound in this complex, RNA polymerase moves along the coding sequence, generating the complementary RNA strand.

(3) **Termination.** The end of the coding DNA sequence is marked by a specific nucleotide sequence. When it arrives at this point of termination, the complex dissociates, releasing the completed RNA strand.

8.1 Initiation of eukaryotic transcription

Most signals that control gene activity act at the stage of initiation. To initiate transcription, the enzymatic machinery of RNA biosynthesis is recruited to a regulatory gene sequence, the promoter. For the synthesis of ribosomal, transfer, and messenger RNA (rRNA, tRNA, and mRNA), eukaryotic genes have different types of promoter recognized by three different RNA polymerases, types I, II, and III. Here the interaction of **RNA polymerase II** with genes encoding mRNA is discussed in more detail since it represents the primary target of signaling cascades. RNA polymerase II consists of 10 subunits and has a molecular mass of about 500 kDa. In its active form, the enzyme is associated with many additional proteins. It is estimated that, for the start of transcription, up to 50 different protein molecules are assembled at the gene promoter.

Eukaryotic **gene promoters** are divided into different functional regions: the core promoter, adjoining the start point of transcription to the 5′-end ("upstream"), and the proximal/distal promoter, located further upstream. In addition to a conventional promoter, certain genes contain alternative promoters located in the coding region (**Sidebar 8.1**). These regulatory sites control the production of truncated protein variants that may act as antagonists of the complete protein (for an example see Section 8.6). Such inhibitory fragments (which resemble dominant-negative mutants) may also be generated by alternative splicing of pre-mRNA (see Sidebar 8.1).

An **enhancer** is a regulatory sequence separated from the corresponding promoter. Enhancers may be found at far distances either upstream or downstream of a gene; nevertheless, they can interact with the gene due to a special folding geometry of the DNA.

8.1.1 Initiation complex

The RNA polymerase complex required for the initiation of transcription is assembled at the core promoter, whereas the proximal/distal promoter and the enhancer interact with gene-specific transcription factors, which "label" the start point and control the progression of RNA synthesis. In most genes the core promoter contains a TATA box [sequence TATA(T)AAG(A)] at 30–50 base pairs (bp) upstream of the starting point of transcription. This motif determines the interaction of the DNA

Sidebar 8.1 Mechanisms to create variability of gene expression Covalent modification of proteins is only one way in which the degree of proteomic complexity is increased in the face of a limited number of genes. Another possibility is the post-transcriptional processing of mRNA. While bacterial genes are composed of noninterrupted coding sequences, the genes of archaea and eukaryotes are assembled in such a way that the coding sequences, called exons, are interrupted by noncoding sequences, called introns. The number of introns is variable; some genes may contain up to 50 introns. The primary transcript, the pre-mRNA, mirrors the intron–exon pattern of the gene. During maturation to functional mRNA, the introns are cut out enzymatically, resulting in splicing of the exons. In the course of this process, the exons may be either combined in new patterns, provided with a new reading frame, or partly lost. Owing to these rearrangements, known as **alternative splicing**, a gene may give rise to a wide variety of protein isoforms exhibiting modified functions. It has been estimated that at least half of all pre-mRNAs in humans undergo alternative splicing. The muscle proteins troponin, tropomyosin, and titin as well as potassium channels are examples of alternative splicing. All exist in many splice variants, which are tissue-specifically distributed and differ in their functional properties (see, for instance, Section 14.4). Splice variants may even exhibit antagonistic effects, thus playing important roles in regulation.

In extreme cases, one gene may supply thousands of protein isoforms, even outdoing the number of genes (for an example see Section 7.4). Therefore, alternative splicing is one of the most efficient mechanisms by which the complexity of the proteome is increased, as compared with the genome. Alternative splicing and exon shuffling between different genes seem to be major reasons for the evolution of interrupted genes.

Splicing of pre-mRNA occurs in the nucleus. Similar to gene transcription and mRNA translation, splicing is under the control of a large number of regulatory proteins (estimated at more than 100) forming spliceosomes with small nuclear RNA. The basic constituents of these organelle-like complexes are small nuclear ribonucleoprotein particles (snRNPs). These cooperate with many regulatory proteins that, like transcription factors, stimulate or inhibit splicing depending on whether they bind to enhancer or silencer sequences in the exons. In their function, such sequences resemble the response elements of gene promoters.

Abundant splice factors are the stimulatory SR proteins [Ser (S)- and Arg (R)-rich proteins] and inhibitory nuclear ribonucleoproteins. Like transcription factors, splice factors are targets of signaling cascades. However, the investigation of these regulatory pathways is still in its infancy. The few cases studied in detail indicate that phosphorylation and dephosphorylation are central regulatory events, involving MAP kinases, calcium/calmodulin-dependent kinases, protein kinase C, cyclin-dependent kinases, and Tyr kinases of the Src type.

It should be mentioned that alternative splicing of pre-mRNA is only one of the powerful mechanisms for generating protein variants from a single gene. Another means is based on **alternative gene promoters**, located at internal intron–exon transitions rather than at the start point of a gene like conventional promoters. The proteins thus produced are amino (N)-terminally truncated forms that functionally may resemble dominant-negative mutants as they are widely used in the laboratory. An example is the generation of ICER, the truncated form and inhibitor of the transcription factor CREB (see Section 8.6). Corresponding sites are also found in certain mRNA species, enabling the generation of truncated protein isoforms by alternative translation initiation. mRNAs allowing such **alternative start codon usage** encode transcription factors (including nuclear receptors) and growth factors.

with the RNA polymerase complex. Additional control elements such as the CAAT box and the GC box are located further upstream. Some genes also contain so-called downstream promoter elements at about 30 bp. Normally such genes lack a TATA box.

In eukaryotes, the selective interaction of RNA polymerase with a promoter is not controlled by a σ factor as in bacteria (see Section 3.6.6). Instead, in a defined sequence, the enzyme forms a multiprotein complex with **general transcription factors** that, in contrast to other transcription factors, are not gene-specific. Their role is to assemble the initiation complex at the transcription start site of a promoter, which is marked by a gene-specific transcription factor.

General transcription factors are grouped in the TFII family. Among the six members complexed with RNA polymerase, TFIID binds directly to promoter DNA at the

Figure 8.2 Schematic representation of the eukaryotic RNA polymerase II complex of transcriptional initiation The core enzyme (black) is shown with the six general transcription factors of the TFII family. The DNA-binding TFIID complex is shown in color.

TATA box and upstream sequences. This factor is not a simple protein but a complex of at least 12 subunits including a TATA box-binding protein and associated proteins. Some of the associated proteins are co-activators of transcription, such as enzymes catalyzing the covalent modification of histones (see Section 8.2). TFIIA and -B stabilize the interaction of TFIID with DNA, while TFIIB and -F enforce the binding of the polymerase to the complex (**Figure 8.2**). To this end TFIIA and -B interact with both DNA and the TATA box-binding protein.

TFIIF can be compared with a bacterial σ factor since it hinders the polymerase from binding to DNA sequences outside of gene promoters. TFIIF facilitates the binding of TFIIH, a large enzyme complex consisting of nine subunits and containing a protein kinase that catalyzes a carboxy (C)-terminal phosphorylation of RNA polymerase II (see Section 8.1.2), as well as helicases, which under ATP hydrolysis unwind the DNA double strands at the start point of transcription.

8.1.2 RNA polymerase phosphorylation triggers the elongation phase

The activity of RNA polymerase II is controlled by signaling cascades. In the C-terminal domain of the large subunit, multiple copies of a heptapeptide with three serine (Ser) residues are found. Two of these residues are phosphorylated at the end of the initiation and during the elongation phase. As a result, the polymerase dissociates from the gene promoter, a process known as promoter clearance or promoter escape. The initiation complex is replaced by an elongation complex, with the phosphorylated C-terminal region of RNA polymerase serving as a scaffold for the recruitment of elongation factors and proteins that promote the maturation of mRNA by splicing, capping of the 5′-end by methylguanosine, and polyadenylation of the 3′-end (see textbooks of molecular biology for details).

Phosphorylation of the polymerase is catalyzed by several protein kinases, which are under the control of exogenous signals (**Figure 8.3**). One of these is a subunit of the general transcription factor TFIIH and is identical with the **CDK-Activating Kinase (CAK)**. In addition to catalyzing the phosphorylation of RNA polymerase, CAK, a complex of the Cyclin-Dependent Kinase CDK7 with the regulatory protein cyclin H is also involved in cell cycle control. Two other kinases that phosphorylate RNA polymerase are **CDK8** (as a complex with cyclin C) and **CDK9** (as a complex with cyclin K or T; for more details about cyclins and CDKs, see Sections 12.2 and 12.3). CDK9 and cyclin T (K) are subunits of the positive transcription elongation factor b. Besides the polymerase, this complex also phosphorylates elongation factors, thus controlling their activities.

8.1.3 Gene-specific transcription factors

The gene- or sequence-specific transcription factors are predominant target proteins of signals regulating gene activity. By interacting with regulatory gene sequences (and in some cases with the polymerase complex), transcription factors either promote or hinder gene transcription.

Figure 8.3 Effects of signal-induced protein phosphorylation on different components of the transcription complex (Left) Kinases of the CDK family existing as complexes with regulatory cyclins catalyze the phosphorylation of RNA polymerase II in the C-terminal domain and of elongation factors. This occurs at the transition from the initiation to the elongation stage and results in a clearance of the promoter from the initiation complex, allowing the formation of the elongation complex. (Right) Transmodulating activities of many gene-specific transcription factors are under the control of various protein kinases.

The DNA sequences of proximal/distal promoters and enhancers specifically binding transcription factors are called **response elements**. A gene is generally controlled by several response elements, each of which interacts with an individual transcription factor; that is, genes are regulated by a combinatorial strategy (see also Section 3.3.2). This enables the cell to modulate the activity of thousands of genes through a relatively small number of transcription factors (approximately 6% of the human genome). The factors are organized on the regulatory gene sequences into complex patterns rather than binding side by side. Additional proteins such as co-activators and co-repressors are integrated into this puzzle. These multiprotein complexes are called **enhanceosomes**.

The role of transcription factors is either to assemble the initiation complex at a distinct gene locus or to prevent such an assembly. They possess a modular structure typical for signal-transducing proteins that consists of at least three domains: a regulatory domain functioning as a receptor of input signals, a DNA-binding domain for specific interaction with a response element, and a trans(cription)-modulating domain that either activates or deactivates gene transcription (**Figure 8.4**). Upon activation by the input signal, the transcription factor binds noncovalently to the corresponding response element through an interaction of amino acid residues with complementary nucleotide residues in the major groove of the DNA double helix. By

Figure 8.4 Mode of action of a gene-specific transcription factor (activating type) In the inactive form, the transcription factor is folded in such a way that the regulatory domain (RD) blocks the DNA-binding domain (DBD) and the transactivating domain (TAD). An input signal (teal circle) induces a conformational change that relieves the blockade. The activated transcription factor binds by its DBD to a response element in a regulatory gene sequence and at the same time recruits by its TAD the initiation complex together with a transcriptional co-activator such as a histone acetyltransferase (see Figure 8.7).

means of their transmodulating (here transactivating) domains, the transcription factors that act as inducers now recruit the RNA polymerase complex as well as transcriptional co-activators. These co-activators are proteins that change the chromatin structure in such a way that transcription is facilitated because the RNA polymerase complex can bind to the promoter region and read the gene. Inhibitory transcription factors known as repressors have an opposite effect: they recruit proteins called co-repressors that catalyze the formation of an inactive chromatin structure or block the access of the RNA polymerase complex to DNA in another way. In a particularly clear manner, transcription factors demonstrate that data processing in the protein network is not simply the result of an interaction between biochemical reactions but critically depends on the assembly of the components at a specific site at the right time.

The active form of a transcription factor is a homo- or heterodimer. Such a structure strengthens the binding to DNA and allows subtle allosteric regulation and efficient suppression of signaling noise. Frequently the DNA-binding domains are involved in dimerization. This dual function is due to specific structural elements such as zinc finger, leucine zipper, and helix–turn–helix motifs (see Section 8.3.2 for examples of such domains; more details may be found in textbooks of molecular and cell biology).

The input signals controlling transcription factor activity consist of interactions with other proteins or low-molecular mass ligands, or of covalent modifications such as phosphorylation (or a combination of both). The controlled translocation between cytoplasm and nucleus also plays an important regulatory role. The lifespan of transcription factors is strictly regulated by degradation along the ubiquitin–proteasome pathway competing with de novo synthesis. Examples are discussed below.

Summary

- mRNA transcription is initiated by a complex of RNA polymerase II with six types of TFII general transcription factors binding at a gene promoter. This complex is a target of signaling reactions, particularly phosphorylations catalyzed by several cyclin/cyclin-dependent kinase (CDK) pairs that induce the replacement of the initiation complex by the elongation complex.

- Gene-specific transcription factors specifically bind to response elements in gene promoters and enhancers, thus labeling individual genes as targets of the RNA polymerase complex and associated proteins (co-activators) required for transcription. In contrast, inhibitory transcription factors (repressors) recruit proteins (co-repressors) that hinder RNA polymerase binding and transcription. Most gene promoters exhibit binding sites for several different transcription factors. Gene-specific transcription factors are the ultimate targets of cellular signal transduction.

8.2 Histones, nucleosomes, and chromatin

Transcription factors are unable to regulate gene transcription on their own. Instead, they rely on the assistance of either co-activators or co-repressors. The reason for this is the special physicochemical structure of the eukaryotic genome. Electrostatic repulsion would make it impossible to package the huge DNA molecules (amounting to a total length of 2 m per cell in humans) in the tiny volume of a nucleus (10 μm diameter). Thus, in eukaryotic cells, strongly basic proteins, the histones, neutralize the negative charge of DNA by forming a highly organized nucleoprotein complex known as chromatin. [Histone-like packing proteins have evolved with archaea; bacteria use other proteins for chromosome formation.] This DNA packaging creates a problem for gene transcription because the strong interactions with histones

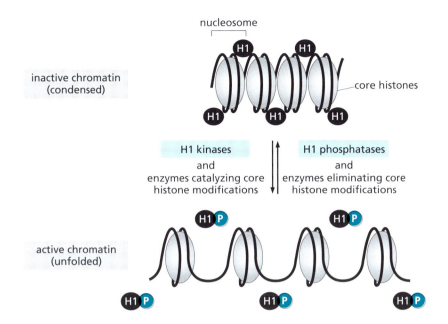

Figure 8.5 Inactive and active chromatin Inactive chromatin has a highly condensed structure due to strong interactions between DNA and the histone core of nucleosomes as well as the binding of histone 1 (H1). At sites that have become labeled by gene-specific stimulatory transcription factors, this structure is unfolded by H1 phosphorylation to such a degree that transcriptional co-activators inducing covalent modifications of the core histones gain access to the core histones of the nucleosomes. Chromatin activation is reversed by H1 phosphatases and enzymes eliminating the covalent modifications of core histones.

hinder access of the RNA polymerase complex to the genes, even when these have been marked by transcription factors. Transcription requires, therefore, a temporally and spatially limited weakening of these interactions. Such a mechanism provides an ideal target for regulatory signals. In fact, in a most ingenious way, nature has killed two birds with one stone by combining the process of DNA packaging with transcriptional control. To this end, the physicochemical properties of chromatin are changed in such a way that transcription is either facilitated or repressed. Therefore, all input signals that modulate gene transcription induce changes of chromatin structure aiming at an opening or closing of genes.

There are five major types of histone: "linker histones" such as histone 1 (H1) and "core histones" such as the histones H2a, H2b, H3, and H4. Core histones are phylogenetically highly conserved proteins characterized by a C-terminal histone fold, a structure that enables interactions with each other and with DNA. These interactions give rise to particular complexes consisting of two copies of each core histone species. In chromatin, this complex is wrapped by about 1.8 windings (approximately 146 bp) of the DNA double helix. This structure, called a **nucleosome**, is the basic building block of chromatin. Individual nucleosomes are connected by a linker DNA sequence comprising about 50 bp; thus, chromatin resembles a pearl necklace (**Figure 8.5**). The linker histone H1 is not a component of nucleosomes but binds to the linker DNA, causing an additional condensation of the chromatin structure by contracting the pearl chain. H1 exists in several variants and exhibits greater phylogenetic variability than the core histones.

8.2.1 Transcriptional co-activators and co-repressors control DNA–histone interactions

The strong interactions between DNA and histones are broken up by covalent modifications that diminish the basic character of the histones and "soften" the compact nucleosomal structure. Such modifications include acetylation, methylation, ADP-ribosylation, ubiquitylation, and phosphorylation. The enzymes catalyzing these reactions function as transcriptional co-activators. Their opponents, the co-repressors, stabilize the histone–DNA complex by reversing these modifications. The work of the histone-modifying enzymes is further supported by translocases that interrupt the histone–DNA bonds by mechanical force (as represented in **Figure 8.6**). Since transcriptional regulation would be impossible without co-activators and co-repressors, gene-specific transcription factors need to recruit histone-modifying

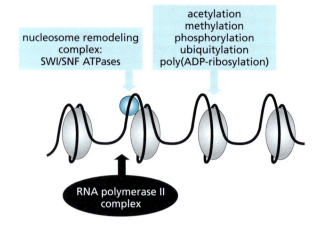

Figure 8.6 Mechanisms of chromatin activation: unfolding of the nucleosomal structure The condensed structure of the nucleosome is weakened by mechanical force exerted by SWI/SNF ATPases (left) and covalent modifications (right). These alterations enable the interaction of DNA with the RNA polymerase complex. Note that histone methylation also may suppress gene transcription (see Figure 8.10).

Figure 8.7 Modulation of transcriptional activity by histone acetylation By acetylation (Ac) of Lys residues of core histones, the compact structure of the inactive nucleosome is weakened, enabling transcription. Therefore, histone acetyltransferases function as co-activators and histone deacetylases as co-repressors. Both enzyme families are recruited to individual gene promoters by stimulatory and inhibitory transcription factors, respectively. Via bromo domains, additional regulatory proteins may become bound to acetylated histones.

enzymes to individual gene promoters in order to clear the path for the RNA polymerase complex. Thus, both transcription factors and cofactors are targets of gene-regulatory signaling cascades.

To induce transcription, the compact structure of inactive chromatin is destabilized in a stepwise manner. In a first step, the linker histone H1 is removed from DNA by phosphorylation (see Figure 8.5). Thus, in transcriptionally active chromatin (called euchromatin), H1 is found to be phosphorylated at many serine and threonine residues. These patches of negative charge weaken the interaction with DNA to such an extent that H1 may become displaced by **HMG proteins** (High-Mobility Group; the name refers to electrophoretic mobility), which belong to the large collection of non-histone proteins that modify the architecture of chromatin and its interactions with other regulatory proteins.

The goal of H1 phosphorylation and HMG protein interactions is to make the nucleosomal core accessible for enzymes catalyzing those covalent modifications that are required to break up the nucleosomal structure itself.

8.2.2 Histone acetyltransferases are co-activators of transcription

The major covalent modification neutralizing the positive charge of core histones is acetylation. For the initiation of transcription, histone acetylation is indispensable, whereas gene repression is always accompanied by histone deacetylation (**Figure 8.7**). The targets of acetylation are the ε-amino groups of conserved lysine (Lys) residues in the N-terminal regions of the four core histones. These regions are exposed at the surface of the nucleosomes and as carriers of positive charge are essential for the interaction with DNA.

The acetyl groups supplied by acetyl-coenzyme A are introduced by **histone acetyltransferases (HAT)**. This enzyme family comprises a large number of members exhibiting different specifications. Quite abundant are **CBP** (CREB-Binding Protein; CREB is a cAMP-activated transcription factor [see Section 8.6]) and the closely related **p300** (protein with a molecular mass of 300 kDa). Both are multipurpose histone acetyltransferases exhibiting numerous regulatory domains (**Figure 8.8**) by which they interact not only with CREB but also with many other transcription factors and control proteins. In fact, CBP/p300 combines the properties of an enzyme with those of a multifunctional scaffold protein, thus recruiting large protein complexes at individual gene promoters and softening the nucleosomal structure by histone acetylation. CBP/p300 and the proteins assembled by these enzymes are targets of numerous signaling cascades.

Apart from being essential partners of stimulatory transcription factors, histone acetyltransferases are also involved in other processes requiring a change of

| steroid hormone receptors | Ets1
HNF4
p65
P73
Pit1
Mdm2
RXR
STAT2
TBP
TFIIF
TFIIEB | BRCA1
CREB
c-Jun
c-Myc
Elk1
HPV E2
Sap1a
SREBP
Tax
YY1 | | c-Fos
dMad
E1A
E2F
Ets1
GATA1
HPV E6
JunB
MyoD
NeuroD | p53
Pit1
PyLT
SMAD
SF1
Tat
TAFIIB
Twist
YY1
high-mobility group proteins |

H₂N — bromo — CD — COOH

other acetyltransferases — acetylated proteins — protein kinases PKA, MAPK, CDK, pp90ᴿˢᴷ, MEKK1, CaMK other acetyltransferases α-tubulin, importin α histones H2A, H2B, H3, H4

chromatin structure, including DNA replication, DNA repair, and genetic recombination. The acetyltransferases acetylate Lys residues of histones selectively rather than randomly, thus generating a characteristic acetylation pattern. Together with other post-translational modifications, this pattern resembles an epigenetic, DNA-independent code (see Section 8.2.8), which can be deciphered by proteins that possess interaction domains for sequences with acetylated Lys residues. A typical recognition motif for acetyl-Lys is the **bromodomain**, a tetra-helical peptide sequence of about 110 amino acids found from yeast to humans in many transcription factors and other chromatin-associated proteins including CBP/p300 (see Figure 8.8). Via bromodomains, complexes of histone acetyltransferases with other proteins are bound to acetylated histones.

Figure 8.8 Histone acetyltransferase CBP/p300: a versatile transcriptional co-activator and scaffold protein Schematic multidomain structure of the 2441-amino-acid protein is shown with the interaction partners identified so far (black, transcription factors and related non-histone proteins of chromatin; teal, other interaction partners). CD, catalytic domain; bromo, bromodomain; gray rectangles, other interaction domains.

8.2.3 Histone deacetylases are co-repressors of transcription

The opponents of histone acetyltransferases are the histone deacetylases (HDAC). As co-repressors, they inhibit or terminate gene transcription, forming complexes with various other proteins such as inhibitory transcription factors. Histone deacetylases also play a key role in gene silencing and the formation of heterochromatin, highly condensed and transcriptionally inactive areas of chromatin arising in the course of tissue differentiation. Such zones are characterized by DNA sequences with methylated CpG residues (see Section 8.2.8), and histone deacetylases are directed to those sites by methyl-CpG-binding adaptor proteins, thus guaranteeing continuous gene silencing.

As inhibitors of gene transcription, histone deacetylases may act as tumor suppressors. An interesting example is **BRCA1** (BReast CAncer gene 1). This enzyme associates with a specific repressor of estrogen-sensitive genes. A loss-of-function mutation of BRCA1 may cause breast and ovarian cancer and is thought to be one reason for the increased incidence of these diseases in certain families. Complexes of repressors with histone deacetylases also attenuate gene transcription activated by other steroid hormone receptors (see Section 8.3).

Another tumor suppressor protein, the **retinoblastoma protein Rb**, forms a complex with HDAC1 and inhibits the transcription of genes required for a proliferating cell to enter the DNA replication phase. Upon phosphorylation of Rb, catalyzed by cell cycle-specific cyclin-dependent protein kinases, the complex dissociates and

Figure 8.9 Histone deacetylation by Sir2-like ADP-ribosyltransferases (sirtuins) The enzymes transfer the acetyl residue (highlighted in color) from histone to NAD⁺ simultaneously with the separation of the nicotinamide residue.

the repressor activity is lost. A deleted *rb* gene is a risk factor for retinoblastoma, a tumor of the eye (for details see Section 12.6).

Exogenous signals regulate the function of histone deacetylases predominantly by phosphorylation. In muscle cells, for instance, calmodulin-dependent protein kinases provoke a translocation of histone deacetylases from the nucleus into the cytoplasm, where they become sequestered by proteins of the 14-3-3 family (see Section 10.1.4). The enzymes return to the nucleus only upon dephosphorylation, catalyzed by signal-controlled phosphatases. Phosphorylation also modulates the enzymatic activity of histone deacetylases and their interactions with other chromatin-associated proteins.

Sirtuins

Conventional histone deacetylases are zinc proteins that catalyze the hydrolysis of the amide bond between Lys and the acetyl residue. Another family of deacetylating enzymes called **Sir2-like histone deacetylases** or **sirtuins** (referring to the homologous yeast enzyme Sir2, Silent mating type information regulator 2) operate by quite a different mechanism: they are ADP-ribosyltransferases that cleave the amide bond by use of nicotinamide adenine dinucleotide (NAD⁺) instead of water, yielding nicotinamide and 2′-*O*-acetyl-ADP-ribose (**Figure 8.9**). Sirtuins have been found from bacteria to humans and may play a key role in gene silencing and cell senescence. In addition to histones, other proteins such as transcription factor p53, α-tubulin, and TATA box binding proteins are deacetylated by sirtuins. Since sirtuins need NAD⁺ as a co-substrate, they are sensors of the metabolic redox state, that is, the energy status of the cell, and they probably coordinate it with repression of transcriptional activity. This may be one of the reasons for the evolution of this costly, energy-consuming mechanism of deacetylation.

Competitors for NAD$^+$ are the poly(ADP-ribosyl)polymerases (PARPs) (see Section 8.2.6). Since these enzymes are involved in gene activation while sirtuins are involved mostly in gene inactivation, a reverse regulation has been proposed. The consumption of NAD$^+$ by PARPs would turn down sirtuin activity and vice versa. Moreover, the reaction product 2′-*O*-acetyl-ADP-ribose might be an intracellular signaling molecule resembling cyclic ADP-ribose, which is generated in an analogous manner (see Section 14.5.3). However, in contrast to cyclic ADP-ribose, 2′-*O*-acetyl- ADP-ribose has not been found to induce a release of Ca^{2+} ions from the endoplasmic reticulum.

The activity of sirtuins was found to correlate with an increase of the lifespan in *Saccharomyces cerevisiae*, *Caenorhabditis elegans*, and *Drosophila melanogaster*, at least under situations of caloric restriction that are monitored by sirtuins. Moreover, genetic studies have prompted some authors to refer to sirtuins as products of **longevity genes**. Whether or not this also holds true for vertebrates, including humans, remains to be shown. Notwithstanding this reservation, pharmaceutical companies have begun to develop drugs that potentiate sirtuin effects. Their prototype is resveratrol, a phenolic compound (3,4′,5-trihydroxystilbene) produced by certain plants to fight parasitic organisms and found in higher doses in grapes and red wine.[*] In cell cultures as well as in *C. elegans* and *Drosophila*, resveratrol increases sirtuin activity and prolongs the lifespan. Whether or not there is a causal relationship between both effects is not entirely clear, because resveratrol also acts as a potent scavenger of reactive oxygen species and free radicals, which are thought to play a key role in aging and cell death.

Another type of longevity genes, the *FOXO* genes, are discussed in Section 4.6.6.

8.2.4 Two faces of histone methylation

The core histones, in particular H3 and H4, are not only acetylated but also methylated in their exposed N-terminal regions. Methylation is catalyzed by histone methyltransferases, demethylation by corresponding demethylases (see Section 2.7). Thus, histone methylation is as reversible as any other post-translational modification. A large variety of histone methyltransferases exhibiting different specificities have been identified. Depending on the enzyme type, either ε-amino groups of Lys residues or guanidino groups of arginine (Arg) residues are methylated one-, two-, and in the case of Lys, three-fold.

Histone methylation has either a stimulatory or an inhibitory effect on transcription. Thus, methylation cooperates with acetylation in reversible gene activation but also plays a key role in gene silencing and the formation of heterochromatin—that is, in the permanent inactivation of certain chromatin areas in the course of tissue differentiation—thereby cooperating with DNA methylation. Which of the two effects is expressed depends on the site of methylation (**Figure 8.10**). As a rule, methylation of Arg residues in H3 is required for *transcriptional activation*. The same holds true for methylation of H3 Lys residues 4, 36, and 79, catalyzed by multiprotein complexes consisting of strictly site-specific methyltransferases. These modifications may hinder access of repressor histone deacetylase complexes and inhibit competitive methylation of Lys9 and Lys27.

In fact, methylation of Lys9 and Lys27 in H3 leads to *gene silencing*. The reaction is catalyzed again by highly specific methyltransferases that generate interaction sites in H3 for a protein HP1, which has been shown to bind to methylated H3 Lys9 via a **chromodomain** (chromatin organization modifier domain), thus triggering heterochromatin condensation. Chromodomains are 30–70 amino acids long and are found in various, mostly transcription-repressing chromatin proteins. Methylation

---

[*]The high concentration of resveratrol in red wine may explain, at least partly, the so-called French paradox: that in France the considerably lower incidence of coronary heart diseases despite a high-fat and high-cholesterol diet correlates with a higher consumption of red wine.

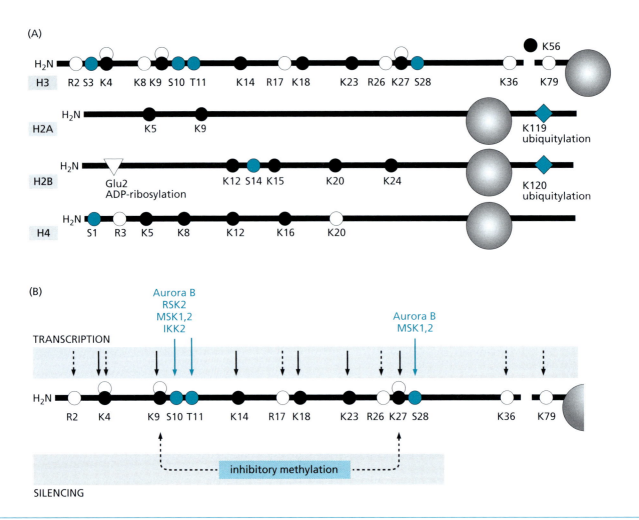

Figure 8.10 Covalent histone modifications (A) Proven modifications of core histones. The black lines stand for the N-terminal regions directly contacting DNA; the larger balls (right) indicate globular domains. Acetylation is symbolized by small black circles, methylation by small white circles, and phosphorylation by small colored circles; ubiquitylation (of H2A and H2B) is shown by a colored rhombus; and poly(ADP-ribosylation) (of H2B) is indicated by a triangle. (B) Association of individual modifications of histone 3 with transcription and gene silencing. Beside methylation of K9 and K27 silencing requires complete dephosphorylation and deacetylation (not shown). Phosphorylation is symbolized by teal arrows (with the corresponding protein kinases), acetylation by black arrows, and methylation by black broken arrows. K, lysine; R, arginine; S, serine; T, threonine. An additional modification found in H3 is a *cis–trans* isomerization of peptidyl–prolyl bonds (not shown). A comprehensive list of histone modifications and the corresponding enzymes is found in Kouzarides, T (2007).

of Lys9 occurs only when Lys14 has *not* been acetylated first, a situation resembling a logical BUT NOT gate. To make the situation even more complex, the acetylation of Lys14 depends on phosphorylation of Ser residue 10, resembling a logical AND gate. A similar pattern of spatially and temporally coordinated acetylations and methylations is seen for H4.

The different covalent modifications of the core histones are mutually dependent. This spectrum of covalent signaling reactions has been investigated in detail for steroid hormone-dependent gene promoters. According to these studies, a steroid hormone receptor acting as a transcription factor recruits the initiation complex including RNA polymerase II to a gene promoter that is associated with H3 acetylated at Lys14. This complex contains acetyltransferases that catalyze the acetylation of additional Lys residues in H3 and other core histones, thus weakening their interactions with DNA. These changes create the conditions for an activating methylation of H3 catalyzed by a methyltransferase also bound in the transcription complex.

Consequently, transcription is accelerated by making room for translocases that catalyze a mechanical rearrangement of nucleosomes (see Section 8.2.7). In addition, mRNA elongation depends on specific histone modifications in those nucleosomes that are located downstream of the polymerase complex. These modifications include a trimethylation of Lys4 and a methylation of additional Lys residues (Lys36 and -79) in H3.

8.2.5 Histone kinases act as teammates of acetyl- and methyltransferases

The basic character of histones is neutralized by phosphorylation more efficiently than by acetylation and methylation. Moreover, selective phosphorylation of individual Ser residues promotes the interaction of histones with acetyl- and methyltransferases. Of particular importance for transcriptional activation is the phosphorylation of H3 at Ser10 since it is, as mentioned, the condition of acetylation at Lys14 and methylation at Lys4. The phosphorylation generates a signal for the recruitment of histone acetyl- and methyltransferases. Corresponding phosphorylation sites are found in the N-terminal regions of H2b and H4 (H2A is not phosphorylated).

The phosphorylation of core histones is catalyzed by protein kinases that also activate a variety of transcription factors. In other words, for gene activation, the decondensation of chromatin through histone phosphorylation/acetylation and the marking of individual genes by transcription factors are strictly coupled events. Protein kinases falling into this category are the mitogen/stress-activated kinases MSK1 and MSK2 and the ribosomal S6 kinase RSK2. These enzymes are under the control of MAP kinase cascades (details in Section 11.5) and as complexes with the cognate MAP kinase isoforms bind to chromatin. Other histone kinases are IKK2, a component of the NFκB cascades where it phosphorylates the inhibitor IκB (see Section 11.7), and the "casein kinase" CK2 (see Section 9.4.5). Both MAP kinase and NFκB cascades are major signaling pathways transducing exogenous signals to the genome.

Extensive histone phosphorylation occurs in the course of cell division. Here the removal of the linker histone H1 is of particular importance since DNA replication and sister chromatid separation are blocked as long as chromatin exists in a condensed form. Therefore, H1 is phosphorylated across the whole genome in two steps in strict correlation with DNA replication and mitosis. The reaction is catalyzed by cyclin-dependent protein kinases such as CDK1 and CDK2 that are key regulators of the cell cycle (see Section 12.3). Core histone phosphorylation is also required for cell division. It is catalyzed, in particular, by the protein kinase Aurora B which controls sister chromatid segregation during mitosis (see Section 12.11).

8.2.6 Poly(ADP-ribosylation) and ubiquitylation of histones

In addition to acetylation, methylation, and phosphorylation, histones are covalently modified by poly(ADP-ribosylation) and ubiquitylation. **Poly(ADP-ribosylation)** is a reversible signaling reaction by which an ADP-ribosyl residue is transferred from NAD^+ to the carboxy groups of aspartate and glutamate residues and enlarged by polymerization (see Section 2.9.2). Targets are, in particular, H1 and H2B (see Figure 8.10). The key enzyme is poly(ADP-ribosyl) polymerase1 (PARP1), found mainly in the nucleus. It contains a DNA-binding zinc finger domain and an automodification domain, the poly(ADP-ribosylation) of which results in auto-inhibition of PARP1, demonstrating control of enzyme activity by negative feedback.

The poly(ADP-ribosyl) residue carries a strong negative charge. Poly(ADP-ribosylation), therefore, weakens the histone–DNA interaction in a particularly effective manner. In fact, PARP1 activity is stimulated by certain gene-regulatory proteins and correlates with high transcriptional activity, for instance, in the puffs of insect chromosomes (see Section 8.3.1). Moreover, PARP1 is connected with DNA repair since it is strongly activated by binding to damaged DNA. The enzyme functions mainly as a transcriptional co-activator, although in some cases it is also involved in transcriptional inhibition. The reverse holds true for poly(ADP ribose) glycohydrolase, the opponent of PARP, which renders the modification reversible.

A post-translational modification of Lys residues by **mono-ubiquitylation** is seen for core histones H2A and H2B and linker histone H1. In contrast to poly-ubiquitylation, this reaction does not trigger proteolytic degradation but correlates with histone methylation and gene silencing, particularly in the course of chromosome condensation during mitosis. The individual modifications clearly depend on each other. In yeast cells, for instance, ubiquitylation of Lys123 in H2B is a prerequisite for methylation of Lys4 in H4. Histone mono-ubiquitylation is a reversible modification catalyzed by highly specific E3 ubiquitin ligases and reversed by de-ubiquitylating isopeptidases. As potential transcriptional co-activators and co-repressors, these enzymes resemble acetyltransferases and deacetylases.

8.2.7 Nucleosome-remodeling complexes as molecular machines of gene regulation

The covalent modifications of histones by acetylation, methylation, phosphorylation, poly(ADP-ribosylation), and ubiquitylation are not sufficient for a complete disruption of the interactions between DNA and histones. They need to be supported by a mechanical separation of DNA from core nucleosomes. This energy-consuming process of nucleosome or chromatin remodeling is catalyzed by at least four families of ATPases, with the **SWI/SNF** family providing the best known example. The name refers to the yeast mutants mating type SWItching and Sucrose Non-Fermenting, where the function of such enzymes was found for the first time. Recognizing acetylated histones, these enzymes are molecular machines that migrate along the nucleosome, pushing the zone of the double-stranded DNA that is detached from the histone core like a "bow wave" (see Figure 8.6), thus giving access for the RNA polymerase complex to interact with the DNA. The mechanism resembles that of motor proteins: upon ATP hydrolysis, the ATPase anchored to both the nucleosomal core and the surrounding DNA undergoes a conformational change resulting in a translocation of the DNA strand (**Figure 8.11**). The abundance of remodeling ATPases (about one machine per 10 nucleosomes) underlines their indispensable role. In fact, nucleosome-remodeling machines are involved not only in gene expression but also in other processes such as DNA repair, gene replication, gene recombination, and chromosome assembly. SWI/SNF proteins are tumor suppressors. Loss-of-function mutations found in about 20% of human tumors belong to the most frequent genetic defects associated with cancer.

SWI/SNF ATPases are multiprotein complexes consisting of a dozen or so subunits. They have a DNA-binding domain and a bromodomain that recognizes acetylated histones. By this means the remodeling complex cooperates with the histone acetyltransferase complex, becoming active only at genes primed for transcription. On

Figure 8.11 Model of ATP-dependent nucleosome remodeling The remodeling ATPase thought to be anchored at the nucleosomal core particle interacts with the surrounding DNA strand at two sites. Owing to the conformational change caused by ATP hydrolysis, the DNA strand is shifted, allowing interaction with the proteins of the transcriptional machinery.

the other hand, some SWI/SNF isoforms also interact with histone deacetylases involved in transcriptional repression. In this case, they are clearing the gene for an interaction with DNA methyltransferases (see Section 8.2.8).

During the elongation phase of transcription, additional measures are required to clear a path for the RNA polymerase complex along the genomic DNA. **FACT** (FAcilitator of Chromatin Transcription), a component of the elongation complex, is a heterodimeric molecular machine or "histone chaperone" that, by means of strongly acidic groups, releases H2A and H2B from the nucleosome, reintegrating them when the RNA polymerase complex has passed. FACT is also involved in DNA repair and replication. **Elongator**, a multiprotein complex with histone acetyltransferase activity, plays a similar role.

8.2.8 Epigenetic long-term memory stabilizes tissue functions

Almost all somatic cells possess the complete genome. Nevertheless, they perform quite different functions depending on the tissue. These tissue-specific patterns of differentiation are maintained over the whole lifespan of the organism. They are passed during cell division in an unchanged form from cell generation to cell generation, thus maintaining tissue individuality and continuity. The differentiation pattern is guaranteed by a set of active genes, whereas the rest of the genome is silenced by stable modifications of chromatin and DNA. This occurs in the course of embryonic tissue development. Tissue-specific gene expression is mirrored by patterns of transcriptionally active euchromatin and transcriptionally inactive (silenced) heterochromatin. Since such patterns are transferred from mother to daughter cells, although they are not encoded in genomic sequences, they are called epigenetic.

Epigenetic heredity depends on covalent modifications of both DNA and histones. A reaction that plays a key role is **DNA methylation**. In mammals the 5′-position of cytidine in the dinucleotide residue CpG is affected almost exclusively. Such CpG sequences are concentrated at distinct sites of the genome, such as gene promoters, forming **CpG islets**. CpG methylation facilitates the formation of highly condensed heterochromatin. In the genome of mature mammalian tissue cells, 70% of all CpG elements are methylated on average, with the methylation pattern changing from tissue to tissue and in the course of embryonic development. As expected, an exceptionally high degree of DNA methylation is found in inactive regions of the genome, whereas nonmethylated CpG elements are concentrated in the regulatory and coding domains of active genes.

DNA methylation hinders the access of stimulatory transcription factors. Moreover, methylated CpG islets are binding sites of histone deacetylase complexes, nucleosome remodeling ATPases, and other transcriptional co-repressors. For complex formation, **methyl-CpG-binding proteins** are required, which act as adaptors binding to DNA via a methyl-CpG-binding domain and to the co-repressors via a transcription-repressing domain (**Figure 8.12**). Thus, in the vicinity of a methylated CpG element the histones become modified in such a way that transcription is rendered impossible. Therefore, the pattern of DNA methylation together with the pattern of histone modifications represents a code of tissue-specific gene expression.

DNA methylation is catalyzed by several DNA methyltransferases controlling de novo methylation in the course of tissue formation and transferring the methylation pattern from the mother DNA to the daughter DNA during cell division. Nucleosome-remodeling ATPases facilitate the access of these enzymes to DNA. Modified histones and methyl-CpG-binding proteins, together with repressors and histone deacetylases with which the DNA methyltransferases form complexes, direct the whole process. This guarantees that the enzymes required for gene silencing stay concentrated near methylated DNA sequences so that the pattern of differentiation is maintained over many cell generations. Disturbances of this epigenetic information transfer leads to decreased DNA methylation, resulting in serious consequences such as cancer.

Figure 8.12 DNA methylation as a signal of developmental gene silencing The chromatin region (teal) includes transcriptionally active (straight line) and transcriptionally inactive (zigzag line) zones. The inactive zone contains an embryonically determined pattern of methylated CpG islets (symbolized as Me-CpG) that, via Me-CpG-binding proteins (symbolized as Me-CpG-BP), interacts with a silencer complex. This complex, containing DNA methyltransferases, histone deacetylases, nucleosomal remodeling complexes, and other inhibitory proteins, reestablishes the pattern of DNA methylation and inhibitory histone modifications after each cell division in a site specific manner. In an active area of chromatin, DNA is not methylated, enabling the binding of a transcription factor (TF) that recruits the transcription complex.

Summary

- In eukaryotic chromatin, DNA is densely packed by histones. H2a, H2b, H3, and H4 combine to form complexes around which DNA is wound (nucleosomes); H1 occupies the linker DNA between the nucleosomes. In these complexes the access of RNA polymerase is hindered.

- For gene transcription, covalent modifications that weaken the basic character of histones and mechanical forces temporarily lift this blockade. Enzymes that catalyze such modifications are transcriptional co-activators, whereas proteins that stabilize histone–DNA interaction are co-repressors.

- Histone modification activating gene transcription include acetylation, phosphorylation, and poly(ADP)-ribosylation.

- Histone modifications repressing gene transcription include de-acetylation, dephosphorylation and ubiquitylation.

- Histone methylation may either activate or repress gene transcription depending on the enzyme type and the methylation site in the histone molecule. Histone demethylases oppose these effects.

- Complete disruption of the DNA–histone complex, as required for transcription, occurs through a mechanical separation of histones and DNA supporting the effects of covalent histone modifications.

- Permanent gene silencing results from specific DNA methylation. The methylated sites are recruitment centers for transcriptional co-repressors.

- The particular pattern of histone modifications and DNA methylation represents a self-renewing epigenetic code of gene activity that guarantees the constancy of tissue differentiation.

8.3 Transcription factors as hormone receptors

Transcription factors are major targets of signaling cascades activated by extracellular signals. As pointed out in Section 2.12, there are the two basic mechanisms of signal transduction connecting the cell periphery with the genome, which have been preserved from prokaryotes (see Section 3.5) to humans.

Figure 8.13 Signaling cascades controlling gene transcription Most extracellular signals interact with receptors at the cell surface. The most abundant types are shown here. Directly or via second messengers, these receptors activate signaling pathways that lead to phosphorylation of transcription factors. More lipophilic signal molecules may penetrate the cell membrane and activate transcription factors by direct binding (right). Signaling pathways not fitting this scheme will be discussed elsewhere. The figure represents an oversimplification, since in reality complex cross talk exists between the individual pathways, ensuring that the activity of a transcription factor is controlled by a distinct pattern of signals transmitted by different receptor types. ANP, atrial natriuretic peptide; ECM, extracellular matrix; TGFβ, transforming growth factor β; G, G-protein; GPCR, G-protein-coupled receptor; GC, guanylate cyclase; InsP₃, inositol 1,4,5-trisphosphate; DAG, diacylglycerol; P, phosphate residue. Note that protein kinase-coupled receptors exist in two forms: receptors with intrinsic (as depicted here) and receptors with associated protein kinase activities.

(1) Extracellular messengers (such as peptide hormones) that are not lipophilic enough to penetrate the cell **membrane interact** with membrane receptors at the cell surface, activating downstream signaling cascades that, in most cases, lead to phosphorylation of transcription factors.

(2) Lipophilic messengers (such as steroid hormones) able to pass the cell membrane are bound by a class of transcription factors known as **nuclear receptors** (**Figure 8.13**). These receptors demonstrate in an exemplary manner how extracellular stimuli influence genetic activity.

8.3.1 The nuclear receptor family

Among the transcription factors, nuclear receptors constitute a major family of structurally related proteins that appear to be restricted to animals. In humans it comprises 48 members. Only about half of them are receptors by definition because they exhibit ligand binding sites for signaling factors. Many of these ligands are still waiting for identification. Their putative receptors are called "orphan receptors." Prototypes of nuclear receptors are the **steroid hormone receptors**. Steroid hormones are of historical significance because they were the first hormones seen to have an effect on genes. This was a breakthrough in our understanding of hormone action comparable with the discovery of transmembrane signal transduction (see Section 4.2). In this context, the pioneering studies on insects should be mentioned. In the course of pupation and molting, puffed regions arise at specific sites of the giant or polytene chromosomes* that are characteristic for such species. Quite early, these "puffs" observed under the light microscope had been interpreted as sites of increased genetic activity.

---

*Giant chromosomes are generated by DNA replication without subsequent cell division. The chromatids do not become separated but stay together, forming a "cable bundle." The number of such "endomitoses" depends on the insect species and on the tissue. For instance, in the salivary glands of *Drosophila*, a preferred experimental model, nine endomitoses yield the 1024-fold enlargement of the haploid chromosome set.

Figure 8.14 Puffing of giant chromosomes (A) Section of a giant chromosome from the salivary gland of *Acricotopus lucidus* (fourth larval stage) in condensed and puffed state (5000-fold magnification). (B) Simplified scheme of a partly puffed giant chromosome; note the four chromatid strands twisted to chromomeres (resembling the strongly dyed zones in A) and unwound in the puff. (C) Development of a puff (shown in color) in the first salivary gland of *Chironomus* larvae 0, 30, and 60 min after injection of the steroid hormone ecdysone. (A, from Mechelke F [1963] In Funktionelle und morphologische Organisation der Zelle [P. Karlson ed] Springer-Verlag; B, from DuPraw EJ [1968] Cell and Molecular Biology, Academic Press; C, from Clever U [1969] *Verh Dtsch Zool Ges* 62:75–92.)

Puffs could be induced experimentally by the steroid ecdysone, the molting and pupation hormone of insects (**Figure 8.14**). Fifty years ago, this observation gave rise to the hypothesis that hormones were transcriptional activators. Radioactively labeled steroid hormones were then found to bind to cytoplasmic proteins—later on identified as receptors—in their target organs, subsequently migrating into the nucleus, where their accumulation correlated with increased mRNA synthesis.

Ecdysone interacts with a nuclear receptor. The same holds true for another prominent insect hormone, namely juvenile hormone, a lipophilic sesquiterpene derivative that as an ecdysone antagonist prevents molting and premature metamorphosis.

Nuclear receptors appear to be restricted to animals, On the other hand, steroid hormones are also found in algae and plants. However, here they do not interact with nuclear receptors but activate other signaling pathways. Thus, in the weed *Arabidopsis thaliana* the multifunctional phytohormone brassinolide, a steroid, binds to a membrane-bound receptor Ser/Thr kinase that indirectly induces gene transcription by activating protein phosphatases. This promotes a dephosphorylation of certain transcription factors that are inactive in the phosphorylated state.

The general mode of action of nuclear receptors is shown in **Figure 8.15**. Most of them shuttle between cytoplasm and nucleus. Upon ligand binding they dimerize and interact with gene regulatory DNA sequences. A comprehensive list of nuclear receptors is found in Evans and Mangelsdorf (2014).

8.3.2 Receptor structure

Nuclear receptors have the characteristic overall structure of transcription factors that is shown in Figure 8.4: a central DNA-binding domain is flanked by a C-terminal regulatory domain interacting with the ligand, and a transactivating domain called AF1 (transcriptional Activation Function 1; see **Figures 8.16** and **8.17**) positioned N-terminally from the DNA-binding domain. In many receptors, a second transactivating domain AF2 is found, overlapping with the ligand-binding domain (see Figure 8.17). A major function of AF1 and AF2 is to recruit transcriptional co-activators such as histone acetyltransferases. The DNA-binding domain, a two-fold zinc

Figure 8.15 Mode of action of nuclear receptors The ligands (colored ellipses) penetrate the cell membrane and bind to their receptors in the cytoplasm. The ligand–receptor complexes oligomerize and accumulate in the nucleus, where they bind to hormone-responsive elements on DNA and act as transcription factors.

finger sequence of 66 amino acids, interacts specifically with hormone-responsive elements in enhancer and promoter regions of genes. It may be assisted in recognizing its cognate response element by the AF1 domain. This is necessary because the response elements of most steroid hormones are identical (see Section 8.3.3). AF1 domains fulfill the requirements of *specific* interaction sites because they are quite heterogeneous, differing from receptor to receptor. To a certain extent they resemble docking domains of protein kinases (see Sections 1.8 and 11.1).

The receptors have, in addition, a nuclear localization sequence located in the center of the molecule within a binding-site of the chaperone heat-shock protein 90 (HSP90). As shown in Section 8.5, HSP90 plays an important role in the activation of steroid hormone receptors.

Like other receptor types, most nuclear receptors exist in various isoforms that frequently are generated after transcription by alternative mRNA splicing or alternative start codon usage for translation. These isoforms are tissue specifically expressed

Figure 8.16 Ligand-activated nuclear receptors of vertebrates Shown are the domain structures of the receptors (amino terminus to the left), the numbers of amino acids (for human receptors), and the major ligands. For an explanation of abbreviations, see Table 8.1. Other details are explained in the text.

Figure 8.17 Two major types of ligand-controlled transcription factors Shown are the domain structures of (A) nuclear receptors and (B) arylhydrocarbon receptor. The characteristic structural elements of the DNA-binding domains, zinc finger and helix–loop–helix, are depicted in a simplified manner. Cys, cysteine residue; AF1 and AF2, activation functions 1 and 2; NLS, nuclear localization sequence. For PAS domain see Section 3.4.2, Sidebar 3.3.

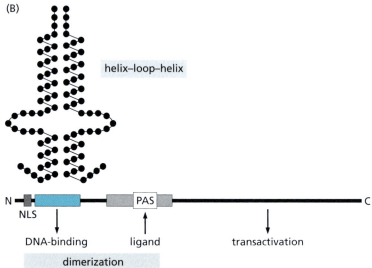

and differ in their properties, thus contributing considerably to the functional versatility of the receptor ligands.

Generally, steroid hormone receptors are active only as homodimers, whereas other nuclear receptors form heterodimers with the 9-*cis*-retinoic acid receptor RXR (**Table 8.1**). The dimerization results in a pronounced cooperative effect. With its response element, a dimer interacts 100 times more strongly and more specifically than a monomer. Some "orphan receptors," however, bind to DNA as monomers.

8.3.3 Hormone response elements

All hormone response elements exhibit a symmetrical structure that fits the receptor dimer and comprises two identically or palindromically arranged hexanucleotide sequences separated by 1–5 nucleotide residues. A palindromic head-to-head arrangement as inverted repeats is typical for steroid hormone response elements. Thus, the estrogen response element has the structure 5'-AGGTCA-nn-n-TGACCT-3' (n stands for any nucleotide). The response elements of other steroid hormone receptors share the common sequence 5'-AGAACA-n-n-n-TGTTCT-3'. To confer specificity, the AF1 domains probably assist the receptors in recognizing their cognate response elements.

Table 8.1 Ligand-dependent transcription factors of vertebrates: a selection

Transcription factor	Dimer	Preferred ligand
Family of nuclear receptors		
Subfamily of steroid hormone receptors		
estrogen receptor ER	ER–ER	17β-estradiol
progesterone receptor PR	PR–PR	progesterone
androgen receptor AR	AR–AR	testosterone
glucocorticoid receptor GR	GR–GR	cortisol
mineralocorticoid receptor MR	MR–MR	aldosterone
Subfamily of non–steroid receptors		
retinoic acid receptor RAR	RAR–RXR	*all-trans*-retinoic acid
retinoid X receptor RXR	RXR–RXR	9-*cis*-retinoic acid
thyroid hormone receptor TR	TR–RXR	3,5,3′-triiodothyronine
vitamin D receptor VDR	VDR–RXR	1,25-dihydroxy-vitamin D_3
peroxisome proliferator–activated receptor		
PPARα	PPAR–RXR	fatty acids, eicosanoids
PPARβ	PPAR–RXR	fatty acids, eicosanoids
PPARγ	PPAR–RXR	fatty acids, eicosanoids
pregnane-X receptor PXR	PXR–RXR	xenobiotics
constitutively activated receptor CAR	CAR–RXR	phenobarbital
farnesoid receptor FXR	FXR–RXR	bile acids
liver-X receptor LXR	LXR–RXR	oxysterols
Family of arylhydrocarbon receptors		
arylhydrocarbon receptor ArH	ArH-ARNT	dioxin, l-kynurenine

In contrast to steroid hormone response elements, the binding sites for nuclear receptors forming heterodimers with RXR are characterized by two parallel or head-to-tail oriented (direct) repeats of the sequence AGGTCA, which differ from receptor to receptor only by the number of connecting nucleotides n (ranging from 1 to 5 according to the so-called 1–5 rule). The structure of the vitamin D response element, for instance, is 5′-AGGTCA-n-n-n-AGGTCA-3′: one of the hexanucleotide sequences binds the complex of RXR and 9-*cis*-retinoic acid and the other one interacts with the liganded vitamin D receptor (**Figure 8.18**).

X-ray analysis has provided deep insight into the receptor–DNA interaction. The DNA-binding site of nuclear receptors contains two α-helices that are oriented by the zinc

Figure 8.18 Dimeric nuclear receptors and their response elements (A) Estrogen receptor (ER; E, estradiol), an example of steroid hormone receptors, binds as a homodimer to a palindromic response element that is arranged as an inverted repeat separated by three arbitrary nucleotide residues n. (B) Non-steroid hormone receptors, here the vitamin D receptor (VDR) as an example, form heterodimers with the 9-cis-retinoic acid receptor (RXR) as a partner (cRA, 9-cis-retinoic acid; VD, 1,25-dihydroxy-vitamin D). The response elements are two parallel oriented identical repeats separated by 1–5 arbitrary nucleotide residues n (n 3 for VDR-RXR).

Figure 8.19 Nuclear receptors: interactions between DNA-binding domains and response elements Shown in a strongly simplified mode are the palindromically arranged DNA-binding domains of (A) the glucocorticoid receptor (GR) homodimer and (B) the parallel binding domains of the 9-*cis*-retinoic acid receptor (RXR)–thyroid hormone receptor (TR) heterodimer in front of the DNA double helix. Note that the α-helices (cylinders) are placed at right angles by the zinc fingers with one helix each dipping into the major groove of the DNA double helix. (C) Molecular model of the glucocorticoid receptor-DNA complex based on X-ray crystallography. The teal balls symbolize the zinc ions.

finger in a rectangular manner. One of the two α-helices dips into the major groove of the DNA double helix while the other α-helix binds at the outside, thus contributing to the specificity of the interaction between receptor and DNA (**Figure 8.19**).

For non-genomic effects of steroid hormones, see **Sidebar 8.2**.

Sidebar 8.2 Non-genomic effects of steroid hormones In textbooks the theory that steroid hormones are direct modulators of gene transcription is frequently treated like a dogma. However, this is a misconception, because steroid hormones can do more. Half a century ago it was observed that certain steroid hormone effects occurred within seconds or minutes, whereas for gene transcription and subsequent protein biosynthesis the cell needed hours. Since then a whole series of these rapid effects have been identified, which are not suppressed by inhibitors of transcription (such as actinomycin D) or translation (such as cycloheximide) and are called, therefore, non-genomic effects. Three of these non-genomic effects are worth examining in more detail (**Figure 8.20**):

- meiosis of amphibian oocytes
- acrosomal reaction of sperm cells
- anesthetizing effect

Maturing **amphibian oocytes** are arrested in the G2 phase of the cell cycle. This blockade is relieved by progesterone, which enables the cells to enter meiosis. The critical step is activation of the cyclinB–CDK1 complex, which induces the transition of the cell from G2 into the mitosis phase. As discussed in Chapter 12, this reaction can occur only when an inhibitory phosphorylation of CDK1, caused by the protein kinase MYT1, is reversed by the protein phosphatase Cdc25. Progesterone seems to affect this regulatory mechanism at two sites: it stimulates a protein kinase (called polo-like kinase) that activates Cdc25, and at the same time, it inactivates MYT1 along the MAP kinase–p90[RSK] pathway (see Section 11.5). The second effect is mediated by the conventional nuclear progesterone receptor, which in its N-terminal region has a proline-rich sequence. Via this sequence, the receptor binds to an SH3 domain of the cytoplasmic Tyr kinase Src thus inducing enzymatic activity (for the mechanism of Src activation, see Section 7.1.2). By means of the scaffold protein Shc, the activated Src stimulates the Ras-GEF mSOS and, thus, the MAP kinase module RAF–MEK–ERK, which transduces mitogenic signals (see Figure 7.12 in Section 7.1.2). In an analogous way, the nuclear estrogen receptor activates the MAP kinase cascade.

The **acrosomal reaction** of sperm is a Ca^{2+}-dependent exocytotic process by which the cell releases lytic enzymes

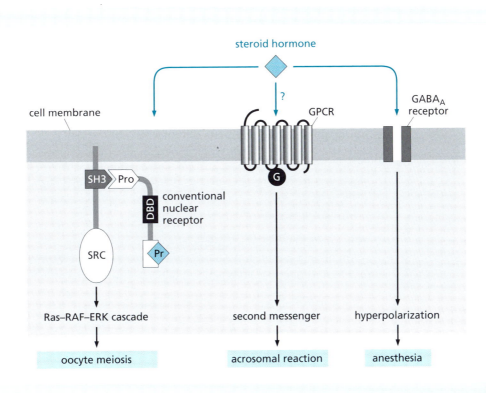

Figure 8.20 Non-genomic effects of steroid hormones (Left) Upon progesterone binding the conventional progesterone nuclear receptor activates, by means of an N-terminal proline-rich domain (Pro), the membrane-anchored tyrosine kinase Src and the Ras–RAF–ERK (MAP kinase) cascade. This signaling pathway plays a role in meiosis of amphibian oocytes. Pr, progesterone; DBD, DNA-binding domain; SH3, Src homology domain 3. (Middle) Steroid hormone controls, via a membrane receptor, the generation of second messengers and induces through Ca^{2+} influx the acrosomal reaction of sperm cells. Whether or not the receptor is G-protein-coupled, as depicted here (GPCR), remains a matter of debate. (Right) Via a γ-aminobutyric acid (GABA$_A$) receptor anion channel, the steroid induces membrane hyperpolarization and thus inhibits neuronal excitation. This effect may result in anesthesia.

to penetrate the zona pellucida (jelly coat) of the egg cell. The response is triggered by oocyte factors. One of these factors, progesterone, induces a Ca^{2+} influx into the sperm cell within seconds. Since the acrosomal reaction is resistant to contraceptive drugs that inhibit the nuclear progesterone receptor, it has been postulated to be mediated by a nonclassical receptor, which has not yet been identified but might be a G-protein-coupled receptor. Such an assumption is supported by the discovery of a GPCR called GPR30 that specifically interacts with estradiol. This receptor is found in the endoplasmic reticulum of a variety of tissue cells and stimulates—via

G-proteins—signaling reactions such as cAMP formation and others.

The **anesthetizing effect of steroid hormones** has been used in human and especially in veterinary medicine for more than 50 years. The rapid effect resembles that of barbiturates. Indeed, like barbiturates, steroids interact specifically with a subunit of the γ-aminobutyric acid (GABA$_A$) receptor anion channel, stimulating the influx of chloride ions and, consequently, hyperpolarization. By this, the excitation of nerve cells is dampened (for more details on GABA receptors see Section 16.3).

8.3.4 Nuclear translocation and activation of gene transcription

Steroid hormone receptors are found in both the cytoplasm and the nucleus. Their active conformation is quite unstable. Therefore, for ligand binding and nuclear translocation they need the assistance of the chaperones HSP70 and HSP90 (see Section 8.5). Upon arrival in the nucleus, the chaperones are replaced by transcriptional co-activators and the nuclear receptor binds to the corresponding response element, marking the gene for the RNA polymerase II complex. Co-activators are first of all histone acetyltransferases such as the enzyme CBP/p300 and more specialized steroid receptor co-activator proteins. As mentioned before, these proteins are both enzymes and scaffolds that assemble additional factors required for transcriptional activation.

The non-steroid nuclear receptors have not been found to interact with chaperones but are converted into the active conformation by other means, probably phosphorylation. Even in the absence of their ligands, most of these receptors are located at the nuclear membrane or in the nucleus where they are bound to chromatin. In this form, they are frequently complexed with transcriptional co-repressors that inhibit the transcription of their target genes but also of other genes. Upon ligand binding, the complex disintegrates and co-activators replace the co-repressors. Most co-repressors are histone deacetylases or at least recruit such enzymes. Examples are the silencing mediator of retinoic acid and the thyroid hormone receptor (SMRT) and the nuclear receptor co-repressor (NCOR).

For steroid hormone receptors, co-repressors have also been described. The most prominent example is the above-mentioned estrogen receptor co-repressor BRCA1, a tumor suppressor protein the deletion of which may cause breast cancer.

Nuclear receptors also interact with other transcription factors such as NFκB and activator protein 1, AP1 (see Chapter 11) forming hetero-oligomers that either promote or attenuate the transcription of thousands of target genes. Such interactions indicate cross talk between various signaling pathways.

The concept of the receptors interacting only with co-repressors or co-activators is an oversimplification. In reality, large multiprotein complexes are assembled at the responsive promoter elements. Apart from the receptors, these complexes contain enzymes that regulate acetylation, methylation, and other forms of covalent modifications of histones, such as nucleosome-remodeling proteins; factors involved in RNA processing and protein degradation; adaptors for binding of the complexes to RNA polymerase; additional transcription factors; and others. Moreover, the transcriptional activity induced by a nuclear receptor is not static; instead it is a process oscillating in periods of seconds to minutes due to a permanent interaction between stimulatory and inhibitory influences causing cycles of receptor binding and release. These may be superimposed with phases of receptor degradation and de novo synthesis controlled by a system of positive and negative feedback loops. An increasing body of evidence indicates that **oscillating transcription** is not restricted to nuclear receptors but might be a common feature of inducible genes. Oscillating activity of transcription factors (and of signal-transducing proteins in general) may provide a way to escape desensitization, thus insuring long-lasting signaling. In other words, the effect induced by a gene-regulatory signal may resemble a stroke of a gong rather than a shot. The idea that, under such conditions, gene transcription represents a frequency-modulated process is easily seen (see also Section 11.6).

8.3.5 Covalent modifications of nuclear receptors

To consider nuclear receptors as transcription factors that are exclusively controlled by ligand binding is another unjustified oversimplification, because in reality their activities are fine-tuned by the data-processing protein network. In this context, post-translational modifications such as phosphorylation, ubiquitylation, and sumoylation play an important role. Each receptor has several phosphorylation sites, particularly in the transactivating domains AF1 and AF2 as well as in the DNA-binding domain. These sites are targets of MAP kinases, cyclin-dependent kinases, protein kinases A, B, and C, and tyrosine kinases that all are components of other signaling pathways. The particular effect depends on which domain of the receptor is phosphorylated. A phosphorylation of the AF domains mostly activates the receptor, for example by reinforcing the affinity for a co-activator, whereas a phosphorylation of the DNA-binding site hinders the binding to the response element by electrostatic repulsion.

Phosphorylation is also important for signal extinction because it labels the receptor for ubiquitylation and subsequent degradation by proteasomes. Under experimental conditions, *activated* nuclear receptors may have a very short lifespan, as they are degraded after one round of transcription (the so-called suicide mechanism; see Section 2.8.2). However, in the living cell, receptor-dependent signaling, once triggered, may escape desensitization by oscillating for a longer period between phases of activity and inactivity.

As co-activators and co-repressors are substrates of protein kinases (see, for instance, Figure 8.7) and other protein-modifying enzymes, highly sophisticated cross talk occurs between different signaling pathways. Therefore, genes controlled by nuclear receptors are, in addition and in parallel, targets of those receptors inducing second messenger formation and protein phosphorylation. Or, in other words: nuclear receptors are embedded in extended signaling networks controlling a wide variety of cellular functions.

Summary

- Nuclear receptors are a family of structurally related transcription factors in animals. Many of them are activated by lipophilic ligands, that are able to penetrate the cell membrane.

- Ligands include steroid hormones, thyroid hormones, dihydroxy-vitamin D_3, retinoids, certain eicosanoids, oxysterols, bile acids, certain fatty acids, and xenobiotics.

- While steroid hormone receptors are homodimers, the other receptors are active only as heterodimers with the retinoid receptor RXR.

- For ligand binding and activation, the nuclear receptors require the assistance of chaperones.

- Binding sites of nuclear receptors on gene promoters consist of two identically or palindromically arranged hexanucleotide sequences separated by 1–5 nucleotide residues. Steroid hormone response elements differ in the hexanucleotide sequences, whereas the other response elements share an AGGTCA hexanucleotide sequence but differ in the number of nucleotide residues linking the hexanucleotides.

- Nuclear receptors, as well as the transcriptional co-modulators they recruit to their effector genes, are targets of a wide variety of signaling reactions that fine-tune the activity and control the lifespan.

8.4 Ligand-controlled transcription factors are xenosensors of the toxic stress response

Some nuclear receptors are strongly activated by foreign compounds (xenobiotics) such as pharmaceutical drugs and environmental poisons and are, therefore, termed xenosensors. Upon heterodimerization with RXR, these receptors evoke the toxic stress response of cells that consists of de novo synthesis of proteins regulating the metabolism, transport, and excretion of xenobiotics but also of endogenous factors such as steroids and bile acids. For the metabolism of these compounds, the large enzyme family of **cytochrome P450-dependent oxygenases** is of central importance. Using molecular oxygen, they oxidize their (mostly lipophilic) substrates to products that, as a rule, are less toxic or biologically inactive and can be rendered water-soluble by conjugation with glucuronic or sulfuric acid and excreted by the kidneys. The nuclear receptors that control genes encoding cytochrome P450 enzymes and other proteins of drug metabolism include PPARα and -β, PXR, CAR, FXR, and LXR.

8.4.1 Peroxisome proliferator-activated receptors

The induction of adipose tissue differentiation is the most prominent among the numerous functions of this receptor type. Peroxisome proliferator-activated receptors (PPARs; frequently pronounced "peepars") exist in three isoforms. The isoform **PPARα** is expressed by tissues with active fat metabolism and regulates the transcription of the corresponding genes. The receptor was discovered as a cellular binding protein of blood fat-lowering drugs such as clofibrate, of the industrial solvent trichloroethylene, and of plasticizers. In animal experiments, these compounds stimulate an increase of peroxisomes in the liver and may promote tumor development. Peroxisomes are organelles that are derived from the endoplasmic reticulum and are found in all eukaryotic cells. They contain enzymes of oxidative lipid and amino acid metabolism together with catalase, which destroys the toxic hydrogen peroxide generated along such pathways.

In contrast to PPARα, the isoforms **PPARβ** and **PPARγ** are not activated by peroxisome proliferators. The ubiquitously expressed PPARβ induces early steps while the more restricted PPARγ activates late steps of fat cell differentiation. In addition, PPARβ promotes the differentiation of skin keratinocytes in the course of wound healing, myelinization of nerve fibers, and placenta development during pregnancy, whereas PPARγ is involved in inflammatory processes and insulin action. This receptor is activated by thiazoledinediones, drugs that as "insulin sensitizers" are employed for the treatment of insulin-resistant type 2 diabetes. Defects of PPAR signaling also play important roles in other disorders such as atherosclerosis and cancer. Endogenous ligands of PPARs are unsaturated fatty acids and their derivatives, particularly certain prostaglandins and other arachidonic acid-derived metabolites such as endocannabinoids (see Section 4.4.5 and Section 16.7).

8.4.2 Receptors PXR, CAR, FXR, and LXR

PXR was originally described as a receptor of pregnane derivatives, but it must not be confused with the progesterone receptor PR. PXR is, in addition, activated by glucocorticoids, drugs, and poisons. It occupies a central position in detoxification and metabolism of both endogenous compounds and xenobiotics. The cytochrome P450 type 3A induced by PXR is of great medical interest since it controls more than 60% of drug metabolism and, in addition, plays a significant role in bile excretion. An adaptive overproduction of cytochrome P450 3A along the PXR pathway is a major reason for drug tolerance or tachyphylaxis.

Another typical xenobiotic receptor is **CAR** (Constitutive Androstane Receptor). The promoters of its target genes are characterized by a phenobarbital response element (phenobarbital is a pharmaceutical drug of the barbiturate family). CAR also binds androstane derivatives, indicating a role in the metabolism of endogenous steroids. It is constitutively active (in the absence of ligands) but ligand binding facilitates its accumulation in the nucleus.

The closely related farnesoid receptor **FXR** is activated by derivatives of farnesol and vitamin A as well as by bile acids. FXR interacts with genes carrying a bile acid response element and encoding proteins that control the biosynthesis, metabolism, and transport of bile acids. Another receptor involved in the regulation of bile acid and cholesterol metabolism is the liver X-receptor **LXR**. Its ligands are oxysterols, which are products of oxidative cholesterol metabolism. The role of this receptor in the metabolism of xenobiotics is disputed.

Agonists of PXR, CAR, FXR, and LXR have been suggested for the treatment of metabolic diseases characterized by elevated blood levels of fatty acids and cholesterol and resulting in obesity, diabetes, and atherosclerosis.

8.4.3 "Dioxin receptor" ArH

A ligand-controlled transcription factor that does not belong to the family of nuclear receptors proper is the **arylhydrocarbon receptor ArH**. It differs from the

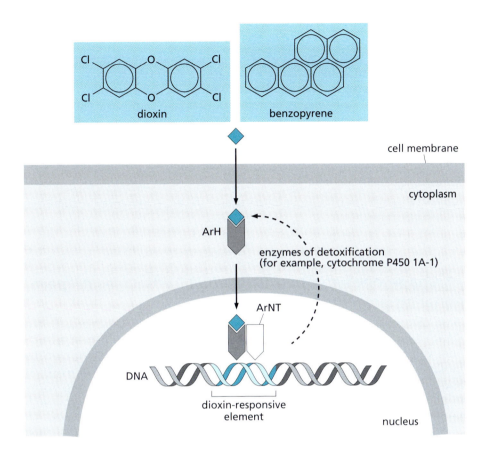

Figure 8.21 Mode of action of the arylhydrocarbon receptor Lipophilic xenobiotics (such as dioxin or benzopyrene, symbolized as a colored rhombus) penetrating the cell membrane are bound by the receptor in the cytosol. Upon translocation into the nucleus and heterodimerization with ArNT, the complex activates the transcription of genes with dioxin-responsive elements, thus inducing the synthesis of detoxifying enzymes. As indicated by the broken arrow, these enzymes also metabolize the xenobiotics by which they are induced.

nuclear receptors by the arrangement and structure of its functional domains (see Figure 8.17). Here, the N-terminal DNA-binding domain has a helix–loop–helix rather than a zinc finger structure, and the ligand-binding site is a PAS domain (see Section 3.4.2). Nevertheless, ArH shares many features with nuclear receptors. For instance, in the cytoplasm, ArH forms complexes with the chaperone HSP90. This results in a more stable conformation and enables ligand binding (see Section 8.5). As a typical xenosensor, ArH interacts with a wide variety of drugs, toxic products including aromatic arylhydrocarbons such as benzopyrene, and food ingredients. L-Kynurenine, a breakdown product of tryptophane has been found to be an endogenous ligand. By far the most potent ArH activator is 2,3,7,8-tetrachlorodibenzo-p-dioxin, briefly called **dioxin** (**Figure 8.21**). This extremely toxic substance, also known as Seveso poison, has gained an unfortunate notoriety: its accidental release by a chemical plant near the Italian village Seveso in 1976 was one of the worst environmental disasters in history.

Upon binding of the ligand, ArH translocates into the nucleus, where it heterodimerizes with the closely related protein ArNT (ArH Nuclear Translocator). [ArNT factors are also involved in the stress response to oxygen deficiency; see Section 8.7.] This complex activates genes carrying a dioxin or xenobiotics response element. Altogether more than 100 genes are directly controlled and an additional 200 genes are indirectly influenced by dioxin, including those for certain cytochrome P450 isoforms. As mentioned in Section 8.4.3, cytochrome P450 normally catalyzes the detoxification of xenobiotics. Under certain circumstances, however, the reaction products may turn out to be even more toxic: for example, in the case of dioxin or polycyclic aromatic hydrocarbons such as benzopyrene. These products are activated in such a way that they covalently bind to DNA, thus causing gene mutations that can result in cancer (benzopyrene is the carcinogenic component of coal tar).

Summary

- Some types of nuclear "non-steroid" receptor interact with lipophilic xenobiotics and, in turn, stimulate the transcription of genes that encode detoxifying enzymes. Their endogenous ligands are lipids such as eicosanoids, certain steroids, farnesol and vitamin A derivatives, bile acids, and cholesterol metabolites.

- The arylhydrocarbon receptor ArH is a ligand-binding transcription factor that differs from nuclear receptors by the nature of its DNA binding domain and its overall domain structure. It is activated by xenobiotics such as carcinogenic hydrocarbons and dioxin and induces the synthesis of enzymes that normally produce less toxic (however, sometimes also more toxic) metabolites.

8.5 Chaperones and peptidyl-prolyl isomerases prepare signaling proteins for work

As we have seen, the chaperones HSP70 and HSP90 are essential for the activation of steroid hormone receptors. This function indicates that these "heat-shock proteins" can do more than just protect other proteins from thermal denaturation. Indeed, more and more client proteins have been found, the functions of which are controlled by chaperones even in the absence of stress situations. These proteins are almost exclusively components of the signal-processing network. They include transcription factors, membrane receptors, protein kinases, G-proteins and the corresponding GDP–GTP exchange factors (GEFs) and GTPase-activating proteins (GAPs), signal-generating enzymes, and ion channels. Thus, stress protection is only one aspect in a much larger functional repertoire of HSPs.

The mode of function of the two chaperones has been studied most thoroughly with steroid hormone receptors as models. These studies have revealed an unexpectedly complex mechanism by which HSP70 and HSP90 change the receptor conformation in such a way that a hydrophobic binding pocket is opened for the ligand (**Figure 8.22**). Since such a conformation is unstable in aqueous solution, it does not appear spontaneously but is induced in an energy-dependent process consuming two molecules of ATP, the hydrolysis of which is catalyzed by the intrinsic ATPase activities of HSP70 and HSP90.

The first ATP hydrolysis promotes the formation of a multiprotein complex and is catalyzed by HSP70 (see Section 2.5.1). The components of the complex are the steroid receptor, HSP70, an HSP90 dimer, an adaptor protein (HSP-complex organizing protein HOP), and the co-chaperone HSP40, which stimulates the ATPase activity of HSP70 and thus functionally resembles a GAP of a G-protein. In this complex HSP90 is activated by ADP–ATP exchange and stabilized by an interaction with another co-chaperone, p23, in such a way that it can bind the steroid hormone receptor. HSP90 then transmits the conformational tension produced by the nucleotide exchange to the receptor, a reaction requiring another ATP hydrolysis catalyzed by HSP90 itself. As a result, HSP90 returns to the ADP-loaded state and the chaperone complex disintegrates.

As indicated by the large number of client proteins, it seems to be a general function of the HSP70/HSP90 pair to generate active but energetically unfavorable conformations of signal-transducing proteins. Thus, the HSP70/HSP90 complex represents an important switching element in the signal-processing protein network.

8.5.1 Peptidyl-prolyl *cis–trans* isomerases

Frequently the chaperones form complexes with other proteins that support their effects on protein conformation. These assistants, known as **immunophilins**,

(A)

(B)

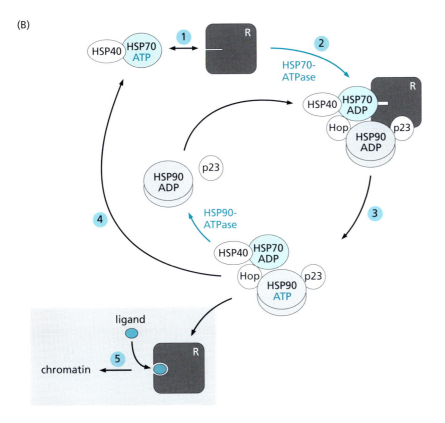

Figure 8.22 Stepwise activation of a steroid hormone receptor by the chaperones HSP70 and HSP90 The principle of the effect is shown in (A). In the cytoplasm most of the receptor molecules exist in an inactive conformation that cannot bind the ligand. Activation is an ATP-consuming process that is catalyzed by a chaperone complex. Ligand binding shifts the equilibrium between inactive and active receptors into the direction of the active form. The details of the reaction are shown in (B). Inactive receptor R with a closed ligand-binding pocket interacts with the ATP-loaded HSP40/HSP70 complex (step 1). This results in a partial opening of the ligand-binding pocket, powered by HSP70-catalyzed ATP hydrolysis, and binding of HSP90 dimer, Hop, and p23 (step 2). Next HSP90 is activated by ADP–ATP exchange and the ligand-binding pocket of the receptor becomes fully opened (step 3). Upon ATP hydrolysis catalyzed by HSP90, the complex dissociates (step 4), enabling the primed receptor to interact with the ligand and with chromatin (step 5). During the disintegration of the complex, HSP70 is assumed to become reloaded with ATP. Upon dissociation or degradation of the ligand, the receptor automatically relapses into the inactive conformation (see panel A); that is, the role of the ligand is to stabilize the structure of the active receptor.

cyclophilins, or FK-binding proteins, were discovered as cellular binding proteins of immunosuppressants such as cyclosporine, FK 506, and rapamycin. They are widely distributed: the human genome encodes about 30 isoforms.

All immunophilins exhibit the activity of peptidyl-prolyl *cis–trans* isomerases (PPIases). These enzymes, found in pro- and eukaryotes, catalyze the isomerization of peptide bonds at the imide nitrogen of prolyl residues (**Figure 8.23**). *Cis* and *trans*

Figure 8.23 Peptidyl-prolyl *cis–trans* isomerase reaction.

conformations of proteins are in equilibrium. However, under cellular conditions the setting of this steady-state is very slow, requiring catalysis. Whether the equilibrium is *cis*- or *trans*-directed depends on the type of substrate protein. Therefore, the result of a PPIase reaction resembles either a *cis* to *trans* or a *trans* to *cis* rearrangement. Since the isomerization causes a pronounced conformational change, PPIases are counted among the (co)chaperones.

Although immunosuppressants inhibit the enzymatic activity of immunophilins, their clinical effects seem to be due primarily to another mechanism: inhibition of the Ca^{2+}-dependent protein phosphatase calcineurin, which activates the NFAT transcription factors that are essential mediators of the immune response (see Section 14.6.4).

In the steroid receptor–HSP90 complex, specific immunophilins contribute to a stabilization of the active receptor conformation. This function is not restricted to nuclear receptors but is described for other transcription factors as well, which become either inhibited or activated. Thus, like the chaperones, immunophilins play an important role in cellular signal processing.

In contrast to cyclophilins and FK-binding proteins, another subfamily of PPIases, the parvulin-like enzymes (the name comes from a bacterial protein), do not react with immunosuppressants. Certain members of this family called **PIN1** (meaning Protein Interacting with NIM-A, a mitotic kinase discussed in Section 12.11) have gained considerable attention as proteins forming logical AND gates with protein kinases: they differ from other PPIases in that they interact exclusively with prolyl residues adjacent to phosphorylated Ser and Thr residues. In other words, PIN1 PPIases operate downstream of those protein kinases that catalyze Pro-directed phosphorylation. Such kinases include cyclin-dependent kinases of cell cycle and transcription control (see Sections 8.1.2 and 12.4), MAP kinases (see Section 11.1), and glycogen synthase kinase 3 (see Section 9.4.4). These enzymes phosphorylate only proteins with a *trans*-prolyl–peptidyl bond, and the same conformation-specificity has also been found for the protein phosphatases PP2A, calcineurin, and the dephosphorylating enzyme of RNA polymerase II (see Section 8.1). Subsequent to phosphorylation or dephosphorylation, PIN1 catalyzes a *trans–cis* rearrangement of the substrate protein by which the stability, activity, and subcellular distribution are changed.

Considering the enormous versatility of kinases and phosphatases, it is not surprising that PIN1 PPIases have a wide variety of substrates. These include (just to mention a few examples) the cell cycle regulators cyclin D1, Cdc25, WEE1, and MYT1, (see Chapter 12); the transcription factors cJun (see Section 11.6.2), NFκB (see Section 11.7), p53 (see Section 12.8.3), NFAT (see Section 14.6.4), and β-catenin (see Section 5.9.2); and RNA polymerase II and some histone deacetylases as well as the protein kinase CK2 (see Section 9.4.5). Phosphorylated NFκB, p53, and β-catenin are protected from proteolytic degradation, and cJun and cyclin D are activated by PIN1-catalyzed isomerization. As discussed in more detail in the following chapters, these stabilizing and activating effects are controlled by additional input signals, rendering PIN1 a component of extended signal-processing networks. Diminished PIN1 activity correlates with neurodegenerative diseases such as Alzheimer's dementia where the *cis–trans* isomerization of certain phosphorylated proteins plays a crucial role in pathogenesis (see Section 13.1.3), whereas over-expression of PIN1 is a common feature of cancer. PIN1 probably stimulates oncogenic signaling by rendering phosphorylated key proteins of mitogenic and anti-apoptotic pathways resistant to degradation along the ubiquitin–proteasome pathway. Examples of such key proteins are β-catenin and NFκB.

Summary

- Activation of nuclear receptors includes the opening of a lipophilic ligand-binding pocket. In the case of steroid hormone receptors, this is achieved by the chaperone ATPases HSP70 and HSP90. Activation of steroid hormone receptors demonstrates a more general effect of these chaperones: control of signaling protein conformation by catalyzing the generation of an active but energetically unfavorable form.

- Enzymatic isomerization of the peptidyl–prolyl bond is a means of conformational change that supports and stabilizes the effects of chaperones. Peptidyl-prolyl *cis–trans* isomerases (PPIases) are found in pro- and eukaryotes and, as immunophilins, bind to immunosuppressive drugs.

- The PIN1 subfamily of PPIases isomerizes only prolyl residues adjacent to phosphorylation sites; that is, it cooperates with proline-directed protein kinases. PIN1 enzymes are involved in cancer, diabetes, and Alzheimer's disease.

8.6 Transcription factors as substrates of protein kinases

Transcription factors undergo a wide variety of post-translational modifications controlling their activity. Examples will be found throughout this book. Here we shall restrict the discussion to phosphorylation, the most versatile and most abundant modification.

Phosphorylation may modulate such diverse features as nuclear translocation, oligomerization, interaction with cofactors, DNA binding, transactivating activity, and lifespan. It may occur in the cytoplasm or, upon translocation of the protein kinase, in the nucleus. In some cases, preformed complexes of transcription factors with corresponding protein kinases and phosphatases as well as other proteins required for transcriptional regulation are associated with chromatin and await an input signal. Their particular construction renders such complexes potential oscillators, indicating a frequency-modulated effect.

An illustrative example of transcription factor phosphorylation is provided by CREB (cAMP Response Element Binding protein). CREB and its isoform CREM (cAMP Response Element Modulator) are both abundant transcription factors, whereas the third member of this family, ATF1 (Activating Transcription Factor 1), is mainly restricted to nerve and glandular cells. Each of the three factors has a leucine zipper domain for dimerization, a DNA-binding domain, and a kinase-inducible domain, the phosphorylation of which results in activation. The DNA-binding domain contains several basic amino acids for the specific recognition of cAMP response elements (CREs) consisting of the palindromic sequence TGACGTCA or half of it (CGTCA). Such sites have been found in the promoters of more than 100 different genes. They interact with the dimeric transcription factors, provided these have been phosphorylated at a Ser residue in the kinase-inducible domain.

Figure 8.24 Activation of transcription factor CREB by protein phosphorylation Via a G_S-protein-coupled receptor, an extracellular signal induces the formation of cAMP (colored rhombus), which activates protein kinase A (PKA), resulting in a dissociation of the catalytic (C) and regulatory (R) subunits. The catalytic subunits of PKA accumulate in the nucleus, where they catalyze the phosphorylation and activation of CREB (P, phosphate residue), thus enabling the transcription factor to interact, as a dimer, with cAMP-responsive elements found in more than 100 genes.

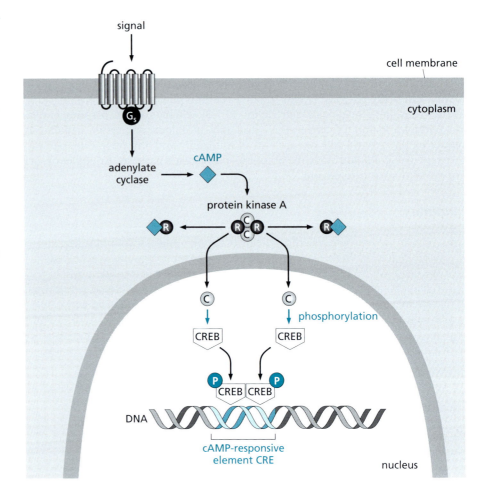

CREB, CREM, and ATF1 exist in many variants generated by alternative splicing of the corresponding premRNAs and via intronic gene promoters. The latter are sequences located in an intron of a gene and induce a truncated form of mRNA. These variants lack the DNA-binding and dimerization domains and may act as competitive inhibitors resembling dominant negative mutants. An example is the truncated CREM variant **ICER** (Inducible cAMP response Element Repressor) that suppresses the transcription of certain CRE-containing genes during cell maturation.

The names CREB and CREM indicate that these factors are regulated by cAMP. However, they do not interact directly with the cyclic nucleotide but become phosphorylated by cAMP activated by protein kinase A (PKA), (**Figure 8.24**). Along this standard pathway of signal transduction, all signals activating adenylate cyclase via G_S protein-coupled receptors (see Section 5.2) or Ca^{2+} (see Section 4.4.1) may influence gene expression. Moreover, CREB and CREM are phosphorylated and activated not only by PKA, but also by various other protein kinases (summarized in **Figure 8.25**). Although these phosphorylations all occur at the same strategic Ser residue, cells are able to distinguish between the different input signals, probably because additional factors direct CREB and CREM to individual genes or, vice versa, protect certain genes from the interaction with CREB and CREM (as a rule, genes are controlled by a collection of different regulatory factors rather than by a single transcription factor). Upon phosphorylation CREB/CREM recruit the initiation complex of transcription to the genes with CREs. To this end, they cooperate with the co-activator CBP/p300 (remember that CBP stands for CREB-Binding Protein), which like CREB/CREM is the target of various protein kinases, or signaling pathways (see Figure 8.7). The phosphorylation signals are extinguished by the phosphatases PP1 and PP2A.

Figure 8.25 Signaling pathways leading to CREB phosphorylation and activation Along the signaling pathways shown, protein kinases (box) are activated, phosphorylating CREB at a Ser residue in the transactivating domain. Ca^{2+} may stimulate certain isoforms of adenylate cyclase as well as Ras-controlled signaling cascades (see Section 10.1). More details are found in Section 4.6.1 (protein kinase A), Section 4.6.6 (protein kinase B), Section 11.6 (protein kinases MSK1 and p90RSK), and Section 14.6.3 (CaM kinases).

cAMP-response elements are found in a wide variety of genes. Particular interest has found their role in nervous tissue. As originally shown by pioneering genetic experiments with *Aplysia* sea slugs and *Drosophila* fruit flies, CREB plays a central role in memory fixation and learning. The factor stimulates the de novo synthesis of proteins that reinforce and stabilize synaptic contacts (synaptic plasticity; see Section 16.4.1). Moreover, CREB promotes the transcription of anti-apoptotic genes such as *bcl-2* and thus prolongs the lifespan of neurons. In such cases, neuronal survival factors such as nerve growth factor and brain-derived neurotrophic factor are major input signals leading to CREB activation. Factors of the CREB family such as ATF1 also control day–night rhythms by inducing the expression of the periodicity genes *per1* and *per2* in a periodical manner (for more details see Section 9.4.5).

As shown in later chapters, the phosphorylation of transcription factors is a theme with many variations. By no means is the target always the transactivating domain, with activation being the outcome. In fact, there are many examples of phosphorylation of the DNA-binding domain, mostly, but not always, resulting in inactivation (see steroid hormone receptors). This means that transcription factors may also be stimulated by dephosphorylation with NFAT providing an example (see Section 14.6.4). Another variant is that both the transcription factor and an accessory protein, for instance an inhibitor, are controlled by phosphorylation. NFκB is a typical example (see Section 11.7).

Summary

The function of most transcription factors is controlled by phosphorylation. An example is provided by cAMP response element binding protein (CREB), which is activated by phosphorylation catalyzed by a variety of protein kinases. It is a target of several signaling cascades. Upon activation, CREB recruits the transcriptional co-activator CBP/p300 to genes containing the cAMP response element.

8.7 The hypoxic stress response

Owing to their multidomain structure, transcription factors may simultaneously transform several input signals into one output signal, working as molecular "nanoneurons." An example is steroid hormone receptors: as discussed previously, they are not only activated by their ligands but their functions are modified by various additional signals such as phosphorylation and other covalent alterations. The hypoxic stress response provides another impressive and relatively well-studied example. This emergency reaction enables the cell to cope with a diminished oxygen supply.

The critical event of the hypoxic stress response is the rescue from inactivation of the **Hypoxia-Inducible transcription Factors (HIFs)**. In fact, when sufficient oxygen is available, the cellular concentration of HIFs is kept at a very low level due to continuous degradation along the ubiquitin–proteasome pathway. This situation is reminiscent of Wnt and hedgehog signaling, where input signals also protect the corresponding transcription factors from destruction (see Section 5.9).

The input signal leading to a stabilization of HIFs is a decline of the cellular oxygen concentration. As a result, the HIF isoforms associate into α/β-heterodimers, which become enriched in the nucleus and stimulate the transcription of more than 70 genes with hypoxia-responsive elements. Such genes give rise to the formation of glucose transporters and of enzymes of anaerobic energy production, which cause the formation of lactic acid from glucose yielding 2 ATP (as compared with more than 30 ATP molecules obtained by oxidative glucose metabolism to CO_2 and water). At the same time, HIFs induce the synthesis of membrane transporters and ion exchangers required to protect the cell from acidification due to lactic acid production.

Other HIF-sensitive genes encode transferrin and its receptor, facilitating the intake of iron ions (required for enzymes of oxidative metabolism), as well as of factors that improve the tissue's blood and oxygen supply. The latter include vasodilatory enzymes such as NO synthase (see Section 16.5) together with factors inducing the formation of blood capillaries (vasculo-endothelial growth factor and its receptor) and of erythrocytes (erythropoietin).

Several covalent signaling reactions control the stability and activity of the HIFs: hydroxylation, acetylation, phosphorylation, ubiquitylation, sumoylation, and S-nitrosylation (see **Figure 8.26**). As expected, HIFs respond very sensitively to changes of the cellular oxygen concentration. The oxygen sensor of the hypoxic stress response is an oxygenase, **prolyl 4-hydroxylase**. This type of enzyme is well known for its important role in collagen biosynthesis, where it stabilizes the triple helix by proline hydroxylation. In the hypoxic stress response, a particular isoform (known as prolyl hydroxylase domain protein, PHD) catalyzes the hydroxylation of two Pro residues in the transactivating domain of HIFα. This unusual covalent signaling reaction occurs under normal oxygen supply and signals a specific ubiquitin ligase (VHL) to mark the transcription factor for degradation. Under oxygen deficiency, Pro hydroxylation does not take place and HIFα (which is continuously produced by the cell) is not recognized by VHL. In contrast to the α subunit, HIFβ (which is identical with the factor ArNT mentioned in Section 8.4.3) is neither hydroxylated nor degraded by the VHL–proteasome system.

A second hydroxylation occurs at an Arg residue in the C-terminal transactivating domain of HIFα. This causes the transcription factor to lose its activity, since

Figure 8.26 Domain structure, interactions, and post-translational modifications of the hypoxia-inducible transcription factor HIF1α Inhibitory modifications are depicted in teal; stimulatory modifications are shown in black. The interaction partners are given in gray boxes. HLH, basic helix–loop–helix motif; TD, transactivating domain; CBP, CREB-binding protein. PAS, see Section 3.4.2.

it cannot interact any longer with the co-activator CBP/p300. The hydroxylation is catalyzed by the oxygenase **FIH** (Factor Inhibiting HIF) and provides an additional on–off switch. Under hypoxic conditions, both the prolyl 4-hydroxylase PHD and, to a lesser extent, the Arg hydroxylase FIH become less active because of a lack of substrate oxygen.

Like many other transcription factors, HIF becomes **acetylated**. This covalent modification, catalyzed by a specific acetyltransferase, takes place at a Lys residue in the transactivating domain. Its effect resembles that of Pro hydroxylation in that HIF is attacked by VHL and is degraded in proteasomes. The signaling pathways linked to HIF acetylation are not yet known.

Another frequent modification of transcription factors is **sumoylation**. Here, an oligomer of the ubiquitin-related peptide SUMO (Small Ubiquitin-like MOdifier) is bound to a Lys residue (see Section 2.8), resulting in inactivation, but not degradation, of HIF. In contrast to acetylation, ubiquitylation, and sumoylation, **phosphorylation** activates HIF. Only MAP kinases and CK2 have been identified as corresponding kinases thus far. An activating signal is also provided by **S-nitrosylation**, which is due to a nonenzymatic reaction of a cysteine residue in the C-terminal transactivating domain with nitric oxide (for more details on this reaction, see Section 16.5).

Hypoxic stress response and cancer

The ubiquitin ligase VHL is a tumor suppressor protein. Its deletion causes **Von Hippel–Lindau syndrome** which is characterized by benign and malignant tumor growth in a variety of organs. The tumorigenic effect is caused by an accumulation of HIF in the affected cells, that results in uncontrolled proliferation due to an oversupply of metabolic energy. However, the inglorious role played by a defective hypoxic stress response is not restricted to this rare hereditary disease. **Angiogenesis**, the supply of a tumor with new blood capillaries, is essential for tumor development since otherwise the rapidly dividing cells would die from starvation. A diminished blood supply leads to hypoxia as it is prevailing in most tumors. As a response, the HIF system is activated, which in turn triggers angiogenesis. Most cancer cells exhibit, in addition, a particular metabolic defect discovered in 1924 by Otto Heinrich Warburg (who was awarded the 1931 Nobel Prize in Physiology or Medicine) and known as the **Warburg effect**: even when the oxygen supply is sufficient, they prefer anaerobic to oxidative glucose metabolism, thus maintaining the angiogenetic stimulus. Moreover, the lactic acid produced from glucose is one reason for the characteristic acidification (acidosis) of tumor cells, which may promote invasive growth by activating acidic hydrolases that degrade the extracellular matrix. These observations prompted Warburg to postulate that cancer is a disease of mitochondrial respiration, a hypothesis that recently has regained attention. Given the central position the HIF system occupies in low-oxygen adaptation and pH control, the Warburg effect may be due to defective regulation of the hypoxic stress response, although the underlying mechanism is unknown. When its role in tumorigenesis is considered, the HIF system provides an important target of future anti-cancer strategies.

Summary

- The hypoxia-inducible transcription factor HIF controls genes that rescue the cell from hypoxic stress.

- HIF provides a typical example of a "molecular neuron," processing a variety of stimulatory and inhibitory input signals. Oxygen inactivates the factor through a rather unusual post-translational modification: protein hydroxylation at Pro and Arg residues. Stabilization of HIF caused by an inactivation of the ubiquitin ligase VHL may lead to cancer.

Further reading

Section 8.1

Gill G (2001) Regulation of the initiation of eukaryotic transcription. *Essays Biochem* 37:33–44.

Kornblihtt AR et al. (2013) Alternative splicing: a pivotal step between eukaryotic transcription and translation. *Nat Rev Mol Cell Biol* 14:153–165.

Shilatifard A, Conaway RC & Conaway JW (2003) The RNA polymerase II elongation complex. *Annu Rev Biochem* 72:693–715.

Slattery M et al. (2014) Absence of a simple code: how transcription factors read the genome. *Trends Biochem Sci* 39:381–399.

Thomas MC & Chiang CM (2006) The general transcription machinery and general cofactors. *Crit Rev Biochem Mol Biol* 41:105–178.

Section 8.2

Biggar KK & Li SSC (2015) Non-histone protein methylation as a regulator of cellular signalling and function. *Nature Rev Mol Cell Biol* 16:5–7.

Gurard-Levin ZA, Quivy JP & Almouzni G (2014) Histone chaperones: assisting histone traffic and nucleosome dxnamics. *Annu Rev Biochem* 83:487–517.

Hottiger MO (2015) Nuclear ADP-ribosylation and its role in chromatin plasticity, cell differentiation, and epigenetics. *Annu Rev Biochem* 84:227–263.

Jeltsch A & Jurkowska RZ (2014) New concepts in DNA methylation. *Trends Biochem Sci* 39:310–318.

Kadoch C & Crabtree GR (2015) Mammalian SWI/SNF chromatin remodeling complexes and cancer: mechanistic insights gained from human genomics. *Sci Adv* 1:e1500447.

Kooistra SM & Helin K (2012) Molecular mechanisms and potential functions of histone demethylases. *Nat Rev Mol Cell Biol* 13:297–314.

Kouzarides T (2007) Chromatin modifications and their function. *Cell* 128:693–705.

Li B, Carey M & Workman JL (2007) The role of chromatin during transcription. *Cell* 128:707–719.

Masliah-Planchon et al. (2015) SWI/SNF chromatin remodeling and human malignancies. *Annu Rev Pathol Mech Disease* 10:145–171.

Molecular mechanisms of histone modification function (2014). *Biochim Biophys Acta* 1839:issue 8.

Patel DJ & Wang Z (2013) Readout of epigenetic modifications. *Annu Rev Biochem* 82:81–118.

Shabazian MD & Grunstein M (2007) Functions of site-specific histone acetylation and deacetylation. *Annu Rev Biochem* 76:75–100.

Shilatifard A (2006) Chromatin modifications by methylation and ubiquitination: implications in the regulation of gene expression. *Annu Rev Biochem* 75:243–269.

Suganuma T & Workman JL (2011) Signals and combinatory functions of histone modifications. *Annu Rev Biochem* 80:473–499.

Suva ML, Riggi N, and Bernstein BE (2013) Epigenetic reprogramming in cancer. *Science* 339:1567– 1570.

Venkatesh S & Workman JL (2015) Histone exchange, chromatin structure, and the regulation of transcription. *Nat Rev Mol Cell Biol* 16:178–189.

Verdin E & Ott M (2015) 50 years of protein acetylation: from gene regulation to epigenetics, matabolism and beyond. *Nat Rev Mol Cell Biol* 16: 258–264.

Whetstine J, ed. (2014) Methylation: a multifaceted modification- looking at transcription and beyond. *Biochim Biophys Acta* 1839:1351-1352.

Section 8.3

Evans RM & Mangelsdorf DJ (2014) Nuclear receptors, RXR, and the Big Bang. *Cell* 147:255–266.

Feng S, Cao Z and Wang X (2013) Role of hydrocarbon receptor in cancer. *Biochim Biophys Acta* 1836:197–210.

Gallastegui N et al. (2015) Advances in our structural understanding of orphan nuclear receptors. *Trends Biochem Sci* 20:25–35.

Losel R & Wehling M (2003) Nongenomic actions of steroid hormones. *Nat Rev Mol Cell Biol* 4:46–56.

Prossnitz ER, Arterburn JB, Smith HO et al. (2008) Estrogen signaling through the transmembrane G protein-coupled receptor GPR30. *Annu Rev Physiol* 70:165–190.

Rastinejad F, Ollendorff V & Polikarpov I (2015) Nuclear receptor full-length architectures: confronting myth and illusion with high resolution. *Trends Biochem Sci* 40:16–24.

Stockinger B et al. (2014) The aryl hydrocarbon receptor: multi-tasking in the immune system. *Annu Rev Immunol* 32:402–432.

Section 8.4

Gao J & Xie W (2012) Targeting xenobiotic receptors PXR and CAR for metabolic diseases. *Trends Pharmacol Sci* 33:552–558.

Grygid-Gorniak B (2014) Peroxisome proliferator-activated receptors and their ligands: nutritional and clinical implications – a review. *Nutr J* 13:17–27.

Tyagi S et al. (2011) The peroxisome proliferator-activated receptor: a family of nuclear receptors, role in various diseases. *J Adv Pharm Technol Res* 2:236–240.

Section 8.5

Guy NC et al. (2015) Function of the Hsp90-binding FKBP immunophilins. *Subcell Biochem* 78:35–68.

Hanes SD (2015) Prolylisomerases in gene transcription. *Biochim Biophys Acta* 1850:2017–2034.

La Montagna R et al. (2012) Pin1 and nuclear receptors: a new language? *J Cell Physiol* 228:1799–1801.

Li J, Soroka J & Buchner J (2012) The Hsp90 chaperone machinery: informational dynamics and regulation by cochaperones. *Biochim Biophys Acta* 1823:624–635.

Litchfield DW et al. (2015) Pin1: intimate involvement with the regulatory protein kinase networks in the global phosphorylation landscape. *Biochim Biophys Acta* 1850:2077–2086.

Ratajczak T, Cluning C & Word BK (2015) Steroid-receptor-assisting immunophilins: a gateway to steroid signalling. *Clin Biochem Rev* 13:3152.

Sanchez ER (2012) Chaperoning steroidal physiology: lessons from mouse genetic models of Hsp90 and its cochaperones. *Biochim Biophys Acta* 1833:722–729.

Schmidtpeter PAM and Schmid FX (2015) Prolyl isomerization and its catalysis in protein folding and protein function. *J Mol Biol* 427:1609–1631.

Section 8.7

Eltzschig HK & Carmeliet P (2011) Hypoxia and inflammation. *N Engl J Med* 364:656–665.

Majmundar AJ, Wong WJ & Simon MC (2010) Hypoxia inducible factors and the response to hypoxic stress. *Mol Cell* 40:294–309.

Van der Heiden MG, Cantley LC & Thompson CB (2009) Understanding the Warburg effect: the metabolic requirements of cell proliferation. *Science* 324:1029–1033.

Yang Y et al. (2013) HIFs, angiogenesis, and cancer. *J Cell Biochem* 114:967–974.

Signals Controlling mRNA Translation

9

Gene expression culminates in the translation of a nucleotide sequence into the amino acid sequence of the complementary polypeptide. This translation occurs at the ribosomes and is controlled by various regulatory proteins or translation factors, which are targets of cellular signaling cascades. In other words, exogenous input signals are controlling protein biosynthesis on both the level of gene transcription and the level of mRNA translation. In fact, the regulation of translation is clearly superior as far as speed is concerned, as is required, for example, in acute stress situations. Moreover, protein biosynthesis is dependent on controlled translation in tissues whose genetic activity is diminished or switched off. A classic example that will be discussed in more detail below is hemoglobin synthesis in the mammalian reticulocyte. Finally, strict control of translation is essential since protein synthesis is by far the most expensive cellular process, consuming—depending on the conditions—up to 50% of the ATP and GTP supply. In emergency situations, protein biosynthesis, therefore, becomes drastically turned down; we have seen this already for the stringent response of bacteria (see Section 3.1).

9.1 Eukaryotic mRNA translation: the essentials[*]

Ribosomes are molecular machines specialized for protein synthesis. The eukaryotic ribosome (80S) is composed of a small (40S) and a large (60S) subunit. These subunits are huge multiprotein complexes: the small subunit is made up of an 18S ribosomal RNA and 33 different proteins, and the large subunit contains 28S, 4.8S, and 2S ribosomal RNA and 49 different proteins.

To start translation, the small subunit recruits an **initiation complex** consisting of at least 11 initiation factors (IF), with some of them being multiprotein complexes themselves (the functions of the individual factors are summarized in **Table 9.1**). This assembly occurs in a stepwise manner. In the beginning, the initiation factors eIF1A (e, eukaryotic) and eIF3 induce the dissociation of the ribosomal subunits so that mRNA can bind to the ribosomal translation apparatus. The initiation complex recognizes mRNA by a 5′-terminal 7-methylguanosine nucleotide residue called the **m⁷GpppX cap**, which binds the heterotrimeric initiation factor eIF4F. The latter is composed of the cap-binding protein eIF4E, the RNA helicase eIF4A (a DEAD-box ATPase breaking up secondary structures in the mRNA molecule that otherwise would hinder translation; see **Sidebar 9.1**), and the scaffold protein eIF4G. In cooperation with eIF3, a putative ATPase (made up of 12 subunits), this complex recruits the 40S ribosomal subunit, which has bound additional initiation factors such as eIF1, eIF1A, and eIF5 as well as a complex of initiator tRNA (in eukaryotes, methionyl-tRNA) and the GTP-loaded GTPase eIF2, a typical G-protein.

*Note that the principles of translation are summarized here only in a condensed form sufficient for an understanding of signal-controlled regulatory processes. More details can be found in textbooks of biochemistry and molecular biology.

Table 9.1 Eukaryotic translation factors

Factor	Function
eIF1	recognition of the start codon AUG
eIF1A	Met-tRNA binding to 40S ribosomal subunit
eIF2	GTPase, methionyl-tRNA binding
eIF2B	GEF of eIF2
eIF3	ATPase(?), association of 40S ribosomal subunit with mRNA
eIF4A ⎫	ATPase, helicase
eIF4E ⎬ eIF4F	binding to m⁷GpppX cap of mRNA
eIF4G ⎭	scaffold protein
eIF4B	activator of helicase eIF4A
eIF4H	activator of helicase eIF4A
eIF5	GAP of eIF2
eIF5B	GTPase, association of ribosomal subunits
eEF1A	GTPase, binding of aminoacyl-tRNA
eEF2	GTPase, translocation of peptidyl-tRNA
eEF1B	GEF of eEF1A
eRF1	recognition of stop codon
eRF3	GTPase, disintegration of translation complex

The 40S ribosomal subunit thus prepared now scans the mRNA under ATP consumption until it has reached the start codon AUG, where it associates with the large subunit. This process and the subsequent disintegration of the initiation complex are targeted (energy-consuming) events that require GTP hydrolysis catalyzed by the two G-proteins eIF2 and eIF5B. In this context, another factor, eIF5, plays the role of an eIF2 GTPase-activating protein (GAP), while eIF2B, as a GDP–GTP exchange factor (GEF), reloads eIF2 with GTP for the next initiation cycle. Therefore, eukaryotic cells are consuming two molecules of GTP per initiation step. In addition, for mRNA scanning, two molecules of ATP are hydrolyzed, probably by the ATPases eIF4A and eIF3.

Initiation does not always depend on an interaction between the 5′-cap of mRNA and eIF4F but may occur at so-called internal ribosomal entry sites (IRES). IRES allow certain viruses such as polio-, hepatitis C, and rhinovirus to produce their proteins independently of cellular protein synthesis. Moreover, such viruses completely block cap-dependent cellular mRNA translation by means of proteases degrading the translation factor eIF4G. IRES are also found in about 5% of cellular RNA, in particular in those species that encode proteins controlling mitosis and cell death. These IRES enable the cell to survive and adapt under conditions where normal

Sidebar 9.1 DEAD-box proteins These highly conserved ATPases found in pro- and eukaryotes got their name from a characteristic D-E-A-D (Asp-Glu-Ala-Asp) sequence in their catalytic domain. Most of them are RNA helicases involved in pre-mRNA splicing, maturation of ribosomal RNA, removal of loops and helices that are hindering mRNA translation, etc. Moreover, the enzymes of a subfamily called SNF2p catalyze the restructuring of nucleosomes in the course of gene transcription (for details see Section 8.2.7). The energy needed for such structural changes is derived from ATP hydrolysis. One should not confuse the DEAD box with the death domain of apoptosis-inducing proteins!

translation is suppressed, particularly under stress such as viral infection. A similar emergency reaction enabling a preferred translation of certain mRNA species in spite of an overall inhibition of protein biosynthesis will be discussed in Section 9.3.2. IRES are controlled by proteins called IRES-transacting factors, which are probably chaperones stabilizing IRES-typical three-dimensional structure of mRNA. The input signals controlling these factors are not yet known.

Like initiation complex formation, peptide chain elongation also requires GTP hydrolysis: elongation factor eEF1A, a G-protein, transfers the aminoacyl-tRNA to the A-site of the ribosome, while another GTPase (eEF2) promotes the translocation of the peptidyl-tRNA. A third elongation factor (eEF1B) plays the role of an eEF1A-GEF. Finally, the termination factor eRF3 (cooperating with another factor eRF1) is a GTPase, too. Thermodynamically the energy obtained from GTP hydrolysis is needed because peptide chain elongation generates a state of higher order or decreased entropy.

Summary

Eukaryotic mRNA translation at ribosomes occurs in initiation, elongation, and termination phases. These phases are controlled by a series of translation factors, some of which are regulatory GTPases. Like transcription factors, the translation factors are targets of signaling pathways.

9.2 Protein release by the endoplasmic reticulum is controlled by G-proteins

Ribosomes are associated with the membrane system of the endoplasmic reticulum (ER). This characteristic structure of eukaryotic cells is connected with the extracellular space; thus its lumen represents external conditions inside the cell. Therefore, the ER membranes are the site where the release of secretory and transmembrane proteins starts. Such proteins have been estimated to amount to about a quarter of all proteins produced by a cell. They are directly synthesized into the ER lumen by the ribosomes attached, and they are subsequently transported to the Golgi apparatus where they become packed into secretory vesicles (for details see Section 10.4.1).

To penetrate the ER membrane, the growing polypeptide chain is equipped with an amino (N)-terminal lipophilic amino acid sequence called signal peptide. This sequence binds a **signal recognition particle (SRP)**, which interacts with an **SRP receptor** at the ER membrane (**Figure 9.1** and **Sidebar 9.2**). The receptor forms a complex with a membrane pore, the **translocon**, through which the growing polypeptide chain becomes threaded. At the same time, a signal peptidase associated with the translocon is removing the signal peptide and further post-translational modifications such as N-glycosylation occur, preventing a re-transport of the protein.

Transportation of the translating ribosome to the translocon and the linearization and subsequent threading through the pore of the polypeptide chain are targeted movements and forced conformational changes requiring energy. This energy is supplied by the hydrolysis of two molecules of GTP. For this purpose, SRP and its receptor have G-protein domains; that is, they are regulatory GTPases. The translating ribosome functions as an SRP-GEF, and the free translocon as a GEF of the SRP receptor: both activate their effector G-proteins by catalyzing GDP–GTP exchange. Therefore, the system becomes active only in the presence of an mRNA-loaded ribosome and a free translocon. The receptor–GTP complex facilitates GTP hydrolysis at the SRP molecule: it plays the role of an SRP-GAP that induces the dissociation of the ribosome–SRP complex when the polypeptide chain has entered the translocon pore. Vice versa, SRP-GTP is a GAP of the SRP receptor. Thus SRP and its receptor are interconnected in a reciprocal regulatory circuit (see Figure 9.1).

Figure 9.1 Protein biosynthesis at the endoplasmic reticulum
(1) The ribosome carrying the growing polypeptide chain (dotted line; the broken arrow symbolizes the signal peptide) activates the G-protein Signal Recognition Particle (SRP) by GDP–GTP exchange (GEF effect). (2) GTP-loaded SRP binds to the signal peptide and recruits the translating ribosome at the GTP-loaded Signal Recognition Particle Receptor (SRPR) and the adjacent translocon in the endoplasmic reticulum (ER) membrane. (3) By means of the signal peptide, the growing polypeptide chain is threaded through the pore of the translocon. At the same time, SRP and SRPR stimulate their GTPase activities in a reciprocal manner (GAP effect). (4) While polypeptide synthesis is progressing, the signal peptidase of the translocon removes the signal peptide and the SRP–SRPR complex disintegrates. Upon termination of translation, the naked translocon reactivates SRPR by GDP–GTP exchange (GEF effect; not shown).

Summary

- The release of newly synthesized protein into the lumen of the ER is a targeted process controlled and promoted by two regulatory GTPases: signal recognition particle (SRP) and SRP receptor, a transmembrane protein associated with a peptide channel (translocon) in the ER membrane.

- At the SRP, GDP–GTP exchange is catalyzed by the ribosome; at the SRP receptor, it is catalyzed by the translocon.

- Homologous GTPases and translocons are also found in prokaryotes where they control peptide transport across the cell membrane. Prokaryotic GTPases are the functional predecessors of eukaryotic G-proteins.

Sidebar 9.2 Signal recognition particles and their receptors Eukaryotic signal recognition particles are ribonucleoprotein complexes consisting of six different protein subunits (according to their molecular weights termed SRP9, -14, -19, -54, -68, and -72) and a 7S RNA forming some sort of a backbone. The subunit SRP54 harbors the GTPase domain of the SRP and contains an unusually high number of methionine residues (so-called M-domain) for the hydrophobic interaction with the signal peptide, as well as a nucleic acid-binding helix–turn–helix motif for the interaction with 7S RNA. The eukaryotic SRP receptor is made up of two subunits, SRPRα and SRPRβ, both exhibiting GTPase domains. SRPRβ also exhibits a transmembrane domain. The basic structure of the translocon is a heterotrimer of the transmembrane proteins Sec61p (with 10 transmembrane domains) and Sec61α as well as Sec61β (with one transmembrane domain each). These proteins assemble forming a membrane pore of 5 nm diameter which is opened by the SRP–receptor complex. The translocon is a component of a large complex with numerous other proteins such as a signal peptidase, an oligosaccharyltransferase, and the chaperone BiP (see Section 9.3.2).

Proteins homologous to SRP, SRP receptor, and the Sec61 complex are found in all pro- and eukaryotes examined thus far. However, the translational machinery of bacteria is considerably simpler than that of eukaryotes. The SRP of *Escherichia coli* consists only of the SRP54-homologous GTPase Ffh (Fifty-four homologue) and a 4.5S RNA, and the SRP receptor of the SRPα-homologous GTPase FtsY. The latter is associated with the inner side of the plasma membrane (prokaryotes lack an endoplasmic reticulum). The basic structure of the prokaryotic translocon widely resembles that of its eukaryotic counterpart.

Figure 9.2 Major signaling pathways controlling translation: an overview Both peptide chain initiation and elongation are stimulated by mitogens (and insulin) and partially inhibited by stress factors and during memory fixation. Shown are the translation factors known to be targeted by such signals and the signaling pathways involved. In addition, other translation factors have been found to be substrates of protein kinases, albeit with less conclusive evidence as far as the physiological role of phosphorylation is concerned. PI3K, phosphatidylinositol 3-kinase; PKB, protein kinase B; GSK3, glycogen synthase kinase 3; eIF, eukaryotic initiation factor; GEF, GDP–GTP exchange factor; MAPK, mitogen-activated protein kinase; MNK, an effector kinase of the mitogenic MAP kinase module; mTOR, protein kinase mammalian target of rapamycin; 4E-BP, eIF4E binding protein; eEF, eukaryotic elongation factor; AMPK, 5'-AMP-dependent protein kinase.

9.3 Signaling cascades controlling translation

Like gene transcription, mRNA translation critically depends on the environmental conditions. As a general rule, translation is up-regulated by survival signals such as growth factors and hormones stimulating cell proliferation and cell function and turned down by stress factors. These signals affect both the initiation and the elongation phase. Major target proteins are the cap-binding complex and the GTPases eIF2 and eEF2. An overview is provided by **Figure 9.2**.

9.3.1 eIF2 kinases are stress-sensitive repressors of protein biosynthesis

To save energy, protein biosynthesis is suppressed in stress situations. This takes place at an early stage of translation, when the initiator tRNA loads the initiation complex, a step catalyzed by the initiation factor eIF2. Under stress (and in neurons also during learning), this GTPase is rendered inactive through phosphorylation of a strategic serine residue catalyzed by an eIF2 kinase. As a result, the interaction of eIF2 with its GEF eIF2B is interrupted and translation is blocked at the initiation stage. This effect, occurring in a few minutes, is by far the most rapid cellular stress response.

In mammals, eIF2 kinase is expressed in four isoforms that differ in cellular functions and mechanisms of regulation (**Figure 9.3**). The classical isoform is the **heme-regulated inhibitor HRI**. Its major task is to adjust protein synthesis, particularly of hemoglobin, to requirements such as the availability of heme, iron, and energy.

Figure 9.3 eIF2 kinases as stress-activated translational repressors The four kinases Heme-Regulated Inhibitor (HRI), Protein Kinase R (PKR), homologue of *Drosophila* protein General Control Non-derepressible 2 (GCN2), and Protein kinase R-related Endoplasmic Reticulum Kinase (PERK) respond to different stress situations and inhibit protein biosynthesis by phosphorylating a specific Ser residue of eukaryotic initiation factor 2 (eIF2).

Hemoglobin synthesis is regulated at the level of mRNA translation—because, at least in mammals, the genome of the mature reticulocyte is silent. HRI, expressed primarily in reticulocytes, binds heme via a specific domain and thus becomes inhibited. As a consequence, eIF2 phosphorylation is suppressed and the rate of translation remains high as long as heme is available. Vice versa, heme deficiency leads to a deblockade of HRI. This holds true also for other stress situations because HRI adjusts translation not only to the availability of heme and iron but also to the physiological situation in general. Such an adaptive response provides an economical advantage and, in addition, protects the chaperone system from an overload that would result in intoxication of the cell by incorrectly folded proteins (see Section 9.3.2).

Another eIF2 kinase isoform is **protein kinase R, PKR** (do not confuse with PKC-related kinase, PRK). It is closely related to HRI, but instead of a heme-binding motif it has an interaction domain for double-stranded RNA (therefore PK<u>R</u>) which activates the enzyme. Double-stranded RNA arises primarily in the course of viral infections. Therefore, PKR is a component of cellular defense: by turning down translation, it suppresses the synthesis of viral proteins and spares the energy supply of the infected cell. Appropriately, the de novo synthesis of PKR is induced by antiviral interferons. Viruses have developed various mechanisms to outmaneuver PKR: for instance, by inhibitor proteins, by double-stranded RNA that is not recognized by the kinase, or by phosphatases that counteract the phosphorylation of eIF2.

The third eIF2 kinase isoform is a dimeric transmembrane protein of the ER called **GCN2** (referring to the homologous *Drosophila* protein General Control Nonderepressible 2). GCN2 is activated by unloaded tRNA and phosphorylates eIF2 at the same Ser residue as HRI and PKR. Since unloaded tRNA signals amino acid shortage, GCN2 helps the cell to adjust the rate of translation to the availability of amino acids (another mechanism aiming at this purpose is discussed in Section 9.4.1). At first glance, this adaptive reaction to starvation seems to resemble the stringent response of bacteria as explained in Section 3.1. It follows, however, another mechanistic route. Amazingly, GCN2 inhibits overall translation but, nevertheless, stimulates the translation of mRNA species encoding those transcription factors that activate the genes of amino acid biosynthesis and of the unfolded-protein response. The mechanism of this apparently paradoxical but physiologically meaningful effect is explained in Section 9.3.2. It seems to play an important role in memory fixation in the brain, where the phosphorylation of eIF2α serves as a molecular switch between short- and long-term memory (see Section 16.4.1).

The fourth eIF2 kinase isoform is **PERK (<u>P</u>KR-related <u>ER</u> <u>K</u>inase)**. Like GCN2, PERK is a transmembrane protein of the ER. The enzyme plays an important role in the so-called unfolded-protein response to be discussed in Section 9.3.2.

The stress-induced inactivation of eIF2 is augmented by an inhibition of eIF2B (the eIF2-GEF) that also is due to phosphorylation. The corresponding kinase is glycogen synthase kinase 3 (GSK3), which becomes activated under stress but inhibited by insulin (see Section 9.4.4). In fact, eIF2B phosphorylation is involved in a typical anabolic insulin response, the increase of protein biosynthesis along the phosphatidylinositol 3-kinase–protein kinase B signaling pathway. Insulin abolishes the effect of GSK3, thus facilitating the dephosphorylation (by omnipresent phosphatases) and reactivation of eIF2B.

9.3.2 Unfolded-protein response or how the endoplasmic reticulum copes with stress

In the endoplasmatic reticulum (ER), a group of chaperones takes care of the correct folding of proteins, since un- or misfolded proteins are not only useless but are toxic and possibly may kill the cell by evoking apoptosis (so-called proteotoxicity).

The accumulation of such defective proteins in the ER may result in what is known as **ER stress**: an overload of the chaperone system. Such a situation may have several causes: overshooting protein synthesis, production of mutated proteins, viral infections, disturbed redox state (for instance, due to a defective blood supply or ischemia), disturbed Ca^{2+} homeostasis (the ER is a major Ca^{2+} store of the cell; see Section 14.5), and toxic substances inhibiting chaperone activity or blocking glycosylation of newly synthesized proteins. When confronted by ER stress, the cell is not helpless but makes use of a set of efficient repair and rescue mechanisms known as unfolded-protein response, which together with DNA repair represents the hard core of the cellular stress defense. The ultimate goal of the unfolded-protein response is to increase the protein folding capacity of the ER by promoting the synthesis of chaperones.

The unfolded-protein response involves the following sequence of events:

- recognition of the defective protein by a chaperone
- stop of counterproductive de novo synthesis of housekeeping proteins
- transcriptional activation of stress-responsive genes (for instance, those encoding chaperones)
- elimination of defective proteins along the ubiquitin–proteasome pathway or, if this turns out to be impossible, apoptosis

ER stress sensor BiP

The major stress sensor of the ER is the **chaperone BiP** (Binding Protein). BiP is an ATPase of the HSP70 type, which, like other members of this family, interacts with an HSP90-type chaperone and various co-chaperones to bring its misfolded client proteins into the native conformation (for a more detailed treatment of chaperones see Section 2.5.1). The energy required for this process comes from ATP hydrolysis.

Under ER stress due to an overloading by client proteins, the stock of BiP is used up, resulting in a shortage of free BiP. This situation calls upon the cell to activate signaling cascades, which turn down translation and thus stop a continuous supply of new proteins to the chaperone system. Simultaneously, a large number of anti-stress genes become activated, which encode (among others) chaperones. These genes carry regulatory elements responding to unfolded-protein response and are recognized by three transcription factors called ATF4, ATF6, and XBP1.

The signaling cascades controlling the activity or level of these transcription factors start from three transmembrane proteins of the ER: the eIF2 kinase **PERK** mentioned above, the serine/threonine-specific protein kinase **IRE1 (inositol-requiring enzyme 1)**, and the precursor protein of the **transcription factor ATF6**. These proteins are "BiP receptors." They are inactivated as long as the level of free BiP is high but become activated when the free BiP level drops. (**Figure 9.4**).

Effects on gene transcription and mRNA translation

When the kinases PERK and IRE1 are cleared from BiP they undergo oligomerization, facilitating activation through trans-autophosphorylation of their activation loops. The phosphorylation and inactivation by PERK of eIF2 has already been mentioned above. It is the most rapid response to ER stress. IRE1 evokes a more delayed effect. Here trans-autophosphorylation results in a stimulation of an intrinsic endoribonuclease activity of IRE1. This activity is used to release the mRNA of the **transcription factor XBP1** from a precursor mRNA. For this purpose, IRE1 cuts out an intron sequence of 26 nucleotides. After ligation of the remaining fragments, a reading frame for XBP1 synthesis arises. This signaling pathway is conserved in all eukaryotes and has been investigated in detail in yeast. Mammals express two IRE isoforms. The important role played by IRE1 is mirrored by gene knockout experiments: mice lacking an active *IRE1α* gene die before birth.

Figure 9.4 Signal transduction of the unfolded-protein response Incorrectly folded proteins accumulating in the lumen of the endoplasmic reticulum sequester the chaperone BiP, thus rescuing the transmembrane protein kinases PERK and IRE1, and the ATF6 precursor from inhibition by BiP. This results in an activation of the transcription factors ATF4, XBP1 and ATF6 controlling the transcription of unfolded-protein response (UPR) genes, as well as in an inhibition of eIF2; that is, in an interruption of protein biosynthesis. The ATF6 precursor translocates from the endoplasmic reticulum to the Golgi apparatus, where it undergoes maturation by limited proteolysis catalyzed by proteases S1P and S2P. Dark gray cylinders, transmembrane domains; light gray cylinders, kinase domains; teal cylinders, endoribonuclease domains; ellipses, active ATF6.

In its function, the **transcription factor ATF6** resembles XBP1, though it is activated more rapidly along quite a different pathway: by release from a precursor protein localized in ER membranes. Under ER stress the ATF6 precursor then migrates from the ER to the Golgi apparatus, where it is cut by two proteases in the luminal domain and the transmembrane domain. This leads to a release of active ATF6 into the cytoplasm, where it migrates into the nucleus and activates a series of genes, including that of XBP1. Thus the activation of ATF6 provides an example of a signal-transducing reaction involving partial proteolysis (more in Section 13.1).

The protein kinase PERK stimulates—in addition to the inhibition of eIF2—a de novo synthesis of the **transcription factor ATF4**, which controls the activity of a third of all genes (including those of XBP1 and ATF6) that are induced by the unfold-ed-protein response. This selective translation is due to a special structure of ATF4 mRNA consisting of several upstream open reading frames in the 5'-region that are not in frame with the ATF4 sequence, the translation of which is thus prevented. This inhibitory effect is relieved by phosphorylation of eIF2, which retards the forma-tion of a ribosomal initiation complex. Consequently, the scanning ribosome skips the open reading frames, becoming active only when it has arrived at the ATF4 se-quence. In addition to its function in the unfolded-protein response, the eIF2–ATF4 system plays an important role in brain neurons, where it regulates memory fixation (see Section 16.4.1).

The eIF2 kinase GCN2 activated upon amino acid deficiency resembles PERK in activating ATF4 translation though inhibiting overall translation. Moreover, at least in insects, GCN2 also stimulates the translation of mRNAs encoding transcription factors that regulate genes of amino acid synthesis. This combined action of the two eIF2 kinases, PERK and GCN2, is known as the **integrated stress response**.

PERK not only is an emergency enzyme but also controls the equilibrium between the rate of translation and the efficacy of chaperones under normal conditions. As an example, the β-cells of the Langerhans islets of the pancreas may be mentioned, where PERK signaling harmonizes the glucose-induced synthesis of pro-insulin with protein folding in the ER. In fact, PERK mutations are one of the various causes of diabetes.

Adaptation and apoptosis

The unfolded-protein response is limited in time because the cell adapts to the stress situation. This may occur, for instance, by providing more chaperones or by the destruction of the misfolded proteins along the ubiquitin–proteasome pathway. For this waste disposal, known as **ER-associated degradation**, the proteins have to be re-translocated through the translocons from the ER into the cytoplasm where the proteasomes are located. Whether or not a protein is recognized by the system seems to depend on its glycosylation status.

Another possibility for cell adaptation is by signal extinction. In fact, one of the un-folded-protein response genes encodes a protein, GADD34, which is a regulatory subunit of the protein phosphatase PP1C directing the enzyme to the phosphorylat-ed eIF2. The degradation of the extremely protease-sensitive transcription factors of the unfolded-protein response also aims at signal extinction.

Like genotoxic stress, ER stress may finally culminate in programmed cell death (apoptosis). To this end, several signaling pathways connect the unfolded-protein response with the apoptotic machinery, which in fact is partly localized in the ER. The ER-bound caspase 12 is activated by IRE1, triggering the caspase cascade (see Section 13.2.1). Moreover, one of the unfolded-protein response-inducible genes produces a protein that strongly stimulates apoptosis. Finally, by interacting with the adaptor protein TRAF2, IRE1 may activate the pro-apoptotic protein kinase ASK1. Being a MAP3 kinase, ASK1 is a component of the JNK–MAP kinase module, which typically becomes activated in stress situations and induces apoptosis (see Section 11.1). Apoptosis is also caused by an overload of the ER with *correctly* folded proteins: for instance, upon viral infection. In this case, however, the NFκB signaling cascade rather than the unfolded-protein response is activated (see Section 11.8).

Cytosolic protein quality control

Protein quality systems such as those evoking the unfolded-protein response are not restricted to the ER but also are found in the cytosol, nucleus, and mitochondria. They are hardworking devices, since it is estimated that about a third of all new-ly synthesized proteins are defective and, thus, have to be eliminated. Such defects are due to unavoidable errors of translation as well as all sorts of stress, first and foremost reactive oxygen species, which are a major risk factor of aerobic life. The cytosolic sensors of damaged or incompletely folded proteins are again BiP-relat-ed chaperones of the HSP70 and HSP90 families. Under stress-free conditions they sequester the transcription factor **HSF1** (Heat-Shock Factor 1), keeping it inactive. Defective proteins intercept the chaperones, and HSF1 can now enter the nucleus to activate (similar to ATF4, ATF6, and XBP1) the transcription of stress-responsive genes.

How does such a system distinguish defective from nascent polypeptide chains? The answer to this question is found probably in a set of ubiquitin ligases (such as the Carboxyterminus of HSP70 Interacting Protein, CHIP), which selectively interact with chaperones bound to misfolded proteins. When the stress has ceased, the same ubiquitin ligases also trigger the destruction of the chaperones, thus reestablishing the normal situation.

9.3.3 Translation factors eIF4E and eIF2B are targets of mitogens and insulin

In addition to eIF2, several other factors controlling the initiation of polypeptide synthesis are targets of regulatory signals transduced by protein phosphorylation. As mentioned in Section 9.3.1 **eIF2B**, the GEF of the GTPase eIF2, is inactivated by glycogen synthase kinase 3. This inhibitory effect augments the suppression of translation by eIF2 phosphorylation. It is overcome by mitogens and insulin, which inhibit GSK3 along the PI3K–PKB pathway (see Section 4.6.6).

eIF4E is the subunit of the cap-binding complex that directly interacts with the m^{7}GpppX sequence of mRNA. Stimulatory hormones and growth factors induce a phosphorylation of the factor, causing the detachment of the cap-binding complex from mRNA and an increase of translational efficiency (Figure 9.2). While the mechanism of this effect is still not clear, the protein kinase catalyzing it has been identified as MNK1, an effector kinase of the mitogenic MAP kinase module (see Section 11.6.2). Genes encoding key proteins of cell cycle progression, such as cyclin D, are activated along this pathway. Through eIF4E phosphorylation, mitogenic signals turn up this effect by stimulating the translation of the corresponding mRNA species (see Section 11.5). Because of these pro-mitogenic effects, eIF4E may promote tumor development. In fact, human tumors frequently show high levels of eIF4E, and in cell culture and transgenic animals, overexpression of this factor facilitates neoplastic transformation and tumorigenesis.

An opponent of eIF4E is the **translational repressor 4E-BP** (elF4E-Binding Protein), which exists in several tissue-specific isoforms. 4E-BP sequesters eIF4E, thus preventing formation of the initiation complex. Exogenous signals that induce eIF4E phosphorylation also inactivate 4E-BP through phosphorylation, which in this case is catalyzed by the protein kinase mTOR. As shown in Section 9.4.1, this mechanism plays a particularly important role in the control of protein synthesis.

9.3.4 eEF2 kinase is a repressor of peptide chain elongation

Apart from the initiation phase, peptide chain elongation is a target of signaling reactions. In this context, the eEF2 kinase plays a key role. This Ser/Thr-specific enzyme, found in all eukaryotic cells, is activated by Ca^{2+} ions interacting with the multifunctional Ca^{2+}-binding protein calmodulin (CaM) that is a subunit of eEF2 kinase. In the literature, eEF2 kinase may also be found, therefore, under the name CaM kinase III, although it does not relate to the CaM kinases proper but belongs to the atypical protein kinases (see Section 2.6.3 for a classification of protein kinases). eEF2 kinase is as specific as eIF2 kinase: the only substrate known is the elongation factor eEF2 which, upon phosphorylation of a Thr residue, loses contact with the ribosome, resulting in premature termination of peptide chain elongation.

The activity of eEF2 kinase is controlled by phosphorylation according to the principle of two reins: depending on the kinase type and the site of phosphorylation, it is either inhibited or activated. Signals driving up the rate of translation *suppress* eEF2 kinase activity along a kinase cascade. Such signals include growth factors as well as insulin and other hormones. The kinase cascade proceeds mainly along the axis PI3K–PKB/Akt–mTOR (see Figure 9.2) and is discussed in more detail in Section 9.4.1. Other kinases inhibiting eEF2 kinase are the ribosomal S6 kinases as well as some MAP kinases.

An *activating* phosphorylation of eEF2 kinase is catalyzed by the 5'-AMP-dependent protein kinase AMPK (see Figure 9.2 and Section 9.4.3). By this mechanism, protein biosynthesis, as a major ATP-consuming process, is suppressed in situations of ATP deficiency such as starvation and hypoxia.

An inhibition of peptide chain elongation also occurs upon phosphorylation of another elongation factor, the GEF eEF1B, catalyzed by the cyclin-dependent kinase CDK1. This reaction has been connected with the suppression of protein synthesis observed at the beginning of mitosis, a phase of the cell cycle that is controlled by CDK1 (see Chapter 12).

Summary

- Protein biosynthesis, one of the most energy-consuming cellular processes, is under tight control by signaling cascades that adjust mRNA translation to cell growth and proliferation and prevent wasted energy in emergency situations.

- Ser/Thr-directed protein kinases HRI, PKR, GCN2, and PERK repress mRNA translation by phosphorylating and inactivating eukaryotic initiation factor 2 (eIF2). These highly specific enzymes are activated in stress situations, such as iron and amino acid deficiency, viral infection, and accumulation of incorrectly folded proteins.

- To protect the cell from damage by proteotoxicity, an ER-based signaling mechanism monitors any accumulation of incorrectly folded proteins. The sensing mechanism consists of sequestering the chaperone BiP resulting in an activation of the protein kinases PERK and IRE1 as well as of the transcription factor ATF6 precursor protein in the ER membrane.

  Consequently, mRNA translation becomes inhibited by the eIF2 kinase PERK and a series of genes are activated that encode (among others) chaperones.

- In the cytosol, defective proteins activate heat-shock transcription factor 1 (HSF1), which also stimulates stress-responsive genes. If these countermeasures turn out to be in vain, apoptosis is induced.

- Mitogens and insulin stimulate initiation factors eIF2B and eIF4E, thus adjusting the rate of protein synthesis to the rate of cell proliferation and cell growth. To this end protein kinase B inactivates an eIF2B-inhibitor (glycogen synthase kinase 3) while eIF4E becomes phosphorylated directly through MAP kinase signaling, in parallel with inactivation of the eIF4E repressor 4E-binding protein (4E-BP).

- The highly specific eukaryotic elongation factor 2 (eEF2) kinase is a repressor of peptide elongation. The enzyme is itself a target of protein phosphorylation, adjusting its activity to the physiological context. Depending on the phosphorylation site, it is either inhibited (by mitogenic signals and insulin) or activated (under stress situations such as starvation).

9.4 Network for adjustment of cell growth to the supply situation

Signaling systems controlling ribosomal protein synthesis fulfill an important task: they adjust the rate of translation and thus the state of vitality and the growth rate of cells to the food and energy supply. This system responds with economic measures to stress factors overstraining those parameters. For this purpose, it makes use primarily of protein phosphorylation, which is catalyzed by kinases serving as sensors of energy and nutrients. Disturbances of this network rank among major causes of metabolic diseases, particularly obesity and diabetes.

9.4.1 TOR kinases are sensors of food supply and control protein biosynthesis

Mammalian target of rapamycin

A key signaling mechanism controlling protein synthesis in response to the state of nutrition has been discovered during studies both on yeast cells and on a particular neoplastic disease.

Hamartomas are benign tumors, which may arise in various tissues. The tumor cells are unusually large and obviously overfed. Hamartomas are induced by the deletion of either two tumor suppressor genes encoding the so-called tuberous sclerosis proteins TSC1 (alias hamartin) and TSC2 (alias tuberin). Both together inhibit a signaling cascade, which adjusts the rates of cell growth and cell proliferation to food and energy supply. In this signaling pathway, a central role is played by the Ser/Thr-specific protein kinase mTOR, an enzyme conserved from yeast to mammals in all eukaryotes examined. mTOR stands for <u>m</u>ammalian (or <u>m</u>echanistic) <u>T</u>arget

Figure 9.5 Activation and effects of the mTOR protein kinase complex
1 By inactivating the GAP TSC2 of the small G-protein Rheb, extracellular signals stimulating the PI3K–PKB signaling cascade prompt Rheb-GTP to activate mTORC1. mTORC1 is also activated by amino acids along a pathway that includes the small G-protein Rag. Rag facilitates the interaction between mTORC1 and Rheb at lysosomal membranes. Also shown is the *activation* of the Rheb-GAP TSC2 by 5′-AMP-dependent protein kinase (AMPK) and by hypoxia. Both result in an inhibition of mTOR signaling and protein synthesis and protects the cell in situations of energy deficiency (see Section 9.4.3). mTORC1 enhances the activity of the protein kinase S6K and represses 4E-BP1 and eEF2 kinase activities, resulting in an increased rate of translation. A stimulatory effect resembling that of PKB has the MAP kinase ERK connecting mTOR signaling with mitogenesis (not shown). The teal dotted line shows the negative feedback of insulin signaling: S6K phosphorylates and inactivates the insulin-specific docking protein IRS. This effect is augmented by overnutrition (leading to increased insulin release) and provides one of the causes of diabetes.

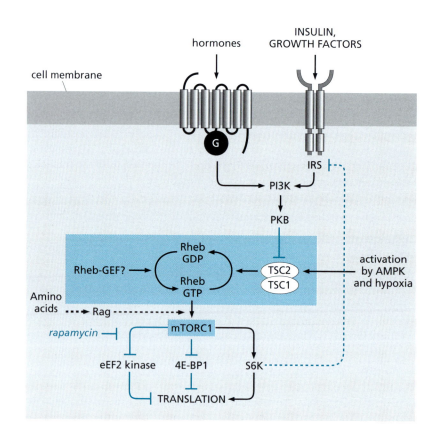

Of Rapamycin. Rapamycin is an immunosuppressive antibiotic from *Streptomyces hygroscopicus* that in cells interacts with the immunophilin FKBP12, a peptidylprolyl *cis–trans* isomerase (see Section 8.5). The complex of rapamycin and FKBP12 is a highly specific inhibitor of mTOR. It reduces both cell size and cell proliferation, mainly by inhibiting ribosomal protein synthesis. One of the reasons for the immunosuppressive effect of rapamycin is seen here: the drug suppresses the activation and proliferation of T-lymphocytes.

Cellular effects of mTOR

The protein kinase mTOR is a key regulator of cell growth. The enzyme (higher vertebrates contain only one *mTor* gene) belongs to the atypical PI3-kinase-like protein kinases. In contrast to PI3-kinase, it does not exhibit lipid kinase activity but is a pure protein kinase. Nevertheless, mTOR connects PI3-kinase signaling with the regulation of protein synthesis (**Figure 9.5**).

mTOR stimulates mRNA translation. Major substrates of the kinase include the ribosomal S6 kinase type p70, 4EBP1, and (indirectly) eEF2 kinase (Figure 9.2). As shown in Section 9.4.2, S6 kinase promotes mRNA translation and cell survival and is activated by mTOR, whereas the translational repressors 4E-BP1 and eEF2 kinase become inactivated.

mTOR also promotes the formation of ribosomes. An important condition of cell growth and proliferation is to adjust the number of ribosomes to the actual physiological requirements. In rapidly growing and proliferating cells, for instance, up to 8000 new ribosomes are produced per minute, whereas in resting or stressed cells, ribosome formation almost comes to a standstill. Ribosome synthesis requires the translation of a special family of mRNAs encoding all ribosomal proteins and several translation factors. The peculiarity of these mRNA species is that they carry a 5′-TOP (Tract Of Pyrimidines), an oligopyrimidine sequence adjacent to the 5′-terminal m^{7}GpppX-cap. When cells are deprived of energy and amino acids, as well as in the absence of mitogenic and growth-promoting signals (or when treated with inhibitory drugs), polysomes (mRNA occupied by actively translating ribosomes) with 5'-TOP-mRNA disintegrate. As a result, formation of ribosomes is blocked and overall protein synthesis comes to a halt, a reaction strongly reminiscent of the stringent response of bacteria (see Section 3.1). Certainly, the mTOR signaling pathway is critically involved in the control of 5'-TOP-mRNA translation. However, the precise mechanism is still a matter of conjecture.

In cells mTOR is integrated in two multiprotein complexes, called **mTORC1** and **mTORC2**. mTORC1 is much more sensitive to rapamycin than mTORC2. The classical functions of mTOR, that is, control of protein synthesis and cell growth, are attributed to mTORC1 while mTORC2 controls the dynamics of the cytoskeleton and, thus, cell shape and cell movements. The mechanisms behind this effect are still not fully understood.

mTORC1 signal transduction

Through its effects on the ribosomal machinery, mTORC1 considerably increases the translational capacity of the cell, provided that the conditions are favorable. To this end, the complex monitors the availability of nutrients (in particular amino acids) and metabolic energy (that is, ATP). The energy status of the cell is measured in two ways. First, since its Michaelis constant (K_m) value for ATP is 5–10 times higher than that of other protein kinases, mTOR becomes active only at a particularly high concentration of intracellular ATP. Second, mTORC1 activity is suppressed by the protein kinase AMPK. AMPK is a cellular energy sensor that becomes active at a low ATP:AMP ratio indicating energy shortage (see Section 9.3.4). Of course, monitoring both nutrient availability and energy status makes sense, since without a sufficient energy supply any activation of protein biosynthesis would be idling (the reader may recall that about 80% of cellular energy is used for cell growth rather than cell proliferation, being consumed primarily by protein synthesis). Thus, by simultaneously responding to two parameters, that is amino acid and ATP concentration, mTORC1 resembles a logical AND gate.

In the signaling cascades leading to mTORC1 activation the Ras-related small G-protein **Rheb** (Ras homology enriched in brain) plays a central role: when loaded with GTP it stimulates mTORC1. Therefore, signals enhancing mTORC1 activity (such as hormones, growth factors, and amino acids) promote the formation of Rheb-GTP either by activating a still ill-defined Rheb-GEF or by inhibiting a Rheb-GAP. Inhibitory signals (such as starvation) have an opposing effect. At this stage, one of the TSC tumor suppressor proteins, **TSC2**, enters the scene: it acts as a Rheb-GAP and has turned out to be a key switch in mTOR signaling (see Figure 9.5).

The function of TSC2 is regulated by the redox status of the cell and by protein phosphorylation. As explained in Section 8.7, oxygen shortage activates the the transcription factor **HIF1α**. HIF1α promotes the de novo synthesis of a protein (REDD1) which prevents TSC2 from sequestration by 14-3-3 proteins (see Section 10.1.4). This effect is augmented by the protein kinase **AMPK**. Being a sensor of ATP shortage AMPK *enhances* the GAP activity of TSC2 by phosphorylating specific Ser and Thr residues. As a result of both effects mTORC1 remains inactive until the oxygen level increases and ATP is produced. When phosphorylated at other sites TSC2 may also be inhibited. Such phosphorylations which result in an activation of mTORC1/2 are catalyzed by the protein kinases **PKB/Akt** (see Section 4.6.6) and **ERK**, the executor kinase of the mitogenic Ras–RAF–MAP kinase module. These are the major pathways along which insulin, growth factors, and other anabolic, growth-promoting, and mitogenic stimuli promote mTORC1 activity and protein synthesis provided that sufficient oxygen, ATP, and amino acids are available. How do amino acids control mTORC1 activity? Here an important aspect of signal processing has to be taken into consideration: intracellular localization. In fact, Rheb is bound to intracellular membranes, in particular those of lysosomes, whereas mTORC1 is dispersed throughout the cytoplasm. To become active it has to assemble with Rheb. For the translocation of mTORC1 to the lysosomal surface, **Rag**, another small G-protein of the Ras superfamily, is required. Rag is activated by a Rag-GEF called "ragulator", which also resides on the lysosome. Ragulator appears to be a sensor of amino acids that accumulate in lysosomes, for instance in the course of autophagy (see Section 9.4.3).

Another putative amino acid sensor of the mTOR system is a class III phosphatidylinositol 3-kinase (known as VPS34). Its mechanism of action is not yet fully understood.

mTOR and disease

In summary, this positive and negative regulation of the mTOR signaling system enables the cell to adjust the effects of anabolic and mitogenic stimuli to the capacity of protein biosynthesis determined by the availability of food (amino acids, sugars) and energy. It is easily understood, therefore, that mTOR is critically involved in the control of anabolic hormone effects and in metabolic diseases. In fact, the function of **insulin** is controlled by negative feedback along the mTOR signaling pathway. To this end, S6 kinase is activated by mTOR inactivates the insulin-specific docking protein IRS1 at several Ser residues, thus blocking signal transduction from the insulin receptor to the PI3K–PKB/Akt pathway (Figure 9.5; more details about IRS1 are found in Section 7.1.3). With an increasing supply of fat and carbohydrates and, as a consequence, a rising insulin level, this inhibitory effect becomes more pronounced, being further reinforced by an adaptive down-modulation of insulin receptors. This may explain why overnutrition is one of the major causes of type 2 **diabetes.** Another cause, lack of physical activity, will be discussed in Section 9.4.3.

As already mentioned, mTOR is inhibited by clinical immunosuppressants such as rapamycin. Moreover, its role as a promoter of diabetes, tumorigenesis, and other growth-related diseases (such as cardiac hypertrophy) makes the mTOR system an attractive target for further inhibitory drugs that are in ongoing clinical trials.

9.4.2 Ribosomal S6 kinases as regulators of cell size and proliferation

In the preceding section, a ribosomal S6 kinase was introduced as a key substrate and effector protein of mTOR. Ribosomal S6 kinase is a collective term for two kinase families. It refers to the substrate protein S6, a component of the small ribosomal subunit, rather than to a location of these enzymes in ribosomes. In fact, the S6 kinases are found at membranes, in the cytoplasm, and in the nucleus, and have functions going far beyond S6 phosphorylation.

Kinases p70^{S6K}

The ribosomal S6 kinase positioned downstream of mTOR belongs to the p70^{S6K} family. Mammalian cells express this type in two isoforms, with S6K1 being the S6 kinase proper. By up-regulating protein synthesis, the enzyme acts as an important positive controller of cell size and body growth. The corresponding gene-knockout animals are, indeed, significantly smaller than wild type. S6 kinase enhances translation by inactivating eEF2 kinase and activating the initiation factor eIF4B. Moreover, its many potential substrates include the insulin receptor substrate IRS1 (see Section 9.4.1 and Figure 9.5) as well as transcription factors and the pro-apoptotic protein Bad, indicating a much wider functional spectrum.

Until recently, phosphorylation of the S6 protein was thought to be essential for the translation of 5′-TOP-RNA required for ribosome synthesis. However, experiments with S6- and S6 kinase-knockout animals have raised doubts about this concept. As a control device for ribosome synthesis, the PI3K–PKB–mTOR cascade is, neverthe-less, the prime candidate. However, the molecular mechanism behind this function is still largely mysterious, and the same holds true for the physiological role of S6 protein and the stimulatory effect of S6 phosphorylation on cell growth.

How is the activity of S6 kinase regulated? As we have seen, it has to be phosphory-lated by mTOR to become active. This phosphorylation takes place in the N-terminal domain containing an mTOR signaling motif and abolishes the intrasterical auto-in-hibition of the enzyme. However, since S6 kinases belong to the AGC kinase family, they also have to be phosphorylated at the activation loop. As for most other AGC kinases, this is catalyzed by the membrane-bound phospholipid-dependent kinase PDK1 (see Section 4.6.4), encountered by S6K upon joining signal-transducing mul-tiprotein complexes that assemble around transmembrane receptors. Thus, in prin-ciple the activation of S6 kinase resembles that of PKB/Akt as shown in Figure 4.43. Most interestingly, PDK1 and mTOR are S6 kinase substrates themselves. While the physiological consequences of mTOR phosphorylation are still elusive, PDK1 phos-phorylation seems to speed up S6K activation by positive feedback.

Kinases p90^{RSK}

The second family of so-called S6 kinases are the ribosomal S6 kinases of the p90 type, p90^{RSK}. To distinguish them from S6Ks, they are called RSKs. Because they are also members of the AGC kinase superfamily, RSKs are related to the p70 S6Ks. For full activation, RSKs (like the p70 S6Ks) have to be phosphorylated by at least two different protein kinases: PDK1 and MAP kinases such as ERK or p38. Therefore, the RSKs are counted among the large group of MAP kinase-activated protein kinas-es (MAPKAPs) and are considered to be important links between the MAP kinase modules and the genome (for more details see Section 11.5). Like S6Ks, RSKs also phosphorylate PDK1 in a positive feedback loop and phosphorylate the S6 protein. However, S6 phosphorylation does not seem to be their primary function. Instead, RSKs act mostly at the level of gene transcription and their preferred substrates are transcription factors such as C/EBP, CREB, SRF, ATF1, and the estrogen receptor, as well as the transcriptional co-activator CBP/p300 and histone 3 (**Figure 9.6**). More-over, RSK species counteract apoptosis by phosphorylating (like S6K) the pro-apop-totic protein Bad (see Section 13.2.5), and they promote cell division by inactivating the cell-cycle-inhibiting protein kinase MYT1 (see Section 12.3). Finally, RSKs acti-vate mTOR signaling, probably by inactivating (like PKB/Akt) the Rheb GAP TSC2. This provides an additional pathway along which growth factors stimulate protein synthesis to cope with cell proliferation.

9.4.3 AMP-dependent protein kinases are sensors of cellular energy status

In the preceding section it was shown that mTOR signaling is inhibited by the Ser/Thr-directed 5′-AMP-activated protein kinase AMPK (not to be confused with cAMP-activated kinases!). This is a key event for translational control because the AMPK family has turned out to be a master regulator of energy metabolism in eu-karyotes (**Figure 9.7**).

As we have repeatedly emphasized, movements, signal processing, and anabolic re-actions such as protein biosynthesis are expensive cell physiological processes, crit-ically depending on the energy status of the cell and specifically on the ATP level. In fact, as long as it is alive, a cell maintains a ratio of ATP to ADP/AMP concentration that is several orders of magnitude above the chemical equilibrium, resembling a battery charged by chemical (ATP) energy (see Section 2.1).

The state of charging is controlled by AMPK: the enzymes of this kinase family act as universal energy sensors, monitoring the ATP to 5′-AMP ratio and also the level of the energy store glycogen. When they recognize a decrease of ATP or glycogen

Figure 9.6 Activation and major functions of ribosomal S6 kinases Double-headed arrows indicate mutual activation of the protein kinases PDK1 and RSK/S6K.

Figure 9.7 5′-AMP-dependent protein kinase AMPK Shown are the three subunits of AMPK, which respond to different input signals. AMPK inhibits ATP-consuming and stimulates ATP-supplying metabolic reactions. The mechanism of activation of the AMPK kinase LKB1 is not clear but may involve protein phosphorylation.

accompanied by a corresponding increase of AMP, AMPKs turn down the ATP-consuming anabolic metabolism (synthesis of macromolecules, fatty acids, cholesterol, etc.) and, instead, stimulate ATP-delivering catabolic reactions (glycolysis, fatty acid oxidation, etc.) and—especially in muscle cells—glucose uptake. This occurs on the levels of transcription, translation, and metabolism.

Activation of AMP-dependent protein kinases

AMPKs are allosterically activated by 5′-AMP. They are heterotrimeric proteins consisting of a catalytic α-subunit and two regulatory subunits labeled β and γ (see Figure 9.7). Humans possess two α-, two β-, and four γ-subunits that are tissue-specifically expressed. In addition, 12 AMPK-related protein kinases with mostly still ill-defined functions have been identified by screening of the human genome.

The activity of AMPK is regulated by an interaction between the three subunits. The α-subunit of AMPK harbors a typical Ser/Thr-specific protein kinase domain. It is activated by a combination of an allosteric effect with phosphorylation of the activation loop, a mechanism that is typical for most protein kinases. Activation loop phosphorylation occurs only in the presence of sufficient 5′-AMP, while in the absence of the nucleotide the kinase is rapidly dephosphorylated and blocked by auto-inhibitory sequences found both in the α- and γ-subunits. This blockade is relieved by a change of the quaternary structure induced when 5′-AMP binds to the γ-subunit. The protein kinase catalyzing activation loop phosphorylation is identical with the tumor suppressor protein LKB1.

ATP and glycogen prevent AMPK activation. In fact, AMPK (like many other protein kinases) combines the properties of a logical AND gate with those of a logical BUT NOT gate: it is activated by phosphorylation *and* binding of AMP *but not* by phosphorylation and an interaction with ATP or glycogen.

Multiple binding sites of 5′-AMP are located in the γ-subunit. At a high ATP:AMP ratio, ATP displaces AMP from these sites, thus blocking the activation of AMPK. A specific binding domain for glycogen is found in the β-subunit, which also serves as an adaptor for the two other subunits. Owing to a pronounced interaction between the subunits, the enzyme is subject to efficient allosteric regulation, showing very steep sigmoidal kinetics and functioning as both a switch and a strong noise filter.

LKB1 is a protein kinase that suppresses tumor growth

The Ser/Thr-specific protein kinase LKB1 (<u>L</u>iver <u>K</u>inase <u>B1</u>, also known as Ser/Thr Kinase 11, STK11), now considered to be the major receiver of input signals controlling AMPK activity, was discovered in the course of a genetic analysis of **Peutz–Jeghers syndrome**. Patients suffering from this relatively rare hereditary disease develop a large number of tumors in the gastrointestinal tract and in other organs that, although they are benign in the beginning, have a pronounced tendency to become malignant. Tumorigenesis was causally related to an *inactivation* of the *LKB1* gene, characterizing LKB1 as a tumor suppressor protein. This was a rather unexpected finding since protein kinases were thought to promote tumor formation through *overactivation* and mutation to hyperactive oncoproteins. LKB1, however, inhibits cell proliferation, for instance by inducing synthesis of the cell cycle inhibitor p21^{WAF1} (see Section 12.4) and the phosphatidylinositol 3-phosphatase PTEN, another abundant tumor suppressor protein (see Section 4.4.3). Such effects easily explain the tumor suppressor function of LKB1. Moreover, since LKB1 is an AMPK activator, its deletion leads to a disequilibrium between anabolic and catabolic metabolism. This situation results in a pathological overproduction of proteins and glycogen, a situation that promotes tumor development. The benign tumors of Peutz–Jeghers syndrome are indeed hamartoma-like with very large and overfed cells. Thus, the down-regulation of LKB1–AMPK signaling has the same consequence as an up-regulation of mTOR signaling: both produce tumors of a similar phenotype. Such an observation is understandable when the antagonism between LKB1–AMPK and mTOR signaling is considered.

LKB1 is only distantly related to other eukaryotic protein kinases. In cells it exists as a complex with several adaptor proteins that is referred to as AMPK kinase (AMPKK). To become active, LKB1 undergoes autophosphorylation. Moreover, it becomes phosphorylated at several Ser and Thr residues by kinases such as PKA, PKC, and ATM (see Section 12.9) which have a stimulatory effect, and ERK which inhibits LKB1. AMPK phosphorylation by LKB1 most probably is controlled by 5′-AMP, which by inducing a conformational change converts AMPK into an LKB1 substrate.

LKB1 is much more than a tumor suppressor, but plays a role in a variety of physiological functions. Thus the enzyme is essential for the establishment of cell polarity and for embryonic development, in particular of the nervous system where it is involved in the control of axon growth. Many of these effects may be mediated by AMPK-related kinases, which, like AMPK, are activated by LKB1. Thus, LKB1 is a master kinase controlling the activity of a whole series of protein kinases, a role reminiscent of that of the lipid-dependent kinase PDK1 with respect to the AGC kinases (see Section 4.6.4).

Autophagy

AMPK is a major inducer of autophagy or autophagocytosis ("self-eating"), with mTORC1 being its opponent. Autophagy is a cytoprotective mechanism found in all eukaryotes. It enables the cell to get rid of superfluous material such as defective organelles and macromolecular complexes that when accumulating would have a damaging effect. The material is degraded and the degradation products (for instance amino acids) are recycled because they are too precious to become wasted. A particularly clear example of recycling is insect metamorphosis. Autophagy is necessary for the maintenance of cells and guarantees an at least temporal survival during starvation, oxygen shortage, mitochondrial damage, infections, and other stress situations. It is essential, since autophagy-deficient animals die shortly after birth. Moreover, the effectiveness of autophagy strongly correlates with longevity in animal models as well as in humans.

For the most abundant form, called macroautophagy, the material to be removed and recycled gets packed into double-membrane vesicles called **autophagosomes**. Autophagosomes grow out of intracellular membranes, in particular the ER, and are transported to the lysosomes where their cargo is broken up. Since "self-eating" is a risky venture, autophagosome formation is tightly regulated by a sophisticated signaling network. The whole process depends on at least 30 autophagy-related genes (*Atg*s). A key protein positioned at the very beginning of autophagosome formation is a Ser/Thr-specific protein kinase. This enzyme is called ATG1 in yeast or UNC-51 in the nematode *Caenorhabditis elegans*. Its mammalian counterpart is known as **ULK (U̲NC-51 L̲ike K̲inase)**. ULK exists in two isoforms, ULK1 and ULK2. The kinase is

Figure 9.8 Induction of autophagy
Upon phosphorylation (P) by protein kinase AMPK, the autophagy kinase ULK1 forms a complex with several regulatory proteins that are phosphorylated. This leads to the formation of autophagosomes and their subsequent fusion with lysosomes. ULK1 activation is inhibited by phosphorylation through protein kinase mTORC1. Vice versa, UMP1-catalyzed protein phosphorylation suppresses AMPK and mTORC1 activity.

a component of a multiprotein complex containing several regulatory and scaffold proteins.

To induce autophagy, ULK has to phosphorylate itself and other proteins of the complex. As a result, additional ATG proteins assemble and the ER membrane is changed in such a way that an autophagosome can develop. An important step is the phosphorylation of the membrane lipid phosphatidylinositol at position 3, because phosphatidylinositol 3-phosphate (PI3P) serves as a platform for ATG proteins with PI3P binding sites. The corresponding enzyme is a class III PI3 kinase known as VPS34. This is one of the proteins that upon ULK activation assemble at the initial complex. As mentioned in Section 9.4.1, VPS34 is also a putative amino acid sensor of the mTORC1 system. The activity of ULK is controlled by environmental conditions. It is up-regulated in shortage situations and dampened during prosperity. The cell's "prosperity sensor" is mTOR whereas the "starvation sensor" is AMPK. In fact, both kinases are the major regulators of ULK activity. Phosphorylation by mTORC1 inhibits ULK and suppresses autophagy whereas phosphorylation by AMPK has the opposite effect, promoting autophagy. Of course, the particular phosphorylation sites on ULK are different. Vice versa, ULK phosphorylates and inhibits both AMPK and mTORC1, thus establishing feedback control (**Figure 9.8**).

The autophagy of damaged mitochondria, called **mitophagy**, is of particular medical interest. Among the initiating events, the accumulation of the Ser/Thr-specific protein kinase **PINK1** in the outer mitochondrial membrane has been investigated in more detail (PINK means PTEN-INduced protein Kinase since it was found to be up-regulated in tumor cells overexpressing the phospholipid phosphatase PTEN, see Section 4.4.3). PINK1 recruits the protein **parkin** (referring to Parkinson's disease), a component of an E3 ubiquitin ligase complex. Upon phosphorylation by PINK1 this complex triggers the proteosomal degradation of mitochondrial proteins. Loss-of-function mutations of PINK1 and parkin lead to an accumulation of damaged mitochondria in neurons, finally resulting in cell death. Such mutations are characteristic for the familial form of Parkinson's disease. Dysregulation of autophagy has also been linked to other neurodegenerative disorders such as Huntington's and Alzheimer's diseases, as well as to chronic inflammatory processes (for instance Crohn's disease) and cancer. By autophagy, tumor cells may cope with the stress of accelerated proliferation and the toxic effects of anticancer drugs. Therefore, the signaling network controlling autophagy has become an attractive target of future clinical interventions. For his pioneering work on the mechanisms of autophagy, Yoshinori Ohsumi was honored with the Nobel prize in Physiology or Medicine in 2016.

AMP-dependent protein kinase signaling in obesity, diabetes, and cancer

AMPK is an inhibitor of mRNA translation: it attenuates mTOR signaling by stimulating the Rheb-GAP TSC2 and suppresses peptide chain elongation by activating eEF2 kinase. But AMPK can do much more. On the metabolic level, it phosphorylates and inhibits key enzymes of anabolic metabolism such as acetyl-CoA carboxylase 1 of fatty acid synthesis, 3-hydroxy-3-methylglutaryl-CoA reductase of cholesterol biosynthesis, glycerophosphate acyltransferase of triglyceride synthesis, and muscle glycogen synthase. Vice versa, AMPK activates proteins of catabolic metabolism such as the glucose transporters GluT1 and GluT4 and the enzyme phosphofructokinase. Translocation of GluT4 to the cell membrane is the key event in the regulation of glucose uptake by muscle and fat cells and a major target of insulin. GluT4 translocation is controlled by small G-proteins of the Rab family and suppressed by the Rab-GAP AS160. This inhibitory effect is abolished by phosphorylation of AS160, catalyzed either by insulin along the PI3K–PKB/Akt pathway or by AMPK. Thus, the effects of insulin and AMPK complement each other, and sufficient activation of AMPK (for instance by physical activity) may save insulin, an effect that is highly important for prevention of diabetes and other metabolic diseases.

AMPK also regulates metabolism at the level of gene transcription by phosphorylating transcription factors. In yeast, for instance, the AMPK-homologous kinase SNF1 stimulates the transcription of genes that are repressed at a high supply of glucose (note the analogy to catabolite repression of prokaryotes), and in vertebrates no less than 370 genes have been identified whose transcription is influenced directly or indirectly by AMPK. In view of this functional versatility, it is conceivable that AMPK is stimulated by a wide variety of physiological and pathological stress factors such as fasting, starvation, oxygen deficiency, ischemia, intoxication, heat shock, hyperosmosis, bodily activity, and that disturbances of AMPK signaling are directly involved in metabolic disorders such as diabetes and obesity.

In this context, two proteohormones activating AMPK signaling deserve special attention. These are the so-called adipocytokines **leptin** and **adiponectin**. Both are released by adipose tissue in response to excessive fat storage and play a significant role in the regulation of body mass. They do this by promoting the hypothalamus to take measures against an overload of the body's energy stores and by suppressing insulin secretion in the pancreas. Leptin, a 16 kDa peptide, has the structure of a cytokine and its receptors belong to the family of tyrosine kinase-coupled cytokine receptors of class I. As such, they induce various cellular reactions along the JAK–STAT and MAP kinase signaling pathways (see Section 7.2.1). However, the mechanism by which leptin activates the LKB1–AMPK axis is still not fully understood. The leptin receptors are expressed in many tissues, in particular, however, in certain cell types of the hypothalamus.

By stimulating hypothalamic functions, leptin suppresses the appetite, reduces the mass of adipose tissue, and activates the overall energy consumption of the body. To this end, it turns off the effects of the hypothalamus hormone neuropeptide Y that evokes the feeling of hunger and promotes anabolic metabolism. Moreover, leptin reduces the sensitivity of sweet-tasting sensory cells, thus rendering sweets less attractive (see Section 15.1). Since sexual maturation depends on the state of nutrition, leptin has been proposed also to act as a triggering signal of puberty. In this respect it again functions as an opponent of neuropeptide Y, which retards sexual maturation by inhibiting gonadotropin release.

Disturbances of the leptin/adiponectin system are a major cause of obesity. In experimental animals, such a condition can be evoked by a loss-of-function mutation of leptin receptor genes, whereas in humans, an overload of the system due to excessive intake of high-energy food may result in leptin resistance due to adaptive down-regulation of the receptors. Since the cellular transduction of the leptin signal (for instance caused by receptor down-regulation) rather than leptin release seems to be impaired, the hope of curing obesity by leptin preparations has turned out to be somewhat premature.

Endogenous antagonists of leptin and adiponectin are the hormones **resistin** from fat cells, **ghrelin** produced by the digestive tract, and, first and foremost, **insulin.** They all activate the anabolic metabolism, resulting in glycogen and fat deposition and an increase of protein synthesis and body mass (for a more detailed discussion of insulin signaling, see Section 7.1.3). It is assumed that leptin/adiponectin and insulin control each other by negative feedback along an "adipoinsulinic axis."

For an obese person this has a serious consequence: the overproduction of leptin/adiponectin going hand in hand with adipocytokinin resistance dramatically diminishes both insulin production and insulin sensitivity.

Such a condition is **type 2 diabetes**, by far the most abundant metabolic disease, and its precursor stage, the so-called metabolic syndrome (in contrast, type 1 diabetes is caused by autoimmune destruction of insulin-producing β-cells in the pancreas). One cause of insulin resistance is considered to be a pathological deposition of fat in skeletal muscles and other organs, which massively impairs carbohydrate metabolism and causes an adaptive down-regulation of insulin receptors. Since skeletal muscle is the tissue with the highest consumption of sugar, this lipotoxic effect has disastrous consequences.

Pathological fat and glycogen deposition is counteracted by the AMPK signaling system as well as by stress hormones. In fact, AMPK is activated by **metformin**, the most widely used antidiabetic drug. Moreover, AMPK is stimulated not only by leptin but also by physical activity and other stress factors. In such situations the (nor)adrenaline level is also increased, speeding up lipolysis and glycogenolysis along the G_S–cAMP–protein kinase A pathway (see Section 5.3 for details). This easily explains why diabetes is promoted by obesity as well as by a lack of exercise, whereas physical activity and a reduction of body mass have a beneficial effect.

Diabetes and obesity clearly increase the risk of certain types of cancer such as mammary, colorectal, and prostate carcinomas, and again a diminished supply of calories as well as physical exercise have been unequivocally shown to exhibit a preventive effect. One of the reasons may be again the activation of AMPKs, which have been shown to suppress the formation of tumor-promoting growth factors such as insulin-like growth factor 1 (see Section 7.1.3).

Diabetes and hexosamine signaling

The strong increase of blood glucose (hyperglycemia) in the course of type 2 diabetes evokes a series of feedback reactions that finally end up in insulin resistance of target cells, in particular in fat, liver, and muscle tissue, and as a consequence in long-term damage of the kidneys, the heart, the eyes, the brain, and the extremities. One reason for this glucose toxicity is the generation of highly reactive sugar metabolites that inactivate or even destroy essential cellular components (see Section 7.3.3). Another reason seems to be hyperactivation of a novel signaling pathway called hexosamine signaling, which was discovered in studies on **glucose-induced insulin resistance**. The key observation was that, in insulin target cells, this dreaded consequence of type 2 diabetes correlates with a characteristic post-translational modification of a wide variety of proteins: covalent binding of *N*-acetylglucosamine (GlcNAc).

Sugar residues bound covalently at hydroxyl or amino groups of proteins are an abundant modification yielding glycoproteins. Most transmembrane proteins, for instance, are glycosylated at the extracellular domains. This is crucial for both their correct positioning and their function. The special thing about the binding of GlcNAc is that it fulfils the conditions of a signaling reaction possibly rivaling protein phosphorylation. Indeed, the reaction is a fully reversible energy-consuming switch that is operated by two enzyme families. These are the O-linked GlcNAc transferases, catalyzing the transfer of GlcNAc from UDP-GlcNAc to Ser and Thr residues of proteins, and the O-linked *N*-acetylglucosaminidases (O-GlcNAcases), which cancel the modification by hydrolysis (**Figure 9.9**). The human genome contains one O-linked GlcNAc transferase gene and one O-linked *N*-acetylglucosaminidase gene. Both give rise to a wide variety of isoenzymes that are generated by alternative pre-mRNA splicing and multiple start codons.

A large number of proteins are modified by hexosamine signaling. They include transcription factors, proteins of RNA translation, cytoskeletal and other structural proteins, signaling proteins, various enzymes, chaperones, proteasomes, and others. Our knowledge of the functional consequences of this posttranslational modification is still rather limited. Depending on the substrate protein, it may block Ser and Thr residues, that is, counteract protein phosphorylation in logical BUT NOT gates, or it may cooperate with phosphorylation in logical AND gates. Blocking of phosphorylation may result in quite different effects including an inhibition of protein degradation along the ubiquitin proteasome pathway, which frequently depends on phosphorylation (in addition, proteasomes are directly inhibited by GlcNAc binding).

What is the present concept on the role of hexosamine signaling in insulin resistance? In insulin target tissues, glucose is sequestered as glycogen and transformed into fat, with hexosamine signaling being a branch of these two major pathways as depicted in Figure 9.9. In fact, hexosamine signaling has a regulatory function since it promotes the release of insulin antagonists, such as leptin and adiponectin, and stimulates fat and glycogen synthesis. In addition, the cell's insulin-dependent glucose transport system becomes gradually inactivated. Persisting hyperglycemia overactivates this feedback control, finally leading to insulin resistance and an impairment of insulin release along the adipoinsulinic axis. This effect is amplified by down-regulation of the insulin receptor and other components of insulin signaling (see Section 7.1.3).

Figure 9.9 Protein modification by *N*-acetylglucosamine (hexosamine signaling) *N*-Acetylglucosamine 1-phosphate is generated from glucose along the pathway shown to the left. It is activated by an O-glycosidic binding of a uridine diphosphate (UDP) residue and transferred to Ser or Thr residues by the enzyme O-*N*-acetylglucosamine transferase (O-GlcNAc transferase, box). The modification is reversed by O-linked *N*-acetyl glucosaminidases (O-GlcNAcase), yielding the nonmodified protein and *N*-acetylglucosamine. Being a typical switch, hexosamine signaling requires energy supplied by the hydrolysis of one UTP molecule.

9.4.4 Glycogen synthase kinase 3 is much more than a suppressor of glycogen deposition

Another signaling enzyme playing a key role in both translational and metabolic regulation is GSK3, an abundant eukaryotic Ser/Thr-specific protein kinase. As shown in Section 9.3.1, it suppresses mRNA translation by phosphorylating and inhibiting the initiation factor eIF2B (see Figure 9.2). Moreover, like several other protein kinases, GSK3 inhibits glycogen synthase by phosphorylation, thus suppressing the anabolic metabolism of glucose (see Section 5.3). However, the activities of GSK are much more extensive. In fact, GSK3 may have more substrate proteins than any other protein kinase indicating its role as a multifunctional biochemical processor with a wide range of effects. In most cases these effects have an inhibitory character.

Unusual mechanism of activation

In mammals GSK3 is expressed in two isoforms, α and β. Although these are closely related to MAP kinases, GSK3 does not require activation loop phosphorylation. Indeed, in contrast to most other protein kinases, GSK3 is constitutively active and, as a rule, becomes *inactivated* by input signals. Such input signals are, first of all, transduced by Ser/Thr phosphorylation of GSK3 at sites outside the activation loop. These phosphorylations are catalyzed by several protein kinases.

A typical example is the inhibition of GSK3β by PKB/Akt (see also Section 4.6.6). As shown in **Figure 9.10**, the inhibitory phosphorylation occurs at a Ser residue in the pseudosubstrate sequence of GSK3 enabling this sequence to cover and block the catalytic domain. This is because the latter recognizes substrate proteins by a corresponding amino acid sequence with a phospho-Ser (or Thr) residue. In other words: to become recognized by GSK3, substrate proteins have to be prephosphorylated by another protein kinase, a reaction called "priming." To reactivate the inhibited GSK3 the pseudosubstrate domain has to be displaced from the catalytic center either by competition with the primed substrate protein or by dephosphorylation.

A protein kinase that frequently has been found to prime GSK3 is CK2 (see Section 9.4.5). Provided a substrate protein has the corresponding primary structure, it may be phosphorylated by GSK3 in cooperation with such a primer kinase in a stepwise manner (see Figure 9.10). For an example (the hierarchical phosphorylation of glycogen synthase), the reader is referred to Section 5.3.1. As far as cellular data processing is concerned, GSK3-catalyzed hierarchical substrate phosphorylation resembles a logical AND gate combined with a digital–analog converter.

Negative key regulator of insulin and mitogenic signaling

The inhibitory effect of GSK3 on glycogen synthase is overcome by the above-mentioned phosphorylation of the pseudosubstrate domain, resulting in a suppression of GSK3 activity. Since PKB/Akt is a major player in this game, all signals activating the PI3K–PKB/Akt pathway are expected to antagonize GSK3 (**Figure 9.11**). This holds true, in particular, for mitogens and insulin. In fact, GSK3 has turned out to

Figure 9.10 "Hierarchical" phosphorylation by and intrasteric inhibition of glycogen synthase kinase 3 (GSK3) (A) Hierarchical phosphorylation of a substrate protein (for example, glycogen synthase) with casein kinase 2 (CK2) as a primer kinase. (B) Interaction of a prephosphorylated (primed) substrate protein with the catalytic domain (gray rectangle) of GSK3 and point of attack of an inhibitory phosphorylation, for instance by protein kinase B (PKB/Akt), in the N-terminal pseudosubstrate sequence (teal rectangle). (C) In inactive GSK3, the phosphorylated pseudosubstrate sequence covers the substrate-binding site, hindering the access of the primed substrate protein.

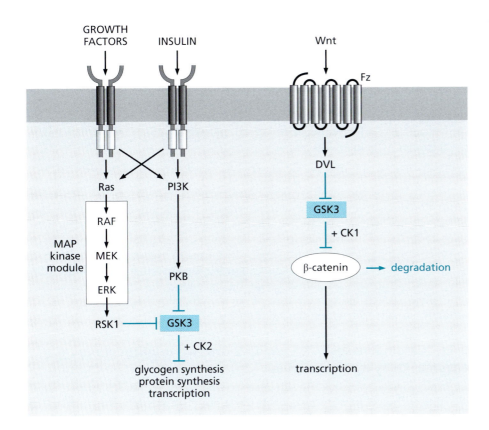

Figure 9.11 Three major pathways leading to an inhibition of glycogen synthase kinase 3 To the left are shown two signaling cascades that originate from Tyr kinase-coupled receptors. The Ras–RAF–ERK pathway results in activation of the protein kinase RSK1, while the phosphatidylinositol 3-kinase (PI3K) pathway results in activation of protein kinase B (PKB)/Akt. Both kinases inhibit GSK3 by phosphorylation of the pseudosubstrate domain, thus rescuing glycogen synthesis, protein synthesis and gene transcription from inhibition. To the right is shown the Wnt pathway, including the receptor Fz and the adaptor protein DVL. In the absence of Wnt, GSK3 (together with the primer kinase casein kinase 1, CK1) inhibits β-catenin activity by inducing proteolytic degradation. In the presence of Wnt, DVL inhibits GSK3, resulting in a stabilization of β-catenin. (For details see Section 5.9.)

be a central switching element in anabolic insulin effects such as glycogen deposition and protein synthesis. As already explained, insulin promotes protein synthesis by rescuing translation from the inhibitory phosphorylation of the initiation factor eIF2B catalyzed by GSK3 (see Figure 9.2 and Section 9.3.3).

In addition, GSK3 is an inhibitory element in mitogenic pathways originating from Tyr kinase-coupled growth factor receptors. In such signaling cascades, GSK3 becomes switched off through pseudosubstrate phosphorylation either by PKB/Akt or by the ribosomal S6 kinase RSK1 that is positioned downstream of the Ras–RAF–ERK cascade (see Figure 9.11).

The inhibitory effect of GSK3 on mitogenic signal transduction is due mainly to a blockade of translation and transcription. Transcription is suppressed by an inhibitory phosphorylation of mitogenic transcription factors such as cJun (see Section 11.6). Another example of a transcription factor that is switched off by GSK3 is β-catenin, which upon phosphorylation becomes a substrate of ubiquitin ligases priming the factor for degradation in proteasomes. Here the activity of GSK3 is controlled along the mitogenic Wnt signaling pathway with casein kinase CK1ε acting as a primer kinase (see Figure 9.11; for details, see Section 5.9). The formation of a complex with β-catenin and two adaptor proteins, axin and APC, probably renders GSK3 resistant to inactivation by PKB and other kinases. The breakdown of this regulatory mechanism is a major cause of colorectal cancer (see Section 5.9).

Glycogen synthase kinase 3 and neurodegenerative diseases

Another aspect of GSK3 function should be mentioned, since it may turn out to be of considerable practical significance. This is the role of GSK3β in neurodegenerative diseases such as **Alzheimer's dementia.** GSK3β has been found to be hyperactive in impaired brain areas and to inactivate by phosphorylation some essential transcription factors besides β-catenin, such as NFAT (see Section 14.6.4) and heat-shock factor HSF1. In this way, the vitality of nerve cells becomes diminished, culminating in apoptosis and necrosis. Potentially even more dangerous seems to be the aberrant phosphorylation of another GSK3 substrate, the **microtubule-associated protein Tau**. By stabilizing microtubule tracks in axons, Tau normally provides the conditions for transport of vesicles filled with neurotransmitters to the synapse. Tau that is rendered insoluble by hyperphosphorylation aggregates into filamentous structures, which are unable to interact with microtubules. As a result, synapses become inactive, finally resulting in cell death. Tau aggregation is regarded as a major risk factor for Alzheimer's disease.

The overactivation of GSK3β in degenerating brain areas seems to be due to an impairment of blood supply, or ischemia. Ischemia is a strong promoter of neurodegeneration since it renders brain neurons sensitive to apoptotic signals. In such cells, GSK3 becomes activated by the β-amyloid peptide Aβ along an ill-defined pathway. Overproduction of Aβ and extracellular deposition of Aβ degradation products as so-called amyloid plaques is considered to be the major trigger of dementia (for details, see Section 13.1.3). These findings nourish the hope that in the future GSK3 inhibitors may become suitable drugs for the prevention and treatment of neurodegenerative diseases, although in view of the multifunctionality of the enzyme, one certainly has to reckon with serious side effects.

9.4.5 "Casein kinases" CK1 and CK2 are two multifunctional enzymes denying their names

As primer kinases of GSK3, the so-called casein kinases CK1 and CK2 were introduced in Section 9.4.4. However, this is by far not their only function. In fact, both represent two groups of Ser/Thr-specific protein kinases that exhibit a functional versatility comparable to that of GSK3. It is a curiosity that, despite their historic name, these kinases do not play a role in the phosphorylation of casein *in vivo* which, instead, is catalyzed by the "genuine casein kinase" G-CK, an atypical protein kinase found, for instance, in the cells of the lactating gland. Therefore, the acronyms CK1 and CK2 are preferred now. The structural similarity between the two enzymes is rather low. With GSK3 they share the property of being constitutively active; they are switched off rather than switched on by input signals.

CK1 and CK2 catalyze the phosphorylation of Ser and Thr residues localized adjacent to acidic amino acids (such as glutamic and aspartic acid) or phosphate. A phosphorylation consensus sequence of CK1 is, for instance, Ser(P)/Thr(P)-x-x-Ser/Thr. Therefore, like GSK3, CK1 and CK2 may phosphorylate proteins in a stepwise manner, cooperating with other protein kinases. By the way, their preference for a negatively charged phosphorylation consensus sequence distinguishes CK1 and CK2 as well as GSK3 from most other protein kinases that prefer positively charged target sequences.

Protein kinase CK1

Mammalian CK1 is expressed in at least eight isoforms, which usually are labeled by Greek letters. Since the enzyme inactivates itself by autophosphorylation, it requires a permanent contact with protein phosphatases to stay active. This situation provides a terminal for input signals that modulate phosphatase activity. CK1 has various functions including that of a primer kinase of AMPK and GSK3 (Figure 9.10). An important example is the control function of CK1 in the hedgehog and Wnt signaling cascades (see Section 5.9). Recently its role in the regulation of **day–night rhythms** has found particular attention.

The cellular and organ functions of all eukaryotes and of certain prokaryotes (such as, for instance, the light-exploiting cyanobacteria) are adjusted to daytime. In vertebrates, the central oscillator controlling day-night rhythms consists of a group of neurons in the hypothalamus, called the nucleus suprachiasmaticus, which regulate subordinated feedback loops in the peripheral organs. Each air passenger travelling

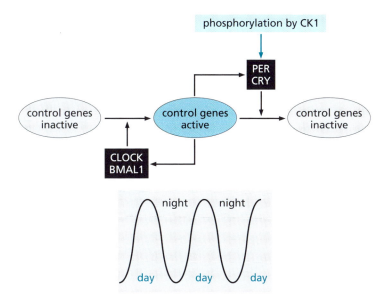

Figure 9.12 Molecular control of day–night rhythm The diurnal rhythm is due to an autonomous oscillating signal system controlling the activity of a large number of genes. These genes encode and become activated by the heterodimeric CLOCK/MAL1 transcription factor (positive feedback loop, left side). Some of the control genes also give rise to the production of the repressor proteins PER and CRY that, upon phosphorylation by casein kinase 1 (CK1), inhibit CLOCK/MAL1; that is, they promote the inactivation of the control genes (negative feedback loop, right side). As a result, the undampened oscillation shown below is generated. CK1 is assumed to be controlled by exogenous signals.

the Atlantic Ocean can tell a thing or two about the resistance of this internal clock! Although genetically fixed, the diurnal or 24-hour rhythm of physiological activities is additionally controlled by exogenous factors such as light, enabling adaptation to changing conditions. To this end, direct contact exists between the photoreceptor cells of the retina and the neurons of the nucleus suprachiasmaticus. In plants, the cryptochromes containing flavin and pterin chromatophores serve as light sensors. Homologous cryptochromes are also found in vertebrate cells. Whether or not they fulfil a similar purpose as the plant cryptochromes is not known.

Rhythmic oscillations such as the day–night rhythm are due to twofold negative feedback or a combination of positive and negative feedback (see Section 1.9). As far as diurnal rhythms are concerned, such regulatory loops control the transcription of a large number of genes including three periodicity genes (*per1–3*) and two cryptochrome genes (*cry*). In the hypothalamus and other organs these genes are activated by the heterodimeric transcription factor **CLOCK/BMAL1**, the activity of which oscillates in a 24-hour rhythm due to a combination of negative and positive feedback. *Positive* feedback results from auto-induction of the genes encoding BMAL1, while *negative* feedback is due to production of the repressor proteins PER and CRY that, upon dimerization, migrate into the nucleus and inhibit CLOCK/BMAL1 (**Figure 9.12**).

Inhibition of the circadian genes correlates with multiple phosphorylation of the PER proteins that is catalyzed in particular by CK1. Studies on transgenic animals have shown this phosphorylation to be indispensable for a precise day–night rhythm. Phosphorylation by CK1 may have more than one effect. Thus, it promotes the ubiquitylation and degradation of PER in proteasomes, but in addition, it might also control the nuclear translocation of PER and its interaction with CLOCK/MAL1. The idea of CK1 as a transducer of exogenous signals modulating the day–night rhythm is easily at hand. However, the input signals controlling CK1 activity are still not known.

Protein kinase CK2

CK2 is a perfect example of a multifunctional protein kinase since more than 100 potential substrate proteins are known. The enzyme is found in all eukaryotes, where it is expressed in several isoforms. As a heterotetramer of two catalytic α-subunits and two regulatory β-subunits, CK2 exhibits a more complex quaternary structure than CK1. CK2 belongs to the very few protein kinases that are constitutively active even in the absence of all intercellular signals known. Activation loop phosphorylation does not seem to be necessary although phosphorylation by other kinases exhibits a modulatory effect on CK2 activity. CK2 is inhibited by acidic molecules such as heparin and is stimulated by basic polyamines. The physiological significance of these reactions is, however, as controversial as the interactions of CK2 with many other proteins.

The essential role of CK2 is emphasized by genetic knockout experiments ending in severe or even fatal developmental defects in such different organisms as yeast, slime molds, nematodes, and mice. However, the large number of potential substrate proteins has impeded most attempts to associate certain cellular events with CK2 activity. Among the putative functions, roles in cell proliferation and apoptosis have been studied in more detail. CK2 is required for an ordered course of the cell cycle and inhibits apoptosis by phosphorylating certain pro-apoptotic proteins such as Bid (see Section 13.2.4). The function of CK2 as a primer kinase of GSK3 has been mentioned in Section 9.4.4, and an inhibitory effect on the transcription factor cJun will be discussed in Section 11.5.

CK2 is enriched in various tumors, where it is assumed to unfold its anti-apoptotic and growth-promoting effects. Therefore, efforts are being made to develop CK2 inhibitors as anti-cancer agents some of which have entered clinical trials.

Summary

- Atypical protein kinase mTOR stimulates ribosomal protein synthesis and cell growth in response to the nutrient and energy supply and to mitogenic and anabolic stimuli. Such signals are transmitted by the small G-protein Rheb via a regulation of its GTPase-activating protein TSC2. TSC2 is a tumor suppressor protein deleted in hamartomas. The stimulatory effect of insulin on mTOR signaling is restricted by negative feedback. An overload of this control system may contribute to diabetes through overnutrition.

- Ribosomal S6 kinases S6K (p70^{S6K}) and RSK (p90^{S6K}) are multifunctional enzymes of the AGC kinase superfamily. While S6Ks as mTOR substrates primarily control protein synthesis and cell size, RSKs are more involved in the regulation of cell proliferation and survival.

- AMP-dependent protein kinase (AMPK) is a cellular master regulator that monitors the energy status of the cell and adjusts energy-consuming anabolic metabolism (inhibited by AMPK) to energy-supplying catabolic metabolism (stimulated by AMPK). In stress situations (starvation) AMPK stimulates autophagy. Activation loop phosphorylation of AMPK is catalyzed by LKB1, a tumor suppressor protein. Disturbances of AMPK signaling are critically involved in metabolic disorders such as obesity and type 2 diabetes.

- Glycogen synthase kinase (GSK) is a constitutively active enzyme that becomes inactivated by input signals. To phosphorylate its many substrates, it frequently cooperates with primer kinases. GSK3 is a major negative regulator of insulin effects and mitogenic signaling and is thought to be critically involved in neurodegenerative diseases such as Alzheimer's dementia.

- "Casein kinases" CK1 and CK2 represent two families of widely distributed Ser/Thr-specific kinases that prefer acidic phosphorylation consensus sequences and are constitutively active. Among the many putative functions of these enzymes, their role as primer kinases of GSK3 and AMPK has gained attention. CK1, in particular, is involved also in the control of diurnal rhythms. Protein kinases that are major signal transducers in the network, which adjust cellular protein synthesis and growth (and proliferation) to the supply of food and energy, are summarized in **Table 9.2**.

Table 9.2 Protein kinases controlling the adjustment of cell growth to the supply situation

Kinase	Activation	Major substrates	Major functions
mTOR	PKB/Akt (or ERK) amino acids, high ATP level	S6K, 4E-BP1, eEF2 kinase (?)	activation of translation increase of cell size
S6K	mTOR (or ERK) PDK1	S6K, Bad, IRS1, eEF2 kinase	activation of translation and cell survival inhibition of insulin signaling
RSK	MAPK PDK1	transcription factors, Bad	cell proliferation and survival
AMPK	5′-AMP LKB1 (physical activity, stress)	metabolic enzymes proteins of mTOR signaling	activation of catabolic and inhibition of anabolic metabolism inhibition of translation
LKB1	autophosphorylation	AMPK and AMPK-related kinases	activation of AMPK and related kinases tumor suppressor
GSK3	constitutively active (inhibition by PKB/Akt, RSK, WnT)	glycogen synthase, RSK, transcription factors, Tau protein, eIF2B	inhibition of insulin signaling, cell proliferation, translation neurodegeneration
CK1	constitutively active	AMPK, GSK3, transcription factors	primer kinase of AMPK and GSK3 control of diurnal rhythms
CK2	constitutively active	GSK3 and many others	primer kinase of GSK3 cell proliferation and survival
ULK	AMPK (inhibition by mTOR)	ATG proteins	formation of autophagosomes

9.5 The signaling network of non-coding RNAs

The vast majority of the human genome is transcribed while only less than 2% is translated into proteins. Hence, 98% of the transcriptome is represented by **non-coding RNAs (ncRNAs)** including housekeeping RNAs (ribosomal RNA, transfer RNA, small nuclear and nucleolar RNA) and regulatory RNAs. Two types of regulatory ncRNA have attracted attention as functional players in signaling pathways: small non-coding RNAs and large non-coding RNAs. The abundance and versatility of these non-coding RNA-species indicates that they are involved in almost any physiological process and that a dysregulation of non-coding RNA-signaling plays a critical role in a variety of diseases, first and foremost cancer.

Small non-coding RNAs control mRNA translation

From yeast to humans, cells possess an efficient mechanism for inactivating or destroying newly synthesized mRNA, thus annulling gene transcription. To this end, mRNA translation is blocked by small pieces of antisense RNA, a process called **RNA interference (RNAi)** or RNA silencing. Functionally it resembles the regulation by small RNA of prokaryotic mRNA translation (see Section 3.2). Mechanistically there are differences, however. While bacteria use single-stranded highly structured RNA molecules of 50–250 nucleotides that are transcribed directly, the corresponding signaling RNAs of eukaryotes are considerably smaller (21–25 nucleotides) and double-stranded. The eukaryotic regulatory sRNAs are termed **microRNA (miRNA)** and **small interfering RNA (siRNA)**. The class of endogenous miRNAs is quite significant: currently 2588 human miRNA sequences are known derived from 1881 primary transcripts.

miRNA is produced in several steps beginning with the transcription of miRNA genes into large primary miRNAs (pri-miRNAs). These primary transcripts are processed by the combined action of two ribonucleases. Still in the nucleus the enzyme

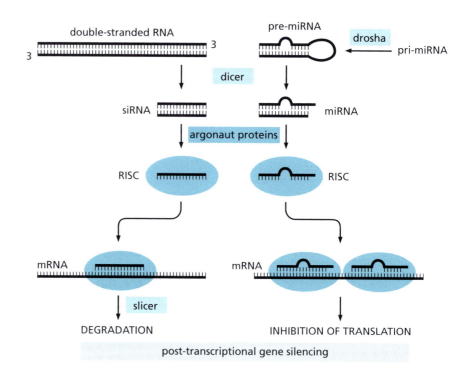

Figure 9.13 RNA interference
Small interfering RNA (siRNA) and micro-RNA (miRNA) are released from high molecular mass double-stranded RNA or hairpin loop RNA (pre-miRNA) by the RNase III "dicer" (pre-miRNA is obtained from high molecular mass pri-miRNA by the RNase "drosha"). Upon dissociation into single strands, si- and miRNA form RNA-Induced Silencing Complexes (RISC) with argonaut proteins (teal ellipse); these complexes are directed to mRNA species with complementary sequences. As a result, those mRNAs are degraded by the endonuclease activity "slicer" of the argonaut protein (left) or are hindered from becoming translated (right).

Drosha, a ribonuclease type II, generates a hairpinlike structure, the precursor miRNA (pre-miRNA). Pre-miRNA is released into the cytoplasm where its double-stranded stem is cleaved by the ribonuclease type III **Dicer** yielding the mature miRNA duplex. For the formation of siRNA, double-stranded RNA—as generated, for instance, in the course of viral infections—is cleaved by Dicer alone independently of Drosha (**Figure 9.13**).

For activation, the double-stranded miRNA and siRNA molecules have to be unwound with the assistance of chaperone proteins, the **Argonauts**, exhibiting (as scaffold) binding domains for miRNA, siRNA, Dicer, and endonuclease activity. One strand of the duplex, the guide strand, stays with the argonaute protein, forming a **RNA-induced silencing complex (RISC)**, whereas the other strand, called passenger strand, is often degraded but can also be activated by RISC incorporation. The single-stranded siRNA or miRNA molecules direct RISC to mRNA molecules with complementary sequences, found primarily in the 3' untranslated regions (3'UTRs). Since the sequence homology between miRNA/siRNA and its target mRNA is incomplete, a single miRNA/siRNA may bind to several different mRNAs. Depending on the degree of complementarity, the mRNA is either degraded by the endonuclease or "slicer" activity of the argonaut protein or its translation is inhibited. This translational blockade can also lead to deadenylation and subsequent destabilization of the mRNA. Dicer and argonaut proteins have an additional function, which seems to be independent of mi- and siRNA formation: at least in yeast, they are required for cell cycle arrest upon genotoxic stress (see Section 12.8).

A special function of RNA interference is the removal of foreign RNA, which is frequently present in a double-stranded form—for example, after viral infection. It has been postulated, therefore, that RNA interference might complete the arsenal of the body's antiviral strategy consisting of antibody formation, production of antiviral cytokines (interferons; see Section 6.3), and inhibition of viral protein synthesis by translation factor eIF2 phosphorylation (see Section 9.3.1).

An additional role is played by miRNA and siRNA in yeast, where both are involved in the formation of transcriptionally inactive heterochromatin. This effect correlates with DNA methylation, and the idea that small non-coding RNAs may form complexes with chromodomain proteins (see Section 8.2.4) and DNA methyltransferases is as attractive as it is speculative.

It is easy to see that RNA interference provides an efficient tool for the investigator to experimentally knock down the transcripts of a selected gene. However, special caution must be taken about the specificity of any observed phenotypes, since individual siRNAs can also target numerous other genes than the one for which they were designed. Hence, either multiple individual siRNAs or one complex pool of many siRNAs must be used for each gene. For the discovery of RNA interference, Andrew Z. Fire and Craig C. Mello were honored by the Nobel Prize in Physiology or Medicine for 2006.

Large non-coding RNAs indicate a highly sophisticated signaling RNA network

In contrast to the rather homogeneous group of small non-coding RNAs, which are similar in size and function, large non-coding RNAs (lncRNAs) are highly heterogenous, consisting of 200 up to more than 1000 nucleotides. They are found in the nucleus and in the cytoplasm, can be mono-exonic or spliced, highly expressed or barely detectable, evolutionarily conserved or primate-specific. The distinction between coding and non-coding RNAs is challenging since short peptides could also be translated from lncRNAs. Since lncRNAs interact with DNA, RNA, and proteins, a plethora of different functions can be invoked for this class of transcripts, suggesting the existence of a **signaling RNA network** that interferes with the signaling protein network. Given the large number of lncRNAs, one may assume that most if not all signaling pathways are involved.

However, the precise function of lncRNAs has been characterized only for a few examples. For instance, the lncRNA XIST is an **epigenetic regulator** and essential for mammalian dosage compensation by silencing the second X chromosome in female cells. Another example is provided by the nuclear lncRNA MALAT1 that regulates the expression of a group of target genes and is involved in lung cancer metastasis. Finally, the cytoplasmic lncRNA CDR1, which is circularized, sequesters and inhibits miRNAs.

As far as formation and inactivation are concerned, lncRNA seems to be subject to the same rules as other RNA species such as mRNA.

Summary

- Non-protein-coding RNAs represent a major fraction of the eukaryotic transcriptome. Many of them have regulatory functions.

- Short regulatory RNA molecules (miRNAs, siRNAs) interact with complementary mRNA sequences thus inhibiting mRNA translation and triggering mRNA degradation. This process, known as RNA interference, enables a post-transcriptional regulation of gene activity.

- Long regulatory RNAs are a heterogeneous group of transcripts probably forming a signaling RNA network with a wide variety of different functions.

Further reading

Section 9.1 and 9.2

Akopian D et al (2013) Signal recognition particle: an essential protein-targeting machine. *Annu Rev Biochem* 82:693–721.

Djebali S et al (2012) Landscape of transcription in human cells. *Nature* 489:101–108.

Kapp LD & Lorsch JR (2004) Molecular mechanics of eukaryotic translation. *Annu Rev Biochem* 73:657–704.

Section 9.3

Bukau B, Weissman J & Horwich A (2006) Molecular chaperones and protein quality control. *Cell* 125:443–451.

Donnelly N et al (2013) The eIF2α kinases: their structures and functions. *Cell Mol Life Sci* 70:3493–3511.

Korrenykh A and Walter P (2012) Structural basis of the unfolded protein response. *Annu Rev Cell Cev Biol* 28:251–277.

O'Flanagan CH & O'Neill C (2014) PINK1 signalling in cancer biology. *Biochim Biophys Acta* 1846:590–598.

Proud CG (2007) Signalling to translation: how signal transduction pathways control the protein synthetic machinery. *Biochem J* 403:217–234.

Section 9.4

Alers S et al (2012) Role of AMPK-mTOR Ulk1/2 in the regulation of autophagy: cross talk, shortcuts, and feedbacks. *Mol Cell Biol* 32:2–11.

Beurel E, Grieco SF and Jope RS (2015) Glycogen synthase kinase-3 (GSK3): regulation, actions, and diseases. *Pharmacol Therap* 148:114–131.

Coll AP, Farooqi IS & O'Rahilly (2007) The hormonal control of food intake. *Cell* 129:251–262.

Crane BR & Young MW (2014) Interactive features of proteins composing eukaryotic circadian clocks. *Annu Rev Biochem* 83:191–219.

Fruhbeck G (2006) Intracellular signaling pathways activated by leptin. *Biochem J* 393:7–20.

Gallego M & Virshup DM (2007) Post-translational modifications regulate the ticking of the circadian clock. *Nat Rev Mol Cell Biol* 8:139–147.

Hardie DG (2015) AMPK: positive and negative regulation, and its role in whole-body energy homeostasis. *Curr Opin Cell Biol* 33:1–7.

Hardie DG and Alessi DR (2013) LKB1 and AMPK and the cancer-metabolism link – ten years after. *BMC Biology* 11:36–47.

Hastings MH, Brancaccio M and Maywood ES (2014) Circadian pacemaking in cells and circuits of the suprachiasmatic nucleus. *J Neuroendocrinol* 26:2–10.

Inoki K, Kim J and Guan KL (2012) AMPK and mTOR in cellular enery homeostasis and drug targets. *Annu Rev Pharmacol Toxicol* 26:2–10.

Jewell JL, Russell RC and Guan KL (2013) Amino acid signalling upstream of mTOR. *Nat Rev Mol Cell Biol* 14:133–139.

Llorens-Martin M et al (2014) GSK-3β, a pivotal kinase in Alzheimer disease. *Front Mol Sci* 7:1–11.

Mariño G et al. (2014) Self consumption: the interplay between autophagy and apoptosis *Nat Rev Mol Cell Biol* 15:81–94.

Richter EA & Hargreaves M (2013) Exercise, Glut 4, and skeletal muscle glucose uptake. *Physiol Rev* 93:993–1017.

Romeo Y, Zhang X and Roux PR (2012) Regulation and function of the RSK family of protein kinases. *Biochem J* 441:553–569.

Russell RC, Yuan HX & Guan KL (2014) Autophagy regulation by nutrient signaling. *Cell Res* 24:42–57.

Russo GL, Russo M & Ungaro P (2013) AMP-activated protein kinase: a target for old drugs against diabetes and cancer. *Biochem Pharmacol* 86:339–350.

Vaidyanathan K & Wells L (2014) Multiple tissue-specific roles for the O-GlcNAc posttranslational modification in the induction of and complications arising from type II diabetes. *J Biol Chem* 289:34466–34471.

Venerando A, Ruzzene M & Pinna LA (2014) Casein kinase: the triple meaning of a misnomer. *Biochem J* 460:141–156.

Section 9.5

Jonas S & Izaurralde E (2014) Towards a molecular understanding of microRNA mediated gene silencing. *Nat Rev Genet* 16:421–433.

Lee JT (2015) Epigenetic regulation by long noncoding RNAs. *Science* 338:1435–1439.

Rinn JL & Chang HY (2012) Genome regulation by long noncoding RNAs. *Annu Rev Biochem* 81:145–166.

Signal Transduction by Small G-Proteins: The Art of Molecular Targeting

10

G-proteins are regulatory GTPases playing a central role in signal processing. One distinguishes between heterotrimeric G-proteins (molecular masses between 85 and 90 kDa), which are almost exclusively coupled to transmembrane receptors, and monomeric ("small") G-proteins with molecular masses between 16 and 36 kDa.

The prototype of small G-proteins is the 21-kDa protein Ras (Ras stands for Rat sarcoma). It has lent the collective term **Ras superfamily** to the whole group. In mammals this family consists of more than 150 isoforms, which are grouped into several subfamilies according to common functional properties (**Table 10.1**). These subfamilies have been found in all eukaryotes so far investigated in this respect.

A main function of small G-proteins is to organize the assembly of highly ordered multiprotein complexes of signal transduction, particularly at membranes. They do this as long as they are "switched on" by GTP binding. As explained in Section 2.4, the energy required for such signal-transducing processes is provided by exergonic reactions occurring in the protein complexes assembled, whereas GTP hydrolysis due to the intrinsic GTPase activity of the G-protein is a switching-off mechanism. However, some small G-proteins, in particular Ran, also use the energy of GTP hydrolysis to facilitate targeted transport processes, frequently against concentration gradients.

As is typical for G-proteins, signal transduction is stimulated by guanine nucleotide exchange factors (GEFs), which catalyze the exchange of bound GDP for GTP, and is terminated by GTP hydrolysis. For the reasons explained in Section 2.4, small G-proteins exhibit only a minute intrinsic GTPase activity. Therefore, the support of GTPase-activating proteins (GAPs) is indispensable. GEFs and GAPs are the major targets of signals controlling the function of small G-proteins. Both are expressed in a large variety of isoforms that connect G-protein signaling with almost any other pathway of signal transduction.

10.1 Ras proteins: generation of order in signal transduction

10.1.1 An abundant oncogene

The small G-protein Ras (p21^{Ras}) was discovered as an oncogenic mutant protein of an RNA virus (retrovirus) causing sarcomas in rats. Later the nonmutated proto-oncogene product Ras was found to represent a highly conserved signaling protein of eukaryotes. Like all retroviral oncogenes, the *ras* proto-oncogene had been taken up by the virus from the host cell genome and undergone oncogenic mutation. For the *ras* gene it was shown for the first time that not only retroviruses but also tumor

Table 10.1 Ras superfamily of mammalian small G-proteins[a]

Ras family		Rho family		Rab family			Arf family	Ran family
Ha-Ras	Rheb	RhoA	RhoE	Rab1A	Rab11A	Rab26	Arf1	Ran
Ki-Ras	kB-Ras1	RhoB	Rnd3	Rab1B	Rab11B	Rab27A	Arf2	
N-Ras	kB-Ras2	RhoC	Rho8	Rab2	Rab12	Rab27B	Arf3	
R-Ras	RagA	RhoG		Rab3A	Rab13	Rab28	Arf4	
M-Ras	RagB	RhoH		Rab3B	Rab14	Rab29	Arf5	
RalA	RagC	TTF		Rab3C	Rab15	Rab30	Arf6	
RalB	RagD	Rac1		Rab3D	Rab16	Rab31	Sar1a	
Rap1A		Rac2		Rab4	Rab17	Rab32	Sar1b	
Rap1B		Rac3		Rab5A	Rab18	Rab33A	Arl1	
Rap2A		Cdc42		Rab5B	Rab19	Rab33B	Arl2	
Rap2B		Rnd1		Rab5C	Rab20		Arl3	
TC21		Rho6		Rab6	Rab21		Arl4	
Rit		Rnd2		Rab7	Rab22		Arl5	
Rin		Rho7		Rab8	Rab23		Arl6	
Rad		Tc10		Rab9	Rab24		Arl7	
Kir/Gem		RhoD		Rab10	Rab25		Ard1	
Signal transduction		**Signal transduction Cytoskeleton**		**Vesicle transport**			**Vesicle formation**	**Nuclear transport Mitosis**

[a]This selection of small G-proteins is grouped in subfamilies according to common structural features and cellular functions. Adapted from Matozaki T et al. (2000) *Cell Signal* 12:515–524.

tissue from humans contained genes mutated to oncogenes. To prove this, NIH-3T3 fibroblasts, an immortal but non-tumorigenic cell line widely distributed in laboratories, was transfected with DNA from a human bladder carcinoma. The result was oncogenic transformation. As a cause, a mutated *ras* gene was identified that encoded a protein where a glycine residue in position 12 of normal Ras was replaced by another amino acid. Since then it has been shown that practically all oncogenic Ras proteins, including the viral ones, suffer from point mutations in positions 12, 13, and 61 that hinder the interaction of Ras with regulatory GTPase-activating factors (see Section 10.1.3). However, in contrast to the situation in some animals, in humans the *ras* oncogene is generated by somatic mutation rather than being introduced by retroviral infection. Being involved in about 30% of all human tumor diseases, *ras* mutations rank among the most frequent causes of cancer, particularly of the killer types such as pancreatic carcinoma (90%), colorectal carcinoma (50%), lung adenocarcinoma (30%), and myeloid leukemia (30%). In view of this situation it is easily conceivable that Ras is one of the best investigated proteins ever. It has turned out to be a central switching device of eukaryotic signal processing. Unfortunately, an effective approach to cancer therapy based on a pharmaceutical intervention in Ras signaling is still missing.

10.1.2 An all-around signal transducer

Mammalian Ras is expressed in 23 isoforms, constituting a Ras subfamily of its own within the Ras superfamily of small G-proteins (see Table 10.1). The prototype is **H-Ras** (referring to a Harvey rat sarcoma virus). Closely related are **K-Ras** (originally found in a Kirsten rat sarcoma virus) and **N-Ras** (N stands for Neuroblastoma). Within the Ras subfamily, these three isotypes are those undergoing most frequently oncogenic mutations.

Quite early, the transforming effect of Ras mutations raised the suspicion that the protein could play a central role in cell proliferation. In fact, cell growth in culture could not be stimulated any longer by serum when the function of Ras had been suppressed, for instance by a monoclonal antibody. However, in the meantime the Ras signal has turned out not only to stimulate but also to inhibit cell proliferation, induce cell differentiation and even cell death depending on the cell type and the physiological context. Such apparently contradictory findings are explained, of course, by the fact that signaling per se is an ambiguous process: the meaning of a signal is allocated by the receiver rather than being an intrinsic property of the signaling factor, which may be used for quite different purposes.

Ras is anchored at the inner side of the plasma membrane by two lipid side chains, a farnesyl and a palmitoyl residue (see Figure 4.8 for an illustration). Without such anchoring the protein is ineffective, because its main function (that it shares with other small G-proteins) is to recruit other signaling proteins to membranes thus assembling large signal transducing complexes in the immediate vicinity of transmembrane receptors.

10.1.3 Upstream of Ras: receptors and ion channels

Like all small G-proteins, Ras has an extremely low GDP–GTP exchange frequency and a minute GTPase activity. It is, therefore, entirely under the control of GEFs and GAPs, both serving as sensors of input signals.

Ras guanine nucleotide exchange factors are sensors of a wealth of input signals

Several types of Ras-GEFs connect Ras with practically all types of membrane receptors and ensure cross talk with other signaling pathways. The most prominent Ras-GEF is **mSOS** (mammalian Son Of Sevenless; for an explanation of the name see **Sidebar 10.1**) that is controlled by Tyr kinase- and G-protein-coupled receptors. Owing to an SH3 domain interaction, mSOS forms a stable complex with the adaptor protein **Grb2** that contacts the autophosphorylated Tyr kinase-coupled receptors either directly (through an SH2-domain) or by means of the scaffold protein **Shc** (**Figure 10.1** and Section 7.1.2). Along the arrestin pathway mSOS is also activated by G-protein-coupled receptors (see Figure 10.1 and Section 5.8). Thus mSOS is the target of a large number of input signals.

mSOS is widely distributed, whereas other Ras-GEFs seem to be more restricted to certain tissues, such as nerves and T-lymphocytes. These GEFs are **GRF** (Guanine nucleotide Releasing Factor) and **GRP** (Guanine nucleotide Releasing Protein). Both contain calcium-binding domains and are activated by Ca^2 ions, GRP in addition by diacylglycerol. They connect Ras with Ca^2 transporting ion channels as well as with the phospholipase C-inositol 1,4,5-trisphosphate ($InsP_3$) pathway releasing Ca^2 and diacylglycerol. As shown in Chapters 14 and 16, calcium-transporting ion channels are controlled by a large number of neurotransmitters as well as by membrane depolarization and action potentials. Indeed, in brain neurons the Ras signaling pathway is involved in the regulation of memory fixation. Thus in mice with a genetic GRF knockout learning capacity and memory are considerably curbed. A specialty of both mSOS and GRF is their integrating

Figure 10.1 Upstream of Ras: signaling pathways leading to an activation of Ras guanine nucleotide exchange factors (GEFs) Activation of Ras-GEFs (colored box) and Ras is mediated by a wide variety of transmembrane receptors. From left to right: a phosphorylated (P) G-protein-coupled receptor activates, via the adaptor protein arrestin, a cytoplasmic tyrosine kinase (Src). By interacting with the adaptor Grb2 (and Shc), Src stimulates the Ras-GEF mSOS. mSOS is also activated by autophosphorylated tyrosine kinase-coupled receptors. The Ras-GEFs GRF and GRP are activated by an increase of the cytoplasmic Ca^2 concentration as brought about by voltage-dependent and receptor-coupled cation channels, or along the G-protein–phospholipase $C\beta$ and $C\varepsilon$ pathways. Moreover, PLCε exhibits Ras-GEF activity on its own. Not shown are signaling pathways that activate Ras through an inhibition of Ras-GAPs.

function: apart from Ras, both GEFs also activate the small G-proteins of the Rho family. This property is due to an additional Rho-GEF domain in the mSOS and GRF molecules (see Section 10.3.1). In fact, as already mentioned, many cellular Ras effects depend on simultaneous activation of Rho-GTPases, providing a logical AND gate (see Section 1.3). A further Ras-GEF, **PLCε**, will be discussed in more detail in Section 10.1.4.

Ras GTPase-activating proteins

At least five isoforms of Ras-GAPs are known for humans. Like Ras-GEFs they are controlled by various input signals. As discussed in Section 2.4.4, a GAP may be considered to be a regulatory subunit of a G-protein providing an arginine residue required for GTP hydrolysis, which thus becomes accelerated by a factor of 10^6. As in Section 10.1.1, this mechanism does not work for Ras mutated to an oncoprotein because, due to the exchange of glycine 12 or 13 by another, larger amino acid residue, access of the Arg finger of GAP to the catalytic domain of Ras is hindered. Thus, the GTPase activity cannot be stimulated any more by GAP (mutation of glutamine 61 has a similar consequence because this residue stabilizes the transition state of GTP hydrolysis). Since the major role of GTP-loaded Ras is to organize signaling protein complexes at the membrane, the mutated Ras resembles a switch blocked in the ON position, or a valve that cannot be closed anymore.

The best-known Ras-GAP is **p120 GAP**. It is activated by Ras in a negative feedback loop but, as indicated by its SH2, SH3, and PH domains, has additional possibilities for interaction that are not fully understood yet. Another Ras-GAP is **neurofibromin**. A loss-of-function mutation of this protein causes the Recklinghausen neurofibromatosis type 1, a hereditary disease characterized by multiple benign (sometimes also malignant) tumors of the nervous tissue appearing predominantly in the skin. A Ras-GAP activated by Ca^2 ions is **CAPRI** (CAlcium-Promoted Ras Inactivator). Its Ca^2-binding sites are two C2 domains enabling Ca^2-dependent binding to membrane phospholipids, thus bringing CAPRI into contact with membrane-bound Ras. Ca^2 ions, therefore, have a dual effect: on the one hand Ras is activated by

Ca^2 -dependent GEFs, while on the other it is inactivated by Ca^2 -dependent CAPRI. How the cell is able to reconcile these opposite effects is not fully understood.

10.1.4 Downstream of Ras: kinases and phospholipases

To perform its organizing effect on signal transduction Ras interacts with effector proteins, recruiting them to the membrane where they come in contact with receptors and other partners. Ras collects such proteins as long as it is in the active GTP-loaded state. The energy required for this organizing process is supplied by complex formation (the binding energies of the interacting partner molecules) and subsequent reactions within the complex rather than by Ras-catalyzed GTP hydrolysis. In fact, the latter only terminates the whole process. Of the various Ras effectors, p120 Ras-GAP has been mentioned already. Other effector proteins studied more thoroughly include protein kinases of the RAF family, phosphatidylinositol 3-kinases, PLCε, and GEFs of the Ras-like Ral proteins.

RAF kinases and the Ras–RAF–ERK cascade

Originally, Raf was discovered as a retroviral oncogene, v-RAF (Rapidly Accelerated Fibrosarcoma), causing rat sarcoma. Later on, the corresponding cellular proto-oncogene was identified. It exists in three isoforms: A-RAF, B-RAF, and C-RAF. The viral Raf oncoprotein is a truncated cellular RAF lacking an autoregulatory amino (N)-terminal domain, which normally inhibits RAF activity intrasterically, resembling a pseudosubstrate motif (see Section 2.6.1). The autoregulatory domain overlaps with two binding sites of Ras-GTP that ensure the membrane recruitment of RAF by Ras (**Figure 10.2**). Membrane-associated RAF is controlled by various input signals inducing a complex pattern of interactions with several protein kinases and scaffold proteins that has not yet been elucidated in every detail.

Interaction with Ras is a necessary but not sufficient condition for RAF activation. RAF has to be phosphorylated, in addition, in the carboxy (C)-terminal region. Two critical phosphorylation sites are a Ser and a Thr residue in the activation loop that probably are autophosphorylated (see Figure 10.2). To fully activate RAF, further phosphorylations at both sides of the active center are necessary: at a Ser residue (catalyzed by PKC) and a Tyr residue (catalyzed by Tyr kinases Src and JAK), and at a C-terminal Ser residue (probably catalyzed by the p21-activated protein kinase, PAK2; see Sections 10.3.4 and 11.3 for more details about this kinase type). Stimulation by PKC of RAF activity and in turn of the mitogenic MAP kinase cascade is one reason for the frequently observed mitogenic effect of endogenous signals or drugs such as the tumor-promoting phorbol esters that activate PKC. PAK2 is stimulated by small G-proteins of the Rho/Rac family (see Section 10.3.2), indicating that

Figure 10.2 Complex regulation of the protein kinase C-RAF The kinase has two binding sites for Ras-GTP in the C-terminal region: a Ras-binding domain (RasBD) and a cysteine-rich domain (CRD). Phosphorylation by protein kinase A or B (PKA or PKB) of serine residues (S) in these domains hinders the interaction with Ras and induces sequestration by 14-3-3 protein; that is, phosphorylation has an inhibitory effect (teal arrows). The kinase domain (gray rectangle), including the activation loop (AL), is localized in the C-terminal region. Phosphorylations of serine (S), threonine (T), and tyrosine (Y) residues in this region have a stimulatory effect on RAF activity (black arrows). The active form of RAF is stabilized by 14-3-3 proteins that recognize a C-terminal phosphoserine residue. Protein kinases found or assumed to catalyze the individual phosphorylations are shown. Auto-P, autophosphorylation.

RAF phosphorylation is one of the mechanisms ensuring the synergism of Ras and Rho/Rac signaling.

RAF is also phosphorylated at Ras-binding sites. This phosphorylation is catalyzed either by protein kinase A (PKA) or by protein kinase B (PKB)/Akt and suppresses RAF activity because it hinders the interaction with Ras, instead enabling a binding of RAF to so-called **14-3-3 proteins** that sequester the inactive kinase. 14-3-3 proteins also promote RAF effects by stabilizing the active form of RAF, provided the kinase had been phosphorylated before at the C-terminal Ser residue, the putative target of PAK2 (see Figure 10.2). This still incomplete picture of RAF regulation indicates that this protein kinase is connected with a wide variety of signaling pathways and provides a key element in Ras signaling. In fact, synthetic RAF inhibitors have proved to be promising anti-cancer drugs.

The only substrate proteins of RAF kinases identified with certainty are the RAF kinases themselves (becoming autophosphorylated) and the MAP2 kinases MEK1 and MEK2. Therefore, RAF fulfils the role of a MAP3 kinase within a MAP kinase module, with the MAP kinase ERK as a signaling outlet (see Sections 2.6.5 and 11.2). This module RAF–MEK–ERK has been called the mitogenic MAP kinase module because it predominantly (but not exclusively) processes mitogenic signals. It is controlled by all input signals of Ras and RAF. The (historically seen) canonical one among these signaling pathways starts from Tyr kinase-coupled growth factor receptors (**Figure 10.3**).

The RAF isoform **B-RAF** (but not yet A-RAF and C-RAF) has been found to be mutated to an oncoprotein in several human malignancies such as malignant melanoma as well as thyroid, colorectal, and ovarian carcinoma. In most cases a valine residue in the activation loop is replaced by a glutamate residue. This mutation creates a negative charge that under normal conditions is introduced by reversible phosphorylation. The result is a constitutively active, nonregulated RAF mutant playing a role comparable to that of mutated Ras.

14-3-3 proteins or how signaling proteins are taken out of circulation

14-3-3 proteins are multifunctional components of cellular signal processing. Their strange name refers to their discovery: in an attempt to isolate proteins from bovine brain by chromatography and electrophoresis, in a fraction numbered 14-3-3, an acidic protein of about 30 kDa was found that later turned out to be representative

Figure 10.3 Activation of the mitogenic MAP kinase module by a receptor tyrosine kinase (RTK): an example of the organizing function of Ras (A) In the absence of an exogenous input signal, the components of the signal-processing protein complex are present in a disordered form in the cytoplasm while Ras is membrane-anchored and in the inactive GDP-loaded form. (B) By inducing autophosphorylation (P) of the receptor, an input signal (such as a growth factor) has triggered the organization of a highly ordered multiprotein complex bound at the surface of a scaffold protein (SP). Grb2 is an adaptor protein linking the activated receptor to the Ras-GEF mSOS that activates Ras and, as a consequence, the MAP kinase module consisting of the three protein kinases RAF, MEK, and ERK. As soon as the activation of the module has become stabilized by phosphorylation and binding of RAF to 14-3-3 proteins (not shown), Ras-GTP is free for a new cycle. The process is terminated by dephosphorylation of the receptor and GTP hydrolysis by a Ras/Ras-GAP complex.

of a whole family of highly conserved eukaryotic proteins. For humans, seven corresponding genes have been identified.

14-3-3 proteins are specialized for forming complexes with other proteins by binding to sequence motifs with phosphorylated Ser residues. In this respect they resemble proteins with SH2 and PTB domains that recognize motifs with phosphorylated Tyr residues.

The interaction of a 14-3-3 protein with its partner is due to a special structure: 14-3-3 proteins associate into homo- and heterodimers, forming a pocket of anti-parallel helices that fits two phosphorylated substrate proteins (or one protein with multiple phosphorylations). In most substrate proteins the 14-3-3-interacting domain is -Arg-Ser-X-X-Ser(P)-X-Pro- or -Arg-X-X-X-Ser(P)-X-Pro-, where Ser(P) is the phosphorylated Ser residue and X is any amino acid. The negative charge rather than the chemical structure of the phosphate group is essential for the interaction; non-phosphorylated proteins are also bound by 14-3-3 proteins provided their interacting motif contains several acidic amino acids such as -Arg-Ser-Glu-Ser-Glu-Glu-. In some cases also a special zinc-finger structure does the job.

In some ways the function of 14-3-3 proteins resembles that of chaperones: they alter the conformation and, thus, the function of a protein, albeit dependent on whether or not this carries a negative charge at a specific site. Consequently, 14-3-3 proteins may either inhibit or promote interactions between proteins (**Figure 10.4**). Inhibition is mostly due to sequestration of proteins for a definite period of time (until the effector protein becomes dephosphorylated, for instance). One example is provided by RAF kinase phosphorylated by PKA; another example is the pro-apoptotic protein Bad that, upon phosphorylation by PKB/Akt, is taken out of circulation by 14-3-3 proteins (this represents one of the pathways along which PKB suppresses programmed cell death; see Sections 4.6.6 and 13.2.5).

Vice versa, 14-3-3 proteins may also facilitate protein interactions in that they act as adaptors, bringing together proteins or determining their intracellular localization, or they protect proteins from being inactivated by dephosphorylation or proteolysis (it should be remembered that ubiquitylation frequently requires a priming

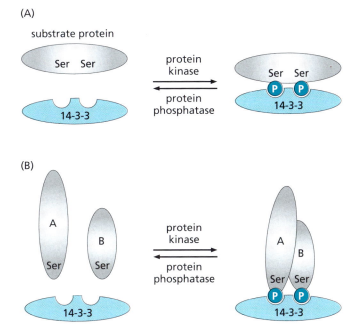

Figure 10.4 Mode of action of 14-3-3 proteins Dimeric 14-3-3 proteins bind other proteins, recognizing them by two phosphorylated serine (Ser) residues (or a corresponding negatively charged motif). (A) Sequestration of a twofold phosphorylated protein is shown, resulting in either inactivation or activation. (B) The adaptor function of 14-3-3 is depicted, bringing together two interacting proteins upon specific Ser phosphorylation.

Sidebar 10.1 Ras and MAP kinase cascades in yeast and nonvertebrates Ras proteins have been found in all eukaryotes from yeast to humans.

In the fission yeast ***Schizosaccharomyces pombe***, RAS1 (the only Ras form expressed in this organism) is a component of a signaling pathway leading from surface receptors to the genome and resembling a Ras–RAF–MAP kinase cascade (**Figure 10.5**). The activating signal is the mating pheromone of the haploid gametes that interacts with a G-protein-coupled receptor, activating the Ras-GEF Ste6 (the name refers to sterile mutants) via the α-subunit of the trimeric G-protein GPA1 (see Section 5.2). The Ras-controlled MAP kinase module consists of the protein kinases Byr2 (a MAP3 kinase functionally resembling a RAF kinase), Byr1 (a MAP2 kinase resembling MEK), and Spk1 (a MAP kinase resembling ERK). The substrates of Spk1 are transcription factors of genes controlling the mating phenotype. Another Ras effector is a GEF of the small G-protein Cdc42 regulat-

ing cytoskeletal dynamics and cell shape. This pathway provides a clear analogy to vertebrate Ras signaling, which also cooperates with a Cdc42 homolog and other small G-proteins affecting the cytoskeleton. In fact, RAS1 of *S. pombe* is quite similar to H-Ras of mammals.

In multicellular nonvertebrates, Ras has various functions. Well known is its role in morphogenetic processes. The classical example of a morphogenetic Ras–RAF–MAP kinase cascade is provided by the development of photoreceptor cells in the *Drosophila* eye. The ommatides of the insect's compound eyes are composed of eight different types of photoreceptor cell inducing each other during embryogenesis. So an R7 cell arises only when its precursor cell receives a specific differentiation signal from an adjacent R8 cell. This R8-derived signal is transmitted by a membrane-bound protein "Boss," a relative of the epidermal growth factor of vertebrates. By a juxtacrine interaction, Boss activates an EGF-receptor-related receptor Tyr kinase

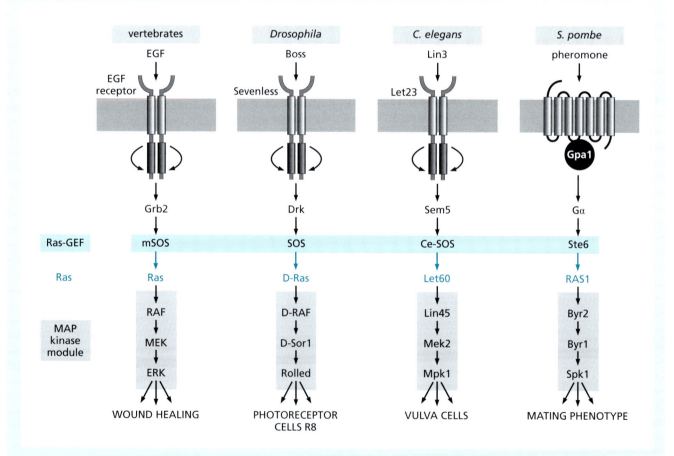

Figure 10.5 Ras-MAP kinase cascades Shown are two morphogenetic pathways in *Drosophila melanogaster* and *Caenorhabditis elegans* compared with growth factor-stimulated signaling in vertebrates and pheromone signaling in *Schizosaccharomyces pombe*. Curved arrows symbolize the autophosphorylation of the receptor tyrosine kinases epidermal growth factor (EGF) receptor, Sevenless, and Let23. Yeast cells do not contain genes encoding receptor tyrosine kinases. Instead they activate the Ras-MAP kinase pathway by means of G-protein-coupled receptors. For other details see text.

"Sevenless" exposed at the surface of R7 precursor cells (the name "Sevenless" refers to a loss-of-function mutation of the receptor resulting in animals lacking R7 cells). Upon autophosphorylation, the receptor interacts with an adaptor protein Drk (resembling the mammalian Grb2) and activates the Ras-GEF "son of sevenless (SOS)." The signaling cascade thus stimulated downstream provides an exact counterpart to the mammalian Ras–RAF–ERK cascade (see Figure 10.5).

Quite a similar signaling cascade controls vulva development in *Caenorhabditis elegans*. As in the *Drosophila* eye, adjacent cells (called anchor cells) induce the differentiation of precursor cells by means of an EGF-related factor called Lin3 (see Figure 10.5). Along analogous signaling pathways, further differentiation processes are controlled in the nematode. In the cases mentioned here, the targets of the MAP kinase cascades always are transcription factors of the Ets family (see Section 11.6.1).

phosphorylation of the substrate protein). An example is provided by RAF kinases phosphorylated by PKC and PAK.

Phosphatidylinositol 3-kinases

These enzymes (PI3Ks) have been treated already in detail in Section 4.4.3. They catalyze the formation of membrane-bound second messengers such as phosphatidyl-inositol 3,4,5-trisphosphate, serving as docking sites for proteins with PH domains. The signal-processing complexes thus assembled promote subsequent reactions that are essential for most Ras effects. Certain (but not all) PI3Ks are activated by a combined action of Tyr kinase-coupled receptors and Ras in that the receptor inactivates an auto-inhibitory subunit p85 and Ras takes care of the correct membrane localization of the catalytic subunit (**Figure 10.6**).

Phospholipase Cε

A particularly interesting and versatile Ras effector is PLCε. The enzyme belongs to those type C phospholipases that hydrolyze phosphatidylinositol 4,5-bisphosphate, yielding the second messengers DAG and InsP$_3$. Thus, the role of PLCε in cellular signal transduction resembles that of PLCβ and -γ (see Section 4.4.4). However, in contrast to the latter, PLCε possesses two Ras-binding domains and a Ras-GEF domain; that is, it has two enzymatic activities (PLC and GEF). Therefore, the phospholipase is both a Ras effector and a Ras activator. The activating effect is amplified by the release of DAG and Ca^2 (via InsP$_3$; see Section 4.4.4) both activating the Ras-GEFs GRP and GRF (see Figures 10.1 and **10.7**). On the other hand, InsP$_3$ is a participant in *negative* feedback control of Ras by releasing the calcium that activates the Ras-GAP CAPRI.

Figure 10.6 Role of Ras in the activation of phosphatidylinositol 3-kinase type α For the activation of PI3Kα, Ras cooperates with a Tyr kinase-coupled receptor such as a growth factor receptor. PI3Kα consists of a regulatory subunit p85 and a catalytic subunit p110 (see Section 4.4.3). On the one hand, p85 inhibits enzymatic activity; on the other hand, it binds via an SH2 domain to the autophosphorylated receptor (P, phosphate group). As a result, the interaction with p110 is interrupted and the Ras-binding domain (RasBD) of p110 becomes exposed (teal arrow), enabling p110 to bind to membrane-anchored Ras-GTP. The association with the membrane is reinforced by a Ca^2–phospholipid-binding C2 domain. Thus the catalytic domain of p110 gains direct contact with its phospholipid substrates such as phosphatidylinositol 4- or 5-phosphate (PIP).

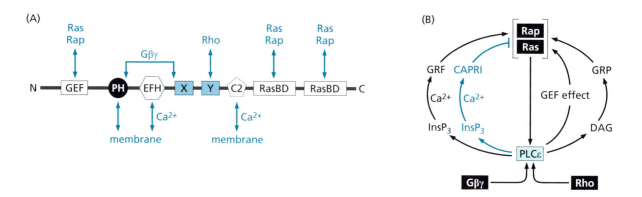

Figure 10.7 Integrating function of phospholipase Cε (A) Domain structure together with major interaction partners of PLCε is shown (GEF, guanine nucleotide exchange factor domain; PH, PIP$_2$-interacting pleckstrin homology domain; EFH (EF-hand) and C2, Ca^2 - and phospholipid-binding domains; RasBD, Ras/Rap-binding domain; X and Y, catalytic domains). (B) PLCε is a Ras and Rap effector and, at the same time, both a positive and negative regulator of Ras and Rap activity. Stimulatory pathways are shown by black arrows; inhibitory pathways are shown in teal. For other details see text.

This is not the end of the story, since PLCε has still more opportunities. It is also an effector and a GEF of the small G-proteins of the Rap type that alternatively are activated along cAMP-controlled pathways. Moreover, PLCε has been reported to become activated by the βγ-subunits of trimeric G-proteins as well as by another small G-protein, Rho, that interacts with a special insert in the catalytic domain found only in PLCε but in no other C-type phospholipase. Most probably, the role of the G-proteins (Ras, Rap, Rho, Gβγ) is to direct the phospholipase to specific sites of the membrane where it becomes anchored through its PH, EF-hand, and C2 domains. As described in Chapter 4, reversible membrane anchoring is a critical step in phospholipase activation since it allows the enzyme to interact with its phospholipid substrates. Thus, the temporal and spatial control of PLCε activity provides an illustrative example of the targeting function of small G-proteins.

Although some PLCε effects still need final proof, at least as far as their physiological meaning is concerned, it is obvious that this phospholipase is able to connect various signaling pathways. The positive and negative feedback loops established by the enzyme might indicate an oscillating, frequency-modulated signaling system.

Summary

- Transducing mitogenic, differentiation-controlling, and pro-apoptotic signals, the small G-protein Ras occupies a central position in cellular signal processing. To perform its major function, recruiting signaling proteins at the membrane, Ras is activated by GTP-binding and anchor at the inner side of the plasma membrane. GTP-loading is catalyzed by various Ras-guanine nucleotide exchange factors (GEFs) that are targets of all kinds of input signals. The extremely low GTPase activity of Ras is up-regulated by GTPase-activating proteins (GAPs) that play an important role in feedback control and as receivers of input signals.

- Major effector proteins of Ras-GTP are RAF Ser/Thr kinases. As MAP3 kinases, they establish the contact between Ras and the mitogenic MAP kinase module RAF–MEK1,2–ERK. RAF activity is regulated by 14-3-3 proteins. These resemble chaperones that modulate protein activity dependent on phosphorylation. A binding of Ser-phosphorylated proteins to 14-3-3 may result either in temporal inactivation by sequestration or in a facilitated interaction with other proteins and protection from inactivation.

- Other important Ras effector proteins are PI3K and phospholipase Cε The latter is both a Ras effector and a Ras activator. It connects Ras signaling with various other signaling pathways, thus promoting signaling cross talk.

- Mutated Ras is an abundant oncoprotein. Since its interaction with GAPs is impaired, oncogenic Ras resembles a switch stuck in the ON position.

10.2 Other G-proteins of the Ras subfamily: an unfinished story

Compared with the Ras proteins proper, particularly H-, K-, and N-Ras, much less is known about the functions of other members of the Ras subfamily. Many of them seem to be involved in the cross talk between pathways controlled by different small G-proteins.

Rap proteins are the closest relatives of Ras. They have been connected with regulation of the actin cytoskeleton and of intercellular adhesion. Indeed, Rap GTPases strengthen cellular contacts, particularly focal adhesions organized by integrins (see Section 7.2.3) and adherens junctions organized by cadherins (see Section 7.3.1). To this end they recruit and activate various proteins engaged in cellular contact.

Rap-GEFs are activated by the second messengers DAG, cAMP, and Ca^2 . One of the Rap-GEFs is, indeed, identical with the Ras-GEF GRP2, exhibiting binding sites for DAG and Ca^2 , while another Rap-GEF binds cAMP. The discovery of this factor, called **EPAC** (Exchange Protein Activated by Cyclic AMP), has broken the rule that in vertebrates only PKA and certain ion channels are effector proteins of cAMP.

Ral (Ras-like) is the common name of a group of small G-proteins (see Table 10.1) connecting Ras with Cdc42 signaling (see Section 10.3). Ral-GEF is stimulated by Ras, and the Ral-GTP produced along this pathway is a negative regulator of Cdc42, exhibiting Cdc42-GAP activity. Only little is known about other functions of Ral although various effector proteins have been found. Deregulation of Ral signaling in tumors nourishes the hope that the protein may become a target of novel anti-cancer drugs.

The abundant GTPases **Rheb** and **Rag** have turned out to control the mTOR signaling pathway regulating ribosomal protein synthesis (see Section 9.4.1), and the G-proteins **Rad** and **Kir/Gem** have been found to inhibit the activity of voltage-dependent Ca^2 channels and to suppress effects of Rho-type GTPases by inhibiting Rho-controlled protein kinases of the ROK family (see Section 10.3). Finally, the Ras-like GTPase **TC21 (R-Ras)** has been identified as an oncoprotein, being able to transform cell lines derived from human tissues. Moreover, it is overexpressed in several types of human cancer, indicating a role in clinical tumor development.

Summary

Only fragmentary information is available about the cellular functions of other members of the Ras subfamily. Some of them such as Rap, Ral, Rad and Kir/Gem seem to cooperate with other small G-proteins in the regulation of the actin cytoskeleton. Rheb and Rag play a key role in the mTOR signaling pathway regulating ribosomal protein synthesis.

10.3 GTPases of the Rho family are master regulators of the actin cytoskeleton and more

Rho-type GTPases constitute a distinct family of small Ras-like G-proteins. They are found in vertebrates, invertebrates, yeast, and plants. In vertebrates this family, named after the protein Rho (Ras homology protein), consists of at least 22 GTPases grouped into several subfamilies. Among these, the **Rho** proteins proper and the **Rac** (an artificial acronym; note that in the literature Rac is used also for PKB/Akt, meaning "related to A- and C-kinases") and **Cdc42** proteins (originally discovered in a Cell-cycle defective mutant 42 of *S. cerevisiae*) are best known. These proteins are not only key players in the regulation of cell shape, cell polarity, and cell motility but are also involved in enzyme regulation as well as in the transcriptional control of various genes. Moreover, they are essential for tumorigenesis caused by oncogenic Ras mutation, although they have not been found to undergo oncogenic mutations on their own. Nevertheless, cell cultures may become neoplastically transformed by an overexpression or overactivation of Rho GTPases.

Among the functions of Rho-type GTPases, their effects on the actin cytoskeleton have been investigated most thoroughly. The cytoskeleton is a highly ordered, although dynamic, structure responsible for the cell shape and all kinds of cellular movements as well as for intracellular vesicle trafficking, endocytosis, and cytokinesis. Moreover, together with the microtubular and intermediate filament skeletons and cellular membranes, it represents a supporting matrix of the data-processing protein network, thus playing a pivotal role in signal transduction (see Section 1.5).

10.3.1 Upstream of Rho, Rac, and Cdc42: all kinds of transmembrane receptors

While Ras proteins are constitutively anchored in membranes, Rho-type GTPases shuttle between cytoplasm and membranes. The membrane anchor is a geranylgeranyl side chain. In the GDP-loaded, inactive Rho protein, this lipid residue is hidden in a complex with a specific **GDI** (GDP Dissociation Inhibitor) protein that renders the GTPase water-soluble and inhibits GDP–GTP exchange. How this inhibitory interaction is relieved is not known. There is no doubt, however, that regulation of the Rho–GDI interaction plays a critical role in Rho signaling. Equally important as receivers of input signals are GEFs and GAPs.

Guanine nucleotide exchange factors and GTPase-activating proteins

More than 70 different GEFs activating Rho proteins have been identified. They constitute a very diverse family of proteins. The large number and high variability of GEFs indicates not only a wide variety of input signals (only a few of which have been identified yet with certainty) but also an involvement of Rho-type GTPases in many cellular processes.

The catalytic domain of Rho-GEFs is also known as DBL homology (DH) domain (referring to an oncogenic Rho-GEF of Diffuse B-cell Lymphoma). Indeed, oncogenic mutations are known for several Rho-GEFs. Interestingly, in addition to their Ras-specific GEF domains, Ras-GEFs such as mSOS and GRF also contain DH domains, indicating again direct cross talk between Ras and Rho signaling.

Most Rho-GEFs have a PH domain C-terminal to the GEF domain. Certainly, the PH domain plays a role in the catalytic process; the mechanism of this effect, however, is still a matter of conjecture. Thus the obvious idea that the PH domain facilitates membrane anchoring of GEFs by binding to 3-phosphorylated phosphatidylinositols (see Section 4.4.5) is not convincingly supported by experimental data. To gain full activity, some Rho-GEFs must be phosphorylated, in addition, at Tyr residues, a reaction catalyzed by Tyr kinases of the Src family (see Section 7.2) and others. Src kinase signaling is induced by Tyr kinase-coupled and heptahelical receptors that are arrestin-coupled (**Figure 10.8**).

Another signaling pathway leading to an activation of Rho-GEFs has the receptors of lysophospholipids, thrombin, endothelin, and thromboxane A_2 as starting points that may couple with trimeric $G_{12,13}$-proteins (see Section 4.3.1). The $G\alpha_{12,13}$-subunits stimulate certain Rho-GEF isoforms directly, while the same Rho-GEFs turn up the GTPase activity of these $G\alpha$-subunits acting as regulators of G-protein signaling in a negative feedback loop (see Section 4.3.2). This regulatory process provides an illustrative example of an interaction between large and small G-proteins.

Figure 10.8 Upstream of Rho, Rac, and Cdc42 To activate a Rho-, Rac-, or Cdc42 GEF, Tyr kinases of the Src family are assumed to cooperate with PI3 kinases. Src kinases are activated by arrestin-coupled heptahelical receptors or by autophosphorylated Tyr kinase-coupled receptors (P, phosphate). The latter also control, together with Ras, the activity of PI3 kinases of the types α, β, and δ, whereas the γ isoform is activated by Gβγ-subunits mainly released from receptor-coupled $G_{i,0}$-proteins. In addition, at least some GEFs are activated by the α-subunits of trimeric $G_{12,13}$-proteins. A direct connection between Tyr kinase-coupled receptors and Rho/Rac/Cdc42 GEFs, indicated by the broken arrow with the question mark, is still a matter of debate. Not shown are pathways including Ras-GEFs with DBL homology domains, along which Rho, Rac, and Cdc42 in principle may become stimulated by nearly all signals that also activate Ras. Also not shown are signaling pathways involving Rho/Rac/Cdc42 GDIs and GAPs, some of which also become activated by Tyr phosphorylation.

Owing to their highly variable and complex activation modes, Rho-GEFs provide perfect examples of logical gates interlinking many different signaling pathways. Considering the biological functions of Rho proteins, such interactions make sense, as for the majority of cell physiological processes (for instance, Tyr kinase-mediated cell proliferation) controlled changes of cell shape and motility are mandatory.

The extremely weak GTPase activity of Rho-type proteins requires stimulation by specific GAPs, which in mammals exist in about 80 isoforms. These are mostly large proteins with various interaction domains enabling intense cross-linking of Rho/Rac/Cdc42 signaling with other signaling pathways. As in the case of Rho-GEFs, input signals break up auto-inhibitory conformations of the GAPs. For this purpose, again the phosphorylation of Tyr residues seems to play an essential role.

Specificity by scaffold proteins

It has been found that individual GEFs and GAPS are coupled with distinct signaling pathways. Considering the large number and diversity of these proteins, how can cells select a specific GAP or GEF for a special purpose? The answer to this mystery is scaffold proteins that direct the GEFs and GAPs to distinct cellular sites and organize the formation of specific complexes with selected regulatory and effector proteins. As a result, scaffold protein A expressed in one cell type may couple Rho-type GTPases with gene expression, whereas scaffold protein B expressed in another cell type brings about a coupling with the actin cytoskeleton. Dictation of specificity by scaffold proteins is turning out to be a fundamental principle of signal processing. We have seen this already for protein kinases (see Section 4.6). In Section 11.4 we shall encounter another example illustrating this principle.

10.3.2 Downstream of Rho, Rac, and Cdc42: control of cell shape and motility

As we have seen, GTPases of the Rho family are controlled by a wide variety of input signals. These correspond to an equally large number of effector proteins, indicating a wide array of functions that are still far from being known in all details. Among these functions, effects on the state of the actin cytoskeleton and its interaction with the motor protein myosin (**Sidebar 10.2**) have been studied most thoroughly, revealing that many proteins involved in the regulation of these processes have interaction domains for the GTPases of the Rho family. In fact, such studies have ranked Rho, Rac, and Cdc42 among the key components of signaling pathways that control the architecture and motility pattern of cells as well as changes of cell shape

Sidebar 10.2 Rho, Rac, and Cdc42 control uptake of extracellular material Rho-type GTPases are typically involved in the uptake of extracellular material by phago- and pinocytosis and clathrin-dependent endocytosis. These processes are due to local changes of cell shape and motility.

Phagocytosis means uptake of larger particles. Some cell types such as macrophages and granulocytes are specialized in eliminating microorganisms by "professional" phagocytosis. They do this by clasping the foreign body with membrane extrusions called pseudopodia, thus forming an organelle, the phagosome. The process is triggered by certain receptors reacting with the complement factor C3b as well as with the Fc portion of immunoglobulins by which the invader has been "decorated." Via the corresponding GEFs, C3b receptors stimulate Rac and Cdc42, while Fc receptors activate Rho. In addition to phagocytes, other cells are qualified to respond by (in this case "nonprofessional") phagocytosis, for instance, in the final phase of programmed cell death (apoptosis) when the remnants of decayed cells are taken up by neighboring cells.

Pinocytosis, understood as the formation of vacuoles filled, for instance, with extracellular fluid or (in the case of antigen-presenting cells) antigens is also controlled by Rac and Cdc42. A special form of pinocytosis is **clathrin-mediated endocytosis** of transmembrane receptors or of material bound to such receptors (see Section 5.8). In the regulation of this process, again GTPases of the Rho family are involved in a way still not entirely elucidated.

Like uptake reactions, the secretory processes of **exocytosis** also depend on reversible changes of the actin cytoskeleton that are controlled by Rho/Rac/Cdc42 proteins which here cooperate with Rab and Arf proteins (see next section).

occurring in the course of mitosis, endocytosis, exocytosis, and pinocytosis. A quick look at the extremely dense packing and branching of the actin fibers filling the cell (**Figure 10.9**) immediately demonstrates that such processes would be impossible without a rapid and continuous reorganization of this cytoskeleton, which has to be looked upon as a highly dynamic structure rather than a stiff scaffold.

Originally the effects of Rho-type proteins were studied by use of fibroblasts grown in cell culture. Upon proper activation, these and some other cell types undergo characteristic morphological changes that have been found to depend on cooperation of Cdc42, Rac, and Rho proteins, with each of them evoking a distinct cellular response (summarized in **Figure 10.10**).

When Cdc42 signaling is stimulated (for instance, along one of the pathways shown in Figure 10.8 or by a constitutively active Cdc42 mutant protein), the cells develop **filopodia**. These are fingerlike membrane extrusions possibly serving as some sort of antennae that help a migrating cell to reconnoiter the terrain. The characteristic cell morphological response to Rac activation consists of the formation of **lamellopodia** and **ruffled membranes**. Lamellopodia are fan-shaped mobile extrusions transiently developing at the front of migrating cells. Both filopodia and lamellopodia are pushed forward by targeted actin polymerization and are filled with either unbranched or branched actin fibers (see Figure 10.9).

The cell responds to a selective stimulation of Rho by formation of **actomyosin stress fibers**. These are bundles of unbranched actin filaments decorated with myosin molecules that traverse the whole cell. Upon stimulation, these fibers contract under ATP consumption, causing changes of the cell shape and all kinds of movements. At the membranes they are anchored by special structures called **focal adhesions**. In a reversible way, focal adhesions sites fix the cell (during migration, particularly the membrane extrusions) with the supporting matrix (*in vitro* the Petri dish, *in vivo* connective tissue structures), thus providing a creeping (amoeboid) movement. In fact, the Rho-activating input signals frequently are derived from the extracellular matrix and are transmitted by integrins, a special type of Tyr kinase-coupled receptors concentrated at the focal adhesion sites. Combining the properties of highly dynamic cell adhesion molecules with those of signal receptors, integrins play an important role in the development and maintenance of tissues (for details, see Section 7.2.3).

Figure 10.9 Electron microscopic picture of the actin cytoskeleton (A) Detail of a lamellopodium of a frog keratinocyte (the bar represents 0.5 µm). (B) Same detail at larger magnification (the bar represents 0.1 µm). (C–E) Actin cytoskeleton of a filopodium. Note the extremely dense packing and high degree of branching of the actin fibers. It must be emphasized that the photographs represent snapshots of an ever-changing situation since the actin structures shown have a lifetime of minutes or less. (From Millard TH, Sharp SJ & Machesky LM [2004] *Biochem J* 380:1–17.)

In a migrating cell, the morphological changes mentioned above occur in a strictly ordered sequence (see Figure 10.10), indicating a high degree of cross talk and overlapping between Cdc42, Rac, and Rho signaling. In fact, sequential activation of Cdc42, Rac, and Rho in a G-protein cascade has been repeatedly described, although the precise mechanism is still somewhat obscure.

In an intact tissue, the patterns of cell motility described here are probably less pronounced than in cell cultures. Nevertheless, cell migration is critical for a wide variety of physiological and pathological processes such as defense reactions, wound healing, and embryonic development as well as tumor invasion and metastasis. Moreover, by controlling the cytoskeleton, intracellular contact sites, and the motility apparatus, Rho-type GTPases dictate the shape and polarity of tissue cells, thus determining tissue morphology and function.

10.3.3 Downstream of Rho: protein kinases controlling dynamics of the actin cytoskeleton

A central question is how Rho, Rac, and Cdc42 signals are transduced to the molecular apparatus controlling the actin cytoskeleton and gene expression. In this section, this question will be answered separately for Rho proteins proper and for Rac and Cdc42 proteins.

Figure 10.10 Coordinated regulation of cell motility by Rho, Rac, and Cdc42: an overview (Top) Three phases of movement of a migrating cell. By actin polymerization, the resting cell (1) develops lamello- and filopodia into the direction of movement (2). These extrusions become anchored to the substrate by new focal adhesions, enabling the cell to contract its rear part (3). (Bottom) Regulation of the motility apparatus and the cytoskeleton in different zones of a cell moving to the right, shown in a schematic and strictly simplified form. (A) Zone of actin polymerization induced by Rac and Cdc42 and resulting in the development of lamello- and filopodia; a re-contraction of these extrusions is prevented by Rac inhibiting myosin light-chain phosphorylation. (B) Zone of transient depolymerization of actin filaments caused by an inactivation of Rho. (C) Zone of contraction; Rho induces the formation of actomyosin stress fibers and enhances the phosphorylation of myosin light chains.

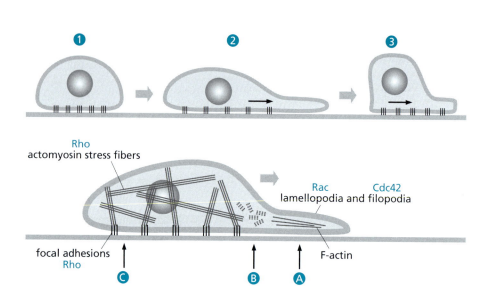

Among the large number of effector proteins of the Rho proteins proper, several inhibitors of actin polymerization and actin–myosin interaction are found. Rho inactivates these proteins, thus promoting the formation and contraction of actomyosin stress fibers. This effect is mediated by several Rho-controlled protein kinases, in particular ROK, LIM domain kinase, CRIK, and PRK.

Among these kinases, the RhO-activated Kinases **ROKα** and **ROKβ** (also known as ROCKs, RhO-Controlled Kinases) are particularly abundant. In addition to an amino-terminal kinase domain, these enzymes contain a Rho-binding domain and a PH domain (for membrane association and auto-inhibition) in the C-terminal part. ROKs belong to the family of AGC kinases and their activation follows the typical scheme explained for PKC-related kinases in Section 4.6.3: the auto-inhibition of the kinase brought about by binding of the C-terminal PH domain to the catalytic domain is relieved by interaction with membrane-bound Rho-GTP, enabling auto-phosphorylation of the activation loop and of a hydrophobic domain located near the C-terminus of the catalytic domain. As a result, the substrate-binding pocket of ROK is opened. [Note that kinase PDK1, which catalyzes activation loop phosphorylation in other AGC kinases (see Section 4.6.4), does not seem to be involved in ROK activation.]

Among the many substrates of ROK, the regulatory light chains of myosin (MLC, myosin light chains) are of particular interest, since they serve as key regulators of actomyosin contraction by hindering the interaction between actin and myosin. This inhibitory activity is abolished by phosphorylation catalyzed by ROKs (or by a Ca^2-dependent MLC kinase, see Section 14.6.2). To augment this effect, ROK simultaneously inactivates the MLC phosphatase (see Section 14.6.2) that would reverse MLC phosphorylation. As a result, Rho increases the level of MLC phosphorylation, thus facilitating contraction of the actomyosin fibers. In parallel, ROK activates other protein kinases called **LIM domain kinases** (LIM stands for the proteins Lin11 of *C. elegans*, Isl1 of rats, and MEC3 of *C. elegans* where the LIM domain was found for the first time). The only substrates of the Ser/Thr-specific LIM domain kinases known are the proteins of the **cofilin/actin depolymerization factor** family. As indicated by their name, these proteins sever F-actin fibers, thus causing a breakdown of the cytoskeleton and providing monomeric actin for further polymerization (see **Sidebar 10.3**). This effect is prevented by LIM domain

Sidebar 10.3 A quick glance at dynamics of the actin cytoskeleton Actin is one of the most abundant and most conservative proteins of the eukaryotic cell. It forms tracks for motor proteins and contractile protein complexes such as actomyosin, and with microtubules and intermediate filaments, it constitutes the cytoskeleton. To be particular, the term cytoskeleton is somewhat misleading because the actin network is not a rigid scaffold but a highly dynamic structure that transforms chemical energy (from ATP) into mechanical energy and undergoes continuous reorganization, thus determining the ever-changing shape and motility of cells.

Actin exists as a monomer (globular or G-actin, molecular mass 42 kDa) and as a polymer (fibrillar or F-actin) and is subject to continuous polymerization and depolymerization depending on the cellular conditions and input signals. G-actin polymerizes spontaneously unless it is hindered by other proteins. In the starting phase, however, spontaneous polymerization is kinetically blocked because actin dimers and trimers are unstable. Once the trimeric state is surpassed, polymerization proceeds rapidly, yielding long filaments. Therefore, the trimers are called nuclei and the process of trimer formation is known as nucleation. In the cell nucleation is controlled (promoted or prevented) by several regulatory proteins, which are targets of signaling pathways. Owing to a head-to-tail arrangement of the asymmetric actin monomers, F-actin filaments show a polarity with pointed and barbed ends.

Although actin binds ATP with high affinity and exhibits ATPase activity, filament formation is a spontaneous exergonic reaction and does not require ATP hydrolysis. Logically, disassembly of the filaments is endergonic and depends on ATP hydrolysis catalyzed by the intrinsic ATPase activity of actin. In other words: actin polymerization has an installed timer. The rates of filament assembly and disassembly are different at both ends of the filament, resulting in a continuous treadmilling of actin monomers from the barbed end, where they become bound as ATP complexes, to the pointed end, where they become released as ADP complexes. As a result, the growing filament appears to move forward and generates pressure that may push forward cell membranes; that is, ATP energy is transformed into mechanical energy. Treadmilling is controlled by regulatory proteins modulating both ADP–ATP exchange at freshly released monomers (such an exchange factor is profilin) and ATPase activity in the polymer (note that the activity of actin is regulated in a similar manner as that of G-proteins). The same holds true for branching and debranching of the filaments, which in addition to treadmilling is essential for the protrusion of membrane processes such as lamello- and filopodia and microvilli. Debranching is driven by ATP hydrolysis, cat-

alyzed by those regulatory proteins (such as Arp2/3) that serve as templates for branching.

In the formation and breakdown of the actin cytoskeleton, a large number of regulatory proteins (more than 50) is involved, exhibiting (among others) the following functions: (Examples of corresponding proteins are given in parentheses; for details the reader is referred to textbooks of cell biology. A complete list of actin regulators is provided by Siripala AD & Welch MD (2007) *Cell* 128:626, 1041.)

- nucleation of new actin fibers, in particular at branching points (Arp2/3)
- nucleation of unbranched actin filaments (Spir, Drf/Dia)
- nucleation of branched actin filaments (WASP, WAVE)
- binding and inhibition of spontaneous aggregation of G-actin (profilin, β-thymosine)
- depolymerization, debranching, severing of the polymer (cofilin, gelsolin)
- capping of the growing end of the polymer (tropomodulin, IQGAP)
- cross-linking of polymers (α-actinin, fascin)
- membrane anchoring (annexin, spectrin, ezrin, moesin, radixin)
- motor protein (myosin)
- cell–extracellular matrix junctions (talin, tensin, dystrophin, vinculin)
- regulation of actin–myosin interaction (troponin, tropomyosin, caldesmon, myosin light chains)

In addition to generating the power for membrane protrusion, actin is involved in still another energy-transforming process: by binding to actin, the ATPase function of myosin is activated. As a result, the actomyosin complex contracts due to a conformational change. In eukaryotes this reaction is the basis of most cellular movements ranging from muscle contraction to amoeboid creeping and changes of the cell shape by stress fiber contraction. Actomyosin is also involved in intracellular cargo transport. Triggered by Ca^2, the interaction between actin and myosin is evoked by all input signals inducing an increase of the cytoplasmic Ca^2 concentration. The classical example is provided by muscle contraction (see Section 14.6.2). In parallel, actin dynamics and actin–myosin interactions are controlled by signals transmitted along the Rho/Rac/Cdc42 pathways. As a rule, the input signals affect regulatory proteins rather than actin or myosin themselves.

Figure 10.11 Effects of protein kinases controlled by Rho Rho is activated by guanine nucleotide exchange factors (GEF), inactivated by GTPase-activating proteins (GAP), and inhibited by GDP dissociation inhibitors (GDI). The effector protein kinases of Rho are shown in the colored box. MLC, myosin light chain; P, phosphate. For details, see text.

kinase-catalyzed phosphorylation, resulting in an increase and stabilization of polymerized actin as required for stress fiber formation. Since the promoting effect of Rho on the formation and contractility of actomyosin stress fibers depends entirely on ROK-catalyzed phosphorylation of MLCs, MLC phosphatase, LIM domain kinase, and some other cytoskeletal proteins, ROKs are considered to be key effectors of Rho (**Figure 10.11**).

As specialized as LIM domain kinase is the Ser/Thr-specific **citron kinase** or **CRIK**, another Rho (and Rac) effector protein (citron is the trivial name of a Rho-interacting protein). Citron kinase is distantly related to ROKs and contributes, in particular, to the formation of a contractile actomyosin ring that separates daughter and mother cells in the course of mitosis.

In contrast to the highly specific kinases LIMK and CRIK, the Protein kinase C-Related protein Kinases **PRK** (see Section 4.6.5) are multifunctional Rho effectors. Effects of these kinases on the actin cytoskeleton, the intermediate filament cytoskeleton, and the microtubule cytoskeleton as well as on cell adhesion and vesicle transport have been repeatedly described, though the molecular details are not entirely known. As discussed in Section 4.6.3, PRKs are stimulated by phospholipid-dependent kinase 1 (PDK1)-catalyzed activation loop phosphorylation. The task of the membrane-anchored Rho-GTP is to bring PRK in contact with membrane-bound PDK1 and to change the PRK conformation in such a way that PDK1 gains access to the catalytic center. To this end Rho-GTP binds to a specific Rho interaction domain of PRK. Thus, in recruiting its effector kinase to the membrane, Rho plays a role resembling that of Ras in the activation of RAF and PI3K.

The enzymes of the PRK family have additional functions, which in most cases have been studied in cell cultures only. The results obtained indicate a role of PRKs as MAP4 kinases of stress-activated p38 MAP kinase modules (see Figure 10.11 and Section 11.3). Isoform PRKα even translocates into the nucleus, possibly modulating gene transcription by phosphorylation of transcription factors.

Another enzyme stimulated by Rho and playing an important role in actin polymerization is **phosphatidylinositol 4-phosphate 5-kinase**. Its product, phosphatidylinositol 4,5-bisphosphate, not only is the precursor of the second messengers DAG and InsP$_3$ but also interacts with actin-binding proteins such as profilin, vinculin, and gelsolin, thus promoting actin polymerization at focal adhesion sites.

The formation of parallel unbranched actin fibers as they are typical for stress fibers is induced by Drf/Dia proteins (formins), which are activated by either Rho or Cdc42 (see Section 10.3.4). Moreover, tension of its own evoked by myosin contraction along the Rho–ROK pathway (see Figure 10.11) seems to be a stimulus for stress fiber formation.

10.3.4 Rac and Cdc42 regulate actin polymerization and generation of reactive oxygen species

Rac- and Cdc42-controlled protein kinases

The number of Cdc42 and Rac effectors rivals that of Rho effectors. In Cdc42 and Rac signaling, three protein kinase families occupy positions resembling that of ROKs and PRKs in Rho signaling (**Figure 10.12**):

- Myotonin-Related Cdc42-binding Kinase **MRCK**

- p21-Activated Kinases **PAK** (p21 refers to the molecular mass of 21 kDa of the Rac protein; see also Section 11.3)

- Mixed-Lineage Kinases **MLK** (treated in more detail in Section 11.2)

As far as their cellular functions are concerned, these kinases have much in common with PRKs and ROKs, again indicating a close relationship and a high degree of overlap between Rho-, Rac-, and Cdc42-dependent signaling. They are recruited by the GTP-loaded small G-proteins to the inner side of the cell membrane where PAKs become phosphorylated and activated by PDK1 while MLK and MRCK undergo autophosphorylation. Experiments showing that in the absence of Cdc42 and Rac certain membrane components such as sphingolipids and gangliosides may stimulate PAK phosphorylation by PDK1 support the concept of a membrane-recruiting effect of Cdc42 and Rac.

Moreover, PAKs, MRCK, and MLKs share several substrate proteins with ROKs and PRKs. Like ROKs, MRCK stimulates actomyosin contraction by inactivating the phosphatase via dephosphorylation of the myosin light chains. Another example of overlapping functions is provided by the stimulatory effects of PRKs, PAKs, and MLKs on stress-activated MAP kinase modules, whose activation results in a wide

Figure 10.12 A selection of pathways downstream of Cdc42 and Rac Both Cdc42 and Rac are selectively activated by guanine nucleotide exchange factors (GEF) and inhibited by GTPase-activating proteins (GAP) and GDP dissociation inhibitors (GDI). Double-headed arrows indicate reciprocal activation of Cdc42, Rac, and Rho in G-protein cascades. Direct effector proteins of Rac and Cdc42 are shown in the colored box (MLKs, mixed-lineage kinases; PAKs, p21-activated protein kinases; MRCK, myotonin-related Cdc42-binding kinase; WASP, Wiscott–Aldrich syndrome proteins; WAVE, WASP/verprolin-homologous proteins). From the many putative pathways, only a few supported by strong experimental evidence are shown. For other details see text.

variety of cellular responses including gene expression (for details, see Sections 11.2 and 11.3).

Like ROKs, MRCK and PAKs stimulate LIM domain kinase, thus promoting the formation of actomyosin stress fibers. However, in contrast to ROKs, PAKs *suppress* myosin light chain phosphorylation and actomyosin contraction. To this end they phosphorylate and inactivate the Ca^2 -dependent myosin light chain kinase. It has been speculated that, through this effect, a retraction of Rac-induced lamellopodia is prevented.

Signaling networks controlling formation and outgrowth of actin fibers

As mentioned in Section 10.3.3, activation of LIM domain kinase hinders the cofilin-induced *breakdown* of actin fibers, thus stabilizing the actin cytoskeleton and promoting the formation of actomyosin stress fibers. In addition, Rho, Rac, and Cdc42 promote the *outgrowth* of new actin fibers. The latter effect is mediated by actin-associated regulatory proteins such as WASP, WAVE, Drf/Dia, Arp2/3, and IQGAP, the activities of which are controlled by the small G-proteins (**Figure 10.13**).

As far as their function is concerned, WAVE, Drf/Dia, and WASP belong to a family of factors that control the formation of new actin fibers in a process called nucleation (see Sidebar 10.3). **WASPs** got their name from **Wiscott–Aldrich syndrome**; this is a severe albeit rare hereditary immune deficiency caused by a loss-of-function mutation of the protein-encoding gene. Consequently, cells of the immune system (T-lymphocytes, macrophages, and monocytes) are restricted in their mobility, rendering them unable to perform movements in response to chemotactic signals.

WASPs are scaffold proteins that assemble functional complexes of nucleation-controlling proteins. As in many other signal-transducing proteins, WASP activity (the ability to bind actin and other proteins) is suppressed by an autoinhibitory conformation that is broken up by an input signal. In the case of WASPs, this is the interaction with membrane-bound GTP-loaded Cdc42. To this end Cdc42 binds to a specific site of the WASP molecule known as the CRIB (Cdc42 and Rac Interactive Binding) or GTPase-binding domain that is found in many Rac/Cdc42 effector proteins. In addition, WASPs have proline-rich domains that interact with SH3 domains of adaptor proteins, especially those of the MAP kinase pathways (such as Grb2 and Nck). Moreover, WASPs are activated by Tyr phosphorylation catalyzed by Src-type kinases. These alternative mechanisms of WASP activation resemble a logical AND gate that links mitogenic stimulation (for instance, by growth factors) with a reorganization of the cytoskeleton. Changes of the cell shape such as rounding-off are, indeed, typical features of cell division.

Figure 10.13 Linear and branched growth of actin fibers induced by Rho-type GTPases Rho or Cdc42 activates Drf/Dia proteins (formins) to induce linear growth at barbed ends, whereas Rac or Cdc42 stimulates WAVE or WASP proteins to form complexes with Arp2/3, thus triggering the outgrowth of a branch from a preexisting actin fiber.

In an analogous way, albeit along another pathway, Rac activates **WAVEs** (WASP/VErprolin-homologous proteins, verprolin is an actin-binding protein of yeast), a group of scaffold proteins related to A-kinase anchor proteins (see Section 4.6.1), which like WASPs assemble nucleation-promoting multiprotein complexes. Since WAVEs do not contain a CRIB domain, Rac interacts with accessory proteins, which in turn transform the WAVE protein proper into an active conformation. WAVEs are enriched particularly in lamellopodia. Like WASPs, the WAVE proteins are under the control of various signaling pathways that either up- or down-regulate their activities. So WASPs and WAVEs may be looked upon as hubs in a network of interactions controlling cell shape and motility.

To induce the outgrowth of actin filaments, WASPs and WAVEs have to form complexes with actin trimers or higher oligomers. Kinetically the formation of such oligomers is highly unfavorable and requires the assistance of regulatory proteins that bypass this obstacle (see Sidebar 10.3). Here the so-called nucleator proteins such as **Arp2/3** (Actin-related protein) enter the stage. Arp2/3, a complex of seven polypeptides, mimics a *stable* actin dimer interacting with both WASP (or WAVE) and actin. In the absence of an input signal, its activity is blocked by an autoinhibitory conformation. When it is activated along the Cdc42–WASP (or Rac–WAVE) pathway, Arp2/3 binds to the side of actin filaments and induces the polymerization of an F-actin side chain (see Figure 10.13), thus playing a key role in the formation of *branched* actin fibers and their outgrowth to a branched lattice (as shown in Figure 10.9). In the cell such branches have a lifetime of only a few minutes before they disintegrate, reflecting the dynamics of the actin cytoskeleton. The process of debranching requires energy provided by ATP hydrolysis, which is catalyzed by the ATPase function of Arp2/3.

In contrast to WAVE and WASP, the **Drf** or **Dia proteins** (Diaphanous-related formins, referring to the homologous *Drosophila* protein Diaphanous that belongs to a family of proteins with formin domains), also called **formins**, do not interact with Arp2/3. Instead they interact with the barbed end of F-actin fibers and induce the outgrowth of parallel bundles of unbranched actin filaments that are found in filopodia and stress fibers (see Figure 10.13). Drf/Dia proteins are abundant in animal cells. They are bound and activated by either Rho or Cdc42.

The outgrowth of actin fibers is also under *negative* control by what is called capping. Capping is understood as a binding of special regulatory proteins at the growing end of the fiber, the further polymerization of which is thus prevented (see Sidebar 10.3). Rac and Cdc42 suppress capping and promote the outgrowth of actin fibers. This effect is thought to be mediated by proteins of the **IQGAP** family (the name refers to the original identification of those proteins as Ras-GAPs, which later turned out to be a misconception), although IQGAPs do not contain CRIB domains. IQGAPs are multipurpose scaffold proteins. They contain various interaction domains that bind other cytoskeletal proteins as well as receptors, protein kinases, and proteins involved in cell adhesion. This indicates a wide variety of IQGAP functions, the physiological relevance of which is still known only fragmentarily. Moreover, it should be mentioned that the role of IQGAPs as Rac/Cdc40 effector proteins is still somewhat controversial since other data indicate a function *upstream* of these G-proteins; for instance, by inhibition of their GTPase activity, thus increasing the concentration of GTP-loaded forms at the leading edge of migrating cells. Probably IQGAPs exhibit both effects depending on the cell type and the physiological context.

Generation of reactive oxygen species

With the control of actin dynamics, the potencies of Rho, Rac, and Cdc42 are by no means exhausted. In fact, cooperating, as mentioned, with Ras along the MAP kinase pathways, they modulate the effects of a large number of proteins and genes.

Among the special functions of Rac, regulation of **NADPH oxidase (NOX)** activity has found particular attention. Using nicotinamide adenine dinucleotide phosphate

(NADPH) as a reductant, this enzyme (or more precisely a whole electron transport chain) catalyzes the reduction of molecular oxygen to superoxide anion radicals and hydrogen peroxide. NOX is a multienzyme complex consisting of Rac, the Rac effector protein p67^{Phox} (Phox stands for Phagocyte NADPH oxidase), the protein p47^{Phox}, and a membrane-bound flavoprotein–cytochrome b complex (gp91^{Phox}). Fulfilling the typical task of a small G-protein, Rac promotes the assembly of this complex at the membrane.

NADPH oxidase is expressed predominantly by phagocytes (monocytes and neutrophilic granulocytes). Committing some sort of chemical warfare, these cell types use the reactive oxygen species produced by NOX as well as even more aggressive derivatives thereof to kill microorganisms taken up by phagocytosis. The reaction proceeds in an explosive manner called oxidative burst and represents an essential component of nonspecific defense and inflammation. NOX isoforms probably serving additional purposes are found in practically every tissue. In fact, the concept of reactive oxygen species providing intracellular signals is gaining steadily increasing support (for details and examples, see Sections 2.2, 7.1.1, and 13.2.5). These signaling pathways are expected, therefore, to be under the supervision of Rac.

10.3.5 Signals controlling microtubule and intermediate filament cytoskeletons

The cytoskeleton of eukaryotic cells consists of three major structural components: actin fibers (microfilaments), microtubules, and intermediate filaments. Although in contrast to the actin filaments they are found to be connected only loosely with small G-proteins, microtubules and intermediate filaments are briefly reviewed here as far as their role in cellular signal processing is concerned. For detailed information the reader may consult textbooks of cell biology.

Microtubules: tracks and scaffolds

Microtubules are polymers of α,β-tubulin dimers. In determining cell shape and polarity, building up the mitotic spindle, and providing tracks for intracellular trafficking, they cooperate with actin filaments. Like the actin cytoskeleton, the microtubule cytoskeleton is a highly dynamic structure undergoing continuous reorganization driven by polymerization and depolymerization. Again these processes are controlled by various regulatory proteins called **MAPs** (Microtubule-Associated Proteins; not to be confused with MAP kinases), which are targets of signaling cascades. These proteins are found in many cell types, but in particular in neurons. One of the neuronal MAPs is Tau, the hyperphosphorylation of which has a bad reputation as one major cause of Alzheimer's dementia (see Chapter 13).

Beside other kinases, those activated by Rho-type GTPases have been observed to catalyze MAP phosphorylation, although as compared with the role of small G-proteins in actin dynamics this field of research is still in its infancy. For instance, microtubules disassemble when certain polymerization-promoting regulatory proteins (stathmins and collapsin response mediator protein 2) become phosphorylated and inactivated along the Rac/Cdc42–PAK and Rho–ROK pathways. Moreover, the Rac/Cdc42 effector protein IQGAP seems to regulate the interaction of microtubules with membrane proteins.

Microtubules serve as tracks for **cargo transport** driven by kinesin and dynein motor proteins. A peculiarity of the microtubule cytoskeleton is that it forms a radial network of filaments with the fast-growing (plus) ends directed to the cell periphery and the slow-growing (minus) ends facing the cell center. Depending on extracellular signals and the dynein and kinesin type, cargo may be transported in either direction.

This process can be easily visualized in melanocytes. In these cells pigment granula, or melanophores, are found either dispersed over the whole cell or concentrated in

Figure 10.14 Signal-controlled cargo transport along microtubule tracks Reversible transport of pigment granules (melanophores) along microtubule tracks (teal lines) in melanocytes is shown. Phosphorylation by the MAP kinase ERK of the motor proteins dynein and kinesin promotes aggregation in the cell center and is inhibited by protein kinase A (PKA). As a consequence, cAMP promotes melanophore dispersion.

the cell center, depending on environmental stimuli. The central aggregation is promoted by phosphorylation of the motor proteins catalyzed by the MAP kinase ERK. As explained in Chapter 11, ERK activity is attenuated by PKA. In fact, an increase of cAMP activating PKA correlates with dispersion of melanophores (**Figure 10.14**).

Owing to the short signaling pathways the redistribution of pigment granula occurs rapidly, explaining the immediate changes of color pattern in the skin of certain animals such as octopuses. Provided these findings can be generalized, microtubule-directed cargo transport must be assumed to be under the control of a wide variety of extracellular signals that activate the Ras–RAF–ERK cascade and cAMP production.

Intermediate filaments: control through phosphorylation

Beside actin-derived microfilaments and tubulin-derived microtubules, intermediate filaments (IFs) are the third cytoskeletal structure found in most vertebrate cells (but not in insects, plants, fungi, and protozoa). They are formed by a wide variety of proteins encoded by more than 70 genes. These proteins assemble into insoluble homo- and heteropolymers, filling the cell with a spongelike structure. Members of the IF protein family are (among others) epithelial keratins, mesenchymal vimentins, muscular desmins, neuronal neurofilaments, and the abundant laminins of the nuclear lamina. In other words, in contrast to the rather conservative chemical composition of microfilaments and microtubules, that of intermediate filaments is highly variable and tissue-dependent.

While actin fibers and microtubules are anchored to focal adhesion sites and adherens junctions, the contact sites connecting the intermediate filaments of adjacent cells are known as **desmosomes**. Moreover, via **hemidesmosomes**, intermediate filaments also anchor cells to connective tissue structures such as the basal lamina of epithelia. Intermediate filaments, therefore, determine the mechanical properties of tissues.

What is known about interactions of the data-processing apparatus with this cytoskeletal structure? A common structural feature of IF proteins is a helical rod module flanked by an N-terminal head and a C-terminal tail domain (**Figure 10.15**). Via a coiled-coil structure of their rod modules, IF proteins form dimers, which then polymerize to filaments. Thus, while the rod module determines the biomechanical properties of the IF protein, the head and tail domains are targets of regulatory signaling pathways. To this end they contain a high number of Ser, Thr, and Tyr phosphorylation sites. A major difference between microfilaments and microtubules on the one hand and intermediate filaments on the other is indeed that polymerization of the former is regulated by ATP and GTP hydrolysis controlled by accessory proteins, while for the latter direct phosphorylation is the control mechanism of choice. In fact, intermediate filaments rank among the most abundant phosphoproteins, and their phosphorylation is a highly dynamic process that in a reversible way controls their grade of polymerization and their intracellular distribution.

Figure 10.15 Schematic structure and sites of post-translational modifications of an intermediate filament protein For details see text.

Parts of the IF cytoskeleton may become polymerized and depolymerized on demand. Depending on the grade of polymerization, such IF proteins shuttle between a soluble cytoplasmic pool and an insoluble cytoskeletal pool, thus to a certain degree resembling actin and tubulin. Multiple phosphorylation promotes depolymerization (or hinders polymerization). It increases during mitosis (when disintegration of the nuclear lamina and of the cytoskeleton is mandatory) and upon tissue injury. In the latter case, phosphorylation certainly is needed to prime cells for cell division aiming at wound repair. In addition, however, intermediate filaments seem to provide a phosphate sink, dampening the activity of pro-apoptotic protein phosphorylation that is stimulated via the MAP kinase stress modules. Moreover, phosphorylation protects IF proteins from degradation by apoptotic caspases or proteasomes. Thus, counterproductive cell death is prevented in the damaged tissue.

Phosphorylation also controls the intracellular compartmentalization of intermediate filaments, for instance in neurons and polarized epithelial cells. As far as the role of IF proteins in cellular stress protection is concerned, an exciting new aspect has been added by studies on wounded nerve cells: this is the role of IF proteins in **intracellular cargo transport**. At the site of damage, MAP kinase cascades (and other signaling pathways) become activated and the MAP kinases migrate into the nucleus to modulate gene expression. When the damage has occurred, for instance in the axon near to the synaptic terminal, the sheer size of a neuron raises the question as to how the large distance (up to 1 meter) between the nerve ending and the nucleus can be bridged in a reasonable time. The most probable mechanism is active transport by motor proteins such as kinesins and dyneins. In fact, in neurons a nonfilamentous form of the IF protein vimentin has been found to serve as a scaffold, anchoring activated (phosphorylated) MAP kinases to the dynein motor and protecting them from inactivation (by dephosphorylation) during their long journey to the nucleus. This special vimentin type is produced from conventional vimentin by limited proteolysis catalyzed by the Ca^2-dependent protease calpain (note that the cytoplasmic Ca^2 level rises at sites of cell damage). The idea is easily at hand that the role of IF proteins in cellular transport is not restricted to damaged neurons but may represent a general mechanism of long-range signaling. The large number of different IF proteins expressed in mammalian cells might provide this process with a high degree of selectivity.

A wide variety of protein kinases have been found to catalyze IF protein phosphorylation. These include PKC, PKB/Akt, RAF, CDK1, CDK5, ERK-type MAP kinases, and (last but not least) ROKs. In addition, Tyr phosphorylation has been observed. Therefore, intermediate filaments are targets of numerous signaling cascades, and pronounced cross talk between the control of intermediate filament dynamics and pathways regulating cell proliferation, cell differentiation, cell shape, and cell motility has to be postulated.

Summary

- Rho, Rac, and Cdc42 couple various cell physiological processes such as proliferation, tissue-specific function, stress responses, and endo- and exocytosis with changes of cell shape, cell motility, and selective gene expression. In this context a major function of Rho/Rac/Cdc42 is to control the dynamics of the actin cytoskeleton.

- Guanine nucleotide exchange factors (GEFs) of Rho, Rac, and Cdc42 proteins constitute a highly diverse group of proteins with various interaction motifs. For activation they require an association with membrane-bound phosphatidylinositol 3-kinase (PI3K) products and, in addition, phosphorylation of Tyr residues catalyzed by Src-type Tyr kinases.

- Effects of Rho on the cytoskeleton are mediated by protein kinases such as Rho-activated kinases (ROKs), and protein kinase C-related kinases (PRKs).

- Like Rho, Cdc42 and Rac control the dynamics of the actin cytoskeleton by activation of protein kinases such as myotonin-related Cdc42-binding kinases (MRCKs) p21-activated kinases (PAKs), and mixed-lineage kinases (MLKs).

- Cdc42 and Rac effector proteins that control the nucleation and polymerization of actin fibers are WASP, WAVE, Drf/Dia, and IQGAP. They play a key role in actin-dependent cellular events such as motility, intracellular vesicle trafficking, exo- and endocytosis, phagocytosis, and outgrowth of membranous structures such as neurites. Rac transduces signals to NADPH oxidase, thus stimulating the generation of reactive oxygen species used either as antimicrobial and pro-inflammatory substances or as messengers in intracellular signaling.

- Another component of the cytoskeleton is microfilaments. The polymerization and depolymerization of these tubulin aggregates is regulated by a series of microtubule-associated proteins, the activities of which are controlled by protein phosphorylation rather than by Rho/Rac/Cdc42 directly.

- Cargo transport along microtubule tracks is regulated by phosphorylation of the associated motor proteins (kinesins and dyneins).

- Intermediate filaments represent the third type of cytoskeletal structure, formed by the spontaneous aggregation of several quite diverse proteins. Their polymerization is controlled predominantly by protein phosphorylation catalyzed by ROKs and other kinases.

10.4 Arf and Rab proteins control vesicle transport

10.4.1 Intracellular traffic of secretory vesicles

A characteristic feature of eukaryotic cells is compartmentalization: the separation by membranes into organelles and other areas of individual biochemical activities. Like the plasma membrane, the intracellular membranes consist of lipid bilayers that are impermeable for hydrophilic substances such as proteins and nucleic acids. Such material is packed, therefore, in membrane-covered vesicles that travel between the different compartments. The principle of this macromolecular trafficking is shown in **Figure 10.16**; a cargo-filled vesicle is released from the membrane of a donor compartment and transported by means of ATP-driven motor proteins along

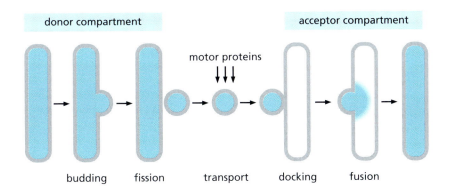

Figure 10.16 Transport of cargo-filled vesicles between cellular compartments

Figure 10.17 Secretion by exocytosis The principle is shown for a glandular cell (left). The secretion is produced at the endoplasmic reticulum, filled into vesicles, and transported via the Golgi apparatus to the cell surface to be released. A cellular structure specialized for secretion, the presynaptic terminal of a chemical synapse, is shown on the right. The secretory vesicles filled with neurotransmitter migrate along cytoskeletal tracks (neurofilaments) from the cell body of the neuron (not shown) to the end of the axonal nerve fiber to become emptied into the synaptic cleft, that is, the gap between pre- and postsynaptic cells. (Adapted from Krstic RV [1976] Ultrastruktur der Saugetierzelle. Springer.)

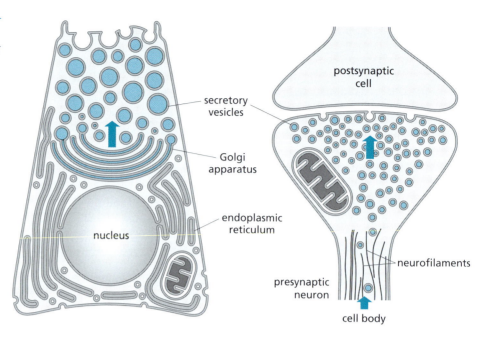

microtubule tracks to an acceptor compartment, where it fuses with the membrane and delivers its content.

This principle holds true for import of extracellular material (**endocytosis**), secretion of intracellular substances (**exocytosis**), and intracellular traffic between different compartments. Endo- and exocytosis are essential for communication between the cell and the environment. By exocytosis, signaling factors such as hormones and neurotransmitters are released and receptors and other proteins are exposed at the cell surface, and the reverse holds true for endocytosis (**Figure 10.17**).

Exocytosis of a new protein starts at the site of biosynthesis (the endoplasmic reticulum), proceeds via the different Golgi compartments (cis-, medial-, and trans-Golgi), and ends with the emptying of the secretory vesicle at the cell surface. For endocytosis, the material bound to receptors at the cell surface assembles in membrane invaginations (coated pits and caveolae), which become separated and released as vesicles into the cytoplasm. The vesicles are taken up by larger organelles, the endosomes, which transfer the material either to lysosomes (to become degraded) or back into the exocytotic cycle (for additional details on endocytosis, see Section 5.8).

To become separated from the donor membrane, the vesicles are covered by a tangle of COat Proteins (COPs), producing a conformational tension that leads to a budding of the membrane. Three types of coat protein complex, called coatomeres, are known, the functions of which depend on the donor compartment:

- **COPI** controls early phases of secretion, in particular at the Golgi complex
- **COPII** is required for vesicle formation at the endoplasmic reticulum
- **clathrin adaptor protein** (see Section 5.8) ensures the endocytotic separation of vesicles from the plasma membrane and vesicle release from the trans-Golgi compartment

The protein coat promotes both the loading and the release of vesicles. If the cargo is membrane proteins, such as receptors, these make contact with coat proteins via their cytoplasmic domains and become concentrated at the site of vesicle that is budding. In contrast, soluble cargo, such as hormones, binds to transmembrane cargo receptors that interact with coat proteins (**Figure 10.18**). Before fusion with

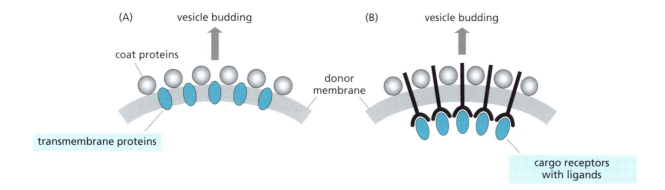

the acceptor membrane, the vesicle has to be uncoated. Vesicle traffic is strictly controlled by the signal-processing protein network. As in most targeted processes, G-proteins are the essential switches, working according to principles conserved from yeast to humans.

10.4.2 Fusion machinery

A vesicle spontaneously fuses with an acceptor membrane when they come in close contact. The surfaces are covered, however, by water molecules, which hinder direct contact between the lipid bilayers. To overcome this obstacle, complementary surface proteins interact with each other, displacing the water coat and pulling the opposite membranes together. This process is strictly controlled to prevent chaotic fusions. In other words, the proteins of the fusion machinery become active only when they receive a proper input signal.

Fusion is a two-step event. In the first step, the vesicle becomes fixed by so-called tethering proteins at the site of fusion, and in the second step, specific protein machinery promotes the fusion of the membranes (**Figure 10.19**).

The first fusion-competent protein identified was found by a classical biochemical approach. It was observed that N-ethylmaleimide (NEM), a customary enzyme

Figure 10.18 Formation of secretory vesicles Cargo molecules (colored ellipses) are concentrated in the donor membrane by interacting with vesicular coat proteins. (A) Transmembrane proteins as cargo. (B) Soluble proteins as cargo; for uptake into the vesicle, they are bound by vesicular target receptors.

Figure 10.19 Steps of vesicle formation and fusion (1) Coat proteins (teal circles) are assembled at the site of vesicle germination; fusion proteins (SNAREs) are symbolized by L-shaped hooks. (2) The vesicle separated from the donor membrane is transported (by motor proteins, not shown) to the target compartment. (3) The vesicle becomes uncoated and recruits tethering proteins (teal ellipses) making contact between vesicle and acceptor membrane. (4) The Ca^2-dependent interaction between complementary fusion proteins (SNAREs) triggers membrane fusion. Binding and dissociation of the coat and tethering proteins is facilitated by small G-proteins of the Arf and Rab families. Complementary binding between v-SNAREs (light gray cylinders) and t-SNAREs (dark gray cylinders) through coiled-coil formation of the cytoplasmic domains is shown schematically at the right.

inhibitor that alkylates SH groups in proteins, prevented the fusion of vesicles with membranes. The effect was traced back to a protein inactivated by the agent. This NEM-Sensitive Factor **NSF**, an ATPase of the AAA protein family (see Section 2.5.2), has turned out to be required for every step of exocytosis. NSF binds to fusing membrane areas, but it needs the support of adaptors, Soluble NSF-Attachment Proteins or **SNAP**s. The SNAPs make contact with the NSF effector proteins called SNAP REceptors or **SNARE**, which are the fusion proteins proper. Mammalian cells express various SNARE isoforms that are differently distributed among tissues and intracellular compartments. Most SNAREs are bound to vesicular and target membranes by a single transmembrane domain. Their cytoplasmic extensions contain helical SNARE motifs that have a strong propensity to develop coiled-coil structures. This mechanism enables SNAREs to form tight complexes with each other. It is generally assumed that, to accomplish fusion, SNAREs bound to vesicle membranes (v-SNAREs) interact with complementary SNAREs on the target membrane (t-SNAREs).

In the course of endo- and exocytosis, more and more proteins of the fusion apparatus inevitably accumulate in the target membrane and must be recycled. This retrograde transport process proceeds along the COPI pathway. To this end the v-SNARE/t-SNARE complexes have to become disintegrated, which is the job of the ATPase NSF, supplying ATP energy for the endergonic complex dissociation.

Neurotransmitter release represents a perfect paradigm of signal-controlled exocytosis

Input signals triggering the SNARE interaction are transmitted by Rab proteins (see Section 10.4.4) and Ca^{2+} ions. This mechanism has been studied most thoroughly for neurotransmitter exocytosis at synapses. The synaptic vesicles filled with neurotransmitter are docked at active zones of the inner side of the presynaptic membrane (see Figure 10.17), a process controlled by the small G-protein Rab3. To be able to fuse and release their content into the synaptic cleft, they require a brief increase of cytoplasmic Ca^{2+} from 0.1 to 1.20 μM triggered by membrane depolarization (or an action potential). In this way, voltage-dependent Ca^{2+} channels in the postsynaptic membrane are opened, allowing Ca^{2+} to flow in from the outside (for details, see Section 14.5.2). These channels are components of the fusion machinery.

The Ca^{2+} ions penetrating the cell bind to **synaptotagmin**, a transmembrane protein of the vesicles. Synaptotagmins (nine mammalian isoforms are known) harbor two C2 domains (see Section 4.6.3), each binding 2–3 Ca^{2+} ions in its cytoplasmic extension (**Figure 10.20**). This results in a conformational change exposing hydrophobic areas of the protein and allowing synaptotagmin with its cytoplasmic domain to dip into the adjacent synaptic membrane, where the C2 domains become fixed by phospholipids. By this means it makes contact with the t-SNARE **syntaxin**, a transmembrane protein, and **SNAP25** (SyNaptosome-Associated Protein of 25 kDa, which must not be confused with the NSF adaptor SNAP), a soluble SNARE protein. Through its two symmetrically arranged SNARE motifs, SNAP25 forms a complex with both syntaxin and the corresponding v-SNARE **synaptobrevin**, thus ensuring membrane fusion. Subsequently the NSF–SNAP complex catalyzes the dissociation of the SNARE complex.

The synaptic proteins are targets of extremely powerful bacterial toxins. The neurotoxins* of **Clostridium botulinum** (cause of poisoning by meat and canned foods) and **Clostridium tetani** (cause of tetanus) rank among the most dangerous biogenic substances. Their average lethal dose (LD_{50}) is less than 0.0001 mg/kg of body weight (for comparison, the LD_{50} is 0.3 mg/kg for cobra toxin and 6.5 mg/kg for sodium

---

* Neurotoxins are only one ingredient of a whole cocktail of *Clostridium* toxins. Other components include ADP-ribosyltransferases, which selectively attack actin, and inhibitors of small G-proteins of the Rho family, which cause actin depolymerization and a breakdown of the cytoskeleton.

Figure 10.20 Model of Ca^2-dependent neurotransmitter release (Top) Simplified domain structure and interaction partners (teal) of synaptotagmin (TM, transmembrane domain). (Bottom) Proposed role of synaptotagmin in vesicle fusion. (1) The vesicle has made contact with the inner side of the presynaptic membrane. (2) Depolarization opens voltage-dependent Ca^2 channels. Ca^2 binds to the C2 domains in the cytoplasmic extensions of synaptotagmin (colored ellipses), which upon a conformational change dip into the cell membrane, reinforcing the interaction between the proteins of the SNARE complex: synaptobrevin, syntaxin, and SNAP25. As a result, membrane fusion is induced (not shown). For the sake of clarity, only two of many synaptotagmin, SNARE, and channel proteins surrounding the site of membrane fusion are depicted. Moreover, additional regulatory proteins not shown here are involved in membrane fusion.

cyanide). The clostridium toxins are zinc-containing proteases that specifically destroy synaptic SNARE proteins.

Neurexin, another t-SNARE-related protein of the presynaptic membrane, is attacked by α-latrotoxin, a poison of the **black widow spider** (*Latrodectus tredecimguttatus*) causing a catastrophic emptying of synaptic vesicles (LD$_{50}$ 0.6 mg/kg). A similar effect is observed upon interaction of the spider poison with latrophilin, a heptahelical receptor with unknown ligand(s) found in presynaptic membranes.

10.4.3 Arf proteins and vesicle formation

To assemble proteins into large complexes at membranes is one of the main functions of small G-proteins. Vesicle formation requires such protein recruitment, in this case of coat proteins. This is managed by the small G-proteins of the **Arf/Sar1 family**. Arf proteins were discovered by their property of increasing the ADP-ribosyltransferase activity of cholera toxin (Arf stands for ADP ribosylation factor). However, this reaction, the physiological significance of which is unknown, has nothing to do with the role of Arf in vesicle formation. Arf proteins have been found in all eukaryotes from yeast (seven isoforms) to humans (15 isoforms). They have been grouped into three subfamilies: Arf proteins proper, Sar (Secretion-associated Ras-like) proteins, and Arf-like (Arl) proteins (see Table 10.1).

The main function of Arf1, -2, -3, and -6 as well as of Sar 1 is to promote vesicle formation by recruiting coat proteins at the site of vesicle budding. The steps of this process are depicted in **Figure 10.21**. The functions of the other Arf proteins are known only incompletely. Some of them seem to be involved in vesicle and membrane trafficking within the Golgi apparatus, probably by promoting the association with tethering proteins.

In the cell, the individual G-proteins have different sites of action. Sar1 controls the formation of vesicles with a COPII coat at the endoplasmic reticulum; and Arf1, of COPI-coated vesicles in the Golgi apparatus. Arf1–Arf3 and probably also Arf6

Figure 10.21 Role of Arf (or Sar1) in vesicle formation (1) Arf-GEF receives an input signal and activates a cytoplasmic Arf protein by GDP–GTP exchange. This causes a conformational change exposing a membrane anchor (black horizontal bar) that was hidden before in a hydrophobic pocket of the Arf-GDP complex. By means of the anchor, the Arf-GTP complex associates with the donor membrane. (2) Arf-GTP recruits the coat proteins and the coated vesicle is released from the donor membrane. (3) A vesicle-bound Arf-GAP (which also may receive input signals) inactivates Arf by GTP hydrolysis, inducing uncoating of the vesicle and the release of Arf-GDP. For the sake of clarity, only one of the many Arf molecules associating with the vesicle is shown.

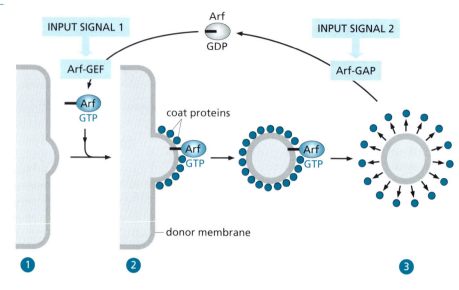

regulate, in addition, the formation of clathrin-coated endocytotic vesicles at the plasma membrane.

The central role played by Arf in secretory processes is underlined by inhibitor experiments using, for instance, a dominant negative mutant Arf protein with a defective GTPase function. Under such conditions the coat proteins cannot be recruited any longer at the donor membrane. Consequently, Golgi membranes are no longer replenished, the Golgi apparatus disintegrates, and secretion and endocytosis are blocked almost completely.

Arf and Sar1 are active only when associated with membranes. For association Arf has two membrane anchors, a tetradecanoyl residue bound N-terminally as an amide and a hydrophobic helix structure, while Sar1 lacks the fatty acid side chain. The membrane anchoring must be reversible, since Arf leaves the donor membrane together with the vesicle and has to be released into the cytoplasm to return and to be used for a new cycle of vesicle formation. This is achieved by a conformational change rendering Arf more hydrophilic: upon GTP hydrolysis a hydrophobic pocket opens in the Arf protein, hiding the lipid anchor.

Guanine nucleotide exchange factors and GTPase-activating proteins controlling Arf activity

Arf is active only in the GTP-loaded form. Signaling inputs are therefore transmitted by GDP–GTP exchange factors. These **Arf-GEFs** (mammalian cells express 14 isoforms) catalyze the nucleotide exchange only when bound to membranes; they are either membrane-integrated or associate upon activation with membranes. Some Arf-GEFs are selectively inhibited by the mold toxin **brefeldin A** (from *Penicillium brefeldianum*) used experimentally for a blockade of intracellular vesicle trafficking.

As is typical for small G-proteins, the GTPase activity of Arf is extremely weak and has to be stimulated by a GAP. Arf-GAPs are a large heterogeneous protein family. The factors exhibit numerous interaction domains linking Arf activity with other signaling pathways. In the course of vesicle formation, Arf-GAPs associate with the vesicular membrane to induce the dissociation of coat proteins by inactivating Arf. To prevent premature uncoating, however, this activity remains inhibited by specific protein interactions until the vesicle has been completely released.

10.4.4 Rab proteins and vesicle trafficking

Like vesicle budding, vesicle targeting and fusion with the acceptor membrane (see Sidebar 10.4) also requires the assembly of proteins at the vesicle membrane. This process is controlled by the small G-proteins of the Rab family that, like Arf proteins, are found throughout the eukaryotic kingdom and are indispensable for exo- and endocytotic vesicle transport.

For pioneering research yeast cells have provided an excellent model. *Saccharomyces cerevisiae* expresses 11 Rab-related G-proteins called YPT proteins (Yeast Protein Transport) and SEC4 (referring to a secretion-defective mutant). The corresponding mammalian homologs were isolated originally from a rat brain cDNA library by polymerase chain reaction with YPT-analogous oligonucleotides. With more than 60 isoforms, these **Rat brain (Rab) proteins** constitute by far the largest subfamily of small G-proteins. They are, of course, not restricted to brain but found in all cell types. Rab and YPT proteins are related to such an extent that they may replace each other in experimental approaches.

Each step of vesicle transport is controlled by Rab (as well as, of course, by Arf) GTPases. Within the Rab family the responsibilities are clearly distributed. Rab1 and Rab2 control, for instance, vesicular traffic from the endoplasmic reticulum to the Golgi apparatus; Rab6, inter-Golgi traffic; Rab8, transport from the trans-Golgi apparatus to the plasma membrane; and Rab4, -5, and -9 regulate endocytosis. Finally, Rab7 is specialized for recycling of down-modulated membrane receptors (see Section 5.8). Moreover, many Rab isoforms exhibit more tissue-specific functions. An example is provided by Rab3 that controls neurotransmitter release at synapses.

Like Arf proteins, Rab GTPases are active only when they are GTP-loaded and membrane-bound, shuttling between cytoplasm and vesicle membranes. Membrane association is achieved by two geranylgeranyl side chains bound to C-terminal cysteine residues as thioethers (more details on lipid anchors are found in Section 4.3.1). This membrane anchoring is controlled by two proteins called **GDI** (GDP Dissociation Inhibitor; such a factor was discussed earlier with regard to the control of Rho proteins, see Section 10.3.1) and **GDF** (GDI Dissociation Factor). By hiding the membrane anchor of Rab in a lipophilic pocket, GDI forms a soluble complex with Rab-GDP that is released into the cytoplasm. In this complex the GDP form of Rab is stabilized (explaining the name GDI). Re-association of Rab with membranes requires the dissociation of the Rab–GDI complex. This is accomplished by GDF

Sidebar 10.4 Lipid messengers control vesicle transport For a long time the question remained open how transport vesicles find the proper target membrane among the many intracellular membranes; that is, how targeted transport is brought about. Now there is a steadily increasing body of evidence indicating that vesicles are targeted by a specific lipid composition of the acceptor membrane. In this context, signaling lipids of the phosphatidylinositol family have turned out to play a key role. As explained in Section 4.4.3, phosphatidylinositols are selectively and reversibly phosphorylated at positions 3, 4, and 5 of the inositol ring. The lipid kinases and phosphatases involved are distributed in such a way that each membrane compartment exhibits a specific pattern of phosphatidylinositol mono-, bis-, and trisphosphates, thus recruiting via corresponding interaction domains individual sets of proteins. Such proteins include the coat proteins COPI and COPII as well as the AP adaptors of clathrin-dependent endocytosis. Since lipid phosphorylation and dephosphorylation are under ever-changing control by input signals, the lipid pattern is highly dynamic, ensuring both the release and the fusion of transport vesicles in a strictly adaptable manner. At the same time, the energy-consuming lipid phosphorylation, together with GTP hydrolysis by Arf and Rab, generates a biochemical gradient that gives coupled processes a definite direction. So the membranes of vesicles released from the Golgi apparatus contain phosphatidylinositol 4-monophosphate [PI(4)P], which upon fusion with the plasma membrane becomes phosphorylated to the 4,5-bisphosphate. An analogous phosphorylation transforming PI(3)P into PI(3,5)P$_2$ is observed during vesicle transport from early to late endosomes. When these phosphorylations are inhibited, vesicular transport breaks down.

Figure 10.22 Control of Rab activity The active Rab–GTP complex (shown in color) is anchored in the vesicle membrane by two geranylgeranyl side chains (horizontal bars) and able to interact with its effector proteins. A Rab GTPase-activating protein (GAP) catalyzes the inactivation to Rab-GDP, which is released from the membrane upon complex formation with GDP dissociation inhibitor (GDI), hiding the membrane anchors of Rab. A signal-activated Rab GDI dissociation factor (GDF) displaces GDI, bringing back Rab to a membrane where it may become activated again by a Rab guanine nucleotide exchange factor (GEF). Most probably, the regulatory proteins GDI, GDF, GEF, and GAP are all controlled by input signals.

targeting the corresponding Rab isoform to a site in the vesicle membrane where it becomes activated by a membrane-bound Rab-GEF (**Figures 10.22** and **10.23**). Among the G-proteins, this sophisticated mechanism of reversible membrane anchoring is restricted to Rab and Rho proteins. It offers additional opportunities for control, because GDIs and GDFs may be targets of input signals.

There is no final answer to the question as to where exactly Rab proteins are involved in vesicle trafficking and whether there is at all a common mode of Rab action. Generally it is assumed that Rab recruits the tethering proteins for the initial contact between vesicle and target compartment. Thus, Rab would play a role comparable to that of Arf in vesicle coating and that of Ras in the formation of signaling protein complexes. However, besides tethering proteins, many additional effector proteins of Rab have been found, exhibiting functions that go far beyond vesicular transport. Such observations indicate further functions of Rab, in particular in the organization and dynamics of the cytoskeleton.

Figure 10.23 Targeting of vesicular transport and induction of fusion by Rab GTPases The figure illustrates a current model which, however, is still partly hypothetical. (1) The vesicle loaded with active Rab-GTP (teal) approaches the acceptor membrane. (2) Rab-GTP recruits tethering proteins (colored ellipses) for the initial contact between vesicle and acceptor membrane. (3) Interaction between the SNARE proteins prepares the membrane for fusion (priming step). (4) Fusion is induced by Ca^2 ions (see Figure 10.20), and the NSF–SNAP complex catalyzes the ATP-consuming dissociation of the SNARE complexes. During steps 3 and 4, Rab is inactivated by a GTPase-activating protein (GAP; probably bound to the acceptor membrane) and released from the vesicle by binding to GDP dissociation inhibitor (GDI) to be reused upon interaction with GDF and activation by guanine nucleotide exchange factor (GEF; probably bound to the vesicular membrane). Upon inactivation of Rab, the tethering proteins are released. For the sake of clarity, only one (or two) of many Rab, SNARE, and tethering proteins involved in the process is shown.

Summary

- Vesicle trafficking between cellular compartments is a targeted process requiring ATP and GTP energy. It proceeds in several steps. To be loaded with cargo and released from donor membranes, vesicles are coated by coat proteins (COPs) and clathrin adaptor proteins that have to be removed before fusion with the target membrane. Tethering proteins enable the initial contact with the target membrane. This contact is reinforced by SNARE proteins found in the vesicle and target membranes. Ca^2 triggers their interaction, which leads to membrane fusion. Subsequently the SNARE complex is dissociated under ATP consumption catalyzed by the AAA protein NSF.

- The exocytotic release of neurotransmitter from neurons provides the best-studied example of an exocytotic process. Here the Ca^2 ions required for membrane fusion enter the cell through voltage-dependent Ca^2 channels and enable a vesicular membrane protein (synaptotagmin) to reinforce the contact between vesicular and cellular membrane. The most powerful neurotoxic poisons, such as botulinum and tetanus toxin, are proteases that destroy the synaptic SNAREs.

- Vesicle formation is regulated by the small G-proteins of the Arf family. Arf-GTP bound to the budding vesicle recruits the coat proteins, while uncoating of the vesicle occurs upon GTP hydrolysis. Both guanine nucleotide exchange factors (Arf-GEFs) and GTPase-activating proteins (Arf-GAPs) are targets of input signals controlling vesicle formation.

- Fusion of secretory vesicles with target membranes is controlled by Rab proteins, the largest family of small G-proteins. Rab-GTP bound to the vesicular membrane recruits the tethering proteins for the initial contact with the target membrane. Upon SNARE interaction, Rab catalyzes GTP hydrolysis which causes its release into the cytoplasm.

10.5 Ran, nuclear transport, and mitosis

Ran (Ras-like nuclear GTPase) is the most abundant among the G-proteins of the Ras superfamily. It differs from the other species in two points.

(1) It does not associate with membranes.

(2) Its major function is to organize the traffic of macromolecules (proteins and nucleic acids) between cytoplasm and nucleus rather than to assemble signal-transducing multiprotein complexes at membranes. GTP hydrolysis provides the energy that gives the transport a defined direction, frequently against concentration gradients.

RNA and ribosomes are exported from the nucleus while chromatin proteins and transcription factors are imported. The traffic proceeds through several thousands of sophisticatedly structured pores in the nuclear membrane that allow passage of more than a million macromolecules per minute.

Most proteins and RNAs cannot travel unaccompanied between cytoplasm and nucleus but have to be carried by transport proteins, the exportins and importins (collectively called β-karyopherins). These carriers recognize their cargo by short amino acid sequences: the Nuclear Localization Sequence (or Signal) NLS and the corresponding Nuclear Export Sequence NES (an additional Nuclear Retention Signal NRS is found in proteins preventing the nuclear export of immature RNA).

Importins and exportins are expressed in several isotypes exhibiting different specificities. In the most abundant "classic" import pathway, cytoplasmic proteins

Figure 10.24 Ran cycle Protein with a nuclear localization sequence [(NLS)protein] is imported into the nucleus; protein with a nuclear export sequence [(NES)protein] is exported from the nucleus. Only export depends on Ran and GTP hydrolysis. NTF2 (nuclear transport factor 2) is a carrier protein for re-import of Ran-GDP into the nucleus; Ran-GEF is a guanine nucleotide exchange factor (also known as regulator of chromosome condensation 1, RCC1); Ran-GAP is a GTPase-activating protein.

carrying a classic NLS bind importin α, an adaptor protein, together with importin β, the carrier protein proper. Probably by interacting with pore proteins (nucleoporins), importin β directs this ternary complex through the nuclear pore. Upon arrival in the nucleoplasm, the complex meets with GTP-loaded Ran that binds to the importins, displacing the cargo protein from the complex. By means of a special exportin (called CAS), Ran-GTP then carries the importins back to the cytoplasm. Since this occurs against a concentration gradient, energy is required; this is supplied by hydrolysis of the Ran-bound GTP, catalyzed by a Ran-GAP bound at the cytoplasmic side of the nuclear membrane. This reaction is supported by Ran-binding proteins acting as adaptors and catalyzing sumoylation (see Section 2.8.3), which is essential for the correct localization of Ran-GAP at the mouth of the nuclear pore. As a consequence of GTP hydrolysis, the Ran–importin complexes dissociate and the transport process is rendered irreversible. Ran-GDP is transported back into the nucleus by the protein nuclear transport factor 2 and becomes recharged with GTP by means of a nuclear Ran-GEF localized in the nucleoplasm.

For the export of other macromolecules, mainly ribonucleoproteins, additional exportins interacting with Ran-GTP are available. The Ran cycle is schematically depicted in **Figure 10.24**. Note that export, not import, is the Ran-controlled step.

Ran-GEF is also known as Regulator of Chromosome Condensation 1 (RCC1), referring to an additional function of Ran: to promote the polymerization of the mitotic spindle. Again the importins α and β are involved in that they sequester spindle assembly factors, which are proteins required for spindle formation. Similar to its role in nuclear transport, Ran-GTP catalyzes the dissociation of such complexes. During mitosis, Ran-GEF (RCC1) is bound to histones H2A and H2B in chromatin, ensuring a high local concentration of Ran-GTP. At the end of mitosis, the chromatin-associated Ran also seems to promote the formation of the new nuclear membrane.

Summary

Cycling between the cytoplasm and the nucleus, Ran, the most abundant small G-protein, controls the targeted transport of macromolecules across the nuclear membrane. The importins promoting the influx of proteins into the nucleus are re-exported against a concentration gradient to the cytoplasm under GTP hydrolysis catalyzed by Ran. Ran-GDP migrates back passively into the nucleus, where it becomes activated by GDP–GTP exchange. Ran is involved also in the formation of the mitotic spindle.

Further reading

Sections 10.1 and 10.2

Cherfils J & Zeyhouf M (2013) Regulation of small GTPases by GEFs, GAPs, and GDIs. *Physiol Rev* 93:269–309.

Gloerich M & Bos JL (2011) Regulatory Rap small G-proteins in space and time. *Trends Cell Biol* 21:615–623.

Kashatus D (2013) Ral GTPases in tumorigenesis: emerging from the shadows. *Exp Cell Res* 319:2337–2342.

Milroy LG, Brunsveld L & Ottmann C (2013) Stabilization and inhibition of protein-protein interactions: the 14–3–3 case story. *ACS Chem Biol* 8:27–35.

Pylayeva-Gupta Y, Grabocka E & Bar-Sagi D (2011) Ras oncogenes: weaving a tumorigenic web. *Nature Rev Cancer* 11:761–774.

Rane CK & Minden A (2014) p21–activated kinases: structure, regulation, and functions. *Small GTPases* 5:e28003.

Samatar AA & Poulikakos PI (2014) Targeting RAS-ERK signalling in cancer: promises and challenges. *Nature Rev Drug Discov* 13:929–942.

Smrcka AV, Brown JH & Holz GG (2012) Role of phospholipase Cε in physiological signaling networks. *Cell Signal* 24:1333–1343.

Zhao J et al. (2011*)* 14–3–3 proteins as potential therapeutic targets. *Semin Cell Dev Biol* 22:705–712.

Section 10.3

Devreotes P & Horwitz AR (2015) Signaling networks that regulate cell migration. *Cold Spring Harb Perspect Biol* 7:a005959.

Janke C & Bulinski JC (2011) Post-translational regulation of the microtubule cytoskeleton; mechanisms and functions. *Nat Rev Mol Cell Biol* 12:773–786.

Julian L & Olson MF (2014) Rho-associated coiled-coil containing kinases (ROCK). Structure, regulation, and functions. *Small GTPases* 5:e29846.

Lawson CD & Burridge K (2014) The on-off relationships of Rho and Rac during integrin-mediated adhesion and cell migration. *Small GTPases* 5:e27958.

Lowery J et al. (2015) Intermediate filaments play a pivotal role in regulating cell architecture and function. *J Biol Chem* 290:17145–17153.

Murali A & Rajalingam K (2014) Small Rho GTPases in the control of cell shape and motility. *Cell Mol Life Sci* 71:1703–1721.

Rotty JD, Wu C & Bear JE (2013) New insights into the regulation and cellular functions of the ARP2/3 complex. *Nat Rev Mol Cell Biol* 14:7–12.

Sadek A & Marshall CJ (2014) Rho GTPases. Masters of cell migration. *Small GTPases* 5:e29710.

Smith JM, Hedman AC & Sacks DB (2013) IQGAPs choreograph cellular signaling from the membrane to the nucleus. *Trends Cell Biol* 25:171–184.

Snider NT & Omary HB (2014) Post-translational modifications of intermediate filament proteins: mechanisms and functions. *Nat Rev Mol Cell Biol* 15:163–177.

Takenawa T & Suetsugu S (2007) The WASP–WAVE protein network: connecting the membrane to the cytoskeleton. *Nat Rev Mol Cell Biol* 8:37–48.

Section 10.4

Ahnert-Hilger G, Münster-Wandowski A & Höltje M (2013) Synaptic vesicle proteins: targets and routes for botulinum toxin. *Curr Top Microbiol Immunol* 364:159–177.

Bhuin T & Roy JK (2014) Rab proteins: the key regulators of intracellular vesicle transport. *Exp Cell Res* 328:1–19.

Hutagalung AH & Novick PJ (2011) Role of Rab GTPases in membrane traffic and cell physiology. *Physiol Rev* 91:119–149.

Jahn R & Fasshauer D (2012) Molecular mechanisms governing exocytosis in synaptic vesicles. *Nature* 490:201–207.

Lowery J et al. (2015) Intermediate filaments play a pivotal role in regulating cell architecture and function. *J Biol Chem* 290:17145–17153.

McNew JA et al. (2013) GTP-dependent membrane fusion. *Annu Rev Cell Dev Biol* 29:529–550.

Mizuno-Yamasaki E, Rivera-Molina F & Novick P (2012) GTPase networks in membrane traffic. *Annu Rev Biochem* 81:637–659.

Stalder D & Antonny B (2013) Arf GTPase regulation through cascade mechanisms and positive feedback loops. *FEBS Lett* 587:2028–2035.

Südhof TC & Rizo J (2011) Synaptic vesicle exocytosis. *Cold Spring Harb Persp Biol* 3:1–14 (a005637).

Swaminathan S (2011) Molecular structures and functional relationships in clostridial neurotoxins. *FEBS J* 278:4467–4485.

Takenawa T & Suetsugu S (2007) The WASP–WAVE protein network: connecting the membrane to the cytoskeleton. *Nat Rev Mol Cell Biol* 8:37–48.

Wu LG et al. (2014) Exocytosis and endocytosis: modes, functions, and coupling mechanisms. *Annu Rev Physiol* 76:301–331.

Section 10.5

Cavazza T & Vernos I (2015) The RanGTP Pathway: From Nucleo-Cytoplasmic Transport to Spindle Assembly and Beyond. *Front in Cell Dev Bio* 3:82.

Güttler T & Görlich D (2011) Ran-dependent nuclear export mediators: a structural perspective. *EMBO J* 30:3457–3474.

Mitogen-activated Protein Kinase and Nuclear Factor κB Modules

11

Eukaryotic cells contain switching elements integrated into modules consisting of several interacting signaling proteins. To a certain extent, such "signalosomes" resemble microprocessors. Even more, one is tempted to refer to them as "molecular ganglions." This means that they collect and bundle the signals transmitted by the "sensory organs" of the cell (the signal receptors at the cell surface) and transduce them to the "cellular brain" (the data-processing molecular network, including the genomic data bank), thus controlling cellular behavior as well as the metabolic apparatus and activity of the genome. Two types of such signaling hub, the MAP kinase modules and the NFκB modules, have been investigated in great detail; they provide especially clear insights into the mode of action of such devices. A major task of these modules is to process mitogenic and stress signals.

11.1 MAP kinase modules are universal relay stations of eukaryotic signal processing

As explained in Section 2.6.5, MAP kinase modules consist of three protein kinases connected in series. Such modules play a key role in cellular data processing ranging from yeast to humans (no such modules have been found yet in prokaryotes). In fact, pioneering research on MAP kinase signaling has been done using non-vertebrates such as nematodes, insects, and—in particular—yeast (**Sidebar 11.1**). These organisms are most suitable for genetic studies such as directed mutations by which the sequence of events can be elucidated (see Section 2.14). With some precautions the results can be extrapolated to mammalian cells since the principles of MAP kinase signaling have been widely preserved in the course of evolution.

What is the secret of this successful model? Most probably it is the special connection of three protein kinases reacting with a strongly sigmoid kinetics. This provides a switch with a powerful noise filter, and in some cases an amplifier, and has—at least theoretically—the properties of an oscillator, allowing the generation of frequency-modulated signals. Moreover, MAP kinase modules are perfect logical gates because they are able to process a large variety of additive, synergistic, and antagonistic signals (see also Section 2.6.5). The three protein kinases interconnected in series are known as MAP kinase (MAPK, MK), MAP2 kinase (MAPKK, MAP2K, MK kinase), and MAP3 kinase (MAPKKK, MAP3K, MKK kinase, MEK) (**Figure 11.1**). Each of these kinases exists in various subtypes that are integrated in individual modules. The core of each module is the pair of MAP kinase and MAP2 kinase, which interact with each other in a highly specific manner. In contrast, the interaction between MAP2 kinases and MAP3 kinases is much less specific: many protein kinases may function as MAP3 kinases, thus transmitting a wide variety of input signals to the module. The same holds true for the MAP kinases themselves, which, admittedly, are firmly wired with the MAP2 kinases, but are not choosy as far as their substrate proteins are concerned. Protein kinases regulating the activity of MAP3 kinases are known as MAP4 kinases.

Figure 11.1 MAP kinase module: signal processor and noise filter (A) Arrangement of the three protein kinases and phosphorylations within the module. The colored and gray boxes symbolize scaffold proteins holding together the module. Upon activation of the module, the strong interaction between MAP2K and MAPK may become weakened, enabling the activated MAPK to translocate, for instance from the cytoplasm into the nucleus. (B) Relationship between input and output signal with (solid line) and without (broken line) modular organization. Note the sharp switch-like response of the module. The area of noise suppression is shown in color.

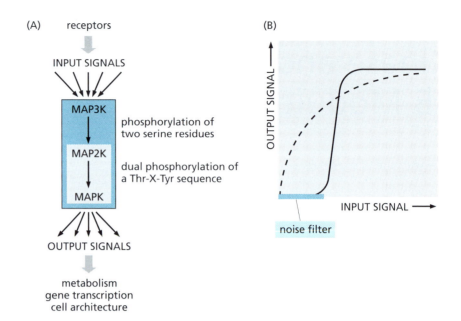

MAP kinases

MAP kinases (mitogen-activated protein kinases) is the collective term for the serine/threonine (Ser/Thr)-specific protein kinases generating the output signals of the MAP kinase modules. The historically based name is somewhat misleading since MAP kinase modules are by no means activated only by mitogenic signals but also are involved in stress response, cell differentiation, and cell death. The mammalian genome encodes 14 MAP kinases and 8 MAP2 kinases. The MAP kinases have been arranged in several subfamilies that are known by historical names such as the following:

- the **ERK family** (extracellular signal-regulated kinases), including isoforms ERK1 and ERK2 (the types ERK3, 4, and 5 constitute separate families).

- the **JNK family** (cJun N-terminal kinases, also called SAPK, stress-activated protein kinases; cJun is a transcription factor), including isoforms JNK1–3, which, owing to alternative splicing, exist in 11 subspecies.

- the **p38 family** (molecular mass of 38 kDa), with four isoforms p38 α, -β, -γ, and -δ.

Each family, probably even each MAP kinase subtype, organizes a module of its own. The ERK modules, in particular, respond to mitogenic signals such as those transmitted by growth factor receptors (and have been called, therefore, mitogenesis modules), whereas the JNK and p38 modules predominantly (but not exclusively) process stress and pro-inflammatory cytokine signals (therefore also called stress modules). MAP kinases catalyze the phosphorylation of Ser and Thr residues adjacent to proline residues and have been called **proline-directed protein kinases**. As mentioned in Section 8.5, such kinases cooperate with Pin1 peptidyl-prolyl *cis–trans* isomerases in modulating the conformation and function of signaling proteins.

Normally MAP kinases are found in the cytoplasm, where they phosphorylate cytoplasmic proteins. However, upon activation they may become translocated into the nucleus to phosphorylate transcription factors (see Section 11.7). The intracellular distribution depends on specific domains in the MAP kinase molecule. ERK, for instance, has both a nuclear localization sequence (NLS) and a cytoplasmic retention sequence (CRS). Via the CRS domain, ERK associates with its MAP2 kinase and is hindered from nuclear translocation unless the module dissociates when activated by an input signal.

Sidebar 11.1 MAP kinase modules of yeast and plants
MAP kinase modules have been found in all eukaryotes studied so far. The genome of the plant *Arabidopsis thaliana* encodes no less than 20 MAP kinases, 10 MAP2 kinases, and 17 putative MAP3 kinases. While cellular functions of the plant modules are widely unknown, the yeast modules belong to the best understood signaling devices of all. The baker's yeast *Saccharomyces cerevisiae* expresses six different modules, and the fission yeast *Schizosaccharomyces pombe* has a similar number. Some of the baker's yeast modules are shown in **Figure 11.2**. As can be seen, these modules are coupled to quite different receptors and signaling devices, while downstream they control enzyme activities, RNA translation, and gene transcription. With the exception of the Sln1 pathway, the stimulation of the MAP kinase modules seems to require the activation of a G-protein such as Cdc42 (Cdc means "cell-cycle defective cell") that helps to anchor the module and the associated MAP4 kinase Ste20 (which resembles mammalian PAK2) at the plasma membrane. Such membrane contact is essential for activation of Ste20 and—in turn—of the MAP3 kinase Ste11. One of the best investigated signaling pathways is the adaptive response that enables baker's yeast to deal with hyperosmotic stress. This occurs when the concentration of sugar and other osmoactive factors in the environment exceeds a critical

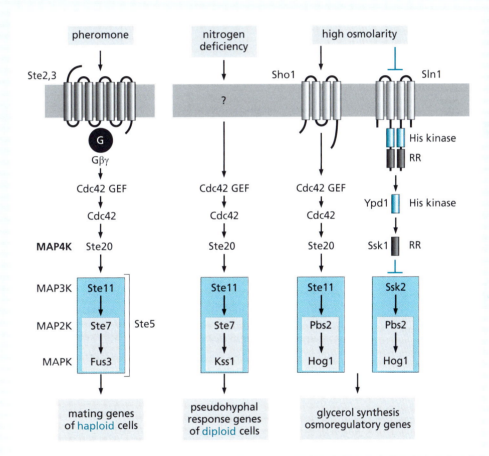

Figure 11.2 Signaling pathways including MAP kinase modules in *Saccharomyces cerevisiae* The MAP kinase modules (colored boxes) are controlled by various exogenous stimuli along quite different signaling pathways, leading to individual cell-physiological effects. Note that the two modules shown to the left differ only in their MAP kinases (Fus3 versus Kss1) addressing different downstream effectors, whereas the two modules shown to the right have different MAP3 kinases (Ste11 versus Ssk2), thus responding to different input signals but generating identical output signals. Various adaptor proteins (not shown) are required for the interaction of the compounds. Cdc42, small G-protein of the Rho family; Cdc42 GEF, guanine nucleotide exchange factor activating Cdc42; Ste2 and Ste3, G-protein-coupled receptors; Sln1, sensor histidine kinase of a two-component system; RR, response regulator of a two-component system; Ste20, Ser/Thr-specific MAP4 kinase; Ste5, scaffold protein of the pheromone module (see Sections 11.4 and 11.5). In the fission yeast *Schizosaccharomyces pombe*, pheromones also interact with a G-protein-coupled receptor and the corresponding MAP kinase module Byr2–Byr1–Spk1 is activated by the Gα-subunit together with a Ras-homologous protein (see Section 10.1.4).

value. To prevent water loss, shrinkage, and cell death, within minutes yeast stimulates the synthesis of glycerol and related compounds by activating corresponding enzymes, and undergoes long-term adaptation by promoting gene expression. These responses are controlled by a MAP kinase cascade, the Hog 1 (high osmolarity glycerol) pathway (see Figure 11.2). The Hog1 module resembles the p38 stress module of vertebrates. In fact, mammalian p38 can restore the hyperosmotic response in yeast cells with genetically deleted Hog1.

Figure 11.3 Three best-known MAP kinase modules of vertebrates In most cases, the modules are activated by different MAP4 kinases (MAP4K) together with small G-proteins of the Ras and Rac/Cdc42 families. By convention the MAP2 kinases frequently are called MKK (MAPK kinases) or MEK (MAPK/ERK kinases). Other abbreviations are MEKK for MEK kinase and MLK for mixed-lineage kinase (see Section 11.3). Further MAP3 kinases are described in the text.

The phosphorylation consensus sequences of MAP kinases are quite abundant: there are hundreds of potential substrate proteins, including protein kinases, transcription factors, and cytoskeletal proteins as well as signal-generating and signal-terminating enzymes. Nevertheless, the individual MAP kinases are surprisingly selective, recognizing only a limited set of target proteins (**Figure 11.3**). This selectivity is due to additional, highly specific protein–protein interactions that considerably restrict the number of enzyme–substrate interactions and are achieved by **docking motifs** (or D-domains) that are not parts of the catalytic substrate binding domain. By means of these motifs, the kinases selectively bind to complementary motifs of individual substrate proteins (**Figures 11.4A** and **11.5**). The principle is reminiscent of a bank safe that is unlocked by two keys. Another tool bringing together MAP kinases and selected target proteins are **scaffold proteins** (**Figure 11.4B**). As explained in Section 11.5, docking domains and scaffold proteins are also used to hold together the whole module.

MAP2 kinases

Each MAP kinase is activated by a firmly attached MAP2 kinase, also called MKK (MAP Kinase kinase), catalyzing in an absolutely specific way the dual phosphorylation of a Thr-X-Tyr sequence in the activation loop. Since these sequences differ

Figure 11.4 Selection and reinforcement of the interaction between MAP kinase and substrate Kinase specificity is due to (A) specific interactions with docking domains or (B) a specific arrangement of the partners by scaffold proteins.

Figure 11.5 Accuracy of MAP kinase–substrate interactions Despite the abundant phosphorylation motif, Ser/Thr- Pro, MAP kinases phosphorylate substrate proteins selectively rather than randomly. This specificity is achieved by domain-interacting motifs (DIM, triangles) of the substrate protein binding to the docking domains of the kinase. (A) Two interacting motifs direct the catalytic domain (CD) of the kinase selectively to one of seven potential phosphorylation sites (black dots; the number is chosen arbitrarily). (B) If the docking motifs are positioned differently, the protein cannot interact with this particular kinase but possibly with another kinase isoform.

in the amino acid X from module to module, each module is equipped with its own specific MAP2 kinases. The Thr-X-Tyr sequences are as follows:

- Thr-Glu-Tyr (TEY) in ERK, phosphorylated by MAP2 kinases MKK1 and MKK2

- Thr-Gly-Tyr (TGY) in p38, phosphorylated by MAP2 kinases MKK3 and MKK6

- Thr-Pro-Tyr (TPY) in JNK, phosphorylated by MAP2 kinases MKK4 and MKK7

MAP2 kinases belong to the small group of dual-specific kinases phosphorylating both Thr and Tyr residues. MAP kinases are their only substrates. As shown below, all MAP2 kinases are activated by MAP3 kinases, which catalyze the phosphorylation of two adjacent Ser and Thr residues in the activation loop. The nomenclature of MAP2 kinases is somewhat confusing (**Table 11.1**).

MAP kinase phosphatases

Antagonists of the MAP2 kinases are the dual-specific MAP kinase phosphatases (DS-MKPs) that dephosphorylate the Thr-X-Tyr sequences of MAP kinases. These enzymes are as selective as MAP2 kinases: one distinguishes between at least 13 isoforms exhibiting different MAP kinase specificities. DS-MKPs are interconnected with MAP kinases in several regulatory loops. Many DS-MKPs are encoded, for instance, by immediate early response genes induced by transcription factors, which

Table 11.1 Nomenclature of MAP2 kinases[a]

MAP2 kinase	Synonyms
MKK1	MEK1 (MAPK/ERK kinase)
MKK2	MEK2
MKK3	SKK1 (SAPK kinase)
MKK4	SKK2, SEK1 (SAPK/ERK kinase), JNKK1 (JNK kinase)
MKK6	SKK3
MKK7	JNKK2

[a]Frequently, but not logically, MEK (mitogen/ extracellular signal-regulated kinase) is used as a collective term for all MAP2 kinases.

are under the control of MAP kinases. Moreover, DS-MKPs require phosphorylated MAP kinases as allosteric activators, and finally, the lifespan of the rather labile DS-MKPs is prolonged by MAP kinase-catalyzed phosphorylation. On the other hand, DS-MKPs function as anchor proteins, sequestering MAP kinases in the nucleus for dephosphorylation. Thus, MAP kinases, MAP2 kinases, and the related phosphatases are interlinked by several positive and negative feedback loops, theoretically indicating an oscillating system.

Summary

- MAP kinase modules comprising three protein kinases interconnected in series are central relay stations of cellular signal transduction that seem to be distributed all over the eukaryotic kingdom. They transform a wide variety of input signals, mostly transmitted by receptors, into a large number of output signals addressing metabolism, gene transcription, mRNA translation, and cell architecture.

- Mitogen-activated protein (MAP) kinases are Ser/Thr-specific protein kinases that are activated by dual phosphorylation of a Thr-X-Tyr sequence. Among the various types, extracellular signal-regulated kinases (ERK), cJun N-terminal kinases, also known as stress-activated protein kinases (JNK/SAPK), and p38 kinases are the most prominent. While ERKs are incorporated in mitogenesis modules, JNK/SAPK and p38 are components of stress modules.

- Upon receiving an upstream signal, the highly specific MAP2 kinases activate MAP kinases by dual phosphorylation of a Thr-X-Tyr sequence. The phosphorylations are reversed selectively by dual-specific MAP kinase phosphatases.

- In principle, MAP kinases are rather nonspecific. However, their specificity and intracellular localization are improved considerably by sequences (docking motifs) and scaffold proteins, enabling more selective interaction with target proteins.

11.2. MAP3 kinases are sensors of MAP kinase modules

To become active, MAP2 kinases have to be phosphorylated at two Ser or Thr residues in the activation loop. A rather large and heterogeneous group of protein kinases, collectively named MAP3 kinases, can do this job. MAP3 kinases are the signal receivers of MAP kinase modules. To this end they dispose of a remarkable variety of interaction domains, thus reflecting the integration of MAP kinase modules in almost any signaling pathway. The two major MAP kinase modules are associated with distinct MAP3 kinase types:

(1) Mitogenesis modules with RAF kinases,

(2) Stress modules with mixed-lineage kinases, MEK kinases, and a heterogenous collection of other kinases.

RAF kinases activate mitogenesis modules

The prototypical RAF kinase was discovered as a retroviral oncoprotein v-Raf (rapidly accelerated fibrosarcoma) isolated from a murine sarcoma virus. This finding represented a key for the investigation of MAP kinase modules. Mammalian RAF kinases are expressed in three isoforms A, B, and C that selectively activate the mitogenic ERK pathway. An important role of RAF is to bring the mitogenesis

MAP3 kinases are sensors of MAP kinase modules   401

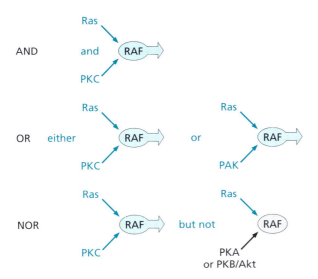

Figure 11.6 MAP3 kinase RAF as a logical gate RAF is activated by both Ras and protein kinase C (PKC), resembling an AND gate, or by Ras and p21-activated kinase (PAK), resembling an OR gate, but not by Ras and protein kinase A (PKA) or protein kinase B (PKB)/Akt, resembling a NOR (or BUT NOT) gate.

modules in contact with those upstream signaling events that lead to an activation of the small G-protein Ras. To this end Ras—in its GTP—loaded form—binds RAF via specific interaction domains. This results in a stimulation of RAF's kinase activity (owing to trans-autophosphorylation of RAF dimers). When a single amino acid of the Ras-binding domain of RAF is substituted, RAF is neither bound to the membrane nor activated. In addition, RAF may interact with the protein kinase PAK2 acting as MAP4 kinase (see Section 11.4). PAK2 is recruited to the receptor complex by additional small G-proteins of the Rho family. Other protein kinases controlling Raf activity are PKC (stimulatory) and PKA and PKB/Akt (inhibitory). Moreover, RAF activity is terminated by feedback control: phosphorylation by ERK disrupts the Ras–RAF complex and releases RAF from the membrane. Thus RAF, like other MAP3 kinases, is the prototype of a logical gate enabling the connected MAP kinase module to perform all operations required for data processing (**Figure 11.6**). The complex regulation of RAF kinases is discussed in more detail in Section 10.1.4.

Mixed-lineage kinases and MEK kinases activate stress modules

What RAF kinases are for the ERK mitogenesis modules, **mixed-lineage kinases** (MLKs) are for the JNK and p38 stress modules. The strange name of this enzyme family is based on an error: owing to structural peculiarities, they were assumed to be a lineage of "mixed-functional" kinases phosphorylating both Ser/Thr and Tyr residues (called dual-specific kinases today). Since then, mixed-lineage kinases have turned out to be bona fide Ser/Thr kinases. Phylogenetically they are distant relatives of the RAF kinases.

In mammalian cells, nine MLK isoforms have been found. Five of them constitute the MLK family proper, while two are known as DL kinases, referring to a characteristic double (D) leucine (L)-zipper domain, and two as ZA kinases, containing a so-called zipper (Z) sterile α (A) motif (**Figure 11.7**). These nine enzymes are components of JNK modules; isoforms MLK3 and DLK in addition are found in p38 modules.

Like RAF kinases, MLKs have to be recruited to the inner side of the cell membrane and to be relieved from intramolecular inhibition to become activated. This is achieved by small G-proteins of the type Cdc42 or Rac, which here play a role resembling that of Ras for RAF activation. Both interact in their GTP-loaded forms with a special binding domain CRIB (Cdc42, Rac Interactive Binding domain;

Figure 11.7 Domain structures of mixed-lineage kinases MLK, mixed-lineage kinase; SH3, Src homology domain 3; LZ, leucine zipper motif; Pro, proline-rich domain interacting with SH3 domains; CRIB, Cdc42, Rac-interactive binding domain; DLK, double leucine-zipper kinase; ZAK, zipper sterile-α kinase; SAM, zipper sterile α motif. For further details, see text.

Figure 11.7). A conformational change is induced and the intrasteric blockade of the kinase activity is cancelled. This mechanism provides one of the pathways along which Cdc42 and Rac stimulate the stress modules (see Figure 11.3; a parallel pathway is described in Section 11.3). CRIB domains are not found in DLKs and ZAKs. These kinases are acivated by signal-controlled dimerization and trans-autophosphorylation (see Figure 11.7)

Currently mixed-lineage kinases are gaining considerable medical interest, since apoptosis is induced in neurons along the MLK–JNK signaling pathway, in particular in the course of **neurodegenerative disorders** such as Alzheimer's dementia, Huntington's and Parkinson's diseases, HIV-induced dementia, and old-age deafness. In fact, relatively specific MLK inhibitors have been found to exhibit neuroprotective effects in animal experiments and are in clinical evaluation.

Another heterogeneous family of MAP3 kinases are the **MEK kinases** or MKK kinases (MEKK, MKKK). In contrast to MLKs and RAF kinases, they are phylogenetic descendants of the yeast enzyme Ste11 (see Figure 11.2), and many of them were indeed isolated by polymerase chain reaction (PCR) with Ste11 probes. There are four closely related isoforms MEKK1–4 (**Figure 11.8**).

Figure 11.8 MEK kinases
(A) Association of the MEK kinases MEKK1–4 with individual MAP kinase core modules and upstream signaling proteins. A broken arrow indicates that the physiological significance of the corresponding interaction is still debated. (B) Domain structures of MEKK1 and MEKK2. MEKK4 has a similar structure as MEKK1 except that a Cdc42, Rac-binding (CRIB) domain replaces the Ras-binding domain (RasBD). MEKK3 closely resembles MEKK2. Pro, proline-rich domain interacting with SH3 domains; PH, pleckstrin homology domain interacting with membrane phospholipids and Gβγ-subunits.

MAP3 kinases are sensors of MAP kinase modules   **403**

- MEKK 1 is the most thoroughly investigated type. Although this enzyme has a Ras-binding domain and may activate the ERK modules in exactly the same way as RAF, it seems to be mainly integrated in JNK modules.

- MEKK1 is also a component of signaling cascades controlled by cytokine receptors that lead to activation of the transcription factor NFκB (see Chapter 6 and Section 11.7).

- MEKK4 has a CRIB domain and is activated by the small G-proteins Cdc42 and Rac. Like MEKK1 it activates stress modules.

- MEKK2 and -3 do not contain G-protein-interacting motifs. Other stimulatory pathways may exist, therefore. They render MAP kinase modules even more versatile than hitherto assumed.

Other MAP3 kinases

Other MAP3 kinases are depicted in **Figure 11.9**. They generally activate the JNK and p38 stress modules, frequently also NFκB signaling (see Section 11.7).

- ASK1, a ubiquitously expressed enzyme, is activated along quite different pathways, either by the Gα-subunit of the large G_{13}-protein or by tumor necrosis factor α (TNFα) and death receptor signaling or by oxidative stress. In fact, by promoting transcription of pro-apoptotic genes, ASK1 plays an important role as downstream mediator of stress- and receptor-transmitted signals that cause inflammation and programmed cell death (for details, see Section 13.2.5).

- TAK1 (transforming growth factor β- activated kinase), is not only integrated in the TGFβ signaling pathway but also a component of signaling cascades originating from TNFα and interleukin-1β receptors, two major pathways for the processing of stress signals. Here specific TAK-binding proteins (TAB1 and -2) serve as adaptors between the kinase and the receptor complexes (see Section 6.3.1).

- TPL2, a TAK1 relative, was identified originally as a proto-oncogene product encoded by a Tumor Progression Locus 2. Its oncogenic mutation causes T-cell lymphomas in rats. In humans some cancer types seem to be promoted whereas others are inhibited by a deregulation of TPL2. Found primarily in macrophages and B lymphocytes, TPL2 seems to play a role in inflammatory defense reactions, in particular since it has been found to be stimulated along signaling pathways activated by bacterial lipopolysaccharides and T lymphocytes.

- Finally, the NFκB-inducing kinase NIK and the TAO kinases (referring to the thousand and one amino acids of the first characterized TAO kinase) remain to be mentioned. The fact that NIK is a component of the NFκB module (see Section 11.7.2) as well as a MAP3 kinase in p38 and JNK modules again indicates intense cross talk between these signaling modules.

Figure 11.9 Signal transduction by MAP3 kinases Apoptotic signal-related kinase 1 (ASK1), transforming growth factor β-activated kinase (TAK1), and kinase "tumor progression locus 2" (TPL2) are depicted. For details, see text. TAB, TAK1-binding protein. Fas is a death receptor inducing apoptosis. CD40 is a member of the TNF receptor family expressed by antigen-presenting cells.

Summary

MAP3 kinases, the signal receivers of MAP kinase modules, constitute a large and heterogeneous group of Ser/Thr-specific protein kinases that catalyze the activation loop phosphorylation of MAP2 kinases. To a certain extent, distinct subfamilies can be assigned to individual modules: Ras-dependent RAF kinases to the ERK modules, and mixed-lineage kinases (MLKs) and other MAP3 kinases to the JNK and p38 modules. In addition, NFκB signaling may be stimulated by MAP3 kinases.

Figure 11.10 Activation of MAP kinase modules by G-proteins cooperating with MAP4 kinases of the Ste20 group The common theme is recruitment of the signal-transducing apparatus to the membrane, where it comes in close contact with receptors of input signals. (A) Activation of the pheromone module in the yeast *S. cerevisiae*. Step 1: activated pheromone receptor induces the release of a Gβγ-subunit from the receptor-coupled trimeric G-protein. Step 2: membrane-anchored Gβγ-subunit recruits the pheromone module to the membrane by interacting with the scaffold protein Ste5 (gray bar). Step 3: MAP4 kinase Ste20, bound to the membrane by the small membrane-anchored G-protein Cdc42 (or by the Gβγ-subunit) and activated by (auto)phosphorylation, catalyzes the phosphorylation of the MAP3 kinase Ste11 and thus activates the pheromone module. (B) Analogous activation of the ERK1 module in mammalian cells; cooperation of the two G-proteins Ras and Rac is assumed. Step 1: activated and autophosphorylated growth factor receptor Tyr kinase stimulates, via the adaptor protein Grb2 and the Ras-GEF mSOS, the small G-protein Ras. Step 2: membrane-anchored active Ras-GTP complex binds the MAP3 kinase RAF, thus recruiting the ERK1 core module (bound to the scaffold protein MP1) to the membrane. Step 3: MAP4 kinase PAK2, bound to the membrane by the small membrane-anchored G-protein Rac and activated by autophosphorylation, phosphorylates RAF, thus activating the ERK1 module. Rac is activated along the Ras signaling pathway or other pathways (see Section 10.3). The guanine nucleotide exchange factors (GEFs) required for Cdc42 and Rac activation are not shown.

11.3 MAP4 kinases and G-proteins: lessons learned from yeast

As we have seen, MAP kinase modules are under the control of a wide variety of input signals, which are received by the MAP3 kinases. Frequently such input signals are transmitted by protein kinases known as MAP4 kinases that phosphorylate and activate MAP3 kinases, thereby cooperating with G-proteins. To this end, the partner molecules must be assembled at the inner side of the cell membrane. In Figure 11.6 this is schematically shown for the pheromone signaling cascade of *S. cerevisiae*, compared with the mammalian RAF–ERK module. Both the yeast and the mammalian modules are recruited to the plasma membrane by membrane-bound G-proteins: a Gβγ subunit in the case of the pheromone pathway, Ras for the RAF–ERK cascade. In addition, the corresponding MAP4 kinase has to be bound to the membrane to become activated. This is accomplished again by G-proteins such as Cdc42 interacting with the kinase Ste20 of the pheromone module, or Rac interacting with the MAP4 kinase PAK2 of the RAF–ERK module. These interactions result in a conformational change of the MAP4 kinase in such a way that an activation loop phosphorylation (by autophosphorylation or by another protein kinase) becomes possible. The activation of the G-proteins requires loading with GTP as catalyzed by guanine nucleotide exchange factors (GEFs). Moreover, G-proteins are inactivated by controlled GTP hydrolysis. Both the GEF and the GTPase activities are the signal receivers proper of the MAP4 kinase–G-protein complex (not shown in **Figure 11.10**). The MAP4 kinase of

MAP4 kinases and G-proteins: lessons learned from yeast   405

the yeast pheromone pathway is Ste 20. Evolutionary descendants of Ste20 are the **p21-Activated protein Kinases, PAKs,** of vertebrates (discussed in Section 10.3.4). From the six PAKs found in humans only PAK2 is an efficient MAP4 kinase, while the other PAKs preferentially phosphorylate proteins controlling the dynamics of the cytoskeleton (see Section 10.3). Characteristic structural features of both PAKs and Ste20 are a carboxy (C)-terminal kinase domain and a domain, such as CRIB, for membrane binding and activation by small G-proteins.

Another family of mammalian protein kinases related to yeast Ste20 are the **germinal-center kinases (GCK).** The strange name is explained by the original discovery of these enzymes in B-cell germinal centers, sites of B-cell maturation. However, these kinases are by no means restricted to lymphoid tissue. Mammalian GCKs exist in 20 isoforms (**Table 11.2** shows a selection). They differ from PAKs in the position of the kinase domain (located in the amino (N)-terminal region) and a pronounced variability of the regulatory domains.

GCKs are predominantly involved in the processing of stress signals. In addition, they mediate signals that induce the differentiation of lymphoid and hematopoietic tissues. Such signals are, for instance, transmitted by cytokines of the TNF family. In fact, a characteristic property of GCKs is that by interacting with TRAF (TNF receptor-associated factor) adaptor proteins they make contact with and are activated by cytokine receptors of the TNFα receptor family (see Section 6.3.1). By adaptors containing SH3 domains such as Nck, Grb2, Crk, etc., some GCKs are also linked to Tyr kinase-coupled receptors transmitting growth factor and cytokine signals (see Section 7.1). Thus, GCKs transmit a series of different exogenous signals to the JNK modules. Whether or not they are bona fide MAP4 kinases or interact more indirectly with the modules is not always clear.

A special member of the GCK family is the **JNK-inhibiting kinase JIK** that competes with MAP2 kinases for JNK. While MAP2 kinases activate JNK, phosphorylation by

Table 11.2 Selection of putative MAP4 kinases of the mammalian GCK family

		Activation	MAPK module
GCK	germinal-center kinases	TNFα	JNK
GCKR	GCK-related kinases	TNFα, UV	JNK
GLK	GCK-like kinases	TNFα	JNK
HPK1	hematopoietic progenitor kinase	TNFα, Tyr kinases	JNK; hematopoietic tissues
NIK[a]	Nck-interacting kinase	TNFα	JNK; binds adaptor Nck
TNIK	TRAF- and Nck-interacting kinase	TNFα	JNK; binds adaptors TRAF and Nck
NRK	NIK-related kinase	TNFα	JNK
SOK1	Ste20-like oxidant stress-activated kinase	oxidative stress	?
KRS1	kinase responsive to stress	extreme stress	?
MST	mammalian Ste 20-like kinases	extreme stress, Fas	JNK, p38
LOK	lymphocyte-oriented kinase	extreme stress	?; in lymphocytes only
JIK	JNK-inhibiting kinase	?	inhibition of JNK

[a]The MAP4 kinase NIK must not be confused with the MAP3 kinase NIK (NFκB-inducing kinase), a signal transmitter in the NFκB system (see Sections 11.2 and 11.7.2).

JIK, occurring at another site of the JNK molecule, has an inhibitory effect. This inhibition is relieved by growth factors such as epidermal growth factor. It has been proposed that epidermal growth factor, although it normally activates only the ERK module, may stimulate the JNK modules indirectly along this pathway.

Other Ste20-homologous mammalian kinases related to both GCKs and MLKs are the above-mentioned MAP3 kinases of the TAO kinase family.

Summary

- The standard mechanism of MAP3 kinase activation is protein phosphorylation catalyzed by a wide variety of MAP4 kinases.

- Most of them are related to the yeast kinase Ste20, which transduces pheromone signals. For activation they require an interaction with small G-proteins of the Rho/Rac/Cdc42 family, which induce membrane attachment of the kinases and activation loop phosphorylation.

- Other MAP4 kinases become activated along cytokine-controlled signaling pathways.

- With the exception of PKC functioning as a MAP4 kinase of the ERK modules, the MAP4 kinases known thus far preferentially stimulate the stress modules.

11.4 Organization of MAP kinase modules by scaffold proteins

Scaffold proteins are molecular sockets that bind several signal-transducing proteins in such a way that the signaling interactions are facilitated and rendered more specific. Somehow, such complexes or "signalosomes" are reminiscent of integrated electronic circuits. For the assembly and function of MAP kinase modules, scaffold proteins play a key role.

The prototype of such a scaffold protein is **Ste5** of the yeast *S. cerevisiae*. It organizes the pheromone-activated Fus3 module (see Figures 11.2 and **11.11**) and is indispensable for the module's function. Quite a similar module, differing from the pheromone-activated module in the MAP kinase only (Kss1 instead of Fus3), controls an entirely different physiological event: the invasive growth of diploid and filamentous yeast cells in response to nitrogen deficiency (see Figure 11.2). This module is not fixed on Ste5 but possibly bound by another, still-elusive scaffold protein. In an additional yeast module, the Hog1 module of osmolarity adaptation, the MAP2 kinase Pbs2 serves as a scaffold protein (see Figure 11.11).

Figure 11.11 Scaffold proteins of MAP kinase modules in *S. cerevisiae* The pheromone module has a separate scaffold protein (Ste5) organizing the ordered interaction of four kinases. In the osmolarity module, the MAP2 kinase Pbs2 serves as a scaffold protein on its own, holding together the whole module. Both scaffold proteins represent standard types. Analogous proteins are found in vertebrate cells. Teal curved arrows symbolize activation steps (phosphorylations).

Table 11.3 Scaffold proteins of mammalian MAP kinase modules[a]

Scaffold protein	MAP kinase module	Remarks
KSR	RAF–MKK1–ERK1	RAF-related pseudokinase
MP1	MKK1–ERK1	
JIP1,2	MLK–MKK7–JNK	interacts with Rho-GEF p190
JSAP1 (JIP3)	MEKK1–MKK4–JNK	
MEKK1	RAF–MKK2–ERK2	interacts with cytoskeleton
MEKK2	MEKK2–MKK7–JNK1	
filamin	TRAF2–MKK4–JNK	interacts with actin filaments
β-arrestin 2	ASK–MKK4–JNK3	activated by angiotensin II

[a]KSR, kinase suppressor of Ras; MP, MEK partner; JIP, JNK-interacting protein; JSAP, JNK/SAPK-associated protein; TRAF2, TNF receptor-associated factor 2, an adaptor protein bringing together the TNF receptor with JNK and NFκB modules (see Section 6.3.1).

The MAP kinase modules of mammalian cells are held together by similar scaffold proteins that, in addition, recruit regulatory proteins such as guanine nucleotide exchange factors (GEFs), the activators of small G-proteins. Inside the cell, such complexes may be transported as cargo by kinesin motor proteins. By this means, probably whole sets of signal-processing proteins are rearranged, resulting in mechanical changes of the "hardware components" aiming at an adaptation of the network's switching pattern. Examples of scaffold proteins are **JIP (JNK interacting protein)** for JNK modules and **KSR** (kinase suppressor of Ras) for ERK modules. KSR was discovered in genetic studies with nematodes and insects as a gene that suppressed the Ras-dependent phenotype. Mammals express 2 isoforms. KSR is a RAF-related pseudokinase (see Section 2.6.3) that, like RAF, interacts with the MAP2 kinase MKK1, at the same time being a substrate of the MAP4 kinase TAK1. When fully phosphorylated, the KSR–MKK1 complex is sequestered by a 14-3-3 protein and cannot organize the Ras-activated ERK modules (explaining its name). Only upon partial dephosphorylation is KSR able to bind RAF and ERK. A collection of scaffold proteins of mammalian MAP kinase modules is shown in **Table 11.3**.

Scaffold proteins increase the specificity or targeting and the efficacy of signal transduction. Moreover, they assemble "their" signal-transducing complexes at distinct cellular sites, shielding them from disturbing influences. This occurs, however, at the expense of signal amplification, which is not possible in a module with a 1:1 ratio of components. For this reason, MAP kinase modules *without* scaffold proteins may also exist, enabling a sequential signal amplification, albeit now at the expense of specificity (**Figure 11.12**).

Summary

MAP kinase modules are frequently organized on scaffold proteins, increasing the specificity and efficiency of signal transduction. However, this is achieved at the expense of signal amplification, which can occur only in the absence of scaffolding.

Figure 11.12 Signal amplification by a MAP kinase module not fixed on a scaffold protein Upon activation by an input signal, MEKK1 phosphorylates many MKK4 molecules, each of which then phosphorylates many JNK molecules (teal arrows; P, phosphate group). The cascade is interrupted only by dephosphorylation of MKK4 and inactivation of MEKK1. An asterisk marks the active MEKK1.

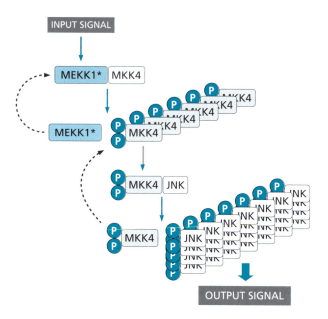

11.5 Downstream of MAP kinase modules: MAP kinase-activated protein kinases

MAP kinases have various substrates, such as the cytoplasmic phospholipase A_2 (the key enzyme of arachidonic acid metabolism), the Ras–GEF mSOS, and the above-mentioned MAP kinase phosphatases, as well as transcription factors that will be discussed below. Moreover, MAP kinases may activate other protein kinases.

In mammalian cells, several structurally related types of such MAP kinase-activated protein kinase (MAPKAP-kinases) have been found:

- ribosomal S6 kinases (RSK), also known as p90^{RSK} or MAPKAP kinases 1

- mitogen- and stress-activated Kinases MSK

- MAPK-iNteracting Kinases MNK

- MK2, -3, and -5 (unfortunately the acronym MK is also used for MAP kinases, which must not be confused with these MAPKAP kinases)

Yeast cells express almost no MAPKAP kinases, indicating that (as expected) their data-processing network is organized more simply.

The widely distributed MAPKAP kinases of the types RSK and MSK are closely related. Both are characterized by two functional kinase domains, a PKC- and a calmodulin kinase-related one (**Figure 11.13**), and both promote survival programs by activating cell proliferation and inhibiting apoptosis. Their major targets are a series of transcription factors such as SRF, CREB, ATF1, NFκB, cJun, and cFos that stimulate immediate early response genes as a rapid reaction to mitogenic signals and stress factors. Another substrate is histone 3. Because it weakens the histone–DNA interactions in nucleosomes, H3 phosphorylation is an important step in transcriptional stimulation. Moreover, RSKs protect cells from apoptosis by inactivating pro-apoptotic proteins such as Bad and stimulating the transcription of anti-apoptotic genes. Finally, RSKs play an important role in the up-regulation of the cell cycle: they inactivate the cell cycle inhibitors p27^{KIP} and protein kinase MYT1. More details on RSK are found in Section 9.4.

The stimulatory effect of RSKs and MSKs on transcription is supported by the MAPKAP kinases of the type MK and MNK in that both promote ribosomal protein synthesis. To this end MNK1 phosphorylates and *activates* the translation factor eIF4E, while MKs arrive at the same result by phosphorylating and *inactivating* the

Figure 11.13 Basic structure of MAP kinase-activated protein kinases (Top) A peculiarity of ribosomal S6 kinases (RSK) and mitogen- and stress-activated kinases (MSK) are two catalytic domains: an N-terminal protein kinase C-related domain 1 and a C-terminal calmodulin-dependent kinase-related domain 2. (Bottom) Other MAPK-interacting kinases (MNK) and MAP kinase-activated protein kinases (MK) contain only the calmodulin-dependent kinase-related domain 2. Activation loops are symbolized by colored bars. Arrows indicate the sites phosphorylated by individual MAP kinases and phospholipid-dependent kinase 1 (PDK1). Auto-P, autophosphorylation; NLS, nuclear localization sequence.

translational repressor eEF2 kinase (see Section 9.3). These reactions are summarized in **Figure 11.14**. They are essential for the cellular stress response.

Summary

MAP kinase-activated protein kinases (MAPKAP kinases) transduce signals received from the MAP kinase modules. One of their major functions is to promote cell proliferation and to prevent apoptosis. To this end, they modulate the activities of regulatory proteins and stimulate mRNA translation and gene transcription. A well-known subfamily of MAPKAP kinases is the ribosomal S6 kinases p90^{RSK}.

Figure 11.14 Effects of MAP kinase-activated protein kinases In response to mitogens and stress factors transmitted by MAP kinase modules, the protein kinases shown in the colored box modulate proliferation and activate survival programs. To this end, they stimulate the transcription of early response and survival genes and inactivate cell cycle inhibitors such as p27^{KIP} and MYT1 but also the cell cycle stimulator CDC25. At the same time, they promote mRNA translation by inactivating the translational inhibitor eEF2 kinase and stimulating the translation factor eIF4E. Inhibition of the Ras-GEF mSOS by p90 RSK provides a negative feedback control of the Ras–RAF–ERK cascade.

11.6 Downstream of MAP kinase modules: transcription factors

As we have seen, MAP kinase modules are central relay stations of signal transduction from the periphery to the genome. In fact, all MAP kinases, from yeast to humans, catalyze the phosphorylation of transcription factors either directly or via MAPKAP kinases, thus controlling the activity of a large number of genes.

Pivotal targets are the immediate early response genes. As the name says, such genes are regulated by signaling cascades directly, without any detour. The transcription factors associated with such genes have (apart from MAP kinase phosphorylation sites with the characteristic Ser/Thr-Pro core) docking domains that render the kinase–substrate interaction specific and direct the phosphorylation to selected sites in the protein molecule. The principle of this mechanism is illustrated in Figure 11.5. MAP kinase-catalyzed phosphorylation may either activate or inactivate transcription factors by regulating one or more of the following features:

- DNA binding
- transactivating or transrepressing efficacy
- interactions with transcriptional co-activators or co-repressors
- nuclear import or export (**Sidebar 11.2**)
- ubiquitin-mediated degradation

Which of these effects predominates depends on the type of transcription factor, the MAP kinase isoform, and the cell-physiological context. In the following, some transcription factors will be introduced that have turned out to be standard substrates of MAP kinases.

11.6.1 Transcription factors of the Ets family

Ets factors are of critical importance for the formation, maintenance, and function of tissues. Their name refers to an erythroblastosis virus E26 (E twenty-six) where one of these factors was found to be encoded by an oncogene. Nevertheless, the Ets family is probably restricted to multicellular organisms. In mammals, it comprises around 30 members grouped into 10 subfamilies.

Sidebar 11.2 What is nuclear translocation? Transcription factors are frequently found in the cytoplasm. They accumulate in the nucleus when they are activated by an input signal. It is said that they are "translocated" or "migrate." Does this mean that they actively move? Certainly not. In fact, the distribution of macromolecules between cytoplasm and nucleus is solely a matter of diffusion. The equilibrium of diffusion is determined first of all by the factor's ability to pass the pores in the nuclear membrane. To this end most—but not all—transcription factors possess amino acid domains that interact with nuclear pore proteins and either facilitate or inhibit the passage. These are called nuclear localization "signals" (NLS) and cytoplasmic retention "signals" (CRS), respectively (see Section 10.5). Through transcription factor activation, a conformational change is induced that either exposes or hides such sequences. Regulatory proteins—such as the IκBs (see Section 11.8)—interacting with the transcription factor may also camouflage such sequences or hold back the factor in another way. On the other hand, such proteins by means of their own NLS may help a factor without NLS to enter the nucleus, or they may block a CRS of the factor. In any case, the equilibrium of diffusion is shifted in either direction.

Once in the nucleus the activated transcription factor binds to DNA, which prevents its rediffusion into the cytoplasm. In certain cases, this effect may be reinforced by a nuclear retention "signal" that hinders the passage of the factor through the nuclear pores. As a consequence, most of the factor accumulates in the nucleus.

Inactivation restores the initial equilibrium of diffusion.

Similar rules hold true for enzymes such as MAP kinases that "migrate" into the nucleus to activate transcription factors. It is clear that the many control points of nuclear localization are targets for a wide variety of regulatory input signals.

Like most transcription factors, Ets factors are activated by input signals that change the conformation from an inactive resting state into an active state. These input signals consist of protein–protein interactions and protein phosphorylation uncovering the DNA-binding domain of Ets. The phosphorylation of Ets factors is catalyzed by individual MAP kinases. It has a stimulatory effect, although there are exceptions to this rule. Moreover, some Ets factors are transcriptional repressors, becoming inactivated by phosphorylation.

Ets factors are structurally and functionally quite diverse. What they all have in common is an Ets domain that exhibits a helix–loop–helix structure serving DNA binding. Within the Ets factor family this domain is highly conserved. Nevertheless, the individual Ets isoforms exhibit specificity for selected genes. This is achieved by protein domains outside the Ets domain as well as by a variety of post-translational modifications. Moreover, Ets factors easily form heterodimers, and complexes with other transcription factors. An example is provided by the **ternary-complex factors (TCFs)**, an Ets subfamily comprising the isoforms Elk1 (Ets-like) and SAP1,2 (Serum response factor- Accessory Proteins). These factors control the transcription of various immediate early response genes by binding to a serum response element located in the gene promoter. (Serum response elements were discovered in experiments with cell cultures, which require serum, a cocktail of growth factors, for proliferation.) To acquire this activity the TCFs have to interact with another transcription factor, the homodimeric **serum-response factor (SRF)**. As a result, a ternary complex consisting of Elk1, SRF, and serum response element is established, which gave rise to the name ternary-complex factor (**Figure 11.15**). This complex is activated by MAP kinase-catalyzed phosphorylation of the Ets component. While Elk1 is a substrate of ERK, JNK, and p38, SAP1 and -2 are phosphorylated by p38 only.

Ets factors are phosphorylated not only by MAP kinases but also by various other protein kinases and, of course, are substrates of protein phosphatases terminating the Ets signal. Moreover, like Ets, SRF also is regulated by protein phosphorylation as catalyzed, for instance, by the MAPKAP kinase RSK. Thus, the control of ternary complex factor activity shows in a particularly clear manner that transcriptional regulation represents a combinatorial process where the resulting effect depends on the local state of excitation of the protein network rather than on a single signaling event.

Figure 11.15 Activation of a ternary-complex factor By MAP kinase-catalyzed phosphorylation (P), the conformation of an Ets factor (here the isoform Elk1) is changed in such a way that the "hidden" Ets domain is exposed, enabling Elk1 to form a complex with the serum response factor SRF (via an SRF interaction domain, SRF-ID) and to bind to a serum response element (SRE) in a corresponding gene promoter. In turn, SRF further unfolds the conformation of Elk1 so that its transactivating domain (TAD) may now recruit the initiation complex of transcription. Helix–loop–helix proteins with inhibitory domains (ID-HLH) may compete with Elk1 for DNA binding, suppressing gene transcription. The oncogenic Ets protein of erythroblastosis virus has a constitutively active conformation and does not require phosphorylation as a stimulatory input signal.

Still, the physiological functions of Ets factors are not completely understood in all details. Relatively well-known is their role in developmental and regenerative processes such as the formation of blood vessels (angiogenesis) and in hematopoiesis as well as in neuronal differentiation. At least in cell cultures, Ets factors are controlled also by proliferative signals transmitted by growth factors and serum and mediated by the ERK1,2 modules. This has been thought to indicate a role in wound repair and explains the tumor-promoting efficacy of Ets, proven by numerous experiments.

11.6.2 Transcription factors of the AP1 family

The second large family of transcription factors directly controlled by MAP kinase-catalyzed phosphorylation is collectively termed AP1 (Activator Proteins 1). Each AP1 factor is a homo- or heterodimer composed of proteins belonging to the subfamilies Jun, Fos, ATF2 (activating transcription factor 2), and MAF (musculo-aponeurotic fibrosarcoma factor). These factors exist in 20 isoforms at least. Dimerization is achieved by leucine zipper motifs. This combinatorial principle gives rise to a large number of transcription factors differing in gene specificity. Jun and Fos were discovered originally as oncoproteins of animal sarcoma and osteosarcoma viruses (vJun and vFos); the cellular proto-oncogene products are differentiated with the prefix c (cJun and cFos).

The efficacy of AP1 factors is regulated by both de novo synthesis and phosphorylation. Fos factors are encoded by typical immediate early response genes, the transcription of which ranks among the most rapid responses of cells to stress and proliferation signals transmitted by MAP kinase modules. Like most eukaryotic gene promoters, the *fos* promoter contains several response elements that interact with different input signals. In other words, the transcription of the *fos* genes is activated along several signaling pathways (**Figure 11.16**) including the Ras–RAF–ERK–TCF cascade, cAMP-controlled pathways, and the JAK–STAT cascade with various cytokine receptors as starting points (see Section 7.2.1). This situation resembles a multifunctional logical gate.

In nonstimulated cells, the cFos partner cJun is phosphorylated in the carboxyterminal domain, preventing its interaction with the gene promoter. This phosphorylation is catalyzed by GSK3β or casein kinase 2 (CK2) (see Section 9.4.5). When GSK3β is inactivated by mitogenic signals along the PI3K–PKB/Akt pathway (see Section 4.6.6), cJun becomes dephosphorylated by omnipresent protein phosphatases. To become active, cJun as well as ATF2 (the two AP1 factors most thoroughly studied in this respect) have to be phosphorylated at two Ser residues in their amino-terminal transactivating domains. These phosphorylations are catalyzed by JNK (for cJun, explaining the name Jun N-terminal kinase) or p38 (for ATF2), and as in the case of Ets, their specificity is enhanced by docking domains. Together with the de novo synthesis of cFos, this represents the major pathway along which stress factors, cytokines, and growth factors change gene expression.

cJun homodimers and cJun/cFos heterodimers preferentially bind to so-called **TPA response elements** in gene promoters. TPA (12-O-tetradecanoylphorbol 13-acetate), a poison from the spurge *Croton tiglium*, has become most prominent as a potent tumor promoter in animal experiments. It mimics the effects of the second messenger

Figure 11.16 cfos gene promoter This promoter provides a striking example of a gene-regulatory sequence with various signaling inputs rendering it an efficient logical gate. Shown are the different response elements with the associated transcription factors (colored box) and protein kinases. The protein kinases represent a selection: as described for CREB in Section 8.6, most transcription factors are substrates of more than one kinase. Protein kinases: JAK, Janus kinase; ERK, extracellular signal-related kinase; JNK, cJun N-terminal kinase; PKA, protein kinase A. Transcription factors: STAT, signal transducer and activator of transcription; TCF, ternary-complex factor; AP1, activator protein 1; CREB, cAMP response element-binding protein. Response elements: SIE, STAT-inducible element; SRE, serum response element; TRE, TPA response element; CRE, cAMP response element; DR, direct repeat. For additional details, see text.

diacylglycerol (see Section 4.6.5) and is a strong activator of genes, though the molecular mechanism of this effect is not entirely clear. The corresponding TPA response element is found also in the c*fos* and many other gene promoters including the c*jun* promoter, enabling cJun to stimulate its own biosynthesis and that of cFos protein in a positive feedback loop.

Different from cJun/cFos, the cJun/ATF heterodimers recognize their target genes by cAMP response elements, which were detected originally as binding sites of the CREB transcription factors. The latter are activated along the cAMP–PKA cascade and other signaling pathways (see Sections 4.6.1 and 8.6). Still other response elements mediate the interaction of cJun/MAF heterodimers with their target genes.

The effects of the individual MAP kinase modules on Ets/TCF as well as on some other transcription factors are summarized in **Figure 11.17**. However, this is not the end of the story, because as we have seen, ERK and p38 (at least) may activate transcription factors also via protein kinases positioned downstream of the modules, such as RSK, MSK, MNK, and MK. This situation and additional secondary reactions

Figure 11.17 Direct control of transcription factor activity by MAP kinase modules To the left and right of the three major MAP kinase modules (colored boxes), a selection of transcription factors phosphorylated by MAP kinases is shown. **MEF2** (myocyte enhancer factor 2) controls genes of muscle differentiation and stimulates the transcription of the c*jun* gene. **CHOP** (CREB-homologous protein) is involved in the genotoxic stress response. Together with other factors, it induces cell cycle arrest at the G1–S transition to provide time for DNA repair (see Section 12.9). **NFAT** (nuclear factor of activated T cells) predominantly controls interleukin genes, playing an important role in inflammatory reactions and in the immune response. Phosphorylation by JNK prevents its translocation from the cytoplasm into the nucleus (an inhibitory effect; Section 14.6.4). **MAX** is a binding partner and activator of the transcription factor **cMyc** that controls many genes for proliferation, apoptosis, and cell differentiation and may undergo oncogenic mutation. Not shown are transcription factor phosphorylations, catalyzed by MAP kinase-activated protein kinases (see Figure 11.13). PI3K, phosphatidylinositol 3-kinase; ATF2, activating transcription factor 2. To the right, phosphorylation of cJun is shown (activation is symbolized by black arrows; inactivation, by teal arrows). As can be seen, cJun works as a logical BUT NOT gate. JNK, cJun N-terminal kinase; GSK3β, glycogen synthase kinase 3β; CK2, casein kinase 2.

Sidebar 11.3 Control of gene transcription by frequency-modulated signals? Although appearing to be quite simple at a first glance, the modulation of gene activity by signals such as phosphorylation harbors a fundamental problem, at least as the situation in a single cell is concerned. In fact, genes respond to input signals in a variable, graded manner. However, the law of mass action describing stochastically graded chemical reactions is not applicable here owing to the small number of reaction partners (two gene copies only). Therefore, variable gene activity should obey other rules than those of random distribution. One possibility certainly consists of the interaction of a gene with more than one stimulatory and inhibitory signal, rendering it a stepwise-regulated logical gate. Another quite appealing and not mutually exclusive mechanism resulting in a graded transcriptional response would be frequency modulation of signals. Indeed, owing to its particular properties and diverse intrinsic feedback loops, the MAP kinase module, in particular, may fulfill the conditions of a chemical oscillator (see Section 1.9). This would mean that the activity of a transcription factor is not controlled according to a rigid ON–OFF scheme but instead by a discrete sequence of phosphorylating and dephosphorylating impulses, the frequencies of which depend on the intensity of the particular input signal. Such a mechanism would enable a graded transcriptional response and may explain why a single input signal, although processed by the same MAP kinase module, may exhibit different or even opposite effects depending on the concentration of the signaling factor and the duration of action. It is conceivable that this might be one reason for the triumphant progress of the MAP kinase modules in evolution. As a restriction, it must be emphasized, however, that currently the concept of frequency-modulated transcriptional regulation is still widely hypothetical, mainly because of a lack of suitable experimental methods.

make plausible the fact that the activities of more than 1000 genes are changed upon stimulation of the Ras signaling pathway. In other words, and to repeat what has been stated above: exogenous signals do not activate linear biochemical reactions but stimulate large areas of the data-processing protein network, thus generating a diffuse excitation pattern similar to that seen in the brain's neuronal network upon sensory stimulation (see Figure 1.13). Moreover, at least theoretically MAP kinase modules fulfill the conditions of oscillating systems. Therefore the activation of cellular protein networks might involve frequency modulation of signaling (**Sidebar 11.3**).

Summary

- Prototypes of transcription factors controlled by MAP kinase-catalyzed phosphorylation are the Ets factors (including ternary-complex factors, TCFs). Upon phosphorylation they form complexes with other transcription factors (such as serum-response factor, SRF). These complexes modulate the transcriptional activity of various immediate early response genes involved in developmental and regenerative processes.

- Transcription factors of the activator protein 1 (AP1) family, comprising a large number of homo- and heterodimers of the proteins cJun, cFos, ATF2, and MAF, are controlled by JNK- and p38-catalyzed phosphorylation. In combination with other transcription factors, they regulate the activities of a wide variety of genes.

11.7 NFκB signaling pathway

A major "spinal cord" of cellular data processing, which rivals the MAP kinase modules in its variedness and complexity, is the signaling pathways that control the NFκB (nuclear factor kappa B) family of transcription factors.[*] These factors rank among the most versatile regulators of gene transcription. In fact, the NFκB cascades have

---

*Originally NFκB was found to bind to the enhancer of the immunoglobulin κ gene in B lymphocytes and was thought to stimulate antibody formation. Today we know that this is not a major function of the factor. Moreover, it is not exclusively nuclear but is found in the cytoplasm as well.

been estimated to connect more than 200 exogenous stimuli with about 500 different genes. By binding to specific DNA sequences (κB elements) in gene promoters and enhancers, NFκB transcription factors can either stimulate or inhibit gene transcription.

Like MAP kinase modules, NFκB signaling is involved in an immense number of physiological processes dealing with organ development, cell proliferation, cell death, and cell differentiation. A major task of the NFκB system is to process signals that control innate and acquired immune defense. Thus it plays a central role in inflammatory reactions (see Section 6.3). To this end, NFκB proteins regulate the transcription of genes encoding pro-inflammatory cytokines and chemokines, cell adhesion molecules, and enzymes such as NO synthase and cyclooxygenase that catalyze the production of inflammatory mediators. Certain cytokines induced along the NFκB pathways, such as TNFα and interleukin 1, activate the de novo synthesis of NFκB through positive feedback. Such a system may easily run out of control, resulting in severe disorders such as rheumatoid arthritis, bronchial asthma, chronic inflammatory bowel disease, multiple sclerosis, and cancer as well as in death by shock. Therefore, NFκB signaling is under strict and expensive control to ensure that the factors become active only when several safety devices have been unlocked.

11.7.1 NFκB module and its components

The basic structure of an NFκB module is shown in **Figure 11.18**. It consists of a protein kinase that senses input signals and a transcription factor that transmits output signals. The transcription factor is composed of a catalytic subunit and a regulatory subunit, which in the absence of an input signal inhibits the catalytic subunit. This inhibition is relieved by phosphorylation and subsequent degradation of the regulatory subunit. In other words: the module works according to the principle of double negation (minus times minus is equal to plus).

The active transcription factors of the NFκB family are homo- or heterodimers of the proteins *cRel*, RelA alias p65, RelB, p50 alias NFκB1, and p52 alias NFκB2 (the term Rel refers to a retroviral oncogene *vRel* derived from cRel and found in avian reticuloendotheliosis virus). In common these proteins have N-terminal **Rel homology domains (RHD)** for dimerization, nuclear targeting, and DNA binding (a closely related Rel similarity domain is also found in NFAT transcription factors that address another set of genes; see Section 14.6.4). Upon activation by input signals, the dimers accumulate in the nucleus and bind to the above-mentioned κB elements in the promoter and enhancer regions of the target genes. The Rel and NFκB

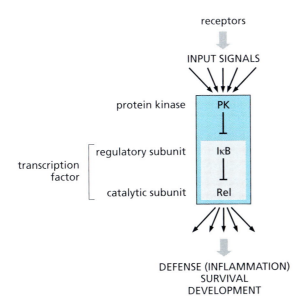

Figure 11.18 NFκB module The module functions according to the principle of double negation. The core of the module is a transcription factor consisting of a regulatory subunit (IκB) and a catalytic subunit (Rel). In the absence of an input signal the regulatory subunit inhibits the catalytic subunit. Input signals activate a protein kinase (PK) that phosphorylates IκB, priming it for degradation. When it is relieved from inhibition, Rel activates a series of genes involved in defense reactions (inflammation), cell survival, and tissue development.

Figure 11.19 Modular structures of NFκB subunits Rel proteins are catalytically active transcription factors that contain a Rel homology domain (RHD) for DNA binding, nuclear translocation (NLS, nuclear localization sequence), dimerization, and interaction with IκB, and a transactivating domain (TAD) for binding of transcriptional co-activators such as CBP/p300. Proteins NFκB1 and NFκB2 are truncated forms of Rel that lack TAD. NFκB1 is released from the protein p100 by limited proteolysis. p100 and p105 contain ankyrin repeats that block the RHD, thus preventing nuclear localization and DNA binding. The same effect is achieved by the inhibitory subunits IκB when they interact with Rel or NFκB proteins.

transcription factors may dimerize to any possible combination except for the RelB homodimer, which has not been found *in vivo*. The 15 dimers thus generated exhibit individual but overlapping gene specificities. Among them, the RelA–p50 heterodimer is the most abundant one, representing the prototypical NFκB protein. While the Rel factors activate transcription, p50 and p52 have an inhibitory effect when they occur as *homodimers* (whereas heterodimers with cRel, RelA, or RelB are stimulatory). The reason for this is that p50 and p52 are truncated forms of Rel that lack a transactivating domain (**Figure 11.19**) and thus are unable to recruit the protein complex of the transcriptional machinery, though via their RHDs they bind like Rel to DNA. Moreover, in the presence of p50–p50 or p52–p52 homodimers, transcriptional co-repressors such as histone deacetylases are assembled at the gene promoter. Thus, p50 and p52 homodimers are powerful competitive inhibitors of NFκB signaling that resemble dominant-negative mutants.

Rel/NFκB-homologous transcription factors are also found in invertebrates. *Drosophila*, for instance, expresses three isoforms called Dif, Dorsal, and Relish. They are essential for certain stages of embryonic development as well as for the defense against pathogenic organisms. Some examples are discussed in more detail in Section 6.3.2. In agreement with its functions in defense and development, the NFκB system probably entered the stage of evolution with the appearance of multicellular eukaryotes, since it is not found in unicellular organisms such as yeast (the lack of NFκB in the nematode *Caenorhabditis elegans* is explained by secondary gene loss).

11.7.2 Signal processing by NFκB modules along canonical and noncanonical pathways

A characteristic feature of Rel and NFκB transcription factors is that in most cell types they shuttle between cytoplasm and nucleus. In the absence of input signals (that is, in nonstimulated cells), they are mainly concentrated in the cytoplasm where they exist as complexes with their inhibitory subunits, the **IκBs** as shown in Figure 11.18. IκBs are expressed in several isoforms. Their characteristic structural elements are ankyrin repeats enabling protein–protein interactions (see Section 13.1). Analogous ankyrin repeats are also found in two other proteins of the NFκB complex, **p100** and **p105**, where they are linked with N-terminal Rel homology domains (see Figure 11.19). By means of their ankyrin repeats IκB and p100/p105 bind to Rel/NFκB proteins which thus are hindered from entering the nucleus and from interacting with DNA. It is easily conceivable that the triggering event in NFκB signaling is the inactivation of the inhibitors. This is achieved by phosphorylation, priming them for destruction along the ubiquitin–proteasome pathway (see Section 2.8.1).

Figure 11.20 Major pathways of NFκB signaling (Left) Canonical NFκB module, including phosphorylation (P) and proteolytic degradation of the inhibitor IκB. The factor Rel thus activated forms a heterodimer, for instance with p50 as shown here, that accumulates in the nucleus. The complex of the protein kinases IKK1 and -2 and IκB assembled by the scaffold protein NEMO has been called a signalosome. (Right) Noncanonical NFκB module. Upon phosphorylation (P), the protein p100 is partially degraded along the ubiquitin–proteasome pathway and releases active p52, which as a heterodimer with RelB accumulates in the nucleus. In a similar way, p50 is released from a precursor protein p105 (not shown). In addition to these major pathways, some atypical pathways of NFκB signaling exist: upon UV irradiation, IκB is phosphorylated by casein kinase CK2 instead of by IKK, releasing the Rel–p50 complex, while under genotoxic stress NEMO forms a complex with the checkpoint kinase ATM (see Section 12.9) that activates the canonical module (not shown). TNFR, tumor necrosis factor receptor; IL1R, interleukin 1 receptor; TLR, Toll-like receptor of bacterial components; BCR and TCR, B- and T-cell antigen receptors; NIK, NFκB-inducing kinase; MHC, major histocompatibility complex.

Phosphorylation is catalyzed by a complex of two Ser/Thr-directed protein kinases, the **IK kinases** IKK1 (alias IKKα) and IKK2 (alias IKKβ). Both are held together by the scaffold protein NEMO (NFκB essential modulator, formerly known as IKKγ although it lacks kinase activity) that at the same time acts as an additional sensor for upstream signals (**Figure 11.20**). The genes activated by Rel/NFκB include that of the IκB isoform IκBα. Thus, like MAP kinases (which induce the production of MAP kinase phosphatases), NFκB also controls its own activity via a negative feedback loop. Another analogy with MAP kinase modules is that the NFκB module is controlled by upstream protein kinases that catalyze activation loop phosphorylation of IKKs (see Section 11.7.3) and that most of these IKK kinases are members of the MAP3 kinase family such as MEKK1, MEKK3, NIK, NAK, and TAK1. This means that frequently both modules are activated simultaneously. Other IKK kinases are the receptor-interacting kinase RIPK1, a key regulator of inflammation and controlled cell death (see Sections 6.3.1 and 13.2.5), and the DNA checkpoint kinase ATM linking the NFκB system with DNA damage repair (see Section 12.8). Moreover, certain protein kinases can replace the IKKs in NFκB activation. For example, casein kinase CK2 is stimulated upon UV irradiation and directly inactivates IκB.

One distinguishes between two pathways of NFκB signaling and the corresponding modules, a canonical one and a noncanonical one (see Figure 11.20). The **canonical module** consists of an IKK1//IKK2/NEMO complex, IκB, and an NFκB-heterodimer. Upon activation by IKK kinases, phosphorylation and degradation of IκB are triggered by IKK2 and the active NFκB-heterodimer is released into the nucleus. The canonical pathway is induced by a wide variety of cytokines, bacterial components, and other factors. The major response is a stimulation of innate immunity, resulting in an inflammatory reaction (see Section 6.3 for details). In the **noncanonical module** NFκB activation is blocked by p100 instead of by IκB. This blockade is overcome by phosphorylation of p100 catalyzed by IKK1. p100 becomes *partly* degraded, releasing NFκB2 (p52) that specifically forms a heterodimer with RelB. The activation of IKK1 is catalyzed rather selectively by the MAP3 kinase NIK (NFκB-Inducing Kinase). The noncanonical pathway starts from certain TNF receptor types and from the T- and B-cell antigen receptors. It plays a role in the regulation of acquired immunity since it leads to B-cell maturation and triggers the development of secondary lymphoid organs such as lymph nodes. IKK1 supports the effect of NFκB in still another way: by phosphorylating the Ser 10 residue in histone 3. As explained in Section 8.2.5, this reaction primes the histone for subsequent acetylation, thus playing a key role in transcriptional activation.

Unlike p100, the IκB isoform p105 is phosphorylated by IKK2 rather than by IKK1 resulting in ubiquitylation and complete degradation. In fact, p50 (NFκB1) is generated directly by translation rather than by partial precursor degradation.

The multiple pathways of NFκB signaling indicate that each dimer addresses an individual set of genes. This considerably increases the versatility of the system.

11.7.3 A complex system of signal processing

What we have seen up to now is nothing but the bare skeleton of NFκB signaling. In reality the NFκB system ranks among the most complex and most interlinked systems of signal processing known. Like the MAP kinase modules, it is connected directly or indirectly with numerous signaling pathways, and we are still far from giving a reliable estimate of the range of this circuitry.

Cross talk with other transcription factors

First of all, the functional repertoire of Rel/NFκB factors is far from exhausted by dimerization with each other. In fact, NFκB binding sites are positioned on gene promoters in such a way that the individual Rel/NFκB monomers or dimers may associate with other transcription factors to form signaling complexes, each of which addresses one or more selected genes and either promotes or suppresses transcription. Such transcription factors include STAT (see Section 7.2.1), nuclear receptors (see Section 8.3), ATF and CREB (see Section 8.6), cJun and cFos (see Section 11.6), C/EBPβ (CCAAT/enhancer-binding protein beta, an abundant transcription factor with special functions in the inflammatory response and immune reactions), and others, which are activated along quite varied signaling pathways. Thus, when stimulated through TNFα or interferon γ, RelA and STAT1 may form a heterodimeric transcription factor that controls genes encoding certain pro-inflammatory mediators, while upon stimulation of the Toll-like receptor TLR4, the pair RelA–IRF3 (interferon-regulated factor 3; see Section 6.3.2) induces the transcription of genes with an interferon response element. The well-known anti-inflammatory effect of glucocorticoids is explained at least partly by a direct interaction with the nuclear glucocorticoid receptor, resulting in sequestration and inhibition of NFκB, an effect which is called transrepression.

Post-translational modifications

The components of the NFκB modules provide a most instructive example of how the activity of signaling proteins is modulated by a large number of post-translational modifications. However, a closer look reveals an extremely complex and somewhat confusing picture that is not yet understood in every detail. Therefore, only some basic facts are described in the following. An overview is found in **Figure 11.21**.

Let us start with the protein NEMO, the scaffold and sensor of the canonical NFκB modules. Here signal transduction to the IKK complex requires ubiquitylation by polyubiquitin chains linked via Lys63. As explained in Section 2.8, this modification does not trigger protein degradation in proteasomes but provides interaction sites for other proteins. The importance of NEMO ubiquitylation is underlined by the fact that different stimuli activate different E3 ubiquitin ligases that catalyze the ubiquitylation of individual Lys residues in NEMO. Vice versa, de-ubiquitylating enzymes may attenuate NFκB signaling. Among the ubiquitin ligases targeting NEMO, we have encountered the TRAFs as transducers of TNFα and interleukin 1 signals (see Section 6.3).

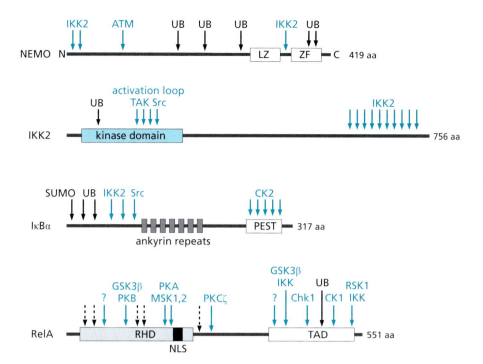

Figure 11.21 Multiple post-translational modifications of selected proteins of the NFκB module Teal arrows symbolize phosphorylation (with the corresponding protein kinases), black arrows symbolize ubiquitylation (UB) or sumoylation (SUMO), and broken black arrows symbolize acetylation. RHD, Rel homology domain; TAD, transactivating domain; LZ, leucine zipper domain; ZF, zinc finger domain; PEST, domain containing the sequence Pro-Glu-Ser-Thr.

The key event of the canonical pathway is activation loop phosphorylation of IKK2 at two Ser residues, and a major function of ubiquitylated NEMO is to establish the contact with the IKK2 kinases that catalyze this reaction. Frequently this requires a cooperation with other poly-ubiquitylated adaptor proteins. An example is given in Section 6.3.3: as shown in Figures 6.9 and 6.10, both the activated interleukin 1 receptor and Toll-like receptors, via the TRAF6 adaptor protein, stimulate the protein kinase TAK1 complexed with the adaptor proteins TAB1 and TAB2. TAK1 is an efficient IKK2 kinase. Its binding to NEMO depends on a Lys63-linked poly-ubiquitylation of the TAB proteins and of TRAF6. Equally complicated is the process by which genotoxic stress signals activate NFκB modules: it includes sumoylation and ubiquitylation of NEMO and its phosphorylation by the stress-activated protein kinase ATM (see Section 12.8), which then phosphorylates IKK2. Interestingly, activation loop phosphorylation of IKK2 can alternatively occur at two Tyr residues instead of at two Ser residues. Another target of Tyr phosphorylation is IκBα, which thus becomes either degraded or separated from NFκB. In both cases NFκB signaling is stimulated. These reactions open a route connecting the NFκB modules with Tyr kinase-coupled receptors.

The signaling sensor of the noncanonical NFκB module is the protein kinase NIK (see Figure 11.20). It catalyzes activation loop phosphorylation of IKK1 independently of NEMO.

Let us now turn to the NFκB transcription factors. Both their nuclear translocation and their transactivating efficacy are regulated by a wide variety of post-translational modifications. Again phosphorylation plays a major role. Upon stimulation by input signals, phosphorylation occurs both in the cytoplasm and in the nucleus. While cytoplasmic phosphorylation controls dimerization and nuclear translocation and primes the transcription factor for DNA binding, nuclear phosphorylation affects DNA binding, transactivating efficacy, and lifetime. Various protein kinases (such as PKA, IKK1 and IKK2, ribosomal S6 kinases, checkpoint kinase Chk1, MSK1, GSK3β, and PKCζ) catalyze Rel phosphorylation, linking NFκB signaling with many other signaling pathways. Phosphorylation may have opposite effects depending on the phosphorylation site: phosphorylation of the transactivating domains generally enhances transcription factor activity, whereas phosphorylation of the DNA binding domain is inhibitory. Moreover, phosphorylation of distinct sites may prime Rel/NFκB proteins for ubiquitylation and subsequent degradation in proteasomes. Such modifications terminate the NFκB signal.

Like many other gene-regulatory proteins, RelA is acetylated at several Lys residues by CBP/p300 acetyltransferases, the canonical transcriptional co-activators. As a result, DNA binding becomes reinforced and acetylated peptide sequences are generated that serve as binding sites for proteins with Bromo domains to form a multiprotein signaling complex. Such proteins, including CBP/p300 (which in this way promotes its interaction with RelA), are involved in transcriptional activation (see Section 8.2.2). Acetylation also hinders the interaction of Rel with IκBα that would promote its export from the nucleus.

Transcriptional co-repressors exhibiting histone deacetylase activity catalyze the deacetylation of Rel and attenuate its effects on gene transcription. However, it must not be overlooked that, like phosphorylation, acetylation may either stimulate or inhibit transcription factor activity, depending on the site of the protein molecule where it occurs. Thus the intensity as well as the duration of the Rel signal is regulated by acetylation in a complex manner.

11.7.4 NFκB signaling: a decision between life and death

In most tissues, NFκB activates genes that promote the survival of cells by encoding anti-apoptotic and pro-mitogenic proteins. A classic example is provided by peripheral B lymphocytes, where along the NFκB pathways antigens stimulate the transcription of survival genes. Signaling starts from the B-cell receptor (see Section 7.2.2), which upon activation interacts with cytoplasmic Tyr kinases such as Btk and Syk. These kinases phosphorylate and stimulate phospholipase Cγ, thus releasing the second messengers diacylglycerol and inositol 1,4,5-trisphosphate. This leads, in turn, to an activation of protein kinase Cβ, which here acts as an IKK kinase, promoting both canonical and noncanonical NFκB signaling. As a final outcome, genes encoding anti-apoptotic proteins such as the caspase inhibitors IAP and FLIP and proteins of the Bcl2 family are activated (for details, see Sections 13.2.4 and 13.2.5).

NFκB proteins also guarantee the survival of immature T lymphocytes whereas in mature T cells they mediate apoptosis by inducing the synthesis of death receptors such as Fas and TRAIL-R, the corresponding ligands, pro-apoptotic Bcl2 proteins, and pro-apoptotic transcription factors such as p53 and cMyc. Since the T-cell receptor resembles the B-cell receptor in many aspects, signal transduction may be similar also.

A pro-apoptotic effect of NFκB is not restricted to mature T cells but also has been observed in other tissues. Frequently it may be an outcome of inflammation, which in its final stages leads to collateral cell death caused by cytotoxic mediators such as reactive oxygen species and pro-apoptotic cytokines released from white blood cells.

NFκB and disease

Cancer is characterized by disequilibrium between cell gain and cell loss due to decreased apoptosis and increased cell proliferation. Thus, an overactivation of NFκB is thought to be oncogenic. This is indeed the case as demonstrated by the retroviral *vRel* oncogene, which causes lymphomas in birds. However, an oncogenic mutation of an NFκB factor has not yet been found in human tumors. Here NFκB signaling plays the role of a powerful tumor-promoting mechanism. A hallmark of tumor promotion is chronic inflammation. Since NFκB is a key mediator of the inflammatory response, constitutive overactivation of NFκB signaling goes hand in hand also with many **inflammatory diseases** such as rheumatoid arthritis, asthma, chronic gastritis, hepatitis, atherosclerosis and multiple sclerosis. Inflammatory diseases such as *Helicobacter pylori*-associated gastritis, Crohn's colitis, and viral-induced hepatitis are preliminary stages of cancer. In fact, an overactivation of NFκB signaling is associated with a wide variety of clinical and experimental neoplasias, whereas pharmacological or genetic inhibition of NFκB signaling has a protective effect.

Tumor promotion by NFκB signaling is easily explained by a constitutive over-stimulation of anti-apoptotic and pro-mitogenic effects. This may occur at any level of the NFκB complex. For instance, a hyperactive IKKε (an IKK isoform found in a subset of NFκB modules) has turned out to be an oncogenic factor in breast cancer. Moreover, along the NFκB-controlled pathways, reactive oxygen species are produced that are important mediators of innate defense, but as endogenous genotoxic agents they may also contribute to the genetic instability of tumor cells (see Section 12.1 for more details).

Apoptosis not only prevents neoplastic growth but also, by removal of infected cells, provides one of the body's defense strategies against pathogenic viruses. Many virus types are able, however, to switch off the apoptotic machinery by overactivating NFκB signaling. Human viruses misusing NFκB signaling are directly or indirectly involved in tumorigenesis. These are **hepatitis C virus, herpes simplex virus, human immunodeficiency virus (HIV)** and **human T-cell leukemia virus (HTLV-1)**. For example, HIV activates NFκB via various signaling proteins in its host cells and harbors an NFκB-responsive element in its own genome.

With the involvement of NFκB signaling in so many diseases, it is clear that strong efforts are being made to apply corresponding inhibitors in the clinic. This has turned out, however, to be a nontrivial task due to the variability, versatility, and redundancy of this signaling system.

Summary

- The transcription factors of the NFκB family participate in the control of a wide variety of genes, particularly those controlling specific and nonspecific defense reactions that result in inflammation. The active forms are homo- or heterodimers of the proteins RelA, RelB, and *cRel* and heterodimers of these proteins with the truncated Rel proteins p50 (NFκB1) and p52 (NFκB2), respectively. Since p50 and p52 lack a transactivating domain, their homodimers are transcriptional repressors.

- Rel factors are kept in the cytoplasm in an inactive state by binding to inhibitory subunits. Upon activation, these subunits become phosphorylated and degraded by proteolysis, enabling the nuclear translocation of Rel factors. The catalytic and inhibitory subunits of the transcription factors and the activating kinases are combined into modules or "signalosomes".

- The signaling versatility of Rel factors is increased by interactions with other transcription factors. Moreover, the activities of NFκB modular components are controlled and fine-tuned by post-translational modifications such as ubiquitylation, phosphorylation, and acetylation, indicating connections with a wide variety of signaling pathways.

- In most cases the NFκB modules process survival (anti-apoptotic) signals. However, depending on cell type and physiological context, cell death may also be induced. This ambivalence plays an important role in the homeostasis of the immune system. An overstimulation of NFκB signaling may promote inflammatory diseases and tumor growth. Several highly pathogenic viruses ensure the survival of their host cells by overstimulating anti-apoptotic NFκB signaling.

Further reading

Section 11.1

Caunt CJ & Keyse SM (2013) Dual-specificity MAP kinase phosphatases (MKPs). Shaping the outcome of MAP kinase specificity. *FEBS J* 280:489–504.

Ferrell JE Jr (1996) Tripping the switch fantastic: how a protein kinase cascade can convert graded inputs into switch-like outputs. *Trends Biochem Sci* 21:460–466.

Futran A, Link AJ & Shvartsman JY (2013) ERK as a model for systems biology of enzyme kinetics in cells. *Curr Biol* 23:R972–R979.

Keshet Y & Seger R (2010) The MAP kinase signaling cascades: a system of hundreds of components regulates a diverse array of physiological functions. *Methods Mol Biol* 661:3–38.

Kyriakis JM & Avruch J (2012) Mammalian MAPK signal transduction pathways activated by stress and inflammation: a 10-year update. *Physiol Rev* 92:689–737.

Raabe T & Rapp UR (2002) KSR: a regulator and scaffold protein of the MAP kinase pathway. *Sci STKE Perspect* 2002/136/pe28.

Shilo BZ (2014) The regulation and function of MAPK pathways in *Drosophila*. *Methods* 68:151–159.

Tanoue T & Nishida E (2003) Molecular recognitions in the MAP kinase cascades. *Cell Signal* 15:455–462.

Tomida T (2015) Visualization of spatial and temporal dynamics of MAPK signalling using fluorescence imaging techniques. *J Physiol Sci* 65:37–49.

Ubersax JA & Ferrell JE Jr (2007) Mechanisms of specificity in protein phosphorylation. *Nat Rev Mol Cell Biol* 8:530–541.

Widmann C, Gibson S, Jarpe MB et al. (1999) Mitogen-activated protein kinase: conservation of a three-kinase module from yeast to human. *Physiol Rev* 79:143–180.

Section 11.2, Section 11.3, Section 11.4, Section 11.5, and Section 11.6

Brown MD & Sacks DB (2009) Protein scaffolds in MAP kinase signalling. *Cell Signal* 21:462–469.

Buchwalter G, Gross C & Wasylyk B (2004) Ets ternary complex transcription factors. *Gene* 324:1–14.

Cargnello M & Roux PP (2011) Activation and function of the MAPKs and their substrates, the MAPK-activated protein kinases. *Microbiol Mol Biol Rev* 75:50–83.

Cuadrado A & Nebreda AR (2010) Mechanisms and functions of p38 MAPK signalling. *Biochem J* 429:403–417.

Furukawa K & Hohmann S (2013) Synthetic biology: lessons from engineering yeast MAPK signalling pathways. *Mol Microbiol* 88:5–19.

Hollenhorst PC, McIntosh LP & Graves BJ (2011) Genomic and biochemical insights into the specificity of ETS transcription factors. *Annu Rev Biochem* 80:437–471.

Keshet Y & Seger R (2010) The MAP kinase signaling cascades: a system of hundreds of components regulates a diverse array of physiological functions. *Methods Mol Biol* 661:3–38.

Kim EK & Choi EJ (2010) Pathological roles of MAPK signaling pathways in human diseases. *Biochim Biophys Acta* 1802:396–405.

Lavoie H & Therrien M (2015) Regulation of RAF protein kinases in ERK signalling. *Nat Rev Mol Cell Biol* 16:281–295.

Rana A et al. (2013) Mixed lineage kinase-c-Jun N-terminal kinase axis: a potential therapeutic target in cancer. *Genes Cancer* 4:334–341.

Section 11.7

Hayden MS & Ghosh S (2012) NFκB, the first quarter-century: remarkable progress and outstanding questions. *Genes Dev* 26:203–234.

Oeckinghaus A, Hayden MS & Ghosh S (2011) Crosstalk in NFκB signalling pathways. *Nat Immunol* 13:695–708.

Regulation of Cell Division

12

Cell proliferation is mandatory for reproduction and for development, repair, and maintenance of tissues. In fact, the apparatus of cell division is an ultimate target of most of the signal-processing reactions explained in the previous chapters. In the following, we shall have a closer look at pathways that connect this apparatus with the cell's signal-processing network. The investigation of these interactions has tremendous practical consequences because defective transduction of signals controlling cell proliferation is a major cause of cancer.

12.1 The cell cycle

Proliferating cells pass through a sequence of phases presented as a cyclic process. This cell cycle is conserved in all eukaryotes. It provides a perfect example of how a highly complex cellular event is controlled temporally and spatially by signaling reactions that are tightly interlinked and feedback-controlled to ensure that the genetic material has been correctly copied and evenly distributed to the daughter cells. Pioneering work on cell cycle regulation was honored by the Nobel Prize in Physiology or Medicine for 2001, awarded to Leland H. Hartwell, R. Timothy Hunt, and Paul M. Nurse.

The production of intact and genetically identical daughter cells presupposes a precise sequence of the cycle phases. For instance, separation of chromosomes must not occur before complete chromosome condensation, which requires DNA replication to be finished to proceed correctly. Therefore, the cell cycle is interrupted at least three times by **checkpoints** (**Figure 12.1**) that give the data-processing network an opportunity to monitor the precision of DNA replication and chromosome

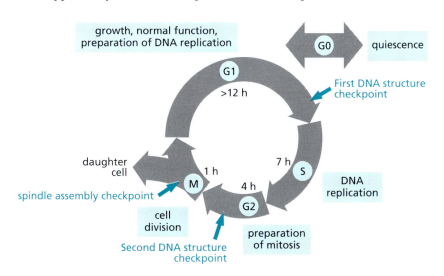

Figure 12.1 Mammalian cell cycle Phases of the cell cycle: S, synthesis phase, period of DNA replication; M, mitosis phase, period of chromosome separation and cytokinesis; G1 and G2, so-called gap phases, where the cell realizes specific programs of differentiation and prepares for DNA replication and mitosis, respectively; G0, resting phase; upon mitogenic stimulation the cell may return to the cell cycle. The three major checkpoints are shown in teal.

Figure 12.2 Cyclin levels and cyclin-dependent kinase activities during the cell cycle (A) Cyclin–CDK activities associated with the individual phases. (B) Oscillation of the cyclin level (teal curve) and the CDK activity (colored bars) depending on the cell cycle phase as represented by the DNA content (black line and ordinate). As an example, the cyclin B–CDK1 pair has been chosen.

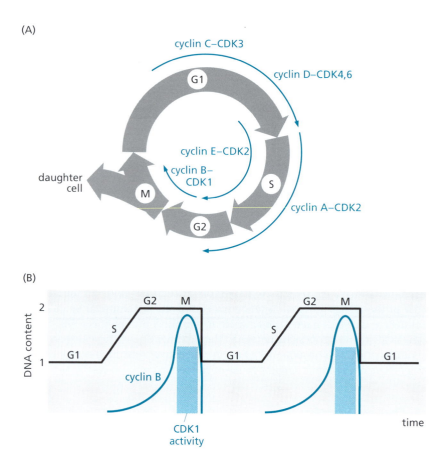

segregation and, if necessary, to launch repair measures or, in the case of irreparable damage, to promote cell death (for details see Section 12.8). By this means, dissemination of harmful mutations is prevented. However, as is typical for data processing, this control is not perfect but may be evaded, in particular in the course of tumorigenesis.

The amazing precision of the cell cycle is guaranteed by control elements meshing like gears. In this mechanism, strictly regulated formation and degradation of regulatory proteins together with precisely timed protein phosphorylation play a key role. The master regulators of cell cycle progression are the **cyclin-dependent protein kinases (CDKs)**. Their importance was discovered originally by studies on yeast mutants with defective or overexpressed *cdc* (*cell division cycle*) genes* and on starfish and amphibian oocytes.

To make a fertilized egg, the cell cycle of oocytes is arrested at the G2–M transition. This arrest could be overcome by progesterone or by a maturation-promoting factor (MPF) isolated from dividing frog cells. MPF was identified as a protein kinase in the late 1980s. A homologous kinase was found to be encoded by the yeast *cdc* gene number 2. Both kinases were originally called Cdc2 but today are known as CDK1. They turned out to be the catalytic subunits of a complex with a regulatory protein called **cyclin B**. Cyclin B appeared in the G2 phase to stimulate CDK1 and then disappeared abruptly at the end of the M phase. In contrast, Cdc2 (CDK1) was always present but was active only in the presence of cyclin B (**Figure 12.2**).

---

* The yeast species *Saccharomyces cerevisiae* and *Schizosaccharomyces pombe* have been extremely valuable models for genetic and biochemical analysis of the cell cycle. Conditional mutations, for instance those that become effective only above a distinct temperature, have enabled the identification of various cell division cycle-controlling genes (cdc) and their products, significantly facilitating the search for homologous vertebrate proteins.

Table 12.1 Cyclin-dependent protein kinases of mammalian cells

Kinase	Regulatory protein	Major function
CDK1	cyclin B	cell cycle control
CDK2	cyclin A and E	cell cycle control
CDK3	cyclin C	cell cycle control
CDK4	cyclin D	cell cycle control
CDK5	p35 and p39	neuronal differentiation
CDK6	cyclin D	cell cycle control
CDK7	cyclin H	CDK-activating kinase
		transcriptional control[a]
CDK8	cyclin C	transcriptional control[a]
		inhibition of cyclin H–CDK 7
CDK9	cyclin K and T	transcriptional control[a]
CDK10	?	transcriptional control
CDK11	cyclin L	transcriptional control[a]
		pre-mRNA splicing

[a]CDK7–9 and -11, together with transcription-associated cyclins H, C, K, L, and T, form complexes with RNA polymerase II. CDK7 (a component of the general transcription factor TFIIH), CDK8 (a component of the RNA polymerase holoenzyme), and CDK9 (a component of an elongation factor) phosphorylate the carboxy (C)-terminal domain of RNA polymerase II, thus participating in regulation of transcriptional elongation (see Section 8.1). The CDK11–cyclin L complex coordinates transcription with RNA splicing (see Section 8.9).

CDK1 and cyclin B are representatives of larger protein families. The human genome, for instance, encodes 20 CDKs and CDK-like proteins, and 29 cyclins or cyclin-like proteins (proteins with an amino acid sequence known as the cyclin box; see below). Only six CDKs and seven cyclins have been shown unequivocally to be involved in cell cycle regulation, while the rest exhibit other or unknown functions (**Table 12.1**).

Summary

- The cell cycle is divided into several phases through which a proliferating cell must pass. The correct sequence of phases depends on the periodic formation and degradation of regulatory proteins known as cyclins.

- At several checkpoints, cell cycle passage is slowed down, giving the data-processing network an opportunity to monitor the precision of DNA replication and chromosome segregation and, if necessary, to launch repair measures or to promote cell death.

12.2 Cyclins: cell cycle regulators and beyond

Cyclins are heterogeneous proteins that share a so-called cyclin box: a domain consisting of a helix bundle that enables them to interact with the corresponding CDKs. The amino acid sequence of the cyclin box distinguishes the cell cycle-associated cyclins A, B, D, and E from the transcription-associated cyclins C, H, K, L, and T and from other cyclins with still-unknown functions (see Table 12.1).

Cell cycle-associated cyclins are periodically synthesized during individual cell cycle phases. They initiate the entry of the cell into a phase and, at the same time, inhibit the transition into the next phase. In other words, as long as a particular cyclin is available, the cell stays in the corresponding phase. Therefore, cyclins are synthesized at the beginning and must be destroyed at the end of a particular phase. Cyclin synthesis is induced by a cyclin–CDK pair from the previous phase of the cell cycle. Cyclin degradation proceeds along the ubiquitin–proteasome pathway and is triggered by ubiquitin ligases that bind to amino (N)-terminal destruction domains (degrons) of cyclins. The cell cycle is dominated by two types of ubiquitin ligases: SCF ubiquitin ligases induce the degradation of G1- and S-phase cyclins, whereas APC/C ubiquitin ligases are more specialized for G2- and M-phase cyclins (more information about ubiquitin ligases is found in Section 2.8).

Summary

- Cyclins are regulatory proteins that activate cyclin-dependent kinases (CDKs). They are periodically produced and degraded in the course of the cell cycle. Each cell cycle phase is associated with and controlled by a distinct cyclin–CDK pair.

- In addition to cell-cycle-associated cyclins, other cyclins are known that, together with corresponding CDKs, participate in the control of gene transcription and pre-mRNA splicing.

12.3 Cyclin-dependent protein kinases: dual control by phosphorylation and dephosphorylation

CDKs are typical eukaryotic Ser/Thr-specific protein kinases. While most of them interact with cyclins, some—such as CDK5—are activated by other proteins (Sidebar 12.1). Moreover, as shown in Table 12.1, CDKs are involved not only in the control of cell division but also in other processes, in particular gene expression. While each cell cycle phase has been assumed to be controlled by a distinct CDK isoform, experiments with genetically manipulated mice have shown that CDK2, -4, and -6 may be dispensable for the basic cell cycle or at least replaceable by other CDKs such as CDK1 in most (but not all) tissues. Such data indicate a certain level of redundancy or fault tolerance in cell cycle control.

Sidebar 12.1 The outsider CDK5 Although closely related to CDK1 and CDK2, the enzyme CDK5 is an outsider of the CDK family, since (despite its name) it is not regulated by cyclins but by other proteins called p35 and p39 that are expressed exclusively in brain. Moreover, CDK5 does not require activation loop phosphorylation and does not respond to MYT1 and WEE1. Most probably, the enzyme does not participate in cell cycle control. Instead, it plays an important role in nonproliferating nerve cells, being essential for **brain development** since it controls the migration and correct positioning of neurons in the embryo.

Exhibiting similar substrate specificity as CDK1, CDK5 phosphorylates many proteins. By controlling cytoskeletal dynamics, the enzyme regulates cell motility, cell shape (such as the outgrowth of neurites), intercellular contacts (for instance at synapses), and neurotransmitter release. CDK5 cooperates with focal adhesion kinases (see Section 7.2.3), which are CDK5 substrates.

Considering the multiple functions, it is conceivable that a deregulation of CDK5 may result in serious consequences. So a hyperactivation of CDK5 due, for instance, to a truncation of the regulatory subunit p35 (to p25) by the Ca^{2+}-dependent protease calpain (see Section 14.6.5), correlates with the progression of neurodegenerative diseases such as **amyotrophic lateral sclerosis** (a fatal destruction of neuromuscular contacts), **Parkinson's disease**, and **Alzheimer's dementia**. A pathological consequence of a hyperactive CDK5 is an excessive phosphorylation of the microtubule-associated protein Tau resulting in massive deformations of the cytoskeleton and finally in cell death (for details about Alzheimer's disease see Section 13.1.3). The reader is reminded that Tau is also phosphorylated by glycogen synthase kinase 3 (GSK3; see Section 9.4.4).

Figure 12.3 Interaction of CDK2 with cyclin A (Left) In the absence of cyclin A, the activation loop (teal) closes the catalytic cleft including the ATP-binding site of CDK2. (Right) Cyclin A (teal) interacts with the PSTAIRE sequence of CDK2 and induces a conformational change that leads to an opening of the catalytic cleft, allowing ATP and substrate protein binding. The active conformation is stabilized by activation loop phosphorylation.

A characteristic structural feature of the cell cycle-associated CDKs is the PSTAIRE (or a closely related) sequence localized in a helix of the ATP-binding site. By means of this sequence, CDK binds to the cyclin box of the corresponding cyclin and is relieved from auto-inhibition. By means of X-ray crystallography this interaction has been studied in detail (**Figure 12.3**). It has been shown that, in the absence of cyclin, the activation loop of CDK buries the ATP binding site and hinders access of the substrate protein to the catalytic center. Owing to a conformational change induced by cyclin binding, the activation loop turns aside, opening the entrance of the catalytic cleft. As in most protein kinases, this open conformation is stabilized by phosphorylation of a strategic threonine residue in the activation loop. This addition of phosphate occurs either via autophosphorylation or via a CDK-activating kinase, which is itself a cyclin H–CDK7 complex. Cell cycle control is a major but not the only function of this kinase type (**Sidebar 12.2**).

At least two cell cycle-associated kinases, CDK1 and CDK2, are arrested in an inactive state by dual phosphorylation of an N-terminal Thr-Tyr sequence catalyzed by two **CDK-inactivating kinases: MYT1** [membrane-associated tyrosine (Y) and threonine (T)-specific kinase], which predominantly phosphorylates the Thr residue, and **WEE1**, a Tyr kinase (the name refers to *WEE1* deletion mutants of fission yeast, characterized by particularly small cells due to an accelerated cell cycle). To activate a CDK, both phosphate groups have to be removed by a highly specialized dual-specific protein phosphatase of the **Cdc25** family (see Section 2.6.4), which in humans comprises three isoforms A, B, and C. For the CDK1–cyclin B complex as an example, these relationships are summarized in Section 12.7.

CDK inhibitors MYT1 and WEE1 and CDK activator Cdc25 are controlled by input signals. In this context, the role of **checkpoint kinases** has been studied in detail. With Cdc25, phosphorylation by checkpoint kinases has an inhibitory effect since it labels Cdc25 for an interaction with 14-3-3 proteins, hindering translocation into the nucleus, and enhances Cdc25 degradation by the ubiquitin–proteasome pathway. On the contrary, MYT1 and WEE1 are stimulated by phosphorylation.

In other words, checkpoint kinases bring the cell cycle to a halt. Since they become active upon DNA damage (see Section 12.8), they provide one of the mechanisms stopping cell proliferation at the DNA structure checkpoints in the late G1 phase and late

---

Sidebar 12.2 CDK-activating kinase: more than a cell cycle regulator The functions of CDK-activating kinase go beyond the control of CDK activity, since as a subunit of the general transcription factor TFIIH, the enzyme also phosphorylates RNA polymerase II. This reaction, providing an example of CDK functions outside the cell cycle, is important for the transition from the initiation into the elongation phase of transcription (see Section 8.1). CDK-activating kinase is activated either by a regulatory protein or by phosphorylation catalyzed by another cyclin–CDK pair, cyclin C–CDK8, which is also involved in transcriptional regulation rather than in cell cycle control.

G2 phase to enable DNA repair or, as a final consequence, to drive an irreparable cell into apoptosis. Another protein kinase inhibiting Cdc25 is the MAP kinase p38-activated kinase 2, which connects checkpoint control with the general stress response.

Summary

- Cyclin-dependent protein kinases (CDKs) are the master regulators of the individual cell cycle phases.

- For activation they require, in addition to cyclin binding, activation loop phosphorylation (either by autophosphorylation or by the cyclin H–CDK7 complex) and dephosphorylation of an N-terminal Thr-Tyr sequence (by the dual-specific phosphatase Cdc25).

- Threonine kinase MYT1 and tyrosine kinase WEE1 catalyze rephosphorylation of this sequence, resulting in CDK inactivation. These kinases and Cdc25 are targets of signaling reactions.

12.4 Cyclin-dependent kinase inhibitors: keeping the cell cycle under control

Strict regulation of CDK activity, as required for precise cell cycle progression, depends not only on inhibitory phosphorylation but also on proteins that specifically antagonize the stimulatory effects of cyclins. These important regulators of cell cycle progression exist in two families, called INK4 (inhibitors of CDK4) and CIP or KIP (CDK- or kinase-inhibiting proteins).

The **INK4** family of CDK inhibitors includes the proteins $p15^{INK4}$, $p16^{INK4}$, $p18^{INK4}$, and $p19^{INK4}$. They specifically inhibit CDK4 and CDK6, hindering the access of cyclin D by means of ankyrin repeats (see Section 13.1). $p16^{INK4}$ is a tumor suppressor protein inactivated, for example, in malignant melanoma. Transcription of the $p15^{INK4}$ gene is induced by transforming growth factor β, and this effect contributes to the strong anti-proliferative effects of TGFβ, especially on epithelial cells. Probably ^{INK4} proteins are also involved in cell aging or senescence.

The **CIP/KIP** family comprises the inhibitor proteins $p21^{KIP}$, $p27^{KIP}$, and $p57^{KIP}$ (found in only a few tissues). They are less selective than the ^{INK4}s and block various CDKs. $p27^{KIP}$ plays a particularly important role in the control of tissue homeostasis: it keeps the cell in a nonproliferative state of quiescence called G0 phase and is a primary target of mitogenic signals (see Section 12.5). In contrast, $p21^{KIP}$ permanently regulates the cell cycle progression of proliferating cells. Its de novo synthesis is strongly induced by the transcription factor **p53**. Since p53 is activated by checkpoint kinases (that is, upon DNA damage), this response provides, in addition to the above-mentioned activation of WEE1 and MYT1 and inactivation of Cdc25, a second mechanism to interrupt the cell cycle at the DNA structure checkpoints and to trigger apoptosis if necessary.

Like cyclins, CDK inhibitors are short-lived proteins that are rapidly ubiquitylated and degraded in proteasomes. To be recognized by ubiquitin ligases (of the SCF type, see Section 2.8.1), the CDK inhibitors have to be phosphorylated. This is accomplished mostly by cyclin–CDK complexes, thus allowing the cell to enter the next phase of the cell cycle. The factors controlling CDK activity are summarized in **Figure 12.4**.

Summary

The activities of cyclin-dependent kinases are specifically controlled by inhibitory proteins comprising the two families INK4 (inhibitors of CDK4) and CIP/KIP (CDK- or kinase-inhibiting proteins). They are short-lived: produced only on demand to be rapidly destroyed afterwards.

ACTIVATION	INHIBITION
E3 ubiquitin ligases (degradation of WEE1) cyclin Cdc25 (Thr/Tyr dephosphorylation) CDK-activating kinase (activation loop phosphorylation)	CDK inhibitors (INK4, CIP/KIP) WEE1, MYT1 (Thr/Tyr phosphorylation) E3 ubiquitin ligases (degradation of cyclin, Cdc25)

CDK

Figure 12.4 Activation and inhibition of cell cycle-associated cyclin-dependent kinases: an overview

12.5 G0 cells, restriction points, and the effect of mitogenic signals

Inhibition of cell proliferation may lead to senescence and cell death. Alternatively, cells may remain in a resting state, the G0 phase, to return after some time or upon demand into the cell cycle. For example, liver cells can leave and re-enter before they finally lose their proliferative capacity by differentiation or senescence.

The G0 blockade can be overcome only by a synergistic effect of several mitogenic signals (such as hormones and growth factors) acting on the cell for several hours until the cell cycle begins to progress independently of exogenous signals. This occurs at the so-called **restriction point** in the late G1 phase. The reason is a sequential activation and repression, respectively, of genes.

In the beginning (0–4 h) the triggering signal, for instance a growth factor, directly controls (via particular signaling cascades) the transcription of immediate early proliferation genes. Later, when the restriction point has been passed, the products of these genes stimulate delayed early genes, the products of which finally turn on late genes in the early S phase. Over 100 early genes encode, for instance, transcription factors such as cJun, cFos, and cMyc, growth factors and their receptors, as well as proteins of the cytoskeleton and the extracellular matrix. Concomitantly with the activation of such genes, the genes of the CDK inhibitors are repressed. Delayed genes encode the G1 cyclins C and D, and the late genes are responsible for events involved in DNA replication.

Up-regulation of G1 cyclins and down-regulation of KIP-type inhibitors trigger the G0–G1 transition

For a cell entering the G0 phase, the cell cycle must be brought to a halt in the G1 phase by CDK inhibitors and a shortage of G1 cyclins C and D. Vice versa, for a resting cell the de novo synthesis of the cyclins and the removal of the inhibitors are the signals to enter the proliferative phase. The cyclins then activate the corresponding CDKs (CDK3, -4, and -6, see **Figure 12.5**) that phosphorylate and stimulate **E2 transcription factors (E2Fs)**. To become active, these factors form heterodimers with the DNA-binding proteins DP1 and DP2, and control the genes of the G0–G1 transition and practically all late genes required for the entry into the S phase and DNA replication (see Figure 12.5). For the de novo synthesis of cyclins the transcription of the cyclin genes is turned on along mitogenic signaling cascades such as the Ras–RAF–Erk MAP kinase and the NFκB cascade. For MAP kinase signaling and cyclin D, as an example the sequence of events is shown in **Figure 12.6**. Mitogenic MAP kinase modules are stimulated by the small G-protein **Ras**. that, in addition, activates the phosphatidylinositol 3-kinase–PKB/Akt pathway (see Section 10.1). Most growth factors as well as many hormones and cytokines activate Ras via their receptors. Ras not only promotes cyclin formation at the level of transcription but also at the level of mRNA translation. To this end the MAP kinase Erk phosphorylates the protein kinase MNK1 which, in turn, phosphorylates and activates the translation factor eIF4E (see Section 9.3.3). To make the story complete, Ras also prevents cyclin degradation. This again occurs via PKB/Akt, which inactivates GSK3β. GSK3β,

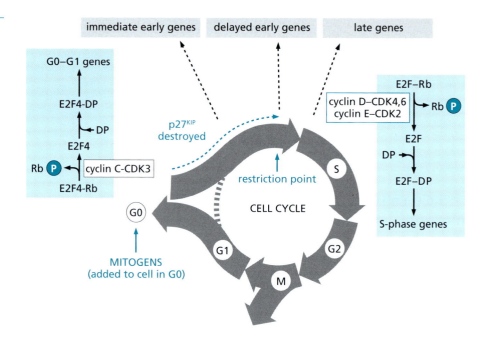

Figure 12.5 Mitogenic stimulation of resting (G0) cells The figure shows three phases of gene expression and the activation of the E2F transcription factors due to phosphorylation of the negative regulator Rb by cyclin C–CDK3 at the G0–G1 transition and cyclin D–CDK4 (or -6) as well as cyclin E–CDK2 at the G1–S transition (colored boxes). The dotted teal line represents the time a mitogen has to be present until the restriction point has been passed.

a ubiquitous protein kinase, primes cyclins for degradation along the ubiquitin-pro-teasome pathway (see Section 9.4.4).

The cell cycle inhibitor p27^{KIP} is responsible for keeping the cell in the G0 phase. Consequently, the synthesis of p27^{KIP} is inhibited and its degradation promoted upon mitogenic stimulation. This is achieved via the Ras–PKB/Akt pathway, leading to phosphorylation and inactivation of the inhibitor and of the Forkhead transcription factors controlling the *p27^{KIP}* gene (see Figure 12.6). Upon phosphorylation, these factors are sequestered in the cytoplasm by 14-3-3 proteins and, thus, are hindered from entering their sites of action in the nucleus (see Section 10.1.4). To reinforce these effects, PKB/Akt stimulates the formation of the protein SKP2. As a regulatory subunit SKP2 renders the ubiquitin ligase SCF, specific for the degradation of p27^{KIP}.

Figure 12.6 Up-regulation of cyclin D and down-regulation of p27^{KIP} through mitogenic signaling cascades The scheme shows signaling pathways activated by growth factors and controlling the de novo synthesis (via gene transcription) of the CDK activator cyclin D and the degradation of the CDK inhibitor p27^{KIP} (via the ubiquitylation–proteasome pathway). These signaling cascades frequently undergo oncogenic mutations. PI3K, phosphatidylinositol 3-kinase; PKB, protein kinase B/Akt; ERK, extracellular signal-regulated kinase; GSK3β, glycogen synthase kinase 3β; SKP2, subunit of the ubiquitin ligase SCF; TF, transcription factor.

To become recognized by SCF, p27^{KIP} must first be "primed": that is, phosphorylated at a distinct Thr residue (see Section 2.8.1). This is catalyzed by either the MAP kinase Erk or the cyclin E–CDK2 complex. In the latter case, a positive feedback loop becomes established since CDK2 activity increases to the same extent as the level of p27^{KIP} drops.

Contact inhibition and the transformed state

Cultivated cells enter the resting phase not only at a growth factor deficiency but also at an increasing population density. This phenomenon, known as contact inhibition, similar to bacterial quorum sensing, is caused by interactions of cell adhesion molecules, such as cadherins (see Section 7.3), and dramatically lowers the cell's sensitivity for proliferative signals.

Tumor cells have little or no tendency to undergo contact inhibition. In addition, they may supply themselves by growth factors acting in an autocrine manner. In fact, a typical sign of the transformed state is that the cells grow in Petri dishes one on top of another to generate a criss-cross pattern, and transformed cells do not need serum as a source of growth factors. Neoplastic transformation requires oncogenic mutations of signal-transducing proteins such as Ras, PKB, growth factor receptors, and transcription factors as well as deletion of tumor suppressor proteins such as the PKB antagonist PTEN or the retinoblastoma protein pRb. Together such disturbances result in excessive production of cyclins D and C, as is found in a wide variety of cancer cells.

Summary

- During the G1 phase, cells may leave the cell cycle, entering a G0 phase. This resting state is associated with a surplus of CDK inhibitors (in particular p27^{KIP}) and a shortage of G1 cyclins (such as cyclin D).

- Vice versa, mitogenic signals trigger a return into the proliferative state by inducing down-regulation of CDK inhibitors and up-regulation of G1 cyclins. The signaling pathways leading to cell proliferation are targets of carcinogens.

12.6 Retinoblastoma proteins: master regulators of the cell cycle

The transcription factors E2F activated at the G0–G1 transition control the genes required for the entry into the S phase and DNA replication (including genes encoding DNA polymerase, factors of the DNA pre-replication complex, ribonucleotide reductase, thymidylate synthase, thymidine kinase, and proteins of DNA repair).

After the G0–G1 checkpoint, the activation of such genes would lead to a chaotic situation. To avoid this, in nonproliferative cells E2F–DP complex is inhibited by the **Rb proteins** p107 or p130. These transcriptional repressors hinder the interaction of E2F with the transcriptional machinery, in particular by recruiting transcriptional co-repressors such as histone deacetylases. Rb proteins become inactivated by multiple phosphorylations catalyzed by cyclin C–CDK3 at the G0–G1 transition and by cyclin D–CDK4 (or CDK6) at the G1–S transition (see Figure 12.5). As a result, E2F is released from the inhibitory complex and the co-repressors are displaced by transcriptional co-activators such as histone acetyltransferases.

The prototype of Rb proteins, providing their name, is the retinoblastoma protein pRb, originally discovered in an eye tumor (**Sidebar 12.3**).

The genes activated by E2F–DP1 complexes include those of cyclins E and A. Together with CDK2, these cyclins control S-phase progression. While cyclin E–CDK2 speeds up the cell cycle via increased Rb phosphorylation (for protein kinase Cdc7's cooperative role with cyclin E–CDK2, **Sidebar 12.4**), cyclin A–CDK2 has a dampening effect in that it inhibits DP1 by phosphorylation. The result is a precisely balanced equilibrium of positive and negative feedback. To avoid a disturbance of S-phase progression by cyclin D, the latter is phosphorylated by glycogen phosphorylase kinase 3β and thus primed for ubiquitylation and proteosomal degradation at the end of the G1 phase. The same happens to cyclin E in the late S phase and to cyclin B in the mitosis phase.

Summary

- Transcription of the genes promoting the progression from G0 phase to and through the phase of DNA replication is stimulated by E2 transcription factors. This effect is opposed by Rb (retinoblastoma) proteins acting as transcriptional repressors.

- Rb proteins are inactivated by phosphorylation catalyzed by cyclin C–CDK3 and cyclin D–CDK4/6.

- Excessive Rb phosphorylation is a hallmark of tumorigenesis. Moreover, the pRb isoform is the epitome of a tumor suppressor protein undergoing carcinogenic loss-of-function mutations.

12.7 Regulation of G2 phase and G2–M transition: precise like clockwork

At the transition from the S to the G2 phase, CDK2 is replaced by CDK1 as a partner of cyclin A. The cyclin A–CDK1 complex then induces the synthesis of cyclin B, which subsequently interacts with CDK1, controlling the transition into the mitosis phase. In an exemplary fashion, the **cyclin B–CDK1 complex** shows the sophisticated control of a CDK by the signal-processing network, ensuring a to-the-point activation in the correct time window.

Cyclin B–CDK1 controls the function of proteins needed by the cell to progress from G2 phase into M phase, and from prophase into metaphase of mitosis. These CDK-regulated proteins include lamin of the nuclear membrane (that disintegrates upon lamin phosphorylation), condensin proteins and histone 1 (the phosphorylation of which initiates chromosome condensation), and nucleolin (its phosphorylation leads to the breakdown of the nucleolus). Moreover, transcription is inhibited by phosphorylation of RNA polymerase II, and the suppression of protein synthesis observed at the G2–M transition is thought to be due to phosphorylation and inactivation by CDK1 of the translation factor eEF1B.

Finally, microtubule-associated proteins required for the formation of the mitotic spindle become phosphorylated, and the actin and intermediary filament cytoskeleton is rearranged, enabling the dividing cell to detach from the support and to round up. At the end of metaphase, CDK1 is inactivated by cyclin degradation, dephosphorylation of the activation loop, and rephosphorylation of the N-terminal Thr-Tyr pair catalyzed by MYT1 and WEE1 (**Figure 12.7**).

However, let us back up in the cell cycle. During late G2 phase, immediately before chromosome condensation, the cell once more slows down cell cycle progression to inspect the integrity of the genetic material. This second DNA structure checkpoint offers a last chance to repair defects in one DNA strand using the intact strand as a template, or to commit suicide in the case of irreparable damage.

Summary

- The progression from G2 phase to mitosis is controlled by cyclin B–CDK1 activated precisely at the entry into M phase.

- CDK1 is inactivated by cyclin degradation, dephosphorylation of the activation loop (which was phosphorylated by CDK-activating kinase), and rephosphorylation of the N-terminal Thr-Tyr pair catalyzed by the protein kinases MYT1 and WEE1.

Figure 12.7 Regulation of cyclin-dependent kinase 1 In the G2 phase, CDK1 forms a complex with newly synthesized cyclin B. At this time point, however, the kinase activity of CDK1 is still suppressed by the phosphorylation (P) of an adjacent Thr-Tyr pair catalyzed by the Ser/Thr kinase MYT1 and the Tyr kinase WEE1. At the G2–M transition, these residues become dephosphorylated by the phosphatase Cdc25, and a Thr residue in the activation loop of CDK1 (teal dot) is phosphorylated by CDK-activating kinase, which is a cyclin H–CDK7 complex. Owing to strictly controlled de novo synthesis of cyclin H, CDK-activating kinase becomes active exactly at this time point. The now active cyclin B–CDK1 complex phosphorylates various proteins required for the G2–M transition and progression of the M-phase. In the telophase, cyclin B is ubiquitylated by the ubiquitin ligase APC/C and degraded in proteasomes. CDK1 is released and is inactivated by dephosphorylation of the activation loop (catalyzed by the protein phosphatase PP2A) and rephosphorylation of the N-terminal Thr-Tyr pair (catalyzed by MYT1 and WEE1). The process includes several feedback loops, which for the sake of clarity are not shown. Cyclin B–CDK1 stimulates Cdc25 and inhibits WEE1, thus increasing the efficacy and punctuality of CDK1 activation by positive feedback. Moreover, the two inhibitory kinases MYT1 and WEE1, as well as their opponent Cdc25, are under the control of upstream signaling pathways. DNA damage and incompletely replicated DNA, respectively, induce the genotoxic stress response, preventing CDK1 activation and interrupting the cell cycle at the late G2 phase (also known as antephase) before chromosome condensation.

12.8 Genotoxic stress response: a matter of life and death

The stability of the genetic material is jeopardized permanently by exogenous influences such as the UV radiation of the sun; ionizing radiation of minerals, nuclear waste, or from space; genotoxic chemicals (which may also be of endogenous origin); and viruses. Moreover, DNA replication does not work without error. All these factors damage DNA and may cause gene mutations, resulting in immediate cell death or in diseases such as cancer. The ability to cope with genotoxic stress is, therefore, a fundamental condition of health and survival.

As already mentioned, somatic cells have a dual strategy: they try to repair the damage or, if this turns out to be impossible, to commit suicide by undergoing apoptosis. This must be done before DNA replication and chromosome segregation: that is, at the DNA structure checkpoints in G1 and G2 phase. At these checkpoints, a sophisticated data-processing network is activated to inspect and evaluate the damage and to launch countermeasures if necessary. In general, the DNA damage response involves three stages:

- recognition of the damage by sensor proteins

- activation of transducer protein kinases

- activation of downstream effector proteins controlling the repair process, cellular stress response, cell cycle progression, and if necessary apoptosis

12.8.1 DNA-dependent protein kinases: gatekeepers of gene stability

The machinery of DNA repair is activated within a few minutes after DNA damage. In this process, transducer protein kinases known as DNA-dependent protein kinases play a key role. They are members of a larger family of eukaryotic Ser/Thr-directed kinases called atypical since they are not closely related to the conventional eukaryotic protein kinases. In addition, many of them exhibit lipid kinase activity (see Section 2.6.3). However, DNA-dependent kinases do not phosphorylate lipids; they are bona fide protein kinases and therefore are subsumed under the collective term **phosphatidylinositol 3-kinase-like kinases (PIKK)**. On the basis of structural variations, PIKKs are divided into the subfamilies TOR, DNA kinases, and ATM/ATR.

The enzymes of the TOR group have been examined already in Section 9.4.1. They are involved in the adjustment of ribosomal protein synthesis and cell growth rather than in DNA repair. In contrast, DNA kinases and ATM/ATR are engaged in genetic recombination and DNA repair.

Enzymes of the **DNA kinase** subfamily transduce signals triggering the repair of DNA double-strand breaks by illegitimate recombination and V(D)J-recombination (which is the mechanism that regulates, through gene rearrangements, the diversity of antibodies and T-cell receptors). The enzyme is attached to DNA by a regulatory subunit consisting of the two proteins Ku70 and Ku80 serving as sensors in that together they recognize free ends of double-stranded DNA (as generated by double-strand breaks) and other structural irregularities of the DNA double helix. As a result the Ku70/80 heterodimer is allosterically activated and stimulates DNA kinase activity. In certain autoimmune diseases, such as lupus erythematosus, Ku70/80 acts as an autoantigen. A role of DNA kinase in genotoxic stress response is still a matter of debate.

Genuine stress enzymes that are activated in emergency situations such as DNA damage are the DNA kinase-homologous protein kinases **ATM** and **ATR**. Their names come from the hereditary disease **ataxia telangiectasia** (ATM stands for Ataxia Telangiectasia Mutated and ATR stands for ATM- and Rad3-related; Rad3 is a DNA-dependent protein kinase of yeast). This disease is characterized by genomic instability leading to brain defects, increased radiation sensitivity, immune deficiency, and a high cancer risk, particularly of leukemia. The causes are homozygous loss-of-function mutations of the genes encoding ATM and ATR. For the integrity of DNA, ATM and ATR are of uppermost importance. Therefore, they have been called guardians at the gateway to genomic stability. While ATM primarily is a sensor of DNA defects caused by ionizing radiation, ATR is more specialized for damage from UV light and inhibitors of DNA replication (for example, cytostatics).

The factors monitoring DNA damage and recruiting ATM and ATR to the sites of damage are large protein complexes. These complexes are multifunctional stress sensors that, in addition to DNA strand breaks and UV-induced defects, recognize a blocked DNA replication fork, and respond to oxidative and nitrosative stress, starvation, and heat shock.

The core components of the complexes are ATR-interacting protein and MRN (for ATM). MRN is a complex of the endo- and exonuclease MRE11 (referring to Meiotic REcombination), the adaptor protein Rad50, and the protein NBS1 (referring to a mutation of NBS1 that causes the **Nijmegen breakage syndrome** characterized by increased radiation sensitivity and cancer incidence). Like ATM, NBS1 is an essential protein since corresponding gene knockout mice are not viable. Additional components of the complexes include Rad and Hus proteins (these terms refer to radiation-sensitive and hydroxyurea-sensitive yeast mutants, where the proteins were discovered; hydroxyurea is an artificial inhibitor of DNA replication), and **BRCA2** (breast cancer protein 2). Together with BRCA1 that also is part of the DNA repair complex (see below and Section 2.8.1), BRCA2 is causally related to the hereditary or familiar form of breast cancer.

When the stress sensor complexes are bound to the site of damage, ATM and ATR become activated by autophosphorylation and assemble substrate and other proteins such as enzymes of DNA repair, histone variant H2AX that recruits additional ATM/ATR substrates, histone deacetylases to stop transcription during DNA repair, transcription factor E2F1 (which is inhibited by ATM but, on the other hand, stimulates transcription of the *atm* gene), Tyr kinase cAbl, and others. A major function of such proteins is to connect the machinery of DNA repair with other signaling pathways the activation of which is essential for a proper cellular stress response. Such pathways include the MAP kinase and NFκB cascades, which occupy a central position in stress adaptation (note that two of the MAK kinases, JNK and p38, had been discovered as stress-activated protein kinases; see Section 11.1). Activation of MAP kinase modules by ATM and ATR possibly involves the Tyr kinase cAbl (see Section 7.1.1), whereas the NFκB complex is activated through phosphorylation of the scaffold protein NEMO (see Section 11.7).

12.8.2 Reversible inhibition of the cell cycle at DNA structure checkpoints

As far as the effects of DNA damage on cell cycle progression are concerned, two ATR/ATM substrates occupy a key position. These are the **checkpoint kinases ChK1** and **ChK2** that transduce the signals leading to cell cycle arrest and apoptosis when they are activated by ATM/ATR.

DNA repair is possible only when the cell cycle is transiently halted. To this end, CDKs are inhibited by the following:

- degradation of cyclins
- activation of the CDK-deactivating protein kinases WEE1 and MYT1
- destruction of the CDK-activating protein phosphatases of the Cdc25 family
- de novo synthesis of the CDK inhibitor p21^{KIP}

Figure 12.8 Events occurring at the DNA structure checkpoints Damaged DNA sequentially activates the protein kinases ATM, ATR, and ChK. By phosphorylation, ChK primes Cdc25 for degradation and activates the inhibitory kinases MYT1 and WEE1 as well as the transcription factor p53, stimulating the synthesis of CDK inhibitor p21^{KIP} and of pro-apoptotic proteins. Cdc25 is also inhibited by the protein kinase MK2 that is activated by the p38 MAP kinase module. Along this pathway, the cell cycle is slowed down by a variety of environmental stress factors including genotoxic radiation.

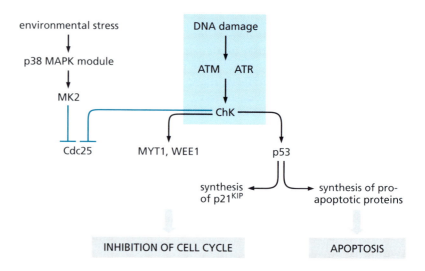

These reactions are controlled by the checkpoint kinases: by phosphorylating specific Ser residues, they activate WEE1 and MYT1, and prime the stimulatory CDK phosphatases of the Cdc25 type for ubiquitylation and proteolytic degradation. To induce p21^{KIP} synthesis, the checkpoint kinases stabilize the corresponding transcription factor p53. This is also the way leading to apoptosis if DNA repair turns out to be impossible (**Figure 12.8**).

Once the integrity of the genome has been successfully examined or reestablished, the blockade of the cell cycle must be lifted. This **checkpoint recovery** is achieved mainly by protein phosphatases that reactivate Cdc25 and inactivate the checkpoint kinases as well as MYT1, WEE1, and p53, and by a ubiquitin transferase that triggers the degradation of proteins of the DNA damage sensor complexes and of WEE1. The type 3 ubiquitin ligase in charge belongs to the SCF family. As is typical for these enzymes, their substrate proteins must be phosphorylated to be recognized (see Section 2.8.1). At least in the G2 phase of the cell cycle, this phosphorylation is catalyzed by the Polo-like kinase Plk1 (see Section 12.11). How Plk1 senses the end of the repair process remains a mystery.

SCF-type ubiquitin ligases are multiprotein complexes. Using alternative substrate-binding subunits they can specifically recognize individual phosphorylation sites in substrate proteins. This may explain why the ubiquitin ligase required for checkpoint recovery has a bivalent function since it is also involved in checkpoint initiation where it triggers the proteosomal *destruction* of the phosphatase Cdc25. Cdc25 is labeled for ubiquitylation by phosphorylation via ChK1 and this phosphorylation is recognized by another substrate-binding subunit of the ubiquitin ligase than that catalyzed in WEE1 by PlK1. In other words, the same ubiquitin ligase may control different or even antagonistic signaling pathways (**Figure 12.9**). This case provides a particularly clear example of the key role played by the cooperation of protein phosphorylation with protein ubiquitylation in cellular signal processing.

12.8.3 Tumor suppressor p53, the "guardian of the genome"

p53 is a multifunctional transcription factor, exhibiting the typical domain structure of such proteins (**Figure 12.10**).

A major function of p53 is to protect the genome from harmful mutations, that is, to prevent genetic instability, justifying the name "guardian of the genome." Owing to this property p53 is a tumor suppressor protein, the loss-of-function mutation of which ranks among the most frequent causes of human neoplasia, involving more than 50% of human cancers. p53 has two close relatives, the transcription factors **p63** and **p73**. They have not yet been associated with tumorigenesis. Instead, they

Figure 12.9 Reversible stop of cell cycle progression at the DNA damage checkpoint in the G2 phase The key role of a specific SCF-type ubiquitin (Ub) ligase exhibiting a bivalent effect is shown. Upon DNA damage, the cell cycle activator CDK phosphatase Cdc25A becomes phosphorylated along the ATR–ChK1 pathway and labeled by ubiquitylation for proteasomal degradation. When the damage has been repaired, a still-unknown signal activates the Polo-like kinase Plk1, which labels the cell cycle inhibitor CDK kinase WEE1 and other DNA damage sensor proteins for ubiquitylation and proteasomal destruction.

suppress mutations leading to embryonic malformations. In the human genome, more than 4000 potential p53-binding sites have been found, leaving open the question as to whether or not all of them are bona fide target genes of p53.

Characteristic effects of p53 are:

- to arrest the cell cycle
- to induce apoptosis
- to promote DNA repair
- to inhibit angiogenesis (particularly in tumors)

For these purposes p53 activates the genes of the cell cycle inhibitor p21^{KIP} and of the pro-apoptotic proteins Bax, Noxa, Puma, APAF1, and Fas (see Section 13.2.5), as well as genes encoding inhibitors of angiogenesis such as thrombospondin and maspin. In addition, p53 induces the synthesis of the phospholipid phosphatase PTEN, a tumor suppressor protein blocking the anti-apoptotic phosphatidylinositol 3-kinase–PKB/Akt signaling pathway (see Section 4.6.6). At the same time, transcription of genes encoding anti-apoptotic and proangiogenic factors is suppressed. These p53-regulated processes are induced either by oncogenic alterations and excessive mitogenic stimulation or by DNA damage (**Figure 12.11**).

Interestingly, cell cycle arrest by p53 is only partly mediated by p21^{KIP} synthesis. Along a parallel pathway, p53 stimulates the transcription of genes encoding specific **microRNA** species, which in turn repress via RNA interference (see Section 9.5) the formation of a series of cell-cycle-promoting factors including CDK4, cyclin E, and E2F.

Figure 12.10 Domain structure, interactions, and modifications of transcription factor p53 Through its DNA-binding domain (DNA-BD), p53 interacts with the regulatory sequences of genes encoding the cell cycle inhibitor p21^{KIP} and pro-apoptotic proteins, and recruits via its transactivating domain (TD) transcriptional co-activators such as the acetyltransferase CBP/p300, which in addition to histones also acetylates the C-terminal region of p53. The transactivating domain is also a binding site of the ubiquitin ligase MDM2 preparing p53 for proteolytic degradation. The ubiquitylation of the C-terminal domain may be competitively inhibited by sumoylation preventing degradation. In another pathway leading to a stabilization of p53, the protein ARF interrupts the interaction with MDM2. OD, oligomerization domain (in the figure this domain is shown to be flanked by nuclear import and export sequences); RD, regulatory domain (containing nuclear import and export sequences). Phosphorylation sites (serine residues) are marked by teal numbers.

Figure 12.11 Activation of the p53 signaling pathway by genotoxic stress and oncogenic mutations DNA damage leads to a stabilization of p53 due to ChK-catalyzed phosphorylation of p53 and inactivation of the ubiquitin ligase MDM2 that targets p53 for degradation. The transcription factor E2F1, rendered hyperactive in response to oncogenic mutations, activates the gene of the MDM2 inhibitor ARF, thus also stabilizing p53. HIPK2, homeodomain-interacting protein kinase 2; P, phosphorylation.

Experimental animals exhibiting excessive p53 activity are more or less protected from getting cancer, but they pay for this benefit by a shortened life span. This may explain why oxidative stress and the loss of telomeres are life-restricting factors since both stimulate p53 signaling. Therefore, an early death may be the price to be paid for protection against cancer, and p53 is under strict control in order to bring the two risks, death by cancer versus death by aging, into a tolerable relationship.

Activation of p53

In normal cells p53 is prone to rapid degradation, so its cellular level is very low. Thus, to activate p53 it must be rescued from destruction. p53 degradation is triggered by the specific **ubiquitin ligase MDM2** (the name refers to mouse double minutes, extrachromosomal pairs of aggregates consisting of amplified genes found in particular in tumor cells) that primes p53 for proteolysis in proteasomes and, in addition, binds to the transactivating domain, thus directly inhibiting the interaction of p53 with the transcriptional machinery. By inducing MDM2 synthesis, p53 promotes its own destruction. MDM2 knockout mice show once more how important it is to keep p53 activity at a low level under normal conditions: already as embryos, these animals die from excessive apoptosis in all tissues unless their p53 gene has been deleted too.

Inactivation of MDM2 is the major purpose of input signals that stimulate p53 activity. Among them, phosphorylation of MDM2 by the protein kinases ATM and cAbl as well as interaction with the protein **ARF** or p14^{ARF} have been studied in more detail. Both interrupt the interaction of MDM2 with p53 thus stabilizing p53 and turning on p53 signaling. These effects explain the tumor suppressor function of MDM2, whereas ARF may mutate to an oncoprotein. [The name ARF indicates that the protein is encoded by an Alternative open Reading Frame of the gene for the CDK inhibitor p16^{INK4A}. ARF should not be confused with the ADP-ribosylation factor Arf, a small G-protein controlling vesicle formation (see Section 10.4.3).

While ATM responds primarily to DNA damage, ARF is rather activated by oncogenic signals (see Figure 12.11) such as overexpression or mutation of β-catenin of the

Wnt signaling pathway (see Section 5.9), mutated Ras, overexpression of the cMyc transcription factor, or deletion of the retinoblastoma gene. As explained above, such alterations lead to strong activation of the transcription factor E2F1, resulting in excessive transcription of the gene encoding ARF.

Regulation of p53 activity

The activity of p53 itself is regulated by post-translational modifications in a complex manner, determining whether apoptosis or another process is stimulated. Again a major role is played by phosphorylation. Targets are several Ser and Thr residues concentrated in the transactivating domain. Their phosphorylation results in p53 activation because it disrupts the interaction with MDM2 (which, in addition, is inactivated by phosphorylation) (see Figure 12.10). Phosphorylation is catalyzed by the checkpoint kinases that (as we have seen) are under control of the kinases ATM and ATR. Moreover, the activity of p53 is also intensified because phosphorylation promotes the interaction with the transcriptional co-activator CBP/p300. In addition to histones, this acetyltransferase also acetylates p53 at several Lys residues in the regulatory domain, thus further stimulating p53 activity (see Figure 12.10). Such acetylation is observed, in particular, after DNA damage. The phosphorylation of a special Ser residue (Ser46 in humans) in the transactivating domain of p53 is a signal triggering apoptosis. p53 thus modified forms a complex with the transcriptional co-activator PML (promyelocytic leukemia protein, a tumor suppressor) and stimulates, among others, the transcription of the gene encoding apoptosis-inducing factor 1, an important mediator of programmed cell death (see Section 13.2.5). The protein kinase catalyzing the pro-apoptotic phosphorylation of p53 is called **HIPK2** (see Figure 12.11). HIPK stands for homeodomain-interacting protein kinase; it was discovered as a co-repressor of genes controlled by transcription factors with homeodomains. This enzyme belongs to the rare dual-specific protein kinases that phosphorylate Ser, Thr, and Tyr residues.

Summary

- At DNA structure checkpoints before DNA replication and chromosome condensation, the cell cycle is slowed down to enable DNA repair or elimination of an irreversibly damaged cell by apoptosis.

- Damage sensors are the DNA-dependent protein kinases ATM and ATR that phosphorylate a large number of proteins required for inhibition of cell cycle progression, DNA repair, and apoptosis.

- Major targets of ATM/ATR are checkpoint kinases, they induce the synthesis of CDK inhibitors and pro-apoptotic proteins via an activation of transcription factor p53.

- p53 is the supervisor of the DNA structure checkpoints. It activates genes encoding cell cycle inhibitors, pro-apoptotic proteins, and proteins inhibiting angiogenesis. To become activated p53 has to be rescued from proteolytic degradation catalyzed by the ubiquitin ligase MDM2. p53 is the tumor suppressor protein most frequently deleted in human cancers.

12.9 Phases of mitosis

If at the second DNA structure checkpoint (G2–M checkpoint) the genome is found to be intact, the cell is allowed to enter the phase of mitosis. As one of the most complex cellular processes, mitosis—the symmetric distribution of genetic material among daughter cells—is controlled by a highly sophisticated signaling apparatus that works with extreme precision and monitors again at several checkpoints. In analogy to other phases of the cell cycle, reversible protein phosphorylation and the periodic formation and degradation of control proteins provide the key regulatory events and the corresponding enzymes—kinases, phosphatases and ubiquitin ligases—are the major targets of control signals.

Mitosis is divided into the following phases:

(1) **Prophase:** chromatin condenses to chromosomes, the centrosomes begin to separate, and the mitotic spindle begins to grow.

(2) **Prometaphase:** the cell rounds up, the nuclear membrane disintegrates, the centrosomes are separated, and the mitotic spindle continues to grow.

(3) **Metaphase:** the chromosomes align at the equatorial plain (metaphase plate) and recruit in their central region (the centromere) the kinetochore protein complex, which anchors the microtubule fibers that link the centrosomes at the spindle poles with the chromosomes.

(4) **Anaphase:** the sister chromatids separate to become daughter chromosomes, which are pulled to the spindle poles by motor proteins and by shortening of the microtubules; and in preparation for the division into daughter cells through cytokinesis, a contractile microfilament ring forms around the cell equator.

(5) **Telophase:** the new nuclear membrane is formed, the chromosomes de-condense, and the daughter cells become separated by contraction of the equatorial microfilament ring.

The central event of mitosis is the separation of the sister chromatids. It requires the formation of the bipolar mitotic spindle consisting of the centrosomes with the attached microtubules and various regulatory and motor proteins. Only a spindle working with flawless accuracy guarantees correct alignment of the chromosomes in metaphase and their proper segregation during anaphase. If half the chromosomes do not end up in each daughter cell one speaks of **aneuploidy**. A correlation exists between the degree of aneuploidy and the malignancy of cancer, and when aneuploidy occurs during meiosis it may lead to developmental disorders such as Down syndrome. To avoid such accidents, the cell cycle is interrupted in late metaphase at a **spindle assembly checkpoint** (**Figure 12.12**) until the spindle–chromosome complex has taken on the correct structure (see below). Only then does anaphase begin.

Centrosomes help to organize the spindle. These small nuclear organelles consist of two centrioles, each serving as microtubule-organizing centers. The microtubules radiating from the centrosomes are anchored at the kinetochores, which are localized in specific chromosomal structures, the centromeres. Kinetochores are large complexes of structural, motor, and regulatory proteins. To form a bipolar spindle, the centrosomes have to duplicate. This occurs in the S phase together with DNA replication (it has been postulated that, before the S phase, centrosome duplication is prevented by a protein "nucleophosmine" that becomes inactivated by cyclin E–CDK2-catalyzed phosphorylation at the G1–S transition). The two daughter centrosomes do not segregate immediately after duplication but are held together until prophase, when they migrate to the cell poles. Disturbances of centrosome duplication are one cause of genetic instability promoting cancer development. These occur, for instance, when DNA damage is not repaired owing to loss of p53 activity, and give rise to multipolar mitotic spindles.

Summary

The central event of mitosis is condensation of the chromosomes and separation of the sister chromatids directed by the centrosomes that generate the mitotic spindle. To avoid aneuploidy and tumor development, correct distribution of the genetic material is controlled at a spindle assembly checkpoint before chromosome segregation.

condensing chromosomes

centrosome

interphase
(including G1,
S and G2 phase)

prophase

chromosome

prometaphase

disintegrating
nuclear membrane

bipolar
spindle

actomyosin ring

cytokinesis

metaphase plate

metaphase

spindle
assembly
checkpoint

formation of new
nuclear membrane

telophase

anaphase

12.10 Ubiquitin ligase APC/C: giving the beat of mitosis

As we have seen, to execute cell division in a flawless manner, a cell must both activate (produce) and inactivate (destroy) regulatory proteins such as CDKs and cyclins at the appropriate time. This holds true also for mitosis, which like other cell cycle phases is controlled by cyclin–CDK complexes. Thus cyclin A–CDK1 is active until prometaphase, when cyclin A is degraded by proteolysis. The complex controls the early phases of chromosome condensation and, together with cyclin B–CDK1, the breakdown of the nuclear envelope. Cyclin B–CDK1 remains active up to late metaphase and regulates chromosome condensation and spindle assembly. CDK activity follows a strict time course: it must be high in the beginning of mitosis and be turned down at the onset of anaphase since otherwise a separation of the sister chromatids is not possible. In fact, the switch from high to low CDK activity brought about by cyclin degradation is a key event that determines the onset of anaphase. The elimination of cyclins (and other proteins regulating mitosis) is triggered by E3 ubiquitin ligases, in particular those of types **SCF** and **APC/C** (anaphase-promoting complex/cyclosome; this notation is used to avoid confusion with the adaptor protein APC of the Wnt signaling pathway [see Section 5.9]), which tag the proteins for degradation in proteasomes. These ligases are essential: without them, the cell cannot divide and arrests at the end of metaphase. SCF ubiquitin ligases preferentially attack cyclins and other proteins involved in regulation of G1 and S phases, whereas APC/C enzymes control later phases of mitosis.

A key event is the separation of sister chromatids in anaphase. To this end, a specific endopeptidase, **separase**, destroys a subunit of the protein cohesin, which is the

Figure 12.12 Mitotic cycle and spindle assembly checkpoint For details, see text.

"glue" holding together the chromatids at the centromeres. To prevent premature separation before anaphase, this enzyme is inhibited by binding to a protein called **securin**. Coincident with cyclin degradation and the onset of anaphase, securin is eliminated with the help of APC/C, rendering separase active. To perform this job, APC/C must be phosphorylated by CDK1–cyclin B. This explains why the activity of the kinase must be high in late metaphase.

Phosphorylation determines the substrate specificity of APC/C in that it controls interactions with various regulatory subunits. In fact, the ligase is a large multi-protein complex consisting of more than a dozen subunits (see Section 2.8.1) that resembles in size other molecular machines such as ribosomes, proteasomes, or chaperonins, thus justifying the name cyclosome.

An important regulatory principle of the cell cycle is that APC/C has the choice between two major substrate-recognizing subunits, known as **Cdc20** and **Cdh1**, which confer different specificities. While Cdc20 mediates the degradation of securin and some prometaphase proteins such as cyclin A, Cdh1 is associated with the elimination of proteins that control the onset of anaphase and later phases of mitosis. These include cyclin B, mitotic protein kinases (see Section 12.11), and subunit Cdc20. The degradation of such proteins is essential for mitotic exit and a clearing of the cell for the subsequent G1 phase. Since APC–Cdh1 also induces the degradation of cyclin A, which is required for DNA replication, the ubiquitin ligase becomes inactivated at the end of the G1 phase by the combined effects of a protein complex containing cyclin A–CDK2. The latter phosphorylates Cdh1, priming it for ubiquitylation by the E3 ligase SCF and degradation in proteasomes. Thus, like cyclins, subunits Cdc20 and Cdh1 are synthesized only during definite periods of the mitotic cycle to become sequestered or degraded afterwards, and the regulation of APC/C activity appears to be as sophisticated as that of CDK activity.

Figure 12.13 Reprogramming of ubiquitin ligase APC/C during the cell cycle At the beginning of mitosis, APC/C is phosphorylated by the CDK1–cyclin B complex. In the phosphorylated state, APC/C binds the subunit Cdc20 and induces the transition from metaphase to anaphase by priming securin for degradation. By this means, the separation of sister chromatids is triggered. In dephosphorylated APC/C, Cdc20 is replaced by the subunit Cdh1, rendering the complex specific for an alternative set of substrates such as cyclin B, Cdc20, and other mitotic proteins. Their degradation triggers cytokinesis and "clears" the cell for the subsequent G1 phase. At the end of G1 phase, an inhibitory protein (EMI1) generated along the E2F pathway suppresses APC/C–Cdh1 activity, and cyclin A–CDK2 phosphorylates Cdh1, priming it for ubiquitylation by the E3 ligase SCF and degradation in proteasomes. The diagram to the left shows in a schematic manner the activities of the two APC/C complexes during different cell cycle phases.

Which of the subunits interacts with APC/C depends on the state of APC/C phosphorylation: APC/C must be phosphorylated at several sites to bind Cdc20 but dephosphorylated to interact with Cdh1 (**Figure 12.13**). APC/C phosphorylation is catalyzed predominantly by CDK1–cyclin B and occurs as long as enough cyclin B is available. It provides the triggering signal for chromosome separation in anaphase. In contrast, APC/C–Cdh1 becomes active when cyclin B disappears, giving protein phosphatases the chance to dephosphorylate APC/C (**Figure 12.14**). By eliminating cyclins and mitotic protein kinases as well as Cdc20, the APC/C–Cdh1 complex speeds up its own formation in an autocatalytic process. This reprogramming of the ubiquitin ligase takes place in late anaphase.

metaphase

cohesin

spindle formation
sister chromatids are held
together by cohesin
APC/C becomes phosphorylated
APC/C–Cdc20 has triggered inactivation
of cohesin and degradation of cyclin B
sister chromatids segregate
transient inactivation of APC/C–Cdc20
at the spindle assembly checkpoint

early anaphase

reprogramming of APC/C by
dephosphorylation and
subunit exchange

late anaphase

APC/C–Cdh1 triggers full degradation
of cyclin B and other mitotic proteins

telophase and
cytokinesis

cyclin B–CDK1 activity

**Figure 12.14 Correlation of
anaphase transit with declining
cyclin B–CDK1 activity and a
change of substrate specificity of
APC/C** For details, see text.

Summary

The E3 ubiquitin ligase anaphase-promoting complex/cyclosome (APC/C) induces two key events of mitosis: in early anaphase, it induces proteolytic degradation of securin, an endogenous inhibitor of separase, the protease that causes sister chromatid separation; and in metaphase, it induces degradation of cyclin B and other proteins, thus enabling anaphase and telophase. The switch in function of APC/C is due to reprogramming by dephosphorylation and an exchange of regulatory subunits.

12.11 Mitotic protein kinases: formation of the mitotic spindle

Mitosis is controlled by a network of protein kinases that includes, in addition to the master regulator CDK1–cyclin B, the following families of highly specialized Ser/Thr-directed kinases:

- Aurora kinases
- Polo-like kinases
- NIM-A-related kinases
- Bub kinases
- Mps kinases

Like cyclins, mitotic kinases are expressed only for a short time interval. Most of them are present from the G2–M transition up to the end of the M phase, when they undergo degradation triggered by the APC/C–Cdh1 complex. Major targets of the mitotic kinases are proteins of centrosomes, of sister chromatids (such as securin), and of the centromere–kinetochore complex.

Mitotic kinases require activation loop phosphorylation. It is not always clear whether this is catalyzed by other kinases or is due to autophosphorylation, particularly because the input signals activating these kinases are known only incompletely. Moreover, the apparently somewhat redundant ways in which these kinases share the job of controlling chromosome segregation and cytokinesis is still not fully understood.

Aurora kinases

Aurora kinases got their name from a *Drosophila* mutant (although initially they were discovered in *Xenopus* eggs). Mammalian cells express three isoforms called Aurora A, B, and C. While isoforms A and B are widely distributed, isoform C is restricted to sperm cells. Aurora kinases are key regulators of cell division that cooperate but differ in substrate specificity and cellular localization. Thus, Aurora A and C are enriched at centrosomes, while Aurora B is enriched at chromosomes. Aurora A, by phosphorylating several centrosomal proteins in the prometaphase, regulates centrosomal functions that are essential for the formation of a bipolar spindle. Therefore, Aurora A controls chromosome segregation.

The same holds true for Aurora B, albeit at chromosomes. Together with some other proteins, Aurora B is a component of a chromosomal passenger complex that stays with the chromosomes up to the end of metaphase, becoming more and more concentrated at the centromeres. During anaphase and telophase, it migrates to the equatorial zone that finally develops into the cleavage furrow of the dividing cell. Inhibitor and knockout experiments show that the complex is essential for spindle formation, correct alignment of chromosomes, spindle assembly quality control, formation of a central spindle during anaphase, and cytokinesis. Substrates of Aurora B are various centromere and kinetochore proteins. Their phosphorylation is required for spindle formation but also loosens the contact between kinetochores and microtubules. In this way, Aurora B might correct defaults in the links between chromosomes and spindle.

A pathological overexpression of Aurora kinases is characteristic for a variety of human tumors. Moreover, in cell culture Aurora overexpression causes aneuploidy and genetic instability resulting in a transformed phenotype. Therefore, Aurora kinases are potential targets of novel anti-cancer drugs that are on clinical trials.

Polo-like kinases

Polo-like kinases (Plk), which owe their name to the "Polo" mutant of *Drosophila*, have been found in all eukaryotes whose mitotic mechanism has been scrutinized. Among the five isoforms of the mammalian cell, at least Plk1 and Plk4 play a significant role as mitotic regulators. Thus, the duplication of centrioles in the S phase critically depends on **Plk4**: inhibition or elimination of Plk4 correlates with severe spindle anomalies, defective embryonic development, and increased cancer susceptibility. Probably, the Plk4 substrates are structural or regulatory proteins required for centriole duplication.

In controlling spindle formation, Plk4 cooperates with the isoform Plk1. Like Plk4 and Aurora kinases, Plk1 is enriched in centrosomes and kinetochores. Major substrates are microtubule-associated proteins, the phosphorylation of which triggers spindle formation at the centrosomes. As explained above, Plk1 is also involved in the recovery from G2 checkpoint control when DNA damage has been successfully repaired. To this end, the kinase phosphorylates key regulatory proteins. As a result CDK1-inhibitors such as the protein kinases MYT1 and WEE1 become inactivated whereas the CDK1-activating phosphatase Cdc25 becomes stimulated. This reaction provides an instructive example of the ambiguity of the phosphorylation signal: while Cdc25 is primed for ubiquitylation and degradation by phosphorylation of a Ser residue in the "ubiquitylation degron" (see Section 2.8.1) catalyzed by checkpoint kinase 1, phosphorylation of another Ser residue by Plk1 has a stimulatory effect.

Like CDK1–cyclin B, Plk1 also may phosphorylate the ubiquitin ligase APC/C, thus promoting the entry into anaphase. Vice versa, APC/C complexed with Cdh1 primes Plk1 for degradation at the end of mitosis.

Plk1 is activated by other protein kinases such as Aurora A.

NIM-A-related kinases

The name of the **NIM-A-related kinases (Nrk)** comes from the NIM (never in mitosis) mutant of the mold *Aspergillus nidulans*, which exhibits a blockade of the G2–M transition. The mold enzyme NIM-A is active only during prophase and metaphase and is degraded afterwards.

Mammals express eleven Nrk isoforms, which here are called Nek1 to Nek11. Nek2 is associated with centrosomes, where it phosphorylates a centrosomal Nek-associated protein (C-Nap1) that (like cohesin) probably adheres centrosome pairs. Since the glue effect is lost upon phosphorylation of C-Nap1, Nek2 can cause centrosome separation and spindle formation. A role in spindle formation is attributed also to the isoforms NEK6, -7, and -9, whereas NEK1, -10, and -11 seem to be involved in the control of the DNA damage response.

Bub and Msp kinases

Additional mitotic protein kinases found in all eukaryotes are associated with the kinetochores. They act as **spindle checkpoint kinases** and are known as **Bub, BubR** (Bub-related), and **Mps** kinases. Bub (budding uninhibited by benzimidazole) refers to yeast mutants that, owing to a defective spindle checkpoint, are unable to interrupt the cell cycle even though they do not form a mitotic spindle in the presence of the microtubule poison benzimidazole. Mps (monopolar spindle) refers to a yeast mutant with disturbed centrosome duplication. At least *in vitro*, Mps kinases exhibit rare dual specificity: the ability to phosphorylate Ser, Thr, and Tyr residues. The yeast enzyme has also been shown to regulate centrosome duplication in G1 phase (in animals this function is attributed to Plk4). These effects indicate that proteins of the spindle checkpoint and of centrosomes might be Mps kinase substrates.

Summary

Apart from cyclin B–CDK1, several other Ser/Thr-directed protein kinases control the passage of the cell through the mitotic cycle. These mitotic protein kinases (Plk, SLK, Nrk, Aurora, Bub, and Mps) are thought to be involved in control of centrosome separation and function, spindle formation, histone phosphorylation, regulation of APC/C activity, and mitotic checkpoints.

12.12 The spindle assembly checkpoint provides a last chance to correct mistakes

During metaphase, the mitotic spindle releases a large number of free microtubules thus trying to "capture" the kinetochores (the microtubule-binding protein complexes associated with sister chromatids). This stochastic process is prone to mistakes. Therefore, a control mechanism is essential to ensure the correct bipolar attachment of sister chromatids (where each pair of sister chromatids is attached by a microtubule to each centrosome). In fact, spindle quality control is extremely important for the well-being of the cell: if it does not work properly, the risk of malignant transformation is strongly increased. The control mechanism becomes active at the spindle assembly checkpoint, where the separation of the chromosomes is delayed until it is guaranteed that each sister chromatid is attached correctly to prevent aneuploidy. To this end, the degradation of securin is interrupted by a transient

Figure 12.15 Mitotic checkpoint complex: a still somewhat speculative model As the key mechanism of anaphase arrest, a transient inactivation of the ubiquitin ligase APC/C–Cdc20 complex is postulated that becomes reversed when the kinetochore has been found by a microtubule or incorrect microtubule–kinetochore binding has been corrected. (A) The motor protein CENP-E (dark gray horizontal bar) recognizes a kinetochore lacking a microtubule and recruits the mitotic checkpoint complex consisting of several protein kinases (shown in color) and the regulatory proteins Mad1 and Mad2. (B) One (or several) of the protein kinases phosphorylates Mad1 (P, phosphate residue), disrupting the interaction with Mad2. Mad2 thus released displaces the regulatory subunit Cdc20 from the ubiquitin ligase APC/C, thus inactivating APC/C. This results in a transient stop of the mitotic cycle. (C) The kinetochore is correctly contacted by a microtubule. Probably owing to a conformational change of CENP-E, the mitotic checkpoint complex dissociates from the kinetochore and is degraded. By dephosphorylation of Mad1, the Mad1–Mad2 complex is reestablished. As a result, APC/C is relieved from inhibition and can now induce the metaphase–anaphase transition. The kinetochore is depicted here as a single entity; in reality, it is a large multiprotein complex. This rather simplified scheme is based on assumptions, some of which are awaiting final experimental proof.

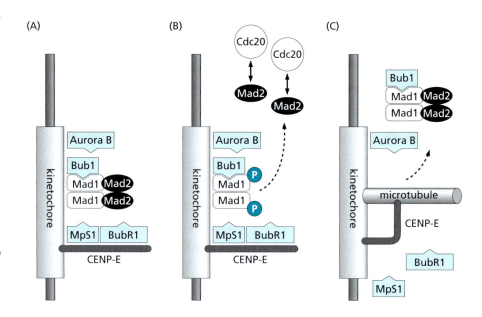

inhibition of the ubiquitin ligase APC/C–Cdc20. The importance of this checkpoint control is emphasized by the fact that when the checkpoint is not passed, the cell submits suicide by apoptosis. In fact, mutations of checkpoint proteins contribute to cancer development.

The checkpoint control is triggered by several proteins, forming a mitotic checkpoint complex. However, it is still not entirely clear how this complex inhibits APC/C–Cdc20 until all chromatids have become precisely aligned and fixed at the spindle. One concept, the mechanism of reversible Cdc20 inactivation, is shown in **Figure 12.15**.

According to this hypothesis, the complex binds to kinetochores that lack a contact with the spindle or whose microtubules are incorrectly aligned in relationship to the spindle pole. The kinesin motor protein **centromeric protein E** (CENP-E) transiently associates with kinetochores in the metaphase and recognizes such defects and recruits the **checkpoint kinase BubR1**, which becomes activated.

A major substrate of BubR1 is a complex of the proteins **Mad1** and **Mad2** (the name refers to mitotic arrest-defective yeast mutants). Upon phosphorylation, this complex displaces subunit Cdc20 from the ubiquitin ligase APC/C, rendering APC/C unable to induce securin degradation and sister chromatid segregation. As a result, the mitotic cycle is blocked in late metaphase. This blockade is overcome only when a microtubule from the spindle pool displaces the BubR1–Mad complex from the kinetochore and a mechanical tension develops between the kinetochore pair of adjacent sister chromatids.

Transient inhibition of APC/C–Cdc20 is only one aspect of spindle quality control. In addition, incorrectly attached microtubules must be released from the kinetochore. This step depends on **Aurora B**-catalyzed phosphorylation of proteins that control the mechanical tension between microtubules and kinetochores (incorrect attachment changes this tension). Aberrant tension is detected by accessory proteins that form the chromosomal passenger complex with Aurora B, as mentioned above. Experiments with mutants and inhibitors have shown that, in addition to Bub and Aurora B kinases, spindle control also requires the kinase Msp. How Msp shares the job with the other kinases remains an open question.

Summary

- At the spindle assembly or anaphase checkpoint, separation of the chromosomes is interrupted until it is guaranteed that all sister chromatids are bound by microtubules from the opposite spindle poles to ensure symmetric segregation of the genetic material among the daughter cells.

- The checkpoint control is due to a sophisticated (not yet fully understood) interaction between proteins that recognize spindle defects and enzymes that catalyze reversible phosphorylation of regulatory proteins to induce a transient stop of the mitotic cycle and allow time for repair measures.

12.13 Cytokinesis: how a new cell is born

Cytokinesis is the final stage of cell division, where the daughter cells form. As far as biochemical control mechanisms are concerned, it is the cell cycle phase least understood. By means of functional genomics and proteomics, several hundred proteins and low-molecular-mass factors have been found to be potentially involved without obtaining a clear picture of their interactions.

The major event of cytokinesis is the contraction of the cell perpendicular to the spindle by a microfilament ring formed in the equatorial plane. Until complete separation, both daughter cells stay connected by the **central spindle** that consists of microtubules not linked to kinetochores and arises when the chromosomes are pulled to the spindle poles. The central spindle is essential for correct progress of cytokinesis since it assembles regulatory proteins in the region of membrane constriction probably through mechanical transport by motor proteins along the microtubule "tracks" of the spindle.

As we have seen, a critical condition of cytokinesis and mitotic exit is to inhibit cyclin B–CDK1 activity, since this activity would suppress continuation of M-phase events such as formation of a new nuclear envelope and cytokinesis. Therefore, the destruction of cyclin B is initiated in late metaphase by the ubiquitin ligase APC/C–Cdh1 complex (see Figure 12.14). Additional control elements are considered to be the protein kinases Aurora B and Plk since their inhibition retards cytokinesis. Moreover, Aurora B dissociates from chromosomes during anaphase and telophase to assemble in the central part of the spindle and in the zone of membrane constriction (midzone), where it probably is bound by specific anchoring proteins. However, for a better understanding of the mechanisms involved, identification of the elusive kinase substrates is mandatory.

The cleavage itself is controlled by several GTPases, which become concentrated in the equatorial plane of the central spindle. Among them, the small G-protein **Rho** (see Section 10.3) promotes the aggregation of actin fibers to the contractile ring. A Rho GTP–GDP exchange factor (GEF) and a Rho GTPase-activating protein (GAP), both regulating Rho activity, are bound by adaptor proteins in the central spindle plane. Rho-GEF can activate Rho only when cyclin B has been degraded, since otherwise it would be inhibited by CDK1-catalyzed phosphorylation. Such an inhibitory effect seems to be supported additionally by activation of Rho-GAP through phosphorylation by Aurora B.

Further GTPases participating in cell division are the **dynamins** (see Section 5.8) and the **septins**. Septins are large proteins found in all eukaryotes except plants (in the human genome, seven septin genes have been identified). When activated by GTP, they aggregate to short filaments associating with the actin fibers of the contractile ring where they might recruit additional proteins required for cytokinesis. Like dynamins, septins are also involved in exocytosis.

Contraction of the microfilament ring depends on a local increase of the cytoplasmic Ca^{2+} concentration activating the Ca^{2+}/calmodulin-regulated **myosin light chain (MLC) kinase**. This enzyme phosphorylates the short myosin chains enabling myosin to interact with actin and thus induce contraction (see Section 14.6.2). The increase in cytoplasmic Ca^{2+} is also associated with destruction of cyclin B and onset of anaphase. In fact, CDK1–cyclin B phosphorylates and activates **MLC phosphatase**, which efficiently counteracts the effect of MLC kinase unless it becomes dephosphorylated. Inactivation of MLC phosphatase is enhanced by Rho activating the protein kinase ROK, which inhibits MLC phosphatase activity (see Section 10.3.3). (Note that the dual control of MLC phosphatase resembles that of Cdc25: depending on the target site, phosphorylation either promotes or inhibits enzyme activity.) As mentioned above, Rho becomes active only upon degradation of cyclin B. So, as noted, the removal of this cyclin initiated by the ubiquitin ligase APC/C provides a critical step in mitosis (see Figure 12.14).

Summary

- Mechanisms regulating cytokinesis, the mechanical formation of daughter cells, are only incompletely understood.

- A critical step is the inactivation of cyclin B–CDK1 through degradation of cyclin B triggered by the ubiquitin ligase APC/C–Cdh1 complex. It provides the condition for cell separation by the contractile microfilament ring the formation and function of which is controlled by G-proteins of the Rho, dynamin, and septin families as well as by myosin light chain phosphorylation.

12.14 Mitotic exit: characterization in yeast

To finish mitosis and to open the gate to a new interphase, the signaling system that initiated M phase must be switched off completely. A significant step is the reprogramming of the ubiquitin ligase APC/C by subunit exchange, which renders the enzyme specific for anaphase and telophase proteins including cyclin B (see Figure 12.13). As shown for *S. cerevisiae*, to this end a signaling complex is activated in late anaphase. The major executor of this mitotic exit network (MEN) is the dual-specific **protein phosphatase Cdc14** that counteracts the mitotic CDKs and promotes the reprogramming of APC/C via replacement of the subunit Cdc20 by the subunit Cdh1 (see Figure 12.13). At least in *S. cerevisiae*, this reaction clears the way for the degradation of Clb2 (the yeast homolog of cyclin B), resulting in inactivation of the kinase Cdc28 (the yeast homolog of CDK1). To reinforce this effect, Cdc14 also activates the CDK inhibitor Sic1 (the yeast homologue of p27^{KIP}).

Punctual activation of Cdc14 is a critical condition of exact regulation of the mitotic exit. In fact, the enzyme is sequestered in the nucleolus and blocked by a protein complex known as Fob1–Net1 until it becomes activated by the MEN signaling system and transiently released into the cytoplasm during anaphase. The MEN complex of yeast is composed of a small Ras-like G-protein Tem1, the associated GEF (Lte1) and GAP (a heterodimer of the proteins Bub2 and Bfa), and the protein kinases Dbf2–Mob1 (**Sidebar 12.5**), Cdc15, and Cdc5 (a Polo-like kinase). These proteins are interconnected to a signaling cascade as depicted in **Figure 12.16**. Since Tem1 and Cdc15 are inactivated by CDK-catalyzed phosphorylation and become reactivated by the phosphatase Cdc14, the mitotic exit network seems to enhance its efficacy by positive feedback.

To get the network in full swing in *late* anaphase, Cdc14 becomes partly activated in *early* anaphase. This is brought about by another protein network called **FEAR** (Cdc fourteen early anaphase release) that involves the inactivation of the Cdc14 inhibitor Fob1–Net1 by the combined action of the protease separase (here called Esp1)

Sidebar 12.5 NDR kinases and the Hippo pathway The protein kinase Dbf2 of the mitotic network of yeast is a member of a larger family of related Ser/Thr-directed protein kinases that probably occur in all eukaryotes. Called NDR kinases (nuclear Dbf2-related) they represent a subfamily of the AGC kinases (see Section 2.6.3) and are activated typically by (auto)phosphorylation of the activation loop and phosphorylation in the C-terminal hydrophobic region. The mammalian enzymes bind, in addition, to a calmodulin-containing Ca^{2+} sensor protein of the S100 family that stimulates the activating (auto) phosphorylation depending on the cellular Ca^{2+} level. The role of NDR kinases in cell cycle control has been studied most thoroughly for yeast. The homologous enzymes of higher eukaryotes seem to fulfill similar purposes.

One of the mammalian NDR kinases has become known as the **tumor suppressor protein LATS** (large tumor suppressor), the deletion of which correlates with tumor growth in multiple organs. LATS is a negative regulator of cell division, and its activity and cellular localization strictly depend on the cell cycle phases. Active LATS displaces cyclin B from CDK1, thus inhibiting the G2–M transition. Moreover, by promoting the synthesis of the pro-apoptotic protein Bax (see Section 13.2.4), LATS facilitates cell death.

LATS is a component of a signaling cascade referred to as **Hippo pathway**. Hippo is a Ser/Thr-directed protein kinase of the Ste20 family (see Section 11.3). The name is based on a "hippopotamus-like" tissue-overgrowth observed in *Drosophila* mutants with an inactive *Hippo* gene. The mammalian homolog is called MST (mammalian Ste20-like). By controlling the equilibrium between cell gain and cell loss, the Hippo pathway plays an important role in the regulation of organ size. As a key event, Hippo signaling *inactivates* the transcriptional co-activator Yap. Yap promotes cell proliferation, overcomes contact inhibition, suppresses cell death, and is mutated to an oncogene in a variety of human cancers. In other words, upon activation of the Hippo pathway, cells slow down proliferation and undergo apoptosis. Yap is inactivated by phosphorylation catalyzed by LATS while LATS is activated by Hippo/MST through phosphorylation. This sequence MST/Hippo–LATS–Yap is under the control of various extracellular signals including stress hormones, lysophosphatidic acid and sphingosin-1-phosphate (the "find-me" signals of phagocytosis, see Section 13.2), and cadherin-type cell contact proteins.

and CDK1 (here called Cdc28). The clockwork-like interaction of several regulatory circuits guarantees flawless temporal coordination between chromosome segregation, cytokinesis, and termination of mitosis.

Cdc14 is a highly conserved protein found from yeast to humans. Moreover, homologs of the other MEN proteins are distributed equally widely. Therefore, one might expect mammalian cells to have as much complexity in regulation of the late mitotic phase as yeast. It must be emphasized, however, that for metazoans the mechanism of mitotic exit differs from that of yeast in a way that is still not fully understood.

Figure 12.16 Mitotic exit network of *S. cerevisiae* The signaling system depicted in a simplified form is composed of the proteins shown in the large gray box. Its activation in late anaphase is probably due to a reorganization of the components resulting in conformational changes. The black solid arrows symbolize activating reactions, while the broken arrows indicate (putative) positive feedback loops based on a dephosphorylation of Lte1 and Cdc15 that are both inhibited by CDK1-catalyzed phosphorylation. These and other inhibitory effects are shown in teal.

Figure 12.17 Signaling network controlling mitosis: a preliminary summary of the state of the art The scheme summarizes the interactions of protein kinases (gray boxes) and of the ubiquitin ligase APC/C as far as they are known at present. MEN, mitotic exit network of yeast.

In **Figure 12.17**, an attempt is made to summarize schematically the complexity of signaling pathways controlling mitosis, with particular emphasis laid on protein phosphorylation. It must not be overlooked, however, that this scheme provides only a preliminary image of the state of the art, since this field of research still is under intense investigation.

Summary

In yeast cells, a mitotic exit network activated in anaphase is thought to provide the conditions for entry into the G1 phase of the cell cycle. A key event is dephosphorylation (by the phosphatase Cdc14) of CDK1 substrate proteins including the ubiquitin ligase APC/C, which are reprogrammed to trigger the elimination of cyclins and other mitotic proteins.

Further reading

Section 12.1

Ferrell JE, Tsai TYC & Yang Q (2011) Modeling the cell cycle: why do certain circuits oscillate? *Cell* 144:874–885.

Harper JV & Brooks G (2004) The mammalian cell cycle: an overview. *Methods Mol Biol* 296:113–153.

Section 12.2, Section 12.3, and Section 12.4

Karlsson-Rosenthal C & Millar JBA (2006) Cdc25: mechanisms of checkpoint inhibition and recovery. *Trends Cell Biol* 16:285–292.

Malumbres M & Barbacid M (2005) Mammalian cyclin-dependent kinases. *Trends Biochem Sci* 30:630–641.

Murray AW (2004) Recycling the cell cycle: cyclins revisited. *Cell* 116:221–234.

Nurse P (2002) Cyclin dependent kinases and cell cycle control (Nobel Lecture). *Chembiochem* 3:596–603.

Sage J (2004) Cyclin C makes an entry into the cell cycle. *Dev Cell* 6:607–616.

Shah K & Lahiri DK (2014) Cdk5 activity in the brain—multiple paths of regulation. *J Cell Sci* 127:2391–2400.

Section 12.8

Awasthi P, Foiani M & Kumar A (2015) ATM and ATR signaling at a glance. *J Cell Sci* 128:4255–4262.

Lee H (2014) Cycling with BRCA2 from DNA repair to mitosis. *Exp Cell Res* 329:78-84.

Meek DW (2015) Regulation of the p53 response and its relationship to cancer. *Biochem J* 469:325–346.

Section 12.9, Section 12.10, Section 12.11, and Section 12.12

Carmena M, Earnshaw WC & Glover DM (2015) The dawn of Aurora kinase research: from fly genetics to the clinic. *Front Cell Develop Biol* 3:1–5.

Cheesemen IM & Desai A (2008) Molecular architecture of the kinetochore-microtubule interface. *Nat Rev Mol Cell Biol* 9:33–46.

Fry AM et al (2012) Cell cycle regulation by the NEK family of protein kinases. *J Cell Sci* 125:4423–4433.

Lew DJ & Burke DJ (2003) The spindle assembly and spindle position checkpoints. *Annu Rev Genet* 37:251–282.

Liu X & Winey M (2012) The MPS1 family of protein kinases. *Annu Rev Biochem* 81:561-585.

London N & Biggins S (2014) Signalling dynamics in the spindle checkpoint response. *Nat Rev Mol Cell Biol* 15:735–747.

Ruchaud S, Carmena M & Earnshaw WC (2007) Chromosomal passengers: conducting cell division. *Nat Rev Mol Cell Biol* 8:798–812.

Sivakumar S & Gorbsky GJ (2015) Spatiotemporal regulation of the anaphase-promoting complex in mitosis. *Nat Rev Mol Cell Biol* 16:82–94.

Teixeira LK & Reed SI (2013) Ubiquitin ligases and cell cycle control. *Annu Rev Biochem* 82:387–414.

Wang G, Jiang Q & Zhang C (2014) The role of mitotic kinases in coupling the centrosome cycle with the assembly of the mitotic spindle. *J Cell Sci* 127:4111–4122.

Zitouni S et al (2014) Polo-like kinasess: structural variations lead to multiple functions. *Nat Rev Mol Cell Biol* 15:433–452.

Sections 12.13 and 12.14

D'Amours D & Amon A (2004) At the interface between signaling and executing anaphase: Cdc14 and the FEAR network. *Genes Dev* 18:2581–2596.

Eggert US, Mitchison TJ & Field CM (2006) Animal cytokinesis: from parts list to mechanisms. *Annu Rev Biochem* 75:543–566.

Rock JM & Amon A (2009) The FEAR network. *Curr Biol* 19:R1063–R1068.

Yu FX, Zhao B & Guan KL (2015) Hippo pathway in organ size control, tissue homeostasis, and cancer. *Cell* 163:811–828.

Signal Transduction by Proteolysis, and Programmed Cell Death

13

Proteolysis plays a most critical role in cellular data processing since it terminates signaling by proteins and peptides in a controlled way. The major player in this process of signal extinction is the ubiquitin–proteasome system. Various examples of this pathway are found throughout this book.

Besides ubiquitin-triggered protein degradation, proteases are involved in signal processing in two additional ways.

(1) Proteases are components of proteolytic switches standing at the top of signaling cascades (see Section 2.11). They are coupled with signal receptors in such a way that, upon activation, they release from proteins peptide fragments that act as second messengers. The energy needed for signal processing stems from hydrolysis of the peptide bond. A perfect example of this signaling principle is the notch system.

(2) Proteases also stand at the bottom of signaling cascades. The most prominent examples are the proteases that, upon activation by death signals, drive cells into suicide.

13.1 Secretase-coupled receptors: generation of peptide second messengers

13.1.1 Notch, a key signal of animal development

The notch system represents a proteolysis-driven mechanism of signal transduction across the cell membrane, or a proteolytic switch, in a particularly clear form. This system enables communication between cells of developing tissues and, therefore, is of central importance for the embryogenesis of metazoans. Mutations of notch proteins cause malformation and cancer.

Notch (the name refers to a *Drosophila* mutant with notched wings) is a dimeric transmembrane receptor that is expressed in four isoforms in mammals. Owing to a specific proteolytic cut, each subunit consists of two parts linked by noncovalent binding at the cell surface. The extracellular domain contains 36 so-called epidermal growth factor (EGF)-like repeats (**Sidebar 13.1**) that are essential for ligand binding. The high variability of notch effects results from post-translational modifications of the protein. Thus, the ligand and cell-type specificity of notch is regulated by the glycosyltransferase "fringe" that attaches *N*-acetylglucosamine residues to the EGF-like repeats (for more details on this post-translational protein modification, see Section 9.4.3).

The notch ligands are called **DSL proteins,** referring to the *Drosophila* mutants Delta and Serrate and the *Caenorhabditis elegans* mutant Lag-2. They also are transmembrane proteins bound at the surface of nearby cells and acting as juxtacrine signals. On the other hand, their extracellular domains, containing (like notch) EGF-like repeats, may be released proteolytically into the extracellular medium. By this "ectodomain shedding" peptides are generated which interact with notch in a paracrine mode.

DSL induces a two-step proteolytic activation of **notch** (**Figure 13.1**). In the first step, known as S2 cleavage, transmembrane proteases of the ADAM family clip the extracellular domain of notch, releasing it from the cell surface in a reaction called **juxtamembrane proteolysis**. Into the transmembrane fragment of notch thus generated, a second membrane-integrated protease, γ-secretase, makes a single cut. By this **regulated intramembrane proteolysis** or S3 cleavage, the cytoplasmic portion of notch, NICD (notch intracellular domain), becomes detached into the cytoplasm, acting as a second messenger in that it binds the NFκB-related transcription factor CSL. As a result, CSL is set free from a complex with transcriptional co-repressors and now can stimulate the expression of several genes involved in developmental processes. NICD and CSL interact with each other through so-called ankyrin repeats (see Sidebar 13.1).

13.1.2 ADAM proteases and γ-secretase

The S2 cleavage of notch is catalyzed by two proteases, ADAM10 and ADAM17. They represent the family of ADAM proteases (with more than 30 isotypes) found in all animal species but not in plants, yeast, and prokaryotes. All are characterized by a single transmembrane domain and an extracellular disintegrin and metalloprotease (ADAM) domain. The active center contains a zinc ion. The disintegrin domain, a sequence of about 70 amino acids, interacts with integrins of neighboring cells and provides ADAM proteases with the properties of **cell adhesion molecules** (originally, this domain was found in certain snake venoms, which impair blood clotting by blocking integrin interactions required for platelet aggregation). This role of ADAM proteases is best exemplified by the sperm-derived protein fertilin (**Figure 13.2** and Section 7.2.3).

Another specific function of ADAM 10 and ADAM 17 is **ectodomain shedding**. This is the proteolytic detachment of peptides and proteins from the extracellular domains of membrane-bound precursor proteins (see Figure 13.2). Ectodomain shedding is not restricted to notch ligands but provides an important step in intercellular communication. Typical examples are provided by the proteins of the EGF, ephrin, and tumor necrosis factor α (TNFα) families that, upon detachment from

Figure 13.1 Notch signaling cascade The transmembrane receptor notch consists of two proteins linked by Ca^{2+}-dependent noncovalent binding. Upon interaction with the membrane-bound (or proteolytically released) ligand DSL notch is cleaved in two steps by (1) ADAM protease and (2) the intramembrane protease γ-secretase. The cytoplasmic fragment NICD (notch intracellular domain) thus generated activates the transcription factor CSL by heterodimerization via ankyrin repeats (see inset; α-helices are symbolized by cylinders, β-sheets by arrows). In reality notch exists as a homodimer rather than (as shown here for the sake of simplicity) as a monomer. The stacked discs represent EGF-like repeats.

Figure 13.2 Activation and mode of action of ADAM proteases The protease activity of ADAM proteins is stimulated by the proteolytic detachment (P1) of an inhibitory prodomain (shown in color) converting pro-ADAM into ADAM (reaction 1). Upon additional activation by phosphorylation of the intracellular domain, ADAM functions as a "sheddase," releasing a signaling protein (here a TNFα trimer, gray circles) from a membrane-bound precursor (reaction 2). By a proteolytic release (P2) of the metalloprotease domain MpD, certain ADAM isoforms may become cell adhesion molecules, such as the sperm-derived fertilin (an ADAM1–ADAM2 heterodimer, reaction 3) that, upon dimerization, interacts with integrins of neighboring cells (left; see Section 7.2.3). Whether the removal of MpD is a general condition of ADAM to act as a cell adhesion molecule or is restricted to certain subtypes is not known. MpD (gray cylinder), metalloprotease domain with an essential zinc ion; Pro (colored cylinder), inhibitory prodomain; DisD (gray ellipse), disintegrin domain; C, cysteine-rich domain; E, EGF-like repeats; T, transmembrane domain.

Figure 13.3 γ-Secretase complex The membrane-bound protease γ-secretase is composed of the catalytic subunit presenilin (with the site of proteolytic autoactivation in the third cytoplasmic loop and the catalytic center shown in teal), the regulatory subunits nicastrin and PEN2, and the multipass transmembrane protein APH1, which probably functions as a scaffold. The extracellular domain of nicastrin is heavily glycosylated, as indicated by the dotted structures.

membrane-bound precursors, mediate paracrine effects. Moreover, the extracellular domains of many transmembrane receptors are released, as we have seen for notch. While notch is activated this way, other receptors may become inactivated. Moreover, upon detachment their extracellular domain may function as a decoy, intercepting the ligand.

Like most proteases, ADAM proteases are under strict control to avoid potentially dangerous proteolysis by mistake. To this end, the enzymes are produced as zymogens that are kept silent by an intramolecular interaction of a cysteine residue in the amino (N)-terminal prodomain with an essential zinc ion in the catalytic sequence. This interaction is known as a cysteine switch. For activation the prodomain is removed by autoproteolysis or by a specific protease referred to as proprotein convertase. In addition, a series of inhibitory proteins known as tissue inhibitors of metalloproteases have been identified. Finally, the cytoplasmic tails of ADAM proteins contain various recognition motifs for intracellular signals that most probably are involved in the regulation of protease and cell adhesion activities. In fact, ADAM-catalyzed ectodomain shedding is induced by a large number of factors that activate protein kinase C, MAP kinase cascades, and other signaling pathways. An important example of this inside-out signaling has been discussed in Section 5.8.3: transactivation of EGF receptors by G-protein-coupled receptors that activate ADAM proteases, releasing EGF from membrane-bound precursor proteins.

Disregulation of ADAM proteases contributes to various severe diseases including asthma, cardiac hypertrophy, and rheumatoid arthritis. Moreover, an overexpression of ADAM10, -12, and -17 correlates with increased aggressiveness and poor prognosis of many cancers. One reason for this seems to be an excessive release by ectodomain shedding of tumor-promoting growth factors such as EGF and its relatives.

The intramembrane protease that catalyzes the S3 cleavage of notch and the release of the transcriptional regulator NICD is called **γ-secretase**. This enzyme is a membrane-bound multiprotein complex consisting of a catalytic subunit known as presenilin and three regulatory transmembrane proteins, called nicastrin, APH1 (originally discovered in a maldeveloped anterior pharynx of *C. elegans*), and PEN2 (presenilin enhancer 2). The catalytic subunit has eight transmembrane domains, with domains 6 and 7 containing the catalytic center (**Figure 13.3**). Like most proteases, γ-secretase exists as an inactive zymogen that, upon demand, is activated by limited autoproteolysis, in this case by a cut in the third cytoplasmic loop of presenilin. In transmembrane domains 6 and 7 this cleavage exposes two catalytically competent aspartate residues that were hidden before inside the protein. These residues contribute to the formation in the membrane of a hydrophilic cavity that is required for substrate binding. In the absence of the regulatory subunits presenilin becomes rapidly degraded. The signaling pathways (including protein phosphorylations and ubiquitylations) along which γ-secretase activity is fine-tuned are still not fully understood. Besides notch, many additional substrates of γ-secretase are known and homologs of γ-secretase are also found in unicellular eukaryotes and plants. This indicates an important—yet mostly elusive—role of these enzymes in cell physiology and cellular signaling.

13.1.3 Protease-mediated signaling and Alzheimer's disease

γ-Secretase is a key enzyme not only of notch signaling but also of Alzheimer's disease. Owing to an ever-increasing life span, this age-dependent dementia has become a medical and social problem of top priority, at least in highly developed countries. Research on the relatively rare hereditary or familial form of Alzheimer's disease, comprising about 5% of all cases of Alzheimer's disease, has led to the characterization of two genes encoding two closely related proteins called **presenilins**. Overactivation of these proteins because of gene mutation is the primary cause of the familial form of the disease. It is now generally accepted that presenilins represent two isoforms of the catalytic subunit of γ-secretase (see Figure 13.3).

What is the role of γ-secretase in Alzheimer's dementia? In the course of the disease, brain neurons die because insoluble protein aggregates accumulate between the cells. As shown in **Figure 13.4**, these **β-amyloid plaques** are generated by a two-step proteolytic cleavage of the membrane protein APP (β-amyloid precursor protein; its mutation also may cause the familial form of the disease). In the first step, a transmembrane protease called **β-secretase** removes the N-terminal extracellular domain of APP, while in a second step, the membrane-bound fragment becomes dissected by γ-secretase into two additional fragments. One of these fragments, APP intracellular domain, closely resembles the notch fragment NICD in that it heterodimerizes with two transcription factors, Fe65 and Tip60, thus controlling a series of genes such as that of the transcription factor NFκB, which becomes repressed. However, in contrast to the notch system, a primary signal inducing the proteolytic cleavage of APP (that is, the APP ligand) is not known.

The second fragment produced by γ-secretase, called **β-amyloid peptide** or **APβ**, is released into the extracellular space, where it may aggregate into the dreaded β-amyloid plaques. An especially aggressive variant, APβ42, with 42 instead of 40 amino

Figure 13.4 Formation of β-amyloid plaques in the course of Alzheimer's disease The transmembrane protein APP (β-amyloid precursor protein) is degraded in two subsequent steps by β- and γ-secretases (reactions 1 and 2). The cytoplasmic fragment AICD (APP intracellular domain) acts as a transcriptional regulator, whereas the extracellular fragment APβ (β-amyloid peptide) may aggregate into β-amyloid plaques, causing neuronal cell death and Alzheimer's dementia. α-Secretase (an ADAM10 protease) cleavage of APP within the APβ-domain (gray cylinder) prevents β-amyloid formation.

acids, predominates in the brain of patients with Alzheimer's disease; because of a more hydrophobic carboxy (C)-terminus, this variant has a greater tendency to oligomerize.

An important difference from notch is that β-secretase is not an ADAM-type protease. In contrast to β-secretase, ADAM proteases (also called α-secretases) have been shown to cleave APP *within* the APβ sequence and thus preclude β-amyloid formation. In fact, experimental overexpression of ADAM10, the major α-secretase in brain, protects mice from amyloid deposition and neurological disorders resembling Alzheimer's disease. An inhibition of β- and γ-secretases appears to be a promising approach for Alzheimer therapy. However, only β-secretase inhibitors are still subject of clinical trials, whereas γ-secretase inhibitors turned out to be too toxic.

In the course of Alzheimer's disease, insoluble protein aggregates are deposited both outside and inside the cell. The intracellular deposits consist of bundled microfibrils called degeneration fibrils or neurofibrillar tangles that lead to deformations of the cytoskeleton and, thus, massively impair secretory processes such as neurotransmitter release as well as cell motility. The production of these aggregates is due to a malfunction of one of the many microtubule-associated proteins (MAPs, not to be confused with MAP kinases) that regulate the formation and degradation of the microtubule cytoskeleton. This MAP species, called **Tau protein**, has been found to be hyperphosphorylated and interlinked by disulfide bonds in neurons of patients with Alzheimer's disease. One Tau kinase responsible for the pathological phosphorylation is identical with the cyclin-dependent kinase CDK5 (see Section 12.3). Under the conditions of Alzheimer dementia and other neurodegenerative diseases, this enzyme is hyperactive since its regulatory subunits p35 and p39 have become inactivated by partial proteolysis. Another Tau kinase involved in Alzheimer's dementia is glycogen synthase kinase 3 (see Section 9.4.4). Such kinases catalyze proline-directed phosphorylation of serine and threonine residues. Interestingly, the effect of Tau (and of APP) depends on the conformation of the peptidyl-prolyl bonds adjacent to the phosphorylation sites: only when these bonds are *cis* do pathological protein aggregates form, whereas the *trans* proteins fulfill normal physiological purposes. One reason for this is that phosphorylated *cis*-Tau is resistant to dephosphorylation by phosphatases such as PP2A that are *trans*-specific. As explained in Section 8.5.1, peptidyl-prolyl bonds adjacent to phosphorylation sites undergo *cis–trans* isomerization catalyzed by the peptidyl-prolyl *cis/trans* isomerase PIN1. This enzyme is down-regulated in neurons of patients with Alzheimer's disease. In contrast, experimental PIN1 overexpression protects animals from amyloid deposition and brain damage. Such findings not only emphasize the important role of enzymes like PIN1 in cellular signal processing but also indicate new opportunities for diagnosis and therapy. Alzheimer's dementia is the prototype of a disease thought to be caused by protein misfolding and accumulation of insoluble protein aggregates. For other examples see **Sidebar 13.2**.

Sidebar 13.2 Protein misfolding disorders and protein deposition diseases Fibrillary protein deposits resulting from protein misfolding and misassembly and resembling the β-amyloid plaques of Alzheimer's disease are found in the final stages of a variety of diseases that are extremely difficult to treat. Like β-amyloid, the aggregates are insoluble, have a steel-like strength, and are widely resistant to proteolytic degradation. They strongly impair cellular functions, finally leading to cell death. Such amyloids [the name refers to a starch (amylose) -like appearance of the aggregates] can derive from quite different proteins, mostly because of an acquired or genetically predetermined misfolding. Since protein misfolding and misassembly normally are prevented by molecular chaperones; the corresponding diseases are certainly due to severe disturbances of the cellular chaperone system such as the unfolded protein response (see Section 9.3.2).

Examples of mutated proteins involved in familial (hereditary) forms of disease are huntingtin, a neuronal protein of unknown function forming aggregates in brain neurons of patients suffering from **Huntington's disease**; superoxide dismutase 1, a redox enzyme found as amyloid in motor neurons of patients afflicted with **amyotrophic lateral sclerosis**; and cystic fibrosis transmembrane conductance regulator (CFTR), a chloride channel found to be aggregated in certain types of **cystic fibrosis**.

Examples of proteins undergoing an acquired misfolding and aggregation are β-amyloid and α-synuclein, a cytoplasmic neuronal protein found to form aggregates (called Lewy bodies) in the degenerating brain areas of patients suffering from **Parkinson's disease**. Moreover, for a century amyloid deposits in the pancreas have been known to be a hallmark of **type 2 diabetes**. They are due to an acquired misfolding of **amylin**, a hormone-like peptide that is co-secreted with insulin, suppressing the emptying of the stomach and producing a feeling of satiety and thirst. As one of the pathogenic factors of diabetes, amylin currently gains considerable interest in basic and applied medical research.

The mechanism of spontaneous aggregation has been studied in detail for the prion protein PrP, providing another example of acquired misassembly. **Prions** (proteinaceous infectious particles) are infectious proteins causing **transmissible spongiform encephalopathies**. The bovine form BSE, better known as mad cow disease, and its human counterpart, **Creutzfeldt–Jacob disease (kuru)**, have become notorious in the past.

The correctly folded cellular protein PrPC is abundant in the central nervous system and in hematopoietic cells, where it is found to be bound to membranes via a glycosylphosphatidylinositol anchor (see Section 4.3.1). Its physiological function is still widely elusive. Under certain conditions it can change its helical structure into a conformation characterized by stacks of β-sheets. This form, called PrPSC (SC stands for scrapie, the corresponding disease in sheep), has a high tendency to form amyloid aggregates. Moreover, PrPSC catalyzes the conformational change of native PrPC; it is an infectious agent. The autocatalytic process, resembling crystallization, starts with a seed of a few aggregated PrPSC molecules, which serve as a nucleus, inducing misfolding and aggregation of PrPC.

In addition to pathological formation, functional amyloid formation is also known. It facilitates the generation of an extracellular matrix in bacterial colonies, mating of yeast gametes, spore formation in fungi, blood clotting and melanin biosynthesis in humans, and protection of eggs of insects and fishes. Functional amyloidogenesis in humans provides a serious obstacle in the development of anti-amyloidogenic drugs.

13.1.4 Other examples of protease-mediated signaling

Notch and APP signaling are representative for other examples of signal transduction by γ-secretase. The enzyme always cleaves a transmembrane protein, releasing the intracellular domain that functions as a modulator of gene transcription. An especially interesting case is provided by the partial proteolysis of **ErbB4** (**Figure 13.5**). A member of the EGF receptor family, this transmembrane protein is a receptor tyrosine kinase (also known as HER4, human epidermal growth factor receptor 4; see Section 7.1.1). The endogenous ligand of ErbB4 is the EGF-related growth factor heregulin. Like other receptor Tyr kinases, ErbB4 controls numerous signaling pathways by means of protein phosphorylation and protein–protein interactions. For ErbB4 it has been shown that a receptor Tyr kinase still possesses additional mechanisms of signal transduction. Thus, the activated (liganded and autophosphorylated) receptor becomes a substrate of γ-secretase. Its cytoplasmic Tyr kinase domain generated by intramembrane proteolysis migrates into the nucleus, where it most probably modulates the activity of transcription factors by phosphorylation or direct interaction. In analogy to notch and APP, a shedding of the extracellular domain precedes the intramembrane proteolysis of ErbB4. The protease catalyzing this step is ADAM17. Vice versa, the ligand binding domain of a receptor may become detached and intercept its diffusible ligand, thus preventing the ligand from interacting with its receptor. Along such a pathway, the activity of growth hormone is attenuated, for instance.

Signaling peptides are also released from proteins integrated in *intracellular* membranes. An example is provided by cholesterol metabolism. Upon a shortage of cholesterol, genes encoding key enzymes of **cholesterol biosynthesis** become activated. In their promoters these genes have a sterol response element (SRE) that interacts with a transcription factor SREBP (SRE binding protein). An inactive

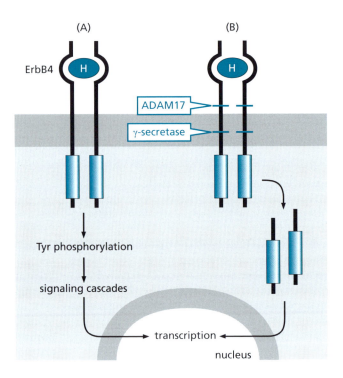

Figure 13.5 Signal transduction by the receptor tyrosine kinase ErbB4/HER4 Upon activation by the ligand heregulin (H), the homodimeric receptor ErbB4 controls gene transcription either (A) along a phosphorylation-dependent pathway or (B) by a proteolytic release of the cytoplasmic domains by ADAM17 and γ-secretase; the released domains then migrate into the nucleus. Colored cylinders symbolize the Tyr kinase domains. For more details, see text.

precursor of this factor is a protein with two membrane-spanning domains, which at sufficient cholesterol supply is stored in the endoplasmic reticulum together with another transmembrane protein acting as a cholesterol sensor. A shortage of cholesterol causes these proteins to translocate to the Golgi membranes, where the active SREBP becomes released by juxta- and intramembrane proteolysis catalyzed by the proteases S1P (site-1 protease) and S2P (site-2 protease) and migrates into the nucleus (**Figure 13.6**). In principle this reaction resembles the activation of notch, with S1P doing the job of ADAM and S2P that of γ-secretase. SREBP activates gene transcription with the consequence that the cholesterol level increases, the process

Figure 13.6 Regulation of cholesterol synthesis Transcription of genes containing a sterol-responsive element (SRE) and encoding enzymes of cholesterol synthesis is stimulated by the dimeric transcription factor SREBP. In the presence of cholesterol, SREBP and a cholesterol sensor protein are sequestered in the endoplasmic reticulum. Shortage of cholesterol causes a translocation of the protein complex to the Golgi apparatus, where SREBP is released from the membrane by the site-1 and site-2 proteases S1P and S2P. Cholesterol inhibits the translocation (teal pistil).

of SREBP activation becomes stopped, and upon phosphorylation by glycogen synthase kinase 3, SREBP (together with key enzymes of cholesterol biosynthesis) becomes degraded along the ubiquitin–proteasome pathway. In other words, cholesterol formation is under strict feedback control. It is easily seen that SREBP signaling is of great clinical interest since defects of this control mechanism may result in a pathological overproduction of cholesterol.

In quite a similar manner, the **transcription factor ATF6** controlling certain chaperone genes is activated. In this case the trigger is an accumulation of incorrectly or incompletely folded proteins in the endoplasmic reticulum. These proteins become rearranged into their native conformations by their chaperones to be newly synthesized. The signaling pathway, called unfolded protein response, is discussed in detail in Section 9.3.2. Taken together, these examples show that controlled intramembrane proteolysis provides a highly efficient and versatile mechanism of signal transduction, the potencies of which cannot yet be estimated down to the last detail.

Summary

- The receptor-dependent release of peptide second messengers acting as transcriptional regulators seems to be a widespread mechanism of signal transduction.

- The prototype of this signaling system is represented by notch, a transmembrane receptor the intracellular domain of which is released as a transcriptional activator upon receptor activation by ligands of the DSL family. The reaction is catalyzed by ADAM proteases and the intramembrane protease γ-secretase.

- Related enzymes play a key role in the pathogenesis of Alzheimer's dementia, where they catalyze the formation of plaque-forming β-amyloid peptide. Another pathogenic process involved in Alzheimer's disease is the deposition of insoluble aggregates of cytoskeletal proteins caused by a hyperactivation of microtubule-associated Tau protein. In plants, the hormones auxin and gibberellic acid induce the proteolytic degradation of transcriptional repressors by activating specific ubiquitin ligases (**Sidebar 13.3**).

Sidebar 13.3 Some plant hormones regulate gene expression along proteolytic pathways The plant hormones auxin and gibberellic acid are potent stimulators of growth and differentiation. To this end they activate discrete sets of genes along a quite direct but rather unusual signaling pathway. The key event consists of a proteolytic degradation of transcriptional repressors known as IAA/AUX proteins in the case of auxin and DELLA domain proteins in the case of gibberellic acid. This occurs along the canonical proteasome pathway controlled by SCF-type E3 ubiquitin ligases. As explained in Section 2.8.1, these enzymes are multiprotein complexes with F-box subunits serving as specific binding sites for the substrate proteins to become ubiquitylated (**Figure 13.7**). In the SCF ubiquitin ligases of plants involved in hormone effects, the F-box subunits interact not only with the substrate proteins such as transcriptional repressors but also with auxin or gibberellic acid. Moreover, the hormones strongly stimulate substrate ubiquitylation. In other words, these enzymes resemble hormone receptors with an intrinsic ubiquitin E3 ligase activity. The *Arabidopsis* genome seems to encode about 700 F-box proteins, which is about 10 times more than in animal cells. Thus, it is easy to imagine that the role of receptor ubiquitin ligases is not restricted to auxin and gibberellic acid but may be a more general mechanism of signal transduction in plants. Indeed, jasmonic acid, a plant wound hormone, may use such a pathway. Other plant hormones interact with transmembrane receptors with His kinase (see Section 3.6.1) or Ser/Thr kinases activities (see Sidebar 6.1).

Figure 13.7 Mechanism of gene activation by the plant hormones auxin and gibberellic acid The hormones stimulate gene expression by triggering proteolytic degradation of transcriptional repressors (R, gray ellipse). To this end they interact with individual F-box subunits of SCF-type E3 ubiquitin ligases (multiprotein complex shown in gray; cullin is an adaptor protein), which specifically promote the ubiquitylation of R catalyzed by the ubiquitin transferase E2 (shown in color). The figure shows a generalized scheme. In reality, each hormone has its own set of E3 ligases and repressors.

13.2 Signal-controlled suicide of cells

A constant tissue mass (as a condition of constant tissue function) is due to a dynamic equilibrium between cell gain and cell loss adjusting the birth rate of cells (amounting to approximately 10^{12} cells per day in adult humans) precisely to the death rate. This equilibrium becomes transiently shifted in the course of developmental and growth processes as well as during tissue repair, with each change being controlled by a complex system of stimulatory and inhibitory signals. Such control mechanisms are permanently disturbed in tumors.

Controlled cell death occurs through suicide. To this end the cell disposes of three mechanisms, apoptosis, necrosis, and autophagy. **Apoptosis** is an everyday mechanism of cell death. It plays a central role in the maintenance and repair of tissues. As a kind of a cleansing operation, apoptosis removes potentially dangerous cells, such as virus-infected or cancer cells, as well as elderly cells that have become superfluous for the organism or, for any possible reason, have escaped from their tissue or have lost their function otherwise. Likewise, apoptosis is important for tissue formation, since in the course of embryonic organ development normally more cells become produced than are finally needed in the mature tissue. The excess dies by apoptosis. An instructive example is provided by the nematode *C. elegans*, a favored model of apoptosis research. During its development to the mature animal consisting of 959 somatic cells, exactly 131 cells undergo apoptosis under the control of 10 different *ced* (cell death) genes. Another example is insect metamorphosis, but vertebrate development may also be considered: for instance, the metamorphosis of amphibians or the formation of hand and foot limbs. A particularly high excess of cells is required for the morphogenesis of the nervous system to establish synaptic contacts. As soon as this has been accomplished, more than 50% of the embryonic neurons become superfluous and die by apoptosis. Especially stricken by death are those cells that failed to made contact with their target tissue, which is a source of survival factors, the neurotrophins. Such a mechanism prevents incorrect "wiring" of the nervous system.

These studies have shown that apoptosis is triggered either by specific death signals or by a lack of survival signals or trophic factors as well as by related stress situations. In its strongest form the hypothesis of trophic factors postulates somatic cells to undergo suicide automatically in the absence of survival signals or when they become neglected.

Apoptosis is characterized by fragmentation of the nucleus and degradation of proteins and nucleic acids. However, the dying cell does not disintegrate but shrinks, finally to becomes digested by phagocytes. These are attracted and activated by **find-me** and **eat-me signals** released from the apoptotic cell. Sphingosine-1-phosphate and lyso-phosphatidylcholine are typical find-me signals that interact with corresponding membrane receptors of the phagocyte, while phosphadidylserine exposed at the surface of the dying cell activates—as an eat-me signal—the engulfment machinery of the phagocyte. Apoptosis generally is a non-inflammatory process. Nevertheless, it can be accompanied by inflammation, in particular in the course of pathological processes.

Necrosis is primarily a defense reaction caused by stress factors. It always evokes inflammation since in contrast to apoptosis, it results in a complete disintegration of cells releasing pro-inflammatory factors.

The term **autophagy** describes the self-eating of cells, where cellular components including organelles are digested by lysosomal enzymes. By providing luxury material for vital metabolic processes, it is a survival strategy in stress situations such as starvation (see Section 9.4.3). Normally, it does not result in cell death.

13.2.1 Caspases are the executors of apoptosis

Apoptosis is due to a destruction of DNA and vital proteins resulting in the decay of the cell. The enzymatic apparatus for the execution of apoptosis is found in a latent form in every cell. Upon demand, it becomes activated by exogenous and endogenous signals that interact either with membrane receptors (death receptors) or mitochondria. The major executors of apoptosis are several proteases, which are interconnected in series to form a proteolytic cascade (**Figure 13.8**). They are called caspases (cysteine/aspartate-specific proteases).

Caspases constitute a family of endopeptidases (13 members in humans) showing a peculiarity: in contrast to most other proteases, they hydrolyze peptide bonds C-terminally to aspartate residues, thus degrading a large number of vital proteins including those responsible for signal processing, cellular structures (for example, the nuclear membrane), DNA repair, and transcriptional regulation. To complete the arsenal of cell degradation, caspases stimulate other hydrolytic enzymes such as the **caspase-activated DNase (CAD)** that cuts DNA between the nucleosomes, yielding fragments of $n \times 190$ nucleotides that can be visualized by gel electrophoresis as a characteristic apoptotic ladder. As a potentially highly dangerous enzyme, CAD is always expressed together with an inhibitor protein, **ICAD**, that forms a complex with CAD. Caspases destroy ICAD, thus releasing the active DNase, which is caspase-resistant.

Potentially as dangerous as CAD are the caspases themselves. Therefore, they are subject to strict control, and several safeguards against accidents are installed in the system. In the absence of input signals, all caspases are present as inactive zymogens or procaspases, and in addition, they are blocked by specific caspase inhibitors (see Section 13.2.4).

When apoptosis is triggered, the caspases start to activate each other by partial proteolysis, forming a **caspase cascade** that to some extent resembles the blood-clotting

Figure 13.8 Two major pathways leading to apoptosis Apoptotic signals either stimulate membrane-bound death receptors along an extrinsic pathway or damage mitochondria along an intrinsic pathway. Both pathways lead to an activation of caspase cascades. The destruction of vital cellular components, such as proteins and DNA, finally results in cell death.

and complement cascades. Along such proteolytic cascades the initial signal becomes amplified enormously.

Like ADAM proteases, procaspases are kept silent by the interaction of the N-terminal prodomain with the catalytic center. For activation, this domain must be removed proteolytically. At the same time, a second proteolytic cut is made in the C-terminal half of the molecule, yielding a small and a large fragment (**Figure 13.9**). The active caspase is a heterotetramer composed of two small and two large fragments.

Caspase activation follows the mode of trans-autoactivation: each enzyme activates the next one. Such a mechanism requires many enzyme molecules to join and is called, therefore, **proximity-induced autoactivation**. This is an important principle of signal transduction, being valid also for other processes such as phosphorylation cascades. The aggregation of the proteins is facilitated by corresponding interaction domains as well as by adaptor and scaffolding proteins.

Aggregation of caspases is induced by pro-apoptotic signals acting on procaspases 2, 8, 9, and 10 (see Figure 13.8). Among the procaspases, only these species have in

Figure 13.9 Domain structure of apoptosis-regulating caspases The initiator or upstream procaspases 8 and 9 as well as the effector or downstream procaspase 3 and an active caspase heterotetramer are shown (large colored rectangle, large catalytic subunit; gray rectangle, small catalytic subunit; arrows, sites of proteolysis). The structure of procaspase 2 resembles that of procaspase 9. DED (small colored rectangle), death effector domain (B).

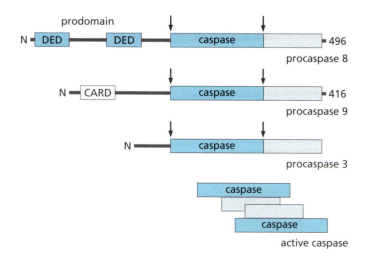

their prodomains sequences such as DED (death effector domain, in procaspases 8 and 10) or CARD (caspase-recruiting domain, in procaspases 2 and 9) capable of interaction with receptor and sensor proteins of pro-apoptotic signals (see Figure 13.9). Called initiator (or upstream) caspases, caspases 2, 8, 9, and 10 are placed, therefore, at the top of the caspase cascade, inducing the activation of the effector caspases 3, 6, and 7 placed downstream. There are other caspases that are not involved in apoptosis but play a role in inflammatory processes (**Sidebar 13.4**).

13.2.2 Receptor-induced apoptosis or the kiss of death

Several intercellular signal proteins are specialized to induce apoptosis. They are counted among the TNF family of cytokines, acting in both a paracrine and a juxtacrine manner. Their transmembrane receptors constitute a subfamily of TNF receptors (see Section 6.3.1). These death receptors are characterized by death domains (DDs; see **Sidebar 13.6**) localized in their cytoplasmic portion.

Thus far, six such receptors have been identified in mammals. Their prototype is **Fas**, also known as CD95 or Apo-1. Fas is strongly expressed in lymphocytes but is also found in other tissues. Its ligand, **FasL**, is a homotrimeric transmembrane protein expressed in particular by cytotoxic T-lymphocytes and natural killer cells, which in this way kill Fas-positive cells.* This mechanism plays an important role for elimination of virus-infected and tumor cells as well as in the autoregulation of the immune system. Autoimmune diseases are prevented and excessive antigen-specific T cells are destroyed in the subsiding phase of an immune reaction, preserving a small population of FasL-resistant memory cells.

As characteristic for TNF receptors, ligand binding results in the recruitment of adaptor proteins which together with other signaling proteins assemble in large signaling complexes. Thus, Fas activates the caspase cascade via the adaptor protein **FADD** (Fas-associated death domain protein) recruited by a homotypical interaction between DD motifs (**Figure 13.10**). FADD is the only protein that contains, in addition to DD, a DED motif (see Sidebar 13.6) that interacts homotypically with the corresponding domain of initiator procaspases 8 or 10, the two caspases that contain DEDs. (Caspase 8 is also known in the older literature as FLICE, Fas-ligand-activated ICE-like protease,

---

* Cytotoxic T cells and natural killer cells may kill virus-infected cells not only along the Fas pathway but also by injecting the content of secretory granules. A component of the secretion, the protein **perforin**, forms channels in the cell membrane, thus causing a breakdown of the membrane potential and the osmotic pressure. Another component, the protease **granzyme B,** triggers (like an initiator caspase) the caspase cascade; it activates, in addition, the pro-apoptotic protein Bid and destroys the DNase inhibitor ICAD.

Figure 13.10 Induction of apoptosis by the death receptor Fas/Apo1 The trimeric receptor is activated by interaction with a trimeric ligand exposed at the surface of a neighboring cell and stimulates, via the adaptor protein FADD, procaspase 8 and, thus, the caspase cascade leading to cell death. The direct downstream substrates of caspase 8 are caspase 3 and the mitochondrial "BH3-only" protein Bid. The inhibitor FLIP, a pseudoprocaspase 8, competes with procaspase 8 for the binding sites of FADD. Note the homotypical, noncovalent interactions between the death domains (black rectangles) and the death effector domains (teal rectangles), and the aggregation of procaspase 8 "chains". Colored cylinders, active center of procaspase 8. To the right, a long and a short FLIP protein is depicted. The gray cylinder represents a pseudocaspase domain.

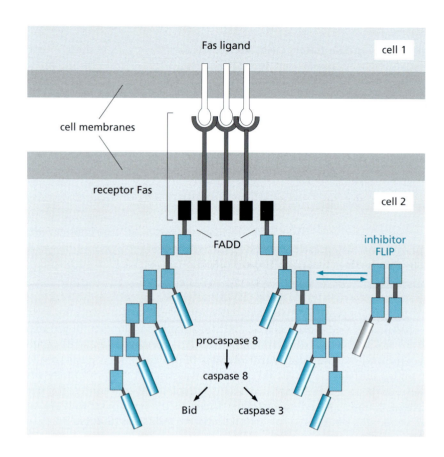

referring to its relationship with ICE [caspase 1]. Caspase 10 is restricted to humans, having developed only recently by a doubling of the caspase 8 gene.)

Both procaspases thus become activated and trigger the caspase cascade. For this purpose a rather large **death-inducing signaling complex (DISC)** of many receptor molecules is formed, assembling enough molecules of procaspase 8 or 10 for proximity-induced autoactivation. DISC formation is facilitated by the clustering of membrane lipids in rafts, with ceramides playing an important role in this process. This may explain, at least partly, the pro-apoptotic effect of such lipids, as mentioned in Section 4.4.5. For certain cell types (type I cells) DISC formation is sufficient to induce apoptosis, whereas in other cell types (type II cells) it has to be amplified by the mitochondria-based mechanism. In the latter case, caspase 8 activates the protein Bid by partial proteolysis (see Section 13.2.5).

The exact stochiometry of DISC is still a matter of dispute. Most frequently, a Fas trimer seems to bind two FADDs, and one FADD may recruit up to nine procaspase 8 molecules (see Figure 13.10). The latter interact with each other via their death execution domains. Such a procaspase "chain" may facilitate proximity-induced autoactivation of caspase 8.

As mentioned above, apoptosis plays an important role in the embryonic development of the nervous system. Here cell death is induced by (among others) the **low-affinity p75-nerve growth factor receptor**; however, this receptor stimulates the de novo synthesis of pro-apoptotic proteins rather than triggering the caspase cascade directly. Ligands of the receptor are neurotrophins such as the nerve growth factor, a protein that does not belong to the TNF family. How can a growth factor, typically a survival signal, induce cell death? The answer lies in the expression of two different receptor types: in contrast to other TNF receptors, the p75-nerve growth factor receptor is able to establish a heteromeric complex with a receptor Tyr kinase, TrkA (see Section 7.1.1). Only via this **high-affinity nerve growth factor receptor** does nerve growth factor function as a survival signal. Opposing effects have been described also for other death signals and receptors. So, depending on the physiological context, Fas may either trigger apoptosis or co-stimulate mitogenic activation of T cells, thus providing an especially clear example of the principle that the meaning of a signal always depends on the functional state of the receiver cell (for mechanistic details on the mitogenic effect of death receptors, see Section 6.3.1).

For some receptors of the TNF receptor family, the pro-apoptotic effect even recedes into the background when compared with other effects. An example is the TNF receptor type 1 that predominantly stimulates the NFκB signaling cascade, a typical survival pathway, rather than apoptosis (see Section 6.3.1). The same holds true for the death receptors DR3 and DR6. Other members of the TNF receptor family contain incomplete or no death domains and are unable to induce apoptosis. They may act similarly to dominant-negative mutants and dampen pro-apoptotic signaling.

13.2.3 The switch between apoptosis, necrosis, and survival

Apoptosis is only one way leading to controlled cell death. Other suicide mechanisms are summarized under the collective term **regulated necrosis**. Like apoptosis, regulated necrosis is a strictly signal-regulated process. However, it does not involve caspase activation and differs from apoptosis also in morphological aspects. Regulated necrosis must be distinguished from accidental necrosis which is due to unspecific injury and seems to proceed in an uncontrolled manner.

Regulated necrosis is primarily a response to cellular stress such as bacterial and viral infection, mitochondrial damage, overproduction of reactive oxygen species, DNA damage, and Ca^{2+} overload. It probably has evolved as a defense strategy against such pathogens, in particular certain viruses, that subvert apoptosis of infected cells by inhibiting caspases. In contrast to apoptosis, regulated necrosis results in membrane disruption. Intracellular components are released that as damage-associated molecular patterns (DAMPs, see Section 6.3.2) provoke an inflammatory response. Like apoptosis, regulated necrosis may be stimulated either by cell surface receptors or along cytoplasmic signaling pathways. A prototype of receptor-dependent necrosis referred to as **necroptosis** is induced by tumor necrosis factor via the TNF receptor 1 (**Figure 13.11**).

Figure 13.11 Survival and suicide mechanisms induced by the TNF receptor 1 Upon interaction with its ligand the trimeric receptor TNFR1 activates a pro-inflammatory survival pathway by assembling a complex I that consists of the adaptor proteins TRADD, TRAF2, RIPK1, and IAP1 and -2 (left). When TRAF2 and RIPK1 are poly-ubiquitylated (Ub, teal dots) by TRAF2, IAP1 and -2, and an additional E3 ubiquitin ligase, LUBAC (not shown), they stimulate MAP kinase (MAPK) and NFκB modules. To induce cellular suicide, complex II (DISC) is formed (right). This requires a de-ubiquitylation of TRAF2 and RIPK1, caused by, for instance, a down-regulation of IAP1 and -2. The recruitment of FADD enables complex II to activate procaspase 8 (procasp 8) and apoptosis, while necrosis is suppressed by caspase 8-catalyzed inactivation of RIPK1. A complex (ripoptosome) of inactive RIPK1, RIPK3, FADD, and procaspase 8 (procasp 8) may be released into the cytoplasm. This complex may contain, in addition, the inhibitor FLIP (not shown). Upon inhibition of caspase 8, the ripoptosome is changed in a necrosome consisting of active (phosphorylated, P) RIPK1, RIPK3, FADD, and MLKL (mixed lineage kinase like protein). Phosphorylation of MLKL results in regulated necrosis. Death domains are symbolized by black rectangles.

The executor of necroptosis is the protein **MLKL** (mixed lineage kinase like protein), which aggregates and causes membrane rupture when phosphorylated. MLKL phosphorylation is catalyzed by the Ser/Thr-kinase RIPK3, provided this enzyme has undergone autophosphorylation, and is aggregated in a filament-like structure. Activation of RIPK3 occurs only in a complex with autophosphorylated RIPK1. In fact, RIPK1 is the master switch for changing between survival, apoptosis, and necroptosis. At the TNF receptor 1 it is bound via death domains to the TRADD–FADD adaptor complex; other death receptors interact directly with RIPK1. When polyubiquitylated by the E3 ubiquitin ligases IAP1 and -2, receptor-bound RIPK1 acts as a scaffold for MAP kinase and NFκB modules, whereas de-ubiquitylated RIPK1 is required for the recruitment of FADD and caspase 8 and apoptosis. This complex, the **ripoptosome**, may be released into the cytoplasm and provides a major platform for an activation of the caspase cascade and apoptosis. In the ripoptosome the necrotic pathway is blocked because the kinase activity of RIPK1 is inhibited by caspase 8. Under stress or upon treatment with SMAC mimetics (see Section 13.2.2) the situation may change because caspase 8 may be inhibited or deleted. Consequently, RIPK1—now rescued from inhibition—phosphorylates itself, recruits RIPK3, and necrosis is induced. The cytoplasmic complex of active kinases RIPK1 and RIPK2, and their substrate protein MLKL, is referred to as the **necrosome**. Along analogous pathways, ripoptosome and necrosome formation is induced by other death receptors, but also by interferon receptors, toll-like receptors TLR3 and TLR4 (see Figure 6.9, Section 6.3.3), and the T-cell receptor. For more on RIP kinases and their function, see **Sidebar 13.5**.

Cell death and autophagy

Autophagy, as explained in Section 9.4.3, differs from apoptosis and necrosis in that it means "self-eating" of cellular components rather than suicide of whole cells. It enables the cell to get rid of defective and outdated material that, when accumulating, could have a damaging effect. The material is degraded and the degradation products are recycled. As a cytoprotective mechanism, autophagy is essential for the health and maintenance of cells and their survival under stress situations such as starvation. Thus, on a first glance, cell death seems to be an opponent of autophagy. Indeed, autophagy prevents cell death and cell death inhibits autophagy. A major reason for the inhibition of cell death by autophagy seems to be the removal of pro-apoptostic proteins and damaged mitochondria (mitophagy, see Section 9.4.3)

Sidebar 13.5 RIP kinases In humans, seven different RIP kinases have been found. They are Ser/Thr-directed and are closely related to the IRAK protein kinases, which transduce interleukin 1 signals (see Section 6.3.2). While RIPK1 and RIPK3 are engaged in signal transduction via death receptors, RIPK2 contains a CARD domain. It transmits signals from cytoplasmic NOD-like receptors which recognize bacterial components and evoke similar pro-inflammatory effects as the TNF receptor 1. This pathway, described in Section 6.3.2, plays an important role in nonspecific defense of pathogenic microorganisms.

RIPK1 is a central switch for the decision between life and death in general, and between different forms of survival and suicide in particular. As such the enzyme plays a key role in developmental processes and defense reactions. RIPK1 acts both as a scaffold (for the induction of cell proliferation and apoptosis) and as a protein kinase (for the induction of necrosis).

RPK1 is essential, since corresponding gene-knockout animals die immediately after birth owing to a severe retardation of lymphatic system development. This is caused by a loss of the scaffold function of RIPK1, because mutant mice with kinase-deficient RIPK1 are viable and their cells respond normally to apoptotic and survival signals, but not to necroptotic stimuli.

The domain structure of RIPK1 is shown in **Figure 13.12**. Besides a kinase domain, the protein has a C-terminal death domain (not found in other RIPK species) for the homotypic interaction with death receptors or the TRADD adaptor of the TNF receptor 1. Via an RHIM domain (RIP homotypic interaction motif), RIPK1 homotypically associates with other RHIM-containing proteins such as TRIF (see Figure

6.10) or RIPK3. To activate the NFκB pathway, RIPK1 must become polyubiquitylated at lysine 377 (see Section 11.7.2). This modification is recognized by the kinase TAK1, an activator of the IKK complex of the NFκB module. For the induction of necroptosis, autophosphorylation of the kinase domain and de-ubiquitylation are required.

Necrostatin, a tryptophan derivative, is a specific inhibitor of RIPK1. It has been employed as a valuable tool in mechanistic studies.

Figure 13.12 Domain structure of RIPK1 Shown are the kinase domain, the interaction domains RHIM and DD, and the sites of phosphorylation, ubiquitylation, and cleavage by caspase 8 (teal arrows). The specific inhibitor necrostatin 1 blocks the kinase domain

which otherwise would induce apoptosis. Vice versa, an overproduction of autophagosomes that cannot be taken up any more by lysosomes may cause regulated necrosis, and the degradation by caspases of autophagy-controlling proteins (ATGs) may suppress autophagy. As yet, there is no convincing evidence for a physiological role of cell death through autophagy in vertebrates. However, in non-vertebrates, such as insects and nematodes, the suppression of autophagy (for instance by deletion of essential autophagy genes) blocks certain steps in larval development that are due to massive tissue degradation. Since an inhibition of apoptosis has no effect, cell loss seems to be due to autophagic cell death.

13.2.4 Apoptosis inhibitors protect cells from suicide

A safeguard against errant apoptosis consists of several endogenous inhibitors of caspase 8/10-induced apoptosis referred to as **FLIP** (Fas ligand-activated ICE-Like inhibitory protein, also known as I-FLICE) and **IAP** (inhibitor of apoptosis proteins). Both FLIP and IAP were originally discovered in certain viruses, which they use to block apoptosis of the infected host cell. Another anti-apoptotic protein is TUCAN (tumor-up-regulated CARD-containing antagonist of caspase nine). By means of a CARD sequence TUCAN competes with caspase 9, the effector caspase of mitochondria-dependent apoptosis (see Section 13.2.5).

The FLIP family

Three human FLIP isoforms are known: a long species (FLIP$_L$) containing an inactive pseudocaspase domain, and two short species (FLIP$_S$). Moreover, these primary transcripts are processed by proteolytic cleavage to other active forms (**Figure 13.13**). Resembling the structure of procaspase 8, all forms contain two DED domains in tandem arrangement that are essential for their function. By homotypic DED–DED interactions FLIP is recruited to death receptors or ripoptosomes, thus competing with procaspase 8. Since many procaspase 8 molecules are recruited to death receptors (see Figure 13.10), the inhibitory effect directly depends on the FLIP concentration. At lower concentrations and when processed by partial proteolysis, FLIP even may activate an pro-inflammatory pathway, the NFκB cascade. To this end, the processed isoforms p22 and p48 bind to and activate the IKK complex. Regulated

Figure 13.13 Anti-apoptotic proteins The domain structures of procaspase 8 and the apoptosis inhibitors FLIP, X-IAP, and IAP1 are shown. The number of amino acids of the human isoforms is shown to the right. The teal arrows indicate the points of proteolytic cleavage of FLIP yielding p22-FLIP and p43-FLIP. BIR, baculovirus IAP repeat; DED (small colored rectangle), death effector domain; CARD, caspase-recruiting domain.

necrosis is suppressed by FLIP$_L$-procaspase 8 heterodimers, while the short FLIP isoforms seem to promote necrosis. As yet, such apparently contradictory effects have hindered a successful clinical application of FLIP antagonists against cancer, although FLIP has been reported to be elevated in many cancers.

The IAP family

IAPs have been found in all kinds of animals and in yeast, but not in plants. In common they have a structural element referred to as BIR domain (Baculovirus IAP repeat). Another domain found in most IAPs is the RING domain conferring E3 ubiquitin ligase activity. Among the 7 human IAP species X-IAP and IAP1 and -2 (also called cIAPs) are the best-investigated isoforms (see Figure 13.13). By means of its BIR domains, X-IAP inhibits caspases 3, 7, and 9 by hindering substrate proteins from access to the active center. In contrast, IAP1 and -2 are only weak caspase inhibitors. They suppress apoptosis in a more indirect way by controlling the formation of signaling complexes at death receptors. As explained for the TNF receptor 1, the E3 ubiquitin ligase activity of IAP1 and -2 is required for the assembly of signaling complexes that transmit survival signals, whereas pathways leading to cell death are blocked. Upon genotoxic stress, for instance caused by cytotoxic anticancer drugs, IAP1 and -2 become down-regulated and cell death is induced.

The cellular opponents of IAPs are the **SMAC** proteins (second mitochondrial activators of caspases). In many tumor types IAPs are overexpressed, protecting the tumor cells from apoptosis by inhibiting caspases and promoting survival pathways. Frequently, IAP overexpression is accompanied by reduced SMAC expression. This situation correlates with cancer progression, poor prognosis, and therapeutic resistance. Novel anticancer drugs have been developed that—as SMAC mimetics—cause the degradation of IAP1 and -2 and suppress the anti-apoptotic effect of X-IAP. Consequently, caspases are rescued from inhibition. Moreover, TNF expression becomes stimulated and TNF signaling is directed to the pro-apoptotic pathway (see Figure 13.11). When combined with other treatments, SMAC mimetics are potent inhibitors of tumor growth and may have reached the clinical stage during the printing of this book.

Another opponent of IAPs is the mitochondrial protease **OMI/HtrA2** (outer mitochondrial membrane-derived inhibitor; HtrA refers to a homologous high temperature requirement protein A of bacteria). It catalyzes the proteolytic degradation of IAPs and mimicks caspase effects.

13.2.5 Mitochondria-induced apoptosis: the consequences of stress and neglect

Cell death is induced not only by endogenous signals reacting with death receptors but also by many other, rather heterogeneous, stimuli. These include DNA damage and further stress effects as well as a lack of survival factors such as mitogenic signals. This type of induced cell death has been called intrinsic apoptosis or "apoptosis by neglect." It represents the counterpart to the receptor-mediated extrinsic apoptosis or "apoptosis by instruction." The primary mediators of intrinsic apoptosis are the mitochondria. In fact, the regulation of programmed cell death is one of their major functions, and for this purpose they are equipped with a sophisticated machinery of interacting proteins.

Bcl2 proteins are key regulators of mitochondria-mediated apoptosis

The key event in mitochondria-dependent apoptosis is a permeabilization of the outer mitochondrial membrane, resulting in a release of pro-apoptotic proteins from within the intramembrane space into the cytosol (**Figure 13.14**). The structural integrity of the outer membrane is controlled by Bcl2 proteins. This protein family, with about 25 members in humans, includes anti- and pro-apoptotic proteins. It has

---

Sidebar 13.6 Death domains Protein aggregation required for the autoactivation of caspases is brought about by homotypical interactions of the sequence motifs DD, DED, CARD, and PYRIN.

DD motif. This death domain proper has been found to be encoded in 33 human genes, including those of receptors and adaptor proteins such as FADD and TRADD. A major function of the DD motif is to recruit multiprotein complexes at death receptors and some other receptors of the TNF receptor family. **FADD** exclusively transduces apoptotic signals, whereas **TRADD** controls two potentially opposing signaling pathways: the pro-apoptotic cascade, when it interacts with FADD, and the mitogenic NFκB cascade, when it interacts with the adaptor protein RIP2 (see Section 6.3.1). Under normal conditions, the receptors coupled with TRADD activate preferentially the NFκB pathway. Only when the latter is blocked may apoptosis be induced.

Alternatively to FADD, another adaptor protein, **RAIDD** (the acronym indicates that RAIDD associates with the adaptor RIP2 and is homologous to the ICE homology protein 1, which is identical with caspase 2) may interact with TRADD. RAIDD is the only protein that exhibits both DD and CARD motifs. Along this pathway, procaspase 2 becomes activated. A further adaptor protein with a DD motif is **MyD88**, which couples the protein kinases of the interleukin 1 receptor-activated kinase family (also DD proteins) and the associated NFκB module with Toll-like receptors of bacterial surface components (details in Section 6.3.3).

Other protein kinases with DD motifs are the **DAP kinases** (death-associated protein kinases). These enzymes induce apoptosis and, therefore, are tumor suppressor proteins. DAP kinases exist in several isoforms. Some of them are similar to myosin light-chain kinase, becoming activated by Ca^{2+}/calmodulin and phosphorylating myosin light chains. This type of DAP kinase is assumed to be involved

in Ca^{2+}-induced apoptosis, probably inducing the typical membrane blebbing of apoptotic cells. Recently it has been observed that DAP kinases also become activated by the pro-apoptotic lipid ceramide.

DED motif. This death execution domain, a bundle of six helices, is found in 13 human genes, including those for the initiator caspases 8 and 10, in FADD, and in the caspase inhibitors FLIP. It mediates the assembly of the DISC aggregates for death receptor-induced apoptosis (see Figure 13.9).

CARD motif. This CAspase-Recruiting Domain is related to the DED motif and has been found in 22 human proteins, including caspases 1, 2, 4, 5, 9, 11, and 12. In the course of mitochondria-induced apoptosis (see Section 13.2.4), APAF1 and procaspase 9 aggregate via CARD motifs.

For nonspecific immunosurveillance, the **NLR proteins** (see Section 6.3.2) resemble the APAF1 proteins. Their CARD motifs enable them to recruit the CARD protein **RIPK2**, thus stimulating the pro-inflammatory NFκB signaling cascade (see Section 6.3.2) and the release of the pro-inflammatory cytokines.

PYRIN motif. Close relatives of the CARD motifs are the PYRIN domains. They are restricted to vertebrates (with 19 human isoforms) and are involved in inflammatory processes in particular. The PYRIN domain (derived from the Greek prefix *pyro* for fire) was initially discovered in the pyrin protein that is produced by white blood cells and keeps the inflammatory process under control by a still ill-defined mechanism. It is found in NOD-like pathogen receptors, which play an important role in pathogen defense (see Section 6.3.3). A mutation of the corresponding gene is the cause of familial Mediterranean fever, a hereditary disease. Poxviruses paralyze the host's immune system by using PYRIN sequences found in their genome. As a result, defense reactions (inflammation) are suppressed and infected cells cannot be removed by cell death.

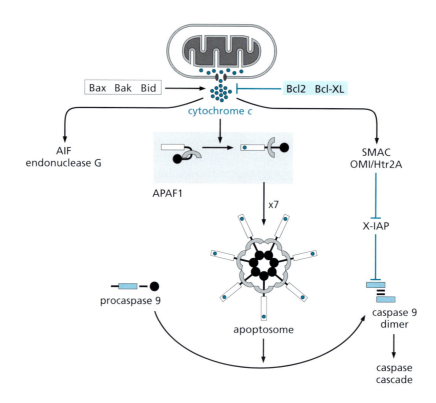

Figure 13.14 Mitochondria-mediated apoptosis Mitochondria permeabilized by the stress-sensitive proteins Bax, Bak, and Bid release several pro-apoptotic proteins. One of them, cytochrome c (colored dots), binds to the scaffold protein APAF1, causing a conformational change by which CARD motifs (black dots) and nucleotide-binding oligomerization domains (NODs, gray curved shapes) become exposed. Upon ATP-hydrolysis (not shown) the activated APAF1 oligomerizes to a wheel-like structure, the apoptosome. This "machine" recruits (via its CARD motifs) procaspase 9 molecules, which undergo proximity-induced autoproteolysis to yield active caspase 9 dimers (catalytic domains are colored). Further pro-apoptotic proteins released from mitochondria are SMAC and OMI/Htr2A, which sequester the caspase inhibitor X-IAP, as well as apoptosis-inducing factor (AIF) and endonuclease G. The anti-apoptotic proteins Bcl2 and Bcl-XL (shown in the colored box) protect the mitochondria from permeabilization.

been named after the B-cell lymphoma gene 2, which by a chromosomal translocation may become a hyperactive oncogene protecting the cell from apoptosis. For the first time this oncogenic mutation has shown that cancer can be due not only to uncontrolled cell proliferation but also to a suppression of natural cell death. As expected, the Bcl2 protein encoded by this gene inhibits apoptosis, and the same holds true for the related Bcl-XL protein, which may also mutate to an oncoprotein.

Pro-apoptotic proteins of the Bcl family are Bax, Bak, Bad, Bid, Bim, Noxa, and Puma. The only structural element common to pro- and anti-apoptotic Bcl2 proteins is the **BH (Bcl2 homology) domain**, which is expressed in four isoforms and through which the proteins interact with each other, forming homo- and hetero-oligomers (**Figure 13.15**). In addition, many Bcl2 proteins have a transmembrane domain for anchoring in the outer mitochondrial membrane. Such a membrane interaction enables Bax, Bak, and Bim oligomers to form pores large enough for proteins to pass. However, normally permeabilization is prevented by the anti-apoptotic proteins Bcl2 and Bcl-XL, which are also membrane-bound and antagonize the effects of Bax, Bak, and Bim.

Pro-apoptotic signals overcome this control by interacting with Bid, Bad, Bim, Noxa, or Puma. Since these proteins contain only the BH isoform BH3, they are called

Figure 13.15 Proteins with BH domains BH, B-cell lymphoma gene 2 (Bcl2) homology; gray rectangles, transmembrane domains. For details, see text.

BH3-only proteins. They transduce apoptosis-inducing input signals downstream to the "death proteins" Bax and Bak, which are activated, and to the "survival proteins" Bcl2 and Bcl-XL, which are competitively inhibited (**Figure 13.16**). Moreover, Bcl2 and Bcl-XL are also turned off interacting with active Bax and Bak (and vice versa). Thus, the competition between pro- and anti-apoptotic proteins finally decides between life and death for a cell.

How is Bax (and Bak) activated by BH3-only proteins? In the cytosol Bax is found in an inactive conformation. Interaction with a BH3-only sensor protein causes a conformational change, thus exposing the transmembrane domain that was hidden before inside the Bax molecule. As a result, Bax oligomerizes and forms membrane pores in mitochondria. In contrast to Bax, Bak is a constitutive component of the outer mitochondrial membrane, which upon stimulation by a BH3-only sensor protein oligomerizes directly to membrane pores.

BH3-only proteins receive their input signals along various pathways that are activated in stress situations. As we have seen, Bid is stimulated by death receptors via caspase 8-catalyzed partial proteolysis. The sensor proteins Noxa and Puma are up-regulated at the level of gene transcription: their genes are turned on by the transcription factor p53, which becomes activated upon DNA damage (Section 12.8.3). Changes of the microtubule cytoskeleton cause a release of active Bim, while a more severe breakdown

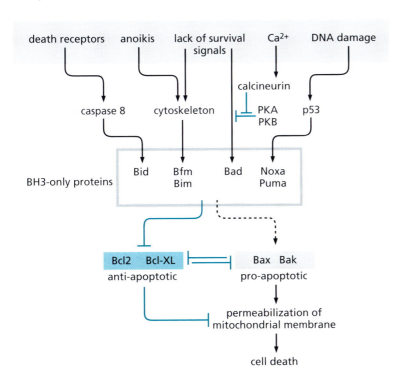

Figure 13.16 Control of apoptosis by proteins of the Bcl2 family Various stress signals interact directly or indirectly with the sensor proteins of the BH3-only subfamily, which then inhibit the anti-apoptotic Bcl2 proteins (colored box), thus leading to an activation of the pro-apoptotic proteins Bax and Bak. Bax and Bak permeabilize (probably together with the BH3-only protein Bid) the outer mitochondrial membrane, thus triggering apoptosis. Bad is inactivated by phosphorylation catalyzed by protein kinase A or B (PKA, PKB). An antagonist of these kinases is the Ca^{2+}-dependent protein phosphatase calcineurin, which mediates apoptosis by Ca^{2+} stress, for instance in the course of stroke and other neurodegenerative diseases (see Sections 14.6.4 and 16.4.2). The transcription factor p53 becomes active upon DNA damage, stimulating the de novo synthesis of Noxa and Puma (as well as of Bax). For additional details, see text.

of the cytoskeleton activates the pro-apoptotic sensor protein, Bfm. Such a situation, called **anoikis** (from the Greek word for "homeless"), occurs when the cell's contact with the extracellular matrix and with neighboring cells is interrupted and, consequently, the cell is in danger of release from the tissue. This protective mechanism against tissue disintegration is repealed in infiltrating and metastasizing tumors.

An important mediator of extracellular signals is Bad. Bad is inhibited by phosphorylation catalyzed by **protein kinase B/Akt (PKB)** as well as by protein kinase A and MAP kinases. PKB prevents apoptosis in still another way: it phosphorylates and inhibits the Forkhead/FOXO transcription factors that promote the transcription of the Bim gene. When these anti-apoptotic effects of PKB are considered, it is not surprising that hyperactive PKB functions as an oncoprotein, Akt, preventing the elimination of tumor cells by Bad- and Bim-mediated apoptosis (see also Section 4.6.6). PKB and the other kinases mentioned above are normally stimulated by a wide variety of growth factors, cytokines, and hormones. Upon a lack of these survival factors, the kinase activity ceases and, consequently, Bad becomes dephosphorylated, now stimulating apoptosis by neglect.

The apoptosome: a molecular killing machine

The permeabilization of the outer mitochondrial membrane results in a release of mitochondrial proteins that induce cell death. One of these proteins is the respiratory enzyme **cytochrome *c*** that is considered to be a major death signal.

Once in the cytoplasm, cytochrome *c* binds to **APAF1** (apoptotic proteases-activating factor), a large chaperone-like protein of the NLR family (see Section 6.3.2). APAF1 contains a CARD motif and a nucleotide-binding oligomerization domain and exhibits ATPase activity. The interaction with cytochrome c induces a conformational change that triggers the exchange of bound dATP (or ATP) by dADP (or ADP) and enables the protein to oligomerize to a wheel-like structure with seven spokes. This complex is called the **apoptosome** (see Figure 13.14). It is closely related to the inflammosome (see Section 6.3.2). Both are molecular machines for caspase activation, the inflammosome for pro-inflammatory caspase 1, the apoptosome for the pro-apoptotic caspase 9. To this end, several procaspase 9 molecules are homotypically bound via their CARD motifs to the CARD motifs of APAF1 to become transformed into active caspase 9 by proximity-induced autoproteolysis. Caspase 9 then triggers the apoptotic caspase cascade. In other words, apoptosomes and caspase 9 play the same role for mitochondria-induced apoptosis as the DISC aggregates and caspase 8 for receptor-mediated apoptosis.

Other proteins released from degenerating mitochondria are a DNA-degrading endonuclease G that cooperates with the caspase-activated DNase CAD, the IAP inhibitors SMAC and OMI/Htr2a already mentioned, as well as an apoptosis-inducing factor (AIF) that triggers apoptosis independently of caspases by promoting chromatin condensation and DNA degradation in a way still not fully understood. AIF is a flavoprotein with NADH oxidase activity and has, like cytochrome c, a physiological function in intact mitochondria, particularly in oxidative phosphorylation and detoxification of reactive oxygen species. AIF seems to play an important role in neuronal cell death (stroke).

Apoptosis by calcium overload

Calcium ions, the most abundant and most versatile second messenger, may promote either cell survival or cell death. Whether or not they kill a cell depends on the dose and duration of the Ca^{2+} signal: low doses and short pulses facilitate cell survival whereas any signal inducing a strong or long-lasting increase of the cytoplasmic Ca^{2+} concentration may trigger apoptosis. The deadly effect of Ca^{2+} ions is assumed to play an important role in stroke, where brain neurons are flooded by Ca^{2+} ions owing to an overactivation of cation channels in the plasma membrane (details in Section 16.4.2).

Calcium causes cell death along various pathways. Thus, the pro-apoptotic effect of Bad that had been suppressed by protein kinases is restored by dephosphorylation. This is catalyzed by several protein phosphatases including **calcineurin**, which contains the calcium-binding subunit calmodulin and is activated by Ca^{2+} (see Section 14.6.4).

However, calcineurin does more: it also activates the NFAT transcription factors (see Section 14.6.4 for details), which in turn stimulate the transcription of death receptor genes. In addition to calcineurin, other Ca^{2+}-dependent enzymes, including proteases such as calpain, endonucleases, and enzymes producing stress signals such as reactive oxygen species and NO, become activated along signaling pathways that lead to an increase of the cytoplasmic calcium concentration. These all may contribute to the pro-apoptotic effect, but—depending on the conditions—also to regulated necrosis.

Finally, a group of Ca^{2+}/calmodulin-dependent Ser/Thr-specific protein kinases known as **DAP kinases** (death-associated protein kinases) should be mentioned (see also Sidebar 13.6). Their activation by various apoptotic signals, including the pro-apoptotic lipid ceramide, evokes characteristic morphological changes of dying cells such as a loss of cell adhesion and membrane blebbing, but also the formation of autophagosomes. DAP kinases are closely related to myosin light-chain kinases, and most of their functions may be explained by effects on the cytoskeleton. DAP kinases are tumor suppressor proteins the deletion of which may promote neoplastic growth.

The expression of DAP kinases is controlled by **p53**. As discussed in detail in Section 12.8.3, this transcription factor controls a wide variety of genes that encode (in addition to DAP kinases) proteins of cell cycle regulation, DNA repair, cell differentiation, and senescence. p53 also stimulates the expression of other pro-apoptotic proteins such as Bax, Noxa, Puma, APF-1, and death receptors. Appropriately, p53 is a tumor suppressor protein because its loss of function protects tumor cells from dying by apoptosis.

Cell death by oxidative stress

The transcription of genes relevant for apoptosis is controlled by still other signals including those transduced by the stress modules of the MAP kinase complexes. These modules may be stimulated by the MAP3 kinase **ASK1** (apoptotic signal-regulated kinase 1; see Section 11.2) with the input signals transmitted by receptors of the TNF receptor family. Moreover, ASK1 also responds to oxidative stress, since in its inactive form it is complexed with the redox sensor thioredoxin (details on redox signaling are found in Section 2.2). Reactive oxygen species act as redox signals by oxidizing the dithiol form of thioredoxin to the disulfide form, thus causing the complex to disintegrate and release active ASK1 (**Figure 13.17**).

Another target of oxidative stress is the NFκB module that processes survival signals by stimulating the transcription of genes for the anti-apoptotic proteins Bcl2, Bcl-XL, IAP, and FLIP. The transcription factor NFκB becomes inactivated by reactive oxygen species owing to an oxidation of essential cysteinyl-thiol groups to intramolecular disulfide groups. The oxidized factor is reduced and reactivated by reduced thioredoxin.

Summary

- Programmed cell death occurs through apoptosis or necrosis. Generally, only necrosis is accompanied by inflammation.

- Apoptosis is induced by specific death signals interacting with transmembrane receptors of the TNF receptor superfamily (extrinsic pathway), and by various stress factors triggering the release of pro-apoptotic proteins such as cytochrome *c* from mitochondria (intrinsic pathway). Alternatively, and depending on the conditions, death receptors may activate survival pathways.

Figure 13.17 Cell death by oxidative stress: role of the oxygen-sensitive protein kinase ASK1 In its inactive form, the homodimeric apoptotic signal-regulated kinase 1 (ASK1) exists as a complex with reduced thioredoxin (TRX). Reactive oxygen species (ROS) oxidize TRX promoting the dissociation of ASK1, which activates itself by trans-autophosphorylation (P). As a MAP3 kinase, ASK1 activates stress-sensitive MAP kinase modules, thus leading to expression of pro-apoptotic genes. MKK, MAP2 kinases; JNK, c-Jun N-terminal kinase.

- Apoptosis is executed by the protease family of caspases and by enzymes attacking chromatin and DNA structure.

- Upon apoptotic stimulation, active caspases are released from inactive precursors (procaspases) by proximity-induced autoproteolysis. This occurs in death-inducing protein complexes (DISCs) recruited at death receptors and in ripoptosomes (extrinsic pathway), or in apoptosomes (intrinsic pathway).

- Apoptosis is controlled by a series of proteins such as the inhibitors FLIP and IAP (extrinsic pathway) and the Bcl2 proteins that either inhibit or stimulate apoptosis (intrinsic pathway). Other pro-apoptotic control proteins are SMAC and OMI/Htr2.

- Necrosis is caspase-independent but proceeds along other pathways leading to a complete disintegration of cells. A major mechanism (necroptosis) involves the protein kinases RIPK1 and RIPK3 that together activate the membrane-disrupting protein MLKL. These kinases are found in DISCs and ripoptosomes. Post-translational modifications of RIPK1 decide as to whether a cell undergoes apoptosis, necrosis or survives. Complexes of the RIPKs and MLKL are referred to as necrosomes.

13.3 Cancer: a disease of signal processing

The relationship between cellular signaling and cancer runs like a thread through this book and deserves to be summarized before we enter chapters on neuronal signaling. Tissue cells are permanently threatened with death from apoptosis, necrosis, senescence, and injury. In fact, their lifespan generally is much shorter, typically weeks to months, than that of the whole organism. To compensate this never-ending cell loss, continuous replacement of cells by division of stem cells is indispensable, the rate of which differs from tissue to tissue. That means that, in the adult organism, the constant mass (and thus the constant function) of a tissue is due to a precisely controlled steady-state between the birth rate and the decay rate of cells. This steady-state is called **tissue homeostasis**. Tissue homeostasis is not maintained automatically but is controlled by an intricate interplay of stimulatory and inhibitory signals that may be both local and systemic. Survival signals such as growth factors and hormones promote cell proliferation, whereas death signals such as the Fas ligand and related cytokines as well as a wide variety of stress factors exert opposite effects. In the state of tissue homeostasis, the effects of death and survival signals compensate each other; but there may be a functional require-

(A)

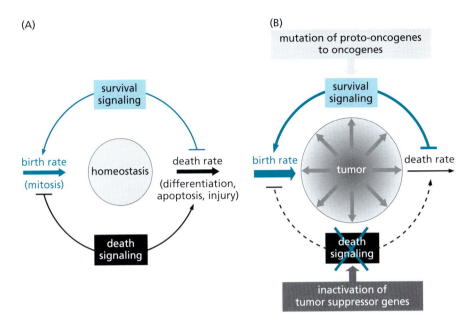

(B)

mutation of proto-oncogenes
to oncogenes

inactivation of
tumor suppressor genes

**Figure 13.18 Tissue homeostasis
and tumor growth** (A) A tissue is
depicted schematically, where the
birth rate of cells is matched by the
death rate because the positive and
negative effects of survival signaling
are precisely balanced out by those
of death signaling and vice versa
(homeostasis). (B) In a tumor, gene
mutations have led to down-regulation
of death signaling and up-regulation
of survival signaling, causing a
permanent disequilibrium of the rates
of cell birth and cell death (arrows
indicate activation; pistils indicate
inhibition).

ment for a *temporary* imbalance: for instance, wound repair or in the course of hormone-regulated physiological cycles.

When the tissue moves away from homeostasis, it either wastes away (atrophy) or hyperproliferates (hyperplasia). Hyperplasia occurs when cells respond more strongly to survival signals than to death signals (**Figure 13.18**). A tumor develops when this disequilibrium becomes irreversible and starts from a single cell. In other words, cancer is a disease of cellular signal processing; a tumor cell misinterprets the communicative signals of the body, thus to some extent resembling a psychotic person. If one prefers the metaphor of the "cellular brain" for typifying the signal-transducing protein network, one may speak of a "cellular psychosis."

Tumor genes

According to current theories a tumor starts with a minute event: a *single* gene mutation that has escaped DNA repair in a *single* cell. That is, most tumors are monoclonal. In most cases this initiating mutation is caused by environmental factors. These are certain viruses, genotoxic chemicals, UV light, and ionizing radiation. Primary targets are genes encoding signal-processing proteins. To emphasize their role in cancer development (oncogenesis), these genes are called **proto-oncogenes** and tumor suppressor genes (some examples are shown in **Figure 13.19**). The proteins encoded by proto-oncogenes are components of networks that process proliferative and survival signals. A permanent overexpression of such proteins facilitates tumor development. Moreover, upon mutation of the corresponding proto-oncogenes to oncogenes, these proteins become hyperactive and control-resistant (this type of gene mutation is called, therefore, a gain-of-function mutation). In contrast, the proteins encoded by **tumor suppressor genes** are integrated in networks that process anti-proliferative, pro-apoptotic, differentiation-promoting, and DNA repair signals. In this case, tumor growth may be stimulated by loss-of-function mutations (gene mutations that result in inactivity of the encoded proteins) or by the generation of dominant negative mutants that compete with tumor suppressor proteins.

Owing to a hereditary predisposition, the sensitivity to tumorigenic factors may be increased considerably (as found in so-called cancer families). When this predisposition is a deletion of one of the two alleles of a tumor suppressor gene, as is frequently the case, such a hereditary effect is easily understood. As long as the second allele is intact, the functional suppressor protein is produced, albeit perhaps at a diminished rate. However, only one more mutagenic hit is needed to delete the

Figure 13.19 Examples of signal-processing proteins the mutation of which may cause cancer

FROM PROTO-ONCOGENES		
ErbB	EGF receptor (receptor tyrosine kinase)	Section 7.1.1
Ras	small G-protein	Section 10.1.1
PKB/Akt	Ser/Thr kinase	Section 4.6.6
Src	Tyr kinase	Section 7.2
cAbl	Tyr kinase	Section 7.1.1
Jun, Fos	transcription factors	Section 11.7
Bcl2	anti-apoptotic protein	Section 13.2.4
Cbl	ubiquitin E3 ligase	Section 7.1.1

FROM TUMOR SUPPRESSOR GENES		
PTEN	PIP-3 phosphatase	Section 7.4.2
APC	inhibitor of β-catenin signaling	Section 5.9.2
Rb	transcriptional repressor	Section 12.6
p53	transcription factor	Section 12.8.3
BRCA1	histone deacetylase related	Section 8.2.3
BRCA2	DNA damage repair complex	Section 12.8.1
SWI/SNF	chromatin remodelling	Section 8.2.7
RolB	phospho-Tyr phosphatase (*Agrobacterium rhizogenes*)	

second allele and fully ruin the cellular brake. Prominent examples of this situation are provided by the retinoblastoma gene, the APC gene of colon tumorigenesis, and the breast cancer genes *BRCA1* and *BRCA2* (see Figure 13.19). Other reasons for a hereditary predisposition to cancer may be phenotypic features (such as poor sun protection in fair-skinned people) as well as inborn defects of DNA repair, of epigenetic chromatin modifications controlling the constancy of tissue differentiation (see Section 8.2.8), of enzymes metabolizing chemical carcinogens (see Section 8.4), of endogenous pathways leading to potentially carcinogenic metabolites, and of other mechanisms that would promote genetic damage.

Multistage carcinogenesis

As mentioned, tumorigenesis is thought to start with a single mutagenic hit generating an "initiated" cell. From this cell a tumor develops, proceeding via a series of clinical stages such as dysplasia (abnormally developing tissue), benign tumor, carcinoma* *in situ*, and, finally, malignant tumor (characterized by destructive growth and metastasis). The appearance of these stages depends on additional gene mutations and chromosomal disorders (**Figure 13.20**). In fact, when a tumor has progressed to the state of malignancy it may have accumulated up to half a dozen such defects. This pathological development is strongly promoted by nongenotoxic influences that, for example, disturb DNA repair and stimulate the metabolic activation of genotoxic carcinogens as well as the blood supply of tumors. Moreover, without nongenotoxic influences, many tumors would not develop at all since the probability is practically zero that a sequence of specific mutations accidentally occurs in a *single* (initiated) cell. This indicates that, besides gene mutations, additional processes are involved in cancer development by which the probability of further mutations to accumulate is increased. The easiest way is to expand the clone of genetically pre-damaged cells by stimulation of cell proliferation and/or inhibition of cell death: the more such cells are available, the higher is the probability of an additional mutagenic hit. Stimuli that induce such a response are called **tumor promoters**.

The clonal expansion of mutated cells finds its explanation in the genetic defect, rendering these cells *less sensitive* for death and *more sensitive* for survival signals. When continuously stimulated in such a way, the pre-malignant cells therefore gain a selective advantage over their non-mutated neighbors, although the latter respond as well to the stimulus.

---

* The term carcinoma (and dysplasia) refers to epithelial tissue only, whereas cancers of connective tissue are collectively called sarcomas. The mechanisms of tumor development are comparable in both cases.

INITIATION   PROMOTION   MALIGNANT PROGRESSION

epithelium

connective
tissue

initiated   preneoplastic benign tumor malignant
tissue   state   (adenoma)   tumor
               (dysplasia)   (carcinoma)

Figure 13.20 Development of colon carcinoma as an example of multistage carcinogenesis Most frequently colon carcinogenesis is initiated by a loss-of-function mutation of the tumor suppressor gene *APC*, causing a deregulation of Wnt signaling. The initiating agent is not known; dietary components are under suspicion. For a minority of patients a *hereditary* defect of *APC* is the cause of the disease adenomatous polyposis of colon that harbors a high cancer risk. In the course of tumor promotion the initiated cell expands to a clone, becoming visible as dysplasia. Certain bile acids are suspected of functioning as endogenous tumor promoters, and the same holds true for chronic irritation such as Crohn's disease. Benign tumors (adenomas or polyps) arising from dyplastic colon epithelium exhibit two more gene mutations, a gain-of-function mutation of the proto-oncogene K-ras and a deletion of the tumor suppressor gene *DCC*, while in the final carcinoma stage the tumor suppressor gene *p53* is deleted in addition. It is assumed that the late mutations occur spontaneously as a result of increasing genetic instability.

It is easily conceivable that tumor promotion plays a critical role in cancer development, since otherwise the disease most probably would not proceed beyond a harmless stage. The classic example of tumor promotion is the permanent activation of cell proliferation by the diterpene ester TPA in experimental mouse skin carcinogenesis (see Section 4.6.5). In this model TPA evokes a hyperproliferative and inflammatory reaction that is closely related to a wound response. Accordingly, for tumor promotion TPA treatment can be replaced by repeated wounding (**Figure 13.21**), and similar observations have been made for other animal models. They indicate that *endogenous* factors may exert tumor-promoting effects. Such factors include "wound hormones" such as growth factors and pro-inflammatory mediators as well as hormones such as prolactin, estradiol, and testosterone, and bile acid metabolites. Moreover, chronic inflammation due to continuous irritation or infections can promote tumor development, and the same holds true for a wide variety of chemicals (so-called non-genotoxic carcinogens) and other stimuli. Thus most tumors develop against a background of chronic irritation. Prominent examples are provided by colon, lung, and skin cancer. Tobacco smoke indeed contains both mutagenic and tumor-promoting components, and for sunlight (causing skin cancer) the short-wave UV component is mutagenic while the long-wave UV component (as used in tanning salons) is a tumor promoter. Clonal expansion is one result of tumor promotion, the other is **genetic instability**. This means that in the course of neoplastic development genetic damage may occur faster than expected from the burden of environmental genotoxic agents. A major cause of genetic instability is an up-regulation of metabolic pathways generating metabolites such as reactive oxygen species, organic free radicals, and other compounds that damage DNA by acting as endogenous mutagens. Moreover, the cellular mechanisms of DNA repair and elimination of mutated cells may be impaired.

The development of genetic instability means that tumorigenesis becomes increasingly independent of external influences. This has been shown using the mouse skin model: while the appearance of premalignant states requires continuous tumor promotion, the development of malignant carcinomas from these states occurs spontaneously without any further treatment although it depends on additional gene mutations (**Figure 13.22**).

Signal transduction therapy of cancer

Conventional anticancer drugs are cytostatics. Although they may preferentially kill cancer cells, they do not spare normal proliferating cells. Thus, this therapeutic approach is limited by strong or even fatal side effects. Currently, the focus is on the development of more specific or targeted therapies, also called "precision medicine." The strategy is to use the information about a patient's tumor mutation signature and the profile of active genes and proteins for diagnosis and treatment. In targeted cancer therapy, drugs are given that interfere with molecular targets in tumor cells themselves but which may also block distinct molecular targets in various cell types of the tumor microenvironment such as blood vessel cells or immune cells that directly or indirectly promote tumor growth, the progression from the benign to the malignant state, and metastasis. To a large extent such approaches are based on the discovery of cancer being a disease of cellular signaling. In fact, strong efforts are being made to develop drugs that correct oncogenic defects of signal transduction.

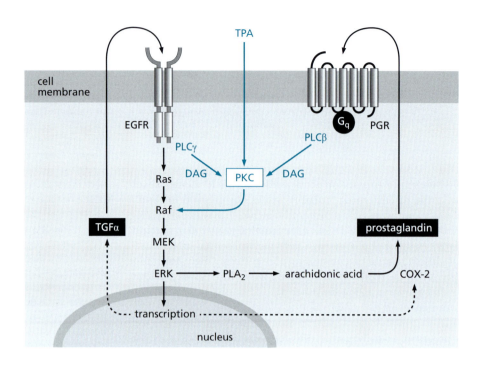

Figure 13.21 How tumor promoters interfere with signaling pathways In the mouse skin model of multistage carcinogenesis, repeated wounding is a tumor-promoting stimulus, while the phorbol ester TPA promotes skin tumor development by interfering with signaling cascades along which the effects of wound hormones are processed. The figure shows two auto-amplifying signaling cascades that, upon wounding, become activated by transforming growth factor α (TGFα): a mitogenic wound hormone, and prostaglandin, a pro-inflammatory mediator. TGFα interacts with the epidermal growth factor receptor (EGFR), a receptor Tyr kinase, and prostaglandin with a G-protein-coupled prostaglandin receptor (PGR). In both cases, different isoforms of phospholipase C (PLC) become activated, releasing diacylglycerol (DAG) which activates PKC (teal arrows). Among the various downstream reactions of PKC activation, the stimulation of the mitogenic MAP kinase cascade, depending on phosphorylation of the kinase RAF, is shown (for details see Section 10.1.4). This reaction enhances the direct effect of TGFα on the MAP kinase cascade, which is due to an interaction between EGFR and the small G-protein Ras. The cascade leads to an activation of numerous genes (see Section 11.7), including that of TGFα, resulting in positive feedback (left). At the same time, the biosynthesis of prostaglandin is stimulated by phosphorylation and activation of phospholipase A_2 (PLA₂), catalyzed by the MAP kinase ERK and transcriptional up-regulation of cyclooxygenase 2 protein (COX-2; for details see Section 4.4.5). As a result, a second positive feedback loop becomes established (right). Both signaling cascades exhibit a pro-mitogenic and anti-apoptotic effect. The metabolism of arachidonic acid is also a potential supply of genotoxic metabolites. In premalignant cells these effects on signal transduction are intensified by an oncogenically mutated Ras protein. Therefore, the conditions of tumor promotion are fulfilled. Both feedback loops may be stimulated by the phorbol ester TPA, rather than by repeated wounding, which specifically mimics the activating effect of diacylglycerol on PKC. The effects of both TPA and DAG critically depend on long-chain fatty acid residues for membrane anchoring and PKC recruitment. DAG and phorbol esters with short acyl groups are inactive, as are nonesterified glycerol and phorbol, respectively.

These drugs are either small molecules or monoclonal antibodies that affect cellular signal processing at different levels. Currently, the major signaling proteins targeted include growth factors, growth factor receptors, and corresponding downstream signal-processing molecules (**Table 13.1**). An example is provided by Tyr kinases that—as we have seen—frequently undergo oncogenic mutations. Synthetic Tyr kinase inhibitors are called tyrphostins (tyrosine phosphorylation inhibitors). Several hundred such compounds have been synthesized, and as yet up to 20 drugs have been registered for clinical use while more are in clinical trials. The first of these drugs is Imatinib, also known as Gleevec, which entered the clinic in May 2001. It has been used successfully for the treatment of chronic myelogenous leukemia. Imatinib/Gleevec inhibits the BCR-Abl Tyr kinase (see Chapter 7, Sidebar 7.1) by blocking the enzyme's ATP-binding site. The activity of BCR-Abl kinase is essential for the survival of leukemia cells. Related drugs inhibit receptor Tyr kinases. For instance, inhibitors of the EGF receptor are used to treat non-small-cell lung cancer. Even though the anti-tumor effects of such drugs may be quite impressive, their long-

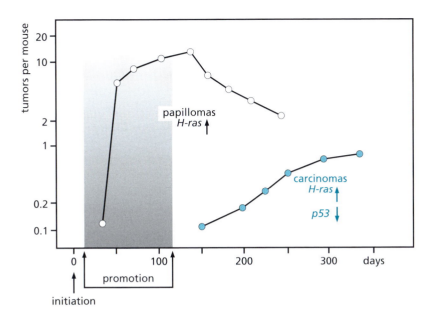

Figure 13.22 Spontaneous malignant progression indicates genetic instability For initiation of mouse skin carcinogenesis, the mutagen 7,12-dimethylbenz(a) anthracene is given at day 0 in a dose that does not cause tumor development by itself. For tumor promotion, the phorbol ester TPA is applied twice a week over 16 weeks. As a result, up to 12 benign epithelial tumors (papillomas) per animal arise. Without any further treatment most of the papillomas regress. These papillomas exhibit a single gain-of-function mutation of the proto-oncogene *H-ras*. Ten percent of papillomas do not regress but spontaneously progress to carcinomas exhibiting at least one more gene mutation, a deletion of the *p53* tumor suppressor gene. Note the logarithmic scale of the ordinate. (Partly based on data from Burns FJ et al. [1976] *Cancer Res* 36:1422–1427).

term use is hampered by side effects and the development of resistance, for instance owing to additional mutations in the kinases' ATP-binding sites or an adaptive over-expression of proteins with an analogous action profile. Efforts are being made to overcome this obstacle, both by developing novel drugs and by novel treatment regimens, for instance by combining the administration of tyrphostins with other anti-cancer drugs such as receptor or kinase antibodies or inhibitors of protein kinases acting downstream of receptor tyrosine kinases. Admittedly, the signal transduction therapy of cancer is still in its infancy. But when the rapidly progressing elucidation of the cell's signal processing network is considered, the approach promises to have great potential for the future. After all, the range of compounds tested in preclinical cancer research is continuously growing.

Table 13.1 A selection of signal-processing proteins as targets of approved anti-cancer drugs.

Growth factors and hormones
EGF, TNF-family factors, VEGF, estrogens, androgens, retinoic acid

Growth factor receptors (receptor Tyr kinases)
EGFR, Kit, Met, PDGFR, Ret, Ros, VEGFR

Other receptors
Smoothened

Non-receptor Tyr kinases
BCR-Abl, Btk (Bruton's kinase), JAK, SRC

Ser/Thr kinases
CDK4, CDK6, MEK1/2, RAF1/2, mTOR

Lipid kinases
PI3-kinase delta

Histone deacetylases (HDAC)

PolyADP-ribose polymerase (PARP)

Proteasomes

Summary

- Cancer results mainly from gene mutations (predominantly caused by environmental factors) that lead to a continuous deregulation of cellular signal processing and, consequently, to an imbalance between the rates of cell gain and cell loss.

- The progression from a single mutated cell to a malignant tumor requires additional genetic and epigenetic alterations and has to be promoted by nonmutagenic endogenous and exogenous stimuli (tumor promoters) that cause a long-lasting stimulation of mutagenic and anti-apoptotic signaling and, in addition, may stimulate the production of genotoxic metabolites, thus creating a state of genetic instability.

- Signaling pathways involved in cancer development increasingly become targets of novel therapeutic approaches.

Further reading

Section 13.1

Bengoechea-Alonso MT & Ericsson J (2007) SREBP in signal transduction: cholesterol metabolism and beyond. *Curr Opin Cell Biol* 19:215–222.

Bray SJ (2006) Notch signalling: a simple pathway becomes complex. *Nat Rev Mol Cell Biol* 7:678–689.

Chow B & McCourt P (2007) Plant hormone receptors: perception is everything. *Genes Dev* 20:1998–2008.

Duffy MJ et al. (2009) The role of ADAMs in disease pathophysiology. *Clin Chim Acta* 403:31–36.

Duggan SP & McCarthy JV (2016) Beyond γ-secretase activity: the multifunctional nature of presenilins in cell signalling pathways. *Cell Sign* 28:1–11.

Hagen G (2015) Auxin signal transduction. *Essays Biochem* 58:1–12.

Knowles TPJ, Vendruscolo M & Dobson CM (2014) The amyloid state and its association with protein misfolding diseases. *Nat Rev Mol Cell Biol* 15:384–396.

Nyren-Erickson EK et al. (2013) A disintegration and metalloproteinase 12 (ADAM12): functions, roles in disease progression, and clinical implications. *Biochim Biophys Acta* 1830:4445–4455.

Prusiner SB (2013) Biology and genetics of prions causing neurodegeneration. *Annu Rev Genet* 47:601–623.

Walker LC & Jucker M (2015) Neurodegenerative diseases: expanding the prion concept. *Annu Rev Neurosci* 38:87–103.

Wang H et al. (2015) The role of notch receptors in transcription regulation. *J Cell Physiol* 230:982–988.

Section 13.2

Ashkenazi A, Yuan J & Wells JA (2014) Regulated cell death. *Meth Enzymol* 544:545.

Biermann M et al. (2013) Surface code - biophysical signals for apoptotic cell clearance. *Phys Biol* 10:065007.

Budhidarmo R & Day CL (2015) IAPs: molecular regulators of cell signalling. *Sem Cell Dev Biol* 39:80–90.

Cabal-Hierro L & Lazo PS (2012) Signal transduction by tumor necrosis factor receptors. *Cell Sign* 24:1297–1305.

Christofferson DE, Li Y & Yuan J (2014) Control of life-or-death decisions by RIP1 Kinase. *Annu Rev Physiol* 76:129–150.

Fulda S (2015) Smac mimetics as IAP antagonists. *Sem Cell Dev Biol* 39:132–138.

Hong SJ, Dawson TM & Dawson VL (2002) Nuclear and mitochondrial conversations in cell death: PARP-1 and AIF signaling. *Trends Pharmacol Sci* 25:159–265.

Larrieu A & Vernoux T (2015) Comparison of plant hormone signalling systems. *Essays Biochem* 58:165–181.

Lavrik IN & Krammer PH (2012) Regulation of CD95/Fas signaling at the DISC. *Cell Death Diff* 19:36–41.

Lord SJ, Rajotte RV, Korbutt ES et al. (2003) Granzyme B: a natural born killer. *Immunol Rev* 193:31–38.

Mariño G et al. (2014) Self consumption: the interplay between autophagy and apoptosis *Nat Rev Mol Cell Biol* 15:81–94.

Moldoveanu T et al. (2015) Many players in BCL2 family affairs. *Trends Biochem Sci* 39:101–111.

Öztürk S, Schleich K & Lavrik IN (2012) Cellular FLICE-like inhibitor proteins (cFLIPs), fine-tuners of live and death decisions. *Exp Cell Res* 318:1324–1331.

Pasparakis M & Vandenabeele P (2015) Necroptosis and its role in inflammation. *Nature* 517:311–320.

Riley JS et al. (2015) DED or alive: assembly and regulation of the death effector domain complexes. *Cell Death Dis* 6:e1866.

Schilling R, Geserick P & Leverkus M (2014) Characterization of the ripoptosome and its components: implications for anti-inflammatory and cancer therapy. *Meth Enzymol* 545:83–102.

Silke J, Rickard JA & Gerlic M (2015) The diverse role of RIPkinases in necroptosis and inflammation. *Nat Immunol* 16:689–697.

Tait SWG & Green DR (2012) Mitochondria and cell signalling *J Cell Sci* 125:807–815.

Vanden Berghe T et al. (2014) Regulated necrosis: the expanding network of non-apoptotic cell death pathways. *Nat Rev Mol Cell Biol* 15:135–144.

White E (2012). Deconvoluting the context-dependent role for autophagy in cancer. *Nat Rev Cancer* 12:401–410.

Yuan S & Akey CW (2013) Apoptosome structure, assembly, and procaspase activation. *Structure* 21:501–515.

Section 13.3

Al-Lazikani B, Banerji U & Workman P (2012) Combinatorial drug therapy for cancer in the post-genomic era. *Nat Biotechnol* 30:1–13.

Hanahan D & Weinberg R (2011) The hallmarks of cancer: The next generation. *Cell* 144:646–674.

Levitzki A (2013) Tyrosine kinase inhibitors: views of selectivity, sensitivity, and clinical performance. *Annu Rev Pharmacol Toxicol* 53:161–185.

Levitzki A & Mishani E (2006) Tyrphostins and other tyrosine kinase inhibitors. *Annu Rev Biochem* 75:93–109.

Sun C and Bernards R (2014) Feedback and redundancy in receptor tyrosine kinase signaling: relevance to cancer therapies. *Trends Bioch Sci* 39:465–474.

Yap TA & Workman P (2012). Exploiting the cancer genome: Strategies for the discovery and clinical development of targeted molecular therapeutics. *Ann Rev Pharmacol Toxicol* 52:549–573.

Signal Transduction by Ions

14

In the previous chapters, mechanisms of signal processing have been discussed that derive their energy from biochemical reactions such as the hydrolysis of energy-rich bonds or redox processes. In parallel, all cell types, from the simplest prokaryotes to the most complex brain neurons, use the electrical energy stored in the membrane potential (like in a charged battery) for data processing and information transfer. The membrane potential is due to a disequilibrium of ions generated and maintained by ATP-consuming ion pumps. To significantly change the potential it is sufficient to transport only a minute proportion of ions across the membrane (see Section 2.10.1). Therefore, most signals using the membrane potential as transducer interact with ion channels rather than with ion pumps. In fact, ion channels are controlled by a wealth of extra- and intracellular stimuli, and are able to regulate almost any cellular function including gene expression. This is achieved by connecting the influx and efflux of ions with the standard pathways of biochemical signal transduction.

Ion channels are transmembrane proteins. They are composed of subunits with several transmembrane domains, mostly α-helices that form a central pore for ion flux. Like most signal-transducing proteins, ion channels exist in at least two conformations, an active one (the channel is open) and an inactive one (the channel is closed). These conformations are in an allosteric equilibrium that provides a signal-operated switch: channel-gating signals stabilize the open form, while channel-closing signals stabilize the closed form. These signals may consist of electricity, controlling voltage-operated ion channels; molecules, interacting with ligand-operated ion channels; and mechanical pressure, acting on mechanosensitive ion channels. Moreover, temperature-sensitive ion channels are known (see Chapter 15). The ions transported by such channels have the following functions:

- H^+: control of pH value
- Na^+: change of membrane potential (depolarization)
- K^+: change of membrane potential (mostly hyperpolarization)
- Cl^-: change of membrane potential (de- or hyperpolarization)
- Ca^{2+}: second messenger, sometimes also change of membrane potential

In addition, monovalent cations and anions determine the osmotic pressure inside the cell. The families of anion and cation channels are strictly separated.

Basic information on the membrane potential and ion channels is found in Section 2.10. Here and in Chapters 15 and 16 we will discuss the structural and chemical features of individual ion channel families with particular emphasis laid on their role in cellular signal transduction. A detailed discussion of (electro) physiological phenomena goes beyond the scope of this book; for that the reader is referred to textbooks of neurosciences and cell biology.

14.1 Cation channels: prototypical structures and gating mechanisms

A typical cation channel is composed of at least two α-subunits that associate with regulatory subunits labelled by additional Greek letters (β, γ, δ, etc.). Each α-subunit contains up to six transmembrane helices or segments and at least one P-loop that dips into the membrane surface. The pore consists of a selectivity filter determining the ion specificity of the channel, a water-filled central cavity for ion enrichment, and a gate that opens or closes in a reversible manner upon signal reception (**Figure 14.1**). While the gate and central cavity are lined by transmembrane segments, the selectivity filter is formed by the P-loops (see Section 2.10.3 for more details on ion selectivity).

One distinguishes between two prototypes of cation channels that have not changed very much from prokaryotes to humans. They differ in the architecture of the subunits (**Figure 14.2**).

(1) Prototype 1: Each subunit contains six transmembrane segments, numbered S1–S6, and one P-loop between S5 and S6. This architecture is characteristic for prokaryotic and eukaryotic voltage-gated cation channels, cyclic nucleotide-gated (CNG) cation channels, transient receptor potential (TRP) channels, the ryanodine and inositol 1,4,5-trisphosphate ($InsP_3$) receptor calcium channels of the endoplasmic reticulum (ER) and the two-pore channels (TPC) of the endolysosomal membranes (must not be confused with the two-P-domain potassium channels K_{2P} of prototype 2).

(2) Prototype 2: Each subunit contains only two transmembrane segments, S5 and S6, and one P-loop in between. This simple and probably more ancient architecture is characteristic for the epithelial Na^+ channel (ENaC) family and for inwardly rectifying potassium channels (K_{ir}). Immediate descendants are the two-P-domain potassium channels (K_{2P}) where two of the subunits are covalently linked.

Figure 14.1 Principles of cation channel structure Cation channels are transmembrane proteins composed of at least two subunits that constitute a central pore. The pore is lined by transmembrane helices and contains a selectivity filter that determines ion specificity, a central cavity, and a gate that reversibly opens or closes in a signal-controlled manner. (A) Molecular model of a channel pore based on structure analysis of the bacterial potassium channel KcsA. Shown are two of the four subunits. Each subunit consists of two tilted transmembrane helices and one P-loop. The P-loops form the selectivity filter (teal) that strips the ions (black balls) from water molecules (teal dots). In the water-filled central pore, the ions become rehydrated. The gate is formed by the inner helices. (B) Space-filling model of a cation channel in closed and open states. Three of the four subunits of a prototypical channel are shown; the positions of the individual pore components may differ from channel type to channel type. Moreover, both the extracellular and intracellular domains may contain ligand binding sites. (A, adapted from fig.cox.miami.edu/~cmallery/150/memb/ion_channels.htm. B, adapted with modifications from Schumacher M & Adelman JP [2002] *Nature* 417:501–502.)

(A)

(B)

bacterial
precursors

K_VAP
NaChBac
MloK1

KcsA
MthK

prototype 1

P   P

voltage
sensor

voltage-operated
cation channels

cyloplasm

CNG
TRP
RyR
InsP$_3$R
TPC

prototype 2

P

ENaC
K$_{ir}$

cyloplasm

P   P

K$_{2P}$

Figure 14.2 Prototypes of cation channels Sodium, potassium, and calcium channels have two prototypes. These are characterized by the structure of the subunits, shown here in schematic form. Nonspecific cation channels coupled with neurotransmitter receptors are distantly related (not shown). CNG, cyclic nucleotide-gated channel; ENaC, epithelial Na$^+$ channel; InsP$_3$R, inositol 3,4,5-trisphosphate receptor; KcsA, K$^+$ channel of *Streptomyces lividans*; K$_{ir}$, inwardly rectifying K$^+$ channel; K$_{2P}$, 2P-domain K$^+$ channel; K$_V$AP, voltage-gated K$^+$ channel of *Aeropyrum pernix*; MloK1, cyclic nucleotide-gated K$^+$ channel of *Mesorhizobium loti*; MthK, Ca^{2+}-controlled K$^+$ channel of *Methanobacterium thermoautotrophicum*; NaChBac, voltage-gated Na$^+$ channel of *Bacillus halodurans*; RyR, ryanodine receptor; TRP, transient receptor potential channel; TPC, two-pore channel. The cylinders represent transmembrane domains. Note that at least two subunits are required for a functional channel.

Most of these ion channels are rather selective, transporting either sodium, potassium, or calcium ions. Other, more nonselective cation channels are distantly related to these prototypes or to other prokaryotic precursors. They are coupled with neurotransmitter receptors and are discussed in Chapter 16. As shown in Section 14.7, anion channels may exhibit a completely different architecture.

Voltage-gated ion channels represent the majority among the nearly 500 different ion channel proteins of humans. They open upon membrane *depolarization* and close by a mechanism of self-termination when the membrane potential reaches a critical value. From prokaryotes to humans, all voltage-gated ion channels are thought to operate according to the same principle on the basis of the characteristic molecular architecture resembling prototype 1 (**Figure 14.3**). In the most ancient species, the voltage-gated potassium channels, the four subunits are separated, whereas in the evolutionarily more modern sodium and calcium channels the four subunits share one large polypeptide chain. Voltage-gated ion channels are ultra-high-speed switches that easily put into the shade technical devices such as transistors. Therefore, they are predestined for rapid signal propagation in the nervous system. To this end they have a structural element that, upon depolarization, changes its position or conformation in such a way that the channel gate opens. The core of this structure is transmembrane segment S4: it carries a strong positive charge due to a clustering of basic arginine residues and is lipophilic enough to move within the lipid bilayer of the membrane. Mutation analyses have confirmed the concept of S4 acting as voltage sensor. Insights into the mechanism of channel gating have been obtained by X-ray analysis of crystallized ion channel proteins (mostly of prokaryotic origin) and other studies on protein structure. The results superseded or at least relativized older hypotheses. Such a putatively outdated theoretical concept, still found in most textbooks, is the sliding helix model. It implies that, upon depolarization, helix S4 makes a screw turn in the direction of the cell surface, resulting in an irislike opening of the channel pore. Detailed studies on the voltage-gated potassium channel K$_V$AP of the archaebacterium *Aeropyrum pernix* indicate that the truth

Figure 14.3 Gating mechanism of a voltage-dependent ion channel (A, B) Molecular model of the archaebacterial potassium channel K_VAP. (A) View from above, showing the four subunits with six transmembrane helices each (numbered S1–S6) and the central pore occupied by an ion (teal dot). (B) Vertical section, showing two of the four subunits and a closed ion gate. Note the perpendicularly bent hairpin loops consisting of the voltage-sensitive S3b and S4 segments (paddles). (C) Model of channel gating. Upon depolarization, the positively charged paddles turn upward, opening the ion gate. (A and B, adapted from Tielemann DP, www.aups.org.au/Proceedings/37/15–27/.)

might be much simpler. They have shown that, in contrast to previous assumptions, the S4 helix does not move vertically. Instead, it forms a hairpinlike structure with a part of segment S3 that is fixed at the basis of the channel protein like a paddle at its axis of rotation. A change of the membrane potential lets the paddle swing within the membrane's interior: in the hyperpolarized state, its position is almost perpendicular to the channel closing the gate, while upon depolarization it swings upward, opening the gate (see Figure 14.3). Certainly, many variations of this basic theme may exist. Thus, in voltage-gated potassium channels of animals, the S4 helix seems to undergo both tilting and sliding.

Voltage-dependent ion channels interact with a wide variety of other signal-transducing components that adjust their effectiveness to the given physiological context. Thus, their activities may be fine-tuned by protein phosphorylation. To this end the cytoplasmic domains of the α-subunits harbor several phosphorylation sites. An example for the consequences of ion channel phosphorylation is found in Section 14.5.2.

In contrast to the voltage-gated channels, most of the other channels of prototype 1 are controlled by intra- or extracellular ligands rather than by the membrane potential (some channels also by both). One reason for voltage insensitivity is an insufficient positive charge of the S4 segment. The ligands interact with extended intra- or extracellular structures of the receptors (for an example, see Section 14.4.1).

Channels of prototype 2 do not contain an S3–S4 paddle or a comparable voltage sensor. Nevertheless, some of them may respond to voltage but in a reversed manner: they are closed upon depolarization. To this end the channel is plugged by positively charged polyamines and Mg^{2+} ions, a mechanism that differs entirely from that of voltage-gated channels (see Section 14.4.2). Other channels of prototype 2 either are consistently open or are controlled by ligands.

For their pioneering work on the structure and function of transmembrane channels Roderick MacKinnon and Peter Agre were awarded the Nobel Prize in Chemistry in 2003.

Summary

- Cation channels are composed of at least two subunits forming a central pore. Two basic subunit structures are distinguished that can be traced back to prokaryotic precursors. The simplest structure consists of two transmembrane segments (helices) lining the pore and a P-loop forming the selectivity filter.

- Voltage-gated channels contain four additional segments. Here a hairpin structure of segment 3 with the positively charged segment 4 (paddle) acts as a voltage sensor that opens or closes the channel depending on the membrane potential.

- Related channels with an insufficiently charged segment 4 are not gated by depolarization. A special channel type is closed upon depolarization owing to a plugging of the pore by positively charged polyamines and Mg^{2+} ions.

14.2 Voltage-gated Na+ channels: masters of the action potential

Sodium ions have two major functions: they control the membrane potential and they determine the osmotic pressure. Accordingly, Na+-selective channels exist in two basic forms for these different purposes: voltage-gated channels for changing the membrane potential and voltage-independent channels for regulating electrolyte transport. As mentioned before, in the voltage-dependent sodium channel the four groups of transmembrane helices characteristic for prototype 1 channels share together as subdomains one polypeptide chain, a 260 kDa protein (**Figure 14.4**).

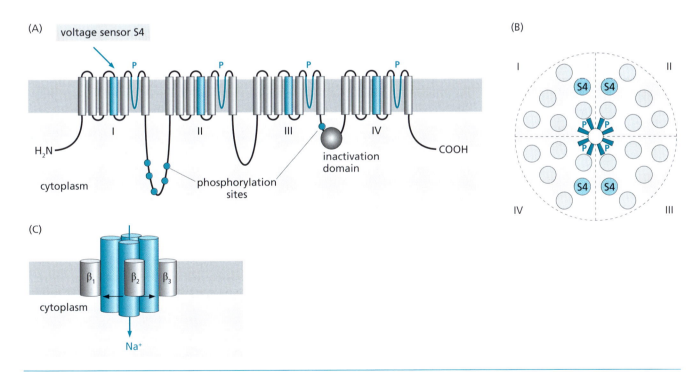

Figure 14.4 Voltage-controlled axonal Na+ channel (A) Transmembrane topography of the α-subunit with the voltage sensors S4 (teal cylinders) and the P-loops (shown in teal). The inactivation domain localized in an intracellular loop between transmembrane subdomains III and IV blocks the gated channel, rendering it transiently refractory (see also Figure 14.11). (B) Arrangement of the transmembrane helices and P-loops (P) seen from above. (C) Model of the subunit architecture of the complete channel, showing the four subdomains of the α-subunit (teal) surrounded by the regulatory transmembrane subunits β1, β2, and β3. (B, adapted from Stevens SF [1991] *Nature* 349:657–658.)

Figure 14.5 Architecture of the axonal Na$^+$ channel (α-subunit) The central pore of the channel has four influx and four efflux openings lined by the S4 voltage sensors, the S6 transmembrane domains, and the P-loops. The model is based on electron microscopy, X-ray diffraction, and nuclear magnetic resonance spectroscopy. (From Catterall WA [2001] *Nature* 409:988–990.)

The structure of this α-subunit, a rather late development of evolution found only in multicellular organisms, has been investigated in detail. With four influx and four efflux openings, it is surprisingly complex (**Figure 14.5**). The pore is formed by the hydrophobic S6 transmembrane helices and, at a central bottleneck, by the selectivity filter of the P-loops, while the S4 voltage sensors mark the boundaries of the influx and efflux openings.

Although the experimental expression of the pore-forming α-subunits is sufficient to generate a functional channel, in most voltage-dependent channels the α-subunits are surrounded by several regulatory subunits (see Figure 14.4). Thus, the α-subunit of the voltage-dependent Na$^+$ channel forms a complex with three different β-subunits, β_1, β_2, and β_3. These are single-pass transmembrane proteins that exhibit structural features of cell adhesion molecules of the immunoglobulin (Ig)-like superfamily (see Section 7.3.1). As such, they probably control the interaction of the channel with the extracellular matrix, adjacent cells (such as glial cells), and the cytoskeleton. Such interactions play an important role in the regulation of channel activity and in the clustering of sodium channels along myelinated axons at the nodes of Ranvier. The saltatory propagation of action potentials along nerve fibers depends on this clustering. The important role of the β-subunits is underlined by the fact that point mutations of β_1 cause hereditary epilepsy. As may be expected from their putative functions, the β-subunits are a late development of evolution, found only in vertebrates.

Owing to the low cytoplasmic Na$^+$ concentration and negative charge inside the cell, the gating of voltage-dependent sodium channels always results in a rapid Na$^+$ influx: the interior of the cell becomes less negative or even positive, or in other words, the cellular battery is discharged by depolarization. But this can also be accomplished by activation of nonspecific cation channels (see Sections 14.5.6 and 14.5.7 and Chapters 15 and 16). A specialty of voltage-gated Na$^+$ channels is signal propagation in the nervous system and in muscle cells, because here each gating event provides a triggering signal for the gating of other voltage-dependent ion channels in the neighborhood. As a result, an action potential migrating across the cell membrane is triggered.

As explained in Section 2.10.2, an action potential is defined as a complete pulse-like depolarization of the cell. This is caused by a rapid discharge and recharge of the membrane battery owing to a controlled influx of Na$^+$ ions followed by an efflux of K$^+$ ions through voltage-operated channels.

To migrate like a wave along neuronal and muscle cell membranes for an action potential two conditions must be fulfilled: the peak of depolarization must be sharp (transient) and it must be propagated. The sharpening is due to a rapid inactivation of voltage-gated sodium channels that terminates the depolarizing phase of the action potential, enabling voltage-gated potassium channels, by hyperpolarization, to reset the membrane potential to the resting value (see Section 2.10.1). For the propagation of action potentials, adjacent voltage-gated Na$^+$ channels cooperate in such a way that the depolarization caused by one channel triggers the gating of the next channel and so on. Such a unidirectional propagation of the action potential, resembling a burning fuse, is possible only when, after activation, the channel becomes refractory for a certain period of time, ensuring that it is not gated again by the depolarization caused by the next channel (**Figure 14.6**). Here again, rapid channel inactivation is necessary. The inactive state is thought to be due to a cytoplasmic globular domain of the α-subunit blocking the channel pore until it has taken on the closed conformation (ball-on-a-chain model, see Section 14.4.1).

An inhibition by drugs and toxic compounds of voltage-dependent sodium channels interrupts impulse propagation in the nervous system. This is the mechanism of **local anesthetics**, such as lidocaine and tetracaine, but also of highly toxic neurotoxins that are applied as experimental probes. Both famous and infamous is

tetrodotoxin, the poison of the puffer fish (**Figure 14.7**). Its half-maximal lethal dose (LD$_{50}$) for humans is 8 µg/kg of body weight (though botulinum neurotoxin, at 0.005 µg/kg, is far more toxic). In the puffer fish, as "Fugu", a particular highlight of Japanese cuisine, the toxin is concentrated in inner organs such as the liver. These have to be removed carefully but not entirely, since the gourmet is said to experience a slight paralysis of the tongue. In spite of intense training and strict examinations of the kitchen staff, about 200 people are said to fall victim to the delicacy per year.

Of comparable toxicity are the secretions of skin glands of South American colored frogs. They contain highly active Na+ channel blockers such as **histrionicotoxin** (see Figure 14.7) and are used as arrow poisons by native tribes. Particularly infamous is **saxitoxin**, the poison of the dinoflagellate *Gonyaulax catenella*. The explosive multiplication of these algae causes the ecologically disastrous red tide. Saxitoxin is a major cause of mussel poisoning.

Investigation of the pharmacological and molecular properties of ion channels has been made possible by the patch-clamp technique (**Sidebar 14.1**).

Summary

- Voltage-gated Na+ channels propagate action potentials in excitable cells. They are complexes consisting of at least four proteins.

- The channel-forming protein, the α-subunit, comprises four groups of six transmembrane helices and one P-loop each, sharing together one large polypeptide chain. It is associated with regulatory transmembrane proteins, the β-subunits.

- After gating, the channel transiently becomes blocked by a cytoplasmic domain of the α-subunit, sharpening action potentials and enabling their targeted propagation.

- Axonal Na+ channels are inhibited by local anesthetics and some powerful animal neurotoxins.

depolarization

closed

open

refractory

Figure 14.6 Unidirectional spreading of a wave of membrane depolarization propagated by cooperating Na+ channels The membrane depolarization caused by channel opening is sufficient to open the next channel. Since, after activation, a channel does not return immediately into the closed state but becomes refractory for a certain period of time, the wave can spread in one direction only (here from left to right). A second wave may arise when the refractory channels have again taken on the closed conformation.

Tetraodon - Fugu fish

tetrodotoxin

Dendrobates histrionicus

histrionicotoxin

Figure 14.7 Toxins, of the puffer (Fugu) fish and of colored frogs, that block the axonal Na+ channel

Sidebar 14.1 Investigation of ion channels by the patch-clamp technique A detailed investigation of the pharmacological and molecular properties of ion channels has become possible by the patch-clamp technique developed by Erwin Neher and Bert Sakmann (awarded the Nobel Prize in Physiology or Medicine for 1991). By means of this method, a tiny piece of the cell membrane (the patch) is isolated that contains a few channels (or even a single one), the conductivity of which may be measured without confounding interactions. For this purpose, a small-diameter glass pipette filled with a salt solution and connected with a measuring instrument is pressed onto the cell surface and a membrane patch (about 1 μm diameter) sealed with the mouth of the pipette by suction. Depending on whether the patch is left on the cell or separated, channel conductivity may be measured in a whole-cell environment or in an isolated preparation (**Figure 14.8**).

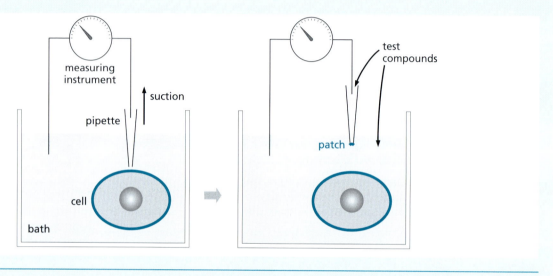

Figure 14.8 Patch-clamp method A small piece (patch) of a cell membrane sealing the tip of a micropipette is separated by suction. An electrical circuit is established by generating a potential difference between bath and pipette. In this circuit, the patch provides resistance. Test compounds added either to the bath or to the salt solution within the pipette may change the conductivity by interacting with the extra- or intracellular domains of the ion channel(s) in the patch.

14.3 The multi-talented Epithelial Na⁺ channels

Epithelial Na⁺ channels (ENaCs) constitute the other large family of sodium-specific ion channels. With two transmembrane helices and a P-loop but lacking a voltage sensor, an ENaC subunit resembles prototype 2 of channel structure. The pore-forming α-subunits are expressed in several isoforms that associate into a heterotetramer to form – together with regulatory β- and γ-subunits – a channel. ENaCs are widely distributed in multicellular organisms. The homologous channels of *Caenorhabditis elegans* are called degenerins; therefore, an ENaC/degenerin superfamily is sometimes cited. Since ENaCs are specifically blocked by the diuretic amiloride, they are also frequently called **amiloride-sensitive sodium channels**. Although some of them manage the directed transport of Na⁺ ions through epithelia, the adjective "epithelial" is somewhat misleading, since channels of this family are found in many other tissues.

Those ENaCs that transport sodium ions through epithelia play a key role in regulating the body's salt and water balance. Many epithelia (in the kidney, intestine, glands, and respiratory tract) take up Na⁺ from the urine, contents of the bowel, and secretions and supply them to the blood. The best-known example is Na⁺ reabsorption in kidney tubules. This process is stimulated by hormones such as vasopressin and aldosterone and turned down by ANP, the natriuretic peptide

(A)

ENaC

membrane

cytoplasm

N

C

(B)

environment

body

water

K⁺ channel

K⁺

K⁺

ATPase

Na⁺

ENaC

Na⁺

Na⁺

water

apical
membrane

basolateral
membrane

Figure 14.9 Mode of action of epithelial Na⁺ channels (ENaC) (A) Subunit of the channel that exhibits two transmembrane helices (shown as cylinders) and a large extracellular loop that dips into the membrane, forming a P-loop (teal). Each channel consists of at least four subunits. (B) Epithelial cell connected to neighboring cells by tight junctions (dark gray bars) that seal the body against the environment. Active ion transport (against a concentration gradient) accomplished by the apical ENa channel (teal) and the basolateral Na⁺/K⁺-dependent ATPase is accompanied by a passive water flux. K⁺ taken up by the cell through the ATPase is extruded again through basolateral K⁺ channels. For chloride, the counterion of Na⁺, a separate transport system exists (see Section 14.7).

hormone from atrial heart cells. The effects of the hormones are partly mediated by ENaCs: while vasopressin and aldosterone induce the de novo synthesis of ENaC proteins and probably activate latent channels, ANP exhibits an opposite effect.

To do their job, ENaCs have to cooperate with ion pumps and K⁺ channels. Directed Na⁺ transport requires an asymmetrical distribution of these components in epithelial cells: the Na⁺/K⁺-dependent ATPase (ion pump) is localized in the basolateral (inner) cell membrane, together with K⁺ channels, and produces an ion gradient promoting the influx of Na⁺ through ENaCs that are components of the apical (outer) membrane (**Figure 14.9**).

Other ENaCs are gated by sensory stimuli and trigger action potentials that are propagated by voltage-gated Na⁺ channels and migrate to the brain. They are found in mechanosensitive and pain receptors of the skin and in certain taste cells (see Chapter 15). Channels of a subfamily called **acid-sensing ion channels (ASICs)** open at an acidic pH value. They respond to sour food but also to tissue acidification resulting from inflammation, injury, and diminished blood supply (for instance, in the course of angina pectoris) and evoke pain sensations. Acid-sensing sodium channels are also expressed in brain neurons. Gene-knockout experiments indicate that they are involved in learning and memory fixation. This concept is supported by observations showing pronounced pH fluctuations and waves of acidification spreading in active brain areas.

Summary

- Epithelial Na⁺ channels (ENaCs) are composed of four subunits with two transmembrane helices and one P-loop each. ENaCs regulate the osmolarity of body fluids by transporting ions through epithelia (for example salt reabsorption in the kidney). In certain sensory cells the voltage-independent gating of ENaCs triggers action potentials.

- Some ENaC isoforms (ASICs) preferentially found in neurons are acid-sensitive, responding to tissue acidification.

14.4 K⁺-selective ion channels are regulators of hyperpolarization and osmotic pressure

K⁺-selective channels constitute the largest and most diverse ion channel family and have a long evolutionary history. Their variabilty and versatility, resulting both from numerous genes and from alternative splicing of pre-mRNA, is amazing. **Figure 14.10** provides an overview of the K⁺ channel families of vertebrates.

Owing to the high intracellular K⁺ concentration, the gating of K⁺ channels mostly results in an efflux of K⁺ ions against the voltage gradient. This exerts a hyperpolarizing effect (the electrical charge inside the cell becomes more negative) that has two major consequences: it lowers the resting potential, thus hindering depolarization, and it terminates depolarization waves such as action potentials. Thus, an activation of K⁺ channels attenuates and blocks depolarization-dependent events such as neurotransmitter release and muscle contraction, whereas vice versa, channel closing or blockade may facilitate depolarization. In addition, K⁺ channels help to control the osmotic pressure of cells.

14.4.1 Voltage-controlled K⁺ channels terminate action potentials

The first voltage-dependent potassium channel discovered was the product of a gene responsible for the Shaker mutation of the *Drosophila* fly (mutant animals exhibit shaking under anesthesia). Related channels were found in all vertebrates. They are key regulators of action potentials: in excitable cells, the voltage-dependent potassium channel K_v is the direct opponent of the voltage-dependent Na⁺ channel. Since it opens upon depolarization, its major effect is to terminate the action potential by negative feedback (Figure 14.10). The K_v channel has prototype 1 architecture and resembles the voltage-dependent Na⁺ channel, albeit with the variation that the four α-subunits are separated rather than joined on a single polypeptide chain. This constructive principle renders K_v channels extremely variable: in mammals, the α-subunits are encoded by more than 50 genes and, according to a unit construction system, the isoforms assemble into homo- and heterotetramers. The different combinations are tissue-specifically expressed and respond individually to changes of

Figure 14.10 K⁺-selective ion channels (A) Model of the prokaryotic KscA channel. The structure consisting of four identical subunits forming a central ion pore (the teal dot symbolizes a potassium ion) is typical for most potassium channels. (Adapted from Wikipedia: Potassium Channel, https://en.wikipedia.org/wiki/Potassium_channel.) (B) The pore-constituting α-subunits (with the numbers of α-subunits forming a channel) of vertebrate potassium channels. From left to right: voltage-controlled channel K_v; voltage- and Ca²⁺-controlled channel BK (Slo); Ca²⁺/CaM (calmodulin)-controlled channel SK; inwardly rectifying channel K_{ir}; G_i-protein-controlled inwardly rectifying channel GIRK interacting with Gβγ; and 2P-domain channel. Each channel subunit exists in various isoforms. Voltage-sensing helices are colored. (C) The influence of K⁺ channels on the action potential: the influx of K⁺ lowers the resting potential and terminates the phase of depolarization, causing repolarization and afterhyperpolarization.

(A) pore α-subunits

cell membrane

β-subunits

(B) pore α-subunits

cell membrane

β-subunits

ball on a chain

the membrane potential. In the vertebrate K_v channel, the pore-forming α-subunits associate with regulatory β-subunits. These are encoded by 12 genes and—owing to alternative mRNA splicing—expressed in a large number of isoforms. The β-subunits of potassium channels share only the name but neither structure nor function with the β-subunits of the voltage-dependent Na^+ channels. In fact, most of them are not membrane-spanning cell adhesion molecules of the Ig-like family but cytoplasmic proteins with a wide variety of functions. As an example, the doorkeeper function may be mentioned: by means of a positively charged globular structure fixed to a mobile polypeptide chain, certain β-subunits reversibly plug the channel pore (ball-on-a-chain model; see **Figure 14.11**). This effect resembles that of the inactivation domain in the sodium channel's α-chain. Depending on the channel type the ion-selective pore can also be blocked through a rearrangement of the membrane-spanning segments of the α-subunit occurring when the membrane potential reaches a critical value. As we have seen, such a short-term desensitization of ion channels is essential for rapid termination as well as for unidirectional spreading of the action potential.

A K_v isoform that has gained considerable medical interest is derived from a <u>h</u>uman <u>E</u>ther-a-go-go-<u>R</u>elated <u>G</u>ene and called hERG. The strange name refers to a *Drosophila* mutant that when anesthetized by ethyl ether starts to shake its legs like a go-go dancer. The major function of hERG is to shape and terminate the action potential that controls the heartbeat. This occurs in two steps: immediately after the peak of depolarization and at the end when the resting potential is to be reached again. In between the channel becomes desensitized for a short period of time. This results in a plateau phase of partial repolarization that is important for supplying the heart muscle with Ca^{2+} (a typical action potential of the heart is shown in Section 14.6.2). Defects of hERG function—caused by point mutations or intoxication (for instance by antipsychotic drugs)—are the reason for life-threatening cardiac arrhythmias. Therefore, the channel is a target of a variety of medical drugs.

Certain K_v channels, particularly those at neuromuscular synapses, are blocked by **dendrotoxins**. These proteins are components of mamba snake venom. They are widely used as tools for the pharmacological investigation of potassium channels. Other K_v channel blockers employed for research are tetraethylammonium salts and 4-aminopyridine. Derivatives of the latter are clinically used for the treatment of multiple sclerosis.

Figure 14.11 Model of the voltage-controlled potassium channel K_v (A) Molecular model showing α-subunits that form the membrane pore (black). At the cytoplasmic side, they are associated with the β-subunits (teal) exhibiting mobile globular domains (ball on a chain) (B) that fit the central pore and, thus, may interrupt the ion flux. The molecular model is based on X-ray crystallography. (B, adapted from Aldrich RW [2001] *Nature* 411:643–644.)

Calcium-sensitive potassium channels

Action potentials induce secretory processes, such as neurotransmitter release at synapses, and trigger muscle contraction. Both responses depend on an increase of the cytoplasmic Ca^{2+} concentration. To this end, voltage-dependent calcium channels in the presynaptic membranes of neurons and in muscle cell membranes become gated by depolarization. This effect is, of course, opposed by K_v channel gating. In addition, the resulting Ca^{2+} influx is regulated by a special group of Ca^{2+}-sensitive potassium channels via negative feedback. These K^+ channels are structurally related to K_v channels; two types are distinguished by their electrophysiological features, BK channels and SK channels.

BK channels (big conductance K^+ channels) function as coincidence detectors resembling logical AND gates: though they are voltage-controlled, gating is not evoked by depolarization alone but requires, in addition, an increase of the cytoplasmic Ca^{2+} level. Unlike K_v channels, the α-subunits of BK channels have a unique seventh transmembrane segment S0 at the amino (N)-terminus and an extended cytoplasmic Ca^{2+} sensor domain (see Figure 14.10). BK channels are expressed in almost any tissue. Owing to alternative mRNA splicing, post-translational modifications, and variable subunit structures they are found in innumerable modifications. The major function of BK channels is to damp excitatory stimuli. Their role in a particular physiological process, the relaxation of smooth muscles, is described in Section 14.6.2. Mammalian cells express two close relatives of the BK channel, Slo2 and Slo3. Together with BK (also called Slo1) they constitute the family of **Slo channels** (the name refers to the *Drosophila* mutant "Slowpoke" lacking an active Slo1 channel). Slo2 is an Na^+-controlled potassium channel found predominantly in brain, while the expression of the acid-sensitive Slo3 seems to be restricted to testis.

SK channels (small conductance K^+ channels) become gated by Ca^{2+} alone, interacting with the calcium sensor protein calmodulin, which is a channel subunit. The α-subunits resemble those of K_v channels, although SK channels are not voltage-controlled since they do not contain a sufficient number of positively charged amino acid residues in the S4 helix to function as a voltage sensor. SK channels are (together with K_v and BK channels) responsible for the afterhyperpolarization phase of an action potential, when the membrane potential falls below the resting potential for a short time (see Figure 14.10).

Calcium-sensitive potassium channels are targets of **apamin**, a component of bee venom, and of the scorpion poison **charybdotoxin**. These peptides block the channels and cause pathological hyperactivation of the nervous system. The consequences are extreme pain and other severe dysfunctions.

14.4.2 Inwardly rectifying K^+ channels set the resting potential

A large group of potassium channels was discovered owing to their apparently strange electrophysiological properties: in contrast to K_v channels, they become gated upon hyperpolarization and closed upon depolarization. In humans the pore-forming α-subunits of these channels are encoded by at least 15 genes, giving rise to a high structural and functional variability. The individual channel isoforms are tissue-specifically expressed, in particular in the central nervous system, the musculature, endo- and exocrine glands, and epithelia.

Hyperpolarization-gated potassium channels belong to channel prototype 2. As such they lack an intrinsic voltage sensor such as the S3–S4 hairpin domain. How can they be voltage-dependent in a reverse manner from K_v channels? The answer is that upon depolarization the cytoplasmic side of the ion gate becomes plugged by positively charged polyamines like spermine and by Mg^{2+} ions.

The gating of K_v channels always results in an efflux of K^+ ions. In contrast, K^+ channels that are gated by hyperpolarization may allow a K^+ *influx*, resulting in depolarization when (depending on the membrane potential and the intracellular K$^+$ concentration) the electrical force of the inward-directed voltage gradient exceeds the chemical force of the outward-directed concentration gradient. This effect, which is essential for setting the resting potential, led to the name inwardly rectifying potassium channels (K$_{ir}$ channels). However, when the potential becomes less negative, such channels begin to propagate K$^+$ efflux, causing hyperpolarization. This would impair or even prevent depolarization-dependent processes. Therefore, the channels must be closed upon depolarization.

K$_{ir}$ channels that regulate the resting potential are consistently open at sufficiently negative potentials (leak channels). They are of utmost importance for the generation and shaping of action potentials: in the initial phase their closure facilitates depolarization, and in the subsiding phase (when the repolarizing K$_v$ channels are already closed again) they open again and complete the repolarization of the cell, thus terminating the action potential (see Figure 14.10). This holds true, in particular, for long-lasting excitations such as those required for heart contraction that critically depend on the correct function of the leak channels.

The manipulation of K$_{ir}$ channels offers a possibility to dampen processes triggered by action potentials: all that has to be done is to keep the channel open by appropriate signals in spite of depolarization. In fact, signal-regulated K$_{ir}$ channels are major negative control devices. This is shown by the following examples.

G$_i$-protein-regulated K$^+$ channels

This channel type is termed K$_{ir}$3 or better known as GIRK (G-protein-gated inwardly rectifying K$^+$ channel). Its depolarization-dependent closure is prevented by Gβγ-subunits that in sufficient concentrations are released from the particularly abundant G$_{i,0}$-proteins and bind to the four channel subunits (**Figure 14.12**). Open GIRK channels lower the resting potential and impair the generation of an action potential. Together with inhibition of cAMP production, this is the effect of all extracellular signals that interact with G$_{i,0}$-protein-coupled receptors, provided the corresponding cell expresses GIRK channels (as is the case in many organs, particularly in brain, heart, and endocrine glands). G$_{i,0}$-coupled receptors activating GIRK channels are widely distributed. In particular, they mediate inhibitory effects of neurotransmitters (see Table 16.1 in Section 16.1).

Thus, this channel type plays an important role in the nervous system and the target organs of neuronal signaling. Again the effects focus on Ca^{2+}-dependent events: as already explained hyperpolarization primarily turns down the influx of Ca^{2+} ions through voltage-operated Ca^{2+} channels. This results in an attenuation of Ca^{2+}-dependent processes such as muscle cell contraction and neurotransmitter secretion. We have already encountered such GIRK effects: the parasympathetic down-regulation of heart activity (see Section 5.5) and the α-adrenergic feedback inhibition of noradrenaline release at sympathetic synapses (see Section 5.3.3). Other examples are found in Chapter 16. Gβγ-subunits also inhibit voltage-operated Ca^{2+} channels directly, thus augmenting the effect of hyperpolarization (see Section 14.5.2).

Figure 14.12 G$_i$-protein-regulated K$^+$ channel GIRK Gβγ-subunits, mainly released from G$_{i,0}$-proteins upon activation of a G-protein-coupled receptor (GPCR), prevent closure of the K$_{ir}$ channel and induce hyperpolarization, while the corresponding Gα$_i$-subunits inhibit adenylate cyclase. As a result, voltage-, Ca^{2+}-, and cAMP-dependent signaling is inhibited.

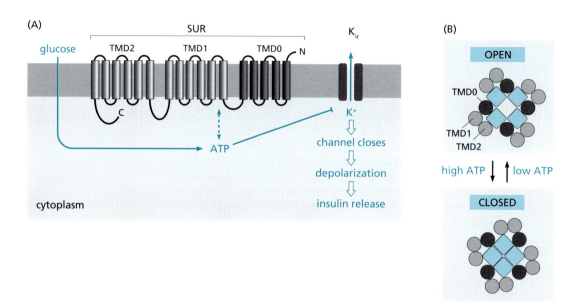

Figure 14.13 Regulation of insulin secretion by the KATP channel
(A) The channel consists of four pore-forming K_{ir} subunits and four regulatory subunits SUR (sulfonylurea receptor). SUR has the structure of an ABC transporter with two groups of six transmembranes each (TMD1 and -2, light gray) and two cytoplasmic ATP/ADP-binding sites. In addition, it contains a group of five N-terminal transmembrane domains TMD0 (dark gray). At low ATP levels, the channel is open and thus inhibits insulin secretion. ATP produced in response to glucose uptake binds to both the channel and SUR. As a result, the channel closes and depolarization and insulin secretion are facilitated. (B) Arrangement of the subunits (Kir colored) as seen from above (based on X-ray crystallography).

$G_{i,0}$-dependent signaling opposes G_s- and $G_{q,11}$-controlled signaling. Vice versa, GIRK channels are inhibited along the $G_{q,11}$-pathway, most probably by protein kinase C-catalyzed phosphorylation.

ATP-regulated potassium channel K_{ATP}

This channel provides a further example of a ligand-controlled K_{ir} channel (**Figure 14.13**). Although closely related to the GIRK channel, it forms a complex with the ATP/ADP-binding transmembrane protein **SUR (sulfonylurea receptor)** rather than with Gβγ-subunits. SUR belongs to the family of ATP-binding cassette (ABC) transporters (see Section 3.3.1). As a specific target of antidiabetic drugs of the sulfonylurea type, this protein is of eminent medical importance. In fact, the K_{ATP}–SUR team participates in the control of insulin secretion by the β-cells of the islets of Langerhans. Here SUR acts as an energy sensor, regulating channel conductance in accordance with the intracellular ATP/ADP ratio that directly depends on the blood sugar level: the cells respond to rising blood sugar by increased uptake of glucose through membrane-bound glucose transporters. This leads to an increased formation of ATP that is bound by both SUR and the channel subunits. As a result, the channel is closed and the resting potential becomes less negative, facilitating membrane depolarization and insulin release. Since the latter is an exocytotic process, it is triggered by Ca^{2+} ions entering the cell through voltage-dependent Ca^{2+} channels that sensitively respond to any change of the membrane potential. A lowering of the blood sugar level causes a fall of the intracellular ATP level and the ATP bound to SUR becomes displaced by ADP. Since under this condition the K_{ATP} channel remains open, the cell becomes hyperpolarized and insulin secretion decreases. By stabilizing the ATP complex of SUR, sulfonylurea drugs promote insulin release from the pancreas.

As we have seen, K_{ATP} channels control the membrane potential depending on the metabolic state of the cell. They do this not only in the pancreas but also in other tissues including heart and vascular smooth muscle, skeletal muscle, and brain. In a specialized set of hypothalamus neurons, a fall of the blood sugar level activates K_{ATP} channels. In turn a neuronal impulse is sent to the pancreatic α-cells that release the sugar-mobilizing hormone glucagon. Another important response to channel gating is the widening of blood vessels because hyperpolarization hinders smooth muscle contraction. The brain and the heart are protected in this way from the harmful consequences of diminished blood supply (resulting in decreased ATP production) in the course of stroke and myocordial infarction. Such effects are

targets of additional pharmaceutical drugs. As a more curious example, minoxidil may be mentioned: by stimulating potassium transport through K$_{ATP}$ channels, it lowers the blood pressure and is said to prevent hair loss by promoting blood circulation in the skin.

14.4.3 2P-domain channels function as background channels, pressure relief valves, temperature sensors, and pain relievers

A third family of K$^+$-selective ion channels is closely related to K$_{ir}$ channels. These channels consist of two α-subunits: each has four transmembrane helices and two P-loops and may be looked upon as two covalently linked subunits of a K$_{ir}$ channel (see Figure 14.10). The P-loop tandem led to the name 2P-domain or K$_{2P}$ channels. Since, like other K$^+$ channels, 2P-domain channels exhibit only one pore, the name twin pore channels (sometimes found in the literature) is misleading.

Fifteen human isoforms have been found. On the basis of structural parameters, they are subdivided into six subfamilies denoted by four-letter acronyms beginning with T (Tandem of P-loops), such as TWIK (T in a Weak Inwardly rectifying K$^+$ channel) or TREK (TWIK-RElated K$^+$ channel). In contrast to K$_{ir}$ channels, K$_{2P}$ channels do not fully close upon depolarization. This qualifies them as background channels, setting the resting potential at different levels and generally dampening excitability. While some are constitutively open (leak potassium channels), others are gated by a variety of physical or chemical stimuli.

Such a physical stimulus is mechanical pressure from inside the cell. Thus, some K$_{2P}$ channels are cellular pressure relief valves, opening when the cell membrane is under tension. Such a situation occurs in a hypotonic environment. To prevent a rupture, the cell has to lower the concentration of osmotically effective molecules, mainly by releasing electrolytes, such as K$^+$, the cytoplasmic concentration of which is particularly high. To retain electric neutrality, pressure-sensitive chloride channels open simultaneously with potassium channels (see Section 14.7.1). As a result, a KCl solution is poured out by the cell. Another physical stimulus is temperature. Temperature-sensitive K$_{2P}$ channels are expressed, for instance, in sensory neurons and in the hypothalamus. They are thought to function as temperature sensors because they close in response to cooling, thus facilitating the generation of action potentials. Certain hypothalamus neurons use this channel type to control body temperature. This effect is mediated by prostaglandin E$_2$ which, by interacting with a G$_s$-protein-coupled receptor, induces cAMP formation in neurons. As a result, protein kinase A becomes activated and phosphorylates a particular K$_{2P}$ channel. Owing to this modification, the channel becomes blocked, rendering the resting potential less negative and facilitating depolarization. This leads to an increased activity of those hypothalamus neurons that cause a rise in body temperature. Since fever develops along the same route, drugs that inhibit prostaglandin synthesis, such as acetylsalicylic acid (aspirin), exhibit an antipyretic effect (see Section 4.4.5).

Other 2P-domain channels are gated by intracellular signaling molecules such as arachidonic acid and lysophospholipids. These lipid mediators are released in the brain in response to disturbances of the blood supply. The corresponding 2P-channels probably participate in neuronal protection from damage caused by stroke: by suppressing a pathological overstimulation of voltage-dependent Ca^{2+} channels, which is a characteristic feature of stroke (see Section 16.4.2), they prevent a fatal flooding of the cell by Ca^{2+} ions. In fact, K$_{2P}$ channels are potential targets for the treatment of stroke, but also of pain and cardiovascular and mental diseases.

Some K$_{2P}$ channels are of great medical interest because they play an important role in the action of **gaseous narcotics** such as chloroform, halothane, and diethyl ether that open the channels, thus hindering the generation of action potentials in those neurons that transmit pain signals to the brain.

Summary

- The highly variable voltage-dependent K^+ channels (K_v) are the counterparts of the voltage-dependent Na^+ channels. Upon gating by depolarization, they cause hyperpolarization, thus terminating action potentials. A subfamily of K_v-related channels is regulated by Ca^{2+} ions and is important for negative feedback control of Ca^{2+}-dependent processes such as smooth muscle contraction and neurotransmitter release.

- Inwardly rectifying potassium channels (K_{ir} channels) are open at negative membrane potentials and become closed upon depolarization. They are important for setting the resting potential and shaping action potentials.

- The K_{ir} channel GIRK, gated by $G\beta\gamma$-subunits released from $G_{i,o}$-coupled receptors, is a key mediator of inhibitory neurotransmitter effects and plays an important role in the negative feedback control of neurotransmitter release at synapses and in the parasympathetic attenuation of heart activity.

- Another K_{ir} channel is under the control of the ABC transporter protein SUR that closes the channel upon ATP binding, thus facilitating events depending on depolarization (such as insulin secretion and smooth muscle contraction), whereas ATP shortage has an opposite effect. SUR is a target of antidiabetic drugs.

- A third family of K^+ channels characterized by a 2P-domain structure also controls the resting potential. In addition, it is specialized for osmotic pressure relief. For this reason, these channels have to cooperate with chloride channels. 2P-domain channels expressed in neurons control body temperature and transmit pain sensations.

14.5 Calcium ions: the most versatile cellular signals

Ca^{2+} ions occupy a central position in cellular data processing. Without any exaggeration, it may be stated that there is practically no cell physiological process that is not influenced and regulated directly or indirectly by calcium. For calcium signaling, cells keep the cytoplasmic calcium concentration at a level that is about 4 orders of magnitude lower than in the extracellular medium. Consequently, gating of Ca^{2+} channels automatically leads to a calcium influx that provides the triggering signal for cellular responses. This gating is induced and modulated by exogenous and intracellular input signals. In most vertebrate cells, calcium influx does not change the membrane potential significantly because the Ca^{2+} concentration is 2–4 orders of magnitude lower than that of sodium, potassium, and chloride ions. Instead, Ca^{2+} ions function as second messengers, being recognized (specifically bound) by effector proteins that, upon interaction with Ca^{2+}, undergo conformational and functional changes, thus producing output signals. This is in clear contrast to the monovalent ions Na^+, K^+, and Cl^-, which primarily control the membrane potential.

14.5.1 Generation and termination of Ca²⁺ signals

As will be shown, cells modulate the Ca^{2+} signal in an almost unlimited manner, with respect to intracellular localization, intensity (or amplitude), and temporal fluctuation (or frequency). Moreover, they are able to generate stationary and spreading Ca^{2+} waves as well as spatial concentration patterns. These sophisticated processes require precise control of both the rise and the decay of Ca^{2+} signals because otherwise the cell would be flooded by Ca^{2+} and become paralyzed and eventually killed. Therefore, we shall first address the question of how the cellular Ca^{2+} level is regulated.

Calcium pumps

Like other second messengers, Ca^{2+} acts as an allosteric modulator of protein functions. In the cytoplasm of nonstimulated cells the concentration of free Ca^{2+} ions is kept at a level of about 0.1 μM, which is too low to significantly influence effector proteins (which respond only at 1–10 μM Ca^{2+}). To achieve this low concentration, Ca^{2+} ions are pumped out of the cell and sequestered in the lumen of the ER, in the mitochondrial matrix, and by cytoplasmic calcium-binding buffer proteins. The concentration of free Ca^{2+} in the extracellular fluid, ER, and mitochondria is about 1 mM, or about 4 orders of magnitude higher than in the cytoplasm. Thus, to generate and maintain the low cytoplasmic Ca^{2+} level, the ions must be transported against a steep concentration gradient. This requires energy derived from ATP hydrolysis, catalyzed by **Ca^{2+}-dependent ATPases**. These enzymes are integrated in the ER and plasma membranes. ATP hydrolysis causes a conformational change of the ATPase that conveys a noncovalently bound Ca^{2+} ion from the inside to the outside of the cell. Thus, these ATPases are calcium pumps that keep the resting cytoplasmic calcium level low and, upon gating of Ca^{2+} channels, ensure a rapid termination of the Ca^{2+} signal, preventing the fatal consequences of calcium overload.

Similar or even stronger pumps are the membrane-bound **Na^+–Ca^{2+} exchangers** or transporters. They compensate the efflux of one Ca^{2+} ion by an influx of three Na^+ ions, thus exhibiting (in contrast to the ATPase) a depolarizing effect. The energy is derived from the Na^+ gradient across the cell membrane; that is, it is ultimately based on the activity of the Na^+/K^+-dependent ATPase generating this gradient. Owing to its depolarizing action, the exchanger not only lowers the cytoplasmic calcium level but also raises the resting potential and facilitates the subsequent generation of an action potential. This effect plays an important role in the periodic firing of synapses and in continuously oscillating heart muscle cells, which both depend on Ca^{2+} influx provoked by depolarization.

An electrogenic Na^+–Ca^{2+} exchanger is also found in mitochondrial membranes. Together with a neutral H^+–Ca^{2+} exchanger, it catalyzes the efflux of mitochondrial calcium into the cytoplasm in a strictly controlled way against a steep charge gradient across the membrane (180 mV, inside negative) that is generated by the respiratory electron transport chain and drives mitochondrial ATP synthesis. Mitochondria contain, in addition, nonselective Ca^{2+} pores that open upon Ca^{2+} overload. These permeability transition pores penetrating the outer and inner mitochondrial membranes are probably also responsible for the release of cytochrome *c* and other pro-apoptotic proteins from mitochondria in the course of programmed cell death (see Section 13.2.5).

While the ER takes up cytoplasmic Ca^{2+} by means of the Ca^{2+}-dependent ATPase SERCA (Sarco/Endoplasmic Reticulum Calcium-dependent ATPase), mitochondria absorb cytoplasmic Ca^{2+} through a special transport protein. This **uniporter** is localized in the inner mitochondrial membrane and allows the influx of Ca^{2+} from the cytoplasm into the mitochondrial matrix. It works only at rather high Ca^{2+} concentrations of 0.01 mM and more and is assumed, therefore, to be positioned in the immediate neighborhood of Ca^{2+} channels in the plasma membrane and the ER, where the Ca^{2+} concentration is particularly high. The pathways and devices controlling the cytoplasmic calcium level are summarized in **Figure 14.14**.

Calcium channels

Calcium channels are the opponents of calcium pumps. The cell membrane contains both ligand- and voltage-controlled types, whereas in the ER, only channels gated by intracellular messengers are found. A third channel type of the plasma membrane, called a store-operated calcium channel, responds to the Ca^{2+} concentration in the lumen of the ER, which beside mitochondria is the major cellular calcium store. Finally, Ca^{2+} is also transported together with Na^+ and K^+ by nonspecific cation channels, particularly those controlled by neurotransmitters (see Chapter 16). The different types of calcium channel are summarized in **Table 14.1**.

Figure 14.14 Regulation of the cytoplasmic Ca²⁺ level Extracellular Ca²⁺ is taken up through voltage- and ligand-dependent membrane channels (teal) that are opposed by a Ca²⁺-dependent ATPase (PMCA, plasma membrane Ca²⁺-dependent ATPase) and an electrogenic Na⁺–Ca²⁺ exchanger (EX$_{Na}$). Intracellular calcium stores are the endoplasmic reticulum (ER) and the mitochondria. The ER membrane contains ligand-controlled Ca²⁺ channels (teal) opposed by a Ca²⁺-dependent ATPase (SERCA, sarco/endoplasmic reticulum Ca²⁺-dependent ATPase). The mitochondria deliver Ca²⁺ to the cytoplasm by an electrogenic Na⁺–Ca²⁺ exchanger (EXN$_a$), an H⁺–Ca²⁺ exchanger (EXH), and permeability transition pores (not shown). Ca²⁺ influx into the mitochondrial matrix occurs through a uniporter (UP) along a charge gradient that is generated by the respiratory electron transport chain. The resting concentrations of Ca²⁺ in the individual compartments are given in teal.

14.5.2 Voltage-operated calcium channels

The plasma membrane contains voltage-operated Ca²⁺ channels (VOCCs) that are opened by depolarization. They have been found in all eukaryotic cell types examined thus far, but not in prokaryotes. Because they connect changes of the membrane potential with Ca²⁺-dependent cellular processes such as secretion and movements,

Table 14.1 Ca²⁺-transporting membrane channels: an overview

In the plasma membrane
voltage-operated Ca²⁺-selective channels (VOCCs)
storage-operated Ca²⁺-selective channels (SOCCs)
arachidonic acid-regulated Ca²⁺-selective channels (ARCCs)
nonselective cation channels of the TRP family
cyclic nucleotide-gated nonselective cation channels
ionotropic neurotransmitter receptors with intrinsic nonselective cation channel activity
In the ER and in the outer nuclear membrane
InsP₃/Ca²⁺-controlled Ca²⁺-selective channels
ryanodine receptor Ca²⁺-selective channels
In other organelles
NAADP-controlled channels
sphingosine 1-phosphate-controlled channels
mitochondrial uniporters
mitochondrial permeability transition pores

they are major players in neurotransmitter release at synapses and in cellular motility, including muscle contraction. In fact, action potentials generated by nerve and muscle cells primarily serve the gating of VOCCs.

Basic structures and types of voltage-operated calcium channel

The molecular architecture of VOCCs is similar to that of the voltage-dependent Na$^+$ channel. The structural core is a large pore-forming protein with 4–6 transmembrane helices and four P-loops. It is called α_1-subunit and is flanked by additional subunits (α_2, β, γ, δ) that exhibit stabilizing and regulatory functions (**Figure 14.15**). The β-subunit, a cytoplasmic protein, controls the exposure of the α_1-subunit at the cell membrane by masking sequences that otherwise would hinder membrane trafficking. In addition, it modulates the electrophysiological properties of the channel and interacts (via an SH3 domain) with other signaling proteins. Thus, it differs entirely from the β-subunits of voltage-dependent Na$^+$- and K$^+$-channels. The role of the δ-subunit, a single-pass transmembrane protein, is that of an adaptor for the extracellular α_2-subunit that modulates the efficacy of the channel. Both subunits are encoded by the same gene and are connected to each other by disulfide bonds. The role of the γ-subunit, a protein with four

Figure 14.15 Voltage-operated Ca^{2+}-channels: basic architecture and major occurrences Shown are the transmembrane architecture (A) and the arrangement of the subunits (B) of an L-type channel. P, P-loops of α_1-subunit. In (A), the S4 voltage sensors are colored. (Adapted from Hall ZW [1992] Molecular Neurobiology. Sunderland, MA: Sinauer Associates.)

transmembrane domains, is not well understood. Whether this subunit is common to all VOCCs or restricted to the cytoskeletal type is an open question. An additional cytoplasmic subunit of most VOCCs is the Ca^{2+}-binding protein calmodulin (CaM). It mediates feedback inhibition of channel gating.

Humans possess 22 genes encoding α_1-subunits. Together with four $\alpha_2\delta$, four β, and eight γ genes, they could assemble theoretically into more than 1000 different combinations. How many of these are realized is an open question, however.

On the basis of pharmacological data, VOCCs are grouped in several subfamilies labeled L, N, Q, R, and T, which differ from each other primarily by the α_1-isoforms.

Apart from being expressed in certain brain regions, **L-type channels** (long-lasting) are typical for skeletal, smooth, and heart muscle cells, where they are involved in stimulus–contraction coupling (see Section 14.6.2). They are known also as **dihydropyridine receptors** in reference to a family of medical drugs that selectively block L-type channels and are used for the treatment of angina pectoris, cardiac arrhythmia, and high blood pressure. L-type channels open gradually when the membrane potential becomes less negative. At maximal depolarization (at the peak of an action potential), they are opened to about 70%. By such a graded response, the digital all-or-nothing signals of action potentials are transformed into analog signals.

VOCCs that induce neurotransmitter release across presynaptic neuronal membranes belong to the **N** (neural), **P**, **Q**, and **R subfamilies**. They respond to moderate to strong potential changes, as are typical for action potentials, and they are targets of highly toxic animal poisons. Examples are ω-conotoxin and ω-agatoxin. **Conotoxins** are a group of peptides found in the venom of marine cone snails (genus *Conus*). They block various ion channels; the ω-form in particular inhibits the N-type VOCC. This blockade exerts an analgesic (pain-relieving) effect much stronger than that of morphine. In fact, synthetic derivatives of ω-conotoxin are used in pain therapy.

The peptide **ω-agatoxin** is a component of the venom of funnel web spiders (family Agelenidae). It blocks P and Q channels. The **T-type channels** (transient) expressed by neurons and other tissues (such as heart muscle) are gated, in contrast to other VOCCs, upon a minute increase of the resting potential and become refractory above –40 mV. Owing to this peculiar property, they play an important role in periodic changes of the membrane potential such as the oscillations occurring in the pacemaker cells of the heart (see Section 14.6.2).

Regulation of voltage-operated calcium channels

The key role of VOCCs in cellular signal transduction and the potentially dangerous effects of Ca^{2+} require precise regulation of channel gating (**Figure 14.16**). Like Na^+ and K^+ channels, VOCCs undergo auto-inactivation, ensuring a sharpening of the Ca^{2+} signal and preventing fatal flooding of the cell by Ca^{2+}. For this voltage-dependent desensitization, a cytoplasmic domain of the α_1-subunit is thought to plug the pore, resembling the ball-on-a-chain mechanism of Na^+ and K^+ channel inactivation. Ca^{2+} itself also blocks the channel by interacting with the calcium sensor protein CaM, which is a separate subunit of the channel noncovalently bound at the carboxy (C)-terminal domain. Transient channel refractoriness is also a condition for unidirectional propagation of calcium waves (see Section 14.5.4).

In contrast, phosphorylation catalyzed by PKA or Ca^{2+}/calmodulin-dependent protein kinase II (CaMKII) facilitates and prolongs channel gating, causing, for instance, sympathetic and stress effects in the heart (see Section 14.6.2). Neuronal calcium channels, in particular, are blocked by $G\beta\gamma$-subunits mainly released from receptor-coupled $G_{i,o}$-proteins. As mentioned in Section 14.4.1, $G\beta\gamma$-subunits, in addition, activate potassium channels of the GIRK type, rendering the resting potential more negative. In other words, $G\beta\gamma$-subunits suppress the generation of a Ca^{2+} signal

Figure 14.16 Regulation of voltage-operated Ca²⁺ channels Resembling voltage-dependent Na⁺ channels, the Ca²⁺ channels undergo desensitization by an auto-inhibitory domain positioned between subdomains I and II. A C-terminal domain interacts with a cytoplasmic complex of Ca²⁺ and the major calcium-binding protein calmodulin (providing negative feedback control of channel permeability). Moreover, Ca²⁺ channels are inhibited by interaction with Gβγ-subunits (mainly derived from $G_{i,o}$-proteins). In the presynaptic neuronal membrane, an interaction with syntaxin/SNAP25 triggers exocytosis of secretory vesicles. The L-type channel couples with ryanodine receptors in skeletal muscle and is activated by C-terminal phosphorylation catalyzed by protein kinase A or CaM kinase II in heart muscle (sympathetic effect; see Section 5.5).

at two points: membrane depolarization and Ca²⁺ influx. This reaction plays a major role in negative feedback control of neurotransmitter release (see Section 16.3). As an example, we have encountered the α_2-adrenergic inhibition of noradrenaline release in sympathetic neurons discussed in Section 5.3.3. Moreover, along these pathways, certain neurotransmitters and sedative drugs attenuate neuronal activity by interacting with $G_{i,o}$-coupled receptors (see Chapter 16).

14.5.3 Calcium channels of the endoplasmic reticulum

The lumen of the ER provides the intracellular calcium store primarily used by the signal-processing apparatus (whereas mitochondria regulate the overall calcium homeostasis of the cell). The calcium store of the ER cooperates with the cellular environment in that it is refilled by extracellular Ca²⁺ (see Section 14.5.5). In fact, the Ca²⁺ concentration in the ER lumen (about 1 mM) resembles that of the extracellular space. The ER membrane contains Ca²⁺-selective ion channels, which at the cytoplasmic side become gated by ligand binding. Two such ion channel families, the **inositol 1,4,5-trisphosphate (InsP₃) receptors** and the **ryanodine receptors**, have been investigated in detail. These families are closely related and consist of four subunits each. Each subunit has six transmembrane domains, a P-loop, and an extended N-terminal (cytoplasmic) domain that includes the ligand-binding site. Thus, in their membrane topology these calcium channels resemble the α-subunit of the voltage-dependent potassium channel as depicted in Figure 14.10. However, compared with the K⁺ channel subunits, their membrane integration is upside-down, with the ligand binding sites facing the cytoplasm. Moreover, they are not voltage-gated because the positive charge of their S4 helices is not high enough. Both receptor types exist in various isoforms that are tissue-specifically expressed.

Strictly speaking, InsP₃ and ryanodine receptors are not only Ca²⁺-selective channels but also Ca²⁺ sensors. In fact, their calcium permeability is regulated by cytoplasmic Ca²⁺ according to a biphasic mode: the channels open at Ca²⁺ concentrations below 300 nM (positive feedback) and close above 300 nM Ca²⁺ (negative feedback). Channel gating by Ca²⁺ is known as **Ca²⁺-induced Ca²⁺ release** from the ER (**Figure 14.17**). This provides the basic mechanism of intracellular calcium signaling.

Figure 14.17 Ca²⁺-induced Ca²⁺ release from the endoplasmic reticulum Upon sensitization by intracellular messengers (cADPR, InsP₃, Ca²⁺), receptor ion channels (InsP₃R and RyR) in the endoplasmic reticulum (ER) membrane release Ca²⁺ into the cytoplasm. (A) Each Ca²⁺ pulse opens the next receptor ion channel, generating a calcium wave that spreads across the ER membrane. (B) Side views of RyR and InsP₃R embedded in the ER membrane . Each channel consists of four identical subunits. (C) Two-dimensional crystal structure ("checkerboard pattern") of tetrameric RyR receptors in the ER of skeletal muscle (seen from above). The teal dots symbolize the central ion pores. (D) Each ion channel becomes desensitized when the Ca²⁺ concentration in its immediate vicinity exceeds a certain value. This mechanism protects the cell from a fatal increase of the Ca²⁺ level and renders a Ca²⁺ wave unidirectional, somewhat resembling an action potential. The models in (B) and (C) are based on cryo-electronmicroscopy. (Partly adapted from van Petegem F [2015] *J Mol Biol* 427:31–53.)

While InsP₃ receptors are found in most tissues, ryanodine receptors are particularly enriched in skeletal and heart muscle where they play a key role in excitation–contraction coupling (see Section 14.6.2). In the sarcoplasmatic reticulum they are arranged in extended quasi-crystalline "checkerboard patterns" (see Figure 14.17). In these arrays the individual channels interact with each other not only via Ca²⁺-induced Ca²⁺ release but also by direct mechanical contact. With up to 5000 amino acid residues per subunit, ryanodine receptors rank among the largest channel proteins. Both ryanodine and InsP₃ receptors contain numerous docking sites for regulatory proteins and for small signaling molecules, as well as many potential phosphorylation sites that could be targeted by a variety of protein kinases (theoretically each ryanodine receptor subunit may exist in more than 10^{42} different phosphorylation states!). Thus, ryanodine and InsP₃ receptors may be considered as huge signal-processing machines.

Ryanodine receptors were discovered as cellular binding proteins of the plant alkaloid and insecticide ryanodine. Their endogenous ligand seems to be **cyclic ADP-ribose (cADPR)**. This compound is released from nicotine amide adenosine dinucleotide (NAD⁺) by the abundant enzyme ADP-ribosyl cyclase (the mammalian form is known as CD38), which, owing to an intrinsic hydrolase activity, also inactivates cADPR (**Figure 14.18**). The signaling pathways leading to cADPR generation are not fully understood.

The endogenous ligand of InsP$_3$ receptors, functionally resembling ADPR, is InsP$_3$, the Ca^{2+}-mobilizing second messenger of a wide variety of extracellular signals (see Chapter 4). The function of both cADPR and InsP$_3$ is to increase the receptors' sensitivity for Ca^{2+}-induced Ca^{2+} release.

Another reaction catalyzed by ADP-ribosyl cyclase is the deamination of NADP⁺. The **nicotinic acid adenine dinucleotide phosphate (NAADP)** thus generated also induces a Ca^{2+} influx into the cytoplasm; however, the source is endolysosomal organelles in addition to the ER, and the corresponding ion channels seem to belong to the two-pore channel (TPC) family (see Figure 14.2). The content of NAD⁺ and NADP⁺ is a measure of the metabolic (redox) state of the cell, which via cADPR and NAADP might be linked with Ca^{2+} signaling.

Figure 14.18 Intracellular messengers inducing Ca^{2+} release from internal stores ADP-ribosyl cyclase catalyzes the formation of cyclic ADP-ribose from NAD⁺ and of NAADP from NADP⁺. Other Ca^{2+}-releasing messengers are inositol 1,4,5-trisphosphate (see Figure 4.18, Section 4.4.4) and sphingosine 1-phosphate (see Figure 4.24, Section 4.4.5).

In certain tissues, a release of Ca^{2+} from intracellular stores is triggered, in addition, by **sphingosine 1-phosphate** cooperating with $InsP_3$. The mechanism of this effect is still widely unknown.

The calcium channels of the ER are interlinked with various signaling pathways. $InsP_3$ receptors are inhibited by phosphorylation catalyzed by Ca^{2+}-dependent protein kinases such as PKC and CaMKII. The same effect is seen for protein kinase G, the activator of which (cGMP) is produced along a pathway involving the Ca^{2+}-dependent synthesis of nitric oxide, NO (see Section 16.5.1). Thus, by attenuating Ca^{2+} release, these kinases are components of negative feedback loops controlling Ca^{2+}-dependent pathways.

Phosphorylation by PKA, in contrast, stimulates $InsP_3$ receptors and ryanodine receptors, and the same holds true for phosphorylation by tyrosine-specific kinases of the Src family. The effect of the cAMP-activated PKA provides another example of complex feedback control, since both the generation and the breakdown of cAMP are either stimulated or inhibited by Ca^{2+}, depending on the adenylate cyclase or phosphodiesterase isoform involved (see Sections 4.4.1 and 4.4.2).

Figure 14.19 Ca^{2+} waves An input signal induces a local accumulation of intracellular messenger molecules (such as $InsP_3$, cADPR, or Ca^{2+}), which at a "hot spot" triggers Ca^{2+}-induced Ca^{2+} release from the endoplasmic reticulum. This effect may be locally restricted, spread as a wave, or develop into a stationary pattern. The frequency of the oscillating Ca^{2+} signal depends on the intensity (A) and the nature (B) of the input signal (upper panel). (Adapted from Smedler E & Uhlén P [2014] *Biochim Biophys Acta* 1840:964–969.)

14.5.4 Waves and patterns, or the beauty of frequency modulation

Like neuronal data processing, cellular data processing is assumed to be dominated by the principles of spatial privacy and temporal transitoriness. This has been convincingly shown for subcellular Ca^{2+} signals, which may be either restricted to very small microdomains or spread as waves across the whole cell or even within a syncytium coupled by gap junctions.

When visualized by specific dyes, local Ca^{2+} signals appear as narrow (1–3 µm) and short-lived (less than 1 s) sparks or as somewhat more extended and longer-lasting puffs and spikes. The smallest sparks develop at the cytoplasmic mouth of Ca^{2+} channels; larger ones and spikes are found at so-called "hot spots": where receptors of extracellular input signals and sensitized $InsP_3$ and ryanodine receptors are concentrated (**Figure 14.19**).

Figure 14.20 Pattern formation A computer-generated pattern in a feedback system with a short-range stimulatory and a long-range inhibitory effect is shown. To the left, the pattern was induced by a single input signal and has developed in a highly regular fashion. To the right, the result is shown for statistically fluctuating input signals inducing an irregular pattern. (From Gierer A [1985] Die Physik, das Leben und die Seele. Munich and Zurich: Piper.)

Since, at higher concentrations, calcium causes apoptosis (see Section 13.2.4), a strict limitation and temporal oscillation of Ca^{2+} signals is primarily a question of life and death. This essential control is guaranteed by a strong inhibitory field consisting of a high Ca^{2+}-buffering capacity of the cytoplasm, which is brought about by the cooperation of Ca^{2+}-binding proteins with the membrane-bound calcium pumps and uniporters. As a result, the diffusion rate of Ca^{2+} in the cytoplasm (which is a semisolid medium due to the high concentration of macromolecules) is reduced to a value of 10 $\mu m^2/s$, or about 2% of the rate in water. Therefore, we are dealing with a system where the stimulatory events (such as Ca^{2+}-induced Ca^{2+} release) are short-range and controlled by local autofeedback as well as by feedback with the much more extended inhibitory field. This provides the conditions for Ca^{2+} signals to be intimate (restricted to very small areas of the cell) or, on the other hand, to oscillate generating waves as well as complex patterns of excitation that may be looked upon as interferences of stationary waves (**Figure 14.20**).

In fact, sparks as well as hot spots are starting points of calcium waves migrating across the ER. Such waves are generated by a cooperation of many channels. The underlying mechanism is Ca^{2+}-induced calcium release. It enables adjacent calcium channels to activate each other (see Figure 14.17). The unidirectional wavelike propagation of Ca^{2+} release requires the activated channel to become refractory for a short period of time. This desensitization is achieved by Ca^{2+}-dependent channel blocking at concentrations above 300 nM (see Figure 14.17). Such a situation resembles the spreading of action potentials (see Figure 14.6). The ER has been called, therefore, an excitable medium, though the chemical calcium waves are much slower than the electrochemical action potentials. However, this apparent drawback is compensated by the short distances within the cell compared with the nervous system.

Like any wave, a calcium wave is subject to modulation of both amplitude and frequency. By frequency-dependent modulation, the efficacy of information processing is increased tremendously (one may consider, for instance, the large difference between a black-and-white photograph [being mainly amplitude-modulated] and a color picture, or between a series of drumbeats and a symphony).

The wavelength of an oscillating calcium signal may fluctuate between milliseconds and several minutes. It directly depends on the nature and the intensity of the input signal: the stronger this is, the more InsP$_3$ receptors (or their relatives) become activated, the more sparks are generated at a hot spot, and the higher is the frequency of the resulting calcium wave.

The Ca^{2+} effector proteins located downstream respond to the different frequencies in an individual manner. In other words, depending on the frequency of the wave, different sets of proteins are addressed by calcium. A prototype of a protein acting as a frequency detector is CaM kinase II, which is discussed in detail in Section 14.6.3.

How to prove frequency modulation of signals

To reiterate, practically all cellular processes are modulated directly or indirectly by Ca^{2+}. Therefore, one may start from the assumption that, via signaling cross talk, the oscillating pattern of Ca^{2+} signals is transduced to other signals, which, vice versa, may modulate calcium signaling. This has been studied in detail for the interaction between Ca^{2+} and cAMP signaling.

Investigation of frequency-modulated signals requires the monitoring of rapid changes of concentrations in living cells. Optical methods have turned out to be particularly suitable since they work almost without delay, enabling an exact localization of signals. Among the most advanced techniques are those employed for an analysis of calcium. For instance, organic dyes that undergo a change in their absorption spectra upon Ca^{2+} binding are used. Such probes are applied by microinjection or as membrane-penetrating derivatives that release the dye upon enzymatic hydrolysis in the cell. More advanced approaches make use of bioluminescence proteins. An example is the protein aequorin from the luminous jellyfish *Aequorea victoria*: it is activated by Ca^{2+} to emit light, the intensity of which depends on the Ca^{2+} concentration. An even better yield of light is obtained with **green fluorescent protein (GFP)**. The fluorescence of GFP is activated by aequorin in the jellyfish, but it can also be excited in the laboratory by UV light (see also Section 5.2). Moreover, GFP has the advantage that derivatives obtained by an exchange of certain amino acid residues or by molecular rearrangements emit light of different colors. However, GFP does not interact directly with Ca^{2+}. Therefore, by means of genetic manipulations, fusion proteins have been constructed that, in addition to GFP, contain CaM as a calcium sensor and the CaM-binding motif of the myosin light-chain kinase (MLC kinase) as an effector domain. In the presence of Ca^{2+}, CaM induces a conformational change of this effector domain, which is transduced to the GFP moiety, changing its fluorescence activity. On the basis of this approach, several Ca^{2+}-indicators have been developed that may be expressed deliberately in cells, enabling a very sensitive assay of temporal and spatial fluctuations of Ca^{2+} concentrations.

An interesting variant of this analytical technique is offered by fusion proteins where the CaM/MLC kinase effector domain pair is flanked on both sides by different-colored GFP derivatives. These so-called chameleons make use of the quantum-mechanical phenomenon of fluorescence resonance energy transfer (FRET; for details see Section 5.2) where one of the chromophores activates the other one to fluoresce, provided they come together very closely. The latter condition is fulfilled only in the presence of Ca^{2+} when the fusion protein changes from a stretched into a folded conformation (**Figure 14.21A**).

Such methods can also be employed to study other signaling molecules. An illustrative example is provided by cAMP (**Figure 14.21B**). Here one uses cAMP-binding PKA as a sensor, making use of a characteristic property of this enzyme: as described

(A) DETECTION OF Ca²⁺

(B) DETECTION OF cAMP

in Section 4.6.1, the catalytic and regulatory domains of PKA are located on separate subunits, which in the absence of cAMP are firmly associated. Binding of cAMP at the regulatory subunits results in a dissociation of the complex. When by genetic manipulations the subunits have become fused with different-colored fluorescent proteins, the concentration of cAMP can be quantitatively assayed by FRET. Since a maximal FRET signal is obtained only when both subunits are completely associated (in the intact complex), cAMP diminishes the FRET signal in a concentration-dependent manner. By employing this and related approaches, a tight coupling between Ca²⁺ and cAMP oscillations has been found that is due, for instance, to the activities of Ca²⁺-controlled adenylate cyclases (see Section 4.4.1) and cAMP phosphodiesterases (see Section 4.4.2). Moreover, cAMP oscillations may occur independently of Ca²⁺ owing to the oscillator-type construction of cAMP response modules as shown in Section 4.6.1.

Since feedback interactions enabling undamped oscillations of output signals are a hallmark of most signal-processing protein complexes, the frequency modulation of signaling certainly is not restricted to Ca²⁺- and cAMP-driven systems but may be assumed to be a (or even *the*) major mode of cellular data processing.

14.5.5 Store-operated calcium channels recharge the cell's calcium battery

Longer-lasting waves and patterns require a continuous replenishment of the Ca²⁺ store in the ER because the calcium-dependent ATPases and ion exchangers of the plasma membrane permanently pump cytoplasmic calcium into the extracellular medium. To replenish the store, the ER is in contact with a distinct class of

Figure 14.21 Assay of Ca²⁺ and cAMP by FRET analysis (A) A fusion protein consisting of calmodulin (CaM) and the CaM-binding domain of myosin light-chain kinase (MLCK) is linked on both sides with green fluorescent protein (GFP) derivatives 1 and 2 emitting different-colored fluorescent light upon excitation by UV light. Binding of Ca²⁺ to CaM induces a conformational change bringing 1 and 2 close together, thus enabling a fluorescence resonance energy transfer (FRET). As a result, the intensity of the fluorescent signal of 1 decreases by the same value as the intensity value for the signal of 2 increases. (B) For a cAMP assay, the catalytic subunits (CU) and the regulatory subunits (RU) of protein kinase A are fused with different-colored GFP derivatives 1 and 2, which in the absence of cAMP are close enough to enable a FRET from 1 to 2. Upon binding of cAMP, the complex dissociates and the FRET signal is diminished.

Ca^{2+} channels in the plasma membrane called **store-operated calcium channels (SOCCs)**, which are opened in response to store depletion in the ER. These channels cooperate with the ER-bound ATPase SERCA, which pumps the Ca^{2+} entering the cell through the SOCCs into the lumen of the ER. Thus, like electrical capacitors, the intracellular calcium stores are discharged and recharged; the replenishment of the stores has therefore been called **capacitative Ca^{2+} influx**. It is essential for any long-lasting or oscillating calcium signal. The nature and mode of operation of SOCCs are still a matter of debate. It seems certain, however, that tissues express different channel types with different mechanisms of action.

As far as the function and the structure are concerned, the channel **CRAC**, a subtype of SOCCs found especially in lymphocytes and some other cell types, has been studied most thoroughly. It causes an electrophysiologically detectable Calcium Release-Activated calcium Current that is highly ion-selective. The CRAC channel is composed of subunits called **Orai**, reminiscent of the mythological gatekeepers of the ancient Greek heaven. The Orai protein is expressed in several isoforms. It consists of four transmembrane helices with the C- and N-termini facing the cytoplasm, but it does not exhibit any structural relationship with other Ca^{2+} channels (**Figure 14.22**). Although in comparison with VOCCs the conductivity of the CRAC channel is small, it is sufficient to guarantee a continuous Ca^{2+} influx and replenishment of the ER store.

The Ca^{2+} load of the ER lumen is monitored by a sensor protein. This protein was discovered in quite another context, as a stromal interaction molecule (**STIM**) and putative tumor suppressor in a splenic stromal cell line. Later STIM was identified as a protein involved in capacitative Ca^{2+} influx. Integrated in the ER membrane by a

Figure 14.22 Model of the capacitative Ca^{2+} influx
(A) Upon Ca^{2+} depletion of the endoplasmic reticulum (ER) lumen, the calciumsensing protein STIM (stromal interaction molecule; shown as gray cylinders) in the ER membrane aggregates opposite to clusters of the calcium-release-activated Ca^{2+} channel (CRAC), activating Ca^{2+} influx from the cellular environment. Ca^{2+} entering the cytoplasm is pumped into the ER lumen by the Ca^{2+}-dependent ATPase SERCA (not shown). (B) STIM, a single-pass transmembrane protein of the ER that exhibits a Ca^{2+}-binding EF-hand domain, is shown to interact via coiled-coil domains with the CRAC subunit Orai, a cell membrane protein with four transmembrane helices. Whether such interaction controls the gating of the CRAC channel is not clear yet.

single transmembrane domain, STIM exhibits a Ca^{2+}-binding EF-hand domain (see Section 14.6.1) that serves as a Ca^{2+} sensor at the luminal side. At the cytoplasmic side STIM has an extended coiled-coil structure and interaction domains that are folded in an inactive conformation at higher Ca^{2+} concentrations in the ER lumen (see Figure 14.22). Upon store depletion, STIM aggregates into patches concentrating opposite assemblies of CRAC channels and other SOCCs in the plasma membrane and the interaction domains become unfolded so that they can contact and gate the channels (see Figure 14.22). The mechanism by which the signal is transduced from STIM to the channels is still not known. Evidence exists for both a direct interaction with channel proteins and a diffusible messenger molecule.

In addition to Orai channels, a series of **TRP channels** (transient receptor potential; see Section 14.5.6) participates in store-dependent Ca^{2+} influx. In fact, almost all TRP channels of the TRPC subfamily seem to function as SOCCs, rivaling or replenishing (probably in a tissue-specific way) the effect of Orai channels. Moreover, like Orai, the TRP channels bind to and are regulated by STIM proteins, indicating that this type of interaction represents a universal signal reporting the charge of the ER battery. Nevertheless, other Ca^{2+}-binding ER proteins such as calreticulin and calsequestrin have been proposed also to monitor store depletion.

Store-controlled channels cooperate with arachidonic acid-controlled calcium channels to regulate the amplitude and frequency of Ca^{2+} waves

As shown by inhibitor experiments, cytoplasmic Ca^{2+} oscillations not only depend on channels like CRAC and on the charge of the ER store but also are regulated by novel ligand-controlled Ca^{2+} channels in the plasma membrane. The gating ligand is arachidonic acid and the channel is called, therefore, **ARCC** (arachidonic acid regulated calcium channel). ARCC plays a crucial role since calcium waves frequently break down when this particular channel is blocked. In the cell, ARCC and SOCC exhibit different functions: while SOCC predominantly controls the amplitude, ARCC regulates the frequency of Ca^{2+} waves. Moreover, Ca^{2+}-dependent adenylate cyclases are activated exclusively by CRAC, probably because they form complexes with this channel.

Arachidonic acid is released from membrane phospholipids by Ca^{2+}-dependent phospholipase A_2 (PLA$_2$), a reaction controlled by a wide variety of extracellular signals (see Section 4.4.5). Its gating effect on ARCC shows that, in addition to being the precursor of eicosanoids, arachidonic acid is a bona fide second messenger (**Figure 14.23**).

Figure 14.23 Arachidonic acid-controlled Ca^{2+} influx A signal (hormone, neurotransmitter, cytokine, or environmental stimulus) stimulates the receptor-dependent release of arachidonic acid by phospholipase A_2 (PLA$_2$). Arachidonic acid opens the Ca^{2+} channel ARCC, and the Ca^{2+} entering the cell induces and modulates Ca^{2+} oscillations generated at the ER according to the mechanism of Ca^{2+}-dependent Ca^{2+} release.

Other ligand-controlled Ca²⁺ channels of the plasma membrane

In contrast to ARCC, which transports exclusively Ca^{2+}, other ligand-controlled "calcium channels" of the plasma membrane are nonspecific cation channels rather than being Ca^{2+}-selective. Nevertheless, some of them exhibit some preference for calcium. Such channels are controlled by intra- and extracellular signals. Prominent examples are the nicotinic acetylcholine receptors, the P2X purine receptors, and the ionotropic glutamate receptor NMDA (*N*-methyl-D-aspartate). These receptor-coupled ion channels are discussed in Chapter 16. For another type of ligand-controlled cation channels, the cyclic nucleotide-gated channels, see Section 14.5.7.

14.5.6 The TRP ion channel family comprises epithelial Ca²⁺ channels and more

As mentioned above, several SOCCs are members of the channel family TRP, which will be briefly reviewed here. The term transient receptor potential refers to the discovery of this channel type in the course of electrophysiological studies on *Drosophila* mutants exhibiting a defect of light adaptation (see Section 15.5).

The TRP family is encoded by no less than 28 genes in humans. They are classified into six subfamilies (a comprehensive list of mammalian TRP channels is provided by Gees et al. (2012)).

- TRPC (canonical)
- TRPV (V refers to the vanilloid receptor explained in Section 15.3)
- TRPM (related to melastatin, which was discovered as a tumor suppressor protein)
- TRPP (related to polycystins, proteins that are mutated in autosomal dominant polycystic kidney disease; see Section 15.1 for details)
- TRPML (related to mucolipin, a protein associated with the disease mucolipidosis)
- TRPA or ANKTM1 (ankyrin-like transmembrane protein 1)

Owing to alternative splicing, the TRP genes give rise to a large number of proteins that associate into homo- and hetero-oligomers. They constitute one of the most versatile ion channel families with manifold functions. In particular, they play a key role in **sensory signal processing** (see Chapter 15).

TRP and TRP-related channels are active in all multicellular eukaryotes examined so far and are found in yeast. They transport cations: some of them in a rather non-specific manner, others with a strong preference for monovalent cations (TRPV5 and 6) or calcium (TRP5 and 6). TRP channels belong to the cation channel prototype 1 (see Figure 14.2). They are composed of four subunits with six transmembrane helices and a P-loop each (**Figure 14.24**). With the exception of the heat sensor TRPV1, the cold sensors TRPA1 and TRPM8, and the taste-sensitive ion channel TRPM5 (see Chapter 15), TRP channels are not voltage-gated because the positive charge of transmembrane domain S4, the voltage sensor in potential-dependent channels, is insufficient.

The mechanism of TRP channel activation is not fully understood yet. Depending on the channel type, obviously more than one signaling pathway leads to channel gating. As far as SOCC function is concerned, an interaction with STIM proteins is a crucial event. Other mediators of channel activation may be diacylglycerol (albeit via a still ill-defined route), cADPR (see Section 14.5.3), and Ca^{2+} ions. Since phosphatidylinositol 4,5-bisphosphate inhibits the opening of many TRP channels, the enzymatic degradation of this membrane-bound phospholipid by phospholipase C (PLC) might provide an additional gating signal.

Figure 14.24 Mode of operation of the epithelial Ca²⁺ channel ECaC (A) One of the four subunits forming a channel. (B) Epithelial cell with an apical ECaC, enabling passive Ca^{2+} influx from the environment (food, urine), and a basolateral ATPase/exchanger complex, pumping Ca^{2+} into the blood. To avoid undesirable second messenger effects, the Ca^{2+} to be transported is transiently bound by the cytoplasmic protein calbindin.

Regardless of whether diacylglycerol production or phospholipid degradation is the triggering signal, most TRP channels seem to be activated by the PLC isoenzymes β and γ that are under the control of G-protein- and Tyr kinase-coupled receptors. Prototypical is the situation in the insect eye. Here the light-sensitive rhodopsin (see Section 15.5) stimulates via a trimeric $G_{q,11}$-protein a β-type PLC, releasing the second messengers diacylglycerol and InsP$_3$. InsP$_3$ triggers the depletion of the Ca^{2+} store in the ER, while in the plasma membrane a Ca^{2+}-selective TRPC channel and a nonspecific TRP-like cation channel (TRPL) are gated. Owing to TRPL gating the cell becomes depolarized, evoking an action potential in the visual neuron. The Ca^{2+} signal generated in parallel by the TRPC channel controls the sensory response by negative feedback and, in addition, serves to refill the intracellular calcium stores. Other examples of TRP channels with known functions are TRPC1, both a SOCC and a putative mechanoreceptor (see Section 15.2), and a corresponding receptor NOMPC (no mechanoreceptor potential channel) of insects, both of which respond to touch stimuli, as well as TRPP1 and 2, which sense fluid currents in kidney cilia. A particular subfamily of sensory TRP channels is the **vanilloid receptors** that belong to the TRPV type. These nonspecific cation channels expressed by sensory neurons respond to temperature stimuli. They are discussed in more detail in Section 15.3.

A channel of special interest is TRPC2, which is expressed in most vertebrates with the exception of primates. The gating of TRPC2 has been found to induce the so-called **acrosomal reaction** in mouse sperm. This Ca^{2+}-dependent exocytotic process is a key event of reproduction: it promotes the release of lytic enzymes that enable the sperm cell to penetrate the zona pellucida wrapping the oocyte. TRPC2 is a Ca^{2+}-transporting channel that is activated by proteins of the zona pellucida along the PLCβ pathway but also by store depletion. The channel is expressed also in the cells of the vomeronasal organ, the sensory organ for (sex) pheromones in vertebrates (see Section 15.4). In primates, the *trpc2* gene has degenerated to an inactive pseudogene and the vomeronasal organ is probably out of function. Whether or not the sperm channel TRPC2 is replaced here by another channel remains an open question.

In vertebrates the **epithelial Ca²⁺ channels (ECaC)**, found particularly in the epithelia of kidney and small intestine, have been identified as the highly Ca^{2+}-selective TRP channels TRPV5 and -6. Their function resembles that of the nonrelated ENaCs: being located in the apical membrane of the epithelial cell, ECaCs are the mediators of vitamin D-controlled Ca^{2+} (re)absorption. In the basolateral membrane two calcium pumps, a Ca^{2+}-dependent ATPase and an Na^+–Ca^{2+} exchanger, convey the Ca^{2+} that has entered the cell through ECaC into the bloodstream (see Figure 14.24). TRPV5 and -6 are leaky channels: they are permanently open. This creates a problem, because a continuous Ca^{2+} influx would lead to undesirable second messenger effects and eventually culminate in cell death. To avoid this, in the cell Ca^{2+} is transiently sequestered by a Ca^{2+}-buffering protein, calbindin, which hands over Ca^{2+} to the ion pumps. The transcription of the genes encoding TRPV5, TRPV6, and calbindin is promoted by calcitriol (1,25-dihydroxyvitamin D_3). The latter is the active form of vitamin D_3 that binds to the vitamin D receptor VDR, a transcription factor (see Section 8.3). Along this route, D vitamins stimulate the uptake of calcium from food and promote bone formation. Vitamin D deficiency is the cause of rickets.

A TRP channel probably functioning as an osmosensor is expressed by heart muscle, liver, and kidney cells and is thought to play an important role in salt and water balance and, thus, in blood pressure control. A similar function is attributed to the channel TRPC4 in endothelial cells. By transporting Ca^{2+} into the cell, this channel triggers the synthesis and release of vasodilatory mediators such as nitric oxide (generated by a Ca^{2+}-dependent NO synthase) and the arachidonic acid derivative prostacyclin (the synthesis of which is triggered by a Ca^{2+}-dependent PLA_2 providing the precursor lipid arachidonic acid). In contrast, Ca^{2+}-selective TRP channels in smooth muscle cells induce vasoconstriction by triggering muscle contraction.

14.5.7 Cyclic nucleotide-gated ion channel family

Another family of ligand-dependent nonspecific cation channels with a high permeability for Ca^{2+} ions is the cyclic nucleotide-gated (CNG) channels. The best-known representatives are the cGMP- or cAMP-controlled ion channels of vertebrate photoreceptor and odorant receptor cells (see Chapter 15). Upon gating, an Na^+- efflux induces depolarization while Ca^{2+} enters the cell along a steep concentration gradient. The ligand specificity of the CNG channels is variable: for instance, the channel in odorant receptor cells does not distinguish between cAMP and cGMP, whereas the channel of photoreceptor cells exhibits a strong preference for cGMP.

CNG channels exist as heterotetramers consisting of two A- and two B-subunits. Each subunit has a cyclic nucleotide binding site as well as six transmembrane helices and a P-loop (**Figure 14.25**); thus the architecture of a CNG channel resembles that of voltage-dependent potassium channels (prototype 1, see Figure 14.2). However, with the exception of the HCN subfamily (discussed in the next few paragraphs) CNG channels are not voltage-controlled because the transmembrane helix S4, which normally serves as a voltage sensor, lacks some of the positively charged amino acid residues required for voltage responsiveness (see Section 14.2). Another component of the CNG channel complex is the Ca^{2+}-binding protein calmodulin. It mediates a blockade of the channel by calcium ions and plays a key role in sensory adaptation (see Section 15.6).

CNG channels and their relatives are widely distributed among eukaryotes. They have been found not only in sensory cells but also in other tissues, where their function remains more or less unknown. An exception may be provided by sperm cells. From experiments on sea urchins, it is known that peptide factors released by the egg stimulate motility, energy metabolism, and the acrosomal reaction of sperm cells. To this end, these peptides interact with a receptor guanylate cyclase on the sperm membrane. The cGMP thus generated induces a Ca^{2+} influx into the cytoplasm, which acts as a trigger of sperm activation. In mammalian sperm cells (unfortunately not yet in sea urchin sperm), the corresponding cGMP-sensitive Ca^{2+} channel has been identified as a member of the CNG family.

A subfamily of CNG channels is the **HCN channels** (hyperpolarization and cyclic nucleotide-gated channels). These voltage-dependent cation channels are unique

Figure 14.25 Gating of a cyclic nucleotide-regulated cation channel by cAMP The model, based on X-ray crystallographic studies, shows the conformational change of ion channel proteins induced by binding of cAMP (teal) to the cyclic nucleotide-binding domains (cNBD). Transmembrane helices S5 and S6, which form the membrane pore, are turning aside and the positively charged bundle of transmembrane helices S1–4 is turning upward, opening the channel. (Adapted with modifications from Rehmann H, Wittinghofer A & Bos JL [2007] *Nat Rev Mol Cell Biol* 8:70.)

in that they are already gated by depolarization at extremely negative resting potentials and are, in addition, activated by cAMP and cGMP. As pacemaker channels, they play a central role in oscillating processes such as the heartbeat (for details see Section 14.6.2).

Summary

- Ca^{2+}-dependent ATPases and Na$^+$/Ca^{2+} exchangers generate and maintain a steep Ca^{2+}-concentration gradient from the cytosol to the extracellular medium, the ER and the mitochondria. Gating of corresponding ion channels and exchangers automatically leads to an influx of Ca^{2+} into the cytoplasm where it interacts with a wide variety of proteins.

- Voltage-operated Ca^{2+} channels (VOCCs) opened by depolarization are found in the plasma membranes of all vertebrate cell types. They are the ultimate targets of action potentials. VOCCs are stimulated by protein kinase A- and CaM kinase II-catalyzed phosphorylation (sympathetic effect) and are inhibited by Ca^{2+} and βγ-subunits of G$_{i,0}$-proteins.

- Two major types of ligand-operated Ca^{2+} channel are found in the ER: inositol 1,4,5-trisphosphate ($InsP_3$) receptors and ryanodine receptors. Both are gated by Ca^{2+}. This Ca^{2+}-induced Ca^{2+} release from the ER into the cytosol is triggered and augmented by the ligands $InsP_3$ and cyclic ADP-ribose (cADPR) and blocked by higher Ca^{2+} concentrations. Other intracellular messengers that induce Ca^{2+} release from internal stores are nicotinic acid adenine dinucleotide phosphate (NAADP) and sphingosine 1-phosphate.

- Ca^{2+}-induced Ca^{2+} release from the ER provides a means for the generation of frequency-modulated Ca^{2+} signals, appearing as both propagating waves and stationary patterns. The wavelength and amplitude depend on the intensity of input signals. Because of the cross talk between individual signal-transducing pathways, the frequency modulation of Ca^{2+} signals can be transferred to other cellular signals.

- Storage-operated calcium channels (SOCCs) in the plasma membrane are connected with Ca^{2+}-sensor proteins in the ER and become gated when the Ca^{2+} level of the internal store falls below a critical value. They are required for a continuous reloading of the internal calcium store, being a condition of long-lasting Ca^{2+} oscillations.

- Intracellular Ca^{2+} oscillations also depend critically on arachidonic acid-controlled calcium channels.

- Physiologically important cation channels with some preference for Ca^{2+} are TRP channels and cyclic nucleotide-gated channels. Most of them are not voltage-sensitive. TRP channels play a key role in sensory signal processing. Depending on the isotype, they are gated along various signaling pathways, most of which include an activation of C-type phospholipases, while cyclic nucleotide-gated channels are activated by cAMP or cGMP. The voltage-sensitive subtype HCN constitutes the pacemaker channels in the heart.

14.6 Downstream of Ca^{2+} signals

As a chemical signal, Ca^{2+}, like other second messengers, interacts with proteins. Such calcium-sensitive proteins are expressed by cells in large numbers. Some of them fulfill the role of buffer proteins: in cooperation with ion channels, ion pumps, and ion exchangers, they control the level of free Ca^{2+} in the cytoplasm and in organelles by binding and releasing Ca^{2+} in a concentration-dependent manner. Examples are calreticulin and calsequestrin in the endo(sarco)plasmic reticulum, and calbindin and parvalbumin in the cytosol.

Most Ca^{2+}-binding proteins, however, are components of signaling cascades. For them Ca^{2+} is an input signal modulating their conformation and function and prompting them to emit an output signal. Some of these effector proteins directly interact with Ca^{2+}, whereas others receive the input signal from a separate Ca^{2+}-sensor protein. Ca^{2+}-effector proteins have Ca^{2+}-recognition domains, with the EF-hand domain being the most abundant one, followed by the C2 and annexin domains. A selection of Ca^{2+}-binding proteins is given in **Table 14.2**.

14.6.1 Calmodulin and other calcium sensor proteins

Proteins with EF-hand domains constitute the largest and most heterogeneous family of calcium-binding proteins. The prototypical **EF-hand protein** is calmodulin (CaM). This universal Ca^{2+}-sensor protein of eukaryotic cells transduces the Ca^{2+} signal to a large number of effector proteins. The mammalian form consists of 148 amino acids and has a molecular mass of 16.7 kDa. Two helix–loop–helix motifs at the N-terminus and two at the C-terminus remind of thumb and forefinger and are denoted as EF-hands (**Figure 14.26**). These are the binding sites of four Ca^{2+} ions

Table 14.2 A selection of Ca^{2+}-binding proteins

Proteins that buffer and sequester Ca^{2+}
calbindin
calsequestrin
calreticulin
parvalbumin

Ca^{2+} sensor proteins
calmodulin
troponin C
S100 proteins
neuronal Ca^{2+}-sensor proteins (recoverin, GCAP)
synaptotagmins

Enzymes
protein kinase C (c- and n-subfamilies)
calpain (protease)
phospholipase A$_2$
phospholipases C
NO synthases (constitutively expressed isoforms eNOS and nNOS)

Receptors and ion channels
InsP$_3$ receptors
ryanodine receptors
Ca^{2+}-activated K$^+$ and Cl$^-$ channels
Ca^{2+}-dependent ATPases

Cytoskeletal proteins
α-actinin
gelsolin
caltractin

Cell adhesion and other proteins
annexins
cadherins

Transcription factors
DREAM

Factors controlling G-protein activity
GRF (a Ras-GEF)
GRP (a Ras-GEF)
CAPRI (a Ras-GAP)

Figure 14.26 Calmodulin Upon binding of Ca^{2+} (teal spheres), the inactive stretched form of calmodulin (left) develops a central helical structure (middle) that, bends into a hydrophobic cavity that fits corresponding domains of partner proteins (teal, right). This rearrangement is supported by hydrophobic interactions between the N- and C-terminal EF-hand domains. EF-hands are helix–loop–helix structures binding one Ca^{2+} ion each at aspartate and glutamate residues in the central loop. The name refers to the arrangement of the two helices E and F to a handlike structure with stretched thumb and forefinger.

that interact with the protein in a cooperative manner. The interaction changes the conformation in such a way that complementary structures of effector proteins fit a hydrophobic pocket of CaM. As a result, both the conformation and the function of the effector protein are modulated. CaM itself is a substrate of numerous protein kinases that change its activity by phosphorylation.

In **Table 14.3**, some enzymes are summarized that contain CaM as a regulatory subunit or the activity of which is controlled by Ca^{2+}/CaM. They are discussed in detail elsewhere.

An isoform of CaM, **troponin C**, is specialized to interact with the troponin-tropomyosin complex of skeletal muscle cells (see Section 14.6.2). The **S100 proteins**

Table 14.3 Ca^{2+}/calmodulin-dependent enzymes: a selection

Enzyme	Major function
CaMKI	regulation of gene transcription (Section 14.6.3)
CaMKII	regulation of synaptic plasticity (Section 14.6.3)
CaMKIV	regulation of gene transcription (Section 14.6.3)
eEF2 kinase (CaMKIII)	eEF2-specific; inhibition of translation (Section 9.3.4)
MLC kinase	M LC-specific; contraction of the actin–myosin complex (Section 14.6.2)
phosphorylase kinase	specific activation of glycogen phosphorylase (Sections 5.3.1 and 14.6.3)
CaMK kinases α and β	activation of CaMKI and -IV (Section 14.6.3)
calcineurin (protein phosphatase 2B)	protein dephosphorylation (Section 14.6.4)
Ca^{2+}-dependent ATPase	Ca^{2+} export (Section 14.5.1)
constitutive NO synthases eNOS and nNOS	synthesis of nitric oxide (Section 16.5.1)
adenylate cyclases 1, 3, and 8	synthesis of cAMP (Section 4.4.1)
cAMP/cGMP phosphodiesterases type 1	degradation of cAMP and cGMP (Section 4.4.2)

(A)   EF-4   EF-3   C   EF-2   N   EF-1

(B)   C   N

(C)   C   N

Figure 14.27 The Ca²⁺-myristoyl switch of recoverin (A) Recoverin in the absence of Ca²⁺. The structure consisting of four EF-hands is folded in such a way that the myristoyl residue (teal) is buried inside the protein, rendering the protein water-soluble. (B, C) Binding of Ca²⁺ (colored circles) at two EF-hands causes an unfolding of the protein, exposing the myristoyl residue that now anchors the protein in lipid bilayers of membranes. (Modified from Weiergräber OH & Koch KW [2005] *Biospektrum* 11:174–175.)

(100% soluble in ammonium sulfate solution) are a group of at least 20 EF-hand proteins that, like CaM, interact with a large number of effector proteins, but in contrast to CaM, they are tissue-specifically expressed. In the clinic they are used as markers for certain tumor types and for brain damage by accidents or stroke. Much remains to be learned about the physiological role of this protein family. Other Ca²⁺-sensor proteins with EF-hands have been found particularly in nerve cells. Well-known examples are recoverin and guanylate cyclase-activating proteins (GCAPs), the expression of which is restricted to retina cells, where they are involved in light adaptation (see Sections 15.5 and 15.6). Both are representatives of the large family of **neuronal Ca²⁺-sensor (NCS) proteins**. An interesting member of this family, the transcriptional repressor DREAM, will be discussed in Section 14.6.4. NCS proteins may shuttle between cytoplasm and membranes. For this reason, most of them contain a tetradecanoyl (myristoyl) residue covalently bound at the N-terminus that functions as a membrane anchor. Depending on the Ca²⁺ signal, this lipid moiety either is buried inside the protein, unable to interact with membranes, or is exposed at the outside (**Figure 14.27**). By means of this **Ca²⁺–myristoyl switch** such proteins can be anchored reversibly at membranes, providing an important condition for a controlled interaction with other signaling proteins (the reader may recall that the intracellular localization of small G-proteins of the Arf family is regulated by an analogous mechanism; see Section 10.4.3).

NCS proteins have various functions including modulation of voltage-dependent calcium and potassium channels, of neurotransmitter receptors, and of pro- and anti-apoptotic effects. As indicated by experiments with gene-manipulated animals, they are involved in learning and memory fixation. In addition, they seem to play a role in psychiatric diseases. Similar functions have been attributed to other Ca²⁺-binding proteins of the EF-hand family.

The second frequent Ca²⁺-binding motif is the **C2** or **Ca²⁺–lipid-binding domain**. As indicated by the name, through this domain proteins are reversibly attached to membranes, where Ca²⁺ interacts electrostatically with acidic phospholipids such as phosphatidylserine. NCS proteins of this family are the **synaptotagmins** that control neurotransmitter release at synapses (as discussed in detail in Section 10.4.2).

In addition, a series of enzymes with C2 domains are known, for instance conventional and new protein kinases C (in fact, C2 stands for conserved domain number 2 of these enzymes), several phospholipases, and the Ras GTPase-activating protein CAPRI (see Section 10.1.3).

The **annexin domain**, the third frequent Ca^{2+}-binding motif, shares with the C2 domain the property of binding proteins via Ca^{2+} ions to acidic membrane phospholipids. It is particularly found in the annexin proteins, which are widely distributed among eukaryotes. By interacting in a Ca^{2+}-dependent manner with membranes, annexins form pseudocrystalline aggregates, thus stabilizing individual membrane regions. Moreover, they function as scaffolds and interact with numerous cytoplasmic proteins, thus constituting signal-transducing multiprotein assemblies involved, for instance, in exo- and endocytosis and in the organization of ion channel complexes.

14.6.2 Muscle contraction

The Ca^{2+} signal plays a key role in stimulus–secretion coupling, particularly in nerve cells (see Section 10.4.2), and stimulus–contraction coupling, particularly in muscle cells. Several different calcium channels cooperate in these processes. This becomes particularly clear for muscle contraction.

Muscle contraction is due to an interaction between the proteins actin and myosin. The extended conformational change thus induced in myosin generates a mechanical tension, resulting in a shift of myosin relative to actin. The energy is derived from ATP hydrolysis catalyzed by myosin, which functions as an actin-stimulated ATPase. In the absence of input signals, the interaction between actin and myosin is blocked by inhibitory proteins such as myosin light chains or tropomyosin (**Figure 14.28**). Ca^{2+} entering the cytoplasm upon stimulation relieves this blockade according to the principle of double negation (inhibition of an inhibitor). In the three major types of muscle—skeletal muscle, smooth muscle, and heart muscle—Ca^{2+} influx and contraction are triggered by membrane depolarization gating VOCCs of the L-type. These channels cooperate with other channels according to a pattern that depends on the muscle type.

Figure 14.28 Stimulation of muscle contraction by Ca^{2+}: the principle Contraction is due to a conformational change and activation of the ATPase myosin triggered by an interaction with F-actin and resulting in a relative shift of the proteins against each other (see textbooks of cell biology for details). This interaction is blocked by the protein tropomyosin and by myosin light chains. Ca^{2+} overcomes this blockade. To this end, it binds in skeletal and heart muscle to the sensor protein troponin C, which, through the adaptor proteins troponin I and troponin T (not shown), interacts with tropomyosin. In other cells, including smooth muscle cells, calmodulin serves as a Ca^{2+} sensor that activates caldesmon and the myosin light-chain kinase (MLCK). While caldesmon, like troponin C, displaces tropomyosin from the actomyosin complex, MLCK inactivates the myosin light chains by phosphorylation. The latter reaction is reversed by a myosin light-chain phosphatase (not shown). In smooth muscle and other tissues, phosphorylation of myosin light chains is essential for contraction and movements, whereas in skeletal muscle it has a modulatory function only. Input signals control the channels and enzymes of cellular calcium homeostasis.

Figure 14.29 Signal transduction in skeletal muscle contraction By interacting with a nicotine receptor cation channel, acetylcholine induces depolarization. Consequently, voltage-operated Ca²⁺ channels (VOCC) of the L-type become gated. Each of these channels activates an associated ryanodine receptor (RyR) in the sarcoplasmic reticulum, both by direct protein–protein interaction and by Ca²⁺ entering the cell through the L-type channel.

Skeletal muscle

The contraction of skeletal muscle cells is triggered by action potentials that are generated by acetylcholine interacting with nicotinic receptor cation channels (see Section 16.2). In turn, L-type Ca^{2+} channels located in the immediate vicinity become gated. These channels are coupled with ryanodine receptors in the adjacent sarcoplasmic reticulum which become activated (**Figure 14.29**). The Ca^{2+} thus released into the cytoplasm is bound by troponin C, which induces a conformational change of tropomyosin. This reaction clears the way for the interaction of actin and myosin, resulting in ATP hydrolysis and contraction. Such an arrangement of interacting proteins is based on a special cellular morphology and enables particularly rapid signal processing [one should bear in mind that more than 100 single contractions per second (100 impulses of signal generation and signal extinction) are taking place in a continuously working muscle].

Smooth muscle

Smooth or vegetative muscles surround the gastrointestinal tract, respiratory tract, blood and lymph vessels, uterus, oviduct, and vagina, as well as the urinary bladder, gall bladder, and many other tissues. They control the state of contraction and thus the function of those organs and are major targets for the signals of the vegetative nervous system.

While skeletal muscle contraction is triggered by an action potential, smooth muscle cells respond to subthreshold depolarizations such as those induced by ligand-operated cation channels, in particular by the ionotropic P2X purine receptor gated by ATP (see Section 16.6). As a result, adjacent L-type channels, which as we have seen gradually respond to depolarization, become partly gated and contraction is induced. This process is supported by additional mechanisms enabling a more subtle control, in particular by the vegetative nervous system. For example, β-adrenergic **sympathetic signals** (as transmitted by noradrenaline) evoke dilatation of most smooth muscles by activating the cAMP–PKA cascade. As a result the Ca^{2+} signal is suppressed, mainly because of phosphorylation and activation of the hyperpolarizing K^+ channel BK and of the Ca^{2+} pumps in the cell membrane and the ER. A similar effect is produced by cGMP released in smooth muscle cells by nitric oxide, the most effective vasodilatory agent of all (see Section 16.5.1).

In contrast, **parasympathetic signals** transmitted by $G_{q,11}$-coupled muscarinic acetylcholine receptors support smooth muscle contraction by triggering $InsP_3$-promoted Ca^{2+} release from the ER and, in turn, activation of ryanodine receptors by Ca^{2+}. On the other hand, Ca^{2+} released from the ER may also evoke smooth muscle dilatation. Such apparently contradictory observations find an explanation in

CONTRACTION   DILATATION

Figure 14.30 Signal transduction in smooth muscle contraction An input signal S1 (left), for instance ATP, activates a receptor cation channel, for instance the P2X purine receptor. By the depolarization thus induced, a voltage-operated Ca^{2+} channel (VOCC) of the L-type is gated and contraction is triggered by Ca^{2+} influx. This mechanism is completed and supported by signals S2 interacting with $G_{q,11}$-coupled receptors and inducing $InsP_3$ release (right). By interacting with its receptor ($InsP_3R$), $InsP_3$ turns up Ca^{2+}-dependent Ca^{2+} release from the sarcoplasmic reticulum. The increase of cytoplasmic Ca^{2+} leads to an activation of BK potassium channels evoking hyperpolarization and muscle dilatation, and depolarization by Ca^{2+}-controlled chloride channels of the types TMEM16A, and bestrophin (collectively called CaCC) supporting contraction. The BK channel is activated, in addition, by depolarization as well as by cAMP- and cGMP-dependent protein phosphorylation (not shown).

interactions of Ca^{2+} with other ion channels such as the Ca^{2+}-activated K^+ channel BK and Ca^{2+}-stimulated chloride channels (**Figure 14.30**; see also Sections 14.4.1 and 14.7.1). While the gating of the BK channel hinders voltage-dependent Ca^{2+} influx by hyperpolarization, the chloride channels evoke contraction by depolarization due to the relatively high chloride concentration in smooth muscle cells. By such positive and negative feedback control, the tone of the smooth muscle cell is fine-tuned.

Ca^{2+} evokes smooth muscle contraction by stimulating the phosphorylation of the myosin light chains that results in a deblockade of the actin–myosin interaction (see Figure 14.28). This reaction is catalyzed by **myosin light-chain kinase (MLC kinase)**, a Ca^{2+}-dependent enzyme with CaM as a regulatory subunit. In addition to MLC kinase, Ca^{2+}/CaM activates the protein caldesmon that displaces tropomyosin from the actin–myosin complex. In other words, in smooth muscle (and other cells), the function of the CaM/caldesmon pair resembles that of troponin C in skeletal muscle.

An alternative mechanism by which input signals control smooth muscle contraction is based on the *dephosphorylation* of the myosin light chains by which their inhibitory effect on the actin–myosin interaction is reestablished. The reaction is catalyzed by a specific **myosin light-chain phosphatase (MLC phosphatase)**. The enzyme is a heterodimer consisting of the phosphoserine/phosphothreonine-directed phosphatase PP1 (see Section 2.6.4) and a myosin phosphatase target subunit (MYPT1). MLC phosphatase is inactivated by protein kinase ROK, which is controlled by the small G-protein Rho (Section 10.3.3; **Figure 14.31**). To this end, ROK phosphorylates and activates a specific inhibitor protein CPI-17 that interrupts the interaction between the phosphatase subunits. At the same time ROK phosphorylates and inactivates myosin light chains, an effect that resembles that of MLC kinase. Along this pathway, contraction can be evoked without a further supply of Ca^{2+}. Stimulation of MLC phosphorylation by Rho/ROK is essential for cytoskeletal rearrangements and motility changes controlled by small G-proteins of the Rho family, not only in smooth muscle but in any other cell types (for details see Section 10.3). A well-known example of a universal Rho-controlled process is cytokinesis.

The reactions required for smooth muscle activation (release of $InsP_3$ and phosphorylation of myosin light chains and MLC phosphatase) are much slower than a voltage-stimulated Ca^{2+} influx and the noncovalent interactions between troponin C and tropomyosin in skeletal muscles. These mechanistic features and anatomical peculiarities explain the relative sluggishness of smooth muscles and their ability to execute long-lasting contractions.

Heart muscle

As in skeletal muscle, in heart muscle the gating of L-type VOCCs triggers Ca^{2+} influx that results in a Ca^{2+}/troponin C-dependent deblockade of the actin–myosin complex and muscle contraction. However, differing from skeletal muscle, here L-type Ca^{2+} channels communicate with ryanodine receptors through Ca^{2+} signals alone rather than by direct protein–protein interaction. By activating a Ca^{2+}-induced Ca^{2+} release at the mouth of the ryanodine receptor, Ca^{2+} entering the cell from outside generates calcium sparks. For this purpose each L-type channel is associated with a battery of four or five ryanodine receptors constituting a structure known as a "calcium synapse" (**Figure 14.32**). Such "synapses" work independently from each other with the result that the stronger the depolarizing input signal, the more "synapses" become activated and the more powerful is the contraction. This mechanism demonstrates in a particularly clear way how, by cooperation of a large number of independent binary units, a digital mode of signal processing based on the binary all-or-nothing code of an action potential is transformed into an analog mode enabling continuous modulation of the amplitude of the output signal (heart muscle contraction). This regulation is superimposed by a frequency modulation due to the positive and negative chronotropic effects of the vegetative nervous system on the pacemaker cells (see below).

Figure 14.31 Control of smooth muscle tone by myosin light-chain phosphorylation Phosphorylation (P) of myosin light chains (LC) results in contraction. Of the two kinases catalyzing this modification, myosin light-chain kinase (MLCK) is activated by an increase and inactivated by a decrease of the cytoplasmic Ca^{2+} concentration, while the kinase ROK is activated by small G-proteins of the Rho family (left). Vice versa, myosin light-chain phosphatase (MLCPase) causes smooth muscle dilatation. The phosphatase is phosphorylated and inactivated by ROK and re-activated by a protein phosphatase (PPase) (right).

Figure 14.32 Signal transduction in heart muscle contraction Through the action potential generated by pacemaker cells, voltage-operated calcium channels (VOCC) of the L-type are gated. Ca^{2+} ions entering the cell stimulate a Ca^{2+}-induced Ca^{2+} release from a battery of cooperating ryanodine receptors (RyR).

Figure 14.33 Generation of pacemaker potentials
(A) Oscillations of the membrane potential in pacemaker cells. In the final phase of each action potential, HCN (hyperpolarization- and cyclic nucleotide-activated) channels become gated, causing slow depolarization (shown in teal). This results in a gradual activation of voltage-operated T-type (T) and L-type (L) Ca^{2+} channels, ryanodine receptors (RyR), and Na^+–Ca^{2+} exchangers (EX), leading finally to a new action potential. The repolarization is initiated by voltage-operated K^+ channels (K_v). (B) Rhythmic fluctuations of the cytoplasmic Ca^{2+} concentration in nonstimulated (solid line) compared with sympathetic- and stress-activated pacemaker cells (dotted line), where the HCN channels are stimulated by cAMP.

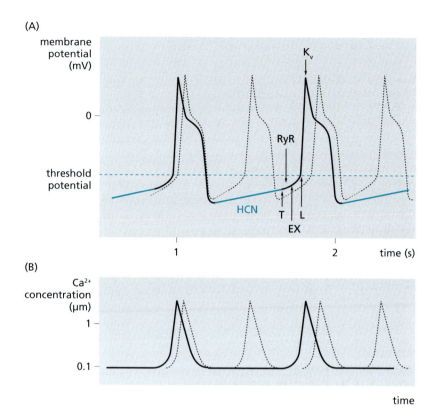

Autonomous oscillations control the heartbeat

The heart muscle is the prototype of a biological oscillator. Its rhythmic activity is brought about by a periodic Ca^2 signal. As a peculiarity, the oscillator and the effector systems are located in different cell types. The periodic signals are generated by a special type of heart muscle cells in the sinoatrial knot called pacemaker cells. As action potentials, they spread across the entire heart muscle, evoking a synchronous contraction of the atrial and ventricular muscle cells, which are electrically coupled by gap junctions. A system of positive and negative feedback loops based on the cooperation of several ion channels renders pacemaker cells autonomous oscillators (**Figure 14.33**).

Each action potential and thus each heartbeat is started by gating a special type of cation channels called pacemaker or **HCN channels** (see Section 14.5.7). These channels are voltage-dependent but, in contrast to most voltage-operated channels, they become gated at an extremely negative potential, as occurs at the end of an action potential. By the influx of Na^+ ions through the HCN channels, the cell becomes slowly depolarized. Beyond a threshold potential, VOCCs, first T-type and later L-type, take over the role of the HCN channels, which more and more become inactive. The influx of Ca^{2+} contributes to depolarization and, at the same time, activates ryanodine receptors and depolarizing Na^+–Ca^{2+} exchangers. Owing to this positive feedback, the Ca^{2+} influx and the depolarization become accelerated, triggering a new action potential when the threshold of about –35 mV is reached. Repolarization occurs in two steps through a gating of voltage-dependent potassium channels of the type hERG (see Section 14.4.1). This gives the action potential the characteristic shape with a plateau phase of partial repolarization between the peak of depolarization and terminal repolarization (see Figure 14.33).

HCN channels belong to the family of cyclic nucleotide-gated ion channels (see Section 14.5.7). In fact, cAMP and cGMP are ligands that specifically activate the channels in addition to depolarization. This has immediate consequences for the heart frequency: upon excitation and stress, sympathetic neurons release noradrenaline

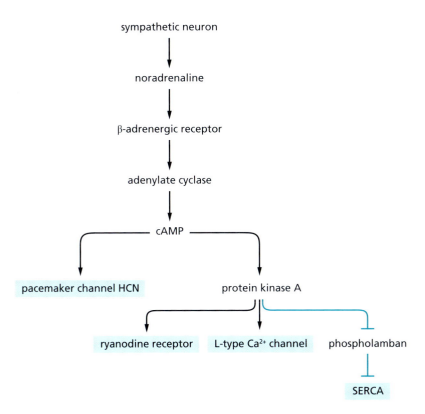

that evokes cAMP formation in pacemaker cells. Through the binding of cAMP to HCN channels, the interval between two subsequent action potentials becomes shortened and the heartbeat increases (see Figure 14.33). This effect is opposed by the parasympathetic signal acetylcholine, which suppresses cAMP formation along a G$_{i,0}$-controlled pathway (see Section 5.5). As a result, the heart calms down.

Under stress, the heart beats faster but also stronger. This amplitude modulation is again caused by cAMP and in this case is mediated by PKA. By means of an adaptor protein AKAP (a-kinase anchoring protein, see Section 4.6), the kinase associates with protein phosphatases and several regulatory proteins and forms a signal-transducing complex with the ryanodine receptor, which becomes sensitized by phosphorylation. Moreover, as mentioned above, PKA phosphorylates and sensitizes the L-type Ca^{2+} channel, thus augmenting the calcium signal.

Another stimulatory pathway leads to the Ca^{2+}-dependent ATPase SERCA. In the sarcoplasmic reticulum of nonstimulated heart muscle cells, this enzyme is inhibited by an interaction with the small transmembrane protein **phospholamban**. Phospholamban again is a substrate of PKA; upon phosphorylation, it loses its ability to interact with SERCA. As a result, cytoplasmic Ca^{2+} is pumped more rapidly into the lumen of the sarcoplasmic reticulum and the "charge" of the calcium battery is continuously kept at a high level. This context is depicted schematically in **Figure 14.34**.

Caffeine enhances the sympathetic effect on the heart by inhibiting phosphodiesterase-catalyzed cAMP degradation and sensitizing ryanodine receptors (for more details see Section 16.9.1). This is the reason for a lively heartbeat following excessive consumption of coffee.

14.6.3 How Ca^{2+} controls protein phosphorylation

Several protein kinases contain CaM as a regulatory subunit that, upon binding of Ca^{2+}, displaces an auto-inhibitory pseudosubstrate domain from the catalytic center, thus activating the kinase. We have already encountered members of this enzyme family: phosphorylase kinase (see Section 5.3.1), eEF2 kinase (also known as CaMKIII; see Section 9.3.4), and MLC kinase. These three kinases are highly specific for their particular substrates. In contrast, other Ca^{2+}/CaM-dependent kinases are less choosy: CaMKI, -II, and -IV and CaMK kinases α and β. They are structurally related to each other and to MLC kinase.

CaMKI is a widely distributed enzyme, whereas its isoform **CaMKIV** seems to be restricted to a very few tissues. Both phosphorylate a large number of proteins. Among the substrates are several transcription factors, such as CREB (cAMP response element binding protein) and its co-activator CBP/p300 (see Section 8.6) as well as the myocyte enhancer factor MEF2, which are activated by phosphorylation. Thus the two kinases play an important role in Ca^{2+}-controlled gene transcription, in particular since they have been shown to shuttle between cytoplasm and nucleus.

For activation of CaMKI and -IV, interaction with Ca^{2+}/CaM is a necessary but insufficient condition. In addition, like most protein kinases, both enzymes must be phosphorylated at a strategic threonine (Thr) residue in the activation loop. This phosphorylation is catalyzed by the **CaMK kinases** in a so-called CaMK cascade. Like CaMKs, CaMK kinases are activated by Ca^{2+}/CaM.

In addition to CaMKs, the CaMK kinases activate other substrates such as protein kinase B (PKB)/Akt. Since PKB/Akt inhibits apoptosis, this is probably a protective mechanism that hinders normal fluctuations of the calcium level from driving cells to death (note that cells undergo programmed suicide when flooded by Ca^{2+} for a *longer* period; see Section 13.2.5).

Calmodulin kinase II: a biochemical frequency detector with memory

By far the best-studied Ca^{2+}/CaM-dependent kinase is CaMKII. The reason for this interest is obvious: this extraordinary enzyme is crucially involved in learning and memory fixation (see Section 16.4.1).

Among the four isoforms of the kinase, CaMKIIα and -β are predominantly expressed in neurons, whereas CaMKIIγ and -δ are found in nearly all tissues. Each of the four isoforms exists in various splice variants. Although closely related to CaMKI and CaMKIV, CaMKII has an additional oligomerization domain in the C-terminal region that enables the enzyme to aggregate into star-shaped hexamers, two of which are stacked (**Figure 14.35**). The outstanding cellular functions of CaMKII critically depend on this quaternary structure, which seems to be unique in the world of protein kinases.

To trigger the activation of CaMKII, an interaction with Ca^{2+}/CaM is necessary. By this means an auto-inhibition through a pseudosubstrate motif is overcome. To stabilize the active conformation, phosphorylation is also required. To this end, the monomer initially activated by Ca^{2+}/CaM phosphorylates the adjacent monomer at a Thr residue in the regulatory domain (provided this subunit has also bound Ca^{2+}/CaM) (see Figure 14.35). In other words, in contrast to CaMKI and CaMKIV, which require the assistance of a CaMK kinase, CaMKII is activated by Ca^{2+}-dependent trans-autophosphorylation proceeding in a stepwise manner. The autophosphorylation has an important consequence: on the one hand it potentiates the affinity of the monomers for Ca^{2+}/CaM several thousandfold; on the other hand it renders the kinase activity autonomous (it does not depend any longer on Ca^{2+}/CaM).

The stronger the Ca^{2+} signal, the more monomers of the complex become occupied by Ca^{2+}/CaM and autophosphorylated until finally the entire complex is activated. Owing to this stepwise stimulation, the CaMKII oligomer functions as a detector, monitoring duration, intensity (amplitude), and frequency of a Ca^{2+} signal and staying active even when the signal has vanished. The duration of this "molecular memory" depends only on the rate of CaMKII dephosphorylation.

Figure 14.35 Activation of CaMKII (A) Six enzyme molecules are shown that aggregate through their C-terminal association domains (dark gray), forming a six-armed starlike structure. Two of these aggregates are stacked (for the sake of clarity, only one aggregate is shown here). In the absence of an input signal, the kinase activity is suppressed as shown for monomers 3–6: a pseudosubstrate motif located in a central regulatory domain (light gray rectangle) blocks the N-terminal catalytical domain (shown in color). In monomers 1 and 2, this blockade has been relieved by Ca²⁺/calmodulin (CaM). The conformational change resulting from this interaction has exposed a strategic Thr residue in the regulatory domain that becomes phosphorylated by the adjacent monomer (teal arrow). This trans-autophosphorylation renders the kinase activity of this monomer autonomous, or Ca²⁺/CaM-independent. With an increasing concentration of Ca²⁺/CaM, the remaining kinase monomers (3–6) also become activated by their neighbors in a stepwise manner. Thus, the degree of phosphorylation and the overall kinase activity of the complex directly depend on the *intensity* of the Ca²⁺ input signal. In the panel below, the activity of the multimeric complex is depicted as a function of the *frequency* of the (oscillating) Ca²⁺ signal (teal). At a low frequency, Ca²⁺/CaM has enough time to dissociate from the complex between the single pulses (left side; the phosphorylation is reversed by protein phosphatases). At higher frequencies this is not possible any longer. Instead, the individual phosphorylations summarize and the overall activity increases in a sigmoidal mode (right side). A similar situation arises when the *amplitude* of the Ca²⁺ signal increases. In this case the teal bars in the figure would become longer (not shown). Note that the *rate* of inactivation (the control value of which is indicated by the dotted lines) slows down with the number of kinase monomers activated since the binding affinity for calmodulin increases with each step. (B) Model of the inactive CaMKII based on X-ray crystallography. One recognizes the two stacked star-shaped oligomers composed of six monomers each are shown from above (upper structure) and from the side (lower structure, pulled apart a little). The association domains are black; the catalytic domains containing the Ca²⁺/CaM binding sites are colored (upper oligomer) or gray (lower oligomer). (B, adapted with modifications from Rosenberg OS, Deindl S, Sung RJ et al. [2005] *Cell* 123:849–860.)

CaMKII comes close to Hans von Foerster's model of a nontrivial calculator: a signal-processing device that is able to learn by adapting its algorithm to the previous operations (see Section 1.6). As shown in Section 16.4.1, this property of CaMKII is highly significant for long-term potentiation of synaptic signal transmission, which is thought to resemble a learning process.

In fact, CaMKII phosphorylates many proteins involved in neurotransmission. These include ryanodine receptors, excitatory glutamate receptors (see Section 16.4), and various transcription factors (such as CREB and the serum response factor), which are activated, as well as the neuronal NO synthase (see Section 16.5.1) and the neuronal Ras-GAP SynGAP, which are both inactivated. To recognize individual substrate proteins and to become targeted to individual cellular sites, CaMKII (like other protein kinases) needs the assistance of various adaptor and scaffold proteins.

14.6.4 How calcium regulates gene transcription

As explained above, Ca^{2+}-dependent protein kinases, primarily CaMKI, -II, and –IV, phosphorylate various transcription factors, thus stimulating the transcription of numerous genes. A well-known example is provided by the transcription factor CREB, which is phosphorylated by CaMK at the same site as by protein kinase Y (see Section 8.6 for details).

NFAT: nuclear factor of activated T cells and more

Another group of transcription factors called **NFAT** is activated by Ca^{2+}-dependent *dephosphorylation* (**Figure 14.36**). The name <u>N</u>uclear <u>F</u>actors of <u>A</u>ctivated <u>T</u>-cells refers to the initial discovery of these proteins. However, NFATs are by no means restricted to T lymphocytes but are found in many cell types, where they control a wide variety of physiological processes. For gene activation, NFATs cooperate with other transcription factors, the type of which depends on the tissue. In T-lymphocytes, for instance, the partners are Jun–Fos dimers of the AP1 transcription factor family that are activated along MAP kinase pathways (see Section 11.6.2). Like most other genes, therefore, NFAT-controlled genes are logical gates that are controlled by the combined effect of at least two different input signals.

Five NFAT isoforms are known and are tissue-specifically expressed. In the absence of Ca^{2+} signals, they are phosphorylated at 10 serine residues at least. Since

Figure 14.36 Calcineurin–NFAT signaling pathway Various signals activating receptors and membrane channels induce an increase of cytoplasmic Ca^{2+} that, upon binding to calmodulin (CaM), activates the protein phosphatase calcineurin. Calcineurin dephosphorylates the transcription factor NFAT (nuclear factor of activated T cells), which thus becomes active and translocates into the nucleus to stimulate, in cooperation with other transcription factors (here AP-1), numerous genes. NFAT is inactivated by phosphorylation catalyzed by kinases such as protein kinase A (PKA) and glycogen synthase kinase 3β (GSK3β). Both DSCR-1 protein, overexpressed in neurons of patients with Down syndrome and immunosuppressive drugs, inhibit calcineurin.

phosphorylation blocks both the DNA-binding domain and the nuclear localization sequence, the factors cannot translocate into the nucleus and interact with DNA unless they are dephosphorylated.

The inhibitory phosphorylation of NFATs is catalyzed by several protein kinases, in particular PKA and glycogen synthase kinase 3β, whereas stimulatory dephosphorylation is induced by Ca^{2+} and catalyzed by the Ser/Thr-directed protein phosphatase PP2B forming a complex with NFAT. In fact, PP2B is the only protein phosphatase controlled by Ca^{2+}/CaM. Because this otherwise widely distributed enzyme is strongly expressed in neurons, it is known also as **calcineurin**. The phosphatase is a heterodimer consisting of a catalytic subunit A and a regulatory subunit B with four EF-hand motifs. In the absence of a calcium signal, the enzymatic activity is suppressed by an auto-inhibitory domain. This blockade is relieved partly by an interaction with Ca^{2+}/CaM and completely by additional binding of Ca^{2+} to the regulatory domain. In other words, like CaMKII, calcineurin is activated by Ca^{2+} in a *stepwise* manner, although according to a different mechanism.

Apart from NFATs, calcineurin dephosphorylates other proteins. Its strong expression in the brain indicates that it participates in learning and memory fixation, a concept that is supported by a large body of evidence (see Section 16.4.1). A permanent (genetically fixed) insufficiency of calcineurin seems to be one among several causes of schizophrenia, and a constitutive down-regulation of calcineurin activity is thought to be involved also in Down syndrome (**Sidebar 14.2**).

In principle, calcineurin and NFAT are activated by every signal inducing an increase of the cytoplasmic calcium concentration. In T lymphocytes, antigens that stimulate Ca^{2+}-release from the ER by activating the T-cell receptor PLCγ–InsP$_3$ cascade transmit such signals. Along this pathway, antigens turn up the transcription of genes controlling T-cell proliferation and the immune response, including genes of the T-cell mitogen interleukin 2 and its receptor; interleukins 3, 4, 5, and 8; interferon γ; Tyr kinase Lck (see Section 7.2.2); death receptor Fas and its ligand (see Section 13.2.3); and prostaglandin H synthase COX-2 (see Section 4.4.5).

Owing to its key function in T-cell activation, the NFAT signaling pathway has found considerable interest in the clinic, in particular since it has turned out to be a target of **immunosuppressive drugs** such as cyclosporine and FK506, which have revolutionized transplantation medicine. These drugs bind to cellular proteins called immunophilins (such as cyclophilin and the FK506-binding protein; for details see Section 8.5) that block the interaction between calcineurin and NFAT, thus suppressing the immune response.

In addition to their effects on the immune system, NFAT transcription factors exhibit other functions. They participate in the development of the central nervous system,

---

Sidebar 14.2 Calcineurin and Down syndrome Affecting 1 out of 700 newborns, Down syndrome is by far the most frequent birth defect caused by chromosomal damage. The disease is due to a tripling of chromosome 21 (trisomy 21), resulting in a pathological overexpression of approximately 300 genes. One of these genes encodes a protein that inhibits calcineurin (and thus NFAT) signaling. It is called calcipressin or DSCR1 (Down syndrome critical region of chromosome 1). Because the synthesis of this inhibitor is induced by NFAT, it seems to be a component of a negative feedback mechanism controlling the calcineurin–NFAT signal. Another gene that maps to a Down syndrome region of chromosome 21 is the unique cell adhesion molecule and receptor DS-CAM (Section 16.1.5). Because these proteins play a central role in signaling pathways controlling embryonic development, continuous overexpression of DSCR1 and DS-CAM may contribute significantly to the organic and mental symptoms of patients with Down syndrome.

cardiovascular system, skin, and skeletal muscles. Moreover, they regulate the adaptation of organ functions to environmental conditions. An example with considerable medical consequences is provided by the hypertrophy of the heart muscle in response to continuous strain (the athletic heart). The underlying mechanism is thought to consist of a chronic activation of postsynaptic Ca^{2+} channels (L-type) in neuromuscular synapses. As a result, the calcineurin–NFAT system becomes overstimulated and activates genes of muscle proteins.

The DREAM of pain research

Currently only one transcription factor is known that *directly* interacts with Ca^{2+}. This is **DREAM**, Downstream Regulatory Element Antagonist. The name of this protein refers to its discovery as a factor bound by a regulatory sequence of the prodynorphin gene located downstream of the transcription start point. DREAM has four EF-hand motifs and belongs to the same family of neuronal Ca^{2+}-sensor proteins as the retinal protein recoverin.

In the absence of a Ca^{2+} signal, the factor inhibits transcription, acting as a gene repressor. Upon binding of Ca^{2+}, DREAM dissociates from the DNA, allowing access by the RNA polymerase complex and transcriptional activators (**Figure 14.37**). In addition to its repressor functions, DREAM seems to have other effects. It has been independently characterized as "potassium channel interacting protein 3," binding to voltage-dependent K^+ channels, and as "calsenilin," interacting with the protein presenilin-2 associated with Alzheimer's disease (see Section 13.1). In both cases, DREAM modulates the functions of its binding partners. Thus, DREAM is a multifunctional transcription factor and in this respect resembles, for instance, p53 and β-catenin.

DREAM has caused some stir in the context of **pain research**. Mice carrying an inactive *dream* gene have been found to be widely insensitive to pain, probably because of an overproduction of **dynorphin** in the spinal cord. This endogenous "opiate" interacts with the $G_{i,0}$-coupled opioid receptor κ (see Section 16.8) and inhibits the generation of action potentials and the transmission of pain signals to the brain, apparently without development of tolerance and addiction (**Figure 14.38**). The name DREAM thus takes on a second meaning, as it may represent a "dream target" in the battle against chronic pain. On the other hand, DREAM may also *enhance* pain by inhibiting neurons that dampen the activity of pain neurons.

In addition to CREB, NFAT, and DREAM, there are many other transcription factors, the activities of which are regulated by Ca^{2+} signals, albeit in an indirect way. An

Figure 14.37 Transcriptional repressor DREAM As a tetramer, DREAM (downstream regulatory element antagonist) binds to downstream regulatory elements (DRE) of certain genes inhibiting transcription (top). Upon binding of Ca^{2+} the tetramer dissociates into two dimers, which leave the DNA giving way to the RNA polymerase II complex (bottom).

overview is given in **Figure 14.39**. The individual signaling pathways are discussed elsewhere in this book.

14.6.5 Calcium and cell death

Uncontrolled flooding of the cytoplasm by Ca^{2+} inevitably kills the cell by causing apoptosis and necrosis. This is the reason for tissue destruction resulting from stroke or myocardial infarction. In these cases Ca^{2+} enters the cell from the outside or is released by damaged mitochondria or by the ER (for a more detailed discussion of apoptosis, see Section 13.2).

The apoptotic signal is transduced by several Ca^{2+}-dependent proteins. One of these, calcineurin, acts as an antagonist of the anti-apoptotic PKB/Akt pathway by dephosphorylating and activating the pro-apoptotic protein Bad. Moreover, NO synthases and oxygenases of arachidonic acid metabolism activated along the Ca^{2+}-dependent PLA_2 pathway catalyze the generation of cytotoxic metabolites such as peroxynitrite and reactive oxygen species that damage mitochondria, thus triggering apoptosis.

Ca^{2+}-dependent proteases participating in apoptotic and necrotic protein degradation are the **calpains** (**Sidebar 14.3**). These enzymes, for instance, destroy anti-apoptotic proteins such as Bcl2 and Bcl-XL and activate pro-apoptotic proteases of the caspase family. More recently, a family of **death-associated protein kinases** has been discovered that promotes apoptosis. Because these Ser/Thr-specific enzymes are closely related to MLC kinase, most of them are activated by Ca^{2+}/CaM. Like cyclin-dependent kinases, they also require dephosphorylation to become fully active (see Section 13.2.5 for more details).

Figure 14.38 Pain relief by Ca²⁺ signals (A) A peripheral pain sensor activates (via the neurotransmitter substance P) a spinal neuron that transmits a high-frequency burst of action potentials (teal circle) as a pain signal to the brain. (B) A spinal interneuron simultaneously activated by substance P decreases the firing frequency of the pain neuron by means of dynorphin. At the level of transcription, the synthesis of dynorphin is inhibited by DREAM (A) and this inhibition is overcome by Ca^{2+} released in the course of an action potential (B). (Modified from Vogt BA [2002] *N Engl J Med.* 347:362–364.)

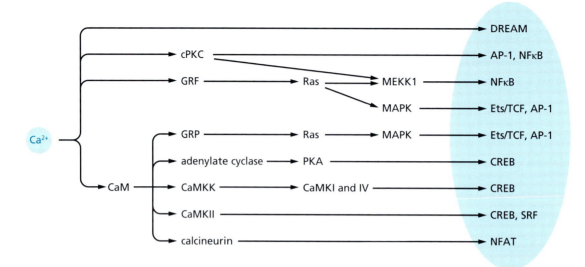

Figure 14.39 Ca²⁺-regulated gene transcription: an overview Shown are signaling pathways leading to an activation of transcription factors or to a removal of repressors. These factors are shown in the oval. For details see text. CaM, calmodulin; cPKC, conventional protein kinase C; GRF, guanine nucleotide-releasing factor, and GRP, guanine nucleotide-releasing protein (both Ras-guanine nucleotide exchange factors; see Section 10.1.3); MEKK1, MAP2 kinase 1; MAPK, mitogen-activated protein kinase; PKA, protein kinase A; CaMKK, calmodulin-activated kinase kinase; CaMK, calmodulin-activated kinase; DREAM, downstream regulatory element antagonist; AP-1, activator protein 1; NFκB, nuclear factor κB; Ets/TCF, erythroblastosis virus E26/ternary-complex factors; CREB, cAMP-response element binding protein; SRF, serum response factor; NFAT, nuclear factor of activated T cells.

Summary

- Ca²⁺-binding domains such as the EF-hand, the C2 motif, and the annexin domain are found in a wide variety of signal-transducing proteins which participate in the control of almost any cellular process.

- Most prominent roles play such proteins in stimulus-secretion coupling (such as neurotransmitter release) and stimulus–contraction coupling (such as muscle contraction).

- In all types of muscle cell, contraction is triggered by voltage-operated Ca²⁺ channels in the plasma membrane and augmented by ryanodine and InsP₃ receptors in the ER. For the graded response of smooth muscle cells both contraction and dilatation are fine-tuned by additional ion channels and exchangers and G-protein-coupled receptors. A key parameter is the phosphorylation status of myosin light chains, controlled by Ca²⁺/CaM-activated MLC kinase (triggering contraction) versus MLC phosphatase (triggering dilatation).

- The channel HCN controlling heart frequency in pacemaker cells belongs to the family of cyclic nucleotide-gated cation channels. Its sensitization by cAMP explains the sympathetic and stress effects on the heart frequency. Through protein kinase A-catalyzed phosphorylation of the ryanodine receptor, the L-type channel, and the inhibitor protein phospholamban, cAMP also increases the power of heart contraction. Caffeine, an inhibitor of cAMP degradation, intensifies heart activity.

- Several protein kinases are activated by Ca²⁺/CaM. These include highly specific enzymes such as glycogen phosphorylase kinase, EF2 kinase, and MLC kinase as well as the less specific CaMKI, -II, and -IV. CaMKI and -IV participate in gene activation. CaMKII is a frequency detector that—on the basis of a unique oligomeric structure—monitors the duration, amplitude, and frequency of oscillating Ca²⁺ signals and thus provides a molecular short-term memory store.

- The activity of numerous transcription factors is regulated *indirectly* by Ca²⁺. An example is provided by the transcription factor NFAT that is activated by dephosphorylation catalyzed by the Ca²⁺/CaM-dependent protein phosphatase PP2B (calcineurin). Calcineurin is inhibited by immunosuppressive drugs. Moreover, impaired calcineurin activity seems to be involved in Down syndrome.

Sidebar 14.3 Calpains: calcium-dependent proteases
Calpains are cysteine proteases with EF-hand domains. They become activated by relatively high Ca^{2+} concentrations and are inhibited by specific proteins, the calpastatins. Since calpains have a wide spectrum of substrate proteins, they have been connected with various cellular processes, although their physiological role is still not fully understood.

An effect with significant consequences for signal transduction consists of the proteolytic removal of auto-inhibitory domains from signal-processing proteins, which thus may become constitutively active. As examples, the continuous activation of protein kinase C by destruction of the pseudosubstrate motif, and of the neuronal protein kinase CDK5 by partial proteolysis of the regulatory subunits p35 or p38 have been mentioned already (see Sections 4.6.3 and 12.3). Though the physiological significance of such effects is still a matter of debate, an increasing body of evidence indicates a role of permanent CDK5 activation in neurodegenerative diseases such as Alzheimer's dementia.

Other processes in which calpains are thought to be involved are the following:

- the cell cycle, where under certain circumstances they support protein degradation (for instance of cyclins) by proteasomes
- cell movements, where they disintegrate contacts between integrins and the cytoskeleton
- synaptic long-term potentiation
- necrosis and certain types of apoptosis

In several diseases, accompanying disturbed Ca^{2+} homeostasis, calpains contribute to functional disorders and cell death. Examples include cataracts, muscular dystrophy, myocardial infarction as well as stroke, multiple sclerosis, Alzheimer's dementia, and other neurodegenerative disorders.

- The only transcription factor known to interact *directly* with Ca^{2+} is the repressor DREAM; it is inactivated by Ca^{2+} binding and may be involved in the processing of pain sensations.
- At higher concentrations Ca^{2+} is a strong inducer of apoptosis by activating Ca^{2+}-dependent enzymes and other proteins that trigger pro-apoptotic signal transduction. Ca^{2+}-induced cell death plays a crucial role in stroke, myocardial infarction, and neurodegenerative diseases such as Alzheimer's dementia.

14.7 Anion channels

Anion channels are found in all organisms from the simplest prokaryotes to mammals (Section 3.6.7). Nevertheless, considering their investigation, they are the Cinderellas among the ion channels. Only recently has this situation changed—in particular because of structural studies—and now anion channels are gaining steadily increasing attention. In addition to chloride, they transport bromide, iodide, nitrate, and other negatively charged ions; however, in animal cells these ions are much less abundant than chloride. Therefore, anion channels are usually called chloride channels. The gating of chloride channels may have either a depolarizing or a hyperpolarizing effect depending on the steepness of the ion gradient across the cell membrane. When the concentration of chloride in the cytoplasm is much lower than in the extracellular space, chloride ions will enter the cell through an open channel despite the negative charge inside. Consequently, the intracellular charge becomes more negative: the cell hyperpolarizes, and events depending on depolarization are impaired. This is the case, for instance, in skeletal muscle cells and in most neurons. In other cell types, such as smooth muscle cells, the concentration gradient is not steep enough to overcome the electric force of the membrane potential. Here the opening of an anion channel results in an efflux of chloride, causing depolarization (see also Section 14.6.2).

Figure 14.40 Families and major functions of chloride channels Helices are symbolized by cylinders. The number of subunits per channel is shown in parentheses. ATP-binding sites are depicted as colored rectangles. ClC, chloride channel; R, regulatory domain; CFTR, cystic fibrosis transmembrane conductance regulator; GABA, γ-aminobutyric acid. For more details, see text.

Six families of chloride channels have been found in mammalian cells (partly illustrated in **Figure 14.40**):

- the ClC (Cl channel) family
- the TMEM16 (transmembrane member 16) family
- the bestrophin family
- the LRRC8 (Leucine-Rich Repeat Containing 8) family
- the CFTR (cystic fibrosis transmembrane conductance regulator) family
- the family of GABA (gamma-aminobutyric acid) and glycine receptor anion channels

The major functions of these channels are control of cell volume and osmolarity, trans-epithelial electrolyte transport, regulation of the intracellular pH value, setting of the resting potential, and (mostly inhibitory) neurotransmission. In this chapter, channels of the ClC and CFTR types will be discussed; the receptor anion channels used for neurotransmission are explained in Section 16.3.

14.7.1 Anion channels and Cl⁻/H⁺ exchangers of type ClC

This ancient channel type is found in prokaryotes, most probably participating in regulation of pH value and cell volume (see Section 3.6.7). Mammals express at least

Figure 14.41 Three-dimensional structure of a chloride channel of the type ClC The figure shows one of the two subunits of the channel. Note that the transmembrane helices (cylinders) are arranged in a twisted and tilted rather than in a parallel and perpendicular manner. The teal sphere in the central pore symbolizes the chloride ion. (Adapted from Jentsch TJ [2002] *Nature* 415:276–277.)

nine different isoforms. Four of them are ion channels while the remaining five isoforms operate as antiporters transporting two negative chloride ions per one positive proton across membranes. This provides a remarkable example of an electrogenic anion/cation exchanger. The overall structure, derived from the bacterial channel, is not related to that of cation channels. Instead, it comprises no less than 18 α-helices; 10–12 of these are transmembrane domains, whereas the others dip into the membrane only partly or not at all (Figure 14.40 and **Figure 14.41**). Each channel is composed of two such subunits and exhibits two pores (see Figure 3.37 in Section 3.6.7).

Most chloride channels/antiporters are gated by low pH, high Cl$^-$ concentration and membrane depolarization. Their voltage-sensitive domain is not related to the S3–S4 voltage sensor of cation channels and the mechanism of voltage-dependent gating differs from that of voltage-gated cation channels in that it depends on the Cl$^-$ and H$^+$ concentrations. The individual ClC types exhibit individual functions and tissue distributions. The channel **ClC1**, for instance, is more or less restricted to skeletal muscle cells, whereas the related channel **ClC2** is broadly expressed. (The first channels of the ClC type to be isolated were those of the electric organ of the electric ray *Torpedo californica*. These so-called ClC0 channels resemble the type ClC1 of skeletal muscles, which are the evolutionary precursors of electric organs.) While ClC2 participates in the regulation of the intracellular pH value and the cell volume, ClC1 controls the resting potential, which in skeletal muscle cells is set mainly by chloride channels rather than by potassium channels as in neurons. Consequently, the channel is gated by depolarization. Defects of the *clc1* gene cause severe neuromuscular diseases such as congenital myotony.

Another highly specialized chloride channel is **ClC-K**, expressed predominantly in the inner ear and kidney. In the kidney, the channel is involved in NaCl reabsorption (**Figure 14.42**). ClC-K forms a complex with a regulatory subunit with two

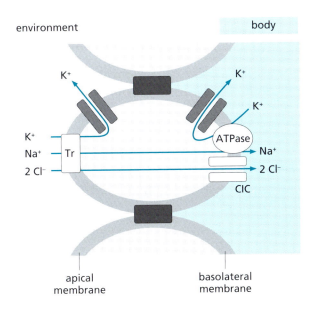

Figure 14.42 NaCl reabsorption in Henle's loop of the kidney By means of an Na$^+$/K$^+$/2 Cl$^-$ co-transporter (Tr), the epithelial cell (shown as an oval) takes up NaCl and KCl from the primary urine (environment). At the basolateral side (body), Na$^+$ is pumped into the blood by a K$^+$/Na$^+$-dependent ATPase with chloride following passively through a ClC channel. Excessive K$^+$ leaves the cell through apical and basolateral potassium channels of the type K$_{ir}$ (Figure 14.10). In an alternative transport system, epithelial cells take up Na$^+$ by Na$^+$ channels (ENaC, Figure 14.9).

transmembrane domains called barttin. This name alludes to **Bartter syndrome**, which is characterized by a severe defect of salt balance. The reason is loss-of-function mutations of the barttin and ClC-K genes, leading to massive salt loss in the kidney and to deafness since the production of endolymph in the inner ear is impaired.

ClC7, a $2Cl^-/H^+$ exchanger, plays a key role in the **resorption of bone material**. This is the task of a phagocyte-like cell type, the osteoclast, which produces acidic hydrolases that dissolve the bone substance. The acidic milieu required is generated by H^+-dependent V-ATPases located in the osteoblast membrane. They pump protons from the intra- to the extracellular space (see Figure 14.40). For the transport of larger amounts of protons, a parallel transport of anions (Cl^-) is established by ClC7 since otherwise the electrical gradient across the osteoclast membrane would become too steep, suppressing ATPase activity. As a result, hydrochloric acid accumulates in the lumen between osteoclast and bone. Bone resorption is of utmost importance for the development of the skeleton and the repair of bone damage. Indeed, a mutation or deletion of the *clc7* gene seems to be a cause of juvenile osteoporosis, a severe developmental disorder characterized by dense and fragile bones. Acidification is also important for the function of certain intracellular organelles such as endosomes, lysosomes, and secretory vesicles, the membranes of which also contain H^+-dependent ATPases and Cl^-/H^+ exchangers such as ClC7.

Some pressure-sensitive ClC types play a role as **cellular pressure relief valves**. In cooperation with 2P-domain potassium channels, they enable the cell to adapt to hypotonic stress by promoting an efflux of KCl (see Section 14.4.3).

14.7.2 Ca^{2+}-and volume-regulated chloride channels

Mammals express 10 **TMEM16** channels. While the functions of most of them are still controversial, two isoforms, TMEM16A and -16B, have been characterized as Ca^{2+}-regulated chloride channels; that is, they require Ca^{2+} ions for opening. TMEM16A and -B share this property with the **bestrophins**, which are expressed in four isoforms. Nevertheless, TMEM16 and bestrophin channels are structurally unrelated to each other and to any other ion channel type. Thus, TMEM16A consists of two subunits with eight transmembrane domains each, whereas in bestrophin 1 the channel is formed by five subunits with four transmembrane domains each.

Both channel families play a key role in a wide variety of physiological processes including sensory signal transduction and other neuronal functions, muscle contraction (for an example, see Figure 14.30), intracellular Ca^{2+} signaling (for instance during apoptosis and cell proliferation), re-structuring of cellular membranes, epithelial chloride secretion, and control of cell volume and osmotic pressure.

A channel especially involved in volume control is **LRRC8**. It is expressed in five isoforms and seems to be the core protein of volume-regulated anion channels. Similar to certain ClC types (see above), LRRC8 responds to changes in osmolarity and to cell swelling, the molecular mechanisms of which are still not fully understood. The physiological importance of TMEM16, bestorphin, and LRRC8 is emphasized by the fact that mutations of the corresponding genes may cause a wide variety of severe diseases.

14.7.3 Chloride channels of type CFTR

The CFTR channel plays an essential role in the hormonal regulation of mucous epithelia. This was discovered during studies on the genetic background of **cystic fibrosis** (or mucoviscidosis), a severe and fatal hereditary disease that is characterized by massive disturbances of water and electrolyte balance. Consequently, mucous membranes suffering from desiccation produce a highly viscous secretion that strongly impairs their function and promotes bacterial infection, in particular in the respiratory and gastrointestinal tracts. The reason is a loss-of-function mutation of the *cftr* gene, resulting in reduced CFTR channel permeability.

Figure 14.43 Domain structure and regulation of the CFTR chloride channel To open the channel, the regulatory domain R must be phosphorylated by either cAMP- or cGMP-dependent kinases and ATP must be hydrolyzed by the intrinsic ATPase domains (shown as colored rectangles). The figure shows two (patho)physiologically important signaling pathways leading to channel gating. AC, adenylate cyclase; PKA, protein kinase A; PKG, protein kinase G; GC, guanylate cyclase.

Among the ion channels, CFTR is unique because it belongs to the family of ABC proteins, which normally transport organic molecules rather than ions (see Section 3.3.1). Like most ABC transporters, CFTR has 12 transmembrane domains arranged in two groups of six each, as well as two ATPase domains in the cytoplasmic region (**Figure 14.43**). These structural elements are completed by a regulatory domain (R in Figure 14.43) that contains various phosphorylation sites and is not found in other human ABC proteins. The regulatory domain is the target of hormonal signals controlling CFTR permeability.

Gating of the channel resembles a logical AND gate because it requires both phosphorylation of the regulatory domain and ATP binding. Upon ATP hydrolysis by the intrinsic ATPase activity (as well as upon dephosphorylation), the channel is closed. This built-in timer function resembles that of G-proteins. Phosphorylation is required because the unphosphorylated regulatory domain blocks the channel pore. It is mainly catalyzed by either cAMP- or cGMP-dependent protein kinases. The cAMP-dependent pathway is activated, for instance, by (nor)adrenaline as part of the systemic stress response; the cGMP-controlled pathway is stimulated by the peptide hormone **guanylin** produced by the small intestine and regulating the electrolyte balance. While (nor)adrenaline interacts with a G_s-protein-coupled β-adrenergic receptor, guanylin stimulates a receptor with intrinsic guanylate cyclase activity (this receptor type is discussed in detail in Section 4.5).

While the inactivation of the CFTR channel causes desiccation of mucous membranes, an overactivation has the opposite result. This explains the effect of some prominent bacterial toxins that attack CFTR-related signal transduction. The G_s-protein is a specific target of **cholera toxin** and **heat-labile enterotoxin** of *Escherichia coli*; they cause a hyperphosphorylation and continuous activation of CFTR (see Section 4.3.3), which leads to a catastrophic loss of salt and water, particularly in the intestinal epithelium. The **heat-stable enterotoxin** of bacteria (such as certain *E. coli* strains) causing diarrhea has a similar effect by overstimulating the guanylin receptor.

Summary

- Anion channels are complex transmembrane proteins with prokaryotic ancestry. They are expressed in six families that are structurally unrelated to each other. Depending on the steepness of the Cl^- gradient, their gating has either a hyperpolarizing or a depolarizing effect. Some of these channels are regulated by calcium ions.

- Anion channels have numerous functions including control of the resting potential, regulation of the osmotic pressure and cell volume, transepithelial electrolyte transport, bone resorption, sensory signal transduction, and muscle contraction.

- CFTR-type channels belonging to the ABC transporter family are essential for the function of mucous membranes. They are gated by a combined effect of ATP hydrolysis and protein phosphorylation, catalyzed by either cAMP- or cGMP-dependent protein kinases. Cholera toxin and heat-stable enterotoxin of colibacteria deregulate the function of the channel, while a loss-of-function mutation causes cystic fibrosis.

Further reading

Section 14.1

Bezanilla F (2008) How membrane proteins sense voltage. *Nat Rev Mol Cell Biol* 9:323–332.

Yu FH, Yarov-Yarovoy V, Gutman GA & Catterall WA (2005) Overview of molecular relationships in the voltage-gated ion channel superfamily. *Pharmacol Rev* 57:387–395.

Zheng J & Trudeau MC (2015) Handbook of ion channels. CRC Press.

Sections 14.2 and 4.3

Catterall WA & Zheng N (2015) Deciphering voltage-gated Na^+ and channels by studying prokaryotic ancestors. *Trends Biochem Sci* 40: 526-534.

Kusche-Vihrog K, Jeggle P & Oberleithner H (2014) The role of EnaC in vascular endothelium. *Pflügers Arch Eur J Physiol* 466:851–859.

Kwong K & Carr MJ (2015) Voltage-gated sodium channels. *Curr Opin Pharmacol* 22:131–139.

Lingueglia E & Lazdunsli M (eds) (2015) Acid-sensing ion channels in the nervous system. *Neuropharmacology* 94:1–118.

Section 14.4

Biel M & Michalakis S (2009) Cyclic nucleotide-gated channels. *Handb Exp Pharmacol* 2009:111–136.

Enyedi P & Czirjak G (2010) Molecular background of leak K^+ currents: two-pore domain potassium channels. *Physiol Rev* 90:550–605.

Gonzalez C et al. (2012) K^+ channels: function-structural overview. *Compr Physiol* 2:2087–2149.

Hibino H et al. (2010) Inwardly rectifying potassium channels: their structure, function and physiological roles. *Physiol Rev* 90:291–366.

Kalia J et al. (2015) From foe to friend: using animal toxins to investigate ion channel function. *J Mol Biol* 427:158–175.

Nichols CG (2006) K_{ATP} channels as molecular sensors of cellular metabolism. *Nature* 440:470–476.

Podda MV & Grassi C (2014) New perspectives in cyclic-nucleotide-mediated functions in the CNS: the emerging role of cyclic nucleotide-gated (CNG) channels. *Pflügers Arch Eur J Physiol* 466:1241–1257.

Sanguinetti MC & Tristani-Firouzi (2006) hERG potassium channels and cardiac arrhythmia. *Nature* 440:463–469.

Tengholm A (2007) Cyclic AMP: swing that message! *Cell Mol Life Sci* 64:382–385.

Yang H, Zhang G & Cui J (2015) BK channels: multiple sensors, one activation gate. *Front Physiol* 6:1–16.

Zylbergold P, Ramakrishnan N & Herbert TE (2010) The role of G proteins in assembly and function of Kir3 inwardly rectifying potassium channels. *Channels* 4:411–421.

Section 14.5

Atlas D (2013) The voltage-gated calcium channel function as the molecular switch of synaptic transmission. *Annu Rev Biochem* 82:607–635.

Berridge JR, Bootman MD & Roderick HL (2003) Calcium signaling: dynamics, homeostasis and remodelling. *Nat Rev Mol Cell Biol* 4:517–529.

De Stefani D, Patron M & Rizzuto R (2015) Structure and function of the mitochondrial calcium uniporter complex. *Biochim Biophys Acta* 1853:2006–2011.

Fahrner M et al. (2013) The Stim1/Orai signaling machinery. *Channels* 7:330–343.

Fedorenko OA et al. (2014) Intracellular calcium channels: inositol-1,4,5-trisphosphate receptors. *Eur J Pharmacol* 739:39–48.

Gees M et al. (2012) TRP channels. *Compr Physiol* 2:563–608.

Guse AH (2015) Calcium mobilizing second messengers derived from NAD. *Biochim Biophys Acta* 1854:1132–1137.

Hofer AM & Brown EM (2003) Extracellular calcium sensing and signaling. *Nat Rev Mol Cell Biol* 4:530–538.

Hogan PG & Rao A (2015) Store-operated calcium entry: mechanisms and modulation. *Biochem Biophys Res Commun* 460:40–49.

Lee HC (2012) Cyclic ADP-ribose and nicotinic acid adenosine dinucleotide phosphate (NAADP) as messengers for calcium mobilization. *J Biol Chem* 287:31633–31640.

Krebs J, Agellov LB & Michalak M (2015) Ca^{2+} homeostasis and endoplasmic reticulum (ER) stress: an integrated view of calcium signaling. *Biochem Biophys Res Commun* 460:114–121.

Parekh AB (2011) Decoding of cytosolic Ca^{2+} oscillations. *Trends Biochem Sci* 36:78–97.

Semedler E & Uhlen P (2014) Frequency modulation of calcium oscillations. *Biochim Biophys Acta* 1840:964–969.

Strehler EE (2015) Plasma membrane calcium ATPases: from generic Ca^{2+} pumps to versatile systems for fine-tuning cellular Ca^{2+}. *Biochim Biophys Res Commun* 460:26–33.

Thompson JL & Shuttleworth TJ (2013) Explaining the unique feature of the ARC channel, a store-independent Orai channel. *Channels* 7:364–374.

Van Petergern F (2015) Ryanodine receptors: allosteric ion channel giants. *J Mol Biol* 427:31–59.

Section 14.6

Ames JB & Lim S (2012) Molecular structure and target regulation of neuronal calcium sensor proteins. *Biochim Biophys Acta* 1820:1205–1213.

Lisman J, Yasuda R & Raghavachari S (2014) Mechanism of CaMKII action in long-term potentiation. *Nat Rev Neurosci* 13:169–182.

Costigan M & Woolf CJ (2002) No DREAM, no pain: closing the spinal gate. *Cell* 108:297–300.

Kortvely E & Gulya K (2004) Calmodulin, and various ways to regulate its activity. *Life Sci* 74:1065–1070.

Lemmon MA (2008) Membrane recognition by phospholipid-binding domains. *Nat Rev Mol Cell Biol* 9:99–111.

Minami T (2014) Calcineurin-NFAT activation and DSCR-1 auto-inhibitory loop: how is homeostasis regulated? *J Biochem* 155:217–226.

Picollo A, Malvezzi M & Accardi A (2015) TMEM16 proteins: unknown structure and confusing functions. *J Mol Biol* 427:94–105.

Section 14.7

Bulley S & Jaggar JH (2014) Cl$^-$ channels in smooth muscle cells. *Pflügers Arch Eur J Physiol* 466:861–872.

Cant N, Pollock N & Ford RC (2014) CFTR structure and cystic fibrosis. *Int J Biochem Cell Biol* 52:15–25.

Orrenius S, Gogvadze V & Zhivotovsky B (2015) Calcium and mitochondria in the regulation of cell death. *Biochem Biophys Res Commun* 460:72–81.

Chin A et al. (2012) V-ATPase in osteoclasts: structure, function and potential inhibitors of bone resorption. *Int J Biochem Cell Biol* 44:1429–1435.

Huang F, Wong X & Jan LY (2012) Calcium-activated chloride channels. *Pharmacol Rev* 64:1–15.

Kunzelmann K (2015) TMEM16, LRRCA8, bestrophin: chloride channels controlled by Ca^{2+} and cell volume. *Trends Biochem Sci* 40: 535–543.

Stauber T, Weinert S & Jentsch TJ (2012) Cell biology and physiology of ClC chloride channels and transporters. *Compr Physiol* 2:1701–1744.

Sensory Signal Processing

15

Sensory cells translate environmental signals into the universal language of the central nervous system: action potentials, the frequency of which is interpreted by the brain as a sensation. To this end they express specific receptors for light, temperature, touch, sound waves, gravity, smell, and taste (or, more precisely, for the corresponding chemicals) as well as, in certain animals, for electricity and magnetism.

In most cases the activation of a sensory receptor results in a depolarization of the sensory cell. This so-called receptor potential induces the release of neurotransmitter at the synaptic contact with the postsynaptic neuron. The neurotransmitter interacts with corresponding receptors at the postsynaptic membrane, causing an action potential in the sensory neuron. The depolarization of sensory cells is due to a gating of cation channels (in some cases a blockade of hyperpolarizing K channels produces the same result). Most of these channels are cation-nonspecific and belong to the transient receptor potential (TRP) and cyclic nucleotide-gated (CNG) channel families (see Sections 14.5.6 and 14.5.7). Depending on the environmental stimulus, the ion channels are receptors themselves or are controlled by other sensory receptors via second messengers (**Figure 15.1**). Sensory cells are highly sensitive. To respond over a large range of signal intensity, they make use of very efficient adaptive mechanisms, thus adjusting their input sensitivity to the strength of the stimulus.

15.1 Taste

A fundamental condition of life is that organisms must be able to check the quality of food to distinguish between edible and inedible and to avoid poisoning. Therefore, taste probably represents the most ancient of all senses. Taste receptors are chemo-receptors that recognize environmental key compounds. In bacteria, for instance, certain amino acids and sugars signal food. We can discriminate between at least five basic taste qualities: sour, salty, sweet, bitter, and "umami." In addition, a sixth sense recognizing "fatty taste" may exist. Other variants of tasting are mediated by smelling and by the sensations "hot" and "cold" (see Section 15.3). In mammals, taste cells are found predominantly in the taste buds of the oral epithelium. Like most other sensory cells they exhibit a bipolar shape: their apical membrane, directed towards the oral cavity, is divided into numerous microvilli and densely occupied by taste receptors, while the basolateral membrane forms a synaptic contact with a sensory neuron (**Figure 15.2**). Each taste quality is thought to have its own cell and receptor type; a simplified overview of taste signal transduction is shown in **Figure 15.3**.

15.1.1 Salty and sour tastes are mediated by ion channels

The **salty taste** is important for the controlled intake of electrolytes. It is caused mainly by sodium chloride, or more precisely by Na ions. The corresponding "receptors" are epithelial Na channels (ENaC family; see Section 14.3) that depolarize the cell

Figure 15.1 Two pathways for the transformation of sensory stimuli into action potentials In sensory cells, environmental signals activate cation channels either directly (left side) or via a second messenger S_2 (right side), the formation of which is mediated by a G-protein-coupled receptor. In both cases the sensory cell depolarizes and releases a neurotransmitter, which via a postsynaptic receptor induces an action potential in the neuron connecting the sensory organ with the brain. Most of the S_2-activated cation channels belong to the TRP family (see Section 14.5.6).

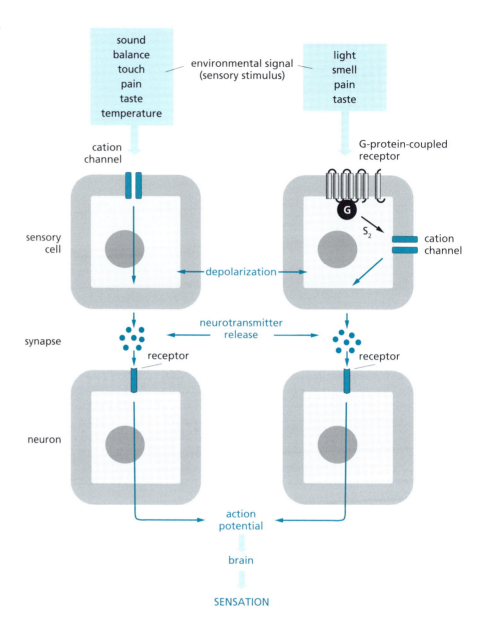

Figure 15.2 Schematic diagram of a taste bud Taste buds are sensory organs at the surface of the tongue. Each taste bud contains about 50 sensory cells (teal) that are in synaptic contact with the fibers of the sensory nerves. The cells' microvilli lining the taste pore are equipped with taste receptors. (Adapted from Alberts B et al. [2014] Molecular Biology of the Cell. New York: Garland Science.)

Figure 15.3 Processing of taste stimuli (A) Salty taste is mediated by a sodium channel of the epithelial Na channel (ENaC) family, sour taste by the proton-sensitive transient receptor potential, polycystin-related (TRPP2) cation channel. (B) The $G_{i,0}$-protein gustducin couples with sweet, umami, and bitter receptors activating phospholipase Cβ via Gβγ-subunits. In turn the ion channel TRPM5 (transient receptor potential, melastatin-related) is gated. Sweet taste may also be mediated by a cyclic nucleotide-gated cation channel CNG_{gust} along the G_s–adenylate cyclase (AC)–cAMP pathway. For other details see text. Note that the figure represents a simplified scheme since in reality each of the five taste qualities is processed by a unique cell type.

via Na influx. As is typical for most ENaCs, their expression is induced by aldosterone. In fact, this steroid hormone, which is secreted by the adrenal cortex as a response to salt deficiency, stimulates adaptive responses that are mediated by ENaCs: the salty taste sensation and Na reabsorption in kidney tubules and other epithelia.

The **sour taste** first of all is a warning signal indicating poisonous chemicals, unripe fruits, or food that has gone bad. It is evoked by protons, which activate depolarizing ion channels. In some species these are ENaCs of the acid-sensing ion channel (ASIC) type (see Section 14.3). Another widespread acid-sensitive cation channel involved in sour taste processing is the P2-type TRP channel (also called PKD2L1). Channels of this type are also known as type 2 **polycystins** (also called PKD2). Together with type 1 polycystins (PKD1) they have been found to be mutated in autosomal dominant polycystic kidney disease (PKD, a severe genetic disorder ending up in renal failure). [Polycystic kidney disease proteins (PKD1 and -2) should not be confused with the entirely unrelated protein kinases D (PKD; see Section 4.6.5).] Type 1 and 2 polycystins have only the name in common. Structurally they differ completely: type 2 polycystins constitute the TRPP2 subfamily of TRP cation channels (see Section 14.5.6), whereas type 1 polycystins are proteins with 11 membrane-spanning domains, an extended extracellular amino (N)-terminus, and a G-protein-activating cytoplasmic domain.

The precise function of PKD1 is unknown except that to constitute an acid-sensitive cation channel, PKD1 and PKD2 must interact via their intracellular carboxy (C)-terminal domains. The heterodimers thus formed have been found to be expressed selectively in distinct taste receptor cells of the mammalian tongue as well as in neurons thought to monitor the acidity of the cerebrospinal fluid. How the channel senses acidity is not clear. The same holds true for its role in polycystic kidney disease. For an example of modern methods for receptor analysis, see **Sidebar 15.1**.

15.1.2 Sweet, umami, and bitter tastes are mediated by G-protein-coupled receptors

The receptors for sweet, umami, and bitter tastes interact with G-proteins rather than with ion channels; they belong to the G-protein-coupled receptor (GPCR)

Sidebar 15.1 Screening for unknown receptors Modern methods for receptor analysis combine bioinformatics and protein expression strategies. By such an approach, for instance, the putative role of polycystins in sour taste processing has been elucidated [Huang AL et al. (2006) *Nature* 442:934–938]. Starting from the assumption that the sour receptor must be a transmembrane protein, around 30,000 computerized mouse open reading frames (indicating expressible gene loci) were scanned for sequences that putatively encode transmembrane segments. In the next step, candidate open reading frames were cross-searched against databases containing mouse expressed sequence tags, a rough method for the identification and quantification of transcribed nucleotide sequences and genes that uses short cDNA-snippets as markers (for details see textbooks of molecular biology). By this means, genes expressed in a wide variety of tissues could be distinguished from those with a very restricted expression pattern; it is assumed that taste receptors would belong to the second category. By use of the polymerase chain reaction (PCR) with cDNA templates prepared from taste papillae mRNA, finally five genes were identified that consistently and specifically were expressed in individual subsets of taste receptor cells, indicating a role in selective taste recognition. One of these genes was found to encode TRPP2. With the help of transgenic techniques, this gene was shown selectively to be expressed in those cells that process sour but not sweet, umami, bitter, and salty tastes. Moreover, animals suffering from a deletion of the *trpp2* gene were unable to respond to sour-tasting chemicals.

superfamily of heptahelical transmembrane proteins. Their activation leads to the generation of second messengers that control corresponding ion channels (see Figure 15.3B). Sweet and umami receptors are closely related, whereas bitter receptors constitute a protein family of their own.

Sweet taste is particularly attractive because it signals energy-rich food (as does the still somewhat enigmatic quality "fatty taste"). Several sweet receptors have been found. They constitute the T1R (Taste 1 Receptors) family, are coupled with G_S-proteins, and respond primarily to sugars but also to some other natural and artificial sweeteners. In the plasma membrane, receptor isoforms combine into heterodimers of the type T1R2–T1R3. Along the G_S-controlled signaling pathway, these dimeric receptors stimulate cAMP production and inhibit (probably through protein kinase A-catalyzed phosphorylation) potassium channels, at the same time gating the cAMP-controlled cation channel CNG_{gust} (see Section 14.5.7). This reaction cascade resulting in depolarization is activated by sugars, whereas artificial sweeteners seem to open the $G_{q,11}$–phospholipase $C\beta$ pathway of signaling. The question as to how this pathway leads to depolarization has not been finally settled. On the one hand, a hyperpolarizing potassium channel has been reported to become blocked by protein kinase C-catalyzed phosphorylation. On the other hand, the cation channel TRPM5 (see Section 14.5.6) has recently moved into the center of interest since it seems to respond to many if not all taste stimuli. Like some other TRP channels, TRPM5 becomes gated by phospholipase C. The effect is mediated primarily by a $G_{i,0}$-protein called gustducin in analogy to the transducin of photoreceptor cells via $G\beta\gamma$-subunits. As explained in Section 14.5.6, the mechanism of TRP channel activation along the phospholipase C pathway is still a matter of debate.

Most interestingly, the adipokinetic hormone **leptin** lowers the sensitivity of sensory cells for signals transducing sweet taste. As explained in Section 9.4.3, leptin prevents excessive weight gain by suppressing the feeling of hunger and by stimulating energy consumption through effects on the hypothalamus, as well as by inhibiting insulin secretion in the pancreas. In parallel, the taste quality of sweetness obviously becomes less attractive. Upon starvation, the secretion of leptin is turned down, resulting in the well-known craving for sweets. Inborn or acquired defects of cellular mechanisms processing leptin signals are some of the main causes of obesity.

The amino acid L-glutamate is a key signal indicating protein-rich food. The taste sensation, typical of chicken soup, is called **umami** after *umai*, the Japanese word for "delicious." Umami receptors belong to the same T1R family as the sweet receptors. However, their functional form is a T1R1–T1R3 heterodimer that, like the T1R2–T1R3 sweet receptors, activates the TRPM5 cation channel. T1 receptors share the selectivity for glutamate with the metabotropic glutamate receptors of the nervous system. Both receptors are active only as heterodimers and quite similar to each other in that they have an extended extracellular domain that contains the ligand binding site, originating probably from a bacterial binding protein for sugars and amino acids (details in Section 16.4).

Bitter taste mostly indicates that a food may be inedible or even toxic. Humans have 26 and mice have 33 genes encoding putative bitter receptors. Like T1 receptors, they are G-protein-coupled, although they belong to another family that is derived from rhodopsin and called T2R (see also Section 5.2). Their great variability reflects the large number of putative ligands with bitter taste. The trimeric G-protein coupled with T2 receptors is gustducin. Upon activation by bitter taste receptors, gustducin appears to stimulate two different types of cation channel: its $G\beta\gamma$-subunit activates phospholipase $C\beta2$ and in turn the TRPM5 cation channel, while the α-subunit stimulates a cAMP/cGMP-specific phosphodiesterase of the PDE1 family that gates a cyclic nucleotide-suppressed cation channel by degrading cyclic nucleotides. Gene knockout experiments indicate that the phospholipase $C\beta2$–TRPM5 pathway is the major route of bitter taste signal processing.

Summary

- Sensory receptors transform environmental stimuli into action potentials that are transmitted to the brain. They are parts of depolarizing ion channels or are G-protein-coupled heptahelical transmembrane proteins that induce channel gating along intracellular signaling pathways.

- There are at least five taste qualities: salty, sour, sweet, bitter, and umami. Salty taste is evoked directly by epithelial sodium channels (ENaCs); sour taste is evoked by various pH-sensitive ion channels of the ASIC and TRPP families. The receptors for sweet, bitter, and umami-tasting chemicals are G-protein-coupled and activate, along cellular signaling cascades, depolarizing cation channels of the transient receptor potential (TRP) and cyclic nucleotide-gated (CNG) superfamilies.

15.2 Mechanical stimuli: touch and sound

Mechanical stress and fluctuations of osmotic pressure are experienced by even the most primitive cells. Therefore, all cells have ion and water channels for osmotic pressure equalization that respond to deformations of the cell membrane. In the MscL channel of *Escherichia coli* we have already seen such a pressure valve (see Section 3.6.7). The corresponding channels of eukaryotic cells belong, however, to another protein family. In fact, **mechanosensitive channels** seem to have developed independently several times in the course of evolution. Accordingly, this receptor family is quite varied, including nonspecific cation channels as well as selective Na , K , and anion channels. In addition, many neurotransmitter receptors (see Chapter 16) respond to mechanical stress. The mechanosensitive ion channels of certain neurons in the hypothalamus, for instance, are sensors of blood osmolarity. When the osmotic pressure rises, they facilitate (through depolarization) the secretion of the neurosecretory hormone vasopressin (also called adiuretin) that inhibits the excretion of water by the kidneys. Particularly sensitive for mechanical stimuli of blood flow and blood pressure are vasoendothelial cells, which respond by production of blood pressure-regulating factors such as prostacyclin, nitric oxide, endothelin, ATP, growth factors, and other signaling substances.

Numerous **sensors of touch and movement** note every muscle and tendon tension and every organ movement. They are concentrated in special sensory organs such as muscle spindles and Ruffini, Meissner, and Pacini bodies of skin.

Figure 15.4 Cellular processing of mechanical stimuli (A) Schematic depiction of a mechanosensory organ in the cuticle of *Drosophila melanogaster*. The organ monitors the slightest air movements. It consists of a flexible bristle that is connected to the dendrite of a sensory neuron (teal). The dendrite contains several mechanosensitive cation channels of the Piezo family and is embedded in a sac filled with endolymph. (B) Model of a mechanosensitive cation channel. The channel is connected with the cytoskeleton and an extracellular anchoring structure in such a way that it becomes gated by any movement. (C) Schematic structure of a subunit of the mechanosensitive channel Piezo 2 (mouse) interwoven with the plasma membrane through 34 transmembrane subunits (gray cylinders). (A, adapted from Gillespie PG and Walker RG [2001] *Nature* 413:194–202. C, adapted from Delman P, Hao J and Rodat-Despoix L [2011] *Nat Rev Neurosci* 12:139–153.)

Most probably, all these cell types possess mechanosensitive ion channels. The touch-sensitive organs in the skin of insects and nematodes have been studied most thoroughly in this respect. In the case of *Drosophila* the channels are connected with a flexible bristle. Any touch, even the slightest air breeze, moves the bristle and generates shearing forces between the bristle shaft and a sensory neuron connected with the bristle. Because the channels in the neuronal membrane are fixed by anchor proteins to the bristle outside and to the actin cytoskeleton inside the cell, they are opened by these forces and the neuron is depolarized, sending an action potential to the brain.

(A) tectorial membrane stereocilia hair cells nerve fibers basilar membrane

(B) ciliar membranes with ion channels actin cytoskeleton elastic connection

Figure 15.5 Processing of mechanical stimuli in the inner ear (A) Anatomical scheme of the organ of Corti. (B). Detail of two neighboring stereocilia. Their mechanosensitive ion channels (teal) are connected with each other and with the actin cytoskeleton by elastic protein fibers (zigzag lines).

Non-specific cation channels supposed to be involved in mammalian mechanosensation include some TRP (see Section 14.6) and ENaC species (see Section 14.3) as well as the **Piezo proteins** (from Greek " piezein," to squeeze). The latter are membrane proteins with up to 36 transmembrane domains and a molecular mass up to 318 kDa (**Figure 15.4**). The functional Piezo channel is tetramer. It is the largest ion channel so far found in the plasma membrane and unrelated to any other channel protein. Mammals express two Piezo proteins. While Piezo 1 is found in many organs including skin, urinary bladder, lung, and red blood cells, Piezo 2 is concentrated in sensory neurons. Protozoa, invertebrates and plants seem to express only one Piezo-like protein.

In the inner ear, highly specialized mechanosensors are found that, as statoacoustic apparatus, respond to sound waves, gravity, and acceleration forces. Acoustic signals are processed by the sensory cells of the cochlea. In reference to their dense bundles of apical stereocilia, these cells are called hair cells. They are located on the basal lamina of the organ of Corti and anchored to the overlying tectorial membrane by the tips of the stereocilia. Sound causes vibrations of the membrane, which are transduced to the stereocilia bundles. Consequently, the mechanosensitive cation channels of stereocilia are opened. Each channel is anchored elastically at both sides: inside the cell to the actin cytoskeleton and outside the cell to the neighboring stereocilium (**Figure 15.5**). The identity of the channels is not entirely clear. Probably they belong to the TRP superfamily.

The closely related sensory cells of the organ of balance are localized in the semicircular canals or labyrinth of the inner ear. They inform the organism about its spatial position and all kinds of acceleration. The inert mass of this system is increased by an incorporation of small limestones (statoconia) in the tectorial membrane. These resemble the gravity stones or statoliths in the organs of balance of many invertebrates.

Summary

Mechanosensors are cation channels firmly anchored to intra- and extracellular structures. They are gated by mechanical forces (exerted by movements, touch, pressure, changes of osmotic pressure, gravity, and sound waves) that induce membrane depolarization.

Figure 15.6 Signal transduction by the vanilloid receptor TRPV1 as compared with cannabinoid receptors The receptor ion channel TRPV1 (transient receptor potential, vanilloid type one, right) is activated by heat, acid, capsaicin, and the endogenous ligand anandamide, an ethanolamide of arachidonic acid. As an endocannabinoid, anandamide also activates the cannabinoid receptors, exerting an opposite effect (left).

15.3 Temperature and pain

Painful pressure, such as that due to an excessively filled urinary bladder or flatulence, is mediated by mechanoreceptors localized at the surface of neighboring neurons. In contrast, another receptor type responds to painful heat and cold. These thermoreceptors have been discovered in nociceptor (pain-sensitive) neurons in the course of studies with natural substances evoking heat, cold, or pain sensations in skin.

A prototype of such a substance is **capsaicin,** the hot ingredient of red pepper (**Figure 15.6**). Capsaicin stimulates a receptor that normally responds to temperatures above 43°C (the pain limit of heat in humans). This receptor, called TRPV1, belongs to the family of the depolarizing TRP cation channels (see Section 14.5.6). Because of the vanilloid-related structure of capsaicin and other hot natural substances, it is also known as **vanilloid receptor 1**. [Interestingly, the TRPV1 receptor of birds does not respond to capsaicin and related chemicals, which allows the animals to consume hot foods such as peppers.] Related receptors exhibit other temperature sensitivities. So TRPV2 becomes active from 52°C, TRPV3 (known as the camphor receptor) from 31°C and TRPV4 from 27°C upward.

The pain limit of cold is around 15°C. The corresponding receptors, called TRPM1, TRPA1, and TRPM8, have been identified by using cooling substances such as menthol or eucalyptol as artificial ligands. They are known as **menthol receptors** and belong to the TRP cation channel family. In contrast to most other receptors of the TRP family, TRPV1 and TRPM8 are weakly voltage-dependent, becoming more readily opened upon depolarization. This effect is modulated by the temperature. If it is assumed that the receptors exist in two conformations, OPEN and CLOSED, the conformation OPEN (for heat-activated receptors) and the conformation CLOSED (for cold-activated receptors) may become stabilized with increasing temperature.

Why does the binding of capsaicin, menthol, and other hot and cold natural products to TRP channels produce a "wrong" temperature sensation? These compounds induce a change of protein conformation. As a result the channel is gated at room temperature instead of at its threshold temperature. In physiology the phenomenon of increased temperature sensitivity is called thermal hyperalgesia.

At least for TRPV1 and TRPV4 receptors an endogenous ligand has been identified. Surprisingly, this substance turned out to be identical with the endocannabinoid **anandamide,** an arachidonic acid derivative (see Figure 15.6). Endocannabinoids are endogenous neurotransmitter-like substances that interact with the cellular receptors for cannabis drugs. However, these cannabinoid receptors are not identical with TRPV ion channels but belong to the G-protein-coupled receptors (see Section 16.7). Their activation has a pain-relieving effect. So, like many other neurotransmitters, endocannabinoids interact with both G-protein-coupled and ion channel-coupled receptors, eliciting opposite responses and dampening their own effects.

Other endogenous stimulants of TRPV receptors are substances that are released from wounded or infected tissue, and provoke the characteristic heat sensation of an inflammatory reaction. Inflammation causes an acidification of the tissue, and the protons thus produced activate the TRPV1 receptor as well as ASIC receptors of the ENaC family (see Section 14.3).

Pro-inflammatory mediators mediating pain sensations include the tissue hormone **bradykinin,** the neuropeptide **substance P** (a member of the so-called tachykinin family), and **prostaglandin E.** They all interact with $G_{q,11}$-protein-coupled heptahelical receptors (BK2-R for bradykinin, TAC1-R for substance P, and EP1-R for prostaglandin E) that activate phospholipase Cβ and, in turn, the TRPV1 channel (see **Figure 15.7**). The formation of prostaglandin E catalyzed by cyclooxygenases is inhibited by nonsteroidal anti-inflammatory drugs (NSAIDs), which include most important pain relievers such as acetylsalicylic acid, better known as aspirin (see Section 4.4.5). Other ingredients of the "inflammatory soup" such as ATP and serotonin facilitate TRPV-dependent depolarization by gating receptor-coupled cation channels (see Table 16.1 in Chapter 16).

Figure 15.7 Cellular processing of temperature and pain signals
Shown are effects on heat and pain receptors (left) and on cold receptors (right). The "inflammatory soup" is a cocktail of endogenous pain-inducing mediators that become locally released upon wounding and inflammation. While anandamide directly interacts with TRPV (transient receptor potential, vanilloid type) channels, serotonin and ATP activate other cation channels, thus facilitating depolarization by TRPV. Substance P, prostaglandin E, and bradykinin bind to $G_{q,11}$-protein-coupled receptors (GPCR), probably activating TRPV channels via phospholipase Cβ (PLCβ) signaling. The same holds true for NGF (nerve growth factor), which interacts with the receptor tyrosine kinase TrkA, thus activating PLCγ. TRPM, transient receptor potential, melastatin-related.

An endogenous factor involved in pain perception is the **nerve growth factor, NGF**. Originally, this tissue hormone was discovered as a neurotrophic protein: an endogenous mediator that promotes the embryonic development of the nervous system, particularly of sensory and sympathetic neurons (for details see Chapter 16.1). This is, in fact, its predominant function in the prenatal phase, whereas in the adult organism NGF is a major pain mediator that becomes strongly expressed in wounded and inflamed tissues. NGF interacts with two quite different types of receptor: a receptor tyrosine kinase of the Trk family (see Section 7.1) and a p75 neurotrophin receptor (p75$_{NTR}$) belonging to the tumor necrosis factor receptor family (see Section 6.3.1). Like other members of this family, p75$_{NTR}$ may induce apoptosis under distinct conditions; for instance, when the NGF-Trk pathway is inhibited or down-regulated.

To evoke pain sensations NGF has to bind to type A Trk receptors that are expressed in sensoric neurons. As a result, TRPV1 ion channels become sensitized. As in the case of bradykinin and other pain mediators, this may occur through an activation of phospholipase C (here of the γ-isoform; see Section 4.4.4). In addition, the expression of the TRPV1 channel and the acid-sensitive cation channel ASIC3 (see Section 14.3) has been found to be induced by TrkA receptors along MAP kinase-dependent signaling cascades stimulating gene transcription.

Since in humans NGF is involved in several painful diseases, which sometimes are resistant to conventional pain relievers such as NSAIDs and opiates (including chronic headaches and arthritis as well as bladder and prostate inflammation), the development of drugs that block NGF signaling has great therapeutic potential.

Summary

- Pain receptors respond to high and low temperature, acid, corrosive and irritating chemicals, mechanical forces, and endogenous factors released in the course of inflammatory processes.

- The receptors are depolarizing cation channels that either interact directly with the stimulants or become gated along intracellular signaling pathways originating from G-protein- or tyrosine kinase-coupled receptors of pro-inflammatory mediators.

- Major players in this game are the vanilloid receptors belonging to the TRPV cation channel family. These are temperature sensors that are sensitized by hot- and cold-tasting food ingredients.

15.4 Smell

Smell is processed by the most versatile chemical sense. Vertebrates have several million olfactory cells that form a neuroepithelium in the nasal mucosa. These sensory neurons are bipolar, with strongly branched dendrites packed with receptors and extending to the surface of the mucosa, and an axon leading to the brain at the opposite pole (**Figure 15.8**).

Owing to the pioneering work of Linda B. Buck and Richard Axel (who were awarded the 2004 Nobel Prize in Physiology or Medicine), we know that smell or odorant receptors constitute by far the largest subfamily of G-protein-coupled receptors and are encoded by one of the largest gene families (composing up to 4% of a mammalian genome).

The nematode *Caenorhabditis elegans* has around 500 genes for putative odorant receptors; these are not related to the vertebrate receptors, though they exhibit the same heptahelical architecture characteristic for G-protein-coupled transmembrane proteins.

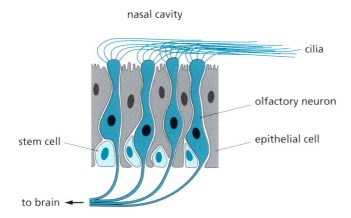

nasal cavity

cilia

olfactory neuron

epithelial cell

stem cell

to brain ◄—

Figure 15.8 Schematic diagram of the olfactory epithelium The diagram shows a section of the olfactory epithelium consisting of sensory neurons (dark teal) and supporting epithelial cells. The neurons have direct contact with the brain. The cilia of the neurons directed into the nasal cavity and bathed in nasal mucus contain the olfactory receptors. The stem cells serve the regeneration of sensory neurons, which have a limited lifespan of a few weeks. (From Alberts B et al. [2014] Molecular Biology of the Cell, New York: Garland Science.)

For most mammals around 1000 odorant receptor genes have been found, 600 of which have regressed to inactive pseudogenes in humans. This number of receptors is sufficient to discriminate between almost countless smell variants, because the nose works similarly to the eye, which distinguishes any number of color varieties. Although each sensory neuron seems to express only one receptor type, an odorant may react with several receptor types or a single receptor may respond to several scents. Therefore, a smell sensation is assumed to result from a combined excitation pattern involving many different receptors and neurons.

As G-protein-coupled receptors, odorant receptors have seven transmembrane domains. The individual receptor types differ from each other primarily in the composition of the transmembrane domains 3, 4, and 5 forming the binding pocket for the ligand (see Section 5.2). The associated olfactory G-protein G_{olf} belongs to the trimeric Gs proteins with its α-subunit activating adenylate cyclase type III (**Figure 15.9**).

The action potential required for signal transduction to the brain is generated by cAMP that binds to and opens nonspecific cation channels of the CNG channel family (see

Figure 15.9 Cellular processing of smell signals in the olfactory system and the vomeronasal organ While the cells of the olfactory system respond to all kinds of odorants (left), the cells of the vomeronasal organ are stimulated specifically by pheromones (right). In both cases signal transduction occurs via specific G protein-coupled receptors. Along the signaling pathways depicted ion channels are activated and an action potential is generated.

OLFACTORY ORGANS

VOMERONASAL ORGAN

odorants

pheromones

GPCR

V1R

V2R

G_{olf}

$G_{i,o}$

$G_{i,o}$

α

βγ

adenylate cyclase

cAMP

Cl⁻

Ca²⁺

PLCβ

cation channel

anion channel

TRP cation channel

ACTION POTENTIAL

ACTION POTENTIAL

Section 14.5.7). This triggering event is accompanied by a tremendous amplification of the signal: a single activated receptor stimulates at least 10 G-protein molecules, one after the other, thus releasing more than 10,000 cAMP molecules, three of which are sufficient to open an ion channel. In addition, olfactory neurons possess another obviously unique mechanism of amplification. In contrast to most other nerve cell types, they are able—by activating the corresponding ion pumps—to generate a steep charge gradient of chloride ions from the inside to the outside. When Ca^2 ions enter the cell through the cAMP-gated cation channels, they stimulate chloride channels, thus inducing an efflux of chloride ions and facilitating depolarization. On the other hand, calcium ions block the cation channels; that is, they exhibit an inhibitory feedback effect in addition to their amplifying effect. The inhibitory effect is important for a rapid adaptation of the sensory apparatus (see Section 15.6).

Many vertebrates have a second olfactory system, the **vomeronasal organ** or Jacobson's organ, in the paranasal sinus (see Figure 15.9). Its sensory neurons respond specifically to pheromones, signal substances that are released by individual organisms and induce species-specific behavior. Best known are the sexual pheromones that attract sexual partners and stimulate the endocrine system to induce mating readiness. The cells of the vomeronasal organ express two families of G-protein-coupled receptors, V1 and V2 receptors, each consisting of approximately 150 receptor isoforms. V1 receptors are closely related to conventional odorant receptors, whereas V2 receptors have an extended extracellular domain, structurally resembling metabotropic glutamate receptors of the central nervous system (see Section 16.4) and T1 receptors that sense the umami taste. These receptor types are coupled with $G_{i,0}$-proteins that activate phospholipase Cβ via their β,γ-subunits. Along this pathway, V1 and V2 receptors open the cation channel TRPC2, thus generating an action potential.

In primates, the vomeronasal organ regresses after birth. Moreover, with one exception, all vomeronasal receptor genes are inactive pseudogenes. Thus, humans obviously recognize sexual attractants through the normal olfactory system in addition to other signals indicating mating readiness.

Odorant receptors are found also in other organs such as lung, heart, kidney etc. The physiological significance of this ectopic expression is obscure with one exception: **sperm cells**. They respond by a chemotactic movement to an attractant released from the egg cell. To this end the sperm cell expresses transmembrane receptors, which closely resemble, or even are identical with, receptors found in the nasal mucosa. The egg pheromone has not yet been identified. However, experiments with artificial substances have shown sperm to be specifically attracted by the scent of the lily of the valley. Moreover, it appears that men who are unable to smell this sort of scent also are infertile, probably because of a deletion of the corresponding receptor gene both in the nasal mucosa and in sperm. These investigations are, of course, of considerable interest for reproductive medicine since they nourish the hope of new therapies for male infertility and of novel approaches to hormone-free contraception.

Summary

- Odorants are environmental chemicals that interact with odorant or smell receptors. These are G-protein-coupled heptahelical transmembrane proteins, constituting by far the largest subfamily of this receptor type.

- There are two groups of odorant receptors: those expressed in olfactory organs and those expressed in the vomeronasal organ. Closely related receptors have been found in sperm and other tissues.

- Through receptor activation, signaling cascades are stimulated, leading to membrane depolarization due to a gating of cation channels and depolarizing anion channels.

15.5 Vision

Among the mechanisms of sensory signaling, visual signal transduction has been investigated most thoroughly. In the lenticular eye—common to all vertebrates—the light-sensitive layer of the retina is composed of two types of photoreceptor cell: rods and cones. The highly light-sensitive rods (several million in the human eye) are specialized for the reception of dim light, while the less sensitive cones (in humans around 6 million) deal with bright daylight and colors (according to the RGB model, colors are processed by three types of cone that are red-, green-, or blue-sensitive). As shown in **Figure 15.10** the light-sensitive membrane surface of a photoreceptor cell is extremely enlarged by invaginations (in cone cells) or separation of up to 1000 stacked discs (in rod cells). These membrane structures contain the light-sensitive pigment rhodopsin (in rods) or the closely related color-specific iodopsins (in cones) forming a highly sensitive light conductor in the outer segment of the cell. The inner segment resembles the cell body containing the organelles and establishing synaptic contacts to the bipolar cells, another type of neurons in the retina. Outer and inner segments are connected by a thin ciliar structure.

In the compound eyes of invertebrates, another type of photoreceptor cell is found. Its light-sensitive membrane is enlarged by a brushlike structure of tightly packed microvilli called a rhabdomere. Again, rhodopsin is the light-sensitive pigment.

The rhodopsin structure is an ancient "invention" of evolution, found even in archaea. However, although they resemble each other in their three-dimensional structures, prokaryotic and eukaryotic rhodopsin probably developed independently. Rhodopsin is composed of the G-protein-coupled heptahelical receptor protein opsin and its ligand, vitamin A aldehyde or retinal. It is inserted as a homodimer into the disc membrane of rod cells (or the membrane invaginations of cone cells). The retinal binding site is localized in a pocket formed by the seven transmembrane domains on the inner side of the disc (that resembles the extracellular space). In the absence of light, retinal exists as an 11-*cis* isomer that, similar to the 13-*cis*-retinal of bacterial rhodopsin (see Section 3.4.5), is bound to a lysine residue (Lys269 in humans) as a protonated Schiff base. Upon illumination, retinal rearranges to the

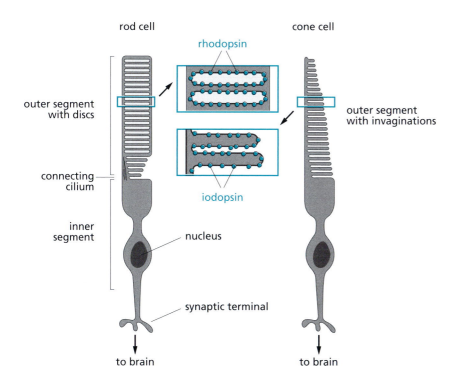

Figure 15.10 Schematic diagram of photoreceptor cells A characteristic structure of photoreceptor cells is the outer segment containing a high number of membrane discs (rod cell, left) or invaginations (cone cell, right) carrying the light-sensitive pigments rhodopsin (rod cell) or iodopsin (cone cell). Upon illumination the membrane potential is changed and via a synaptic terminal a signal is sent to the brain. (Adapted from Nicholls JG, Martin AR, and Wallace BG [1992] From Neuron to Brain. Sunderland, MA: Sinauer Associates.)

Figure 15.11 Processing of light signals in a rod cell of the vertebrate retina Light activates, via rhodopsin and the α-subunit of the G-protein transducin (G_t), phosphodiesterase 6 (PDE6), which hydrolyzes cGMP. As a result, the cyclic nucleotide gated (CNG) cation channel becomes closed and the cell hyperpolarizes. More details are found in the text.

Figure 15.11 Processing of light signals in a rod cell of the vertebrate retina Light activates, via rhodopsin and the α-subunit of the G-protein transducin (G_t), phosphodiesterase 6 (PDE6), which hydrolyzes cGMP. As a result, the cyclic nucleotide gated (CNG) cation channel becomes closed and the cell hyperpolarizes. More details are found in the text.

all-trans configuration. As a result, the covalent Schiff base bond is hydrolyzed and retinal finally dissociates from opsin, diffusing into the disc lumen where it becomes enzymatically re-isomerized.

Hydrolysis of the covalent bond between opsin and retinal causes an allosteric change of opsin conformation: the transmembrane helices are shifted in such a way that the active form (called metarhodopsin II) can interact with the trimeric G-protein transducin (G_t) on the cytoplasmic side of the disc membrane (**Figure 15.11**; see also Section 5.1). Transducin belongs to the $G_{i,0}$-protein family but differs from the other members because it interacts with a unique effector protein, a cGMP-specific phosphodiesterase of the type PDE6 (see Section 4.4.2). By promoting the dissociation of an inhibitory subunit (the γ-subunit of PDE6), the α-subunit of transducin activates this enzyme.

PDE6 catalyzes the hydrolysis and inactivation of cyclic GMP, the key signal of light processing in photoreceptor cells. cGMP opens nonspecific cation channels in the cell's outer segment, thus inducing depolarization. Like the channels of odor-sensing cells, these cation channels are members of the CNG channel family (see Section 14.5.7). Rod cells express the channel isoform CNG1, while cone cells express the channel isoform CNG6. In dark-adapted rods, the constitutively active guanylate cyclases E and F produce cGMP in a concentration sufficient for maximal gating of the CNG1 channels. Both cyclases are receptorlike transmembrane enzymes. Their ligands, if they exist at all, are unknown.

As a consequence of light-induced cGMP degradation, the CNG ion channels are closed and the membrane *hyperpolarizes*. This is because the membrane potential of a photoreceptor cell depends on an equilibrium between two opposing processes: the influx of cations through cGMP-gated ion channels and the removal of those ions by a constitutively active K⁺–Na⁺–Ca²⁺ exchanger in the membrane. In contrast, light may also result in a *depolarization* of the photoreceptor cell in nonvertebrates.

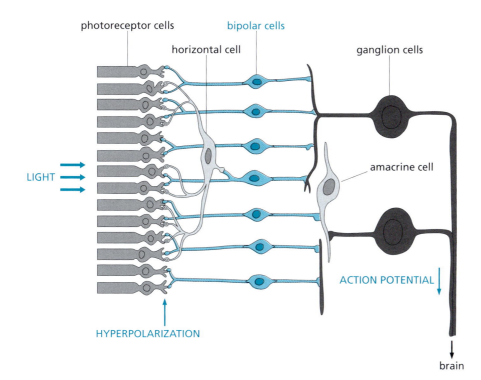

photoreceptor cells

bipolar cells

horizontal cell

ganglion cells

LIGHT

amacrine cell

ACTION POTENTIAL

HYPERPOLARIZATION

brain

Figure 15.12 Wiring of the vertebrate retina The photoreceptor cells transmit their signals via intermediate bipolar cells to the ganglion cells of the visual nerve. Horizontal and amacrine cells make horizontal connections, serving to fine-tune the response. (Partly adapted from Hubel D [1995] Eye, Brain, and Vision. New York: W. H. Freeman.)

The system has an enormous amplification rate. At complete darkness, a single photon has been found to activate up to 500 molecules of transducin, resulting in the degradation of 250,000 molecules of cGMP. Under this condition half-maximal hyperpolarization is induced by only 30 photons. Without adaptation, such an extreme sensitivity is incompatible with efficiency, or the ability to respond over a wide range of signal intensity. At least as impressive, therefore, is the system's ability to adapt, enabling the cell to process between one and 1 million photons per second. The mechanisms of adaptation will be explained in Section 15.6.

Upon illumination, a photoreceptor cell transmits an action potential to the brain. How is this possible, although light induces *hyperpolarization*? The answer is found in the principle of double negation ("minus times minus is equal to plus") and a corresponding neuronal wiring in the retina with several cell types communicating with each other (**Figure 15.12**). In the dark, the photoreceptor cells are depolarized and release the neurotransmitter glutamate at their synaptic contacts with the bipolar cells. These cells exist in two different types. One type expresses *inhibitory* glutamate receptors, rendering the cell inactive upon interaction with glutamate released from the photoreceptor cells. When, upon illumination, the photoreceptor cells are hyperpolarized and the glutamate release ceases, these bipolar cells are relieved from inhibition; they depolarize and release their own neurotransmitter, which again is glutamate. As a result, an action potential is evoked in the ganglion cell connected downstream and equipped with excitatory glutamate receptors. Since these ganglion cells respond to a switching on of light, they are called ON cells. The other type of bipolar cells expresses *excitatory* glutamate receptors. These cells are, therefore, stimulated by the glutamate released from photoreceptor cells and are depolarized in the dark. They induce an action potential in connected ganglion cells, which is switched off upon illumination (OFF cells; see **Figure 15.13**).

The photoreceptor cells of the compound eyes of invertebrates process light signals along another pathway (**Figure 15.14**). In the *Drosophila* eye, retinal is replaced by 3-hydroxyretinal, which also undergoes a *cis–trans* rearrangement upon illumination. However, in contrast to vertebrates, the *all-trans* form remains bound to opsin and becomes re-isomerized to the 11-*cis* form by longer-wavelength (red) light

Figure 15.13 Light processing in the vertebrate retina In the dark, the photoreceptor cell is depolarized, releasing the neurotransmitter glutamate (Glu). Glutamate inhibits a bipolar cell type expressing an inhibitory Glu receptor and activates another bipolar cell type expressing an excitatory Glu receptor. As a consequence, an OFF ganglion cell is activated (black arrows). Upon illumination, the situation becomes reversed: the glutamate release of the hyperpolarized photoreceptor is diminished with opposite effects on the two types of bipolar cells and the activation of an ON ganglion cell (teal arrows).

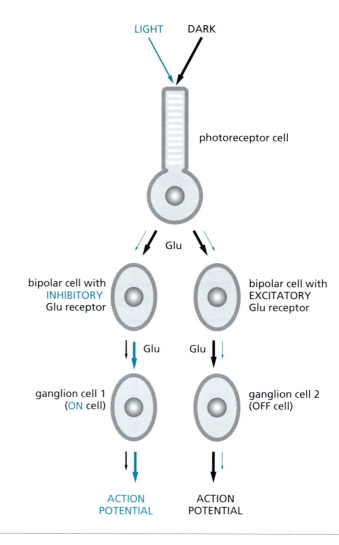

Figure 15.14 Cellular processing of light signals in vertebrate and insect eyes In the vertebrate eye, the light-induced degradation of cGMP along the retinal–transducin (G_t)–cGMP phosphodiesterase pathway leads to a closing of CNG cation channels and hyperpolarization. In the insect eye, light opens TRP cation channels along the 3-OH-retinal–$G_{q,11}$–phospholipase C pathway and induces depolarization. The mechanism of TRP channel gating is still not entirely clear (shown by the question mark). Teal bars show inhibition of G-proteins by cGMP phosphodiesterase, by regulator of G-protein signaling type 9 (RGS-9), and by phospholipase Cβ4, acting as GTPase-activating proteins. More details are found in the text. CNG, cyclic nucleotide gated; TRP, transient receptor potential.

alone, without an enzyme. This allows an unparalleled temporal discrimination of light signals, up to 10 times faster than in the vertebrate eye and corresponding with the high-speed lifestyle of insects.

In invertebrate eyes, light always induces a *depolarization* of the photoreceptor cell. In *Drosophila* this occurs via gating of a nonspecific TRP-like cation channel (TRPL). Simultaneously a Ca^2-selective TRP channel is opened that is essential for adaptation (see Section 15.6). Both these channels are unrelated to the CNG channels of the vertebrate eye; they belong to the TRP ion channel family, to which they have given the name (see Section 14.5.6). Like most other TRP channels they are regulated along the phospholipase C signaling pathway. To this end, light-activated metarhodopsin stimulates a trimeric $G_{q,11}$-protein that activates phospholipase Cβ type 4 (also called NorpA for *Drosophila*) resulting in the release of diacylglycerol (DAG) and inositol trisphosphate. However, the signal proper gating the TRPL channel is still unknown.

Summary

- Light receptors are heptahelical proteins of the rhodopsin family incorporated in the disc membrane of retina cells and coupled to transducin.

- Their ligand is retinal; upon illumination, it undergoes a *cis–trans* rearrangement resulting (in vertebrates) in activation of a phosphodiesterase that degrades the second messenger cGMP. This leads to hyperpolarization due to closing of cGMP-gated cation channels.

- As a consequence, inhibitory neurons positioned between the sensory cell and the visual nerve are switched off, enabling the generation of an action potential in the visual nerve.

- In nonvertebrates (insects) a $G_{q,11}$-protein-controlled signaling cascade directly leads to gating of depolarizing cation channels in the sensory cell.

15.6 Sensory adaptation

The more sensitive a signal-processing device, the more efficient it is. That means that already low signal intensities evoke maximal activation. To respond with always-maximal sensitivity over a wide range of intensity the system must adapt. It must become insensitive for a signal just processed before to be able to respond maximally to a stronger signal (the apparatus is said to be "reset to zero"). Adaptation thus resembles a learning process that is indispensable for "intelligent" signal processing. We have already discussed this phenomenon for bacterial chemotaxis as an example (see Section 3.4.4).

A second condition for outstanding performance is rapid temporal discrimination or dissolution of signals. As far as visual signal processing is concerned, this is immediately conceivable: the transmission of a signal from rhodopsin to the visual nerve lasts only about 150 μs. Therefore, an image disappears as soon as we close our eyes and does not become blurred by eye movements, provided these are not too fast.

On the cellular level, adaptation and temporal discrimination are controlled by highly efficient mechanisms of negative feedback and signal extinction. These will be discussed in more detail for visual signal processing as an example.

The signaling apparatus of a photoreceptor cell adapts at four levels at least: receptor, G-protein, guanylate cyclase, and CNG ion channel. The key signal of sensory adaptation is Ca^2 (**Figure 15.15**). It enters the cell through the open cation channels (in the dark), approaching a cytosolic concentration of 300–500 nM. Upon illumination, this value drops to 50–100 nM because the ion channels are closed whereas ion exchangers and Ca^2 pumps remain active. As a signal of dark adaptation,

Figure 15.15 Key function of Ca²⁺ ions in sensory adaptation Ca^2 entering the cell through a cation channel that is gated by sensory stimuli, either directly or via a signaling cascade (involving second messengers) inhibits signal transduction at several sites. GPCR, G-protein-coupled receptor.

calcium sensitizes the system for light by activating rhodopsin and counteracting all events that lead to an increase of the cGMP concentration and to depolarization. The primary point of attack is the CNG channel itself becoming blocked by a Ca^2 – calmodulin complex. In parallel, Ca^2 turns down the activity of guanylate cyclase by binding to and inhibiting two guanylate cyclase-activating proteins.

Rhodopsin becomes insensitive upon illumination (**Figure 15.16**). The adaptive mechanism is closely related to the adaptation of β-adrenergic signal transduction (see Section 5.7). When activated by light, rhodopsin, by means of the β,γ- subunit of transducin, recruits a rhodopsin kinase from the family of GPCR kinases (see Section 5.7). Upon phosphorylation at several Ser and Thr residues in the C-terminal domain, rhodopsin binds the protein arrestin that interrupts the interaction with transducin. At high Ca^2 concentrations (in the dark), rhodopsin kinase becomes inhibited by the Ca^2-sensor protein recoverin and rhodopsin is reactivated by dephosphorylation (catalyzed by permanently active protein phosphatases). Recoverin belongs to the EF-hand superfamily of Ca^2-binding proteins and is closely related to the guanylate cyclase-activating proteins (for more details about this protein type, see Section 14.6.1). The interaction of recoverin with Ca^2 uncovers a covalently bound myristoyl side chain normally hidden inside the protein molecule. By means of this fatty acid anchor, recoverin becomes fixed in the disc membrane in the immediate vicinity of its target, the rhodopsin/rhodopsin kinase complex (Ca^2 –myristoyl switch, see Section 14.6.1).

Transducin undergoes auto-inactivation due to its GTPase activity. However, for a rapid temporal discrimination of light signals this reaction, lasting 10–20 s, is by far too slow. Therefore, it is stimulated enormously by a GTPase-activating protein complex consisting of the inhibitory γ-subunit of the effector enzyme PDE6 and the protein RGS9-1. The latter, a member of the family of Regulators of G-protein Signaling (see Section 4.3.2), stimulates the GTPase activity of transducin only as a complex with another protein called $G\beta_5$. It is an unusual G-protein subunit found only in nervous tissue and showing, in contrast to other Gβ-subunits, little affinity for Gγ-subunits.

plasma membrane

disc membrane

disc lumen

rhodopsin

rhodopsin kinase

arrestin

recoverin

RGS9

G_t

$G\beta_5$

PDE6

Ca^{2+}

CNG channel

GCAP

guanylate cyclase

Figure 15.16 Adaptation of light-sensory signal processing in the vertebrate eye Adaptation is due to a modulation of guanylate cyclase and phosphodiesterase PDE6 activities that regulate the intracellular cGMP level. Guanylate cyclase is activated by two guanylate cyclase-activating proteins (GCAP). PDE6 is activated by rhodopsin via transducin (G_t) and becomes inactive when rhodopsin is inhibited by phosphorylation through rhodopsin kinase and when the interaction with transducin (G_t) is interrupted by arrestin (recruited by phosphorylated rhodopsin) and the GTPase-activating proteins PDE6 and RGS9 (as a complex with the G-protein Gβ5). These reactions are under the control of Ca^2 ions entering the cell through CNG cation channels. In the dark, when the CNG channels are open, Ca^2 inactivates GCAP and activates PDE6 by inhibiting rhodopsin kinase through an activation of the kinase inhibitor recoverin. As a result rhodopsin activity is increased, the cGMP level drops, and more CNG channels are closed: the system becomes more light-sensitive (dark adaptation). Upon illumination the channels close, the Ca^2 influx is interrupted, and guanylate cyclase and rhodopsin kinase are activated. As a result the rhodopsin activity decreases, cGMP level increases, and more CNG channels are opened: the system becomes less light-sensitive (light adaptation). Ca^2 controls its influx by inhibiting the CNG channel. Teal, inhibitory effects; black, stimulatory effects. CNG, cyclic nucleotide gated.

In the cell membrane, the components of visual signal transduction and adaptation are assembled into multiprotein complexes that are, in addition, fixed by anchor proteins. Such an arrangement guarantees a high temporal discrimination of signals. In an especially impressive manner, this principle is realized in the ultrafast insect eye. Here a scaffolding protein, like a circuitboard, assembles the switching elements into a "transducisome" (or "signalplex") resembling a microprocessor. However, in contrast to conventional technical devices, the biological system is by no means firmly wired but rearranges its components depending on the actual requirements such as adaptation. For example, in strongly illuminated rods, transducin migrates from the outer into the inner cell segment within a few minutes, returning when the light becomes dimmed, while the inhibitor protein arrestin translocates in the opposite direction.

As far as adaptation is concerned, other sensory systems keep up with the visual complex. The sensitivity and adaptability of the sense of smell is legendary. The human nose, though by no means representing the most sensitive olfactory organ in the animal kingdom, is able to recognize only 4 mg of methyl sulfide in an air volume resembling 100 large sports halls but loses this ability quickly through adaptation (this also limits the effectiveness of perfumes). The adaptive mechanisms of olfactory neurons resemble those of photoreceptor cells: Ca^2–calmodulin blocks the cAMP-controlled CNG channel, specific protein kinases phosphorylate and inactivate the odorant receptors, and a specific RGS protein (RGS2) stimulates the GTPase activity of the G-protein G_{olf}.

In the sensory cells of the inner ear, around 30 different proteins are involved in signal processing and adaptation. Again, Ca^2 is the primary adaptive signal that blocks the mechanosensitive TRP channels just as it inhibits the CNG channels in the eye and the nose. Moreover, for a slow adaptation to the sound volume, Ca^2 activates a special type of myosin, thus reducing the mechanical tension between the channel protein and anchoring structures.

Summary

Adaptation enables sensory cells to respond over a wide range of signal intensity. It is a complex process involving feedback loops at every level of signal transduction. A master regulator of adaptation is Ca^2, which invades the cell through cation channels that have become gated in the course of sensory signal processing.

Further reading

Section 15.1

Bagriantsev SN, Gracheva EO & Gallagher PG (2014) Piezo proteins: regulators of mechanosensation and other cellular processes. *J Biol Chem* 289:31673–31681.

Dulak C & Grothe B (eds) (2004) Sensory systems. *Curr Opin Neurobiol* 14:403–518.

Li F (2013) Taste perception: from the tongue to the testis. *Molec Human Reprod* 19:349–360.

Liman ER, Zhang YV & Montell C (2014) Peripheral coding of taste. *Neuron* 81:984–1000.

Nilius B & Honoré E (2012) Sensing pressure with ion channels. *Trends Neurosci* 35:477–486.

Section 15.2

Delmas P & Coste B (2013) Mechano-gated ion channels in sensory systems. *Cell* 155:278–284.

Section 15.3

Battista N et al. (2012) The endocannabinoid system: an overview. *Front Behav Neurosci* 6:1–7.

Hefti FF, Rosenthal A, Walicke PA et al. (2006) Novel class of pain drugs based on antagonism of NGF. *Trends Pharmacol Sci* 27:85–91.

Julius D (2013) TRP channels and pain. *Annu Rev Cell Dev Biol* 29:355–384.

Palkar R, Lippoldt EK & McKemy DD (2015) The molecular and cellular basis of thermosensation in mammals. *Curr Opin Neurobiol* 34:14–19.

Vriens J, Nilius B & Voets T (2014) Peripheral thermosensation in mammals. *Nature Rev Neurosci* 15:573–589.

Section 15.4

Chamero P, Leinders-Zufall T & Zufall F (2012) From genes to social communication: molecular sensing by the vomeronasal organ. *Trends Neurosci* 35:593–606.

Ihara S, Yoshikawa K & Touhara K (2013) Chemosensory signals and their receptors in the olfactory neural system. *Neurosci* 254:45–60.

Jiang Y & Matsunami H (2015) Mammalian odorant receptors: functional evolution and variation. *Curr Opin Neurobiol* 34:54–60.

Kang NN & Koo JH (2012) Olfactory receptors in non-chemosensory tissues. *BMB Rep* 45:612–622.

Liberles SD (2014) Mammalian pheromones. *Annu Rev Physiol* 76:151–175.

Spehr M, Schwane K, Riffell JA et al. (2006) Odorant receptors and olfactory-like signaling mechanisms in mammalian sperm. *Mol Cell Endocrinol* 250:128–136.

Section 15.5 and 15.6

Burns ME & Arshavsky VY (2005) Beyond counting photons: trials and trends in vertebrate visual transduction. *Neuron* 48:387–401.

Field GD & Chichilnisky EJ (2007) Information processing in the primate retina; circuitry and coding. *Annu Rev Neurosci* 30:1–30.

Hardic RC & Juusolo M (2015) Phototransduction in *Drosophila*. *Curr Opin Neurobiol* 34:37–45.

Sharma RK & Duda T (2014) Membrane guanylate cyclase, multimodal transduction machine: history, present, and future direction. *Front Mol Neurosci* 7:1–20.

Signaling at Synapses: Neurotransmitters and their Receptors

16

This chapter provides a brief overview of neurotransmission, with emphasis laid on signaling mechanisms at synapses. In fact, neuronal signaling adds some new aspects to the principles discussed in previous chapters. This holds true, in particular, for ion channel-coupled neurotransmitter receptors, which are only distantly related to the ion channel families discussed in Chapter 14 but, like those, can be traced back to bacterial precursors. For complementary information about further questions of development, anatomy, (electro)physiology, and pathology, the reader may consult textbooks of neuroscience.

The chemical vehicles of neurotransmission are the neurotransmitters. These are defined as signaling molecules that are released by neurons to interact at specific intercellular contact sites, the synapses, in either para- or autocrine mode with membrane receptors of the post- and presynaptic cell. The concept of chemical signal transmission at synapses was put forward by Otto Loewi in Graz, Austria, in 1921 after he had succeeded in slowing down the beat of a frog heart in tissue culture not only by an electrical stimulation of the parasympathetic nervus vagus but also by a *Vagusstoff* (vagus substance) found in the culture medium of a heart electrically stimulated before. Together with Sir Henry Hallett Dale, Otto Loewi was awarded the Nobel Prize in Physiology or Medicine for 1936. Later the *Vagusstoff* was identified as acetylcholine, which still is the prototype of a neurotransmitter.

The term neurotransmitter comprises simple molecules such as acetylcholine, lipid derivatives, amino acids, and biogenic amines derived from amino acids by decarboxylation (such as dopamine, histamine, serotonin, and noradrenaline) as well as peptides and proteins. There is no clear boundary between neurotransmitters and what are called neuromodulators, even though the latter are understood commonly as signaling factors that modulate the release and effects of neurotransmitters.

16.1 Neurons and synapses

Neurons, the prototypes of excitable cells, have developed to perfection the basic cellular property of data processing, which is to "calculate" an output signal from a large number of input signals and to store acquired information in a data-processing network. The communicative cellular structures are the synapses, where neurons and their target cells come up to within a distance of 20–40 nm. As a rule, a neuron has many more synapses for reception than for transmission of signals. The receptive or input synapses are concentrated on the neurites and the cell body, while the transmitter or output synapses are found at the axons, ending in so-called presynaptic boutons. With up to 10^4 contact sites per cell, the data-processing potency of a (brain) neuron reaches an astronomical value. The total number of synapses in the adult human brain has been estimated at 10^{14}, surpassing the number of stars in the galaxy for a factor of 1000. The concept of synaptic contacts was introduced by the Spanish neurophysiologist Santiago Ramon y Cajal (Nobel Prize in Physiology or Medicine 1906) and his British colleague Sir Charles Scott Sherrington (Nobel Prize in Physiology or Medicine 1932), two founders of modern neuroscience.

16.1.1 Mode of action of synapses

The input signals received by a neuron are deciphered and bundled to generate an action potential running down the axonal membrane to the presynaptic bouton, where in most neurons it triggers the release of neurotransmitters (**Figure 16.1**). [Note that in some tissues, such as hypothalamus, pituitary gland, and adrenal medulla, neurons release neurotransmitters as hormone-like neurosecretions into the bloodstream; that is, they function as endocrine glands.] Neurotransmitters diffuse across the synaptic cleft and interact with receptors of the postsynaptic cell (for an illustration see Figure 10.17 in Section 10.4.1). When such receptors are coupled with ion channels, they evoke a change of the postsynaptic membrane potential; when they are coupled with signal-transducing enzymes such as G-proteins, they trigger the generation of second messengers.

The distinct signaling functions of the active pre- and postsynaptic zones are mirrored by characteristic morphological features (Figure 16.1 and **Figure 16.2**). The presynaptic bouton, primarily acting as signaling device, is filled with exocytotic vesicles containing the neurotransmitter. Those vesicles that are primed for release assemble on a protein meshwork. This **presynaptic web** lines the inner side of the presynaptic membrane and contains the protein machinery of exocytosis. The postsynaptic membrane, being a zone of signal reception, contains clusters of receptors, ion channels, scaffold

Figure 16.1 Morphological basis of neuronal signal transmission An idealized neuron is shown (left). Its dendrites are occupied by a large number of synapses receiving input signals, the processing of which results in an action potential. The latter is propagated along the axon to the presynaptic bouton of the axonal synapse where it triggers the release of neurotransmitter resembling the output signal. The basic architecture of the presynaptic terminal of a chemical synapse is shown in the inset to the right. Secretory vesicles filled with neurotransmitter (shown in color) have been transported along cytoskeletal fibers (neurofilaments and neurotubules) from the cell body to the presynaptic bouton where those primed for release become assembled on a meshwork of proteins (presynaptic web). Upon depolarization of the presynaptic membrane (arrival of an action potential), the vesicles are emptied into the synaptic cleft. The mechanism of this exocytotic process is explained in more detail in Section 10.4.2. (Adapted in part from Krstic RV [1978] Die Gewebe des Menschen und der Säugetiere. Heidelberg: Springer.)

proteins, and other constituents of the signal-processing protein network. Under the electron microscope, this zone is visible as **postsynaptic density** (see Figure 16.2). Finally, pre- and postsynaptic membranes are interconnected by a dense web of interacting proteins operating as glue and signal transducers. Thus, the conventional idea of the synaptic cleft as an empty gap is not confirmed by the reality.

The combination of electrical with chemical signaling ensures that in chemical synapses the signal is transmitted only in one direction (unidirectional signaling). Nevertheless, chemical synapses also process retrograde signals, in particular to control neurotransmitter release by negative feedback. For an example, the reader may recall the autocrine self-regulation of α-adrenergic sympathetic signaling as explained in Section 5.3.3.

As with any exocytotic process, neurotransmitter release is strictly calcium-dependent. A key event of synaptic signal transmission is, therefore, the gating of voltage-dependent Ca^{2+} channels. This is brought about by depolarization, through the action potential running down the axonal membrane. In fact, in neurons the major role of action potentials is to trigger the influx of Ca^{2+} ions into the presynaptic terminals. Since Ca^{2+} influx and neurotransmitter release are stochastic rather than all-or-nothing events, synapses release small "packages" (resembling vesicles) of neurotransmitter even in the absence of an input signal, evoking postsynaptic miniature potentials that are either excitatory or inhibitory depending on the type of neurotransmitter and receptor. In other words, as long as a neuron is alive, its synapses never rest. Any excitatory or inhibitory effect of signaling is superimposed upon this apparently chaotic background activity.

For signal termination the neurotransmitter is removed rapidly from the synaptic cleft by presynaptic endocytosis, better known as recapture, or (rather an exception) by enzymatic degradation (see Sections 16.2 and 16.9).

Synapses that communicate via neurotransmitters are called chemical synapses. In vertebrate nervous systems they are the predominant type. Another type of structure is the **electrical synapse**. Here the gap between the cells is bridged by a particular type of membrane channel. Although these **gap junctions** allow ions and other low-molecular electrolytes to pass (**Sidebar 16.1**), they are structurally unrelated to voltage- and receptor-dependent ion channels. In contrast to chemical synapses, electrical synapses work almost without delay, transmitting signals in *both* directions (bidirectional signaling). Their predominant function is to synchronize large collectives

Figure 16.2 Synaptic contact zone The figure based on electron microscopy shows the synaptic contact zone between a presynaptic (pre) and a postsynaptic (post) neuron of the mouse cerebellar cortex. Note the granulofilamentous material in the synaptic cleft (asterisk), the postsynaptic density (PD), and the secretory vesicles filled with neurotransmitter. (Adapted from Spacek J, Atlas of Ultratructural Neurocytology [https://sites.cns.utexas.edu/synapseweb/atlas].)

Sidebar 16.1 Intercellular communication by gap junctions Gap junctions are large membrane channels bridging the gap between adjacent cells. Each cell provides half of the channel called a connexon. A connexon consists of six protein subunits. These subunits, known as connexins, are built in the membrane by four transmembrane domains each. Human cells express about 20 connexin isoforms in a tissue-specific pattern. The central channel pore is 1.2 nm wide and penetrable for molecules up to 1000 Da, enabling an electrical and metabolic coupling of cells and an exchange of small hydrophilic signaling molecules such as cAMP, cGMP, Ca^{2+}, etc. The connexon pore opens or closes by an irislike shift of the subunits. This gating is signal-controlled and is regulated by the membrane potential and intracellular signaling reactions. Thus, protein phosphorylation may either open or close the channel depending on the protein kinase and the connexin and the intramolecular site to be phosphorylated. Signals that trigger an immediate closing are an acidification of the cytoplasm or a high Ca^{2+} concentration, as they both are typical of damaged cells. As a consequence, those cells become sealed. By means of gap junctions, tissue cells may become organized as a syncytium and respond synchronously. Well-known examples are the heart muscle, smooth muscles, and certain regions of the brain. In the course of evolution, gap junctions must have arisen independently at least twice, since the junctional proteins of vertebrates and invertebrates are entirely unrelated. Their equivalents in plants are the plasmadesmata that connect the endoplasmic reticula of adjacent cells, penetrating the cell walls and transporting macromolecules such as proteins.

of cells and to trigger very rapid responses such as flight reflexes and the processing of sensory stimuli in the brain. However, compared with chemical synapses, their capabilities to filter and to modulate signals are rather limited.

16.1.2 Axon growth: how signals regulate a developmental process

Tissue formation depends on directed growth and targeted migration of cells in combination with the induction of specific genetic programs of differentiation. In these processes, each individual step is controlled by intercellular signals acting as morphogenetic factors. The formation of synaptic contacts provides a typical and particularly exciting example. Here a key event—the outgrowth of axons—is controlled and guided toward a target by a wide variety of factors produced by surrounding and target cells. We have already encountered some of these factors in previous chapters. Correct targets emit attractive signals, while wrong targets emit repulsive signals. The signal-processing apparatus of the growing neuron translates such signals primarily into cytoskeletal changes required for the protrusion or retraction of the nerve fiber. The corresponding receptors are concentrated at the tip of the developing axon, called the growth cone. This is a highly dynamic structure that continuously explores its environment.

Axon guidance proteins

The factors dominating targeted axon growth are both diffusible and membrane-bound proteins. Such axon guidance proteins are expressed in numerous isoforms by the target cells. Most of their receptors are single-pass transmembrane proteins of the immunoglobulin-like cell adhesion molecule (Ig-CAM) family (see Section 7.3). Prototypical axon guidance proteins are the **netrins**. These are either secreted into the extracellular matrix or are membrane-bound via glycophosphatidyinositol anchors. Depending on the receptor type, netrins either have an attractive or a repulsive effect on axon growth. The receptor type evoking attraction is called **DCC** (Deleted in Colon Carcinoma) since it was originally discovered as tumor suppressor protein deleted in large bowel cancer (indicating that netrins must have additional functions beside their role in axon guidance!). The repulsive effect of netrins is transduced by the receptor **UNC5** (referring to the mutant "UNCoordinated 5" of *Caenorhabditis elegans*).

With more than 30 isoforms, the **semaphorins** are the largest family of axon-guiding proteins. Their major function is that of repellents that prevent axons from approaching a wrong target (because they provoke a collapse of the axonal growth cone, semaphorins are known also as collapsins). Semaphorins are found as either single-pass transmembrane proteins, or anchored in the membrane through a glycosylphosphatidylinositol residue, or in a diffusible form. Two types of semaphorin receptor, the **neuropilins** and the **plexins**, are known. Both are expressed in a high number of isoforms.

Other families of repulsive axon guidance proteins are the **RGMs** (R̲epulsive G̲uidance M̲olecules) and the **slit proteins** (referring to the *Drosophila* mutant "slit"). Like certain netrins and semaphorins they are membrane-bound through glycophosphatidylinositol anchors. The corresponding receptors are called **neurogenins** for RGMs and **Robos** (referring to the *Drosophila* mutant "roundabout") for slit proteins. RGMs play a specific role in the development of the retina, while slit proteins prevent crossover growing of developing nerve fibers. An important role in axon guidance is also played by the **ephrins** which function both as attractants and repellents. They are treated in detail in Section 7.1.4.

A common feature of the receptors of the axon-guiding factors is that they transduce the signal to the Rho–Rac–Cdc42 machinery as well as to the integrin complex, the two major control systems of the cytoskeleton. This occurs either through direct interactions or via cytoplasmic Tyr kinases that associate with the receptors or, as in the case of ephrin receptors, are integrated in the receptor molecule. An additional mechanism is used by plexins: their cytoplasmic region contains a GTPase-activating protein domain, which catalyzes the inactivation of Rho/Rac/Cdc42 G-proteins and thus induces a retraction of cytoskeletal filaments.

Neurotrophins

Neurotrophins are hormone-like growth factors. They are survival signals that promote the proliferation of immature neurons and protect them from cell death, which otherwise would occur automatically. To this end, neurotrophins stimulate mitogenic and anti-apoptotic pathways such as the MAP kinase and phosphatidylinositol 3-kinase–protein kinase B/Akt cascades (see Section 4.6.6). Typical neurotrophins are **glial cell-derived neurotrophic factor GDNF** and **nerve growth factor (NGF)**. Both interact with receptor tyrosine kinases (see Section 7.1.1): GDNF with Ret (and in addition with neuronal cell adhesion molecules), and NGF with Trk, the high-affinity NGF receptor.

16.1.3 Formation of synapses

At sites where a growing axon meets its target cell, synapses are formed. Like axon guidance, this process is controlled by stimulatory and inhibitory signals that are released from the interacting cells. Many of these factors are identical with those regulating axon growth. Therefore, the signaling events involved in both axon guidance and synaptogenesis overlap to a large degree. Synapse formation is a matter of mutual giving and receiving because the axon is not only guided by factors of the target cell but also, vice versa, transmits signals that organize the functional architecture of the postsynaptic target membrane. This is best exemplified by the events occurring during the formation of a neuromuscular synapse

Agrin and neuregulin: signals coordinating the development of neuromuscular synapses

Studies on neuromuscular synapses have deepened considerably our understanding of developmental processes in the nervous system. Here the key protein of the postsynaptic membrane is the nicotinic acetylcholine receptor. When axons of motor neurons approach muscle fibers, the receptors scattered across the cell surface assemble into large aggregates at the postsynaptic membrane zone in a process called synaptic targeting of receptors. Simultaneously, gene transcription is selectively turned on in those nuclei of the multinuclear muscle cell that are positioned in the immediate vicinity of the developing synapse. The corresponding genes encode proteins that reinforce the synaptic contact.

The major signal inducing neuromuscular synapse formation is transmitted by **agrin**. This heparan sulfate proteoglycan of about 2000 amino acids is expressed in a large number of splice variants, ensuring a high degree of specificity of neuromuscular contacts. It is released from the approaching axon and stored in the extracellular matrix filling the synaptic cleft (**Figure 16.3**). The agrin receptor of the muscle cell is the receptor Tyr kinase **MusK** (Muscle-specific Kinase), which is found only in skeletal muscle. Both MusK and agrin are essential proteins; mice with the corresponding gene knockouts die immediately after birth due to a failure of the respiratory system.

Agrin activates MusK in the conventional way, by stimulating Tyr-directed trans-autophosphorylation within the receptor dimer (see Section 7.1). One major substrate of active MusK is the nicotinic receptor, the aggregation of which at the site of synapse formation depends on phosphorylation. This response probably requires the recruitment at the receptor of a large signal-transducing protein complex that is organized by the scaffold protein **rapsyn**. Gene knockout experiments have shown rapsyn to be as essential for life as agrin and MusK.

As is typical for a receptor, the Tyr kinase MusK activates, in addition, signaling cascades, which turn on the transcription of genes encoding synaptic proteins such as the

Figure 16.3 Formation of a neuromuscular synapse The tip of a nerve fiber (left) approaching the skeletal muscle cell (right) produces an extracellular matrix (ECM), which is impregnated with the protein agrin. By interacting with the receptor Tyr kinase MusK (muscle-specific kinase), agrin induces the aggregation of nicotinic receptors in the postsynaptic muscle cell membrane opposite to the nerve fiber, and via signaling cascades such as Ras–RAF–ERK, the transcription of various genes encoding synaptic proteins is induced. The teal cylinders in the dimeric MusK protein symbolize the Tyr kinase domains.

nicotinic receptor, MusK, and acetylcholinesterase. Such genes are characterized by a synapse-specific response element called N-box that interacts with the transcription factor GABP (Guanosine- and Adenosine-Binding Protein). Being a member of the Ets family of transcription factors, GABP is a target of the Ras–RAF–ERK signaling cascade (see Section 11.6.1). This cascade is activated by MusK.

Another neuronal signal cooperating with agrin is the epidermal growth factor (EGF)-like growth factor **neuregulin**. Its receptor is the receptor Tyr kinase ErbB, the EGF receptor proper, that like MusK evokes the whole program of downstream responses (including activation of the Ras–RAF–ERK cascade) that is, as described in Chapter 7, typical for this receptor family. There is every reason to assume that related events also control the formation of synapses between the vegetative nervous system and its target tissues.

Synapse formation in the central nervous system

The wiring of the human brain ranks among the most fantastic phenomena in nature: within a relatively short period, more than 1 million kilometers of axons and dendrites interconnecting 10^{11} neurons by 10^{14} synapses are established. As explained earlier, these growth processes driven by the dynamics of the actin cytoskeleton (see Section 10.3) are guided by numerous diffusible and membrane-bound signaling molecules. In the central nervous system, such factors are derived from both neurons and the surrounding glial cells.

To generate an intraneuronal synapse, the dendrite of the target neuron stimulates the approaching axon to develop a presynaptic bouton equipped with the signaling machinery, while vice versa, the axon prompts the dendrite to protrude a dendritic spine, which is derived from a filopodium, and to develop a postsynaptic density. This is a rapid event lasting less than an hour, whereas the maturation of the initial contact zone to a fully active synapse takes considerably more time.

For intraneuronal synapse formation, diffusible proteins seem to play a similar role as agrins and neuregulins do for neuromuscular synapses. Examples are provided by

immunoglobulin-like domains

homo and heterotypic interactions ligands

fibronectin-like domains

transmembrane domain

cell membrane

protein kinases

cytoplasm

signal transduction

Figure 16.4 Prototypical structure and mechanism of signaling of Ig-CAMs controlling the wiring of the nervous system The neuronal cell adhesion molecules of the immunoglobulin-like family (such as N-CAMs, DS-CAMs, Syn-CAMs, and DCCs) are characterized by series of immunoglobulin-like and fibronectin-like motifs in the extracellular domain and a single transmembrane domain. Upon *cis* and *trans* interactions with each other or other cell adhesion molecules and soluble or membrane-bound ligands, respectively, they trigger intracellular signaling cascades, probably by recruiting cytoplasmic protein kinases or by *cis*-interactions with receptor Tyr kinases.

ephrin B1, probably acting in a proteolytically detached form (see Section 7.1.4), and neuronal pentraxins, a family of secreted synaptic proteins. Both are required for the clustering of postsynaptic neurotransmitter receptors.

A key role in synapse formation is played by **neural cell adhesion molecules (N-CAMs)**, a subfamily of Ig-CAMs (see Section 7.3). In the receptors DCC, UNC5, Robo, and neogenin we have already encountered members of this family (see Section 16.1.2). N-CAMs share common structural features that include a single transmembrane domain, clusters of Ig-like and fibronectin-like domains in the extracellular part, and cytoplasmic signaling domains. A prototypical structure is shown in **Figure 16.4**. N-CAMs combine two functions. On the one hand they are signal transducing entities promoting axonal growth and synaptic differentiation, on the other hand they glue together pre- and postsynaptic membranes. To this end they interact with each other (homotypically) or—like receptors—with other membrane-bound or soluble molecules. Upon activation N-CAMs recruit various signaling proteins, in particular cytoplasmic and membrane-bound Tyr kinases. In fact, N-CAMs may be considered as Tyr kinase-associated receptors that assemble, like other receptors of this type, large signal-transducing protein complexes at their cytoplasmic domains. On the cellular and molecular level such interactions have widespread consequences, reaching from direct effects on the cytoskeleton to gene activation mediated by MAP kinase cascades, NFκB modules, and nuclear receptors.

By alternative splicing of pre-mRNA (see Section 8.1) some N-CAMs may be generated in an astronomically high number of variants, providing a kind of cell recognition code that ensures a highly selective interlinking of nerve cells. A striking example of biochemical combinatorics is provided by **DS-CAM**, a large membrane protein of about 1500 amino acids. The term DS stands for Down Syndrome, because the *DS-CAM* gene maps to a Down syndrome region of chromosome 21 (see Section 14.6.4). Alternative splicing of the pre-mRNA of the *Drosophila* homolog (Ds-cam) has been estimated to yield about 38,000 isoforms that differ from each other in their extracellular domains. The number of Ds-cam variants thus outdoes the total number of *Drosophila* genes by a factor of 2.5, and it is assumed that individual groups of neurons (or even single individual nerve cells) each express a unique pattern of Ds-cams. In fact, Ds-cam interactions have been found to be highly selective, allowing cells to communicate with an extreme degree of privacy. The situation is assumed to be quite similar for mammalian DS-CAMs.

Nerve fibers expressing identical DS-CAM patterns repel each other. Therefore, DS-CAMs are thought to prevent short cuts such as synaptic contacts between neurites and their cell of origin and to cause a repulsion of dendritic branches of the same neuron, thus allowing for correct outgrowth of neurite trees. To this end DS-CAMs interact with each other. Via adaptor proteins (such as Nck in mammals and DOCK in *Drosophila*) they activate signaling cascades such as the PAK–Rac/Cdc42 sequence, which controls the architecture and function of the actin cytoskeleton (see Section 10.3). The existence of heterotypically interacting DS-CAM ligands has been postulated but not proven.

There are other "glues" involved in neuronal differentiation that do not belong to the N-CAM family. These proteins are the **ephrins** (see Section 7.1.4) and the **neuroligins**. Neuroligins are bound in the postsynaptic (dendritic) membrane. Their receptors are presynaptic (axonal) transmembrane proteins, the **neurexins**. The mechanism of neuroligin–neurexin signaling closely resembles that of N-CAMs in that upon activation large complexes of signaling proteins including protein kinases are recruited at the cytoplasmic domains. The signals thus transduced are essential for the formation of presynaptic boutons including the organization of the machinery for neurotransmitter secretion. Like DS-CAMs, neuroligins and neurexins are expressed in a large number of splice variants. The variations in the extracellular domains of these isoforms represent a splice code that renders the interactions highly specific, determining, for instance, whether a synapse will have a stimulatory or an inhibitory effect. Some neurexin variants also interact with diffusible peptide factors called neurexophilins. Neurexins are the major target of α-latrotoxin, a component of the poison of the black widow spider (*Latrodectus tredecimguttatus*).

Other cell adhesion molecules expressed by nerve cells include cadherins, protocadherins, and integrins. They are important for stabilization of synaptic contacts rather than for synapse formation.

16.1.4 Ionotropic and metabotropic neurotransmitter receptors

The key proteins of signal processing in chemical synapses are the neurotransmitter receptors. Their functional organization is decisively determined by interacting CAMs. One distinguishes between ionotropic and metabotropic receptors. An ionotropic receptor is a part of an ion channel, whereas metabotropic receptors are coupled with trimeric G-proteins; they belong to the large family of heptahelical transmembrane proteins introduced in Chapter 4 (**Figure 16.5**).

Ionotropic receptors are found predominantly (but not exclusively) in the postsynaptic membrane and provide rapid paracrine signal transmission occurring within milliseconds or less. The considerably slower metabotropic receptors are expressed both pre- and postsynaptically and mediate auto- and paracrine neurotransmitter effects such as the autocrine regulation of neurotransmitter release (**Figure 16.6**). It should be noted that in interneuronal signaling the principle of double negation (stimulation by inhibition of an inhibitory effect) is particularly abundant. Thus, an

Figure 16.5 Neurotransmitter receptors: the basic types Receptor ion channels (ionotropic receptors) and G-protein-coupled receptors (metabotropic receptors) are shown. Depending on the intracellular Cl⁻ concentration, an opening of chloride channels may either hyper- or depolarize the cell. $G_{i,0}$-protein-coupled receptors may exhibit either an inhibitory or (along the Gβγ–phospholipase-Cβ pathway) a stimulatory effect.

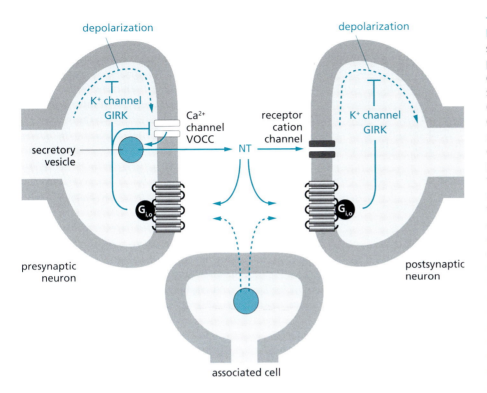

Figure 16.6 Modulation of synaptic signal transmission by pre- and postsynaptic receptors As an example, an excitatory synapse is shown where upon depolarization (teal broken line) the neurotransmitter (NT) released by the presynaptic neuron opens a postsynaptic receptor cation channel, thus depolarizing the postsynaptic cell. Simultaneously, the transmitter interacts with pre- and postsynaptic metabotropic receptors, which for the sake of clarity are assumed here to be $G_{i,0}$-coupled only. In reality, all kinds of G-protein couplings exist. Activation of these receptors leads to a gating of the hyperpolarizing G_i-protein-regulated K^+ channel (GIRK) and to a blockade of presynaptic voltage-dependent Ca^{2+}-channels (VOCC), thus impeding depolarization and neurotransmitter release by negative feedback (G_s- and $G_{q,11}$-coupled receptors may, vice versa, amplify synaptic signaling by positive feedback). Pre- and postsynaptic receptors may also interact with signal molecules released from associated cells. These cells may be neurons or glial cells, which become stimulated by the neurotransmitter or by other means. To modulate synaptic signaling, neurotransmitters and signals from associated cells may react also with pre- and postsynaptic receptor ion channels (ionotropic receptors; not shown)

excitatory neurotransmitter may have an inhibitory effect when (via a presynaptic receptor) it facilitates the release of an inhibitory transmitter, and an inhibitory transmitter would enhance neurotransmission when it acts on an inhibitory neuron.

As shown in **Table 16.1**, acetylcholine, serotonin, γ-aminobutyric acid (GABA), glutamate, ATP, and anandamide interact both with ionotropic and with metabotropic receptors, whereas for the other neurotransmitters only metabotropic receptors are

Table 16.1 Neurotransmitter receptors: an overview

Transmitter	Ionotropic receptors		Metabotropic receptors coupled with		
	Anion channels	Cation channels	G_s	$G_{q,11}$	$G_{i,0}$
acetylcholine		Nicotinic α_{1-10}		M_1, M_3, M_5	M_2, M_4
noradrenaline			β_{1-3}	$\alpha_{1A,B,D}$	$\alpha_{2A,B,C}$
dopamine			D_1, D_5		D_2, D_3, D_4
histamine			H_2	H_1	H_3, H_4
serotonin (5-OH-tryptamine)		$5\text{-}HT_3$	$5\text{-}HT_{4-7}$	$5\text{-}HT_2$	$5\text{-}HT_1$
ATP, etc.		P2X		$P2Y_{1,2,4,6,11}$	$P2Y_{12}$
opioid					MOP, DOP, KOP, NOP
anandamide (cannabinoid)		TRPV			CB_1, CB_2
glutamate		NMDA, AMPA, kainate		$mGlu_{1,5}$	$mGlu_{2-4,6-8}$
GABA	$GABA_A$				$GABA_B$
glycine	GR				
peptide transmitters				metabolic receptors only	

known, and for glycine only an ionotropic receptor has been found. The effect a neurotransmitter receptor has on postsynaptic signaling depends on the type of effector protein. Thus, ionotropic receptors coupled with cation channels and metabotropic receptors coupled with G_s- or $G_{q,11}$-proteins are stimulatory, whereas a coupling with anion channels or $G_{i,0}$-proteins mostly (but not always) has an inhibitory effect. In this chapter some emphasis is laid on ionotropic receptors and those metabotropic receptors that hitherto have not been mentioned. The function of other metabotropic receptors (in particular those for acetylcholine and noradrenaline) has been discussed in some detail in Chapter 5.

Fundamental insights into synaptic signal transduction are due to the work of Arvid Carlsson, Paul Greengard, and Eric R. Kandel, who were awarded the Nobel Prize in Physiology or Medicine in 2000.

Summary

- Synapses are signal-transducing contact sites between neurons and their target cells. Development of synapses is controlled by paracrine and juxtacrine signaling between the approaching neuron and the target cell. The signals are emitted by both pre- and postsynaptic cells. Juxtacrine signals are transmitted by cell membrane-bound proteins combining the properties of receptors with those of cell adhesion molecules (CAMs). Some of these proteins are expressed in astronomically high numbers of splice variants, ensuring specific interconnections.

- Synapses exist in two types: electrical synapses that allow a direct exchange of molecules up to 1 kDa between adjacent cells, and chemical synapses that are bridged by neurotransmitters. Neurotransmitters are released from the presynaptic cell by depolarization-induced exocytosis. They interact with receptors on the pre- or postsynaptic membrane.

- Neurotransmitter receptors are coupled with either ion channels (ionotropic receptors) or trimeric G-proteins (metabotropic receptors). They are integrated in both pre- and postsynaptic membranes as components of positive and negative feedback loops that control synaptic activity. While cation channel- and G_s- or $G_{q,11}$-coupled receptors exert stimulatory effects, anion channel- and $G_{i,0}$-coupled receptors mostly transmit inhibitory signals.

16.2 Acetylcholine receptors

The classical example of a neurotransmitter interacting with both types of receptor is acetylcholine: its muscarinic receptors are of the metabotropic type, while its nicotinic receptors are of the ionotropic type. Since the muscarinic receptors have been discussed in Section 5.4, here we shall confine the discussion to the nicotinic receptors.

16.2.1 Nicotinic receptor and the family of cysteine-loop receptors are prototypes of ligand-gated ion channels

In 1906, the British physiologist J. N. Langley in Cambridge had found that nicotine caused skeletal muscle contraction and that this response could be prevented by curare, the arrow poison of native South Americans. This led Langley to postulate "receptive substances" binding pharmacologically active compounds in the corresponding target tissues. In an attempt to explain the specific effects of antigens, Paul Ehrlich in Frankfurt had put forward a similar concept of cellular receptors in his famous *Seitenketten-Theorie* some years before. Not until 20 years later was acetylcholine identified as the endogenous ligand of nicotine-binding sites at neuromuscular synapses. The "nicotinic" acetylcholine receptors, called nAchR, were found to be concentrated in the membrane of the postsynaptic muscular cell, where upon ligand binding they triggered an action potential, inducing muscle contraction. Today we know that the mammalian nicotinic receptor is expressed in more than 10 isoforms: according to their ligand-binding α-subunits (each receptor is composed of five variable subunits), these are called nAchRα$_1$–α$_{10}$. Providing an impressive example of receptor multiplicity, they are tissue-specifically expressed. Only the isoform nAchRα$_1$ is found in skeletal muscle cells, whereas the other isoforms are restricted to individual neurons in the ganglions of the vegetative nervous system (see Section 4.1.2) and in the central nervous system. In the brain a major function of acetylcholine is to augment neurotransmitter release by interacting with presynaptic nicotinic receptors, thus facilitating depolarization and calcium influx, an effect that explains some of the addictive effects of nicotine (see Section 16.9).

Isolation from electrical organs

The α$_1$-nicotinic receptor was the first receptor to be isolated and purified to homogeneity, mainly by J.-P. Changeux in the laboratory of Jacques Monod in Paris. This success was made possible by nature providing an extraordinarily rich source of the receptor in the electrical organs, which in certain fish species may come to 20% of body weight. These organs have evolved from skeletal muscles; however, in their cells, called electrocytes, nicotinic receptors serve the generation of electrical rather than of mechanical energy. To produce a high voltage, the synapses and the receptors are densely packed. So, in the electrical organ of the torpedo ray (*Torpedo californica*), about 50% of the postsynaptic cell surface consists of nicotinic receptors corresponding to 15,000 receptor molecules per square micrometer and a total receptor-packed surface area of 0.07 m^2 per animal (**Figure 16.7**). To obtain the receptor in pure form, only a 300-fold enrichment was necessary. This provides an extremely favorable starting situation compared with the efforts necessary to isolate other receptors such as the uterine estrogen receptor (100,000-fold enrichment) or the insulin receptor of liver (500,000-fold enrichment). Another advantage consisted of the ability to label the receptor with **α-bungarotoxin**, enabling the investigator to follow up the purification procedure (for this purpose bungarotoxin had been labeled before with radioactive iodine). This peptide from the venom of the Taiwanese snake *Bungarus multicinctus* binds strongly and highly specifically to the receptor, causing a complete blockade of neuromuscular signaling. Related α-neurotoxins are also found in the venom of cobras, mambas, and coral and sea snakes (the acetylcholine receptors of these animals are insensitive to the toxins).

To isolate the nicotinic receptor, a novel approach was introduced that has since become a standard method of protein purification. This approach is **affinity chromatography**, where a ligand (here curare) covalently bound to the chromatographic matrix selectively holds back the protein until it is eluted from the column by an excess of the same or another ligand. To prove the identity of the purified receptor, a biological assay was performed. Upon injection of the material, rabbits developed the symptoms of a severe muscular weakness resembling **myasthenia gravis**, an autoimmune disease of humans caused by an inhibitory antibody directed against muscular nicotinic receptors (it is instructive to compare this classical approach of receptor isolation with present methods as described in Section 15.1, Sidebar 15.1).

The ion channel function of the isolated receptor was demonstrated by incorporating the material into artificial lipid vesicles filled with a solution of radioactively labeled NaCl or KCl. Provided the preparation contained active receptors, acetylcholine could induce an ion flux across the vesicular membrane by gating these receptors.

Structure and properties

Nicotinic receptors are composed of five subunits, which are closely related isoforms of a basic polypeptide structure with four transmembrane helices. The subunits are arranged like a rosette and exhibit two acetylcholine-binding sites at the

Figure 16.7 Nicotinic acetylcholine receptor cation channel of the skeletal muscle type (A) Left: Californian torpedo ray with the electrical organs composed of columns of electrocytes. Right: pentameric receptor with the acetylcholine-binding sites (teal encircled "A") seen from above. (B) Vertical view and section of the receptor with the negatively charged zones determining the cation selectivity of the central channel pore. According to the ion distribution, K$^+$ leaves and Na$^+$ and Ca^{2+} enter the cell through the channel gated by acetylcholine binding. (A, from Hall ZW [1992] Molecular Neurobiology. Sunderland, MA: Sinauer.)

interface between the extracellular amino (N)-terminal extensions and the transmembrane helices of the α-subunits (see Figure 16.7). At the cytoplasmic side, they are anchored at the cytoskeleton.

The subunits are tissue-specifically expressed in various isoforms. Depending on the isoform composition, one distinguishes between nicotinic receptors of the muscle, ganglion, and brain types. Each type comprises several receptor variants exhibiting different properties such as toxin sensitivity. α-Bungarotoxin inhibits only the muscle type, whereas the related κ-bungarotoxin inhibits only the brain type. The receptors of the electrical organ belong (as would be expected) to the muscle type, consisting of the subunits α$_1$, α$_2$, β, γ, and δ.

All nicotinic acetylcholine receptors have turned out to be nonspecific cation channels permeable for Na$^+$, K$^+$, and Ca^{2+}. Gating causes membrane depolarization and Ca^{2+} influx and facilitates the generation of action potentials. Both are essential for muscle contraction. The ion pore is formed by five helices (one per subunit) rather than by P-loops as in more specific cation channels (see Section 14.1). Since P-loops control cation selectivity, their absence may explain the nonselectivity of the receptor ion channels. Novel approaches provide a clear picture of the receptor structure.

CLOSED   OPEN

M2   M2

cell
membrane

cytoplasm

hydrophobic interactions

+ Ach

Figure 16.8 Gating of the nicotinic receptor cation channel For the sake of clarity, only two of the five receptor subunits are shown. In the closed conformation, hydrophobic interactions between the M2 helices hinder the ion flux. The binding of acetylcholine (Ach) induces a conformational change involving a minimal irislike twist of the subunits. As a result, the hydrophobic interactions in the pore become disrupted. (Modified from Unwin N [2003] *FEBS Lett* 555:91–95.)

A breakthrough was the X-ray crystallographic analysis of an acetylcholine-binding protein from the freshwater snail *Lymnaea stagnalis*. In its pentameric configuration this protein is very closely related to the nicotinic receptor and can be used as a template for nicotinic receptor modelling.

Such studies show that, in the absence of the ligand, the central ion pore of the channel is blocked by hydrophobic interactions. The binding of two acetylcholine molecules induces a minimal twist of the subunits, relieving the hydrophobic blockade and opening the ion channel like an iris aperture (**Figure 16.8**).

Like voltage-controlled Na^+ and Ca^{2+} channels (see Sections 14.2 and 14.5.2), the nicotinic receptor cation channel exists in three states: closed, open, and desensitized (**Figure 16.9**). Desensitization is caused by long-lasting or high-frequency acetylcholine pulses. Hyperpolarization, Ca^{2+} ions, and phosphorylation of the cytoplasmic receptor domains by protein kinases A and C as well as by Tyr kinases of the Src family facilitate desensitization. The desensitizing protein kinases are activated

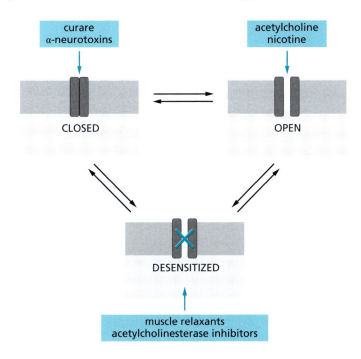

curare
α-neurotoxins

acetylcholine
nicotine

CLOSED   OPEN

DESENSITIZED

muscle relaxants
acetylcholinesterase inhibitors

Figure 16.9 Functional states of the nicotinic receptor The receptor exists in the three states, "closed", "open," and "desensitized," which are stabilized by the agents indicated.

ligand	receptor	
acetylcholine	nAchR	cation
serotonin	5-HT₃R	channels
zinc ions	ZAC	
GABA	type A	anion
glycine	glycine R	channels

Figure 16.10 Basic structure and membrane topology of an ionotropic receptor subunit of the nicotinic receptor type A single receptor subunit is shown; each receptor is composed of five subunits. Depending on the ligand (inset), a receptor is coupled with either a cation or an anion channel. The cylinders symbolize the transmembrane domains, with the domain lining the ion pore shown in color. A characteristic structural feature of this receptor type is an extracellular N-terminal Cys loop formed by a disulfide bridge between two Cys residues. This loop stabilizes the conformation of the ligand-binding site. The large cytoplasmic loop between transmembrane helices 3 and 4 contains several Ser, Thr, and Tyr residues, the phosphorylation of which modulates the function of the ion channel. GABA, γ-aminobutyric acid.

by neuropeptides and the neuromodulator adenosine interacting with G_s-coupled receptors, as well as by $G_{q,11}$-coupled muscarinic acetylcholine receptors. These reactions, which provide an illustrative example of the various interactions (cross talk) between ionotropic and metabotropic receptors, represent a kind of molecular short-term memory ensuring a continuous fine-tuning and adaptation of receptor function.

Metabolically stable synthetic derivatives of acetylcholine cause long-term desensitization of the nicotinic receptors. In the clinic, they are used to immobilize an area for surgery. Such drugs share this effect with inhibitors of acetylcholinesterase, the enzyme that degrades acetylcholine to choline and acetate in the synaptic cleft. Acetylcholinesterase inhibitors include highly toxic insecticides (like E605) and nerve gases for chemical warfare. In lower doses, such agents are also employed clinically for the treatment of myasthenia gravis. The inhibitory effect exerted by these compounds on the neuromuscular junction rivals that of curare and snake venoms.

Other receptors of the nicotinic receptor type

A structure resembling that of the nicotinic acetylcholine receptor is exhibited by the ionotropic receptors of the neurotransmitters serotonin (5-hydroxytryptamine), GABA, and glycine as well as by a still somewhat enigmatic receptor ZAC (Zinc-Activated Channel) of zinc ions (**Figure 16.10**). Together with the nicotinic receptors, these receptors constitute the family of **Cys-loop receptors** (named after a characteristic structural element; see Figure 16.10) with at least 40 members in humans. In contrast, the ionotropic glutamate and purine nucleotide receptors are members of other ion channel families. Genetic studies indicate that Cys-loop receptors join the large family of ion channels with prokaryotic ancestry.

As shown in Figure 16.10, Cys-loop receptors transport either cations or anions. The ion selectivity is determined by rings of charged amino acid residues surrounding the pore: positive charge repels cations and attracts anions, while negative charge has the opposite effect. Cys-loop receptors are purely ligand-gated since they lack a structure acting as a voltage sensor. A multimeric structure with at least two ligand binding sites, as found for nicotinic receptors (and other Cys-loop channels), allows positive cooperativity of the subunits (see Section 1.3). As repeatedly shown in previous chapters, such arrangement provides a powerful noise filter and enables rapid, switch-like channel gating.

Summary

- Together with the closely related ionotropic receptors of serotonin, γ-aminobutyric acid (GABA), glycine, and zinc ions, the ionotropic (nicotinic) acetylcholine receptor constitutes the Cys-loop receptor family.

- These receptors consist of five subunits with four transmembrane domains each forming a ligand-operated nonspecific cation (acetylcholine, serotonin, zinc) or anion channel (GABA, glycine).

- Nicotinic receptors exist in various isoforms differing in subunit composition. They are particularly abundant in electrical organs, skeletal muscle, ganglions of the vegetative nervous system, and certain brain areas.

16.3 γ-Aminobutyric acid and glycine receptors are the masters of inhibitory neurotransmission

The ionotropic receptors of glycine and GABA belong to the same protein family as nicotinic receptors. In contrast to those, however, they are coupled with anion channels, mainly transporting Cl⁻ ions. As explained in Section 14.7, such channels may cause either hyperpolarization or depolarization, depending on the intracellular chloride concentration. In most neurons, this concentration is kept at a low level by exchanger proteins (such as, for instance, a K^+–Cl^- exchanger) in the cell membrane. This ensures a hyperpolarizing influx of chloride ions through glycine and GABA receptor channels. As a result, the generation of action potentials as well as neurotransmitter release at synapses is impeded. (In neonatal neurons, the chloride concentration may be so high that GABA acts as a depolarizing excitatory neurotransmitter. Even in the adult brain, the polarity of GABA receptors may be reversed by changing the intracellular chloride concentration and thus be adapted to the particular conditions.) In fact, GABA and glycine are the most abundant and putatively most important inhibitory neurotransmitters of the central nervous system. It has been estimated that up to 20% of all brain neurons are glycinergic or GABAergic.

16.3.1 Ionotropic GABA$_A$ receptors

One distinguishes between ionotropic GABA$_A$ and metabotropic GABA$_B$ receptors (**Figure 16.11**). As relatives of the nicotinic receptor, GABA$_A$ receptor anion channels differ entirely from other anion channels (see Section 14.7), and within the Cys-loop family they show the most complex architecture. Their pentameric structure arises from an assembly of different subunits expressed in seven isoforms. In the brain, dozens of individual combinations have been found that differ in their physiological and pharmacological properties. The major function of these and the glycine receptors is to dampen the excitability of neurons. Appropriately, GABA agonists mostly exhibit anticonvulsant, sedating, and narcotic effects, whereas antagonists cause hyperactivity, convulsions, and panic attacks. Both effects may be deadly.

Examples of GABA agonists are ethyl alcohol, the narcotic halothane (which also activates hyperpolarizing K_{2P} potassium channels; see Section 14.4.3), barbiturates known as sleep inducers, and benzodiazepines used as fear relievers, whereas avermectin (an insecticide) and picrotoxin (a plant poison used for fishing in southeast Asia) represent the analeptic (fear- and convulsion-inducing) GABA antagonists. In the clinic, analeptics are used against alcohol and barbiturate intoxications as well as for the treatment of psychotic diseases.

Many of the agonists and antagonists interact with specific domains of the receptor that are not identical with the GABA-binding site. Therefore, they function as

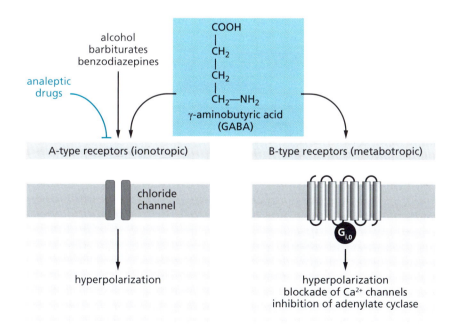

Figure 16.11 Ionotropic and metabotropic GABA receptors
(Left) The chloride channel of the ionotropic A-type receptor is gated by γ-aminobutyric acid (GABA), ethyl alcohol, barbiturates, and benzodiazepines and blocked by analeptics (fear- and convulsion-inducing drugs). (Right) Metabotropic B-type receptors are exclusively coupled with $G_{i,0}$-proteins.

Figure 16.12 Fear-relieving effect of GABA neurons: an example of presynaptic modulation Shown is an excitatory (fear-inducing) brain neuron releasing, for instance the neurotransmitters noradrenaline, serotonin, or glutamate. A presynaptically associated GABA neuron (shown in color) inhibits transmitter release by inducing hyperpolarization. This effect is augmented by alcohol or benzodiazepines.

allosteric modulators of GABA binding. The most prominent example is provided by the **benzodiazepines** that bind to the α_2-subunit and enhance the interaction of the receptor with GABA. It has been speculated that these drugs mimic the effect of an endogenous ligand that remains elusive, however (the pacifying effect of mother's milk, for instance, has been suggested to be due to such a factor). Diazepam and valium, the prototypes of fear-relieving benzodiazepines or tranquilizers, rank among the most successful medical drugs. However, the hope that they are devoid of addictive effects has not been satisfied. In addition, in several cases synergistic effects with alcohol and barbiturates resulted in fatal consequences.

GABA neurons not only inhibit the transmission of fear signals (**Figure 16.12**) but also dampen the excitability of neurons that are engaged in **memory fixation**. In fact, the forgetting of unnecessary information, being one of the essential conditions of successful learning, seems to be primarily under the control of GABA. As everybody knows, memory and learning capability are massively impaired in states of fear and other types of stress. Probably this is due to an activation of GABA neurons (perhaps resulting from a release of the elusive endogenous benzodiazepines?) aiming at a compensatory control of such situations. Vice versa, GABA antagonists have been found to increase learning capabilities.

In the course of learning and memory fixation, the conductivity of the synapses involved is changed. This phenomenon, called synaptic plasticity, is observed in both excitatory and inhibitory neurons and will be discussed in Section 16.4.1.

16.3.2 Metabotropic GABA$_B$ receptors

These receptors are separated into two subfamilies B1 and B2, which both are expressed in numerous splice variants and exposed in the presynaptic as well as in the postsynaptic membrane. A peculiarity of the GABA$_B$ receptors is that, owing to special structural parameters, they are active only as B1/B2 heterodimers (see also Section 5.2).

Although characterized by seven transmembrane domains, the GABA$_B$ receptors differ from the other G-protein-coupled receptors (GPCRs) to such an extent that they were classified as a separate subfamily that comprises, in addition, the metabotropic glutamate receptors, the T1R taste receptors, and the Ca^{2+} receptors. A characteristic structural feature of this receptor type is an extended extracellular domain that binds the ligand between two jaws like a Venus flytrap (**Figure 16.13**). The binding proteins of amino acids in the periplasm of Gram-negative bacteria have a

Figure 16.13 Metabotropic GABA$_B$ receptors Metabotropic GABA receptors heterodimerize by an interaction of the cytoplasmic C-terminal domains forming a "coiled coil" structure. The ligand is bound by extended extracellular structures reminiscent of a Venus flytrap. Quite a similar structure is exhibited by the metabotropic glutamate receptors (see Figure 16.15). For comparison, a conventional G-protein-coupled receptor (GPCR) of the rhodopsin family is shown. In contrast to the GABA$_B$ receptor, this receptor binds its ligand in the pocket formed by the seven transmembrane helices.

similar shape and it is assumed that the extracellular domains of the $GABA_B$ receptors and their relatives evolved by a fusion of these prokaryotic precursor proteins with the heptahelical structure of GPCRs (see also Section 3.6.7).

As discussed in Section 5.2, conventional GPCRs of the rhodopsin family bind their ligand in the central hollow of the rosette formed by the seven transmembrane domains. This results in deformation of the transmembrane structure, leading to activation of the associated G-protein. For $GABA_B$ receptors and their relatives with the ligand-binding site positioned outside of the central hollow such a mechanism has to be ruled out, of course. Here the transmembrane signal is transduced by a mutual shift of the two monomers. Therefore, these receptors require dimerization.

Since $GABA_B$ receptors are coupled exclusively with $G_{i,0}$-proteins, they primarily inhibit synaptic signaling. Nevertheless, they may also stimulate nervous transmission if they are exposed in the presynaptic membrane of *inhibitory* neurons, thus dampening neurotransmitter release by negative feedback. The principle of such an inhibition of an inhibition resembling the arithmetical instruction "minus times minus is equal to plus" is depicted in Figure 16.6. Moreover, depending on the cell type and the receptor isoform, an activation of phospholipase Cβ signaling by the Gβγ-subunit of the $G_{i,0}$-protein may lead to a release of diacylglycerol (DAG) and inositol 1,4,5-trisphosphate ($InsP_3$)/Ca^{2+}. This mostly stimulatory pathway plays a role in long-term potentiation of ionotropic $GABA_A$ synapses (see Section 16.4.1).

16.3.3 Glycine receptors

For the neurotransmitter glycine, only ionotropic receptors have been found. Although their pentameric structure (**Figure 16.14**) resembles that of $GABA_A$ receptors, it is composed of α- or α- and β-subunits only. These subunits exist in various isotypes. Thus, several isoforms of Gly receptors differing in their subunit composition are expressed, depending on the developmental stage and the type of nervous tissue. Gly receptors have two ligand-binding sites. Drugs blocking these sites are highly toxic. An example is provided by strychnine, the alkaloid of the nux vomica plant (*Strychnos*) that is used as rat poison. A hereditary disease linked with defects of glycine neurotransmission is hyperexplexia, also known as "**stiff baby syndrome**." This motorneuronal disorder is characterized by an exaggerated startle reflex of newborns.

Depending on the intracellular chloride concentration in neurons, glycine, like GABA, may function as either an inhibitory or an excitatory transmitter. A particularly high chloride level is found, for instance, in the cells of embryonic nervous tissue. Here an activation of the Gly receptor ion channel leads to an *efflux* of chloride ions resulting in depolarization, whereas in adult neurons the chloride level is much lower so that the hyperpolarizing effect of glycine predominates. Moreover, glycine is a co-activator of excitatory ionotropic glutamate receptors of the NMDA type (see Section 16.4). At this opportunity, the reader may be reminded once more of the principle that signals do not have a meaning per se but trigger preformed responses of the receiver. This principle is illustrated in a particularly clear manner by glycine.

Figure 16.14 Structure of a heteropentameric glycine receptor The receptor is shown from above without ligand. Note the central chloride ion channel. Based on X-ray crystallography. (Adapted from Dutertre S, Becker CM and Betz H [2012] *J Biol Chem* 287: 40216–40223.)

Summary

- The neurotransmitters γ-aminobutyric acid (GABA) and glycine interact with anion channel-coupled ionotropic receptors. Their major effect consists of a dampening of synaptic signal transmission. They play an important role in learning processes and memory fixation.

- Ionotropic $GABA_A$ receptors are activated by various drugs including alcohol, barbiturates and benzodiazepines, while the glycine receptor is blocked by strychnine.

- For GABA, but not for glycine, metabotropic receptors coupling with $G_{i,0}$ proteins are also known. These $GABA_B$ receptors are heterodimers that differ from most other G-protein-coupled receptors by an extracellular "Venus flytrap" structure for ligand binding, which is assumed to be of prokaryotic origin.

16.4 Glutamate receptors are favorites of molecular brain research

In the brain, the excitatory neurotransmitters glutamate and aspartate are the major opponents of GABA and glycine. Like GABA, they interact with ionotropic as well as with metabotropic receptors (**Figure 16.15**). In mammals the ionotropic glutamate receptors are encoded by 18 genes giving rise to a large number of isoforms and splice variants. They are grouped in three families according to their exogenous ligands: **NMDA receptors** (named for the synthetic agonist N-Methyl-D-Aspartate), **AMPA receptors** (named for α-Amino-3-hydroxy-5-Methyl-4-isoxazole-Propionic Acid, another artificial agonist), and **kainate receptors** (kainic acid is a neurotoxin of a sea plant with a glutamate-like structure). NMDA receptors are expressed in seven isoforms; AMPA receptors, in four isoforms; and kainate receptors, in five isoforms (the residual two genes encode individual channel subunits).

The ionotropic glutamate receptors are a separate family that, being unrelated to other neurotransmitter receptors, is traced back to the bacterial glutamate receptor GluR0. As explained in Section 3.6.7, this receptor is thought to have evolved by a fusion of a periplasmic glutamate-binding protein with a prokaryotic K^+ channel of the type KcsA integrated upside-down into the membrane. This also is the basic structure of the mammalian glutamate receptor ion channels; however, these subunits contain an additional carboxy (C)-terminal transmembrane helix (**Figure 16.16**). The ion channel is composed of four subunits, mostly different isoforms, with one ligand binding site each.

Figure 16.15 Ionotropic and metabotropic glutamate receptors For the metabotropic receptors, the number of isoforms is given in parentheses. Details are found in the text. NMDA, N-methyl-D-aspartate; AMPA, α-amino-3-hydroxy-5-methyl-4-isoxazolepropionic acid; R, receptor.

Figure 16.16 Glutamate receptors and their bacterial precursors
(A) Shown in a schematic form are the subunits of the K$^+$ channel KcsA of *Streptomyces lividans*; the bacterial glutamate receptor GluR0 (a fusion product of a periplasmic glutamate-binding protein with a KcsA-like potassium channel integrated upside-down into the membrane); and the mammalian ionotropic glutamate receptor, exhibiting an additional transmembrane helix M4. Each ion channel consists of four subunits. In addition, the heptahelical structure of the dimeric metabotropic glutamate receptors is depicted. A characteristic feature that ionotropic glutamate receptors share with the metabotropic glutamate and GABA receptors is the Venus flytrap structure (teal) derived from prokaryotic precursor proteins, such as GluR0. The extended amino-terminal domain (ATD) of the ionotropic receptors is important for receptor oligomerization. The teal balls represent the ligands. The P-loops lend some cation specificity to the channels (see text). (B) Structure of an ionotropic glutamate receptor (iGluR) compared with a Cys-loop receptor (nicotinic acetylcholine receptor nAChR). Based on X-ray crystallography. (Adapted from Yelshanskaya MV, Li M and Sobolevsky AI [2014] *Science* 345:1070–1074.)

Via their cytoplasmic domains, ionotropic glutamate receptors interact with a large number of intracellular proteins to form extended signal-transducing complexes. The functions of these accessory proteins are still far from being understood in all details.

The ionotropic glutamate receptors exclusively function as depolarizing cation channels. AMPA and kainate receptors primarily transport Na$^+$ and K$^+$. Owing to a distinct amino acid composition of the ion pore, AMPA receptors are impermeable for Ca^{2+}. In animal experiments, mutations causing Ca^{2+} permeability result in epilepsy and premature death. Analogous defects may be involved in amyotrophic lateral sclerosis, a progressive degeneration of motor neurons ending in fatal muscle weakness. In contrast to AMPA and kainate receptors NMDA receptors exhibit a strong preference for Ca^{2+} ions. An additional peculiarity of NMDA receptors is that, to become fully active, they require membrane depolarization and glycine or D-serine as a second ligand (D-serine is derived from enzymatic isomerization of L-serine). Membrane depolarization is required to remove Mg^{2+} ions that plug the NMDA receptor ion channel. This is achieved by the gating of AMPA receptors. Thus, for excitation of the postsynaptic cell, both receptor types share the job: AMPA receptors take care of depolarization and NMDA receptors are responsible for Ca^{2+} influx. We shall see below that, because of this teamwork, this receptor pair is especially suited to translate neuronal signals into long-lasting changes of synaptic conductivity experienced as short- and long-term memory.

Kainate receptors are found in both pre- and postsynaptic membranes. Although they share many properties with AMPA receptors, their contribution to synaptic firing is much smaller. They probably have modulatory functions: by causing partial but long-lasting (also called tonic) depolarization, they facilitate neurotransmitter release by the presynaptic neuron and promote AMPA receptor-dependent depolarization in the postsynaptic cell. In distinct brain regions these effects have a significant influence on synaptic plasticity.

Like GABA receptors, ionotropic glutamate receptors are targets of many drugs and toxins. Examples are provided by phencyclidine, a widely used hallucinogenic drug ("angel dust"), and ketamine, an anesthetic in veterinary medicine, which both block NMDA receptors. The excitatory functions of glutamate receptors are also inhibited by ethyl alcohol and barbiturates. By this means, the stimulatory effects of the same drugs on opposing inhibitory GABA receptors are augmented.

In addition to the receptor ion channels, at least eight different **metabotropic glutamate receptors** (mGlu) have been identified, six of which couple with $G_{i,0}$-proteins and two with $G_{q,11}$-proteins. The $G_{q,11}$-coupled receptors are found primarily in postsynaptic membranes. By stimulating the release of DAG and InsP$_3$, they activate Ca^{2+}-dependent proteins such as protein kinase C and calmodulin-activated kinases and, thus, complete and amplify the Ca^{2+}-mediated effects of NMDA receptor gating. In contrast, the $G_{i,0}$-coupled receptors are concentrated in presynaptic membranes and glial cells. They suppress cAMP formation, inhibit voltage-gated Ca^{2+} channels, and activate GIRK-type potassium channels and thus control glutamatergic signal transduction by negative feedback. Structurally, the metabotropic glutamate receptors are closely related to the metabotropic GABA receptors, exhibiting a Venus flytrap structure for ligand binding. It appears as though, for the individual receptor types, this structure is derived from different prokaryotic precursor proteins.

16.4.1 Data storage by synaptic plasticity

Glutamate receptors are a preferred object of neurobiological research since they are involved directly in learning and memory fixation. According to prevailing theories, learning is associated with the generation of new and the reinforcement of existing synaptic contacts (signaling becomes strengthened in frequently used neurons) as well as with the production of new neurons. The same holds true for the reverse situation: synaptic contacts used less frequently are shut down or eliminated by ubiquitin-controlled protein degradation, in extreme cases even through apoptosis of the postsynaptic cell. Both effects are realized as remembering and forgetting. In other words, like contacts between proteins, the interconnections between neurons are not irreversibly fixed but flexible and adaptable, changing in response to the actual conditions. Owing to this plasticity, synapses are able to adapt to endogenous and exogenous stimuli by long-lasting alterations of conductivity. This holds true for both presynaptic functions (in particular transmitter release) and the signal-transducing network, enabling the postsynaptic cell to adjust the efficacy of its metabolic and genetic apparatus to the input signals.

Long-term potentiation and depression: a matter of protein synthesis

In brain preparations, synaptic plasticity can be measured as **long-term potentiation** and **long-term depression**. These are defined as an activity-dependent increase or decrease of synaptic conductivity that arises very rapidly and may last for many hours, in the case of long-term memory even lifelong. In the beginning, these effects are associated with post-translational modifications of signal-processing proteins induced along signaling cascades originating from neurotransmitter receptors. These modifications are an equivalent of short-term memory. The final establishment of long-term potentiation or depression, thought to resemble long-term memory fixation, occurs after a few hours and depends on de novo synthesis of synaptic proteins, requiring a modulation of both mRNA translation and gene expression. Finally, once fixed, long-term memory is independent of protein synthesis, indicating that the changes of synaptic conductivity, once established, remain stable.

Among the many transcription factors controlling genes involved in memory fixation, the cAMP response element-binding protein **CREB** (called "the master of memory genes") plays an outstanding role. As explained in Section 8.6, its activity is controlled by various protein kinases and phosphatases acting downstream of neurotransmitter receptors. However, considering the logistic problem created by the large distances between the synapses and the nucleus, neurons cannot rely solely on transcription but must also use mRNA translation to control long-term potentiation or depression. To this end, synapses contain a stock of ribosomes and more than 100 species of stable mRNA enabling a local protein synthesis on demand, which has been shown to occur even in dendrites surgically dissected from the cell body. Such local protein synthesis allows for selective and specific modification of individual synapses for the purpose of rapid adaptation to incoming stimuli.

The role of translation factors and microRNAs

Master switches controlling synaptic protein synthesis are two translation factors, the initiation factor **eIF2** and the elongation factor **eEF2**. Both are critically involved in learning and memory fixation. As we have seen in Section 9.3, these factors are inactivated by protein phosphorylation catalyzed by specific eIF2- and eEF2- kinases. As a result, general mRNA translation is suppressed, whereas the synthesis of certain proteins is stimulated due to a special structure of the cognate mRNA species. One of these proteins selectively produced upon eIF2 phosphorylation is the transcription factor **ATF4** (see Section 9.3.2). It is key regulator of stress-sensitive genes, but at the same time it acts as an antagonist of CREB-dependent gene transcription and suppresses memory fixation. In brain neurons the induction of long-term potentiation or depression correlates with a dephosphorylation of eIF2, resulting in an increase of overall translation accompanied by a suppression of ATF4 synthesis. Consequently, CREB is relieved from inhibition. Experiments with gene-manipulated animals and pharmacological inhibitors demonstrate a causal relationship between the eIF2–ATF4 complex and learning. In neurons eIF2 phosphorylation is controlled by the eIF2 kinases PKR, GCN2, and PERK (see Section 9.3.1), and by phosphoprotein phosphatase 1C equipped with specific regulatory subunits (see Section 2.6.4). However, the most important question, along which signaling pathways the enzymes of eIF2 phosphorylation are regulated in neurons, remains open. More is known about the regulation of eEF2 phosphorylation since the eEF2 kinase is activated by calcium influx as well as by cAMP-controlled phosphorylation. As we have seen, both are pathways typically controlled by neurotransmitters.

Another potent mechanism of translational control is the suppression by **microRNA** of mRNA translation (RNA interference, see Section 9.5). MicroRNAs (miRNAs) are strongly expressed in neurons and are concentrated—together with Dicer and Argonaut proteins—in postsynaptic densities. In fact, miRNAs seem to be critically involved in brain development and the regulation of synaptic plasticity. Thus, in experimental animals the expression profiles of selected miRNAs depend on the state of development, and a genetic knockout of such miRNA species leads to severe neurodegenerative disorders. A major neuronal pathway to control the formation of individual miRNA species is via CREB signaling. Upon stimulation of this signaling pathway the turnover rate of miRNAs increases dramatically. Vice versa, CREB synthesis is under the control of specific miRNA species.

The role of scaffold proteins

The proteins modified and produced in the course of memory fixation assemble in the **postsynaptic densities**, which (as explained in Section 16.1) are characteristic of permanently active synapses. These aggregates are visible under the electron microscope (see Figure 16.2). They are composed of signal-transducing proteins, including neurotransmitter receptors and various protein kinases, and are organized and anchored to the cytoskeleton and the cell membrane by adaptor and scaffold proteins.

As an example of scaffold proteins, the **shank proteins** may be mentioned. Through their SH3 domains, ankyrin repeats, PDZ domains, and proline-rich motifs, they interact with a large number of other proteins, thus connecting various signal transducing pathways. Moreover, by means of SAM domains (referring to a Sterile Alpha-helical Motif of yeast), they polymerize into helical fibers that interact with each other to form large sheets beneath the synaptic contact zone. These interactions are promoted by zinc ions binding to the SAM domains. Since in many neurons presynaptic vesicles have been found to contain both neurotransmitter (particularly glutamate) and Zn^{2+}, it has been speculated that Zn^{2+} may enter the postsynaptic cell through ionotropic neurotransmitter receptors and induce the organization of the shank lattice as the structural backbone of the postsynaptic density. Whether such an effect has anything to do with the above-mentioned ionotropic zinc receptors is not known.

Another scaffold protein found in postsynaptic densities is **PICK1** (Protein Interacting with C-Kinase 1). Via PDZ and other interaction domains, it organizes large signal-transducing complexes that may contain glutamate, dopamine, acetylcholine, GABA and serotonin receptors, regulators of G-protein signaling (that control the GTPase activity of G-proteins coupled with metabotropic neurotransmitter receptors), and the SNARE complex protein SNAP25, as well as an atypical PKC isoform. Currently, PICK1 is gaining much interest because genetic defects of this protein strongly correlate with neurological disorders such as schizophrenia, Parkinson's disease, epilepsy, chronic pain, and drug abuse.

The role of glutamate receptors

As we have seen, learning and memory fixation correlate with specific changes of the postsynaptic proteome due to post-translational modifications and effects on mRNA translation and gene transcription. These lead to remodeling of synaptic contacts, measured as long-term potentiation and depression. The question arises how a simple electrochemical nerve impulse is interpreted by the intracellular network of signal processing in such a way that it can evoke such far-reaching changes.

Neuronal signals are transmitted by waves of depolarization and by neurotransmitters released in small packages or quanta. Such signals are frequency-modulated. Therefore, neurons must possess proteins able to read the information encoded in the frequency of the input signal and to transform this information into output signals that are perceivable for the data-processing protein network, thus changing synaptic plasticity. This condition is fulfilled perfectly by **NMDA receptors** because they are able to read frequency-modulated membrane potentials. To this end NMDA receptors have to cooperate with **AMPA receptors**, which transform the neurotransmitter signal received from the presynaptic neuron into postsynaptic voltage pulses that the NMDA receptors then translate into pulses of Ca^{2+} influx. These pulses are read by intracellular frequency detectors such as **calmodulin kinase II** (CaMKII), an outstanding monitor of oscillating calcium signals that controls the function of numerous synaptic proteins including the receptors themselves (see Section 14.6.3).

How do NMDA receptors read the voltage signals emitted from AMPA receptors? The answer is found in the special type of voltage sensor with which the receptor is equipped. This sensor is a channel-blocking Mg^{2+} ion rather than a special transmembrane domain as in other voltage-controlled ion channels. To enable ligand-induced channel gating, this Mg^{2+} ion has to be removed by depolarization. The higher the frequency of the voltage pulses emitted by the AMPA receptor, the more NMDA

presynaptic
neuron

AMPA-R

postsynaptic
neuron

depolarization

Glu

Gly

Ca^{2+}

Ca^{2+} controlled
proteins:
CaM kinases, PKC,
calcineurin,
NO synthase

NMDA-R

cGMP

Mg^{2+}

NO

retrograde amplification

receptors are activated. Consequently, NMDA receptor activity critically depends on the frequency with which the synapse is used. However, as we have seen, voltage alone is not sufficient to gate NMDA receptors; it must cooperate with two ligands, glutamate and glycine or D-serine. In other words, NMDA receptors are equipped with three safety locks; they represent a special type of logical AND gates.

As already mentioned, NMDA receptors predominantly allow the passage of **Ca^{2+} ions**. These are the key regulators of synaptic plasticity. For long-term potentiation, they strengthen the synaptic contact by positive feedback and render it long-lasting by activating a series of Ca^{2+}-dependent signaling proteins such as the above-mentioned CaMKII and other Ca^{2+}-activated enzymes. One of these is the neuronal NO synthase nNOS. It catalyzes the formation of nitric oxide, which functions as a retrograde messenger, stimulating the presynaptic release of glutamate (**Figure 16.17**; for more details see Section 16.5.1). A major task of Ca^{2+}-dependent protein kinases such as CaMKII is to increase the activity and promote de novo synthesis and membrane localization of AMPA and NMDA receptors. While phosphorylation promotes receptor expression required for long-term potentiation, dephosphorylation has the opposite effect, causing long-term depression. In this system a major opponent of kinases is the protein phosphatase calcineurin. Since this enzyme is Ca^{2+}-activated, Ca^{2+} is a signal for both long-term potentiation and long-term depression. Apparently, it is a matter of NMDA receptor efficacy which of the two responses gets a chance: a highly active receptor evokes a strong Ca^{2+} influx that causes long-term potentiation via CaMKII, whereas the weak Ca^{2+} signal induced by a less active receptor activates only calcineurin, thus causing long-term depression. In fact, the number and subunit composition of NMDA and particularly AMPA receptors depend on the frequency with which the synapse is used and is determined by the steady-state between up- and down-regulation of the receptors. This equilibrium is highly dynamic, ensuring a permanent turnover of ionotropic receptors in the neuronal cell membrane. So the cell is able to adjust its receptor capacity very rapidly to the particular circumstances. To avoid overshooting, several negative feedback loops are built into the system. They control the Ca^{2+} influx through NMDA receptors; this influx is suppressed by dephosphorylation and N-nitrosylation (by NO) of the receptor.

Depending on the conditions, metabotropic glutamate receptors (as well as other neurotransmitter receptors) augment or dampen the processes leading to long-term potentiation or depression in that they activate G-protein-dependent signaling cascades. For instance, protein kinase C, stimulated along $G_{q,11}$- and $G_{i,0}$-controlled pathways

Figure 16.17 Mode of NMDA receptor activation Shown in schematic form is a central nervous glutamate-operated synapse. For gating of the postsynaptic NMDA receptor cation channel, three signals have to cooperate: these are transmitted by the two ligands glutamate (Glu) and glycine (Gly) and by membrane depolarization caused by glutamate-activated AMPA receptor ion channels. The depolarization removes Mg^{2+} that blocks the NMDA channel. The NMDA receptor preferentially transports Ca^{2+} ions, thus modulating the functions of postsynaptic Ca^{2+}-dependent proteins. One response is the production of nitric oxide (NO), which exhibits a dual effect as an inhibitor of NMDA receptor activity and a retrograde amplifier of presynaptic transmitter release. The latter reaction is mediated by cyclic GMP (Section 4.6.2).

(see Section 4.6.3), facilitates the incorporation of NMDA receptors from secretory vesicles into the plasma membrane but also promotes the formation of endocannabinoids that, as retrograde messengers, diffuse from the post- to the presynaptic cell and inhibit presynaptic glutamate release by activating $G_{i,0}$-coupled receptors (see Section 16.7). Vice versa, NO synthase, CaMKs, and calcineurin are stimulated along the $G_{q,11}$–phospholipase C–InsP$_3$–Ca^{2+} pathway (see Section 14.6.4).

The molecular mechanisms of long-term potentiation and depression are not restricted to glutamatergic neurons but hold true also for synapses that use other neurotransmitters. The important role played by GABAergic neurons in synaptic plasticity has been mentioned already. Studies that are more detailed have shown that here again the increase of the cytoplasmic Ca^{2+} level in the postsynaptic cell provides a key signal. This signal may be generated by GABA$_A$ receptors, provided these exhibit a depolarizing effect, thus facilitating the gating of voltage-dependent Ca^{2+} channels. However, even in hyperpolarizing GABA$_A$ neurons, a Ca^{2+} signal may arise, in this case through metabotropic receptors (of GABA or of other neurotransmitters) or NMDA receptors activated simultaneously. The downstream effects of Ca^{2+} resemble those discussed for glutamate receptors.

For the manipulation of G-protein-coupled metabotropic receptors in new optogenetic technologies, see **Sidebar 16.2**.

Sidebar 16.2 Optogenetics: controlling neuronal signaling with light A novel method that is going to revolutionize neuroscience is optogenetics, a technique that combines methods of optics with those of genetics. By means of this approach the dream of cell biologists to follow up signaling events in a single cell or even in a subcellular compartment such as a synapse in real time has come a step closer to fulfilment.

Optogenetics uses light as a non-invasive input signal. The cellular targets of such light signals are photosensory proteins. As described in Section 3.4.5, most of these are of prokaryotic or non-vertebrate origin. Nevertheless, they may remain active when expressed in mammalian cells by means of gene transfection. This offers the possibility of using the unique properties of such proteins for manipulating signaling processes with a hitherto unequalled high degree of temporal and spatial precision. Moreover, the properties of photosensory proteins can be specifically altered by genetic and chemical engineering, so that a large optogenetics toolbox has become available.

Prominent examples of such unique tools are light-gated ion channels and ion transporters such as channelrhodopsins of green algae and archaebacterial rhodopsins. **Channelrhodopsins** (see Section 3.4.5) are non-specific cation channels (although exhibiting the structure of a heptahelical receptor). When expressed in neurons they mostly cause depolarization in a millisecond scale upon illumination with blue-green light. By genetic engineering such as targeted gene mutation they can be transformed into light-gated anion channels inducing hyperpolarization. Other hyperpolarizing optogenetic probes are **halorho-**dopsins, the light-sensitive chloride pumps of halobacteria (see Section 3.4.5). Individual channelrhodopsins and halorhodopsins are activated by light of a specific wavelength, for instance depolarizing channelrhodopsins by blue light and hyperpolarizing halorhodopsins by orange light. Thus, the activity of a neuron expressing both rhodopsin types can be stimulated or depressed *ad libitum* by changing the color of the light.

When light-sensitive domains of photosensory proteins are fused with signal-transducing domains of other proteins, optogenetic tools become available, the applicability of which is not only for neuroscience but also for studies of cellular signaling in general. Examples are provided by chimeric receptors (**opto-XRs**) that respond to light rather than to their cognate ligands. They are obtained by replacing the extracellular ligand binding site of G-protein-coupled metabotropic receptors by the light-sensitive extracellular domain of mammalian rhodopsins. Illumination of such constructs results in an activation of G-protein-dependent signaling. More examples of the applicability of optogenetics in and beyond neurosciences are found in Section 5.2.1, Sidebar 5.1.

Optogenetic techniques are not restricted to single cells but are also applicable for intact tissues and living animals. Thus, individual collectives of neurons can be precisely stimulated or inhibited *in vivo*. This opens promising future aspects for medicine. For instance, in gene-manipulated experimental animals expressing channel- and halorhodopsins in discrete neurons, pathological processes such as epileptic seizures can be evoked or dampened by illuminating the corresponding brain area with either blue or orange light using minimally invasive glass-fiber optics.

Figure 16.18 Activation of anti-apoptotic signaling by the NMDA receptor Ca^{2+} passing the cell membrane through the NMDA receptor cation channel (NMDA-R) stimulates signaling cascades that lead to inactivation of pro-apoptotic proteins (Bad, caspase 8) and to expression of genes encoding anti-apoptotic factors (Bcl2, IAP, neurotrophins). Details are found in the text. GRF, guanine nucleotide releasing factor; CaMKK, calmodulin kinase kinase; CaMKIV, calmodulin kinase IV; PI3K, phosphatidylinositol 3-kinase; PKB, protein kinase B; CREB, cAMP response element binding protein; CBP, CREB binding protein; IAP, inhibitors of apoptotic proteins.

16.4.2 A decision between life and death of brain neurons

As we have seen, NMDA receptors are protected against accidental failure. This is for good reason, since hyperactivation would result in a flooding of neurons by Ca^{2+} ions, causing cell death. Indeed, such a fatal event is a characteristic feature of stroke, brain injury, and neurodegenerative disorders such as Alzheimer's and Huntington's diseases. A permanent inactivation of glutamate receptors is, on the other hand, just as fatal because nonstimulated brain neurons also are doomed to die.

How do glutamate receptors ensure neuronal survival? Neurons belong to those cell types that automatically undergo apoptosis when survival factors such as neurotrophic growth factors are withheld; these are essential, in particular, during embryonic development, and neurotransmitters like glutamate are required throughout life. The apoptotic response is prevented by Ca^{2+} in lower concentrations entering the cytoplasm through NMDA receptors or other membrane channels (**Figure 16.18**). The signaling cascades that mediate this survival effect include the Ras–PI3K–PKB pathway, which leads to inactivation of the pro-apoptotic protein Bad (see Section 4.6.6). Here Ras is activated by the Ca^{2+}-dependent guanine nucleotide releasing factor GRF that is expressed preferentially in neurons (see Section 10.1.3). PKB is stimulated, in addition, by Ca^{2+}-dependent calmodulin kinase kinase (CaMKK; see Section 14.6.3).

Another survival pathway leads to the transcription of genes encoding anti-apoptotic proteins such as Bcl2, Bcl-XL, and the caspase inhibitor IAP (see Section 13.2.2) as well as neurotrophic growth factors. One of the transcription factors required for this response is CREB, which (like its co-activator CBP/p300) is activated by phosphorylation catalyzed by CaMKIV, a major transducer of gene-regulatory Ca^{2+} signals (more details about calmodulin kinases are found in Section 14.6.3).

While the survival of excitatory glutamate neurons depends on a continuous normal activation of the glutamate receptors, hyperactivation causes cell death. Such a neurotoxic response is triggered by deficiencies of blood flow (ischemia) and oxygen supply (hypoxia) as well as by injury and bleeding. An infamous clinical outcome of these disorders is stroke. The key mediator is again Ca^{2+}, which at higher concentrations induces apoptosis rather than promoting cell survival (see Section 14.6.5). Some of the relevant signaling pathways are shown in **Figure 16.19**.

Owing to positive feedback, these processes build up in a catastrophic manner. A particularly fatal role is played by excessively produced nitric oxide that, acting as a retrograde signal, potentiates glutamate release and, in high concentrations, exhibits a strong cytotoxic effect of its own by irreversibly damaging proteins and nucleic acids. How this occurs is explained in Section 16.5.

Figure 16.19 Neuronal cell death due to hyperactivation of NMDA receptors: a molecular mechanism of stroke Ischemia and hypoxia (as well as injury) result in an increased firing of a presynaptic glutamatergic neuron (left) resulting in a hyperactivation of NMDA-R and a flooding of the postsynaptic neuron by Ca^{2+} ions (right). As a result, metabolic pathways (here exemplified by the arachidonic acid metabolism triggered by Ca^{2+}-dependent phospholipase A_2) are running out of control, generating (as byproducts) reactive oxygen species such as superoxide anion radicals. The latter react with nitric oxide, NO (produced by Ca^{2+}-dependent neuronal NO synthase, nNOS), to produce the highly cytotoxic peroxynitrite. In addition, Ca^{2+} causes cell death by damaging mitochondria and activating proteases (calpain) and endonucleases. The process is amplified by positive feedback with NO that, as a retrograde signal, via a release of cGMP augments glutamate release.

Is neuronal cell death an irreversible fate? For about 100 years the concept of the adult brain as incapable of regeneration was upheld dogmatically. At present, however, a steadily increasing body of experimental evidence indicates the existence of neuronal stem cells in the brain, which do proliferate depending on the environmental demands or in response to injury. This discovery of what is called "adult neurogenesis" has revolutionized our ideas of the brain's capabilities and has added a new aspect to the concept of neuronal plasticity, which now is considered to rest on three pillars: modulation of the strength of synaptic contacts, of the number of these contacts, and of the production of new neurons.

Summary

- Glutamate and aspartate are the major excitatory neurotransmitters in brain. Their ionotropic receptors are coupled with semispecific cation channels, with the NMDA receptor exhibiting some preference for Ca^{2+}. The metabotropic glutamate receptors are either $G_{i,0}$- or $G_{q,11}$-coupled. Both ionotropic and metabotropic receptors are characterized by an extracellular Venus flytrap structure for ligand binding.

- Glutamate receptors are critically involved in learning and memory fixation.

- Learning and memory fixation depend on synaptic protein synthesis leading to a strengthening or weakening of synaptic contacts (synaptic plasticity). Synaptic protein synthesis is mainly controlled at the level of mRNA translation.

- The expression and activity of the ionotropic glutamate receptors is regulated by Ca^{2+} influx as well as by signals interacting with metabotropic glutamate and other receptors.

- The continuous use of glutamate receptors keeps neurons alive by inhibiting apoptosis. However, overstimulation of these receptors causes a pathological increase of the Ca^{2+} concentration in the postsynaptic neuron leading to cell death, an effect that is amplified through nitric oxide. This signal-induced neuronal degeneration is triggered by ischemia and hypoxia and plays a key role in stroke.

16.5 Nitric oxide: a Janus-faced signal molecule*

As we have seen, NO is an important auto- and paracrine signaling molecule in the brain. Its biological significance was discovered, however, by studies on parasympathetic blood pressure regulation. For this work, the Nobel Prize in Physiology or Medicine for 1998 was awarded to Robert F. Furchgott, Louis J. Ignarro, and Ferid Murad.

The key event of peripheral blood pressure regulation is a relaxation of vascular smooth muscle cells induced by acetylcholine and mediated by the $G_{q,11}$-coupled muscarinic M1 receptor, which activates the phospholipase $C\beta$–$InsP_3$-Ca^{2+} pathway (see Section 5.4). This seems to be in contradiction to the fact that Ca^{2+} triggers muscle contraction rather than relaxation. However, vascular smooth muscle cells do not express M1 receptors and thus cannot respond directly to acetylcholine. Instead, M1 receptors are found in the adjacent endothelial cells where the Ca^{2+} released by $InsP_3$ stimulates the synthesis of an **endothelium-derived relaxing factor (EDRF)** that, as a paracrine signaling molecule, triggers relaxation of the surrounding vascular smooth muscle cells. EDRF has turned out to be identical with the gas nitric oxide (**Figure 16.20**). This discovery was quite a surprise since at that time (in the 1980s) only one gaseous signaling molecule was known, namely, the plant hormone ethylene. The identification of EDRF as NO offered an elegant explanation of the therapeutic effects of drugs like nitroglycerol used against a constriction of coronary vessels (angina pectoris): the metabolization of such compounds results in the formation of NO.

Today it is generally accepted that NO production is not a specialty of endothelial and neuronal cells. Instead, all tissues possess the corresponding enzymatic apparatus. In other words, NO is a universal intercellular messenger. NO arises from oxidation of the guanidino group of arginine (**Figure 16.21**). This reaction requires nicotinamide adenine dinucleotide phosphate (NADPH) and molecular oxygen and is catalyzed by specific oxygenases, the NO synthases (NOS), with flavin mononucleotide, flavin adenine dinucleotide, and tetrahydrobiopterin as redox cofactors.

---

* Janus, the Roman god of doors and gates, was depicted with two heads looking in opposite directions. The term "Janus-faced" is generally used to indicate a conflicting nature. See also Janus kinase (see Section 7.2.1).

Figure 16.20 Nitric oxide: the endothelium-derived relaxing factor (EDRF) In endothelial cells activated by the parasympathetic transmitter acetylcholine, Ca^{2+} stimulates the endothelial NO synthase eNOS. The NO thus generated diffuses into the surrounding vascular smooth muscle cells and stimulates the soluble guanylate cyclase. In the next step, cGMP-dependent protein kinase causes a relaxation of the muscle cell by inactivating the inositol 1,4,5-trisphosphate ($InsP_3$) receptor / Ca^{2+} channel of the endoplasmic reticulum and activating the myosin light chain (MLC) phosphatase. As a result, the blood pressure falls. In both cell types, bacterial endotoxin (interacting with the Toll-like receptor TLR4, see Section 6.3.2) induces the de novo synthesis of the inducible NOS isoform iNOS. This highly active enzyme dramatically amplifies the NO–cGMP signal, thus causing a catastrophic fall of blood pressure known as endotoxin shock (see Section 6.3.2). The release of cytokines additionally induced by endotoxin strengthens this fatal response.

Figure 16.21 Biosynthesis of nitric oxide NO is produced by a two-step oxidation of arginine (above). The enzymes catalyzing this reaction are shown in the box. Nitric oxide synthase (NOS) isoforms are as follows: i, inducible; c, constitutively expressed; n, neuronal; e, endothelial.

Various NOS isoforms are known, which are grouped into three families. The oxygenases of the NOS1 and NOS3 families are constitutively expressed housekeeping enzymes and are collectively called constitutive or cNOS. NOS1 is also known as neuronal or nNOS because it is particularly abundant in the central nervous system, while NOS3 has been called endothelial or eNOS because it was originally identified as the EDRF-producing enzyme. NOS1 and NOS3 need Ca^{2+} and the calcium-binding protein calmodulin, and are activated by any signal causing an elevation of the cytoplasmic Ca^{2+} level (such as acetylcholine in endothelial cells). The enzymes of the NOS2 family, in contrast, are Ca^{2+}-independent and become expressed in most tissues only upon demand, particularly in emergency and stress situations like infections, injury, and inflammation that require an increased output of NO. Thus they are called inducible or iNOS. Inducers of iNOS are pro-inflammatory cytokines and bacterial substances, in particular the endotoxic lipopolysaccharides. iNOS enzymes are much more active than cNOS. They produce NO in cytotoxic doses, mainly for defense purposes. However, excessive NO synthesis, such as that caused by endotoxin, may trigger a catastrophic fall of blood pressure, cumulating in shock and death.

At body temperature NO is a gas that easily penetrates lipid membranes, "soaking" the surrounding tissue cells and acting both as a signaling molecule and as a defensive compound. However, owing to a metabolic lifetime of only a few seconds, its range is quite narrow unless it becomes stabilized by particular proteins. Thus, in most tissues NO acts as a *local mediator*. An illustrative example is the nervous system: here NO produced by nNOS acts both as a neurotransmitter and as a neuromodulator. Its effect as a retrograde messenger in the brain has been mentioned already. In the vegetative nervous system it is responsible for so-called Non-Adrenergic, Non-Cholinergic (NANC) effects. Examples of such effects are smooth muscle relaxation in distinct organs such as gastrointestinal tract, urethra, and penis (where it causes erection; see Section 4.4.2).

NO is both an endogenous messenger and a toxic gas. Therefore, its effects critically depend on its tissue concentration (**Figure 16.22**). At low concentrations, as generated by cNOS, NO functions as a signaling molecule that specifically interacts with a cytoplasmic receptor, **soluble guanylate cyclase** (sGC). This is a hemoprotein that binds the NO (which is a free radical) at the iron atom of the heme group (see Section 4.4.2 for more information). Upon activation by NO, the enzyme produces cGMP, which in turn stimulates protein kinases and ion channels and modulates

Figure 16.22 Concentration-dependent effects of NO At low and moderate concentrations, NO activates soluble guanylate cyclase (sGC) and upon oxidation nitrosylates proteins. At high concentrations, NO exhibits strong cytotoxic effects by its own and by combining with reactive oxygen species (here superoxide anion radical $O_2{}^{.-}$) to highly aggressive intermediates such as peroxynitrite. PKG, cGMP-dependent protein kinase.

the activity of cyclic nucleotide phosphodiesterases (see Sections 4.4.2, 4.6.2, and 14.5.7). Along these signaling pathways, for instance, synaptic plasticity and blood pressure are regulated. Moreover, NO thus ensures the blood supply of the working skeletal muscle, since Ca^{2+} released from the sarcoplasmic reticulum during contraction activates cNOS, and the NO thus produced induces a dilatation of the surrounding blood vessels.

In higher concentrations, NO still exhibits signaling effects: it acts as a post-translational modulator by nitrosylating SH groups of proteins (see also Section 2.2). For this nonenzymatic **S-nitrosylation**, NO is mostly oxidized to the chemically more active N_2O_3. S-nitrosylation is reversible and either inhibits or activates the functions of proteins carrying essential SH and S–S groups. Since S-nitrosylated proteins are more stable than NO, they may serve as transport vehicles, releasing NO at target sites.

As a signaling reaction, S-nitrosylation may be highly specific. A well-known example is provided by the NMDA receptor, which is inactivated by S-nitrosylation of a single SH group. Since, on the other hand, nNOS is stimulated by the Ca^{2+} influx through the NMDA receptor ion channel, S-nitrosylation probably is a means of preventing dangerous hyperactivation of the receptor (see Figure 16.17). This feedback control regulates itself by S-nitrosylation of cNOS, which also becomes inhibited. Thus, in physiological concentrations NO exhibits a neuroprotective effect. Such an action is supported by an inhibition of apoptosis because S-nitrosylation inactivates the initiator caspases 3 and 9 (see Section 13.2.1) as well as the pro-apoptotic protein kinase ASK1 (see Section 13.2) but activates Ras, a central relay station of survival pathways. Moreover, NO induces the expression of heme oxygenase 1, a neuroprotective enzyme (for more on the carbon monoxide generated in the course of oxidative hemoglobin degradation catalyzed by heme oxygenase, see **Sidebar 16.3**).

At even higher concentrations, the cytotoxic effects of NO dominate. An overproduction of NO is due to an induction of iNOS expression, mostly in the course of an inflammatory reaction. As **nitrosative stress** it may have fatal consequences, since as a free radical NO inactivates essential hemoproteins such as cytochrome oxidase (thus inhibiting the mitochondrial respiratory chain) and iron–sulfur proteins. In addition, NO is transformed into highly aggressive products upon reaction with reactive oxygen species. The latter are produced by **oxidative stress**, which is another typical feature of inflammatory reactions. Particularly aggressive is peroxynitrite, which arises from a reaction between NO and superoxide anion radicals. Such agents damage proteins and nucleic acids with the result that the cell eventually may be killed. In a kind of kamikaze attack, cells of innate defense, such as granulocytes and macrophages, make use of NO toxicity for chemical warfare against microorganisms. Since collateral damage of adjacent tissue cells is unavoidable, inflammation is generally accompanied by tissue necrosis. Inflammatory reactions are typical features of **neurodegenerative disorders** such as stroke, Alzheimer's dementia, and Huntington's and Parkinson's diseases. In fact, in the morbid brain areas iNOS is strongly expressed and the NO level is pathologically high. In parallel, an overexpression of enzymes provoking oxidative stress is seen. This holds true in particular for the prostaglandin H synthase cyclooxygenase 2 (COX-2), which like iNOS is induced in emergency

---

Sidebar 16.3 Carbon monoxide Carbon monoxide generated in the course of oxidative hemoglobin degradation catalyzed by heme oxygenase (see textbooks of biochemistry) is a para- and autocrine signaling molecule sharing some functions with NO.

The heme oxygenases exist in at least two subfamilies: HO-1 and HO-2. HO-2 is constitutively expressed, in particular in the liver, testes, blood vessels, and the central nervous system. HO-1 is, in contrast, an emergency enzyme expressed only in stress situations such as inflammation, heavy metal intoxication, oxygen deficiency, impaired blood supply, and heat. Thus, heme oxygenases resemble NO synthases and cyclooxygenases (see Section 4.4.5), which are also expressed both as constitutive housekeeping enzymes for everyday demands and as inducible emergency enzymes dealing with stress. Although it is a gas, CO (in contrast to NO) is not a free radical; therefore, CO is much more stable and less aggressive.

The toxic effect of CO does not go beyond a blockade of Fe^{2+}-containing hemoproteins such as hemoglobin, an effect that has no consequence considering the low physiological CO concentrations. Moreover, soluble guanylate cyclase (sGC), despite being a hemoprotein, is activated rather than inhibited by CO. Although the sGC-activating efficacy of CO amounts to only 5% of that of NO, it results in comparable effects such as lowering of blood pressure (due to a relaxation of vascular smooth muscles) and inhibition of platelet aggregation. Stress-induced CO production probably provides a protective mechanism against tissue damage as caused, for instance, by inflammation. Thus CO turns down the release of pro-inflammatory cytokines (such as tumor necrosis factor α and interleukin 1β) by leukocytes while at the same time inducing synthesis of the anti-inflammatory interleukin 10. Moreover, regulatory cross talk seems to exist between NO synthases and heme oxygenases, ensuring that the cytotoxic effects of NO do not go off course. In fact, NO and its aggressive metabolites belong to the most effective inducers of HO-1, whereas, vice versa, CO inhibits in many (but not all) tissues the activity and expression of iNOS. The role of CO in the brain remains a matter of debate.

---

situations (see Section 4.4.5). Thus, iNOS and COX-2 induction certainly are critical events in neurodegeneration, and considerable efforts are being made to develop drugs by which NO signaling and COX-2 activity can be manipulated. However, these approaches are hampered by the Janus-faced role of NO as a neuroprotective and a neurotoxic signal, and the difficulty of overcoming the blood–brain barrier. This does not devalue preventive measures that generally lower the pressure of nitrosative and (even more so) of oxidative stress. Such results may be achieved by both novel drugs such as antioxidants and COX-2 inhibitors and an adjustment of lifestyle.

Although the destructive role of nitrosative and oxidative stress is unquestioned, it is only one among various brain defects involved in neurodegenerative diseases. Others are a flooding of cells with calcium ions (resulting in apoptosis; see Sections 14.6.5 and 16.4.2), the generation of advanced glycation end products (see Section 7.3.3), and intra- and extracellular protein deposits such as β-amyloid and Tau aggregates (see Section 13.1.3). It is still not entirely clear how these processes interact with each other to bring about the overall pathological changes.

Summary

- NO gas is an abundant paracrine mediator with strictly local effects. It is produced by NO synthases that exist in two forms: constitutively expressed, Ca^{2+}-dependent enzymes and stress-induced, Ca^{2+}-independent enzymes.

- Depending on the concentration, NO functions either as a signaling molecule or as a cytotoxic agent involved in defense reactions and cell death.

- The signaling effects are due to activation of soluble guanylate cyclase, generating the second messenger cGMP, or to reversible post-translational modification of proteins by nitrosylation of sulfhydryl groups.

- Overstimulation of NO formation plays an infamous role in life-threatening pathological events such as stroke and endotoxin shock.

16.6 Receptors of purine and pyrimidine nucleotides: the ATP signal

Purine and pyrimidine nucleotides such as ATP, ADP, UTP, and UDP are not only components of energy metabolism but also neurotransmitters, which are released at purinergic and pyrimidinergic synapses. Like other transmitters, they interact with both ionotropic and metabotropic receptors. Moreover, certain cell types, such as endothelial cells, have been found to contain a cell surface ATPase that directly synthesizes ATP into the extracellular space, where it acts as a paracrine signal.

The **ionotropic P2X receptors** are found in all tissues. They are highly specific for extracellular ATP. P2X receptors are coupled to nonspecific cation channels that cause depolarization when opened. The channel architecture is rather simple, resembling that of the mechanosensitive channel MscL of *Mycobacterium tuberculosis*, which probably is the evolutionary precursor of P2X receptors (see Section 3.6.7). Like this channel, P2X is composed of three subunits with two transmembrane domains each (**Figure 16.23**). The human genome encodes seven different P2X subunits; together with various splice variants, these combine to yield a large number of receptor isoforms. Among many P2X functions, their effect on some forms of smooth muscles has been studied in more detail. Acting (like NO) as a non-adrenergic, non-cholinergic signal of the vegetative nervous system (see Section 16.5.1), here ATP cooperates with the sympathetic transmitter noradrenaline to trigger muscle contraction (see Section 14.6.2). P2X receptors are also quite abundant in the central nervous system, where they enhance neurotransmitter release by facilitating presynaptic depolarization.

The **metabotropic P2Y receptors** also are widely distributed. In contrast to P2X, they are activated not only by ATP but also by ADP, UTP, and UDP; they are both purinergic and pyrimidinergic receptors. In mammals, they are expressed in at least 12 isoforms differing in ligand specificity. Some P2Y receptors expressed in presynaptic membranes are coupled with $G_{i,0}$-proteins. Among other effects, they suppress the release of their own nucleotide ligands by negative feedback, reminiscent of the regulation of α-adrenergic effects (see Section 5.3.3).

Figure 16.23 Ionotropic and metabotropic nucleotide receptors of the P2 type The inset (left) shows the subunit of the ionotropic P2X receptor cation channel with two transmembrane domains and an extended extracellular loop of about 280 amino acids that is stabilized by several disulfide bonds (not shown). Probably the ion channel consists of three subunits and is lined by all transmembrane subunits. For other details, see text.

Other P2Y receptors are coupled with $G_{q,11}$-proteins; along the phospholipase Cβ–InsP$_3$–Ca^{2+} axis, they induce platelet aggregation (a characteristic ADP effect) and NO release.

For more on adenosine receptors (P1 receptors), see **Sidebar 16.4**.

Sidebar 16.4 Adenosine receptors Like nucleotides, adenosine functions as an intercellular signaling molecule. Since it is not stored in synaptic vesicles and not released by depolarization but by special membrane transporters, it is considered to be a neuromodulator rather than a neurotransmitter proper. In addition, extracellular ATP is degraded by ectonucleotidases to adenosine. In contrast to the P2 nucleotide receptors, the adenosine receptors are called P1 receptors. They exclusively belong to the G-protein-coupled metabotropic type, interacting either with $G_{i,0}$- or G_s-proteins (**Figure 16.24**).

A major function of adenosine is to protect tissues from damage. In fact, a strong increase of adenosine in the extracellular medium is an emergency signal indicating an acute oxygen deficiency (hypoxia, ischemia) as is characteristic for stroke, angina pectoris, and myocardial infarction. Via the $G_{i,0}$-coupled receptors A$_1$ and A$_3$, adenosine exhibits

various effects that prevent (at least partly) stress-induced tissue destruction. These effects include increased blood supply (caused by vasodilatation), diminished blood clotting (due to an inhibition of platelet aggregation), and a sedating effect (resulting from presynaptic inhibition of excitatory neurotransmitter release). Vice versa, a dysregulation of the adenosine system correlates with the induction of seizures, particularly in epilepsy. Moreover, there is evidence that activation of the A$_1$ receptor is involved in the generation of sleep and that the stimulating effect of caffeine is due at least partly to a blockade of this receptor (see also Section 16.9.1). The strong pain indicating myocardial infarction is also triggered by adenosine, here probably through a stimulation of A$_2$ receptors. An overactivation of A$_2$ receptors is also seen in neurodegenerative disorders such as Alzheimer's, Parkinson's, and Huntington's diseases as well as in drug addiction, indicating a role of this receptor type in pathogenesis.

Figure 16.24 P1 adenosine receptors The isoforms are called A$_1$, A$_2$, and A$_3$. Extracellular adenosine is generated from ATP by ectonucleotidases (or released from the cell through special membrane transporters) and inactivated mainly by adenosine kinase.

Summary

- The purine nucleotides ATP and ADP as well as the pyrimidine nucleotides UTP and UDP are paracrine mediators, acting as neurotransmitters.

- The corresponding receptors are the ionotropic P2X receptors, coupled with cation channels and exclusively stimulated by ATP, and the metabotropic P2Y receptors, coupled with $G_{i,0}$- and $G_{q,11}$-proteins and stimulated by each of the four nucleotides. P2X receptors play an important role in smooth muscle contraction and as enhancers of neurotransmitter release in the brain.

16.7 Cannabinoid and vanilloid receptors

The search for a cellular binding site of the marijuana ingredient tetrahydrocannabinol led to the identification of two **cannabinoid receptors, CB1 and CB2,** in the 1980s. CB1 is expressed by many tissues but particularly in the brain, whereas CB2 seems to be restricted to immune cells. Both receptors are coupled with $G_{i,0}$-proteins. Along this pathway, cannabinoids inhibit adenylate cyclase and voltage-dependent Ca^{2+} channels and activate GIRK channels, causing membrane hyperpolarization. Thus, they suppress neurotransmitter release, dampen neuronal excitability, and mediate inhibitory effects such as sedation and pain relief. In addition, cannabis drugs transiently induce euphoria, rendering them particularly attractive. This action is probably due to a pre- or postsynaptic inhibition of inhibitory neurons (see Section 16.9). Artificial CB1 receptor blockers are employed, on the other hand, as appetite suppressants in the therapy of obesity. Of course, tetrahydrocannabinol is not the actual ligand of CB receptors. Instead, endogenous cannabinoids have been identified that are derivatives of polyunsaturated fatty acids. Their prototype is arachidonylethanolamide or **anandamide** (the name refers to the Sanskrit word *ananda* for bless), which has been found in various areas of the brain but also in other tissues. The major function of endocannabinoids is the negative feedback control of synaptic activity: upon their release from depolarized postsynaptic neurons, they act as retrograde signals to inhibit presynaptic neurotransmitter release according to the mechanism shown in Figure 16.6. In certain brain regions this effect causes long-term depression of synaptic conductivity. Thus, endocannabinoids are involved in the regulation of synaptic plasticity. Endocannabinoids also play an important role in the embryonic development of the central nervous system. This may explain why excessive marijuana smoking by expectant mothers frequently causes pathological problems in their children, such as deficits of cognition, motion, and social behavior.

In addition to metabotropic CB receptors, **ionotropic receptors of the transient receptor potential (TRP) family** also interact with anandamide (see Section 15.3). These TRPV or vanilloid receptors are coupled to nonspecific cation channels that actually respond to heat and acid. They are sensitized by anandamide as well as by certain exogenous compounds such as the vanillin-related capsaicin, the hot ingredient of chili pepper (for details see Section 15.3). Thus, like many other neurotransmitters, anandamide exerts a dual effect depending on the receptor type: via TRPV it enhances pain, whereas via CB it acts as a pain reliever (**Figure 16.25**).

As lipophilic molecules, endocannabinoids are also able to pass the plasma membrane and to interact with nuclear receptors of the **PPAR** family (see Section 8.4.1). Along this pathway they modulate gene expression.

Summary

- Endocannabinoids such as anandamide are paracrine mediators interacting with $G_{i,0}$-coupled metabotropic cannabinoid receptors (which also bind cannabis drugs) and cation channel-coupled ionotropic vanilloid (or TRPV) receptors (which are also activated by capsaicin from chili pepper).

Figure 16.25 Endocannabinoid receptors The metabotropic endocannabinoid (CB) receptor with its exogenous ligand tetrahydrocannabinol is shown to the left, the ionotropic vanilloid receptor TRPV with its exogenous ligand capsaicin is shown to the right.

● The activation of vanilloid receptors found predominantly in sensory cells has a pain-evoking effect, whereas the cannabinoid receptors controlling neurotransmitter release mediate pain-relieving effects.

16.8 Opioid receptors

Morphine, the alkaloid of poppy isolated by Friedrich Sertürner in 1803, is the prototype of an opiate. The pain-relieving effect of these drugs has been known since early antiquity and is still unsurpassed. On the other hand, opiates are the epitome of narcotics. Rather early, the stereospecific effect of very low doses of morphine was taken as an indication of an endogenous receptor. Since 1973, three opioid receptors labeled by the Greek letters δ, κ, and μ have been identified. Since the discovery of the corresponding endogenous peptide ligands, the terms DOP, KOP, and MOP (Delta, Kappa, and Mu Opioid Peptide receptors) are now preferred. An additional opioid receptor found more recently is NOP, the nociceptive (pain-related) opioid peptide receptor. All receptors are metabotropic: that is, coupled with $G_{i,0}$-proteins (**Figure 16.26**); ionotropic opioid receptors have not been found.

Along the $G_{i,0}$-controlled signaling pathways, the opioid receptors trigger the usual effects such as inhibition of adenylate cyclase and voltage-dependent Ca^{2+} channels, hyperpolarization by gating the GIRK-type potassium channels, and (via $G\beta\gamma$-subunits) activation of phospholipase $C\beta$ and PI3K. The endogenous ligands are small peptides: β-endorphin (31 amino acids) and enkephalins (5–8 amino acids) interact with the receptor DOP; β-endorphin also interacts with MOP; dynorphins (8–17 amino acids) bind to KOP; and nociceptin (or orphanin, 17 amino acids) binds to NOP. Morphine and heroin prefer MOP as receptor (the label μM refers to morphine).

The opioid peptides are released from larger precursor proteins by controlled proteolysis. They all contain the basic enkephalin sequence Tyr-Gly-Gly-Phe-Met/Leu or (as in the case of nociceptin) Phe-Gly-Gly-Phe-Met/Leu. To a certain extent, this

Figure 16.26 Opioid receptors All opioid receptors known are $G_{i,0}$-protein-coupled. Examples of endogenous ligands with the preferred isoforms of opioid peptide receptors, OP, shown in parentheses are depicted in the box; examples of exogenous ligands are shown to the left and to the right.

sequence is frozen in the complex ring structure of morphine since the aromatic ring and other structural elements exactly fit the receptors' binding sites for the aromatic amino acids Phe and Tyr.

As neurotransmitters, opioid peptides inhibit interneuronal signaling post- or pre-synaptically according to the scheme shown in Figure 16.6. Depending on whether the opioid neuron is connected with an excitatory or an inhibitory neuron, the opioid either exerts a sedating and pain-relieving effect or causes euphoria.

Pain therapy with morphines is confronted by two problems: tolerance and addiction. Owing to tolerance, the dose of the drug has to be increased permanently to overcome endogenous receptor desensitization or tachyphylaxis, while addiction means that a withdrawal of the drug results in severe sickness that can be prevented only by new drug administration or a time-consuming and strenuous therapy. As shown in the following section, both symptoms may be understood as the results of a dramatic and long-lasting adaptation of neuronal signal processing resembling long-term memory storage.

Summary

Endogenous opiates such as enkephalins, endorphins, dynorphins, and nociceptins are paracrine peptide mediators and neurotransmitters that interact with $G_{i,0}$-coupled metabotropic receptors (ionotropic receptors are not known) and are also activated by morphine and other opium-derived drugs. Opioid receptors play a key role in pain therapy and drug addiction.

16.9 Narcotics and drug addiction

Drug addiction is due to a misdirected adaptation of signaling events in the brain. In fact, the major targets of narcotics and psychotropic drugs are synaptic signal transmission and neuronal signal processing. Some examples are listed in **Figure 16.27**. For some examples of cultural uses of narcotics past and present, see **Sidebar 16.5**.

Figure 16.27 Drug effects on the level of signal transduction: an overview

STIMULATION	
caffeine	inhibition of cAMP–phosphodiesterase stimulation of ryanodine receptors inhibition of adenosine receptors
nicotine	stimulation of nicotinic receptors
cocaine amphetamines ecstasy	inhibit the recapture of noradrenaline, serotonin, and dopamine

SEDATION, EUPHORIA	
alcohol, barbiturates	inhibition of NMDA receptor stimulation of GABA receptors
cannabis, opiates	presynaptic inhibition via $G_{i,0}$

FEAR RELIEF	
benzodiazepines	stimulation of GABA receptors

HALLUCINATIONS	
mescaline psilocybin LSD ecstasy	stimulation of serotonin, noradrenaline, and dopamine receptors

16.9.1 Nicotine, caffeine, and alcohol

First, we shall have a look at the everyday drugs nicotine, caffeine, and alcohol. The **nicotine** effects experienced by smokers are primarily due to an activation of ionotropic acetylcholine receptors in the central nervous system. These particular neurons stimulate dopaminergic and other signaling pathways of the limbic system. This so-called **rewarding center** of the brain processes emotions, in particular pleasant feelings. Adaptation due to receptor desensitization leads to tolerance and addiction; that is, to the chain-smoker.

Because of its adenine-related structure, **caffeine** targets several cellular proteins that normally interact with adenosine and its derivatives. Caffeine inactivates the inhibitory adenosine receptor A_1 and cAMP phosphodiesterase and sensitizes ryanodine receptors, probably by mimicking the effect of cyclic ADP-ribose. Along the

Sidebar 16.5 Drug applications, past and present Heroin is a synthetic acetylation product of morphine. Compared with the latter, it penetrates the blood–brain barrier more rapidly. At the end of the 19th century, heroin was introduced by the Bayer Company as cough medicine. Its addictive potential was realized only 25 years later. Cough medicines developed later frequently contained the much less dangerous **codeine**, a methylated morphine derivative. The substitute drug **methadone** was developed as a pain reliever during World War II to make Germany independent from opium import. Like morphine, methadone causes addiction. However, because (compared with heroin or morphine) it is much cheaper, exhibits a longer-lasting effect, and may be taken orally, crimes committed in the pursuit of drug acquisition and the risk of infection (due to nonsterile hypodermic syringes) are both considerably diminished.

Until the end of the 19th century, cocaine, the classic drug of native South Americans, was considered to be a harmless stimulant and was used as an ingredient of refreshing beverages such as Coca-Cola and Vin Mariani. Sigmund Freud had prescribed cocaine against morphine addiction until one of his friends fell into an incurable cocaine psychosis. When such harmful effects became known, the cocaine in Coca-Cola was replaced by caffeine.

Figure 16.28 Molecular effects of caffeine Caffeine (trimethylxanthine) has an adenine-related structure explaining many of its effects that are due to an increase of the cytoplasmic cAMP and Ca^{2+} levels. Caffeine inhibits the A_1 adenosine receptor, thus relieving voltage-dependent Ca^{2+} channels (and adenylate cyclase) from $G_{i,0}$-mediated inhibition, and cAMP–phosphodiesterase, thus stabilizing cAMP and (as a consequence) activating protein kinase A. By phosphorylating phospholamban (Section 14.6.2), protein kinase A stimulates the Ca^{2+} pump SERCA of the sarcoplasmic reticulum, resulting in a storage-activated Ca^{2+} influx (Section 14.5.5). The activation of ryanodine receptors leads to an additional Ca^{2+} influx from the endoplasmic reticulum (Section 14.5.3). As a result, heart activity and synaptic signaling become stimulated, resulting in the typical caffeine effects such as increased heartbeat and alertness. For the sake of clarity, other effects of cAMP and Ca^{2+} that are also influenced by caffeine are not shown.

corresponding signaling pathways, the cytoplasmic levels of cAMP and Ca^{2+} become elevated, resulting in a reinforcement of sympathetic actions and of central nervous processes with the well-known consequences such as increased heartbeat, desire to urinate, and alertness (**Figure 16.28**).

Ethyl alcohol stimulates inhibitory $GABA_A$ and blocks excitatory NMDA receptors, sharing these molecular targets with barbiturates. In principle, the physiological and psychological consequences are similar to those caused by cannabis drugs and opiates, both of which exert inhibitory effects via $G_{i,0}$-coupled metabotropic receptors. [Mice with a genetic knockout of the MOP receptor do not respond to alcohol, cannabis, and nicotine, indicating that the target neurons of such drugs release endogenous opioids that, like morphine, activate the MOP receptor.] Therefore, all these drugs (alcohol, barbiturates, cannabis, and opiates) act as sedatives and narcotics. On the other hand, they strongly induce euphoria. Indeed, this dual action renders such drugs particularly attractive. It finds an explanation in the principle of double negation; namely, an inhibitory signal has a stimulatory effect when it suppresses another inhibitory signal. This is shown for opiates in **Figure 16.29**.

16.9.2 Hallucinogenic and stimulatory drugs and antipsychotics

Like caffeine and alcohol, many other drugs and narcotics interact with more than one cellular protein and receptor. Typical multifunctional drugs are the **hallucinogens** depicted in **Figure 16.30**. Owing to their chemical structure, they may fit receptors of dopamine, noradrenaline, and serotonin, neurotransmitters primarily involved in the emotional processing of sensory impressions in the limbic system. In addition, they play a key role in psychotic diseases such as schizophrenia.

A major target of hallucinogens are the **serotonin (or 5-hydroxytryptamine) receptors** that in mammals are expressed in no less than 14 isoforms, emphasizing their importance in brain physiology. While the ionotropic serotonin receptor 5-HT3 is coupled with a nonspecific cation channel resembling the nicotinic acetylcholine receptor, the other receptors are metabotropic, interacting with G_s-, $G_{q,11}$-,

Figure 16.29 Dual effect of opioids Although the $G_{i,0}$-coupled opioid receptors primarily transmit inhibitory signals, opioids may have both sedative and stimulatory effects. (A) For sedation and pain relief, peripheral neurons are hindered by associated opioid neurons from releasing noradrenaline or substance P that, via $G_{q,11}$-coupled receptors, would trigger the release by further neurons of noradrenaline and dopamine, stimulating the waking and pain center in the brain. Substance P is a small peptide (11 amino acids) of the tachykinin family that transmits pain signals from the periphery to the spinal cord (see also Figure 14.38). It interacts with a $G_{q,11}$-coupled receptor NK1. (B) The stimulatory effect of opioids is due, at least partly, to an inactivation of GABAergic neurons that normally turn down the activity of those dopaminergic neurons, which stimulate the rewarding center in the brain. G-proteins are symbolized by black circles.

(A) SEDATION

(B) EUPHORIA

and $G_{i,0}$-proteins. Serotonin receptors are involved in many central nervous processes. Synthetic 5-HT1 agonists are clinically used as antimigraine drugs. They induce vascular constriction in the brain, thus preventing a migraine attack, which is accompanied by vascular dilatation. Serotonin (or tryptaminergic) signaling is also stimulated by some antidepressants.

Stimulants such as **amphetamines**, **cocaine**, and **ecstasy** interact with dopaminergic, adrenergic, and tryptaminergic neurons by inhibiting synaptic signal extinction rather than by receptor binding. To this end, they inactivate presynaptic neurotransmitter transporters, thus preventing transmitter recapture (**Sidebar 16.6**). This effect is augmented by a simultaneous stimulation of neurotransmitter release. As a result, the postsynaptic cell becomes overstimulated.

Many drugs evoke symptoms of acute psychosis. Such observations support the concept that paranoid, schizoid, and manic–depressive diseases are caused by genetically fixed or acquired disturbances of cellular signal processing. Indeed, most **antipsychotics** interact with neurotransmitter receptors, in particular with the D2 dopamine receptor, and with transport proteins catalyzing the presynaptic re-uptake of transmitters (synaptic signal extinction; see Sidebar 16.4). Clinical experience yielded a dopamine theory of schizophrenia, maintaining a causal relationship between this most frequent psychosis and a hyperfunction of dopaminergic neurons. However, the appearance of the disease is so complex that such monocausal explanations are regarded as outdated today. Observations showing dopamine receptor blockers to be useful for treating both psychoses and opiate addiction nevertheless emphasize the close relationship between drug diseases and mental illness.

16.9.3 Drug addiction represents an adaptive memory effect

Drug addiction is a phenomenon that in its complexity rivals psychotic diseases, resisting any simple explanations. For psychotic diseases, genetic predisposition is involved to a certain degree. The same holds true for drug vulnerability whereas drug addiction in most cases is an acquired behavior. The pressure to avoid withdrawal ecstasy   mescaline (peyote cactus)   psilocybin (magic mushroom)   LSD

dopamine   noradrenaline   serotonin (5-hydroxytryptamine)

symptoms certainly is a major reason for physical dependency but is only one of several causes of addiction. As least as important is the urge to satisfy the desire for pleasure, resulting in psychological dependency. Depending on the type of drug, either component may predominate.

The major source of pleasant sensations is the previously mentioned rewarding center of the limbic system. As is well-known, this brain area is activated not only by drugs but also by a wide variety of other stimuli, the effects of which are mediated by endogenous "drugs": that is, signaling molecules such as neurotransmitters (**Figure 16.31**).

Figure 16.30 Hallucinogenic drugs The prominent hallucinogens shown have transmitter-related structures. Therefore, they are potent activators of metabotropic dopamine, noradrenaline, and serotonin receptors. These receptors dominate signal transmission in brain areas involved in the processing of emotions and sensual impressions. LSD, lysergic acid diethylamide; P, phosphate.

Sidebar 16.6 Signal extinction by neurotransmitter transporters Synaptic signaling is terminated through a re-uptake of neurotransmitter by the pre- and postsynaptic neuron and adjacent glial cells. An exception is acetylcholine, which is hydrolyzed by acetylcholinesterase in the synaptic cleft. The re-uptake is managed by extremely efficient and specific transporter proteins within neuronal and glial cell membranes. One distinguishes between monoamine transporters for dopamine, noradrenaline, and serotonin and transporters for GABA, glutamate, glycine, and other transmitters. Most of them are characterized by 12 transmembrane domains with the C- and N-termini extending into the cytoplasm. Thus, at first glance, they seem to resemble ABC transporters (see Section 3.3.1). However, neurotransmitter transporters catalyze a secondary active transport against a concentration gradient powered by a coupled ion flux (of Na^+, Cl^-, K^+, and H^+) rather than by ATP hydrolysis. Glutamate and monoamine transporters can be traced back to bacterial glutamate and leucine transporters, respectively, which have been crystallized and analyzed in detail. Like receptors, transporters rapidly become up- and down-regulated by exo- and endocytosis depending on the situation and the demands. Moreover, their activities are controlled by post-translational modifications such as protein phosphorylation.

Release and re-uptake of neurotransmitters are interlinked. So, a blockade of transporters by cocaine, amphetamines, or ecstasy is accompanied by increased neurotransmitter secretion into the synaptic cleft. This leads to overstimulation of the particular synapses, possibly culminating in threatening psychotic symptoms. In fact, pathological malfunctions of neurotransmitter transporters belong to the molecular causes of many psychoses, and vice versa, a series of antipsychotic medical drugs, in particular antidepressants and antiepileptics, directly inhibit presynaptic reuptake. Neurotransmitter transporters are also found in the membranes of synaptic vesicles storing transmitters. Like the transporters of the presynaptic membrane, they exhibit 12 transmembrane domains, although the types are not closely related.

Figure 16.31 Addiction: a consequence of adaptation The brain's rewarding system may be excited by a wide variety of stimuli. Any permanent strain evokes adaptive responses of the signal-processing apparatus in the neurons involved. Such an adaptation may lead to long-term desensitization. This phenomenon yields two consequences: the dose of the stimulus has to be increased continuously in order to obtain the effect desired, and withdrawal results in pathological hyperactivity with strongly negative sensations. Both the desire for positive feelings (pleasure) and the attempt to avoid negative feelings (withdrawal symptoms) may lead to abuse and addiction, finally becoming firmly established by long-term potentiation.

Figure 16.32 Tolerance and addiction: an adaptive process explained for alcohol as an example This strongly simplified scheme only serves to make clear drug addiction as a result of cellular and molecular adaptation. In the sober condition, excitatory (black) and inhibitory (teal) signals compensate each other. Alcohol activates (among others) the ionotropic GABA_A receptors, causing an increase of the inhibitory signal in the drunken state. Owing to adaptation, this effect becomes matched by a corresponding increase of excitatory signaling, here exemplified by an expression of additional NMDA and serotonin 5-HT_3 receptors (many other possibilities are conceivable). Upon withdrawal of the drug, the inhibitory signal cannot balance out any longer the excitatory signal, resulting in cellular hyperactivity felt as withdrawal sickness.

As is typical for cells, the neurons of the rewarding center are extremely adaptive and rapidly become desensitized upon prolonged stimulation. Consequently, the dose of the stimulus has to be increased more and more to produce the pleasant sensations desired. Thus, to understand and treat addictive diseases, one must know the cellular and molecular mechanisms of adaptation (shown as a strongly simplified scheme in **Figure 16.32**). Although these investigations by no means are concluded, the data available strongly indicate that addiction is due to long-term adaptation and conditioning of cellular signal processing and that drug dependency and memory fixation might be two sides of the same coin.

That adaptation occurs at both the cellular and the molecular level was shown in 1975. For these pioneering experiments, a neuroblastoma/glioma cell line was used. The cells responded to morphine treatment as expected by a diminished cAMP synthesis, because adenylate cyclase was inhibited along the MOP–$G_{i,o}$ signaling pathway. When the treatment was repeated several times, an adaptive de novo synthesis of adenylate cyclase was observed, settling down the cAMP level at normal values. To lower the level again, the morphine dose had to be increased continuously. In other words, single cells were rendered drug-dependent. Upon withdrawal of the drug, the cells responded by an overproduction of cAMP, that is, they exhibited some sort of withdrawal symptoms. This process is shown in **Figure 16.33**. It may be asked why the cells did not adapt by a down-regulation of MOP receptors. An answer may be found in a unique property of this receptor: namely, that adaptive endocytosis is induced only by the endogenous ligand and not by morphine.

Drugs such as opiates, cocaine, and alcohol strongly up-regulate cAMP signaling in the limbic system, a process called cAMP superactivation. Several transmitters such as glutamate, serotonin, noradrenaline, and dopamine control cAMP production in their target neurons. In the rewarding system **dopamine-induced signaling** plays a key role. It is stimulated by practically all addictive drugs, either directly or through

NORMAL ⟶ DRUGGED ⟶ ADAPTED ⟶ HYPERACTIVE

adaptive mechanisms. Five different metabotropic dopamine receptors are known, two of which (D1 and D5) are coupled with G_s-proteins and the others with $G_{i,0}$-proteins. Drug addiction seems to be accompanied by almost irreversible changes of dopaminergic signal transmission. Such a **long-term adaptation** is particularly feared: it is the reason that, after successful withdrawal treatments, even a single contact with the drug may be sufficient to reconstitute dependency. This phenomenon is thought to be based on mechanisms resembling that of long-term memory fixation. In fact, in both cases similar changes of synaptic signal transmission caused by altered gene expression are observed. These include (apart from the above-mentioned cAMP superactivation) an overexpression of NMDA and AMPA receptors, neurotrophins, NO synthase, G-proteins, and tyrosine hydroxylase, the key enzyme of dopamine synthesis. Long-term memory fixation is thought to be due to stable modifications of neuronal signal transduction, caused by both modulation of synaptic strength and formation of new synapses (see Section 16.4.1). Such alterations require protein synthesis and—in the case of long-term memory—are expected to be almost irreversible.

An increasing body of evidence indicates that **epigenetic processes** play a key role in memory fixation and drug addiction. As explained in Chapter 8, epigenetics means long-lasting—sometimes lifelong or even trans-generational—alterations of the genetic readout without a change of DNA sequence. Such effects are due to alterations of chromatin structure caused by histone modifications, by DNA methylation, and by non-coding RNA signaling. In the neurons of the rewarding center of experimental animals such epigenetic reactions are stimulated by addictive drugs. Corresponding alterations are found in brain samples taken from drug-addicted accident or suicide victims. Moreover, in experimental animals, manipulations of histone modifications induce typical drug effects and symptoms of addiction.

Among the transcription factors involved in drug addiction, CREB and δfosB have been investigated in more detail. **CREB**, a substrate of protein kinase A (see Section 8.6), is activated in the course of cAMP superactivation. The reader may recall the central role this factor plays in long-term potentiation of synaptic signaling (synaptic plasticity). **δFosB**, a truncated splice variant of the transcription factor FosB, is (in contrast to other Fos-type factors) metabolically more stable, being able to stimulate adaptive gene transcription even long after the drug has been withdrawn. The reason for δFosB stability is the truncation of the C-terminal region and protein phosphorylation, which both render δFosB resistant to proteolytic attack. In the dopaminergic neurons of the limbic system, the formation of the factor is induced by all sorts of drugs and other addictive stimuli but also by nonspecific stress. The physiological function of δFosB might be to render neurons able to cope with long-lasting stress by an adaptive behavior. For survival it would be a tremendous advantage to remember this trained ability for life. On the other hand, chronic stress and traumatic experience frequently lead to severe psychiatric disorders such as major depression and post-traumatic stress syndrome. As in the case of drug addiction, these chronic diseases correlate with epigenetic alterations in the corresponding brain areas.

Figure 16.33 Drug-addicted cells From left to right: when stimulated by an input signal (not shown), the normal cell starts to produce cAMP owing to an activated adenylate cyclase (teal spheres). A drug (shown as black semicircles) blocks signal transduction and interrupts cAMP synthesis. The cell adapts to the new situation by expressing additional adenylate cyclase molecules, thus reestablishing the original state of signaling. Removal of the drug results in an overproduction of cAMP. This withdrawal symptom can be cured only by new drug administration.

Summary

- Nicotine interacts with ionotropic ("nicotinic") acetylcholine receptors both in the brain and in peripheral organs.

- Caffeine stimulates sympathetic effects as well as the central nervous system by inhibiting cAMP degradation, blocking sedative adenosine receptors, and activating ryanodine receptors (causing an increase of cytoplasmic Ca^{2+}).

- The sedative effects of ethyl alcohol are due to a blockade of excitatory glutamate receptors and an activation of inhibitory GABA receptors. Stimulatory effects of alcohol (and other sedative drugs such as opiates) are caused by an inhibition of inhibitory neurons.

- Most hallucinogenic drugs activate neurotransmitter receptors involved in the processing of sensory emotions such as dopamine, serotonine, and noradrenaline receptors. Stimulants such as cocaine exert their effects mainly by inhibiting the presynaptic reuptake of dopamine, serotonin, and noradrenaline (synaptic signal extinction). The endogenous targets of many antipsychotics resemble those of hallucinogenic and stimulatory drugs.

- Drug addiction is characterized by a continuous stimulation of receptor-controlled signaling mechanisms. As a regulatory consequence, opposing signaling is overstimulated and becomes dominant upon drug withdrawal. The resulting adaptation is a long-term effect resembling long-term memory fixation. Epigenetic processes play a key role in drug addiction.

Further reading

Section 16.1

Bemben MA et al. (2015) The cellular and molecular landscape of neuroligins. *Trends Neurosci* 38:496–505.

Benson DL & Huntley GW (2012) Synapse adhesion: a dynamic equilibrium conferring stability and flexibility. *Curr Opin Neurobiol* 22:397–404.

Blockus H & Chédotal A (2014) The multifaceted roles of Slits and Robos in cortical circuits: from proliferation to axon guidance and neurological diseases. *Curr Opin Neurobiol* 27:82–88.

Emes RD & Grant SGN (2012) Evolution of synapse complexity and diversity. *Annu Rev Neurosci* 35:111–131.

Focant MC & Hermans E (2013) Protein interacting with C kinase and neurological disorders. *Synapse* 67:532–540.

Karopouli E & Kolodkin AL (2014) Semaphorins and the dynamic regulation of synapse assembly, refinement, and function. *Curr Opin Neurobiol* 27:1–7.

Klein R & Kaina A (2014) Ephrin signalling in the developing nervous system. *Curr Opin Neurobiol* 27:16–24.

Krueger DD et al. (2012) The role of neurexins and neuroligins in the formation, maturation, and function of vertebrate synapses. *Curr Opin Neurobiol* 22:412–422.

Nadim F & Bucher D (2014) Neuromodulation of neurons and synapses. *Curr Opin Neurobiol* 29:48–56.

O'Brien J (2014) The ever-changing electrical synapse. *Curr Opin Neurobiol* 29:64–72.

Van Bottum EY, Brignani S & Pasterkamp RJ (2015) Axon guidance proteins in neurological disorders. *Lancet Neurobiol* 14:532–546.

Yang X et al. (2014) Intercellular protein-protein interactions at synapses. *Protein Cell* 5:420–444.

Section 16.2 and 16.3

Davies PA, Wang W, Hales TG et al. (2003) A novel class of ligand-gated ion channels is activated by Zn^{2+}. *J Biol Chem* 278:712–717.

Dutertre S, Becker CM & Betz H (2012) Inhibitory glycine receptors: an update. *J Biol Chem* 287:40216–40223.

Harris RA & Mihic SJ (2004) Alcohol and inhibitory receptors: unexpected specificity from a nonspecific drug. *Proc Natl Acad Sci USA* 101:2–3.

Hogg RC, Raggenbass M & Bertrand D (2003) Nicotinic acetylcholine receptors: from structure to brain function. *Rev Physiol Biochem Pharmacol* 147:1–46.

Zhang J et al. (2013) The structural mechanisms of Cys-loop receptor desensitization. *Mol Neurobiol* 48:97–108.

Section 16.4

Fenno l, Yizhar O & Deisseroth K (2011) The development and application of optogenetics. *Annu Rev Neurosci* 34:389–412.

Hara MR & Snyder SH (2007) Cell signaling and neuronal death. *Annu Rev Pharmacol Toxicol* 47:117–141.

Ohno M (2014) Roles of eIF2α kinases in the pathogenesis of Alzheimer's disease. *Front Mol Neurosci* 7:1–7.

Karakas E, Regan MC & Furukawa H (2015) Emerging structural insights into the function of ionotropic glutamate receptors. *Trends Biochem Sci* 40:328–337.

Taha E et al. (2013) The role of eEF2 pathway in learning and synaptic plasticity. *Neurobiol Learn Mem* 105:100–106.

Tischer D & Weiner OD (2014) Illuminating cell signalling with optogenetic tools. *Nat Rev Mol Cell Biol* 15:551–558.

Trinh MA & Klann E (2013) Translational control by eIF2α kinases in long-lasting synaptic plasticity and long-term memory. *Neurobiol Learn Mem* 105:93–99.

Wang W, Kwon EJ & Tsai LH (2012) MicroRNAs in learning, memory, and neurological diseases. *Learn Mem* 19:359–368.

Yin T & Wu YI (2013) Guiding lights: recent developments in optogenetic control of biochemical signals. *Pflügers Arch Eur J Physiol* 465:397–408.

Section 16.5 and 16.6

Calabrese V, Mancuso C, Calvani M et al. (2007) Nitric oxide in the central nervous system: neuroprotection versus neurotoxicity. *Nat Rev Neurosci* 8:766–775.

Derbyshire ER & Marletta MA (2012) Structure and regulation of soluble guanylate cyclase. *Annu Rev Biochem* 81:533–559.

Jiang R, Taly A, & Grutter T (2013) Moving through the gate in ATP-activated P2X receptors. *Trends Biochem Sci* 38:20–29.

Puchalowicz K et al. (2014) P2X and P2Y receptors – role in the pathophysiology of the nervous system. *Int J Mol Sci* 15:23672–23704.

Rochette L et al. (2013) Carbon monoxide: mechanisms of action and potential clinical implications. *Pharmacol Therap* 137:133–152.

Sperlagh B & Illes P (2014) P2X7 receptor: an emerging target in central nervous diseases. *Trends Pharmacol Sci* 35:537–548.

Section 16.7 and 16.8

Katona I & Freund TF (2012) Multiple functions of endocannabinoid signaling in the brain. *Annu Rev Neurosci* 35:529–558.

Ohno-Shosaku T & Kano M (2014) Endocannabinoid-mediated retrograde modulation of synaptic transmission. *Curr Opin Neurobiol* 29:1–8.

Sheth S et al. (2014) Adenosine receptors: expression, function and regulation. *Int J Mol Sci* 15:2024–2052.

Snyder SH & Paternak GW (2003) Historical review: opioid receptors. *Trends Pharmacol Sci* 24:198–205.

Waldhoer M, Bartlett SF & Whistler JL (2004) Opioid receptors. *Annu Rev Biochem* 73:953–990.

Section 16.9

DeBiasi M & Dani JA (2011) Reward, addiction, withdrawal to nicotine. *Annu Rev Neurosci* 34:105–130.

Fields HL & Margolis EB (2015) Understanding opioid reward. *Trends Neurosci* 38:217–225.

Klengel T et al. (2014) The role of DNA methylation in stress-related psychiatric disorders. *Neuropharmacology* 80:115–132.

Nestler EJ (2014) Epigenetic mechanisms of drug addiction. *Neuropharmacology* 76:259–268.

Nestler EJ (2004) Historical review: molecular and cellular mechanisms of opiate and cocaine addiction. *Trends Pharmacol Sci* 25:210–218.

Peña CJ et al. (2014) Epigenetic signaling in psychiatric disorders. *J Mol Biol* 426:3389–3412.

Ruffle JK (2014) Molecular biology of addiction: what's all the (Δ)FosB about? *Am J Drug Alcohol Abuse* 40:428–437.

Sitte HH & Freissmuth M (2015) Amphetamines, new psychoactive drugs and the monoamine transporter cycle. *Trends Pharmacol Sci* 36:41–50.

Snyder SH (1986) Drugs and the Brain. New York: Scientific American Books.

Putting Together the Pieces: The Approach of Systems Biology

17

17.1 The whole is greater than the sum of its parts

A central theme of this book is that a signal, even when it interacts with only one receptive molecule, in a cell evokes a complex excitation pattern rather than opening a molecular one-way street. We have tried to make this clear by introducing the metaphor of the "brain of the cell."

A signal-induced excitation pattern resembles a data-processing apparatus at work. Such an apparatus is based on a system or network of interacting switching elements. As discussed in Chapter 1, societies, organisms, cells, proteins, and genes represent biological systems at different scales, while computers and other machines are technical systems.

Since, in the past, the lack of suitable methods prevented any scientifically sound access to the problems of complex biological systems, one was satisfied by a hunter–gatherer approach; and this, as we have seen, with tremendous success. It is, however, a fundamental property of complex systems to develop in an unpredictable way emergent functions that are not found in the individual components. This makes it impossible to predict the outcome of biological systems, for instance mitosis, the behavior of a flock of birds, or a Bach cantata, even if we were to possess a complete list of all biochemical reactions or all neuronal interactions at work. Moreover, even much simpler biological systems are still far too complex to be intuitively understood. Instead, mathematical models for evaluating their dynamic properties and providing a quantitative analysis are required. This is the task of systems biology, a new field in the life sciences. A brief outline of principles, current methods, and inherent problems of this approach is given in this chapter.

17.2 Systems biology: origin and focus

By combining the traditional reductionist methodology with a holistic approach, systems biology aims at an understanding of complex systems in a quantitative and predictable way. Systems biology has its origin in both theoretical biology and engineering science. Already in the early 1960s, the Austrian-American biologist Ludwig von Bertalanffy formulated a "general system theory" emphasizing that all living systems share "the common properties of being composed of inter-linked components, in which case they might share similarities in detailed structure and control design." Fifteen years earlier the American mathematician Norbert Wiener, the founding father of cybernetics, had developed a theory of control mechanisms stressing the importance of negative feedback for maintaining stability of technical systems.

Rapid progress in cell and molecular biology, culminating in the characterization of macromolecular structures as well as in the development of recombinant and high-throughput techniques that allow the genome-wide census of components and interactions (genomics, proteomics, transcriptomics, metabolomics, and other "omics"), has provided a basis for the experimental analysis of biological systems. This analysis may proceed in two directions: from the parts to the whole (bottom-up) or from the whole to the parts (top-down). In the **bottom-up approach** the emergent properties of a system from detailed knowledge of subsystems and molecular interactions are deduced. The **top-down approach** is used to identify the molecular networks and mechanisms that are thought to induce emergent properties of the system. This approach is feasible when one possesses a list of components generated—for instance, by an "omics" approach—but lacks a detailed knowledge of their functions and is therefore data-driven. Since both the bottom-up and top-down approaches necessarily complement each other, it has been proposed to combine them in what has been called a "middle-out approach." However, this is hardly more than semantics, saying what has been known for centuries: namely, that the progress of science, including biology, depends on a continuous interplay between inductive and deductive approaches.

A further point deserves to be addressed. The theories of evolution, developmental biology, genetics, and non-equilibrium thermodynamics say that life is the result of a unique historical process. This means that biological systems must have an extremely high degree of individuality in their details, a fact that from the beginning excludes any too far-reaching generalization of statements, whereas on the other hand it calls for a strict standardization of experimental approaches.

Summary

New techniques allowing system-wide studies of cellular components and of the dynamics of their interactions provide the conditions for a systematic and quantitative analysis of biochemical networks. This is achieved by a combination of deductive (bottom-up) and inductive (top-down) approaches.

17.3 System structure: basic network topologies and properties

Newly developed methods of molecular cell biology have proven to be very useful in bottom-up approaches. A typical example is gene-editing such as the CRISPR/Cas9 technique, which allows genomic DNA to be cut at any desired location. Thus, a single component of a molecular network can be altered and the impact of this perturbation on the behavior of the other components can be studied. Systems biology tries to improve the bottom-up approach by generating quantitative data.

Top-down biology has received a strong impetus with the introduction of the "omics" based on the development of high-throughput techniques, which allow the genome-wide identification of biomolecules, their expression patterns, and their interactions (an example is the identification of the sour-taste receptor described in Section 15.1.1). Even though the data may still be incomplete and afflicted by errors, these approaches are used to construct networks of interactions.

A network consists of **nodes** that are interconnected by **links** or edges. In biological networks, nodes may symbolize genes or proteins, and links are the interactions between them. The type of interaction can be undirected or directed, respectively. Protein–protein interactions, for example, are usually represented by undirected networks, whereas transcription, signal transduction, and most metabolic reactions are targeted events depicted by directed networks composed of nodes connected by incoming and outgoing links.

Today, most networks still resemble street maps with the interactions described only qualitatively. For the future, a quantitative assessment is indispensable: this would also take into account the strengths of interactions (resembling traffic types and densities). It is easily seen that this goal can be achieved only by a combination of top-down and bottom-up systems biology.

However, assembly of networks requires considerable effort and a combined application of different experimental and computational techniques. The construction of protein interaction networks focusing on the assembly of protein complexes may give an idea of the problems inherent in such an approach, particularly where the faithful proof of interactions is concerned. In the beginning such networks were based on data obtained by using the yeast two-hybrid method. However, the value of this technique is limited by false-positive results (see Section 5.2). Later on, high-affinity purification followed by mass spectrometry was employed to characterize protein complexes. Recently, we have witnessed many advances in quantitative mass spectrometry. An example is provided by the proteomic ruler approach, which facilitates a proteome-wide quantification of proteins. Here the protein signals obtained by mass spectrometry are converted into the number of protein copies per cell by comparing them with the histone signals that are proportional to the amount of DNA and thus to the number of cells. Methods for the analysis of protein–protein interactions such as Förster resonance energy transfer (FRET) are described in Section 5.2. Furthermore, recent advances in high-resolution fluorescence microscopy allow a more precise localization of interaction partners *in vivo*. A disadvantage of such approaches is that proteins have to be modified by fluorescent tags and exogenously expressed. Furthermore, binding partners in substoichiometric concentrations are not detected. Nevertheless, when combined with biochemical data, the data sets generated by these techniques provide useful information about the global topology of protein–protein interaction maps.

17.3.1 Network parameters and nomenclature

Like a map, a network is described theoretically by a set of parameters:

- the **degree**, k (connectivity), gives the number of links connected to a node; in directed networks one distinguishes between incoming and outgoing degrees
- the **degree distribution**, $P(k)$, describes the probability that a node has k links
- the **distance** is the shortest path length between two nodes
- the **diameter** is the maximum distance between any two nodes in a network
- the **clustering coefficient** gives the percentage of existing links in the neighborhood of a node
- the **betweenness** is understood as the fraction of shortest paths between all pairs of nodes that pass through one node

Networks with a small diameter, or short paths between the nodes, are frequently termed "small-world" networks, referring to a social study where such properties were discovered for the first time. Neural and metabolic networks, for instance, may exhibit a small-world topology, indicating a minimal transition time between the different states. In such networks a local effect spreads rapidly.

The theory predicts a few principles that dominate most networks despite their obvious diversity. A major principle concerns the quality of the nodes, which allows a distinction between two types of network organization. In a **random network**, on average each node has a similar number of links, whereas in a **scale-free network**, nodes with many connections, called hubs, are surrounded by a large number of nodes with only one or a few connections (**Figure 17.1**). The analysis of biological networks reveals that most of them are organized in a scale-free rather than in a random manner. For a protein network, a scale-free topology means that most proteins interact with only a few partners, whereas a few proteins exhibit a high number of interactions. The advantage of a scale-free organization is obvious: it renders the system more robust because a loss of one of the many non-hub nodes is less disruptive

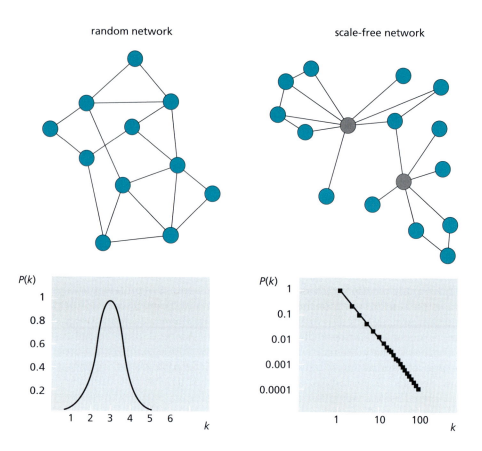

Figure 17.1 Two types of network Networks are described by the number of nodes, the number of links connected to a node (degree *k*), and the probability *P(k)* of connections with a given *k* value (degree distribution). For random networks, *P* shows a bell-shaped Gaussian distribution when plotted against *k*, indicating that on average each node has a similar number of links (in our example, 2–4). Scale-free networks show a small number of high-degree nodes and a power-law distribution of links connected to a node that results in a straight line when *P* is plotted against *k* on a log–log scale. Regular nodes are indicated in teal and high-degree nodes (hubs) are shown in gray.

than in a random network. On the other hand, for example in cancer cells, it can be advantageous to specifically target hubs to inhibit cell proliferation.

A characteristic feature of both street maps and biological networks is a hierarchical organization: lower-degree networks are the components of higher networks. Moreover, even within a network some kind of hierarchy is found, becoming visible in the graphs as groups of highly interconnected nodes, known as clusters or modules. Nevertheless, network topology and organization resemble each other at each level.

Despite their complexity, biological networks contain a limited set of recurring regulatory elements or network motifs, representing interactions that facilitate the progress and control of biological functions, for instance by positive or negative feedback loops. Examples of such subgraphs are shown in **Figure 17.2**. One easily recognizes that such network motifs resemble the biochemical signaling patterns described in Section 1.9.

17.3.2 Signal-processing networks

Traditionally, signal-processing systems have been analyzed by the bottom-up approach. Only recently the top-down approach has entered the field. To handle the problem of unwieldy data sets, which are typical for the top-down approach, one divides the network into several subnetworks. Of course, this is an artificial subdivision, since in reality the subnetworks are fully interconnected in a hierarchically organized super-network. One of the signaling subnetworks describes the interactions of transcription factors and genes. These **gene-regulatory networks** are considered to be slow because the input signals that change the activity of transcription factors operate in a matter of seconds, whereas the output that corresponds to an alteration of gene expression and the accumulation of protein products may develop within hours. Despite variable quality of the data upon which the networks are based (obtained by experiments with model organisms and by large-scale analysis

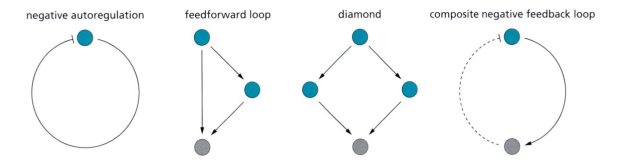

of transcription factor binding sites), interesting and meaningful results have been obtained. These allow predictions that afterward can be verified. An example is the global identification of targets of factors controlling embryonic stem cell maintenance, which has suggested novel pathways of stem cell self-renewal.

Much faster are **signal transduction networks**, which operate on the time scale of seconds to minutes. In these networks, signaling proteins represent the nodes and the edges reflect direct non-covalent interactions or covalent post-translational modifications such as protein phosphorylation.

The most rapid signal processing is observed in **neuronal networks**, where the response time is in the range of milliseconds. In these networks, nodes symbolize neurons and edges represent synaptic connections.

17.3.3 Metabolic networks

A major function of signal processing networks is to control **metabolic networks**, which combine information on proteins (enzymes) and metabolites and focus on the mass flux of chemical compounds such as amino acids, sugars, lipids, and energy required for biochemical reactions. Unlike signaling networks, a main characteristic of metabolic networks is the conservation of mass. For example, one molecule of glucose in glycolysis will produce exactly two molecules of pyruvate. Therefore, the flux, which is the rate of flow of the molecules per enzyme through a linear network, is the same for each step at steady-state equilibrium. These networks are analyzed using flux balance analysis based on the assumption that all metabolites exist in steady-state concentrations and that an organism controls the flux of metabolites to optimize biomass production. Computerized flux models have been successfully used to predict the phenotype of mutant yeast strains grown under different media conditions. These studies reveal that most nonessential enzymes in yeast are required for growth under certain specialized conditions and that isoenzymes or parallel pathways compensate only a small subset of these enzymes.

Summary

- A major goal of current systems biology is to construct maps of interactions to infer functionality. In such networks, components are symbolized by nodes and links indicate their interactions. Networks are described by a set of parameters that characterize the quality of nodes and links.

- Most biological networks are scale-free: a limited number of nodes with many links (called hubs) are connected with a large number of nodes with only a few links. Compared with a random network (in which, on average, each node has the same number of links), a scale-free network is more resistant to disturbances, but hubs may be suitable targets for interventions.

- For the sake of clarity, signal-processing networks may be subdivided in gene-regulatory networks, signal-transduction networks and neuronal networks. They all communicate with metabolic networks.

Figure 17.2 Recurring motifs in signaling networks A frequent network motif observed in signaling networks is negative autoregulation. Feedforward loops occur in gene regulatory and signal transduction networks, whereas a diamond relationship is observed only in signal transduction networks and can be combined in multiple layers to perform functions on multiple input signals. Composite network motifs combine different types of interaction. Most common is a composite negative feedback loop in which one arm represents a transcriptional interaction (arrow) and the other arm represents an inhibitory protein–protein interaction (broken line). The separation of time scales between the slow transcription arm and the faster protein–protein arm helps to stabilize the dynamics of the composite loop.

Figure 17.3 Systems biology approaches consist of an iterative process between laboratory experiments and mathematical modeling In a collaboration between experimenters and theoreticians, contradictory issues in biology are answered in an iterative cycle of quantitative data generation, mathematical modeling, *in silico* predictions, experimental validation, and design of new experiments. Advancement of research in computational science, analytical methods, technologies for measurements, and improved high-throughput methodologies will help to gradually transform biological research into a more systematic and hypothesis-driven science, as represented by this iterative cycle process. (Adapted from Kitano H [2002] *Science* 295:1662–1664.)

17.4 The iterative cycle: laboratory experiments and mathematical model development

Scientific work may be described as an interplay between imagination and experiment. On the basis of available knowledge, potentially contradictory hypotheses are formulated and experiments are designed to test them. The predictions thus made are, in turn, scrutinized experimentally, leading to modified hypotheses. In an iterative cycle this approach is repeated until the hypothesis becomes as congruent as possible with reality. For reasons already discussed, in systems biology, mathematical models are utilized to formulate working hypotheses. (**Figure 17.3**). Thus, by combining experimental data generation and mathematical modeling mainly based on computer simulations, an analysis of complex biological systems becomes possible, which is not feasible with the traditional methods of molecular cell biology. To reach this goal, biologists must become familiar with mathematical approaches, while computer scientists must make some effort to arrive at an understanding of biological problems. Meanwhile, considerable progress has been made to standardize experimental data generation. Moreover, user-friendly software is being developed that biologists can use to become familiar with analytical and numerical methods and with statistical evaluations. On the other hand, theoreticians are encouraged to tailor the development of software tools to specific questions. Some future aims of such approaches are the prediction of effective therapeutic strategies and the development of synthetic systems with desired properties obtained by artificially modifying an existing system or constructing a new one (synthetic biology).

Early investigators who applied a mathematical approach were the neurophysiologists A. L. Hodgkin and A. F. Huxley (1952), who put forward a model of the neuronal action potential, and Denis Nobel (1960), who developed the first computer model of a beating heart.

Summary

- The task of systems biology is investigation of the structure and dynamics of complex networks. The goal is to explain the emergence of network properties that cannot be deduced from the properties of the single components.

- The approach combines the generation of quantitative experimental data with mathematical modeling and computer simulations, requiring a collaboration of experimenters and theoreticians and standardization of approaches.

17.5 Problems of quantitative data generation and mathematical model development

Since a system is more than an assembly of its components and of the linkages between them, its properties cannot be understood solely on the basis of qualitative data. Rather, it is essential to consider the quantitative strength of the interactions and their dynamic behavior. Otherwise the construction of networks would be a descriptive rather than an explanatory venture.

To arrive at an in-depth understanding of a biological system, the following must be known:

- its **structure** involving the components, their mode of interaction, and mechanisms regulating these interactions

- its **dynamics**, or the behavior over time and the role of dynamic motifs such as gradual response, oscillation, adaptation, feedback control, and switch-like behavior

- its **control** by key mechanisms based on conserved design principles and dominating the state of the biological network in a dose–response manner

Although many kinetic parameters have been determined, especially in biochemistry, it is still difficult to link these measurements of the isolated enzyme to the situation *in vivo*. Thus, most of the available data are hardly applicable to infer mechanisms of signal processing. The descriptions used to be qualitative, and in many cases the temporal resolution of the experimental techniques was not high enough to obtain detailed information on process dynamics. However, there is increasing effort in developing standardized workflows that facilitate quantitative data generation of proteomics and transcriptomics at the cell population level. Further, quantitative mass spectrometry approaches now enable the absolute quantification of proteins and the degree of phosphorylation can be quantified by utilizing peptide standards of known concentrations. These developments are important prerequisites in establishing mathematical models that connect signal transduction networks with gene-regulatory networks and that allow quantitative predictions.

A major challenge of quantitative systems biology is that several key properties of cellular signaling can be studied only in individual cells. An example is provided by switch-like responses (**Figure 17.4**); another by oscillatory processes, which in a cell population are masked unless the cells are synchronized (**Figure 17.5**). Therefore, strong efforts are being made to improve the analysis at the single cell level, ranging from imaging and cytometric techniques to RNA or DNA sequencing.

Finally, our knowledge of signaling pathways and gene regulatory networks is based mainly on artificial systems, in particular on cell lines that considerably deviate from normality as far as control mechanisms and genetic stability are concerned. Therefore, to yield reproducible results for model building, it is mandatory to standardize the cell type, cell culture conditions, and experimental techniques.

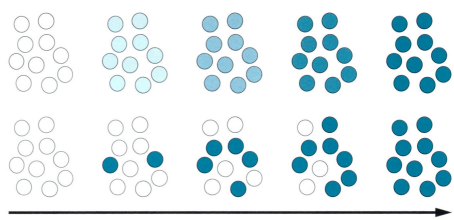

Figure 17.4 Gradual versus switch-like response Increasing stimulus concentrations may trigger a gradually increasing effect (top row), such as the expression of a distinct protein in a cell population. Alternatively, an all-or-nothing effect may be switched on in an increasing number of cells (bottom row). Upon averaging over the cell population, the two response types cannot be distinguished from each other. However, at the single cell level, the differences become evident.

increasing stimulus

Summary

Without quantitative data evaluation, networks remain entirely descriptive. Therefore, the progress of systems biology critically relies on quantitative data generation and computer-assisted data normalization and management for model calibration to elucidate structure, dynamics, and control mechanisms of biological systems.

17.6 Mathematical modeling of a signaling pathway

As stated above, systems biology is based on the combination of laboratory with computer work. Mathematical models of biological systems are abstractions aiming at the discovery of more generic principles. If a data set is sparse, statistical methods can be applied to assess errors and uncertainty. How much data do you need to appropriately describe a biological system? This depends on the time scales of the processes and the model itself. There are mathematical criteria to evaluate how informative a model structure is, given the experimental data. If the model is not informative, its structure has to be refined and the number of parameters has to be reduced. The dynamic behavior and design principles of biological systems can be assessed through simulations by kinetic (reaction) models of covalent modifications, intermolecular association, and intracellular localization, which are calibrated on the basis of experimental data of high quality. Reaction equations describe

Figure 17.5 Oscillatory response Upon averaging over the cell population, the concentration of a protein of interest appears constant if the cells are not in phase. However, detection at the single-cell level reveals oscillatory behavior.

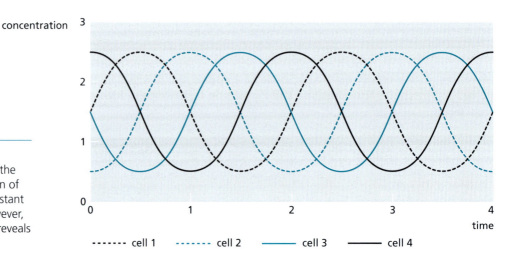

------- cell 1   ------- cell 2   ———— cell 3   ———— cell 4

(A) EpoR

(B) EpoR

(C)

(1) unphosphorylated STAT5

$$\dot{x}_1(t) = -k_1 x_1 \text{EpoR}_A$$

(2) tyrosine-phosphorylated STAT5 monomer

$$\dot{x}_2(t) = -k_2 x_2^2 + k_1 x_1 \text{EpoR}_A$$

(3) tyrosine-phosphorylated STAT5 dimer

$$\dot{x}_3(t) = -k_3 x_3 + \tfrac{1}{2} k_2 x_2^2$$

(4) tyrosine-phosphorylated STAT5 dimer in the nucleus

$$\dot{x}_4(t) = \qquad + k_3 x_3$$

(1) unphosphorylated STAT5

$$\dot{x}_1(t) = -k_1 x_1 \text{EpoR}_A + 2 k_4 x_3^\tau$$

(2) tyrosine-phosphorylated STAT5 monomer

$$\dot{x}_2(t) = k_2 x_2^2 + k_1 x_1 \text{EpoR}_A$$

(3) tyrosine-phosphorylated STAT5 dimer

$$\dot{x}_3(t) = -k_3 x_3 + \tfrac{1}{2} k_2 x_2^2$$

(4) tyrosine-phosphorylated STAT5 dimer in the nucleus

$$\dot{x}_4(t) = -k_4 x_4^\tau + k_3 x_3$$

Figure 17.6 Mathematical treatment of two conflicting hypotheses on signal transduction in the JAK–STAT cascade
(A) Linear signal transmission by a feedforward cascade from the cell surface to the nucleus and signal termination by STAT5 degradation in the nucleus (one-way model). (B) For signal transductions by iterative cycles, STAT5 is dephosphorylated in the nucleus and returns to the cytoplasm to re-enter subsequent activation cycles (cycling model). (C) The STAT5 populations described by coupled ordinary differential equations are labeled in A and B by the concentrations x_1, x_2, x_3, and x_4. The dynamic parameters are indicated by rate constants k_1, k_2, k_3, and k_4 and exponent τ, the delay term introduced for the cycling model. The term $\dot{x}_4(t)$ is equal to the differential quotient dx/dt. EpoR, erythropoietin receptor; P, phosphate residue; JAK, Janus kinase; STAT, signal transducer and activator of transcription.

processes such as catalysis and assembly, and the parameters represent concentrations, binding affinities, and reaction rates. A **deterministic** representation considers bulk concentrations of pathway components, not individual molecules, and assumes the cell to be a well-stirred reactor. **Stochastic** models are applied for reactants that are present in small numbers and include effects arising from random fluctuations around the average behavior. **Boolean** models are used to identify network interactions for large systems.

In most present-day mathematical models of signaling pathways, changes in the concentration of reactants over time are described by coupled ordinary differential equations (ODEs). The model structure is chosen to evaluate the internal structure (intermediary steps) lying between the input and output of the system. Overall orientation is given by maps that qualitatively depict the connections between individual components (**Figure 17.6**). Traditionally, such maps have been used to illustrate signaling pathways without providing information about their dynamic behavior. They are found throughout this book. Such illustrations can be translated into a set of ODEs.

For this purpose, some basic facts about the kinetics of molecular interactions are taken into account. For instance, in a first approximation, the non-covalent dimerization or oligomerization of proteins frequently observed in signaling pathways can be considered to follow the law of mass action (**Sidebar 17.1**), resulting in a **linear response** curve when the concentration r of complexes is plotted against the concentration of individual proteins.

Enzymatic reactions such as phosphorylation/dephosphorylation or protein synthesis/degradation in the simplest case exhibit the **hyperbolic response** curves of Michaelis–Menten kinetics (see Sidebar 17.1). **Sigmoidal response** curves indicating positive cooperativity finally are the result of allosteric interactions of proteins or protein complexes with multiple subunits or domains (see Section 1.3). This enables the proteins to maximally respond to small changes in ligand concentration, resulting in a switchlike behavior, while on the other hand it provides an effective noise filter. A well-known example for allosteric regulation is the binding of cyclic AMP to the inhibitory subunit of protein kinase A (see Section 4.6.1). Sigmoidal kinetics are characterized by the Hill coefficient ≥ 1 (see Sidebar 17.1). More complex response curves resulting from positive feedback or a combination of positive and negative feedback are found in Section 1.9.

In Figure 17.6 the strategy of model development is explained with the help of a distinct signal transduction event, the JAK–STAT pathway. As shown in Section 7.2.1, the transcription factor STAT5 is recruited to an activated cytokine receptor (in our

---

Sidebar 17.1 Kinetics of protein interactions

Law of mass action kinetics: When a reversible chemical reaction A + B ⇌ C + D is in equilibrium, the law of mass action defines an equilibrium constant as $K_{eq} = k_f/k_b = $ [C][D]/[A][B], with [A] and [B] and [C] and [D] being the concentrations of the initial products A and B and the end products C and D. For a reversible chemical reaction, the forward reaction rate and the backward reaction rate are proportional to the concentration of the reaction partners. The equilibrium constant K_{eq} is related to the change of free energy ΔG (see Sidebar 2.1) of the reaction by $\Delta G = -RT \ln K_{eq}$, with R denoting the gas constant and T the absolute temperature.

Michaelis–Menten kinetics: An approximation to mass action kinetics is used for enzyme-catalyzed reactions. Maximal reaction velocity is reached when the enzyme is saturated with its substrate. If the reaction rate is plotted against the substrate concentration, a hyperbolic curve is obtained. The reversible formation of an intermediary enzyme–substrate complex reaching an equilibrium is assumed. Lim-

itations are that, in signal-transduction networks, intermediary complexes do not reach a steady-state or reactions are tightly coupled to other processes. The Michaelis–Menten equation for an enzymatic reaction with one substrate is $v_0 = V_{max}S/(K_M + S)$, where S is the substrate concentration, K_M is the Michaelis constant, V_{max} is the maximal velocity, and v_0 is the initial rate of the enzymatic reaction. If $S = K_M$ then $v_0 = \frac{1}{2}V_{max}$. The smaller K_M is, the higher is the enzyme–substrate affinity.

Hill coefficient: For protein complexes consisting of several subunits that allosterically cooperate with each other, a sigmoidal curve is obtained when the reaction rate is plotted against the substrate concentration. This cooperative behavior is described by the Hill function, which is an extension of the Michaelis–Menten equation: $v_0 = V_{max}S^h/(K_M + S^h)$. Being proportional to the steepness of the sigmoidal function (that is, the deviation from the normal hyperbolic Michaelis–Menten function), the exponent h (Hill coefficient) is a measure of the strength of the cooperative effect.

example the erythropoietin receptor EpoR) to become phosphorylated by the receptor-associated tyrosine kinase JAK2. Upon phosphorylation STAT5 dimerizes and then migrates into the nucleus, where it binds to the promoter sequences of target genes. Signal termination may be assumed to be due either to dephosphorylation, allowing STAT5 to recycle into the cytoplasm (cycling model), or to ubiquitylation and proteolytic degradation of STAT5 (one-way model). To decide between these two possibilities, a set of coupled ODEs is formulated that describes the changes over time in the concentration of unphosphorylated as well as of Tyr-phosphorylated monomeric and dimeric STAT5, both in the cytoplasm and in the nucleus. To capture the cycling behavior, the model is extended by a delay term accounting for the time it takes the dephosphorylated STAT5 to return to the cytoplasm.

In the beginning of the investigation, the ligand-induced phosphorylation of EpoR is measured and used as the input function for the mathematical model. In the next steps, the kinetic rates k_1, k_2, k_3, and k_4 are estimated to obtain an optimal fit between the model and the experimental data. To this end, the ODEs are solved numerically for different sets of k_{1-4} values by employing a computational routine program called **parameter estimation**. The difference between the solution of the ODEs and the experimental data is calculated using the mean square distance, resulting in a χ^2 value that provides a measure for the "goodness-of-fit." The estimated k_{1-4} values are changed stepwise until the χ^2 value reaches a minimum, indicating the optimal fit of the experimental data. Additional parameters used for an optimization of the models describe the experiment-specific measurement scale and background as well as the quality of the individual experiments. For JAK–STAT signaling, different sets of parameters are obtained for the cycling model and for the one-way model; the obtained minimal χ^2 value is smaller for the cycling model than for the one-way model. Therefore, only for the cycling model of the JAK–STAT pathway can a set of parameters be found to obtain a good fit, whereas the one-way model cannot describe the experimental data (**Figure 17.7**).

The aim of the procedure is to establish a minimal model describing the experimental data. If an extended model is developed, it can be tested whether, by removing distinct steps, the calculated χ^2 value significantly increases. For example, for the JAK–STAT model it was shown that the dimerization of phosphorylated STAT5 proteins occurs very quickly. Therefore, the model could be simplified, by assuming phosphorylation and dimerization to occur in one step. Moreover, steps with k values close to zero may be neglected, and when the turnover of proteins is much slower than protein phosphorylation the total amount of proteins can be assumed to remain constant in the course of the experiment. Such minimization of the number of parameters is critical to avoid over-fitting.

Owing to the limited amount and quality of experimental measurements, it is not assured that the parameters can be estimated unambiguously. To evaluate how well the parameters are determined, profile-likelihood analysis can be performed, which gives a measure of the identifiability of parameters. Identifiable parameters are an important prerequisite for reliable predictions of the behavior of systems and are therefore critical for obtaining insights into key properties. The reliability of the predictions can be determined by calculating their uncertainty. If the parameters are not accurately determined given the available data and model structure, the model predictions may be associated with a large uncertainty.

The dynamic behavior of many signaling pathways is determined by feedback loops. Negative feedback loops result in adaptation and robustness, and a combination of negative and positive feedback loops can lead to an oscillatory response and pattern formation (see Section 1.9). These features render biological systems nonlinear. Therefore the impact of perturbations is very difficult to predict and computational methods are required to disentangle the complexity. To systematically identify parameters that critically determine the dynamic behavior of a system and to predict

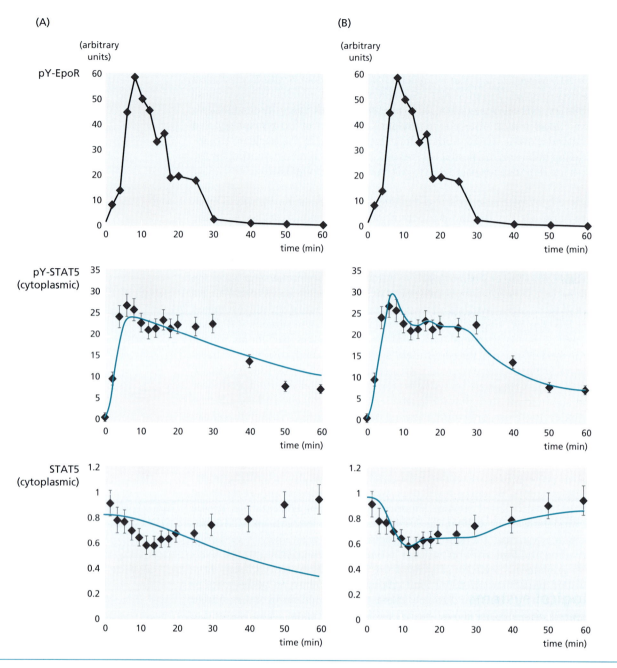

Figure 17.7 Mathematical modeling of the dynamic behavior of STAT5 The time course of ligand-induced Tyr phosphorylation (pY) of the erythropoietin receptor (EpoR) and signal transducer and activator of transcription 5 (STAT5), and the total amount of STAT5 in the cytoplasm, were determined by quantitative immunoblotting given in arbitrary units (gray diamonds). Linear interpolation of EpoR phosphorylation is indicated in black and used as input function for the mathematical models. The mathematical fit achieved with the one-way cascade model (A) and the cycling model (B) is represented in teal. (Adapted from Swameye I, Müller TG, Timmer J et al. [2003] *Proc Natl Acad Sci USA* 100:1028–1033.)

the consequence that changes have on signaling output, a **sensitivity analysis** can be performed. This mathematical approach reveals which concentrations and rate constants have the largest influence on the overall behavior of the system. Usually, sensitivity analysis can be applied only if most parameters can be identified. In the mathematical model of JAK–STAT signaling, the parameters of nuclear export and import turn out to be most sensitive to perturbation, a prediction that is confirmed experimentally. These investigations reveal rapid nucleocytoplasmic cycling as a general design principle for signaling cascades involving latent transcription factors such as STAT. This hypothesis has also been found to hold true for SMAD factors (see Section 6.2.2).

A related method called metabolic control analysis has been proved to be particularly useful for determining rate-limiting steps in metabolic networks. It was shown, for instance, that in such networks a single step is not rate-limiting, but rather many components contribute simultaneously to the control of the network. This means that, as might have been expected, control is not the property of single components but a network property.

The steps of model development for analysing signaling pathways can be summarized as follows:

- **Mathematical model:** sets of coupled ordinary equations (ODEs) describing the dynamic behavior of biological system variables (x) by differential equations
- **Model establishment:** translation of prior knowledge derived from pathway cartoons into a list of reactants and reactions
- **Model calibration:** use of high-quality, quantitative data to estimate model parameters and to ensure a high degree of identifiable parameters
- **Model validation:** generating model predictions that can be experimentally tested
- **Systems analysis:** inferring systems properties by model predictions and by utilizing sensitivity analysis

Summary

- Mathematical modeling and computer simulations of signaling pathways are based on experimental data and take into account the basic laws of (bio)-chemical reaction kinetics.
- To arrive at the best fit between experimental data and model simulation, parameters have to be estimated. Sensitivity analysis is performed to determine those parameters that dominate the overall behavior of a system.

17.7 Synthetic and predictive biology: the improvement of biological systems

An in-depth investigation of the structure and dynamics of biological systems is expected to have far-reaching practical consequences, for instance in medicine, agriculture, and environmental protection. This aspect is illustrated with the help of two examples: systems robustness and drug discovery strategies.

17.7.1 Systems robustness

Perturbation insensitivity or **robustness** is a characteristic property of biological systems that is essential for performance and survival. As shown in this book, it is particularly critical for cellular signal processing (see Section 1.9). Robustness is due to the following:

(1) **Adaptation**, or the system's ability to cope with perturbations over a wide range while remaining sensitive. As a rule, adaptation is achieved by negative feedback control or by a combination of positive and negative feedback (see Section 1.9).

(2) **Parameter insensitivity**, or the ability to cope with changes in internal kinetic parameters. This property protects the system against signaling noise. Parameter insensitivity is achieved by sigmoidal kinetics, as found for proteins consisting of allosterically interacting subunits or subdomains (see Section 1.3).

(3) **Fault tolerance**, which allows the system to continue operating, albeit at lower performance, even when some of its components fail. Fault tolerance requires stability of components and processes, which mostly is due to a redundancy (backup) of regulatory proteins and pathways. Temporal and spatial separation of redundant components and pathways (modularity) prevents a failure from spreading over the whole system.

In engineering science, robustness indicates a superior design. In fact, the approaches used to achieve robustness resemble each other in living and engineered systems. Therefore, in principle, systems biologists can employ the methods used by engineers to analyze and improve the robustness of biological networks. Synthetic biology creates new functional devices, systems, and organisms based on standardized biological building blocks. By modification and recombination of existing genetic elements, new genetic parts, such as plasmids, genetic elements, or artificial chromosomes, are created. The synthetic biology approach has been made possible by two recent developments: an increasing number of available and mechanistically understood genetic building blocks and progress in mathematical modeling.

17.7.2 Improvement of drug development

Diseases can be understood as perturbations of biological systems ranging from the molecular up to the social level. Consequently, mathematical models integrating various signal-transduction cascades are expected to provide clinical researchers with novel mechanism-based drug discovery strategies. In fact, under the conditions mentioned above, mathematical models may predict the therapeutic and side effects of drugs in a more reliable and cost-effective way than traditional methods.

Many diseases, particularly the most complex ones such as diabetes and cancer, are caused and promoted by multiple defects of cellular signal processing (see, for instance, Section 13.3). Like a signal that induces an extended excitation pattern, a defect of signaling is also expected to evoke in cells a widespread perturbation of the signal-processing network (after all, and to stress it again: most signaling cascades are complex, highly interconnected systems with multiple feedback loops rather than linear pathways). Much effort in cancer research has been dedicated to examine the effects of altered protein function due to oncogenic defects caused by somatic gene mutations and to suggest targets for intervention. The most serious obstacles to this approach are intolerable side effects and the development of drug resistance. A prominent case illustrating this dilemma is found in Section 7.1.1. An example where mathematical modeling has led to new insights is provided by the serine/threonine kinase RAF (see Section 10.1.4), which is constitutively activated in many tumors due to oncogenic mutations. Nevertheless, compared with normal cells, the activity of the MAP kinase ERK positioned downstream of RAF is increased only moderately. Mathematical modeling has revealed that half-maximal inhibition of the mutated RAF by a potential anticancer drug would lead to some decrease of ERK activity but might entirely abrogate RAF and ERK activity in healthy cells, thus suggesting that the mutated RAF is a poor drug target since it has lost part of its control (of course, the prediction of the model has to be verified experimentally). An alternative strategy would be to interfere with another node (protein) of the data-processing subsystem to which RAF belongs. However, this would require a detailed knowledge of systems dynamics and of the specific impact of feedback loops or cross talk with other signaling pathways.

To this end, a given process has to be analyzed both on a time scale and a spatial scale. Signaling pathways operate on a time scale of minutes to hours, gene regulatory networks on the order of hours, and phenotypic changes require hours to days; and a perturbation in a single cell not only inflicts the behavior of this single cell, but

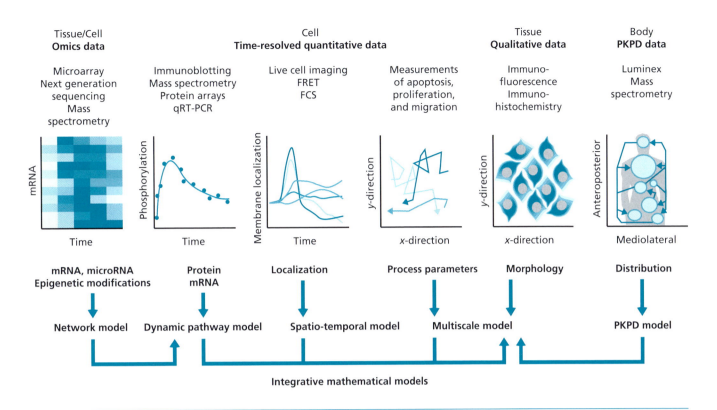

Figure 17.8 Integrative mathematical modeling across the scales Typical methods to generate experimental data at the cellular, tissue, or body scales are represented. Data sets and the corresponding modeling approaches are shown. By linking results of different mathematical models, integrative mathematical models can be established.

also of neighbor cells, the tissue, the organ, and the whole body. Therefore, to design new therapeutic strategies it is mandatory to investigate networks and systems at different levels of complexity rather than just individual components.

As for single cells, mathematical modeling is used to describe drug effects at the tissue level. For example, on the basis of confocal images a model was developed for the regeneration of the liver lobule after intoxication. To assess effects at the scale of the whole body, mathematical models of pharmacokinetics (PK) and pharmacodynamics (PD) are employed. Pharmacokinetics describes what the body does to the drug, such as uptake, metabolism, transport, and secretion; pharmacodynamics describes what the drug does to the body including the desired effect (efficacy), but also unwanted side effects (toxicity). Such models can be extended to take into account more complex parameters such as differences between individuals and those that are specific for an organism, for example the blood volume, or for a certain drug, for example the molecular mass. An essential part of these models is the parameters describing the exchange rates from the arterial blood to the tissue and from the tissue to the venous blood. Such an approach allows extrapolation of the model for an individual with different blood flow or body weight, or adaptation of the model to another compound with known physicochemical properties. Currently, efforts are underway to establish multi-scale models (**Figure 17.8**) that link different approaches so as to maximize the insights that can be obtained and to improve prediction of treatment outcome.

Thus, the potential benefits of systems biology are considerable and it is evident that predictive drug design is one of its major practical goals. Systems biology has become an indispensable tool to uncover otherwise masked aspects of biological systems. However, there is still a long way to go, particularly as far as the highly complex systems of cellular data processing, the "brain of the cell," are concerned in the context of the entire organism, where different temporal and spatial scales have to be considered.

Summary

- To a certain extent, methods used by engineers to analyze and improve a technical system can also be employed in systems biology. Thus, in both cases the insensitivity of systems to disturbances (robustness) is based on the same properties: adaptation, parameter insensitivity (for instance, noise insensitivity), and fault tolerance (achieved by redundancy).

- The analysis of signal-processing systems is expected to have far-reaching practical consequences such as optimized drug design. This holds true particularly for the treatment of highly complex diseases of signaling such as cancer.

Further reading

Albeck JG, MacBeath G, White FM et al. (2006) Collecting and organizing systematic sets of protein data. *Nat Rev Mol Cell Biol* 7:803–812.

Aldridge BB, Burke JM, Lauffenburger DA et al. (2006) Physicochemical modelling of cell signalling pathways. *Nat Cell Biol* 8:1195–1203.

Alon U (ed) (2007) An Introduction to Systems Biology. Design Principles of Biological Circuits, pp 1–295. Chapman & Hall/CRC Mathematical and Computational Biological Series. Chapman & Hall.

Fu P (2006) A perspective of synthetic biology: assembling building blocks for novel functions. *Biotechnol J* 1:690–699.

Hahn B, Böhm M, Raia V et al. (2011) One-source peptide/phosphopeptide standards for accurate phosphorylation degree determination. *Proteomics* 11:490–494.

Hoehme S, Brulport M, Bauer A et al. (2010) Prediction and validation of cell alignment along microvessels as order principle to restore tissue architecture in liver regeneration. *Proc Natl Acad Sci USA* 107:10371–10376.

Hornberg JJ & Westerhoff HV (2006) Oncogenes are to lose control on signaling following mutation: should we aim off target? *Mol Biotechnol* 34:109–116.

Kremling A & Saez-Rodriguez J (2007) Systems biology—an engineering perspective. *J Biotechnol* 129:329–351.

Lienert F, Lohmueller JJ, Garg A et al. (2014) Synthetic biology in mammalian cells: next generation research tools and therapeutics. *Nat Rev Mol Cell Biol* 15:95–107.

Raue A, Steiert B, Schelker M et al. (2015) Data2Dynamics: a modeling environment tailored to parameter estimation in dynamical systems. *Bioinformatics* 31:3558–3560.

Schilling M, Maiwald T, Bohl S et al. (2005) Computational processing and error reduction strategies for standardized quantitative data in biological networks. *FEBS J* 272:6400–6411.

Snider J, Kotlyar M, Saraon P et al. (2015) Fundamentals of protein interaction network mapping. *Mol Syst Biol* 11:848.

Wiśniewski JR, Hein MY, Cox J et al. (2014) A "proteomic ruler" for protein copy number and concentration estimation without spike-in standards. *Mol Cell Proteomics* 13:3497–3506.

Wiener N (1961) Cybernetics or Control and Communication in the Animal and the Machine. MIT Press.

Zhu X, Gerstein M & Snyder M (2007) Getting connected: analysis and principles of biological networks. *Genes Dev* 21:1010–1024.

Index